"Walled Up to Heaven"

HARVARD SEMITIC MUSEUM PUBLICATIONS

Lawrence E. Stager, General Editor
Michael D. Coogan, Director of Publications

STUDIES IN THE ARCHAEOLOGY AND HISTORY OF THE LEVANT

edited by
Lawrence E. Stager

1. *The Archaeology of Jordan and Beyond: Essays in Honor of James A. Sauer*, edited by Lawrence E. Stager, Joseph A. Greene, and Michael D. Coogan
2. *The House of the Father as Fact and Symbol: Patrimonialism in Ugarit and the Ancient Near East*, by J. David Schloen
3. *Canaanites, Chronologies, and Connections: The Relationship of Middle Bronze Age IIA Canaan to Middle Kingdom Egypt*, by Susan L. Cohen
4. *"Walled Up to Heaven": The Evolution of Middle Bronze Age Fortification Strategies in the Levant*, by Aaron A. Burke

"WALLED UP TO HEAVEN"

THE EVOLUTION OF MIDDLE BRONZE AGE FORTIFICATION STRATEGIES IN THE LEVANT

by

Aaron A. Burke

EISENBRAUNS
Winona Lake, Indiana
2008

“Walled Up to Heaven”:
The Evolution of Middle Bronze Age Fortification Strategies in the Levant

by
Aaron A. Burke

Printed in the United States of America

Library of Congress Cataloging-in-Publication Data

Burke, Aaron A.
Walled up to heaven : the evolution of Middle Bronze Age fortification strategies in the Levant / by Aaron A. Burke.
p. cm. — (Studies in the archaeology and history of the Levant ; 4)
Includes bibliographical references and index.
ISBN 978-1-57506-927-2 (hbk. : alk. paper)
1. Fortification—Middle East—History—To 1500. 2. Military architecture—Middle East—History—To 1500. 3. Siege warfare—Middle East—History—To 1500. 4. Middle East—Antiquities. 5. Bronze age—Middle East. I. Title.
UG432.M628B87 2008
725′.18—dc22

2008004947

The paper used in this publication meets the minimum requirements of the American National Standard for Information Sciences—Permanence of Paper for Printed Library Materials, ANSI Z39.48-1984.♾™

For Katherine

CONTENTS

Figures ix
Tables xi
Preface xiii
General Abbreviations xv
Bibliographic Abbreviations xvii

PART ONE
AN INTRODUCTION TO MIDDLE BRONZE AGE FORTIFICATIONS

1. RESEARCHING MIDDLE BRONZE AGE FORTIFICATIONS 3
   - A. Previous Scholarship 3
   - B. Levantine Fortifications in Their Regional Context 15
   - C. Methodology 17
   - D. Geography and Chronology 18
   - E. Moving Forward 20

PART TWO
THE EVOLUTION OF WARFARE AND DEFENSES IN THE LEVANT DURING THE BRONZE AGE (CA. 2500–1500 B.C.)

2. WEAPONRY AND THE ART OF WAR (CA. 2500–1500 B.C.) 25
   - A. Levantine Textual Traditions on Warfare 25
   - B. The Emergence of Siege Warfare in the Ancient Near East 27
   - C. The Role of the Chariot in Bronze Age Warfare 29
   - D. The Weapons and Tactics of Siege Warfare 31
   - E. Mortuary Evidence for Siege Warfare 44
   - F. The Military Context for Middle Bronze Age Fortification Strategies 45

3. MIDDLE BRONZE AGE FORTIFICATIONS 47
   - A. Architectural Elements 48
   - B. Building Materials 73
   - C. Fortifications and the Landscape 77
   - D. Architectural Evolution 78

PART THREE
MIDDLE BRONZE AGE FORTIFICATIONS & THEIR CULTURAL CONTEXT

4. A HISTORY OF WARFARE IN THE LEVANT (CA. 2500–1460 B.C.) 87
   - A. Early Bronze Age Warfare (ca. 2500–1900 B.C.) 88
   - B. Middle Bronze Age Warfare (ca. 1900–1460 B.C.) 94
   - C. Conclusions 101

5. DEFENSIVE STRATEGIES & POLITICAL ORGANIZATION 103
   - A. Bronze Age Settlement Patterns 103
   - B. Defensive Settlement Networks 122
   - C. A Case Study in Bronze Age Polities: The Kingdom of Ashkelon 125

6. THE SOCIOECONOMIC IMPACT OF FORTIFICATION CONSTRUCTION 141
   - A. The Organization of Labor in the Levant 141
   - B. Construction Rates and Resource Requirements 143
   - C. Estimating Labor Consumption 146
   - D. Scenarios for the Construction of Fortifications 153
   - E. Social Complexity and the Construction of Fortifications 155

7. CONCLUSIONS 159

PART FOUR
A CATALOG OF MIDDLE BRONZE AGE FORTIFIED SETTLEMENTS IN THE LEVANT AND MESOPOTAMIA, AND THEIR FORERUNNERS

GENERAL NOTES 165
1. NORTHWESTERN MESOPOTAMIA (CA. 3000–1500 B.C.) 167
2. THE LEVANT (CA. 2400–1500 B.C.) 185
   A. The Northern Levant: Western Syria, Lebanon, and the Hatay 186
   B. The Southern Levant: Israel, Jordan, and the Palestinian Territories 228
3. EGYPT 320

Bibliography 321
Index 351

FIGURES

Figure 1. Third-millennium B.C. bows. 33
Figure 2. Third-millennium B.C. shield and double-curved reflexed composite bow from Mari. 35
Figure 3. Ranges of Bronze Age weapons. 37
Figure 4. Reconstruction of a battering ram of the Neo-Assyrian period. 39
Figure 5. Chest cavity with arrow lodged in vertebrae from Ugarit. 44
Figure 6. Cross-section of freestanding and supplemental ramparts. 51
Figure 7. Section of layered fill in rampart at Gezer employing "sandwich" technique. 53
Figure 8. So-called "Syrian gate" of the mortuary temple of Ramesses III. 59
Figure 9. Illustration of the use of terraces and "anchor walls" between fortification walls and gates. 63
Figure 10. Fields of fire for settlements with and without towers. 65
Figure 11. Sites identified as Kranzhügeln. 79
Figure 12. Examples of main MB–LB settlement types. 123
Figure 13. The MB IIB–C kingdom of Ashkelon. 130
Figure 14. Late Bronze Age "City-States" of the southern coastal plain represented as polygons. 137
Figure 15. Sites in northwestern Mesopotamia discussed in catalog 1. 167
Figure 16. Plan of Tell Beydar with excavation areas. 169
Figure 17. Plan of Tell Chuera. 170
Figure 18. Plan of Mari. 172
Figure 19. Fortifications of Mari Ville I to III. 174
Figure 20. Plan of Tell es-Sweyhat showing course of EB IV fortifications. 176
Figure 21. Plan of Terqa and excavation areas. 177
Figure 22. Reconstruction of Terqa's third-millennium defensive system. 178
Figure 23. Plan of Tell Bi'a. 179
Figure 24. Plan of Tell Munbaqa. 182
Figure 25. Sites in the northern Levant discussed in catalog 2.A. 186
Figure 26. Plan of Afis. 188
Figure 27. Alalaḫ Level VII showing location of earthen rampart. 189
Figure 28. Plan of fortifications excavated at Biruta. 191
Figure 29. Tell el-Burak fortress. 192
Figure 30. Plan of Byblos. 193
Figure 31. West section through northern fortifications of Byblos. 195
Figure 32. Plan and section of NE Gate and earthen rampart at Byblos. 195
Figure 33. Northern ramparts at Byblos. 196
Figure 34. NW Gate with double staircase at Byblos. 197
Figure 35. Plan of Deir Khabiye. 198
Figure 36. Plan of Ebla. 200
Figure 37. Damascus Gate in Area A at Ebla. 201
Figure 38. Euphrates Gate in Area BB at Ebla. 202
Figure 39. Bastion in Area M at Ebla. 202
Figure 40. Bastion in Area V at Ebla. 203
Figure 41. Bastion in Area AA at Ebla. 203
Figure 42. Ebla's outlying rampart. 204
Figure 43. Reconstructed plan of Middle Bronze Age Ḫalab. 208
Figure 44. Plan of Qatna. 214
Figure 45. Fosse and southern ramparts at Qatna. 215
Figure 46. Plan of Tell Sefinat-Nouh. 218
Figure 47. Plan of aṣ-Ṣour. 220
Figure 48. Plan of Tuqan. 221
Figure 49. Gate in Area A. 221
Figure 50. Gate in Area F. 222

Figure 51. Towers in Area A. 222
Figure 52. Glacis at Ugarit. 223
Figure 53. Plan of Ugarit (Ras Shamra). 224
Figure 54. Postern gate at Ugarit. 225
Figure 55. Plan of Umm el-Marra. 226
Figure 56. Sites in the southern Levant discussed in catalog 2.B. 228
Figure 57. Plan of Achzib. 230
Figure 58. Plan of Tell el-ʿAjjul. 231
Figure 59. Plan of Tel Akko. 234
Figure 60. Plan of Aphek. 236
Figure 61. Plan of Tel Ashkelon. 238
Figure 62. Phase 14 gate. 239
Figure 63. Phase 13C gate. 239
Figure 64. Phase 13B–A gate. 241
Figure 65. Phase 12 gate. 242
Figure 66. Phase 11 gate. 243
Figure 67. Tell Beit Mirsim Stratum F fortifications. 244
Figure 68. Tell Beit Mirsim Stratum E fortifications. 245
Figure 69. Plan of Tel Burga. 250
Figure 70. Tel Dan and excavation areas showing MB I–II features. 251
Figure 71. East section of rampart at Tel Dan in Area A-B. 252
Figure 72. Plan of six-pier gate in Area K at Tel Dan. 252
Figure 73. Tel Dan Area K section. 252
Figure 74. Plan of Ekron. 256
Figure 75. Fortifications of Tell el-Farʿah North. 257
Figure 76. Fortifications of Gezer. 261
Figure 77. Ramparts of Tel Haror. 264
Figure 78. Plan of Hazor. 266
Figure 79. North section of fortifications in Areas D and B. 273
Figure 80. Plan of Jemmeh. 274
Figure 81. Plan of Jericho. 277
Figure 82. Schematic north section of Kenyon's Trench I and Area C. 279
Figure 83. Schematic east section of Kenyon's Trench II. 280
Figure 84. Schematic section of rampart in Kenyon's Trench III and Area A. 282
Figure 85. Plan of Kabri. 284
Figure 86. Plan of Lachish. 287
Figure 87. Plan of Masos. 289
Figure 88. Gate and wall of Stratum XIIIA in Area AA. 291
Figure 89. Wall of Stratum XIII in Area BB. 292
Figure 90. Gate and wall of Stratum XII in Area AA. 293
Figure 91. Area BB Strata XIIIA and XII walls. 294
Figure 92. The evolution of Tel Michal's embankments. 297
Figure 93. Schematic of section through Tel Michal's embankments. 297
Figure 94. Plan of Tel Poleg. 302
Figure 95. Plan of Shechem and its Middle Bronze Age fortifications. 307
Figure 96. Plan of Shiloh showing Middle Bronze Age fortifications. 311
Figure 97. Plan of Shimron. 312
Figure 98. Plan and profile of Timnah. 314
Figure 99. Plan of Yavneh-Yam. 315
Figure 100. Plan of Tel Zeror. 318

TABLES

Table 1.	Terminology and chronology for the periods discussed in this work.	19
Table 2.	Middle Bronze Age stratigraphy in the southern Levant and VLC dates.	20
Table 3.	Stratigraphic correlations between Middle Bronze Age sites in the Levant and Mesopotamia.	20
Table 4.	Major developments in warfare from the third to mid-second millennium.	46
Table 5.	Slopes of earthen ramparts from EB IV to LB I.	52
Table 6.	Dimensions of Middle Bronze Age fosses and their estimated volumes.	57
Table 7.	Width and composition of solid fortification walls from ca. 1900 to 1500 B.C.	62
Table 8.	Dimensions of Middle Bronze Age towers and bastions.	67
Table 9.	Dimensions of six-pier gates.	69
Table 10.	Dimensions of four-pier gates.	70
Table 11.	Dimensions of mudbricks used in EB IV and Middle Bronze Age fortifications.	76
Table 12.	Major developments in defensive architecture from ca. 3000 to 1200 B.C.	84
Table 13.	Akkadian kings and the Levantine settlements they claim to have pacified.	90
Table 14.	Relevant historical events between ca. 2500 and 1460 B.C.	97
Table 15.	Initial phases of fortification of Middle Bronze Age settlements in the southern Levant.	99
Table 16.	Terminology for settlement types in Old Babylonian, Nuzi, and Ugaritic texts.	110
Table 17.	Correlations for stratigraphy of Middle Bronze Age settlements in the southern coastal plain.	128
Table 18.	Primary settlements in the kingdom of Ashkelon at the end of the MB IIC.	129
Table 19.	Historical evolution of Middle Bronze Age districts around the district of Ashkelon.	138
Table 20.	Previous estimates of the labor required in the construction of earthen ramparts.	144
Table 21.	Mudbrick production rates per person per day according to various sources.	146
Table 22.	Revised estimates of the labor and earth required for earthen ramparts.	147
Table 23.	Estimated labor required in the construction of fortification walls.	152
Table 24.	Labor requirements for Middle Bronze Age fortification construction.	152
Table 25.	Pertinent data for estimating the available workforce for sites listed in Table 24.	152
Table 26.	Duration of fortification construction employing corvée or hired laborers.	152
Table 27.	Numbers of soldiers required to build fortifications within two to eight months.	152
Table 28.	Estimated length of ramparts at Ebla according to sections identified by Matthiae.	199
Table 29.	Calculated volume of ramparts at Tel Dan.	254
Table 30.	The Middle Bronze Age stratigraphy of Gezer.	261
Table 31.	Dimensions of Gezer towers.	262
Table 32.	Dimensions of Haror's earthen ramparts.	265
Table 33.	Stratigraphic correlations between the upper and lower towns at Hazor.	266
Table 34.	Summary of Middle Bronze Age fortifications of Hazor.	270
Table 35.	Dimensions and volume of Hazor's earthen ramparts.	270
Table 36.	Jericho feature correlations with rampart phases according to Kenyon.	278
Table 37.	Revised phasing of the fortifications of Jericho by Italian expedition.	278

PREFACE

The people is greater and taller than we; the cities are great and walled up to heaven!
Deuteronomy 1:28 (KJV)

Although there has been no shortage of excavated material available to facilitate a study of Middle Bronze Age fortifications in the Levant, there has been to date no attempt to provide a synthesis on the subject. Little consideration has also been given to fortification strategies of this period other than those related to the construction and function of earthen ramparts. The present work constitutes an attempt to review archaeological and historical evidence in order to answer many of the questions that have been raised concerning the unique fortification strategies of the Middle Bronze Age, and their relationship to social and political complexity in the Levant. It seeks, therefore, to address not only the architectural but also the military-historical, socio-economic, and political significance of Middle Bronze Age fortification strategies.

This study is a revised and expanded version of my Ph.D dissertation, *The Architecture of Defense* (Burke 2004), the inspiration for which can be traced back to discussions regarding the preparation of the *National Geographic* painting of the Middle Bronze Age gateway at Ashkelon (see Gore 2001:72). Out of this process came the renewed realization by both Lawrence Stager and David Schloen that the issue of Middle Bronze Age fortifications in the Levant remained inadequately explored. My interest in ancient warfare and the Middle Bronze Age in the Levant left me with no hesitations about the project. Little did I realize the scope required by the project as I traced reference after reference to Middle Bronze Age fortifications not only in the Levant, but throughout the Near East. I also could not have anticipated the many aspects of Middle Bronze Age culture upon which the issue of fortifications impinged. In addition to architectural and military history these also included ethnic identity, and social and political complexity.

The title of this book is adopted from the description, attributed by the author of Deuteronomy to the Israelite spies, of the fortifications of Canaanite towns on the eve of the Israelites' entry into the promised land. While there is no evident connection between this biblical tradition and Middle Bronze Age fortification systems, there is even less evidence for the fortifications of Late Bronze Age towns in the southern Levant and the data available do not reveal an impressive network of fortified settlements such as those during the Middle Bronze Age. It is curious, therefore, that the biblical tradition regarding fortified Canaanite cities finds what superficially appears to be its historical antecedent in the walled cities of Middle Bronze Age Canaan. It was for this reason that Yigael Yadin surmised that the wall of Jericho, which was reportedly brought down by Yahweh before the Israelites, might be identified with the Middle Bronze Age fortifications discovered at the site. Nevertheless, Yadin's hypothesis never received support owing largely to the no less than three hundred year gap (from ca. 1530 to 1250 B.C.) between the late Middle Bronze Age wall and the appearance of the Israelites.

For his countless hours of gracious advising—in and out of the field—his mentoring and friendship I would like to thank David Schloen. I would like to thank Lawrence E. Stager for allowing me to use the excavation records of the Leon Levy Expedition to Ashkelon and for his insightful critique of this work. McGuire Gibson's keen observations regarding Mesopotamian sites were likewise indispensable in the preparation of this work. Without the help of these three individuals the original dissertation would not have assumed the shape and scope that it finally did.

I would like to thank the Ryerson Fellowship Committee at the University of Chicago for funding of travel related to this research from August through October of 2002. It facilitated research in Syria, Lebanon, and Egypt, where I was able to meet with a number of scholars whose suggestions and observations have been incorporated into this work in various ways.

Completion of this manuscript was facilitated by two quarters of course release by the University of California at Los Angeles for which I am extremely grateful. I am also grateful for comments and suggestions from a number of my colleagues in the Department of Near Eastern Languages and Cultures and the Cotsen Institute of Archaeology at UCLA. I would like to thank William Schniedewind and Cale Johnson for their helpful philological comments on chapter 5. Bob Englund and Elizabeth Carter also graciously answered various questions regarding Mesopotamian archaeology and philology for which I am especially grateful. I would also like to thank Charles Stanish, Sarah Morris, and Monica Smith for discussions that fed into this work in various ways. Finally, thanks are also owed to Lord Colin Renfrew, whose stimulating lectures as a visiting scholar at the Cotsen Institute of Archaeology during the winter of 2006 prompted reconsideration of the impact of Indo-European studies on the early study of Middle Bronze Age fortifications. This resulted in a reappraisal of scholarly influences upon William Foxwell Albright and Kathleen Kenyon regarding the origin of the Amorites and Middle Bronze Age fortifications, which are discussed in chapter 1.

It is impossible to adequately acknowledge the many individuals who assisted my research during my travels, while at The University of Chicago, and at UCLA. Nevertheless, a number of individuals deserve special thanks. I am especially grateful for many years of assistance provided by Charles Jones who was at the time of my residence in Chicago, the head archivist of the Oriental Institute's library. Israel Eph'al, whom I consider a mentor and friend, was especially instrumental in setting the tone of my research at a very early stage during his stay in Chicago from 2001 to 2002. Fellow students at the Oriental Institute were also very encouraging and supportive in the years of work required to complete this manuscript. Among them I especially thank Adam Miglio, Jonathan Tenney, and Jacob Lauinger for their assistance. I also wish to thank Kyle Keimer and George Pierce for their helpful comments regarding the final appearance of the manuscript.

My greatest thanks are owed to my patient wife, Katherine Strange Burke, who attended all of my dinnertime workshops on Middle Bronze Age fortifications and related topics. In addition to her support and graciousness, I thank her for her practiced eye in proofreading, but must admit that all mistakes herein remain *mea culpa*.

Finally, I would like to thank Lawrence Stager, editor of the SAHL series, and Michael Coogan for their suggestions and assistance with the publication of this volume.

Aaron A. Burke
Los Angeles, California December 2007

GENERAL ABBREVIATIONS

Abb.	Abbildungen (Ger.)
Akk.	Akkadian
approx.	approximate
Ar.	Arabic
AS	Atchana Survey
ASL	above sea level
avg.	average
BA	Bronze Age
bot.	bottom
cf.	*confer*
cm	centimeter(s)
cm^2	square centimeters
constr.	construction
d.	depth
deg.	degrees
dia.	diameter
DN	divine name
Dyn.	Dynasty
EB	Early Bronze Age
ED	Early Dynastic Period
esp.	especially
est.	estimate
F.	Feature
fig. (*pl.* figs.)	figure
FIP	First Intermediate Period
Fr.	French
g	gram(s)
Gen	Genesis
Gk.	Greek
GN	geographical name
Ger.	German
h.	height
ha	hectares
Heb.	Hebrew
ill.	illustration
Josh	Joshua
kg	kilogram
km	kilometer(s)
km^2	square kilometers
l.	length
L.	Locus *or* Layer
LB	Late Bronze Age
LC	low chronology
LF.	Layer-Feature
ln. (*pl.* lns.)	line(s)
lit.	literally
m	meter(s)
m^2	square meters
m^3	cubic meters
M.	*murale* (Italian), *mur* (French)
MB	Middle Bronze Age
MBab	Middle Babylonian
MC	middle chronology
mo. (*pl.* mos.)	month(s)
n.	note
Neh	Nehemiah
Num	Numbers
OB	Old Babylonian
OK	Old Kingdom
Per.	Period(s)
Ph.	phase
pl. (pls.)	plate
PHM	patrimonial household model
PN	personal name
pop.	population
Ps	Psalms
Refs.	References
2 Sam	2 Samuel
sect.	section
Sects.	Sections
sing.	singular
SIP	Second Intermediate Period
Str.	Stratum *or* Strata
Sq.	Square
Sum.	Sumerian
supp(s).	supplement(s)
T.	Tell (Ar.) *or* Tel (Heb.)
Taf.	Tafel (Ger.)
TN	toponym
Ug.	Ugaritic
VLC	very low chronology
w.	width
W.	Wall

GENERAL ABBREVIATIONS

BIBLIOGRAPHIC ABBREVIATIONS

AAS	*Annales archéologiques de Syrie*
AASOR	Annual of the American Schools of Oriental Research
ÄAT	Ägypten und Altes Testament
ABD	*The Anchor Bible Dictionary*, 6 vols. ed. David N. Freedman (New York, 1992)
ADAJ	*Annual of the Department of Antiquities of Jordan*
AbrN	*Abr-Nahrain*
AEL	*Ancient Egyptian Literature*, 3 vols. M. Lichtheim (Berkley, 1973–80)
AfO	*Archiv für Orientforschung*
AHW	*Akkadisches Handwörterbuch*, ed. W. von Soden. 3 vols. (Wiesbaden, 1965–1981)
AJA	*American Journal of Archaeology*
Akk	*Akkadica*
AkkSup	*Akkadica Supplementum*
ANEP	*Ancient Near East in Pictures Relating to the Old Testament*, 2nd ed., ed. J. B. Pritchard (Princeton, 1969a)
ANET	*Ancient Near Eastern Texts Relating to the Old Testament*, 3rd ed., ed. J. B. Pritchard (Princeton, 1969b)
AnSt	*Anatolian Studies*
AOAT	Alter Orient und Altes Testament
ARAM	*ARAM Periodical*
ARET	Archivi reali di Ebla, Testi
ARM	Archives royales de Mari
ASI	*Archaeological Survey of Israel*, Israel Antiquities Authority, Jerusalem
AT	Alalaḫ texts. See *The Alalakh Tablets*, Donald J. Wiseman (London, 1953)
ÄuL	*Ägypten und Levante*
AW	*Antike Welt*
BA	*Biblical Archaeologist*
BAAL	*Bulletin d'Archéologie et d'Architecture Libanaises*
BAH	Bibliothèque archéologique et historique
BAR	*Biblical Archaeology Review*
BASOR	*Bulletin of the American Schools of Oriental Research*
BDB	*The Brown-Driver-Briggs Hebrew and English Lexicon.* ed. F. Brown et al. (Peabody, MA, 1906)
Berytus	*Berytus Archaeological Studies*
BiMes	Bibliotheca Mesopotamica
BIHL	*Bulletin of the Institute of Archaeology*, London
BM	*Baghdader Mitteilungen*
BMB	*Bulletin du Musée de Beyrouth*
CAD	*The Assyrian Dictionary of the Oriental Institute of the University of Chicago*
CCEM	Contributions to the Chronology of the Eastern Mediterranean
CHANE	Culture and History of the Ancient Near East
COS	*Context of Scripture*, 3 vols., ed. William H. Hallo (Leiden, 1997–2002)
CRAIBL	*Comptes rendus de l'Académie des inscriptions et belles-lettres*
DM	*Damascus Mitteilungen*
DULAT	*A Dictionary of the Ugaritic Language in the Alphabetic Tradition*, G. del Olmo Lete and J. Sanmartín (Leiden, 2003)
EA	el-Amarna text (as trans. by Moran 1992)
EAEHL	*Encyclopedia of Archaeological Excavations in the Holy Land*, ed. M. Avi-Yonah (1975)
EI	*Eretz-Israel: Archaeological, Historical and Geographical Studies*
Epitome	Vegetius Renatus, *Epitome of Military Science*. Trans. by N. P. Milner (Liverpool, 1993)
ESI	*Excavations and Surveys in Israel*
HA(—ESI)	*Hadashot Arkhaeologiyot—Excavations and Surveys in Israel*
HALOT	*The Hebrew and Aramaic Lexicon of the Old Testament*, L. Koehler and W. Baumgartner (Leiden, 1994)
HdO	Handbuch der Orientalistik

IEJ	*Israel Exploration Journal*
JAOS	*Journal of the American Oriental Society*
JARCE	*Journal of the American Research Center in Egypt*
JCS	*Journal of Cuneiform Studies*
JEA	*Journal of Egyptian Archaeology*
JESHO	*Journal of the Economic and Social History of the Orient*
JHS	*Journal of Hellenic Studies*
JNES	*Journal of Near Eastern Studies*
JPOS	*Journal of the Palestine Oriental Society*
JSOR	*Journal of the Society of Oriental Research*
JSOT	*Journal for the Study of the Old Testament*
JSSEA	*Journal of the Society for the Study of Egyptian Antiquities*
MARI	*Mari: annales de recherches interdisciplinaires*
MDOG	*Mitteilungen der deutschen Orient-Gesellschaft*
NABU	*Nouvelles assyriologiques brèves et utilitares*
NEA	*Near Eastern Archaeology*
NEAEHL	*New Encyclopedia of Archaeological Excavations in the Holy Land*, ed. E. Stern (New York, 1993)
NRSV	*The New Oxford Annotated Bible.* New Revised Standard Version (New York, 1991)
OE	*Orient-Express*
OEANE	*Oxford Encyclopedia of Archaeology in the Near East*, ed. E. Meyers (New York, 1997)
OIP	Oriental Institute Publications
OrAnt	*Oriens Antiquus*
Pal	*Paléorient*
PEFQS	*Palestine Exploration Fund Quarterly Statement*
PEQ	*Palestine Exploration Quarterly*
PIHANS	Publications de l'Institut historique-archéologique néerlandais de Stamboul
PRU 6	*Le Palais royal d'Ugarit*, vol. 6, ed. J. Nougayrol (Paris, 1970)
RA	*Revue d'assyriologie et d'archéologie orientale*
RB	*Revue Biblique*
RGTC 12/2	*Répertoire Géographique des Textes Cunéiformes, Band 12/2: Die Orts- und Gewässernamen der Texte aus Syrien im 2. Jt. v. Chr.*, J. A. B. Marín (Wiesbaden, 2001)
RlA	*Reallexikon der Assyriologie*, ed. G. Ebeling et al. (Berlin, 1932–)
RIMA 1	*Royal Inscriptions of Mesopotamia, Assyrian Periods*, vol. 1: Assyrian Rulers of the Third and Second Millennia B.C. (to 1115 B.C.). A. K. Grayson (Toronto, 1987)
RIME 2	*Royal Inscriptions of Mesopotamia: Early Periods*, vol. 2: Sargonic and Gutian Periods (2334–2113 B.C.). D. Frayne (Toronto, 1993)
RIME 3/2	*Royal Inscriptions of Mesopotamia: Early Periods*, vol. 3/2: Ur III Period (2112–2004 B.C.). D. Frayne (Toronto, 1997)
RIME 4	*Royal Inscriptions of Mesopotamia: Early Periods*, vol. 4: Old Babylonian Period (2003–1595 B.C.). D. Frayne (Toronto, 1990)
RS	Ras Shamra (Ugarit) text
SAAB	*State Archives of Assyria Bulletin*
SAHL	Studies in the Archaeology and History of the Levant
SAOC	Studies in Ancient Oriental Civilization
SARI	*Sumerian and Akkadian Royal Inscriptions I: Presargonic Inscriptions*, Jerrold S. Cooper (New Haven, 1986)
SHAJ	Studies in the History and Archaeology of Jordan
Siegecraft	Aeneas Tacticus, *How to Survive under Siege*, Trans. by D. Whitehead (Oxford, 1990).
SMA	Studies in Mediterranean Archaeology
SMS	Syro-Mesopotamian Studies
StEb	*Studi Eblaiti*
TA	*Tel Aviv*
TAVO	*Tübinger Atlas des Vorderen Orients*
TM	Tell Mardikh (Ebla) text
UF	*Ugarit-Forschungen*

UVST	*Ugaritic Vocabulary in Syllabic Transcription*, J. Huehnergard (Atlanta, 1987)
VT	*Vetus Testamentum*
WoAr	*World Archaeology*
WVDOG	Wissenschaftliche Veröffentlichung der Deutschen Orient-Gesellschaft
ZDPV	*Zeitschrift des Deutschen Palästina-Vereins*

PART ONE

An Introduction to Middle Bronze Age Fortifications

1. Researching Middle Bronze Age Fortifications

> If all the city walls and gates of the Middle Bronze Age which have been discovered were adequately published, it would be possible to describe the evolution of the art of fortification in considerable detail.
>
> William Foxwell Albright (1949:87)

THE architecture of the Levant during the Middle Bronze Age is in large part characterized by the remains of massive earthen fortifications that are still visible across the landscape. Like the pyramids of Egypt, the remains of these features—their earthen ramparts, revetment walls, glacis, fosses, towers, and mudbrick fortification walls—attest to the enormous quantity of material and labor that was required for their construction. The impressive size of these features is demonstrated by their survival for nearly four thousand years.

Because fortification systems are the most common form of monumental construction in the Levant during the Middle Bronze Age, they have received considerable attention. Scholars have attempted to establish their origin, cultural affiliation, geographic distribution, and function, as well as the terminology that should be employed to describe them. By examining contributions that have already been made it is possible to identify the questions that remain to be addressed in the study of Middle Bronze Age fortifications. Eliezer Oren, for example, has already posed some of these questions:

> Important information can be provided by key questions: How long did the construction of these fortifications take? Were they constructed rapidly to encounter an impending military threat, or gradually in stages? Were they designed to comply principally with the immediate needs of an existing population, or do they reflect long-term planning for a larger future population? Perhaps most intriguing of all is the identification of the labor force involved in the actual building operations: Were the defenses built by the local population of the settlement or by a task force recruited specifically for this purpose from elsewhere? (1997b:256f.).

However, other questions remain. In light of alternate hypotheses regarding the function of Middle Bronze Age ramparts it has become necessary to demonstrate that such structures must be identified as fortifications. And, once such an issue is addressed a number of questions follow: if they are fortifications, why were they built? What characteristics distinguish them from Early or Late Bronze Age fortifications? And how were they built? Can they be classified into distinct types? Who was responsible for initiating their construction? How long were they intended to function? Where and with which groups did this type of fortification originate? How far did such fortification techniques spread? And, ultimately, did they achieve the purpose for which they were constructed?

In this chapter previous attempts to address such questions are reviewed. Only against the backdrop of the history of research of this topic is it possible to understand how trends in Near Eastern archaeology have influenced, and sometimes distorted, our evolving interpretations of the role of Levantine Middle Bronze Age fortifications. Within the historical and regional scope defined by previous research, the precise geographical and chronological scope of this study, its methodology, and its organization are also outlined.

A. Previous Scholarship

Numerous, though brief, discussions of Middle Bronze Age fortifications furnish fundamental insights into the development of thought regarding the function and origin of these features (e.g., Albright 1960; B. Mazar 1968; 1970; G. E. Wright 1971; Kenyon 1973; Dever 1976; 1977:84–89; Kenyon 1979:148–79; Dever 1987; A. Mazar 1990:174–231; Dever 1992b; Kempinski 1992d; Ilan 1995:esp. 316f.; Joffe 1997; Pitard 1998). Nevertheless, many more observations on the subject are embedded in preliminary reports scattered throughout various journals and a limited number of final reports. In contrast to the quantity of literature available for the southern Levant, summary discussions of Middle Bronze Age fortifications in the northern Levant are almost nonexistent (more recently, see Akkermans and Schwartz 2003). However, pivotal contributions to the Middle Bronze Age archaeology of the northern Levant include the excavations of Alalaḫ by Sir Leonard Woolley, Byblos by Maurice Dunand, and Ugarit by Claude Schaeffer.

Despite the large number of contributions on this subject, only a handful of scholars have introduced unique concepts or approaches to the study of Middle Bronze Age fortifications. This review, therefore, addresses these contributions by examining them in light of themes that have shaped the nature of research on Middle Bronze Age fortifications since William Foxwell Albright's work. These broader themes include their association with chariot-driving Hyksos and Hurrians (1920s to 1960s), functional explanations for their development (1940s to 1960s), their architectural and stylistic classification (late 1960s through 1970s), and recent endeavors to place them within the broader socioeconomic context of the Middle Bronze Age in the Levant (from late 1980s on).

1. The Indo-Europeans, the Hurrians, and the Hyksos

Nearly all of the earliest discussions concerning the origins of the massive earthen fortifications of the Middle Bronze Age in the southern Levant have sought connections with cultures to the north and east. Most often these associations were with Indo-Europeans, specifically the Hurrians with whom the Hyksos were once identified. This fixation was undoubtedly due in large part to the influence of W. F. Albright's earliest treatment of the topic where he stated that "there is no escape from the conclusion that the Hyksos came from a land of tumuli and earthen ramparts, that is, from the plains of Eurasia" (1922:123).

Despite a dearth of evidence for a direct connection between the Hyksos and such distant regions, Albright was led to seek sites and architectural parallels in "Russia and Central Asia." Yet Albright's bold assertion did not prevent his recognition of closer parallels for these fortifications. This particular aspect of Albright's conclusions, which acknowledged that "there was a gradual extension of the Mesopotamian art of fortification through Syria, toward the south, perhaps under the influence of fresh Amorite energy" (ibid., 136), has prevailed in scholarship to the present day. Had Albright realized the implications of his own observation he probably would not have continued to advance a "Central Asian" origin for rampart style fortifications (1926:252ff.). Nevertheless, he did so and called particular attention to a handful of massive enclosures discovered in Turkestan (modern Turkmenistan) at Ghiaur Kalah, which Raphael Pumpelly explored (1908:187ff.), and at Kirk Tepe. Although these enclosures, which measured 1,800 m by 2,100 m at Ghiaur Kalah and more than 300 m square at Kirk Tepe, featured thick walls of sun-dried mudbrick construction, preliminary consideration of the dates for their construction did not, even at the time of Albright's writing, support comparisons with ramparted sites in the Levant. The bulk of the evidence from Ghiaur Kalah, as suggested by coins and ceramics, were Seleucid and later in date.

Albright's attempt to draw parallels between these distant regions was not, however, without criticism. Harry R. Hall, for one, did not agree with Albright's suggestion and favored instead a less radical conclusion:

> Northern Mesopotamia is just as much a land of "tumuli and earthen ramparts" as is Transcaspia, and the Hyksos, if they are to be connected with these forts, need not have come from farther north than North Syria and Mesopotamia, where indeed the knowledge of earthen walls and mounds is more likely to have passed to Transcaspia than the reverse (H. R. Hall 1923:171, n. 1).

Nevertheless, Hall's comments were relegated to a footnote and perhaps for this reason elicited no response except from Albright (1926:252, n. 9).

Still, as G. R. H. Wright noted in his discussion of Albright's remarks between 1922 and 1932, Albright continued to maintain that people of Indo-European or Indo-Aryan stock were responsible for the introduction of this type of fortification to the Levant (1969:24f.). He remained convinced that the connections between these regions provided the most plausible explanation for the diffusion of rampart defenses into the Levant. Furthermore, William M. Flinders Petrie's conclusions based on his work at Tell el-Yehudiyeh (1908:3–10), as summarized by Albright, that "earthwork ramparts were intended to protect chariotry, which could not find any room at all in a typical Canaanite fortress or town" only reified the perceived cultural relationships (1935:224). Albright, therefore, continued to advance what he thought to be the evident progression of "rectangular earthworks" from "Central Europe or Transcaspia."

The picture that Albright painted from the 1920s through the 1930s was one in which chariot-borne Hyksos warriors armed with the composite bow descended on the Near East during the first half of the second millennium, bringing with them the Anatolian practice of using "sloping revetments in fortress walls" (1935:230). This hypothesis gained a stronger foothold as later studies such as Robert M. Engberg's *The Hyksos Reconsidered* asserted a Hurrian or Indo-

Iranian ethnic identity for the Hyksos (1939). This viewpoint remained unchanged until it was widely recognized that the material culture of the ruling elite at Avaris (Tell ed-Dab'a), the capital of the Fifteenth Dynasty, was southern Levantine (or Amorite) in character and not Hurrian.

In the broader context of European scholarship during the first half of the twentieth century, it is difficult to fault scholars who were attracted to what might be called the Indo-European model. Kathleen Kenyon, for example, found herself working closely with Mortimer Wheeler, whose thoughts on Indo-European migrations were well known:

> The Aryan invasion of the Land of the Seven Rivers, the Punjab and its neighbours, constantly assumes the form of an onslaught upon the walled cities of the aborigines. For these cities the term used in the *Rig-veda* is *pur*, meaning a 'rampart,' 'fort' or 'stronghold'. One is called 'broad' and 'wide'. Sometimes strongholds are referred to metaphorically as 'of metal'. Autumnal forts are also named: this may refer to forts in that season being occupied against Aryan attack or against inundations caused by overflowing rivers. Forts 'with a hundred walls' are mentioned. The citadel may have been made of stone; alternatively, the use of mud-bricks is perhaps alluded to by the epithet *ama* ('raw,' 'unbaked') . . . The recent excavations of Harappa may be thought to have changed the picture. Here we have a highly evolved civilization of essentially non-Aryan type, now known to have dominated the river-system of north-western India at a time not distant from the likely period of the earlier Aryan invasions of that region. What destroyed this firmly settled civilization? Climatic, economic, political deterioration may have weakened it, but its ultimate extinction is more likely to have been completed by deliberate and large-scale destruction (Wheeler 1947).

Wheeler's work, therefore, appears to have provided support for the suggestion that these features should be identified with an Indo-European migration. This is particularly apparent in the "Amorite Hypothesis" (see discussion below), a conceptual model best expressed by Kathleen Kenyon in which waves of armed Amorites invaded the Levant during the late third millennium (Kenyon 1966).

2. Chariots and Battering Rams

At the time of Albright's initial appraisal of Middle Bronze Age fortifications very few excavated sites in the Levant could be included in his discussion. Excavations at Tell Beit Mirsim (1926–32), Tell el-Far'ah South (1928–29), and Jericho (1907–09, 1930–36) in Palestine, as well as preliminary exploration of a handful of sites in Syria, and Tell el-Yehudiyeh (1905–06) in Egypt formed the corpus for comparison. However, in later years other excavations provided additional data necessary to test Albright's hypothesis. New projects included Tell el-'Ajjul (1930–34, 1938), Lachish (1932–38), Jericho (renewed from 1952–58), Hazor (1955–58), and Shechem (1956–73) in the southern Levant, Qatna (published 1935), Ugarit (since 1929), and Alalaḫ (1937–49) in the north, as well as Mari (beginning in 1933) in Mesopotamia. All of these projects contributed to the ongoing development of ideas regarding Middle Bronze Age fortifications in the Levant.

John Garstang's popular monograph on Jericho (1940) was the earliest publication from the excavations of these sites that addressed the functional aspects of earthen rampart and wall construction without reference to chariot riders from the north. There Garstang summarized the excavation of Jericho's rampart and revetment wall while making the following novel observations:

> While the whole conception of this new rampart was majestic, its construction was equally imposing. Solidly built of large, rough-hewn stones, laid and set with surprising regularity, it was provided with special features for defensive purposes. All around, about half way up its slope, a "string course" of stone projected slightly, in order to cause a missile launched from above to bound outward against the attackers . . . Nearer the bottom, again, where the size of the building stones was proportionately larger, one course in particular was constructed of enormous blocks each more than a ton in weight. This girdle was clearly designed to resist the battering ram (Garstang and Garstang 1940:91f.).

Nevertheless, Garstang's suggestion did not gain widespread popularity among archaeologists until Yigael Yadin repeated it and reiterated the ineffectiveness of chariots against fortification walls (see discussion below). Yet from this statement it is clear that Garstang was the first to identify the battering ram as one of the reasons for this style of fortification.

Despite Garstang's insightful observation, Albright's interpretation regarding Middle Bronze Age fortifications held sway. The Indo-European model propagated by Albright and Kenyon remained almost unchanged until Albright revisited the question in the first edition of *The Archaeology of Palestine* (1949). There he suggested more specifically that the "battered wall and glacis types of construction came from

Asia Minor to Palestine" (ibid., 89).[1] This was illustrated according to him by the earlier dates of the Middle Bronze Age fortifications at sites such as Alalaḫ and Ugarit in Syria (which it should be noted were within the northern sphere of Levantine culture and not part of Anatolia), although the fortifications of Ugarit would prove to be later than he was aware.

Albright's most insightful observations on the subject as a result of his continued thought on the subject pertain, however, to the characteristic number of gateways within the gate complexes of these settlements. These, Albright noted, citing examples from Mari, Shechem, and Megiddo, usually consisted of two or three doorways that were "flanked by a pair of massive piers" each of which was "the same size and . . . symmetrically disposed in two parallel alignments" (1949:89). He had identified the four and six-pier gate types of the Middle Bronze Age (see pp. 68ff.), though he continued to interpret these characteristics as "closely connected with the introduction of horse-drawn chariots" since the narrowest gates "were wide enough for a single chariot, and the widest ones known were spacious enough to permit two chariots to pass" (ibid., 89f.). Albright's identification of these features with the chariot reveals how deeply rooted was the assumption that the Hyksos, who were viewed as dominating the southern Levant during the Middle Bronze Age, were essentially Hurrian, and that by virtue of their descent as Indo-Europeans were the carriers of a unique tradition of fortification construction and chariot-based warfare.

In her early work Kathleen Kenyon also subscribed to Albright's hypothesis, although she was familiar with Garstang's work since she excavated Jericho from 1952 to 1958. Nevertheless, Garstang's thoughts appear to have left no impression on her and so she took the notion of chariot-based warfare and fortification systems one step further:

> This new type of constructing defences indicates a new type of warfare, for it is axiomatic in military science that new defensive methods are devised in reply to new methods of attack . . . and it seems very probable that the strengthening of the walls by these sloping defences is in fact against chariotry, which would thereby be prevented from rushing up to the foot of the walls (Kenyon 1952:71).

It was only following the work of Yigael Yadin that Kenyon's notion was revealed as flawed. Yadin observed that the "new methods of attack" to which Kenyon referred were, in fact, the direct result of the introduction of the battering ram:

> We must therefore conclude that the glacis of both types were built in order to protect the city walls against the enemy's attempts to break *through* them, as they eliminated at the same time the "dead area" so essential to this method of attack and . . . thickened the most vulnerable parts of the fortifications, i.e., the lower parts.
> These conclusions should have led us to the *prima facie* assumption that the new weapon which endangered the old method of fortification with vertical walls and against which a new kind of protection had to be invented, was some sort of *battering-ram* (Yadin 1955b:29f.).

Yadin was critical, therefore, of the idea that chariot forces would have been deployed against rampart fortifications since "the characteristic features of the chariots were essentially different from those of modern tanks," and thus "could not have been used for any purpose connected with direct attack on a fortified city" (ibid., 28). Although he argued that the fear of having walls breached, instead of their being scaled or undermined, was the primary motivation for the new type of defense, the strength of Yadin's argument came from his incorporation of textual evidence from Mari, which attested to the use of the battering ram (for discussion of the battering ram and siege pole, see pp. 39f.).[2] The extent to which Yadin's assertion was accepted is evident in Kenyon's eventual abandonment of her interpretation and subsequent adoption of Yadin's for Jericho's defenses (1979:165f.).

3. The Amorites and Hyksos

In the same article in which Yadin had suggested that the introduction of the battering ram was the main reason for the construction of Middle Bronze Age rampart fortifications, he also proposed the first clear typology for these features (1955b:25ff.). Yadin distinguished two main types of fortifications with different functions, although they employed similar construction techniques. The first type he defined as "fortified camps" or "enclosures," terminology that had already been used by Albright and others (see discussion of freestanding ramparts on pp. 48f.). These camps were usually large rectangular enclosures surrounded by huge earthen ramparts described as built of *terre pisée* (Fr. "beaten earth"). (For a discussion of why this term

[1]In a footnote in his publication of T. Beit Mirsim Albright acknowledged that his Indo-Iranian origin for rampart fortifications could not be proved (1938b:28f., n. 2).

[2]Yadin cited ARM I 131 and 138, and II 7, *The Siege of Uršu*, and illustrations of siege warfare from the Beni Hasan tombs (1955b:30f.).

should not be applied to earthen ramparts, see p. 50) Their function Yadin asserted, following Albright, was to protect chariot forces from attack. Nevertheless, excavations such as those at Hazor would later demonstrate that the large enclosed areas within the ramparts were, in fact, lower towns and not merely "enclosures" for chariotry.

The second type of fortification, which Albright referred to as "beaten-earth fortified towns," were employed to defend existing tells (see discussion of supplemental ramparts on pp. 49f.). Two predominant construction techniques were identified with this type: the *terre pisée* technique, adopted from the techniques used in rampart construction, and the construction of battered stone walls (ibid., 26). (The stone walls were, however, not a sole means of defense but merely revetment walls located at the foot of the rampart; see p. 55) These two techniques were thought "to add further protection to the already existing fortifications by thickening their lower parts and exposed slopes" (ibid., 28). While both of these techniques were sometimes used together as at Jericho and Lachish, Yadin observed that most often only one technique was employed.

Yadin suggested that inhabitants of "beaten-earth fortified towns" borrowed techniques such as *terre pisée* construction from the inhabitants of "fortified camps" (1955b:26f.). He also concluded that the introduction of battered stone walls indicated that *terre pisée* construction "proved inadequate in the process of developing the weapon against which the glacis were built." Unfortunately, because Yadin was convinced that such fortification types were unknown during the MB IIA, his typology was only relevant for the MB IIB–C. Despite his thorough treatment of warfare in *The Art of Warfare in Biblical Lands* (1963) and his excavations at Hazor, Yadin never modified this typology.

The nuances in terminology employed to describe Middle Bronze Age fortifications were and remain significant. Years after Yadin's typology had been published, Kenyon, for example, intentionally avoided use of terms such as *glacis* and *fosse* because they were borrowed from the study of medieval fortifications. She expressed concern that they created a false certainty about the function of these elements of the fortifications (1966:67). Nevertheless, she did not abandon the use of the equally evocative term *rampart* to refer to the mounded earthwork defenses surrounding Qatna, Khan Sheikhoun, Ugarit, Kar-Kemiš, Hazor, Gerisa, Jericho, 'Ajjul, and Lachish (Kenyon 1966:67–72).[3] Eventually, however, she abandoned this line of argument and freely employed the term *fosse* (for Hazor, see 1973:100), though the appropriate terminology for these features would only be agreed upon following Jacob Kaplan's work (1975), as discussed below. Kenyon also asserted along with Albright that one could "with some certainty ascribe to the Hyksos" this type of defense. While an overview of these various terms may seem pedantic, arguments over terminology reveal the need for and gradual emergence of a glossary for these features, their function, and their historical contexts (for terminology suggested in this work, see chapter 3).

4. Exogenous Origins Deconstructed

Kenyon's publication of *Amorites and Canaanites* (1966) marked a milestone in attempts to identify the cultural origin of Middle Bronze Age fortification techniques. In this work she posited two distinct phases of Amorite ingress into the Levant, which correspond with the start of both the EB IV and MB IIA, and specific destruction levels found at sites throughout the Levant. Although many scholars had more or less adhered to an invasion model even before Kenyon's publication, none had ever articulated the argument in such a nuanced manner. Yadin's suggestion regarding foreigners from the north living alongside the native population can be viewed, for example, as an early argument of this type (see previous section). However, Yadin's argument associated the introduction of "fortified camps" with the arrival of the Hyksos at the start of the MB IIB. Together Yadin and Kenyon's notions accounted for three separate invasions, each of which was accompanied by cultural changes: at the start of both the EB IV and the MB IIA, and at the start of the MB IIB–C with the emergence of a new style of fortifications associated with the Hyksos. (It had not yet been demonstrated that the Hyksos were of Amorite extraction.) Beginning with Kenyon's work, therefore, the various rampart style fortification techniques attested were being increasingly associated with an Amorite-Hyksos cultural *koiné* of exogenous origin.

It was not, however, until Peter Parr's article, *The Origin of the Rampart Fortifications of Middle Bronze Age Palestine and Syria* (1968), that the issue of the origin of rampart fortifications was addressed separately. Parr challenged what he viewed as the

[3]Kenyon, like Albright and others, wrongly identified T. el-Yehudiyeh as possessing earthen rampart fortifications (1966:70). Nevertheless, her conclusions concerning the eventual spread of Amorite culture into Egypt are no less significant given what we know today from excavations at T. el-Yehudiyeh, Mashkuta, and Avaris.

baseless identification of these features with exogenous groups.[4] While he noted that Albright had seemingly changed his opinion on their origin, no longer suggesting an "Indo-Aryan" origin for fortifications though maintaining that they originated in the north, he desired to rectify the influence that Albright's speculation about their northern origin continued to have upon scholars like Kenyon and Yadin. He wanted, for example, to address the misconception propagated by Yadin that the techniques used in later rampart-fortified settlements were adopted by the inhabitants of older tell settlements since direct evidence to support this suggestion had never been presented.

Parr felt that Yadin's suggestion that *terre pisée* techniques preceded battered stone walls was "highly dubious" (1968:20). Although Yadin's argument was evidently based on excavations by Albright at Tell Beit Mirsim, further clarification of the Middle Bronze Age defenses at Lachish and Jericho has since demonstrated that the chronology proposed by Yadin for these features was incorrect (see pp. 11f.). Furthermore, Parr observed that the *terre pisée* technique was, in fact, often used contemporaneously with the battered stone glacis and that "conditions and requirements no doubt dictated the type [of defense] which individual cities adopted" (ibid., 21). A site-by-site analysis of a limited number of Early and Middle Bronze Age fortifications in the southern Levant and an examination of Anatolian forebears all appeared to support his reevaluation and suggested several new conclusions. For example, Parr concluded that Middle Bronze Age defenses were not radically different from those that preceded them during the Early Bronze Age, though during the Middle Bronze Age they were built on a much larger scale.

> The Early Bronze Age architects had seen their cities rising ever higher on their autogenous mounds, and had been faced with the problem of how to deal with the vulnerable slopes . . . Their natural surfaces, perhaps baked hard in the summer heat, would have been scoured and scarred by winter storms, when veritable gullies would have been cut into them by run-off waters . . . The result would have been slopes which, even in time of peace would have weakened the foundations of the walls of their crest, and which in time of war would have provided a would-be attacker with an easy approach, where every irregularity gave cover. The architects of the Early Bronze Age in Palestine had met with this problem in a variety of ways, each of which was designed to consolidate and regularize the embankment. Their Middle Bronze Age successors, faced with the same problems, adopted much the same remedies (Parr 1968:43).

Parr also remained unconvinced that the introduction of the battering ram served as the main motivation for the emergence of earthen rampart fortifications, as this weapon was already in existence during the Early Bronze Age, a fact confirmed by Piotr Steinkeller (1987). Parr, for example, cited the Egyptian illustration from a third-millennium tomb at Deshashe as proof of its early existence. Parr concluded, therefore, that:

> It is unlikely that the designers of the Middle Bronze Age fortifications had any single method of attack in mind when they erected their steeply inclined ramparts; they merely knew from experience that by so doing they were making all the more difficult the task of any enemy seeking to scale, to undermine, or to breach their city walls. It is by an evolutionary interpretation of its origin that we can best understand the function of the MB glacis (Parr 1972:43).

What remained unclear, however, was the chronological distribution of Middle Bronze Age fortified settlements between the northern Levant and the south. Concerning this Parr noted some time later that "the evidence does not permit a refined or trustworthy chronology" (1972:44). Despite evident gaps in Parr's list of Middle Bronze Age sites in the Levant and the absence of any consideration of possible Mesopotamian origins for these features, his treatment was limited; detailed information concerning some of the most important sites had not yet been made available. Nevertheless, while his conclusions may now be viewed as tentative, they have served as the basis of subsequent discussions regarding Middle Bronze Age fortifications.

Parr did not address in his work the commonly held view, since the pioneering work of Albright, that Tell el-Yehudiyeh and nearby Heliopolis in the Egyptian Delta featured "Hyksos" style fortifications during the Middle Bronze Age. Yet in a nearly contemporaneous publication G. R. H. Wright addressed this question. He conclusively demonstrated that earthen embankments at both of these sites were

---

[4]Parr's corpus included the following sites in the southern Levant: Jericho (T. es-Sultan), Lachish (T. ed-Duweir), T. el-Far'ah South, Shechem (T. Balatah), Hazor (T. el-Qedah), T. el-'Ajjul (Sharuhen?), Ta'anach, Megiddo (T. el-Mutesellim), T. Poleg, T. el-Far'ah North, and Ai (et-Tell); and in the northern Levant Alalaḫ (T. Atchana), Kar-Kemiš (Jerablus), and Ebla (T. Mardikh). Conspicuously absent from his catalog were Byblos, Ugarit, Mari, and a group of third-millennium sites identified as Kranzhügeln in eastern Syria, for which publications were available at the time of his writing.

neither constructed during the Middle Bronze Age nor as fortifications (1968). They were, he argued, essentially the remnants of a type of ancient Egyptian construction, the topography or signature landscape of which was similar following centuries of erosion. He would later add that Egyptologists commonly identified such features as "Sanctuary Mounds or Sanctuary Enclosures" (1969:33), and, ultimately, his observations emphasized the lack of data from the Egyptian Delta in support of an Egyptian origin for this type of fortification.

In addition to the relevance of such Egyptian sites Wright addressed archaeological evidence from Iran, which Albright had suggested demonstrated that rampart fortifications were introduced from this region to the Levant (1969). Wright's conclusions were clear, if guarded:

> . . . geographically speaking reference has been made to features in Persia, Transcaspia and the Eurasian plains. However, in the three sites specified from these regions, neither published information nor inspection reveals anything beyond the square form of the enclosure and the fact that the enceinte is of mud or mud brick, which serves to relate them closely to Syrian or Egyptian sites.
> That is, the archaeological facts here brought together do not suffice to determine that there was one type of earthen walled enclosure common to Transcaspia, Persia, Syria and Egypt. Still less do they serve to establish that this enclosure was a fortified camp built by a certain group of people of a defined ethnic origin (Aryan) who played a certain role at a given historical juncture (G. R. H. Wright 1969:33).

Despite these remarks, Wright cautioned that an investigation was required of "a sufficient number of examples [of fortifications], sufficiently closely to determine if a distinct historical or structural category" was relevant (ibid., 33). Nevertheless, no attempt to connect rampart fortifications in the Levant with ancient Iran has been made since Wright's treatment. This is due, no doubt, to the inadequate evidence available from Iran, but also to the recognition of a multiplicity of reasons for the construction of such fortifications.

5. Amorite Fortifications Redux

A number of scholars challenged the assertion by Parr and Wright that rampart fortifications were constructed by indigenous groups in direct response to threats posed by new developments in siege weaponry. Paolo Marrassini (1971), for example, in a little-known article entitled *Sui «campi fortificati» nell'età di Mari* challenged the evidence presented by Parr by calling attention to archaeological evidence from Syria and Mesopotamia. According to Marrassini this evidence, predominantly from the late third and early second millennia, suggests a relationship between fortifications in Syria and Mesopotamia that had been overlooked by Parr and Wright. He notes that while abundant Old Babylonian texts appeared to support the identification of a tradition of earthen rampart construction in Mesopotamia, the corresponding archaeological evidence derived primarily from the southern Levant. Marrassini also suggests that a chronological and stylistic connection exists between the late third-millennium fortifications of Byblos and those of northern Mesopotamian sites like Tell er-Rimah and Tell Khoshi (ibid., 122). Nevertheless, his own assertions were never adequately tested.

With a slightly different approach Jacob Kaplan, director of the excavations at Jaffa (1955–74) and Yavneh-Yam (1966–1969), also sought to address the issues raised by Parr by establishing a typology of Middle Bronze Age fortifications (1975).[5] Kaplan distinguished two main types of earthen ramparts by their relationship to walls, namely the "wall rampart" and the "freestanding rampart." He defined the wall rampart as "a girdle of earth deposited in front of the wall" (as at Tel Poleg) for the purpose of protecting it against erosion and "enemy attempts to undermine it." The freestanding rampart, on the other hand, Kaplan concluded, "originated with the embankments of the ancient irrigation networks" like those in southern Mesopotamia and featured no wall superstructure (ibid., 2).[6] Kaplan's concern was primarily with the internal composition of these fortifications rather than their overall form or their effectiveness.

Although often the terms *glacis* and *rampart* had been used interchangeably, Kaplan's observation facilitated the distinction of these separate morphological features of rampart fortifications. Kaplan noted that the term glacis should be employed to refer specifically to the "rampart slope or other escarpment, including the slope of the tell" built for the specific purpose of serving as an impediment to approach. This observation came, therefore, as the study of Middle Bronze Age fortifications increasingly struggled to properly classify and identify their features. (For a discussion of these terms, see pp. 48ff.)

[5]This article is a translation of a lecture that Kaplan presented in 1971 and later published in Hebrew (1972b).

[6]Kaplan's early thoughts on this are revealed in an article where he attempted to identify connections between Middle Bronze Age material culture in Palestine and that of Mesopotamia (see 1971:esp. 306).

Terminology aside, Kaplan unequivocally clarified the association between the construction of freestanding ramparts and the arrival of the Amorites:

> The appearance of the freestanding rampart is linked by us with the beginning of the spread of the Amorite tribes, throughout Syria and Palestine. Our theory is that the Amorites were the first to build earth rampart enclosures which in plan resemble Roman military encampments, and that these fortifications were inspired by similar fortifications in Mesopotamia (Kaplan 1975:9).

Kaplan suggested, building upon an argument advanced by Jean Bottéro (see Posener et al. 1971), that the inspiration for rampart construction could be traced back to the building of *Muriq-Tidnim* (lit., "keeping Tidnum [Amorites] at a distance"), which was built by the Ur III king Šu-Sin (ca. 1941 B.C.). Kaplan postulated that the wall was probably built upon "parts of existing levees of the main irrigation canals and only added the needed links between them to complete their continuous defence line" (1975:12). While more evidence is needed to test this hypothesis, Kaplan's claim that the builders of Shechem's triple gate (dated to MB IIC) used the triple gate of the court of Nannar in the sacred enclosure at Ur as a model for its construction was extremely conjectural, given the disparity in time and distance between their constructions, as Barbara Gregori has demonstrated (1986:96f.). Despite his many novel ideas, the impact of Kaplan's contributions on this subject has been limited by a lack of publication of his excavations at Middle Bronze Age sites such as 'Avdon, Yavneh-Yam, and Jaffa.

By the early 1980s, according to William G. Dever, three main interpretations of "fortified city-states" of the Amorites in the southern Levant during the Middle Bronze Age could be identified (1985:73). These included that they functioned (1) as "a defense against the more powerful Amorite/Canaanite city-states to the north in Syria," (2) as "strongly-defended regional centers in the internecine war . . . among petty princes of the 'Hyksos' who ruled Palestine itself," or (3) that they were "base-camps from which to launch Asiatic campaigns into Egypt, as well as 'back-up systems' in case these were repelled."[7] Dever himself suggested that the third option best accounted for the archaeological data, despite its limitations.

6. The Role of Fortifications

During the early years of excavation in the southern Levant, Middle Bronze Age fortifications were regarded as having served to repel either chariotry (Kenyon 1952:71) or the battering ram (Garstang and Garstang 1940:91f.; Yadin 1955a). However, following Parr's work (see esp. 1968:43) a widely accepted consensus emerged that the entire complex of fortifications, including the fosse, rampart, wall, and gates, were part of a multifaceted defensive strategy. The strategy was intended to defend against each of the means known for entering towns as observed by Yadin (1963:17): over, through, or under walls. Suggestions by various scholars have sought, therefore, to shed light on how ramparts served more than a single defensive purpose.

Among many considerations was the recognition that the glacis may have functioned to deter sappers from undermining town walls (Lagarce 1984:n. 2; Stager 1991b).[8] Similarly, Jonathan Tubb suggested that rampart fortifications were intended to keep archers and javelin throwers at a distance (1985:193f.). Others have more specifically noted the threat posed by archers who could fire burning arrows into a town (Miller et al. 1986:191). Added to this growing list of defensive considerations was the essential possibility that glacis and ramparts were not primarily military innovations, but rather served simply to protect the slopes of the settlement against erosion (Parr 1968:43; Pennells 1983; Lagarce 1984:159). (All of these ideas are discussed in chapter 2.)

Out of the multiplicity of opinions regarding the purpose of Middle Bronze Age fortifications, the contributions of one scholar in particular have shaped much of the debate. Building upon his experience from the excavation of the Middle Bronze Age gate and rampart at Ashkelon during the 1990s, Lawrence Stager formulated several hypotheses concerning Middle Bronze Age fortifications. One suggestion was that the fosse was integrated into Middle Bronze Age rampart design for the explicit purpose of preventing sapping of the town wall by attacking forces (1991b). Stager argues that the fosse at the foot of the rampart guaranteed that the sappers had to undertake to dig through bedrock before gaining any cover from the arrow-firing defenders on the walls above the rampart.

[7] Although Dever's identification of Middle Bronze Age settlements in the southern Levant as "city-states" may have inclined many scholars toward either of the first two interpretations, Dever did not provide a detailed defense of his characterization (for a discussion regarding whether to call them city-states or kingdoms, see pp. 119ff.).

[8] In his examination of defenses constructed at Ras Ibn Hani and Ras Shamra Lagarce observes, "Son rôle militaire est de protéger le pied des murs contre l'action des béliers et des sapeurs, et aussi de permettre à un défenseur posté au sommet de l'édifice de faire ricocher sur l'assaillant un projectile pondéreux lâche verticalement" (1984:n. 2).

Elsewhere Stager addresses the subject of Middle Bronze Age ramparts and posits that sites with "free-standing ramparts and dry moats . . . derived ultimately from third-millennium settlements in the Baliḫ and Ḫabur river basins" at settlements referred to as Kranzhügeln or "wreath-shaped mounds" (1999:237). Although Stager never developed the notion in detail, he does therein make two important observations. First, he notes that the general shape of Kranzhügeln was characteristic of Middle Bronze Age rampart fortifications. Second, he demonstrates that the technology used in the construction of early examples of mudbrick vaulted arches originated in northern Mesopotamia as early as ca. 2400 B.C. and was only later incorporated into Middle Bronze Age gate structures. Stager's argument drew support from a general consensus that both urbanism and fortification systems were reintroduced at the start of the Middle Bronze Age from the northern Levant and Mesopotamia following a deurbanized interlude in the southern Levant during the EB IV (for a discussion of the chronology of fortification development, see pp. 78ff.).

7. History, Chronology, and Typologies

Since Kaplan's efforts a number of other scholars have also attempted to place the development of Middle Bronze Age fortifications in the Levant within their historical context. Joe Seger sought to provide one such framework in a review of the MB IIC fortifications of Shechem and Gezer (1975). Seger's analysis of these two southern sites, which experienced different building programs but appeared to follow a "common pattern of city development" (ibid., 42), led him to suggest a three-stage development of late Middle Bronze Age fortifications in this region. The first stage consisted of "an era of gradual settlement and assimilation" from about the late MB IIA through MB IIB (ibid., 45). During the first part of the second stage towns underwent rapid "redevelopment and expansion" as a result of the establishment of the Hyksos in Egypt [early MB IIC]. This was followed by "refortification and energetic development, in response to the resurgence of Egyptian power" with the rise of Dynasties XVII and XVIII toward the end of the MB IIC. The final phase of development occurred thereafter at the start of the Late Bronze Age with the "gradual dissolution of Hyksos power" and the fall of their centers in northern Egypt.[9] Nevertheless, precise correlations between the archaeological evidence from other excavated sites and Seger's historical development remain wanting.

A very significant debate concerning the chronology of Middle Bronze Age fortifications emerged from the publication of the preliminary reports on the excavations of Aphek (1972–1976) by Moshe Kochavi in the mid-1970s (1974; 1975). The revelation that the fortifications of Aphek dated to the MB IIA incited Yadin (1978) to reply that Aphek could not be dated to the MB IIA since the fortifications were clearly of MB IIB–C style.[10] Yadin's argument was, of course, based on the conviction that MB IIA sites were not fortified and, furthermore, that most of the large, walled settlements of the MB IIB–C (like Hazor, Yadin 1972b) had not been occupied during the MB IIA (i.e., Jericho, Hazor, Yadin 1978:2ff.). This conviction was further supported, he argued, since Stratum XII at Megiddo dated to the MB IIB–C (ibid., 2), rather than to MB IIA as it is dated today (see A. Mazar 1990:196). Nevertheless, it is evident that Yadin's arguments were premature since they were based on the limited information available for the MB IIA date for Aphek's fortifications, which had at that time only been presented in preliminary reports. Moshe Kochavi, Pirhiya Beck, and Ram Gophna (1979) noted that four sites in the coastal plain with MB IIA foundations also featured fortifications: Aphek, Poleg, Burga, and Zeror. This evidence convincingly refuted Yadin's rigid MB IIB–C date for all massive Middle Bronze Age fortification systems and revealed, therefore, that the earliest rampart fortifications in the southern Levant were indeed of MB IIA date.

Interestingly, in the same issue of *ZDPV* in which Yadin's 1978 article appeared an essay addressing the Middle Bronze Age fortifications of Megiddo argued for their construction during the MB IIA. On the basis of a comparison of the fortifications of Megiddo XIIIA and Egyptian Middle Kingdom cataract forts, the limited evidence for Egyptian material culture discovered at Megiddo, and the evidence for the raising of cattle in Retenu, Amos Harif (1978) suggested that Megiddo functioned as an "Egyptian stronghold" in the southern Levant during the early Middle Bronze Age. However, the crux of Harif's argument rested upon the identification of what was known as "Wall Street" in Area BB. Since this identification has been abandoned the direct parallel Harif sug-

[9]Seger's emphasis of the break between the first and second phases in this development served to reify the position of those who have noted the cultural differences between the MB IIB and IIC.

[10]An earlier version of this article was published in Hebrew (Yadin 1977).

gested with Middle Kingdom forts in Upper Egypt can no longer be supported.

8. Evidence from the Northern Levant

Although parallels had been drawn between large fortified Middle Bronze Age settlements in the southern Levant and those in the north, insufficient publication of the excavations of Alalaḫ, Qatna, Byblos, and Ebla meant that the relationship between these regions remained mostly conjectural. It was not until 1955 that Sir Leonard Woolley provided the first published results of excavations of late Middle and Late Bronze Age defenses of Alalaḫ (1936–39, 1946–49) that such data first became available (Woolley 1955). Until that time scholars had merely cited what amounted to no more than reconnaissance of the ramparts of Qatna, where du Mesnil du Buisson had excavated from 1926 to 1929 (1935a). Similarly, while preliminary reports were made available during the 1950s following excavation of the fortifications of Byblos (see pp. 192ff.), the French mission has not provided final reports for these excavations with plans and section drawings since Dunand's death.

Until recently equally meager information had been provided by the Italian expedition that has been excavating the fortifications at Ebla (Tell Mardikh; see pp. 198ff.). Although excavations began in 1964, only in 1994 did the mission continue with intensive exploration of the site's fortifications in what Matthiae has since referred to as the sixth phase of archaeological exploration of Ebla (1997c:3). By the time of the publication of *Ebla: un impero ritrovato* (Matthiae 1977a), the mission had already excavated the remains of two of Ebla's four Middle Bronze Age gates (in Areas A and L), and Fortress M, a bastion built into the eastern ramparts north of the southeast gate. Not surprisingly, following the English translation of this work (Matthiae 1980b), archaeologists began once again to consider the extent to which such features were indeed northern Levantine in origin.

> The earthwork fortifications of Ebla are perhaps the most imposing example of a type of city wall, well attested and much discussed . . . it is a fact that several important tells of Northern Syria had fortifications like these during the Middle Bronze on a particularly grand scale. It was certainly the case with Qatna to the south and Karkemish to the north, to name only two of the big cities in which owing to their large surface area the circuit of the ramparts on the surface remains very obvious. But in many small cities the same defensive system must have been used. The date of the ramparts of Ebla, which as we have seen, must have been erected about 1950 B.C., is a sure indication that fortifications of this type were characteristic of the first emergence of the great urban culture of the Archaic Old Syrian Period (1980b:199f.).

Several studies in particular owe much of their value to the results of the archaeological work at Ebla. Among these is an important contribution on triple-entry or Syrian-style gates of the Middle Bronze Age in the Levant and northern Mesopotamia by Barbara Gregori (1986). Her findings along with those summarized in a work on town gates by Ze'ev Herzog (1986:esp. pp. 37–88), published in the same year, have been widely accepted as the definitive typologies of Middle Bronze Age gateways (for current discussion, see pp. 68f.). Additionally, Gregori's conclusions (1986:95), which were based upon the established MB IIA dates provided for the gates at Ebla and Tell Tuqan excavated by Matthiae, also substantiated Parr's (1968) hypothesis concerning the northern origin of rampart fortifications. More recent treatments such as *The Urban Landscape of Old Syrian Ebla* by Frances Pinnock (2001) have also made it possible to further understand defenses within the context of large urban Middle Bronze Age centers such as Ebla.

9. Post-processual Approaches

Although the observations and hypotheses discussed above reveal the widespread consideration that Middle Bronze Age fortifications received during the 1970s and 1980s, toward the end of this period a new school of scholarship emerged. These scholars were primarily motivated by a desire to expose the relationship between the construction of fortifications and the society and economy of the Levant during the Middle Bronze Age. Work by these scholars, therefore, focused specifically upon the socioeconomic context of these structures while drawing upon the results of more recent excavations at sites such as Tel Michal (1977–80) and Shiloh (1981–4). Publication of these new hypotheses were mostly confined to a single issue of *Tel Aviv* (1992, vol. 19:2) in articles by Israel Finkelstein (1992), Shlomo Bunimovitz (1992), David Ussishkin (1992),[11] and Ram Gophna (1992a),[12] but also include more recent reassessments of previous excavations of fortification systems by Finkelstein (2002).

The views of this school centered upon a reaction against traditional, functionalist interpretations like

[11] For original article in Hebrew, see Ussishkin (1991).

[12] This is a revised version of his original article on the Middle Bronze Age rampart at Tel Poran published in Hebrew (1977).

those discussed above, which viewed these types of earthworks as fortifications primarily constructed with defense in mind. These scholars remained unconvinced that earthen rampart fortifications even featured fortification walls such that they should be identified as defensive systems. The inauguration of this seemingly new approach can, however, be traced back twenty years to Kaplan's original work on the typology of Middle Bronze Age fortifications (see discussion above), wherein he had suggested that walls were only present with "wall ramparts" but not atop "freestanding ramparts" (1975). Such thinking may also have influenced Rivka Gonen's assertions regarding the limited evidence for Late Bronze Age fortifications (1984). Nevertheless, as will be discussed below, such conclusions have become increasingly tenuous in light of a growing body of archaeological evidence.

Israel Finkelstein has sought to incorporate three particular aspects that had not been previously addressed into his study of Middle Bronze Age fortifications (1992). These included (1) the integration of excavation data for Middle Bronze Age fortifications with survey information relating to settlement patterns, (2) the differentiation of socioeconomic and political conditions at sites in different regions (e.g., between the highlands and lowlands in Israel), and (3) comparisons with political formations and the public building projects in which such polities were involved in earlier and later periods. Though he intended that these lines of evidence would demonstrate the "various purposes" for which Middle Bronze Age earthworks were constructed, Finkelstein's proposed approach contrasts markedly with the very terse discussion he provides of relevant discoveries at highland sites (e.g., Shiloh, Shechem, Hebron, and Beth-El), as well as those in the lowlands (e.g., Dan, Akko, Kabri, Ashkelon, Yavneh-Yam, Mevorakh, and Jericho), and his brief presentation of the survey data. Yet, since many of the sites discussed by Finkelstein lacked final reports (e.g., Ashkelon, Beth-El, Dan, Hebron, Kabri, and Yavneh-Yam) and because he omitted detailed data from other published sites (e.g., Hazor), Finkelstein's observations can only be regarded as a preliminary hypothesis.

Shlomo Bunimovitz (1992), who like Finkelstein has also sought to understand Middle Bronze Age fortifications as a sociocultural phenomenon, also downplays the defensive aspects of these fortification types. Although he has left open the possibility that they may have functioned as defenses, he suggests that their diffusion was part of a process of "competitive emulation between neighbouring tribal groups in the country during MB IIA . . . as a vehicle to enhance social solidarity within the various groups and to reinforce the tribal leaders' status" (ibid., 228). Bunimovitz's study was the result of a desire to return to an understanding of "the cultural mechanism by which the idea of building earthen ramparts around settlements was transferred to Syria from Palestine" (ibid., 223). This statement reveals the recognition that such features occurred earlier in Syria even if interpretations of their function and significance vary. Nevertheless, the "cultural mechanism" Bunimovitz identified was a social phenomenon since he had accepted Dever's observation that "formal and functional considerations" of the type familiar from earlier studies were "too general and simplistic" to explain the basis for the construction of Middle Bronze Age ramparts.

Bunimovitz's study was built upon earlier work by Herzog, Biran, and more recently Finkelstein, each of whom considered the amount of labor required in the construction of these features. In Bunimovitz's opinion, such studies demonstrated that a "strong central government capable of organizing and deploying masses of manpower" would have been required to build such fortifications (1992:224). Still, Bunimovitz's own work assumed a priori that the figures for labor provided in the works of earlier scholars were reliable. (For an assessment of such figures and further discussion of Bunimovitz's conclusions, see chapter 5.)[13]

Within the framework of this socioeconomically focused approach, Ze'ev Herzog constructs a typology of Middle Bronze Age settlements that incorporates evidence for Middle Bronze Age fortifications (1997a:268ff.). He develops this classification based on the premise that MB IIA urbanism was "not sud-

[13]Bunimovitz, like Herzog and Finkelstein, accepts the conclusion that city walls were not built atop Middle Bronze Age ramparts (1992:227). He argues that "it is far easier and quicker to dump earth than to engage hundreds of skilled workers and stone-masons in quarrying, transporting and laying stones for wall foundations, and preparing the enormous quantity of bricks required to build their superstructure" (ibid., 224), although such efforts were employed in the construction of core walls within ramparts, as demonstrated in chapter 3. This argument, which relies upon the typology proposed by Jacob Kaplan, seems contradictory since the main thrust of Bunimovitz's argument is that earthen ramparts, "monumental architecture" built as products of "conspicuous consumption," were intended "to enhance social prestige and power" (ibid., 225). Yet if one of the main goals of constructing ramparts was conspicuous consumption, and building mudbrick walls required even more labor—and everywhere else in the Near East massive mudbrick structures were regarded as evidence of royal power—why should the construction of earthen ramparts, without mudbrick walls, have been preferred?

denly introduced" to the southern Levant by "an urban population from Syria," but that the development was endogenous and was the result of a takeover "by an ascendant political-military ruling class." Herzog characterizes the earliest phase of settlement, the MB IIA as "not urban in character" suggesting, instead, that the data for sites founded and fortified in this period support the interpretation that farmers and pastoralists who comprised the "autochthonous population" (presumably a reference to the EB IV population), reintroduced town life as they moved onto "raised mounds to profit from the natural protection afforded by their steep slopes" (ibid., 268f.). Nevertheless, his interpretation overlooks a number of large sites founded in this period, such as Burga, Shimron, and Ashkelon, which feature no prior occupation (i.e., EB IV) and provide no basis on which to posit a clear connection between their Middle Bronze Age population and an indigenous EB IV population. While Herzog acknowledges that it is "problematic to pinpoint the exact circumstances that generated this process of urbanization," he ascribes this complex process to "a new type of elite," who were responsible for the "erection of urban institutions" (ibid., 269). Although not explicit in his argument, presumably these elites were Amorites.

From this theoretical framework, Herzog identifies five "structural types" of Middle Bronze Age towns based on their topography, plans, and fortifications. These five types included (1) *rampart cities* "covering more than 10 ha," (2) *fortified cities* protected "by walls and glacis, usually 3 to 7 ha in size," (3) *extended cities* "located on mounds enlarged by earthen ramparts," (4) *embankment cities* "located on a plain or low hill and fortified by a city wall and outer glacis," and (5) *girdled cities* "fortified during the MB IIA and succeeded by unfortified cities in the MB IIB" (ibid., 269). Unfortunately, two major problems with this typology are immediately apparent.[14] First, while Herzog recognized that during the MB IIA "occupation was not urban in character," his typology was, nevertheless, intended for the entire Middle Bronze Age. Yet, several of the types of sites he identified only existed during the MB IIB–C. Furthermore, the typology is also restricted to the southern Levant and makes no attempt to account for developments in the northern Levant.

The second problem with Herzog's typology is that his criteria are insufficiently exclusive, making it possible that a site could be included in more than one classification. This pertains specifically to the use of size as a criterion for the first and second site types, even though it is possible that sites of a similar size could be included among the last three categories he identifies. For example, Timnah, which Herzog classifies as an "extended city" (1997a:160ff.), could also be identified as a "rampart city" in miniature, built in the classic style with a core wall and built-up rampart like Hazor, which Herzog does include in this category. Also, several of the so-called "rampart cities" that Herzog identifies, such as Hazor, Haror, and Tel Dan (1997a:269), should probably be identified as "extended cities," as these sites featured some occupation prior to the construction of their MB IIB–C ramparts, though the nature of this occupation has yet to be clarified. Finally, Herzog's criteria for "MB cities" does not address unfortified settlements and, therefore, this typology does not address under what conditions sites were fortified. In short, the validity of Herzog's typology is undermined by mixed categories and an inconsistent reliance on chronological data.

Two important but incorrect assumptions are inherent in the approaches reviewed above. The first of these assumptions is that a lack of evidence exists for fortification walls at sites that have been considered characteristic of Middle Bronze Age fortification strategies in the southern Levant (see pp. 63f.). The second assumption is that the purpose of these earthworks must have been more than, if not other than, defense since the construction of ramparts was an extremely labor-intensive process. (However, as demonstrated in chapter 5, various parallels for the construction of earthwork fortifications reveal that such construction projects required only a fraction of the labor previously suggested.) Nevertheless, these basic assumptions have led to the inaccurate but now widely accepted conclusion, as exemplified by Herzog's statement:

> It has been irrefutably demonstrated that in not even a single instance did earthen ramparts carry or incorporate a fortification wall. (1997a:132f.)

Despite the absence of any comprehensive treatment of Middle Bronze Age fortifications, such assertions have also begun to find acceptance among archaeologists working in Syria. Based on his recent excavation of large sections of the fortifications of Ebla, Paolo Matthiae, for example, recently suggested that functional explanations advanced by Kenyon (1952:71), Yadin (1955b) and Stager (1991b) do not adequately account for the discoveries at Ebla since 1995 (2000:n. 77). Unfortunately, Matthiae does not provide details that demonstrate how this is so, while noting that the premise of the Tel Aviv school's position should not be dismissed since it accounts for socioeconomic and ideological considerations:

[14]Dag Oredsson also observes this problem (2000:51f.).

> Toutefois, il faut retenir les récentes théories plus complexes qui considèrent les remparts dans un cadre socio-économique et idéologique plus élaboré (ibid., 601, n. 77).

In a similar attempt to focus on socioeconomic aspects, Harvey Weiss has recently suggested that "massive city walls were constructed to control labor" during the Akkadian period at sites such as Šubat-Enlil [Tell Leilan] (2000:85). However, no direct correlation between the presence of Akkadian imperial power and the widespread attempt to control labor by limiting the mobility of urbanites has been demonstrated. Nor has it been demonstrated to what extent such walls effectively kept people within these settlements throughout this period.

Despite evident flaws among various attempts to address the socioeconomic concerns relating to Middle Bronze Age fortifications, only Amihai Mazar has challenged these perspectives, notably in his publication of the Middle Bronze Age fortifications at Timnah (1997b). Although he agrees that Middle Bronze Age fortifications may have been understood as symbols of power, he has adhered to a "practical reason for the erection of the ramparts, beyond mere propaganda and the demonstration of power" (ibid., 250f.). He disagrees specifically with the details of this dubious interpretation, which require one to believe that the core mudbrick walls found within the ramparts did not project above the rampart at sites such as Hazor, Kabri, and Shiloh. He also cites Ashkelon as a prime example of the incorporation of various defensive strategies into the construction of the rampart (see pp. 237ff.). As for arguments citing the perceived disadvantages of Middle Bronze Age defenses, Mazar cautiously notes that the perception of town planners during the Middle Bronze Age may have been quite different from our own, and it is even possible that their perceptions were based on "erroneous principles."

B. Levantine Fortifications in Their Regional Context

For comparative purposes a few observations are in order regarding fortification strategies employed throughout the eastern Mediterranean and the Near East during the Bronze Age. A review of this evidence serves to illustrate, first, that two neighboring regions are of particular relevance to the study of Levantine fortifications during the Middle Bronze Age: Egypt and northern Mesopotamia, and second, that the tradition of Levantine Middle Bronze Age fortification strategies affected the nature of Late Bronze Age defensive strategies among the Levant's neighbors. To support these observations it is necessary to provide a brief summary of contemporaneous fortification strategies in neighboring regions.

As this review of previous scholarship on Middle Bronze Age fortification strategies in the Levant demonstrates, there is little evidence, if any, for seeking contemporaneous and related exemplars of rampart style fortifications as far away as Iran. Similarly, limited exploration of Middle Bronze Age settlements in Greece and the Aegean has yielded no comparative data. In fact, the earliest Greek fortifications attested are those at Late Bronze Age Mycenae. On the island of Cyprus, where evidence regarding Middle Bronze Age fortifications have been examined by Michel Fortin (1981; 1983; 1989; 1995), no relationship with Levantine fortifications is evident there either, nor has any connection ever been supported. While little work has been done on the subject in Anatolia (and a few Middle Bronze Age sites have been explored), notable works include Parr's aforementioned discussion (1968), Rudolf Naumann's treatment of Anatolian architecture (1971), and a more recent German project, which has undertaken the reconstruction of a considerable part of the fortifications of the Late Bronze Age Hittite capital, Hattuša. Nevertheless, evidence for Anatolian fortifications dating to the Early and Middle Bronze Ages is sparse and no data suggest a unique or native tradition of fortification construction during the Bronze Age. During the Late Bronze Age, however, it may have been the case that the Hittites co-opted earlier Levantine fortification traditions since they would have been familiar with them in the northern Levant (see pp. 83f.) and southern Anatolia (e.g., Kültepe-Kaneš).

Although there also is no basis for seeking Egyptian influence in the development of Levantine fortification strategies during the Middle Bronze Age, Middle Kingdom fortresses in Upper Egypt do provide a useful source for comparative data regarding the construction of massive mudbrick fortifications. Salvage excavations in the 1950s and 1960s in the Sudan (under what is now Lake Nasser) between the second and third cataracts of the Nile, exposed the remains of these fortresses before they were completely submerged. Among these forts, from north to south, were Buhen (Emery et al. 1979), Kor (Vercoutter 1955; H. S. Smith 1966), Mirgissa (Dunham 1967; Vercoutter 1970), Askut (Badawy 1964; 1965; 1966), Shalfak and Uronarti (Dunham 1967), Semna and Kumma (Dunham and Janssen 1960), the Semna south fort (Vercoutter 1966; Žabkar 1982), and Serra (Knudstad 1966). Although most of these forts were already known (see Clarke 1916) and some sites such as Buhen and Mirgissa had been initially explored by George A. Reisner (1960), Jean Vercoutter, David Randall-Maciver and Leonard Woolley (1911), the data from these salvage excavations resulted in intensive consideration of Egyptian penetration into Nubia (Lawrence 1965; Emery 1965; Vila 1970; Kemp 1972; 1983; 1991; Trigger 1982; S. T. Smith 1995).

Although no attempts have been made to compare these Egyptian fortresses with Middle Bronze Age settlements in the Levant, Barry Kemp identifies two distinct fortification strategies in Egypt that are methodologically relevant to the present study (1983:130ff.; 1991:166ff.). These two strategies correspond with the two main phases of Egyptian expansion into Nubia during the Middle Kingdom. Kemp observes that an earlier set of forts, which were built during the reign of Sesostris I (1953–1908 B.C.), were established on the plain along the Nile with regular plans and therefore featured a unique combination of defensive features surrounding each fort on all sides. These features included a fosse, low parapet walls with apsidal towers and loopholes, and thick fortification walls with towers and bastions (1983:130f.). Although the initial plans of these settlements were rather regular, modifications, maintenance, and improvements during the Middle Kingdom resulted in a variety of final plans at these forts. Among the initial settlements were Ikkur, Kubban, Aniba, Buhen, and perhaps slightly later, Mirgissa. In addition to the "plain type" of forts, as Kemp identifies them, a second group were constructed during the reign of Sesostris III (1872–1853 B.C.; ibid., 131f.). These fortifications, Kemp notes, were irregular in plan since they were usually built very close to the shores of the Nile or even upon islands in the Nile. They included the fortified sites along the Semna gorge: Semna, Kumma, the Semna south fort, Uronarti, Serra, Shalfak, and Askut.

While Kemp's typology serves as an excellent example of the possibility of observing a diachronic development among Middle Bronze Age fortifications, of particular interest to Levantine archaeologists is the exceptional state of preservation of these fortresses. As a result of the extremely arid climate plastered mudbrick walls (up to a height of 11 m), fosses, towers, gates, and loopholes are all preserved to an extent unknown in the Levant, Mesopotamia, or Anatolia. These features provide evidence that can clarify issues regarding the construction, appearance, and preservation of Levantine fortifications of the Middle Bronze Age (see chapter 3).

Another region of significance to this study is Mesopotamia, although the study of fortifications from the late third through the first half of the second millennium B.C. in this region is limited. Although excavation in eastern Syria has provided a wealth of information in recent decades concerning fortified sites in northern Mesopotamia during the third millennium, evidence from contemporaneous sites in southern Mesopotamia during both the third and second millennia is sparse. Sites where outer walls of these early periods have been explored only include, for example, Larsa, Nippur (Fisher 1905; Gibson et al. 1983), Sippar (Tell Abu Habbah), and Tell ed-Der (de Meyer et al. 1971; Paepe et al. 1978; Gasche and Paepe 1980).

Limited exploration of Old Babylonian fortifications is the result of several factors, however, some of which have already been noted by Ruth Opificius (1964). First, despite interest in the rise of urbanism in southern Mesopotamia, few research projects have expressly focused upon the exploration of defensive systems. Instead, focus has often been on other monumental features such as temples, palaces, canals, and irrigation systems, which certainly played an important part in the evolution of urbanism. Second, widely held assumptions regarding Mesopotamian fortifications appear to be largely the result of intensive but limited exploration of the rampart-like embankments at both Sippar (Tell Abu Habbah) and Tell ed-Der. However, only limited evidence of an original fortification wall at the base of the embankment at Tell ed-Der was encountered. Nevertheless, the excavators of the site confidently concluded that their embankments were first and foremost "raised in order to protect the living site from the increasing ravages of fluviatile waters" (Paepe et al. 1978:33). It is probably because these particular results have gone unchallenged, despite their having been obtained at only one site, that there has been little impetus for further exploration of comparable features at other southern Mesopotamian settlements. The third reason for such limited evidence is that fortifications in this region have been either heavily damaged or gradually buried as a result of the meandering of and alluviation from the Euphrates and Tigris Rivers (e.g., Mari and Terqa). Finally, the lack of information for fortifications in this region is also, undoubtedly, related to the modern political circumstances in Iraq, which have understandably limited the number of excavations in the region over more than fifteen years. Collectively these factors have limited our understanding of Old Babylonian defensive features in southern Mesopotamia.

Despite these limitations, excavations of smaller fortified settlements in Mesopotamia have been published in recent years, which do improve our understanding of Old Babylonian fortresses. Excavations include those by the Oriental Institute at Dūr-Samsuiluna located at Khafajah Mound B in the Diyala (Delougaz 1990) and Haradum (Khirbet ed-Diniye) excavated by Christine Kepinski-Lecomte (1992). These sites permit further consideration of the time required in the construction of the fortifications, which were of similar size to those encountered in the Levant (see chapter 5).

In contrast to the lack of information from southern Mesopotamia, the north, particularly in eastern Syria, has yielded considerable evidence for the evolution of urbanism and the development of fortification systems during the Early and Middle Bronze Ages along the middle and upper Euphrates, and along the Ḫabur and Baliḫ River basins. First and foremost, numerous reports have been published for sites such as Terqa (Tell 'Ashara), Tuttul (Tell Bi'a), Yakaltum (Tell Munbaqa), Tell es-Sweyhat, Selenkahiye, Tell Beydar (Nabada?), Urkeš (Tell Mozan), Šubat-Enlil (Tell Leilan), and Tell Chuera (see catalog 1). In addition to these, a few surveys of the evolution of fortification systems have been published (e.g., Oates 1985; Margueron 2000). The literature concerning the role that Kranzhügeln played in the evolution of urbanism and fortification systems in this region must also be considered (e.g., Moortgat-Correns 1972; Lyonnet 1998; Kouchoukos 1999). Nevertheless, no consensus has yet emerged regarding the origin and ultimate influence that the Kranzhügeln of the third millennium B.C. might have had upon later fortification systems (see Stager's suggestion on pp. 10f.). Since this region is of greatest cultural relevance to the Levant during the Middle Bronze Age and it provides such a rich catalog of sites (see catalog 1), considerable attention will be given throughout this work to the role that this region plays in our understanding of the evolution and development of Middle Bronze Age fortification systems in the Levant (see esp. p. 78).

C. Methodology

In seeking to address many of the concerns mentioned in the preceding sections, the approach of this work, while multifaceted, remains straightforward. In short, it comprises an effort to contextualize the consideration of Middle Bronze Age fortification strategies in the Levant by addressing a series of topics relevant to the history and archaeology of the period. For this reason, in addition to a general description of fortification systems and their construction (chapter 3), which might be identified as the core of this work, it is necessary to address various other aspects of warfare and society during the period between ca. 2500 and 1500 B.C. For example, in order to consider the particular types of weapons and strategies against which these fortification systems may have been built, the evidence related to siege warfare in the Levant during the Early and Middle Bronze Ages is reviewed (chapter 2). Based upon the data collected in the construction of the catalog of fortified Middle Bronze Age sites in the Levant (see catalog 2) a thorough discussion of fortification strategies in this period is also provided (chapter 3). Analysis of the fortifications of these sites also led to an assessment of the evolution of fortification strategies (chapter 3.A). With an archaeologically grounded understanding of their development the question is set in its broader historical context in order to establish what if any historical circumstances account for the development of Middle Bronze Age fortifications (chapter 4). The military and political role played by fortified settlements within the Levantine landscape during the Middle Bronze Age are also assessed (chapter 5). Finally, the socioeconomic role of these fortification construction projects has been assessed by means of various lines of evidence for their duration and the resources that would have been required in their construction (chapter 6).

The core of this work remains the primary data upon which the conclusions are based, namely the catalog of all fortified Middle Bronze Age sites in the northern and southern Levant (see catalog 2). It is assembled from preliminary and final reports, personal communications with scholars, and on occasion visits to the sites themselves. Realizing that contemporary sites in northern Mesopotamia and the Egyptian Delta are also relevant to the question of origins, catalogs of relevant sites in these regions were added (catalogs 1 and 3, respectively). All catalog entries consist of a summary of the elements of the fortifications of these sites noting major defensive features, both natural and artificial, and their characteristics. The purpose of this catalog (and the volume as a whole) is, first, to facilitate the comparison of similar features at different sites; second, to provide a resource for future scholarship; and finally to encourage the adoption of a standard lexicon for the analysis of fortification systems, especially those of the Middle Bronze Age.

D. Geography and Chronology

Two critical issues that must be addressed at the outset of this work are its geographical and chronological scope, which are, of course, inextricable. The geographical focus of this research is the Levant for the reason that for quite some time earthen rampart fortifications have been considered characteristic of its Middle Bronze Age settlements (see chapter 3). However, treating the Levant as a single cultural sphere is also important since its inhabitants shared many cultural traits during the Middle Bronze Age. For the purposes of this work, therefore, the Levant is frequently discussed in terms of the regional divisions most often identified. These are the northern Levant including Lebanon, western Syria, and the ʿAmuq Valley in Turkey, and, the southern Levant including Israel, the Palestinian territories, and Jordan. Since the regions neighboring the Levant, such as the Egyptian Delta, Anatolia, and northwestern Mesopotamia, may have played a role in the transmission of fortification strategies during the Middle Bronze Age or just prior to it, findings from these regions are discussed when they bear directly upon the question of the evolution of Middle Bronze Age fortifications in the Levant.

In order to best understand the historical development of Middle Bronze Age fortifications it is also necessary to consider a broader temporal scope. Therefore, evidence for fortification strategies during the Early Bronze Age, especially from the EB IV in the northern Levant is discussed. As demonstrated in this work, evidence from fortified centers in northwestern Mesopotamia reveals that the tradition of fortification construction in this region during the Early Bronze Age related directly to that of the northern Levant during the late third millennium (see p. 78). Since there is no evidence supporting a relationship between Early Bronze Age fortifications in the southern Levant and Middle Bronze Age earthen rampart fortifications, the earlier fortifications are discussed only briefly and primarily for comparative purposes (see pp. 92f.).

While some reference is made to historical and archaeological data from the Late Bronze Age (ca. 1530–1200 B.C.), such data obviously do not bear upon the question of the development of Middle Bronze Age fortifications. Evidence from the Late Bronze Age can only address the extent to which these fortifications were able to survive, which is important but is nevertheless not of concern to the present work. In the end, although Middle Bronze Age fortifications are the focus of this work, the conclusions presented here represent the first comprehensive approach toward understanding the evolution of fortifications in the Levant during the Middle Bronze Age by addressing evidence from the period from ca. 2500 to 1500 B.C.

1. Relative Chronology

The geographic divide between the archaeology of the northern and southern Levant, discussed above, has unfortunately resulted in two distinctly, though related, chronological schema for the period in question (see last two columns of Table 1). However, because the present work seeks to address the Middle Bronze Age across the entirety of the Levant, some compromise must be reached with regard to chronology and terminology. Since there is a dearth of fortified sites in the southern Levant during the EB IV, use of the tripartite division proposed by William Dever (1980) is unnecessary for discussion of EB IV fortified settlements. Rather, when data from the EB IV that are germane to this work are discussed the divisions common in the northern Levant are adopted (i.e., EB IV A or B).

The Middle Bronze Age is identified using the traditional terminology for the southern Levant (e.g., MB IIA–C).[15] However, it is recognized that in general no MB IIC phase, which is identified with the Hyksos in the south, is identified with archaeological assemblages in the northern Levant. When archaeologists have offered more precise subdivisions of these phases in their discussions of fortifications, their dates and/or terminology are included in parentheses. For example, on a number of occasions it has been possible to employ the fourfold division of the MB IIA sequence based on the phases of Tel Aphek (see Cohen 2002a).

In most cases the original archaeological phasing of Middle Bronze Age fortifications that was provided by the excavators has been retained.[16] This was done for several reasons. First, it remains the case that too few ceramic assemblages from fortification systems have been adequately published to permit a

[15] The adoption of the MB IIA–C terminology is a convention of the SAHL series.

[16] Since no north Levantine sites have employed divisions beyond MB I (MB IIA) and MB II (MB IIB–C) in their discussion of fortifications, it has been unnecessary to employ more specific terminology (e.g., Nigro 2002b). Similarly, in the southern Levant a more precise phasing of fortifications outside of the major subdivisions of the Middle Bronze Age chronology beyond the terms normally employed (i.e., MB IIA, MB IIB, and MB IIC) has not been possible.

Table 1. Terminology and chronology for the periods discussed in this work.

<table>
<tr><th>This Work</th><th>Approx. Dates</th><th>Egypt</th><th>South (Old)</th><th>South (New)</th><th>North</th></tr>
<tr><td>EB IVA</td><td>2400/2300–2200</td><td rowspan="2">FIP:Dyn. VII–X
2136–2023</td><td rowspan="4">Intermediate BA
or MB I</td><td>EB IVA</td><td>EB IVa</td></tr>
<tr><td rowspan="3">EB IVB</td><td rowspan="3">2200–1900</td><td>EB IVB</td><td rowspan="3">EB IVb</td></tr>
<tr><td>Dyn. XI
2023–1973</td><td rowspan="2">EB IVC</td></tr>
<tr><td rowspan="2">Dyn. XII
1973–1795</td></tr>
<tr><td rowspan="2">MB IIA</td><td rowspan="2">1900–1700</td><td rowspan="2">MB IIA</td><td rowspan="2">MB I</td><td>MB Ia</td></tr>
<tr><td rowspan="2">Dyn. XIII/XIV
1795–1631</td><td>MB Ib</td></tr>
<tr><td rowspan="2">MB IIB</td><td rowspan="2">1700–1600</td><td rowspan="2">MB IIB</td><td rowspan="2">MB II</td><td rowspan="2">MB IIa</td></tr>
<tr><td rowspan="2">“Hyksos”
(Dyn. XV)
1640–1540</td></tr>
<tr><td>MB IIC</td><td>1600–1530</td><td>MB IIC</td><td>MB III</td><td>MB IIb</td></tr>
</table>

Note: The different start dates for the EB IV signify the use of different chronologies for the north and south, respectively.

reassessment of the basic framework provide. Second, there are too few published ceramic sequences for the northern Levant (particularly north of Hama) by which it is possible to suggest a more precise dating of defensive systems. Third, since no comprehensive regional overview of the question of fortifications has been attempted, one of the purposes of the present work is to set forth an initial framework for the study of Middle Bronze Age defensive strategies in the Levant. Therefore, once the ceramic sequences from the northern and southern Levant are more thoroughly correlated, the relative sequencing established here can be revised to provide even more precise historical dates for the events and building phases under consideration. Thus, a wholesale reconsideration of Middle Bronze Age chronology has not been undertaken in this work.

2. The Very Low Chronology

The dates cited in this work are based on the most recent version of very low chronology (hereafter VLC) for the Levant during the Middle Bronze Age, which relies upon synchronization of the VLC recently advanced for both Mesopotamia (Gasche et al. 1998) and Egypt (Bietak 2002). The reader should be aware, however, that this work is dependent upon chronological data provided by other studies. It does not, according to the author's understanding, provide new data that enable a revision or clarification of the chronology for this period. This section merely explains, therefore, the basis for the adoption of the VLC.

While only recently scholars have begun to realize the merits of the low chronology (hereafter LC) if not the VLC, W. F. Albright was among the earliest advocates of the LC.[17] The most recent data presented in favor of the VLC for the Levant are, nevertheless, the result of contributions to the SCIEM 2000 project.[18] These are outgrowths of Manfred Bietak's objective of establishing the essential bases for synchronizations between Egyptian and Levantine chronologies among others (see Bietak 1984). Among relevant contributions for the present work is Lawrence Stager's chronology for the Middle Bronze Age remains from Ashkelon (Stager 2002), which is based on excavation of the site's MB IIA–B gate (Voss 2002). Because of the extensive inclusion of a wide range of data, it can be asserted that Bietak's efforts have resulted in a broad consensus among archaeologists working in Egypt and the Levant regarding the VLC. Nevertheless, an explanation of the basic merits of adopting the VLC remains necessary.

First, it can be observed that Levantine chronology is at the mercy of Mesopotamian and Egyptian chronologies, since only these enable synchronisms between precisely dated astronomical events and mention of such events within the reigns of kings who belong to more or less complete sequences of interrelated dynasties. Second, it becomes apparent, especially working with chronology from the Levantine perspective, that whether a high, middle, low, or very low chronology is adopted, an equivalent choice must be made for both Mesopotamian and Egyptian chronologies. Thus, if the VLC is adopted for the Levant because of the considerations regard-

[17] Albright long ago advocated the adoption of the LC for Mesopotamia and the Levant (1964; 1965; 1966; 1973); for a summary of it, see Dever (1992a:fig. 1, p. 3), who also provides an excellent bibliography for Levantine chronology during the second millennium. Yadin was also inclined to follow the LC (1972b:108)

[18] The Synchronisation of Civilisations in the Eastern Mediterranean in the Second Millennium B.C. (e.g., Bietak 2002).

Table 2. Middle Bronze Age stratigraphy in the southern Levant and VLC dates.

Approx. Start	Phase	ed-Da'ba (Avaris)	Aphek	Hazor	Megiddo	Ifshar	Ashkelon
ca. 1900	MB IIA1		Pre-Palace	—	—	A–C	—
ca. 1850	MB IIA2	H	Palace I/ Pre-Palace II	—	XIVA XIII	D	Ph. 14 (Gate 1) ↓
ca. 1800	MB IIA3	G/4	Palace II	—	↓	E–H	
ca. 1775	MB IIA4	G/1–3 F	Post-Palace II	Pre-XVII	XII	—	Ph. 13 (Gate 2) Ph. 12 (Gate 3)
ca. 1700	MB IIB	E	Palace III	XVII XVI	XI	M	Ph. 11 (Gate 4) Ph. 10
ca. 1600	MB IIC	D/3–D/2		↓	X		↓
ca. 1530	LB I	D/1–C/2	Palace IV	XV	IX	P	Ph. 9

ing Egyptian chronology, then some version of the VLC will be necessitated for Mesopotamian dates (as predicted by A. Mazar 1990:190). This observation is predicated upon the Middle Bronze Age stratigraphy of Hazor (Ben-Tor 2004). On the one hand, Hazor's phased cultural assemblages share similarities with those of Tell ed-Dab'a, thus establishing strong links with Egyptian chronology (Table 2). On the other hand, references to Hazor in the Mari archive mean that Hazor's earliest MB IIB settlement (Stratum XVII), which is most likely the one identified in the texts (contra Ben-Tor 2004:54), must have been founded prior to Mari's destruction (Table 3). Thus, the central date in question is that of the destruction of Mari, which occurred in 1664 B.C. according to the VLC (cp. MC 1764 B.C. or LC 1696 B.C.). Nevertheless, as demonstrated by Gasche et al. this date is dependent upon the date of the datums used in Mesopotamian chronology, among them the Ur III lunar eclipses (1998:78–92). These establish dates of almost complete certainty for two major events: the death of Šulgi (1953 B.C.) and the fall of Ur (1911 B.C.). Synchronisms with the VLC for Egypt are, however, based on dates presented in Kenneth Kitchen's recent work (2000); these are used in this work. Nevertheless, the fundamental structure of the argument presented in this work, the relative sequence provided for the events discussed (see Table 14), and the evolution of rampart fortifications described are unaffected by the choice of chronology.

Table 3. Stratigraphic correlations between Middle Bronze Age sites in the Levant and Mesopotamia and VLC dates.

Period	Hazor	Mari	Approx. Start
MB IIA1	—	Ville III: *Šakkanaku Period*	1900
IIA2/3	—	*Amorite Period*	1850/1800
IIA4	Pre-XVII		1775
MB IIB	XVII	ARCHIVE Destroyed 1664	1700
MB IIC	XVI	—	1600
LB IA	XV	—	1530

E. Moving Forward

The preceding review of the study of Middle Bronze Age fortifications in the Levant and comparative studies of fortifications in neighboring regions reveals a number of areas ripe for further investigation. A particularly glaring deficiency after eighty years of scholarship on this question remains the complete lack of a broad synthesis of Middle Bronze Age fortification strategies in the Levant, which is now possible due to the sheer number of published archaeological reports. While studies have focused upon the identifications of glacis, ramparts, walls, and fosses, or upon the absence of certain features in the archaeological record, not since Parr's work (1968) has a site-by-site treatment of the data been attempted, such that today one is confronted with the task of constructing an enormous catalog. Absent are comparative studies of building techniques and materials used at these sites, which might suggest the direct relationships between sites throughout the Levant or that might even enable the identification of individual polities (see chapter 5). Only a thorough examination of the planning and construction requirements of Middle Bronze Age fortifications can help identify the work force involved in the construction of these fortifications (see chapter 6).

A more detailed study can contribute not only to a greater understanding of the fortifications themselves (see chapter 3) but also of social complexity during the Middle Bronze Age. With regard to the relationship between the construction of fortification systems and political organization we may consider Eliezer Oren's remarks as illustrative:

> The construction of these elaborate defense systems, perhaps more than any other aspect of the material culture [of the Middle Bronze Age], could help to elucidate the mechanism of state formation in southern Canaan (1997b:256).

That many such questions have remained unaddressed reveals that scholars have only scratched the surface with respect to comparative approaches that provide useful analogies for the circumstances characteristic of the Levant during the Middle Bronze Age, particularly since textual sources are lacking. Unfortunately, modern boundaries continue limit scholarship concerned with questions of multi-regional scope.

In the many years since the first theories were advanced and subsequent excavations of Middle Bronze fortifications in the Levant were published, no single work has sought to comprehensively study the various issues involved in what inevitably must be a detailed inquiry into architectural, social, political, economic, military, and historical aspects of the Middle Bronze Age. The present study attempts to remedy this lacuna by incorporating archaeological data from the Levant and comparative studies from contemporaneous fortified settlements in neighboring regions. It is hoped that this work will serve as an important reference for future archaeological work by providing excavators with a collection of comparanda in the face of a vast number of relevant reports and articles. Perhaps it will also encourage the use of a standardized terminology that might facilitate future study of this topic. To this end it is hoped that future publication of the remains of Middle Bronze Age (and other) fortification systems will attempt to consider the questions addressed here.

PART TWO

THE EVOLUTION OF WARFARE AND DEFENSES IN THE LEVANT DURING THE BRONZE AGE (CA. 2500–1500 B.C.)

2. WEAPONRY AND THE ART OF WAR (CA. 2500–1500 B.C.)

> Generally in war the best policy is to take a state intact;
> to ruin it is inferior to this.
> To capture the enemy's army is better than to destroy it;
> to take intact a battalion, a company or a five-man squad
> is better than to destroy them.
> For to win one hundred victories in one hundred battles is not the acme of skill.
> Thus, what is of supreme importance in war is to attack the enemy's strategy;
> next best is to disrupt his alliances:
> the next best is to attack his army.
> The worst policy is to attack cities.
> Attack cities only when there is no alternative.
>
> Sun Tzu, *Art of War* (pp. 77f.)
> China (fourth century B.C.)

In the past the study of Early and Middle Bronze Age warfare has been mostly limited to analyses of weapons recovered from archaeological contexts (e.g., Deshayes 1960; Salonen 1965; Korfmann 1972; Gonen 1975; Philip 1989; Miron 1992) and brief syntheses on the evolution of fortifications (Parr 1960; Kempinski 1992b; 1992c). Only a few works have attempted a synthesis of the various lines of available evidence (Yadin 1963; 1970), and these works are now in need of revision. Nor has any prior study of Levantine weapons or fortifications attempted to account for the development of military technology and administration in this period in Egypt and Mesopotamia, although military innovations in these regions influenced the evolution of Levantine warfare. Such evidence includes not only artifacts but also iconographic and textual data. Egyptian tomb paintings and Mesopotamian stelae, for example, provide excellent depictions of military technology (Schulman 1964; 1982), and Egyptian and Akkadian texts provide much needed descriptions of military engagements (A. E. Glock 1968; Sasson 1969). However, to fully appreciate the significance of the development of fortification strategies during the latter part of the third and the first half of the second millennium (ca. 2500–1500 B.C.) in the Levant, it is necessary to examine the development of warfare throughout this period.

This chapter, therefore, draws together archaeological, textual, and iconographic evidence in order to describe the general characteristics of ancient Near Eastern, but more specifically Levantine, warfare from the mid-third through the mid-second millennium. These data are employed to demonstrate that siege warfare was the most common type of military engagement throughout this period, reaching its climax during the Middle Bronze Age. By considering the most frequently attested weapons employed in siege warfare it is possible to assess the function and effectiveness of these weapons. Mortuary evidence likewise supports not only the conclusion that siege warfare had become the dominant mode of warfare during this period, but also poignantly illustrates how these weapons were employed. Thus, through these diverse lines of inquiry it is possible to assess the role played by the development of weaponry in the evolution of warfare in the Levant from the period of the Ebla empire through the expulsion of the Hyksos.

A. Levantine Textual Traditions on Warfare

While no traditions in Near Eastern literature can be identified as manuals on the art of war comparable to that of Sun Tzu, proverbs and literary sources indicate that there was considerable reflection on the nature of warfare and its impact on society. Such a tradition is hinted at, for example, in an Old Babylonian letter from Mari:

> In order to defeat the enemy you think up plans of attack and keep maneuvering for position against him. But the enemy will also, against you, try to plot and to maneuver for position, much as wrestlers use tricks against each other. Truly, (it is) as an old proverb which goes: *A bitch in her battling for food gave birth to lame puppies.* As for you, do not do the same,

> provided that the enemy will not ambush you in (your) residences (ARM I 5:4–16 as trans. by A. E. Glock 1968:171).

As vague as it may be, the proverb imbedded in this letter suggests that wisdom regarding warfare existed, though perhaps mostly in proverbial form during the Middle Bronze Age.

The Lamentation over the Destruction of Ur is perhaps the best known Mesopotamian literary account describing the character of siege warfare at the start of the second millennium (see p. 91). Ur, which was destroyed at the hands of the Elamites in 1911 B.C., was according to this tradition the scene of a horrific massacre following its defeat:

> By its walls, as far as they extend in circumference, laments were uttered,
> At the . . . , the weapon makes all cower before it,
> In Ur, the large axes wreak havoc before them,
> The spear, "the might of battle" is hurled straight (to its mark),
> The large bows, the throw-stick, the sling are all devouring,
> The "head" arrows filled their bodies like heavy rain,
> The large stones that of themselves (strike) afar, crush the bones,
> Daily the evil wind brings them back against the (people of the) city,
> Ur that relied on its lions, was given over to carnage,
> Its people were turned over the power of the enemy (*ANET*, p. 618).

Most of the remaining textual sources from Mesopotamia are of a historical or commemorative nature, and thus for the Middle Bronze Age this remains our only detailed "eyewitness" account of the downfall of a major city following a siege.

The *Tale of Sinuhe* (*AEL* 1, pp. 222–235) is perhaps the most frequently cited literary tradition relating to the art of war in the Levant during the Middle Bronze Age. References to the weapons employed by Sinuhe and his opponent, which included the dagger, battle-ax, and missiles (presumably the spear or javelin), are in keeping with the types of weapons characteristic of the MB IIA (see discussions below). However, the tradition of heroic battles between national "champions," though they probably occurred throughout much of the Bronze Age, is not particularly instructive about the extent of warfare in the Levant and its effects on society as a whole. Mention of Sinuhe's thwarting of Asiatics attacks against other Asiatic rulers (i.e., *ḥk3w ḫ3swt*) are equally lacking in details, except to suggest that infighting among these rulers was common during the Middle Bronze Age.

Far more illuminating traditions regarding warfare are preserved in Ugaritic literature dated to the Late Bronze Age. The *Kirta Epic*, for example, illustrates a number of details regarding siege warfare (trans. Pardee 1997a:334f.):

> Like grasshoppers you will invade the field,
> like locusts the edges of the steppe-land.
> Go a day, a second,
> a third, a fourth day,
> a fifth, a sixth day
> Then at sunset on the seventh (day)
> you will arrive at ʾUdmu the great (city),
> at ʾUdmu the well-watered (city).
> Occupy the cities,
> invest the towns.
> The women gathering wood will flee from the fields,
> from the threshing-floors the women gathering straw;
> The women drawing water will flee from the spring,
> from the fountain the women filling (jugs).
> Then for a day, a second,
> a third, a fourth day,
> a fifth, a sixth day,
> Do not shoot your arrows at the city,
> your sling-stones at the towers.
> Then at sunset on the seventh (day of the siege)
> Pabil the king (of that city) will not (be able to) sleep
> For the sound of the roaring of his bulls,
> for the noise of the braying of his donkeys,
> for the lowing of his plough-oxen,
> the howling of his hunting dogs (*COS* 1.102).

From this passage several observations regarding Bronze Age warfare can be made. First, it illustrates that during the mid-second millennium long-distance campaigns remained as common as they had been during the first half of the second millennium (see chapter 4). Second, it confirms that a region's inhabitants routinely fled from unfortified settlements in the countryside to walled towns for refuge during an invasion. Third, it reveals at the least an occasional adherence to a strategy of prolonged siege—without a direct assault—with the intention of wearing down those besieged. Finally, it also provides a glimpse into the real discomforts that siege warfare inflicted on a population (and their animals).

Another Ugaritic text describes the specific rituals associated with climbing to the "top of the wall" to offer prayers during a siege. Identified as a *Prayer for a City Under Siege* (RS 24.266), it is the only Ugaritic prayer preserved. Following the proscriptions for sacrifice are the words of a prayer that were recited (trans. Pardee 1997b:284–85):

O Baʿlu, if you drive the strong one from our gate,
the warrior from our walls,
A bull, O Baʿlu, we shall sanctify,
a vow, (O) Baʿlu, we shall fulfill;
a firstborn, (O) Baʿlu, we shall sanctify,
a *htp*-offering, (O) Baʿlu, we shall fulfill,
a *ʿšrt*-feast, (O) Baʿlu, we shall offer;
To the sanctuary, (O) Baʿlu, we shall ascend,
th(at) path, (O) Baʿlu, we shall take
(*COS* 1.88).

Thus, in addition to Bronze Age Mesopotamian traditions, Levantine traditions from the Late Bronze Age also provide a vivid depiction of warfare during the second millennium.

Comparisons between ancient Near Eastern literary traditions that relate to the art of warfare, however limited they are, and literary traditions from other regions of the ancient world suggest that fundamental similarities in the nature of warfare can be identified between the Bronze Age Near East and slightly later periods in other regions. These additional sources are useful for the light they can shed on details related to Bronze Age warfare in the Near East, which are not made explicit in the limited sources available. For example, the tradition ascribed to Sun Tzu (1963), a Chinese master tactician during the Warring States Period (ca. 450–300 B.C.), suggests that China experienced tactical concerns that were comparable to those of the pre-classical Near East. Observations include that the chariot was common in open battles, the composite bow dominated engagements, siege techniques were virtually identical (and lacked machinery), and cavalry had not yet been introduced. In light of our understanding of warfare as waged by Near Eastern powers during the Bronze and Iron Ages and in the absence of such Near Eastern treatises, Sun Tzu's statements uniquely illustrate the nature of the art of war during the Bronze Age as addressed in this chapter.

Similarly, the specific nature of ancient Near Eastern siege warfare has often been usefully illuminated by classical authors (e.g., Schulman 1964; and Eph'al 1996; see esp. Kern 1999). Specific works such as Aeneas Tacticus's *Siegecraft* (1990), dated no later than the mid-fourth century B.C., and Vegetius Renatus's *Epitome of the Military Science* (1993), dated to the fourth century A.D. clarify how certain techniques were intended to function. Despite the use of such sources, in what follows caution is taken to avoid propagating anachronistic comparisons that are often the result of a poor understanding of the historical development of weaponry and tactics. When necessary, therefore, such anachronisms are directly addressed.

B. The Emergence of Siege Warfare in the Ancient Near East

In the ancient world military engagements may be characterized broadly as one of two main types: the open or pitched battle, and the siege. We possess historical records of a number of important pitched battles in the Bronze and Iron Age Levant, such as the Battle of Megiddo (ca. 1458 B.C.), the Battle of Kadesh (ca. 1275 B.C.), and the Battle of Qarqar (853 B.C.). Smaller-scale ambushes and skirmishes are also attested in historical, including biblical, sources, though they are often distinguished from pitched battles because they occurred outside the main theater of operation. Regardless of scale, land-based military engagements other than sieges all shared in common the role played by natural topography. The key distinction, therefore, between the pitched battle and the siege was the impact of manmade fortifications on the conduct of warfare.

Yet, among textual sources from the Levant, Egypt, and Mesopotamia there are no clear references to large-scale open battles prior to the Battle of Megiddo during the Late Bronze Age. While absence of evidence does not demonstrate that such military engagements never occurred, it does suggest that they were perhaps less frequent than during the Iron Age and later. So, what evidence is there for open battles in the ancient Near East before the end of the Middle Bronze Age (ca. 1530 B.C.)?

The archives of the town of Mari on the middle Euphrates River (ca. 1700 B.C.) are one of the best sources of information concerning military and political activities throughout northern Mesopotamia and the northern Levant in the years before Mari's destruction by Ḫammurapi of Babylon (e.g., A. E. Glock 1968). Stephanie Dalley (1984:146) has, for example, noted that "no detailed accounts of pitched battles" have been found in the Mari documents, although small-scale ambushes are occasionally mentioned:

> When a large detachment went to ambush an enemy expedition, they could not find a suitable place to lie in wait, so that detachment returned without success, and the enemy expedition continued unchecked—it was not ambushed. Now, let a small detachment go and ambush the enemy expedition, and let them take informers (ARM II 22:6–15 as trans. by Dalley 1984:150).

Major siege campaigns are, however, very well attested among Old Babylonian sources such as the

Mari texts. The following text serves as an excellent example.

> When I had captured the towns of Tarrum, Ḫatka and Šunḫum, I approached Ḫurara and surrounded that town. I set up a tower (*dimtam*) and a battering ram (*yašibam*) against it, and in seven days I captured that town (ARM I 131:5–16 as trans. by Dalley 1984:146).

Since the Mari texts provide substantial evidence for the conduct of siege warfare during the Middle Bronze Age, a lack of references to open battles among these thousands of texts is revealing concerning the nature of warfare at this time. This picture is also commensurate with early textual evidence from southern Mesopotamia that suggests that siege warfare was on the one hand a common military operation while on the other pitched battles were rare. Akkadian kings in the second half of the third millennium (ca. 2237–2060 B.C.) undertook various battles to consolidate their empire. In the many inscriptions that record these engagements it is said, for example, that a particular king was victorious over a given town in battle, conquered that town, and destroyed its walls. One notable example refers to Sargon's defeat of Uruk:

> [Sargon] . . . conquered the city of Uruk and destroyed its walls. He was [victorious] over Uruk in battle, [conquered the city], . . . (*RIME* 2.1.1.1, lns. 12–22).

Identical phrases were employed in the same text to describe Sargon's battles against other cities such as Ur, Eninmar, and Umma.

Although it is possible that what is meant in such texts was that opposing armies first fought in open battle outside the town and the victorious invaders ruthlessly destroyed the town, the juxtaposition of references to "battle" and that "he conquered the city" more probably means that the battles referred to were sieges. This is, in effect, a poetic parallelism that depicts the nature of the battle. This interpretation is corroborated by another inscription of Sargon in which a reference to his victory in thirty-four battles is followed by the phrase "he destroyed their (city) walls" (*RIME* 2.1.1.11 lns. 5–6). However, an account by Sargon's grandson, Naram-Sin's, siege of Armānum and Ebla, two towns in the northern Levant (*RIME* 2.1.4.26), provides the clearest evidence that these battles were sieges (see pp. 90f.). The evidence for the destruction of Ebla's Palace G, which I shall discuss in greater detail in chapter 4, corroborates this account.

Later in the third millennium there are numerous references to sieges of towns among the year-names of the kings of the Ur III Dynasty, from ca. 2018 to 1911 B.C. (Sigrist and Gomi 1991:319ff.).[1] At least twelve different towns were besieged during a sixty-nine year span; while several were besieged more than once, one unfortunate town underwent siege at least eleven times. Smaller towns of lesser strategic significance were probably also besieged during this period but did not warrant such recognition. It is particularly noteworthy that in none of the year-names that identify military engagements is there a commemoration of a single pitched battle. Thus, it has been suggested that open field battles were uncommon through the end of the third millennium.

From Ur III sources we learn of Mesopotamian rulers' use of massive fortifications projects intended to protect their lands against a growing threat from Amorites to the northwest. A great wall called *Muriq-Tidnim* (lit., "that which keeps the Tidnum [Amorites] at a distance") was to be constructed in the fifth year of the reign of Šu-Sin. This wall and similar walls, such as *The Wall of the Land* built during the thirty-seventh year of Šulgi's reign, indicate a Mesopotamian military strategy in which attacking forces were not met in open battle but were supposed to be contained by a defensive line, even in the open countryside. Such fortifications may have been intended to predetermine the location of battles with invading troops, who would have been forced to engage in a type of siege warfare. While the Ur III Empire could muster an army without equal to execute punitive or preemptive campaigns against its enemies, it appears to have preferred to build what was, no doubt, a costly defensive perimeter that could be defended like a town's wall against enemy attack. The so-called *Walls of the Ruler* in Egypt, first attested in the Middle Kingdom *Tale of Sinuhe* (*AEL* 1, p. 224, ln. 15ff.), may be evidence that a similar military strategy was adopted in Egypt during an only slightly later period. According to Sinuhe these walls—or what might simply have been a series of fortresses along the coastal highway—were intended "to repel the Asiatics and to crush the Sand-farers" (ibid.).

While there were many cultural differences between Mesopotamian and Egyptian practices in the conduct of warfare, the evidence for open battles is equally scarce in contemporary Egyptian sources from the Sixth Dynasty (ca. 2362–2176 B.C.) and the First Intermediate Period (ca. 2136–2023 B.C.). Yet there is considerable evidence that siege warfare was the most common form of military engagement. The relevant Egyptian sources for this period are not tex-

[1] Year-names within a king's reign commemorated important events such as the construction of major canals, fortification walls, or temples, as well as important battles.

tual but iconographic and therefore include tomb reliefs and mortuary complexes where battle scenes were depicted. These have been thoroughly studied by Alan R. Schulman (1982:esp. 165f.), who discussed two siege scenes dated to the Sixth Dynasty, and two Fifth Dynasty fragments, which according to Schulman may depict pitched battles though this identification is far from certain.

Fragmentary battle scenes dated to the early First Intermediate Period (FIP) do not provide evidence of "assault on a fortified town," according to Schulman (ibid., 179); but there is a late FIP depiction of an Egyptian siege against a town occupied by Asiatics in the tomb of Intef (Theban Tomb 386), which is dated to the reign of Mentuhotep II. This provides clear evidence of Egyptian siege warfare even in this period of political decentralization (see Jaroš-Deckert 1984:pls. 1, 3, and 17). During the end of the FIP and the Middle Kingdom numerous reliefs from tombs in Beni Hasan depict siege warfare (see tombs 2, 14, 15, and 17, Newberry 1893:pls. 14, 47, 5, and 15), while only one reference to a pitched battle, Merikare's battle (see *AEL* 1, p. 102, ln. 70ff.), is made and its historicity remains in question.

As Schulman has observed, all but one of the tombs with illustrations of battle scenes during the later Middle Kingdom depict "a full-scale assault on a walled fortification" (1982:168). The remainder of the Middle Bronze Age in Egypt, during the Second Intermediate Period, witnessed an increased frequency in siege warfare culminating in the siege of Avaris, the capital of the Semitic "Hyksos" rulers of Lower Egypt, as well as the later Egyptian siege of Sharuhen in the southern Levant, where the Hyksos forces fled after they were expelled from Egypt. In summary, although a few relief fragments suggest that open battles were a part of Egyptian warfare during the third and early second millennia B.C., numerous examples of siege scenes suggests that this mode of warfare was dominant in Egypt as it was in Mesopotamia during this period.

The picture provided by textual, iconographic, and archaeological sources dating to the late third millennium and first half of the second millennium is highly suggestive that an emphasis was placed on siege warfare and thus there existed a widespread reliance on fortified settlements. The dependence on fortified settlements during this period has its obvious roots in the recognition of the strategic value of fortified positions. As articulated by Carl von Clausewitz these include the function of fortresses as (1) secure depots, (2) protection for large and prosperous towns, (3) real barriers (i.e., along roads), (4) tactical points of support, (5) staging points, (6) refuge for weak or defeated units, (7) a shield against enemy attack, (8) protection for army camps (i.e., in the surrounding territory), (9) cover for an unoccupied province, (10) support for a general insurrection, and (11) defense of geographical features such as rivers and mountainous regions (1976:393–99).

C. The Role of the Chariot in Bronze Age Warfare

> Dust spurting upward in high straight columns indicates the approach of chariots. When it hangs low and is widespread infantry is approaching . . . When light chariots first go out and take position on the flanks the enemy is forming for battle.
>
> Sun Tzu, *Art of War* (1963:119–20)

Before delving deeper into a discussion of siege warfare it is necessary to address the role of the chariot since it has played an important part in the interpretation of Bronze Age warfare (e.g., Drews 1988) and yet its role remains poorly understood. Although there is no direct evidence for the involvement of chariots in siege operations, references to chariots in late Middle Bronze Age texts, their depiction in seals and sealings, and ceramic models of them from the start of the Middle Bronze Age warrant consideration of their military role during this period. The emergence of the chariot and its presumed use in battle during the Middle Bronze Age also suggests that warfare consisted of more than the siege of cities during the first half of the second millennium.

The historical development of the chariot in the Near East and the role that the horse played in this development are thoroughly treated in the works of Mary Littauer and Joost H. Crouwel (1979; Littauer 2002) and P. R. S. Moorey (1986). Although "battle cars," "straddle cars," and "platform cars," as Littauer and Crouwel have identified them (1979:15–47), were all in use before ca. 1900 B.C., it is incorrect to view these vehicles as chariots. The earliest evidence for the true chariot dates, therefore, to the Middle Bronze Age (see Littauer and Crouwel 1979:48–72):

> This period sees the evolution and general adoption in the Near East of a light, two-wheeled, horse-drawn vehicle with spoked wheels: the *chariot*. It appears either as a flat cart with open railing (in Anatolia and

southern Mesopotamia), or (in Syria) as a gradual modification of the "platform car": disk wheels are replaced by spoked ones, the high front screen by a low one of equal height at front and sides; the seat is removed, permitting rapid access from the rear, and the high, arching pole is reduced to a lower more mildly curving one. Evidence points to a lengthened axle, providing a wider wheel base, which increased stability and eventually permitted a crew of two to stand abreast (an important improvement for military use) [Littauer and Crouwel 1979:71].

The chariot, although certainly used in sport-hunting, as indicated by Syrian seals, was also put to use in "warfare, in cult and, probably, as a parade and a status vehicle" (ibid., 72). While such functions are, of course, well attested in Late Bronze Age seals, such as those from Alalaḫ,[2] Ugarit, and Tell Abu Hawam (for references, see Collon 1982:130), they are also attested in a number of seals and sealings with clear Middle Bronze Age dates.[3] Nevertheless, the emergence of the chariot and its presumed use in battle during the Middle Bronze Age also suggests that warfare consisted of more than the siege of cities during the first half of the second millennium.

Of central importance to the development of the chariot is the issue of the domestication of equids in the Near East, since it has been established that the equids, which drew these chariots, were not themselves ridden at this time. However, as Moorey has demonstrated, the arrival of horses and their use with chariotry during the second millennium was not, as has been traditionally assumed, "a sudden intrusive innovation from the north or from the east" (contra Drews 1988), but instead part of a local development:

> In general, texts reveal that horses were coming to Mari, and to the king of Assyria, at this time [i.e., during the Middle Bronze Age] from the west, from inland Syria and from Anatolia, not from the east. At this stage, in the early and mature Middle Bronze Age, there seems to be no hard case for an exclusive Hurrian or Indo-European involvement in breeding, in training, or in the dissemination of the arts of horsemanship and chariotry (Moorey 1986:198).

Nevertheless, the extent to which chariots were employed in warfare as early as the Middle Bronze Age remains in question.

With respect to the chronological and spatial distribution of the chariot in the Levant during the Middle Bronze Age, it is noteworthy that ceramic chariot models are known from the northern Levant and northern Mesopotamia but are seemingly unattested in the southern Levant. Chariot models dated to the Middle Bronze Age have been found in the northern Levant[4] at Ugarit,[5] Tell Mašin,[6] Qatna,[7] and Hama.[8] No substantive evidence for the use of the chariot during the Middle Bronze Age in the southern Levant has yet been adduced.

Evidence for the adoption of the chariot in Egypt by at least the end of the Middle Bronze Age has also recently come to light from the excavation of the monuments of Ahmose at Abydos (Harvey 2001). This evidence confirms that the chariot was already in use in the southern Levant, whence it was probably introduced to Egypt during the Hyksos period. The use of the chariot in Egypt during this period ultimately resulted in the full incorporation of chariotry in pitched battles during the New Kingdom, when its use in warfare is first attested.[9]

In light of the scarcity of evidence for open battles during the Middle Bronze Age—in contrast to sieges, in which chariots played little part—chariots do not appear to have been a major element of Middle Bronze Age armies. This probably explains why references to chariots in Middle Bronze Age texts are extremely rare and why the presence of chariots must often be inferred from references to the management

[2]AT/48/32, though poorly preserved and of unknown date and provenance, shows a man in a chariot being followed by three men in procession (see Woolley 1955:266, no. 136; and Collon 1982:51, no. 17). AT/46/213 is from Level V and, therefore, probably of Late Bronze Age date (Woolley 1955, no. 44; Collon 1982:129f., no. 19). Dominique Collon suggests that this seal, although thought to be an "inept copy of a well-known subject," dates to the thirteenth century B.C. (1982:130).

[3]Since the publication of Littauer and Crouwel's work a number of additional seal impressions from Syria dated to the late third and early second millennia depicting chariots have been published. In addition to seal nos. 892 to 895 in Buchanan (1966:174f., pl. 56); see also no. 502 (1981:193), and no. 82 dated ca. 1700 B.C. (A. Glock 1988). All of these seals are without provenance. Other seals depicting chariots include those from Babylon such as seal 3 on tablet no. 22 published by Hugo Figulla (1967:pl. 14), which has been dated to the eighteenth century B.C.

[4]An unprovenanced but well-preserved example of a covered chariot of the EB IV to MB I period is in the collection of the American University of Beirut (Baramki 1966:pl. 2 and p. 23). Although it could be from a Lebanese coastal site like Byblos, chariot models are not attested in the site's material culture.

[5]See no. 11 in fig. 97 in *Ugaritica II* (Schaeffer 1949:231).

[6]See fig. 118 in du Mesnil du Buisson (1935b).

[7]See fig. 12 in du Mesnil du Buisson (1935a:60).

[8]See fig. 17:1 (Hama H) in Harald Ingholt (1940), and figs. 64 (Hama J), 110 (Hama H), 132, and 139 in Ejnar Fugmann (1958).

[9]Six New Kingdom chariots have been recovered from the tomb of Tut'ankhamun (Littauer and Crouwel 1985).

of equids. Textual references to the use of chariots in military operations during the Middle Bronze Age are limited to Hittite sources, as Moorey noted (1986:204). Furthermore, these sources mention the use of chariots only for patrolling and enforcing blockades. Only later, during the Late Bronze Age, were tactics mastered that allowed chariots to be employed en masse in open battle, with a devastating effect against enemy infantry.

Although it has been suggested that even in earlier periods chariots were of some use in battle because horses could trample enemy foot-soldiers and disrupt their ranks, as Littauer and Crouwel rightly note, horses tend to avoid trampling humans. They know their physical limitations and instinctively avoid injury unless sufficiently blinkered:

> Indeed, the use of wagons in a charge against each other or against closely ranked infantry seems extremely doubtful. Equids are reluctant to step on yielding bodies because they instinctively know that the restricted articulation of their limbs, evolved for fast forward movement on the level steppe, makes them vulnerable to joint injuries. Moreover, attempts on their part to avoid the bodies by leaping over them could result in damage to their necks as well as to parts of the draught system. Hence the motif of the prostrate enemy "beneath the hooves . . . must be interpreted as merely symbolic of victory, while in reality such bodies would be lying on the hither or further side of the teams (Littauer and Crouwel 1979:32f.).

Consequently, it is now widely recognized that the true effectiveness of the chariot in battle was its capacity to provide a mobile platform for the use of the bow, particularly the composite bow. The introduction of the chariot ca. 1900 B.C. coincided with the widening use of the composite bow, which was replacing the simple bow in military contexts. The chariot provided a means by which an archer's effectiveness could be maximized against troops massed for battle in the open field. In hilly terrain or around settlements built next to wadis or surrounded by walls—especially walls built on ramparts—the chariot was of no use.

For this reason, although ill-informed and misleading comparisons have been made in the past between the chariot and the tank (e.g., Kenyon 1952:71), as noted in the previous chapter many scholars since Yigael Yadin (1955b) have concluded that chariots had no role in attacks against fortified complexes (see pp. 5f.). The chariot's role in warfare—particularly evident during the Late Bronze Age—is better compared to the fighter plane, which serves as a fast and maneuverable firing platform (Moorey 1986:203). The military role of the chariot during the Middle Bronze Age was most likely limited to providing "protection for an army on the march and in combat; to blockade at times of siege; and to transport archers and other elite troops" (ibid., 203f.).

D. The Weapons and Tactics of Siege Warfare

> When I wish to battle, my enemy, even though protected by high walls and deep moats, cannot help but engage me, for I attack a position he must succour.
>
> Sun Tzu, *Art of War* (1963:97)

The approach taken in this discussion of the weapons and tactics of siege warfare during the Bronze Age developed from consideration of both classical sources on the art of warfare and more recent treatments of warfare.[10] It centers on consideration of the weapons used by attacking forces based upon the range at which they were employed during the course of a siege. As part of this approach I have noted the technical improvements these weapons underwent between ca. 2300 and 1500 B.C., as well as the elements of fortification systems that they were intended to overcome. The following section, therefore, begins with weapons of the greatest range and arrives ultimately at a discussion of weapons used in hand-to-hand combat if a siege led to the successful penetration of a town (see Figure 3 below). The context for such an approach is established by the vignette from the *Kirta Epic* discussed in the previous section.

1. Long-Range Weapons

Of all the weapons available to the Bronze Age warrior, the bow and sling provided the greatest range and firepower. These two long-range weapons played a significant role in siege warfare because they made it possible to suppress the efforts of defenders from the top of the walls or during sallies from their gates. In this section we will consider the crucial role that the sling, the simple and composite bows, and various types of shields played during siege warfare from the Early through the Late Bronze Ages.

[10]Much is owed to John Keegan's *The Face of Battle* (1976) and Kimberly Kagan's *The Eye of Command* (2006).

a. The Sling

Although the sling was of considerable importance beginning as early as the eighth millennium B.C., direct material evidence of its existence is limited since it was constructed from biodegradable materials, mostly leather. Manfred Korfmann observed that the absence of clearly identifiable sling stones or bullets has, in some cases, led to the sling being entirely overlooked (1973:38).[11] This omission has been further compounded because the "naturally formed missiles" were also used by slingers and "may never be recognized" (ibid.). Such objects might be no more than river pebbles or roughly shaped, fist-sized stones, rather than well-worked, symmetrically rounded stone projectiles that are more easily identified. Nevertheless, reliefs and textual references inform us of the importance of slingers in the military ranks of Near Eastern armies during the Late Bronze and Iron Ages.

Even though identifying the sling in the Early or Middle Bronze Ages in the Levant is equally problematic, there is some notable archaeological evidence for the projectiles used with it in the Levant and northern Mesopotamia during these periods. For example, numerous clay sling bullets of five different shapes have been recovered from the late third-millennium site of Tell es-Sweyhat in Syria (Stout 1977). These "stockpiles" were found in structures adjacent to the western wall of the upper town. That such projectiles could be effective was demonstrated during experiments conducted by the Sweyhat expedition using a local slinger. This informal experiment determined that sun-dried, clay sling bullets could be effectively used within 100 m of the target (compared to 200 m for sling stones) and sometimes even remained intact after impact (ibid., 65).

Similar archaeological evidence comes from a large cache of baked clay sling stones found in court 22 of the temple of Dagan at Mari (Parrot 1964:9–11). Other sling bullets referred to as "catapult projectiles" (Matthiae 1982b:125) were recovered from a storeroom of the Middle Bronze Age Western Palace in Area Q at Ebla.[12] Since there is no evidence for the catapult before the classical period,[13] these artifacts must, in fact, be sling stones. Their size, though not provided by Paolo Matthiae (1977b:152, n. 9), has been compared to a ballista stone (no. 2379) from a group found in the Iron Age Phrygischer Tempel at Boğazköy published by Rainer Boehmer (1972:227 and pl. 96, nos. 2379–83), which had a diameter of 13.7 cm. Matthiae states that such stones "étaient employés, pendant les sièges, pour créer ou élargir des brêches dans les ramparts" (Matthiae 1977b:152, n. 9).[14] Such a suggestion seeks to compare the effectiveness of slingers with the successful employment of ballistae against walls, as for example during Alexander's siege of Tyre.[15] This suggestion is, however, anachronistic and ignores two important facts: on the one hand, no propulsion machines existed during this period with which to fire such projectiles, and on the other hand, slingers were unable to breach walls.

It is likely, therefore, that slingers employed an array of different sized projectiles, including large stones, against enemy soldiers. Using larger stones would have meant, of course, that the slinger sacrificed the speed and accuracy of a smaller projectile, but would have gained increased mass. The effect of this would have been an increased probability of inflicting casualties on the enemy despite a loss in accuracy. The choice between differently weighted projectiles was also available to archers, who used arrowheads of different weights (see discussion of the bow below).

To this archaeological evidence we can also add iconographic representations as well as textual references to the sling. A depiction of an Egyptian slinger in the Beni-Hasan reliefs from Egypt affirms the contemporaneous use of the sling in Middle Bronze Age warfare in Egypt (Newberry 1894:pl. 15).[16] In the Mari texts the sling is referred to as *waṣpum* (see ARM IX 102:18; *AHW* 3, p. 1475) and the sling stone as *aban waṣpim* (see ARM II 127:7; Sasson 1969:26). Although it is unfortunate that references to slings in the Mari texts are limited to requests for a specific number of the weapon, Margueron also identified

---

[11]For detailed treatments of the sling and its distribution, see works by Korfmann (for prehistoric times, see 1972; and subsequent periods, see 1973).

[12]These were found in storeroom L.2887, located below a staircase in square DfV8ii (Matthiae 1982b:n. 23).

[13]It is noteworthy that the belly bow (Gk. *gastraphetēs*) and its successor the catapult (Gk. *katapeltikon*) described by Diodorus Siculus (14.41–42.1) were products of the classical period. There is no evidence to suggest that anything comparable to this device existed prior to the fifth century B.C. at the earliest. For further discussion of such devices, see works by Eric W. Marsden (1969; 1971).

[14]Although he cites Yadin, Matthiae's assertion finds no support in Yadin's treatment of the methods of warfare during the Middle Bronze Age (1963:69–71).

[15]His was the first historically attested use of the stone-throwing catapult (Diodorus Siculus 17.42.7; 17.45.2).

[16]The iconographic evidence is corroborated by a large number of clay and stone sling projectiles recovered from the Middle Kingdom fortress of Buhen (Emery et al. 1979:95ff., 130, and pl. 45).

sling stones during excavation of areas around the ramparts at Mari (1982:30), thus corroborating their use as alluded to in the Mari texts.

Despite the relatively limited archaeological evidence for the use of the sling in the Levant during the third millennium and the first half of the second millennium, its role in Bronze Age warfare should not be underestimated. As Korfmann has observed (1972:37), even a relatively novice slinger could propel a missile more than 200 m at a speed in excess of 100 km per hour (ibid., 40). Such a missile was considered lethal even without breaking the skin or penetrating a soldier's armor (if he was fortunate enough to have had armor) as noted by Vegetius Renatus (*Epitome*, Bk. I.16, pp. 15f.). The lethal nature of the injuries was due to the internal damage that the projectile could inflict.

As Yadin has also noted, "the function of the sling was often complementary to that of the bow," and when "they were used in battle, the slingmen always served close to archery units" (Yadin 1963:9f.). This is evident in the Lachish reliefs from the Neo-Assyrian period, where the slingers are always depicted behind the archers (see Ussishkin 1982:78, 80, 82, and 84). It was probably also advantageous to position the slingers behind the archers since it made a flatter trajectory possible and thus would have increased accuracy. The Lachish reliefs also seem to confirm, as Korfmann suggests, that the slingers' effective range of 200 m was actually greater than that of the bowmen (see Figure 3 below).

b. *The Simple and Composite Bows*

Although very little archaeological evidence is available for the use of the bow, iconographic evidence contributes to our understanding of the bow's development from the mid-third millennium through the mid-second millennium. Yigael Yadin argued that the composite bow was the main long-range weapon of the Bronze Age throughout this period from as early as ca. 2300 B.C. (1963:62). We may begin, therefore, by considering the development of the composite bow from the simple bow during the second half of the third millennium.

Two types of simple or self bows are known. The first type, which was usually longer, ca. 1.5 to 1.7 m, is referred to as a *single-arc bow* (ibid., 63).[17] Its unwieldy size was the result of its being made of a single piece of wood that had to be long enough to provide the necessary tension to cast an arrow (Figure 1.A). The second type of simple bow was the *double convex bow* (Figure 1.B), which is also known as the *sinew-reinforced bow* (Bergman et al. 1988:660). It is called double-convex because of its form, featuring a slight curvature, such that when strung the arms of the bow bent slightly forward of the handle before bending backward toward the string; this is known as recurve. This improvement was made possible by the addition of sinew to the back of the bow's wooden arms in order to prevent it from breaking (Miller et al. 1986:179). This increased the bow's tension while allowing the bow to be made smaller.

Despite this increase in the bow's tensile capacity, this innovation still fell considerably short of yielding the power available with the next major development in bow technology, the *composite bow* (Figure 1.C). It was achieved by adding horn to the other side of the double convex (sinew-reinforced) bow (Bergman et al. 1988:660f.). Because of the tensile capacity of the composite bow, when unstrung the arms of the bow bent backward, a state known as reflex.

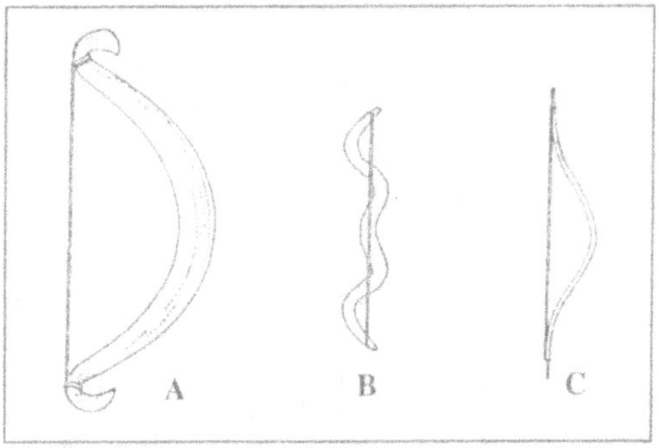

Figure 1. Third-millennium B.C. *bows (after Yadin 1972a:fig. 1): A) Single-arc bow on the Uruk stele; B) Double-convex bow from Mari scene; C) Composite bow shown on Naram Sin's stele.*

The earliest description of the composite bow is found in the Ugaritic *'Aqhatu Legend*, which dates to the end of the Late Bronze Age. Nevertheless, it provides a roughly contemporaneous description of the materials used in the construction of this weapon:

> I'll vow ash wood from Lebanon,
> I'll vow sinews from wild bulls,
> I'll vow horns from rams,
> Tendons from the hocks of a bull,
> I'll vow reeds from GL'IL.
> Give (these) to Kôtaru-wa-Hasīsu,
> And he'll make a bow (*qšt*) for ʿAnatu,

[17]The principle here is the same as that employed in the English long bow, also a simple bow. Although the composite bow was a technology lost to Europe in the medieval period, it remained in use in the Near East. The composite bow was therefore often identified as the *short bow* since it was shorter than typical European bows (Bradbury 1985:12ff.).

Arrows for the sister-in-law of Liʾmu
(*COS* 1.103, pp. 346f.).

The Ugaritic term for the bow was *qaštu* (for vocalization, see *UVST*, p. 175), which was a cognate of Akkadian, *qaštu* (*CAD* Q, pp. 147ff.; also ^{GIŠ}PAN). A second Akkadian term, *tilpānu* (*AHW* 3, p. 1359) appears to have been a direct loanword from Sumerian. It is not certain that any terminology existed to distinguish between the simple and composite bows.

Eric Marsden's work on *Greek and Roman Artillery* (1969) provides an excellent technical description of the composite bow that highlights the value of each of the component parts:

> The [composite] bow contains three principal layers. The central strip of wood, though its relatively slight elasticity may contribute something to the total power of the bow, serves mainly as a base to which the other more resilient materials can be attached. On the back of the wooden layer, that is on the side facing away from the archer, the bowyer fastens a band of animal sinew. On the inner side he glues a series of pieces of horn. When the bow is bent the sinew is stretched and tries very hard to contract to its original size; the horn on the other hand, is compressed and tries to expand to its original shape. The sum of these combined forces is considerable (ibid., 9).

From this description it is evident that both horn and sinew were crucial in the construction of the composite bow, and that the wood, the original element of the bow, served only as a foundation for these materials. Similar descriptions of the components of composite bows from the medieval period testify to the enduring dominance of this weapon,[18] which was in use in the Near East and North Africa for almost four thousand years. Furthermore, Wallace McLeod observes that "the oriental bow used on the fringes of the Greco-Roman world underwent no startling improvement between 700 B.C. and A.D. 700" (1965:3), and its ultimate abandonment was due entirely to the introduction of gunpowder.

The range of the composite bow has been determined from textual evidence, most of which comes from the classical period (McLeod 1965; 1972). McLeod concludes that:

> bowmen were quite accurate up to 50 to 60 metres; that their effective range extended at least 160 to 175 metres, but not as far as 350 to 450 metres; and that 500 metres was an exceptional flight shot (ibid., 8).

Comparative evidence gathered by McLeod from later bow competitions supports that these figures accurately reflect the average range of the composite bow (see Figure 3 below). He also notes that although ranges of 450 m or more were achievable from an elevated position or with favorable winds, shots at such a distance were of minimal effect and wildly inaccurate.[19]

Experiments conducted by C. A. Bergman et al. with a replica of an Egyptian composite bow demonstrate the power and effective range of this bow type (1988). Using arrowheads of four different weights these experiments demonstrated that under controlled conditions the composite bow is capable of casting projectiles weighing between 25 and 90 g a distance of 221 to 109 m respectively. These figures can be used therefore for establishing the effective range of the composite bow depending upon the weight of the arrowhead, thus between 100 and 200 m (see Figure 3 below).

As Miller et al. point out, the tactical advantage of the composite bow lay not only in its greater range but in its increased power, which permitted the firing of projectile points of different weights (1986:189ff.). As they observe, the best evidence for the use of the composite bow in the ancient Near East is derived from the many arrowheads that have been found, which were of varied weight and design because archers "could use heavy arrows at short range to pierce armour, or lighter arrows to harass the enemy at long range" (ibid.). Use of the former type of arrowhead is suggested by depictions of Amenhotep II and other New Kingdom pharaohs firing arrows (probably of bronze) through thick copper targets with the composite bow (*ANEP*, p. 137, no. 390).[20] The latter use of the composite bow for rapid firing of lighter arrows has also been quantified through modern experimentation, which has demonstrated that an expert archer probably could have fired approximately thirty arrows in three minutes (Miller et al. 1986:188).

In northern Mesopotamia, use of the simple bow was probably eclipsed by the composite bow as early as the EB III. This is indicated by a scene discovered in the pre-Sargonic palace at Mari (ca. 2300 B.C.) that illustrates the use of the double-curved reflexed composite bow that Yadin referred to as the "Scythian bow" (Figures 1.B and 2), because it was previously

[18] An anonymous 353-page Arabic manuscript on archery from the Garrett Collection at Princeton University Library discusses the construction of the composite bow (Faris and Elmer 1945). It is dated to the fifteenth century A.D. by the paper on which it was written (ibid., vi).

[19] Yadin's figures of 275 to 300 m for the effective range of the composite bow and 550 to 700 m for its maximal range are therefore unreliable.

[20] For a more complete list of sources for pharaohs firing arrows at copper targets with composite bows, see p. 37, n. 3 in McLeod (1970).

identified with the Scythians of the late Iron Age (1972a:91). Evidence for the composite bow also includes several depictions on fragments of Akkadian stelae. On one of these stele Naram-Sin is depicted as a god, holding a composite bow (see Figure 1.C) and battle-ax in his left arm (Yadin 1963:150). In another broken stele from Lagaš dated to the reign of either Sargon or Naram-Sin, a soldier is shown employing the composite bow in combat (Yadin 1963:151).

Figure 2. Third-millennium B.C. shield and double-curved reflexed composite bow from Mari (after Yadin 1972a:fig. 1).

The primary evidence for the presence of the composite bow among the inhabitants of the Levant during the Middle Bronze Age consists of an illustration from Tomb 3 at Beni Hasan in Middle Egypt, which depicts an Asiatic caravan from the southern Levant "guarded" by a male bearing a composite bow, a quiver, and an ax (Newberry 1893:pls. 30f.). Unfortunately, either a lack of preservation of bows or a cultural predisposition against their inclusion in tombs of the so-called "warrior burial" type (see Philip 1989:145f.) has meant that no Bronze Age archaeological remains of the composite bow from the Levant have been encountered.

Secondary archaeological evidence for the use of the bow consists of arrowheads and the casualties they inflicted during the Middle and Late Bronze Ages (see p. 44). Metal arrowheads of different weights (OB *šiltāḫum*, *CAD* Š/2, pp. 448–51), ranging from 8 to 40 g. (1–5 shekels) and as much as 48 g. (6 shekels) are attested in the Mari documents (see ARM I 38 and XVIII 5), as noted by Stephanie Dalley (1984:148). Thus professional archers appear to have selected arrowheads of different weights depending upon their objectives. Despite such evidence, it has been nearly impossible to distinguish metal arrowheads from spearheads in the archaeological record of the Levant. A discussion of arrowheads is entirely missing, for example, from Graham Philip's catalog of bronze weapons from the Early and Middle Bronze Ages (1989).

Kültepe-Kaneš in central Anatolia provides clear examples of Middle Bronze Age metal arrowheads. Numerous broad-headed, barbed arrowheads of various sizes were recovered from the Level Ib settlement (nos. 54–68 in Erkanal 1977:52, pls. 17f.), as well as Level II (no. 49, p. 51). These vary in size from 4.12 to 8.6 cm in length and 1 to 2.4 cm in width (weights not provided). Examples of similar arrowheads were also recovered from Late Bronze Age Anatolian sites such as Alishar Höyük, Alaça Höyük, Dündartepe, and Hattuša (ibid., 51–54, pls. 17f.).

Several "chisel-ended" arrowheads have also been found in the defensive fosse around the Middle Bronze Age site of Tell Hadidi on the Upper Euphrates (Miller 1983). The find-spot of these arrowheads suggests that they were used in an assault on the settlement at Hadidi during the early second millennium. They are also depicted in use by Middle Kingdom Egyptian troops during the Middle Bronze Age (Newberry 1893:pls. 13f.). Perhaps those targeted with such arrows were intended to be taken as prisoners or as slaves since projectiles of this type are thought to have been used to maim rather than kill.

Despite the abundance of evidence for arrowheads dated to the Middle and Late Bronze Ages from Anatolia and Mesopotamia, evidence for the use of either the simple or composite bow in the Levant during the Middle Bronze Age is ambiguous. Based on comparative evidence from these regions, the Levantine data suggest, in fact, that many of the smallest so-called spear or lance points should be reclassified as arrowheads (see p. 42). If this is indeed the case, then the shape of arrowheads used in the Levant during this period was essentially limited to those forms common also to thrusting weapons, thus making their identification much more difficult. It is also possible that arrowheads in this region were influenced by Egyptian types, such as the ebony points (see pp. 44ff.), which would not be preserved in the Levant.

Yigael Yadin suggested that the bow was primarily a weapon of siege warfare (1963:64), although it is difficult to verify this assertion since large-scale open battles during the Middle Bronze Age are neither depicted in Egyptian reliefs nor attested in the Mari texts. However, given that the recorded confrontations of the Middle Bronze Age nearly all involved sieges, Yadin's assertion is probably correct. His hypothesis may be supported by Jonathan Tubb's (1985:193) observation that arrowheads were less common during the early part of the Middle Bronze Age in the southern Levant than they were during the latter part of the Middle and Late Bronze Age, by which time siege warfare had reached its apex. Whatever the case may be, the gradual proliferation of the composite bow from the EB IV to the end of the Middle Bronze Age would have drastically affected the conduct of siege warfare. The increased range afforded by this type of bow meant that archers could harass a town's defenders at a distance of more than 200 m or set a town ablaze within about 40 m of the wall (see Figure 3).[21]

Although the early use of the composite bow in warfare remains to be clarified, it is generally acknowledged that it had its greatest impact when it began to be used in conjunction with the light horse-drawn chariot some time after ca. 1900 B.C. (Miller et al. 1986:182; Moorey 1986:208). Prior to the Middle Bronze Age, neither simple nor composite bows were used in combination with equid-drawn vehicles. The limited evidence for the adoption of the chariot during the Middle Bronze Age, discussed above, suggests that it took some time before the composite bow was widely used with chariotry. In Egypt, for example, techniques for the use of the composite bow with the chariot in pitched battles were not fully developed until the start of the Late Bronze Age.

The spread of the composite bow through the southern Levant to Egypt appears, therefore, to have been a rather slow process. The primary bow in Egypt during the Middle Bronze Age, for instance, remained the double-convex simple bow (for illustrations, see Yadin 1963:162f.). Because the composite bow is only attested in Egypt from the Late Bronze Age on[22]—that is, after the expulsion of the Hyksos—the weapon is thought by many scholars to have contributed to the Hyksos overthrow of the Egyptians. While this part of the composite bow's history remains enigmatic, it is clear that the Egyptians obtained this technology from eastern lands, presumably shortly before the expulsion of the Hyksos. New Kingdom texts such as *The Autobiography of Ahmose son of Abana* record the early use of the chariot and bow in Egypt—probably the composite bow—in battle (*AEL* 2, pp. 12–14).

c. *The Shield*

It is unfortunate that no artifactual evidence is available for the shield during the late third and early second millennium in the Levant. This is no doubt because shields, like slings and bows, were constructed from organic materials such as leather, wood, or reeds, as is suggested by Egyptian depictions of Asiatic shields in the tomb of Intef, dated to the reign of Mentuhotep II (Jaroš-Deckert 1984:pl. 7). In an elaborate siege scene the Asiatic shields are depicted with an ox-hide shape, which may indicate that they were constructed with wooden or wicker frames and covered with the hide of an animal. As a defensive device the shield's size and shape involved trade-offs between effective protection of the soldier and the ease with which the shield could be manipulated (Yadin 1963:14).

While the shield is often considered a defensive device, it played a significant supporting role for attacking archers, since the degree to which they were protected directly affected their effectiveness. In Mesopotamia full-length shields seem to antedate their use in Egypt by several hundred years. Such a shield, featuring a top that curved back over the archer, protected the archer and shield-bearer from the defender's fire (see Figure 2). Prior to the discovery of the Mari scene, the full-length shield was thought to be a Neo-Assyrian innovation. Further evidence for the use of a large shield in Mesopotamia during the late third millennium B.C. comes from the "Stele of the Vultures," which depicts a "phalanx" of soldiers from Lagaš bearing rectangular shields and spears (ibid., 134f.).

Although the shield was in use even before the third millennium in Egypt (e.g., Hierakonpolis tomb 100, Yadin 1963:117), its first widespread employment appears to have begun during the Middle Kingdom when two types of Egyptian shields were used. The full-length shield with a rounded top and flaring wide bottom was used by stationary soldiers (see Newberry and Griffith 1895:pl. 29), while shields of similar shape but half as tall were carried by infantry (see Newberry 1894:pl. 15). It is clear from these depictions that the shield's primary purpose was to protect against enemy projectiles since no shield-bearing Egyptians are depicted in hand-to-hand combat but shield-bearing soldiers are always depicted in scenes showing enemy archers.

---

[21] Through experimentation R. Miller et al. (1986:191) have determined that a flame arrow could not have remained lit if fired from a composite bow at full draw, but could be fired at half draw, achieving a height of 30 m and a distance of 40 m.

[22] See bows from Tut'ankhamun's tomb (McLeod 1970).

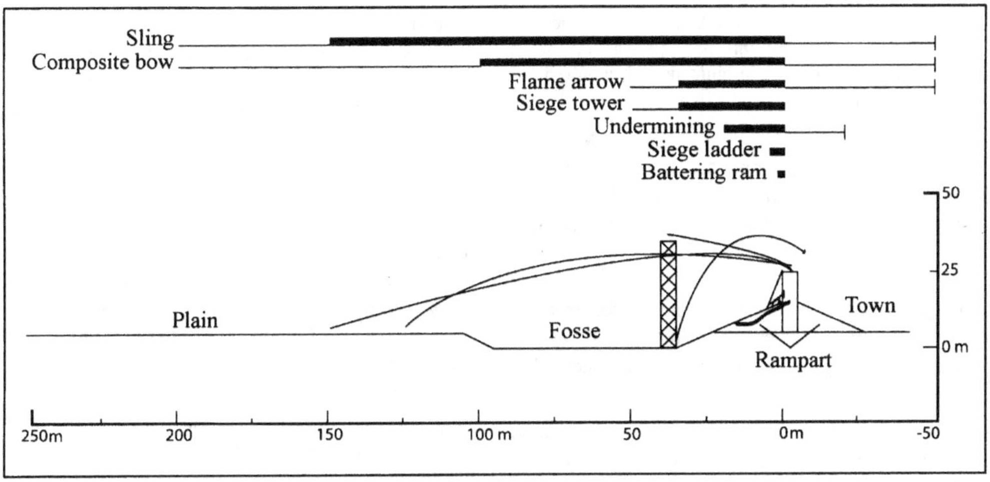

Figure 3. Ranges of Bronze Age weapons: maximal (thin lines) versus effective ranges (thick bars). Dimensions of the fortifications are based on Qatna (Tell Mishrife).

It is noteworthy that none of the Asiatics depicted in the Middle Kingdom Beni Hasan reliefs carry shields, even though they are shown carrying axes and spears. Furthermore, no archaeological evidence of the shield has thus far been unearthed in the Levant for either the Middle or Late Bronze Ages. Therefore, aside from the pictorial evidence that shows small ox-hide shields in use by Asiatics at the start of the Middle Bronze Age in Egypt, the earliest evidence for the form and construction of the shield in the Levant is derived from New Kingdom reliefs (Yadin 1963:83f.). These shields appear to be small, light, and rectangular, and were clearly intended for personal protection in hand-to-hand combat. Egyptian shields during this period were of similar dimensions having evolved from their Middle Kingdom predecessors (ibid.).

d. The Sail

While the sail has received little attention, evidence suggests that the sail was employed by Canaanites[23] during the Late Bronze Age as a countermeasure to enemy projectiles that were fired into towns (Burke forthcoming). Several Egyptian New Kingdom reliefs of Canaanite towns show what appears to be a ship's sail flying above the city in the midst of a siege (see siege of Dapur in Wreszinksi 1935:pls. 78 and 109). In one case it is pierced by arrows while in another by javelins. Were it not for Aeneas Tacticus's work this feature may have remained completely enigmatic. He advocated the use of sails to "offer protection against missiles" (*Siegecraft* §32.1 and §32.9–10), in addition to building wooden towers and erecting sand-filled baskets, stones, and bricks (see *Siegecraft* §32.2).

2. Siege Machinery

Besieging armies would have employed slingers and archers to harass a town's defenders or even to set a town ablaze, supported by chariot-based archers who tracked down those attempting to escape. However, projectiles alone did not provide access into a town during a siege. As Yadin noted, four main modes existed by which a town's defenses could be overcome (1963:17). These included the escalade (i.e., climbing over the walls), going through the gate or a breach in the fortification walls, tunneling under the defenses, or employing a ruse.

Historical evidence, especially from the classical period (see Kern 1999), demonstrates that successful entry to a town during a siege by scaling, undermining, breaching, and/or the use of a ruse almost always guaranteed the victory of the attacking force. This is because the town's defenders relied almost entirely upon these fortifications, which constituted the most important element of their defensive efforts. Such success was usually possible by means of the simultaneous employment of several of these tactics at different points around a town's defenses, which wore down the defenders and prevented them from concentrating their efforts against a single approach. This is evidenced in the following Mari text:

[23]In this work the term Amorite is used to identify the Semitic inhabitants of the Levant during the Middle and Late Bronze Ages. The term Canaanite is reserved for the Amorite inhabitants of the region identified as Canaan in historical texts from the Late Bronze Age onward.

> As soon as I had approached the town of [Qîrḫadat], I set up a tower (*dimtam*) and made its wall fall down by tunneling [*pilši*], and in eight days I captured the town of Qîrdaḫat [*sic*] (ARM I 135:4–13 as trans. by Dalley 1984:146).

In this case both siege towers and tunneling were used, and we may suppose that siege ladders, although not mentioned, were also part of this strategy.

a. The Siege Ramp

A common siege technique involved the construction of a siege ramp out of earth against a wall (*eperu*, *CAD* E, pp. 187f.). The technique is again attested in the Mari texts:

> The town of Nilimmar that Išme-Dagan besieged Išme-Dagan seized. As long as the earth(en ramp) [*epiru*] did not reach the height of the top of the city (wall) [*mulē qaqqad ālim*] he could not conquer the town, he could not seize (it). But when the earth(en ramp) [*epiru*] reached the height of the top of the city (wall) [*qaqqad ālim*] he conquered this (city) [ARM I 4:9–11, trans. in *CAD* E, p. 188].

The goal, therefore, was on certain occasions to build a gently sloping ramp up which troops or perhaps a battering ram could be moved. While such efforts were sometimes successful, they may often have been thwarted by similar earthworks constructed within the town by its inhabitants. Such earthworks were built against the interior face of the wall and were intended to buttress the area that was about to be battered by the ram.

b. The Siege Tower

The Mari archives identify siege towers among the devices employed in Middle Bronze Age siege warfare in Mesopotamia (ARM I 131:10ff., II 7:8ff., XIII 146:15ff.).[24] These are described using the Old Babylonian term ^{GIŠ}*dimtum* (*CAD* D, pp. 144ff.). Unfortunately, the lack of detail in these references and absence of archaeological evidence for these towers makes it difficult to reconstruct them and to determine their precise function. The Mari texts suggest that such towers often needed to be built on-site in contrast to battering rams, which could be delivered to the town by means of wagons or boats, as in the following Mari text.

> My lord wrote to me about sending downstream to Mari some ropes to go around siege towers [*dimātim*] and a battering ram [*yašibim*] (ARM XIV 45:5–7 as trans. by Dalley 1984:145):

Iconographic evidence for the siege tower is nevertheless provided by Middle Kingdom Egyptian reliefs from the Tomb of Intef (no. 386) in Egypt (see discussion in Gaballa 1976:38f.). In this case the open tower built of timbers was wheeled up against the wall of an Asiatic town (ibid., fig. 3). Although siege towers potentially functioned as large, protected ladders as depicted in this Egyptian relief, the fosse and sloping ramparts of Middle Bronze Age fortifications were probably successful in preventing this. For this reason, siege towers most likely functioned as freestanding firing platforms situated at the foot of the rampart, from which archers would provide suppressing fire to protect sapping, tunneling, breaching, and escalade operations. If a siege tower was tall enough, archers of the besieging force could fire down upon soldiers on the battlements or within the town. This function is noted by Vegetius Renatus (*Epitome*, Book IV.17). As Figure 3 demonstrates, the height required for a siege tower to be used in this fashion against the most formidable of Middle Bronze Age defenses, such as those found at Qatna, would have been about 30 m. Even if such a tower could not have been pushed up the rampart toward the walls, its military value can be readily understood since archers, especially when armed with composite bows, would have been as little as 30 to 50 m from the town wall.

c. Breaching, Tunneling, and Sapping

Reliefs from an Old Kingdom Egyptian tomb belonging to Iny at Deshashe appear to illustrate the act of breaking through a town wall. Two individuals are depicted in this scene using crowbar-like implements on the corner of a schematically portrayed fortress as a ladder is raised against the wall behind them (see Schulman 1964:15). This third millennium B.C. depiction is probably typical of the most basic means of penetrating a fortified wall in antiquity, if the defenders were sufficiently distracted. But classical sources indicate that such attempts usually faced stiff resistance and quite frequently failed.

When using ladders to scale the town's defenses proved ineffective and simply breaking through the wall at ground level was impossible, besiegers could attempt to undermine the fortifications by means of tunnels. This was done in the hope that the wall would collapse into the excavated cavity (Sasson 1969:184). Such breaches may have been described using the Old Babylonian term *pilšum* (*CAD* P, pp. 378ff.), which occurs in the Mari texts and refers either to an above-ground breach in a town's wall or possibly to a tunnel used for undermining the wall or

[24]See also discussion by Sasson (1969:33 and n. 182) and Salonen (1965:26f.).

entering the town.[25] Unfortunately, there is still no clear archaeological evidence for sapping tunnels dated to the Middle Bronze Age, though this may be wholly due to limited exploration of town defenses. Nevertheless, such an identification has been suggested for the so-called "sappers' tunnel" at Lachish, although there is no evidence that this feature was excavated before the Iron Age (on its date see S. Yeivin 1951).[26] Middle Bronze Age tunnels identified at Hazor and Tell el-'Ajjul, though located in different locations relative to the defenses, are similarly difficult to identify.

d. The Battering Ram and Siege Pole

Scaling ladders, tunnels, and earthen ramps were rather basic techniques of siege warfare that required considerable energy and time but probably remained largely unaltered from the third millennium B.C. until medieval times. Several new techniques, however, were invented ca. 2300 B.C. which were intended to improve the effectiveness of a besieging army. The first such device is the battering ram. Although the role played by the battering ram in Neo-Assyrian conquests of the first millennium B.C. is well known, the importance of the battering ram during the late third and second millennia is not widely recognized.

Although no physical evidence of battering rams or their accoutrements is known between ca. 2300 and 1500 B.C., the Ebla texts confirm that the battering ram was in use near the start of this period. Piotr Steinkeller (1987) notes that four references in the Ebla texts comprise the earliest evidence of the battering ram.[27] These references include the logogram GUD.SI.DILI, which in Sumerian means the "one-horned bull," though the Eblaite term remains unknown. Of these four references in the Ebla texts only one involved the actual use of the battering ram, while the other three references allude to the maintenance of the weapon. The Old Babylonian term for the battering ram, *yašibum* or *yašubum* (*CAD* A/2, pp. 428f.),[28] is attested in numerous texts from Mari (ARM I 131:12; II 7:12; V 2:13; VI 63:6; VII 16:2, 63:2, 69:3; XIII 146:16; XIV 45). Unfortunately, these references are brief and give little insight into how the device actually functioned.

In order to understand the battering ram's function we must rely upon evidence from the Iron Age. A reconstruction of a Neo-Assyrian battering ram based on terminology derived from texts and comparative evidence from classical sources has been made by Jo Anne Scurlock (1989). This reconstruction is suggestive of how such rams may have functioned during the EB IV and Middle Bronze Age (Figure 4). It is thought that the device featured a beam or tree trunk that was hung from a frame by four chains and fitted with a metal point or "tooth" that served as the ram's head.

Figure 4. Reconstruction of a battering ram of the Neo-Assyrian period (after Scurlock 1989:131).

In Egypt during the Middle Kingdom the battering ram appears to have assumed a slightly different form. As illustrated in Tomb no. 2 at Beni Hasan, dated to 1910 B.C., three Egyptian soldiers are shown wielding a long, skinny pole against a fortification wall (see Newberry 1893:pl. 14). This device appears to have functioned to pick apart the bricks of the fortification rather than to batter them. It seems appropriate, therefore, to identify this instrument not as a battering ram but as a "siege pole."

Although later versions of the battering ram, especially those used in Greece, featured some type of protection for those maneuvering the ram, it is possible that individuals were often protected only by large shields such as those depicted in a third-millennium illustration from Mari (see Figure 3). However, Egyptian reliefs also illustrate a covering over the men using the siege pole, which is identified as a *mantelet* (Newberry 1894:pl. 5), the French word for canopy. It is impossible to infer from these depic-

[25]See ARM I 135:9, I 118:12, and III 37:17, as well as discussion by Erkki Salonen (1965:36f.).

[26]For a photo of the tunnel, see pl. 6 in *Lachish IV* (Tufnell 1958); for the course of the tunnel, see pl. 90. Its identification as a "sappers' tunnel" is dubious for whatever period because the tunnel parallels the line of the defenses for more than 20 m and never made any turn beneath the line of the city wall.

[27]See Sollberger (1980:figs. 27a vii 20–viii 6, and ix 3–12 in same figure); also ARET 4:2 viii 5–11, 4:13 rev. 13 vi 12–16.

[28]See also discussions by Jean-Robert Kupper (1951:125f). and Salonen (1965:29–31).

tions what materials were used to construct the mantelet, but it must have been sufficiently thick to stop arrows and other projectiles, and prepared against the use of fire by the town's defenders.

e. Escalade: Siege Ladders

Despite the wide array of siege tactics available, scaling ladders—the oldest and crudest means of mounting a town's walls—were probably the most common siege device. Although the main disadvantage to using the siege ladder would have been the accompanying casualties for the attacking force, once the soldiers were atop the wall they were in an advantageous position, owing to the elevated ground from which they could begin to overpower the town's remaining defenders.

Although no direct references to ladders are known, ladders were probably specially produced in different lengths to attack various locations around a town. In unforested regions they would have been borne on carts or boats to the site of the siege since obtaining timber of sufficient length for building them would have been difficult. Our only direct evidence for the use of the scaling ladder comes from Egypt. One example from a tomb at Deshashe (Dyn. V) shows a ladder being moved into position to scale the wall of an Asiatic town. This town is illustrated as a stylized oval enclosure, as if it provided an aerial view of the town (Yadin 1963:146). In another example, on a wall painting from the tomb of Kaemheset at Saqqara (Dyn. VI), a ladder with wheels at the bottom is depicted being used by soldiers to scale the wall of a town and tear it down with axes (for an interpretation of its use, see Yadin 1963:147). It is likely, however, that the relatively steep slopes of the ramparts around many Middle Bronze Age settlements in the Levant and northern Mesopotamia made it difficult to use ladders with wheels, if not also regular scaling ladders.

f. Countermeasures against Siege Machinery

In light of the siege techniques discussed above, it is necessary to consider the types of countermeasures that were employed to thwart a siege. Aeneas Tacticus, for example, described a variety of countermeasures that could be employed against siege devices during the classical period, even including the battering ram (*Siegecraft* §32.3–7). Aeneas also noted that fire was commonly employed against siege towers (§32.8–§33.2), which was equally effective against mantelets. Nevertheless, he also observed what could be done to protect wooden structures in the town (or similarly the siege machines themselves) against fire (§33.3–34.2), as might have been set by using flaming arrows. Countermeasures against siege ladders (§36.1–2) and tunneling (§37.1–5) were also described in detail in *Siegecraft*.

With an awareness of such literature, attempts have been made to identify the features of Middle Bronze Age fortifications against which siege techniques were intended to defend. Lawrence Stager, for example, suggests that the fosses built outside the ramparts of Middle Bronze Age Levantine towns like Ashkelon and the great thickness of the earthen ramparts were intended to prevent sapping or tunneling efforts.

> While the city was under siege, a team of excavators from the attacking army would begin their tunnel at some distance from the fortification line they wished to undermine. Their object was to cause the fortifications to collapse or to sneak beneath them and then to surface inside the city, usually at night, to launch a surprise attack. It might take days, even weeks, for the "moles" to reach their objective. Once under the fortifications they might widen the tunnel in order to collapse the defense works above, or if that failed, to stoke the widened tunnel with combustibles, which would then be burned in order to precipitate collapse, while assault troops penetrated the breach above ground (Stager 1991b).

Stager argues that the amount of earth that needed to be moved because of the earthen rampart provided defenders with critical time to "spot the sappers and trap them or smoke them out" (ibid.).

This argument, which is inspired by accounts from the classical period, does not, however, account for the geometry of Middle Bronze Age ramparts. For example, let us consider the labor involved in the undermining of a fortification wall accompanied by an earthen rampart of the largest size attested, which featured a 30° slope. Assuming that a tunnel was begun 50 m away from base of the foundation of the fortification wall, a tunnel dug just below but parallel to the rampart's surface (i.e., along the longest line between the tunnel's entrance and the wall's foundation) would have been about 55 m. The distance would, in fact, be only 5 m more than if the same tunnel were dug from a distance of 50 m toward a fortification wall that lacked an earthen rampart. The rampart only increases the time required to accomplish the same task by ten percent, which would at the most constitute a matter of only a couple of days more labor. Furthermore, the difference in time decreases for ramparts of the same size when the tunnel's incline is in any way reduced or if the tunnel is begun anywhere up the slope of the rampart from the base. These figures are, of course, a maximal estimate of the effort required by the attacking force to dig tunnels. Nevertheless, they demonstrate that ramparts did not buy defenders significant amounts of time or strategic advantage against tunneling efforts.

Several other factors also argue against the value of the rampart and fosse in thwarting efforts to sap walls. To begin with the size of the rampart is irrelevant to the sappers' efforts. This is the case because the nature of the sappers' work and their strategy are not dictated by a rampart's size. As I argue above, if sappers could begin a tunnel on a 30° slope 30 m away from the wall, then we have no reason to believe that under the same circumstances they would have begun any farther from the wall on a larger rampart (e.g., 50 m wide) with a similar slope. On the slopes of the rampart the greatest threat to the attacking force would have been from arrows and missiles fired from the wall. Realizing that at 50 m the composite bow—not to mention the sling—would have been within its most effective range, the difference between the threat posed by these weapons at a distance of between 30 and 50 m was irrelevant to the efforts of the sappers.

This scenario is also supported by the realization that sappers could be protected with large shields, like those discussed above (Figure 2). Devices of various designs, as described by Aeneas, could be put in place to protect the workers (see *Siegecraft* §37.8f.). Though scouting parties could be sent out by the defenders to thwart these efforts, it must be remembered that to do so always jeopardized the town's security by requiring a gate to be opened—an action discouraged during a siege (see *Siegecraft* §28.1–7, §38.8).

Finally, in a number of instances, as demonstrated in the next chapter, it is evident that artificial earthen ramparts were probably less dense than the soil and surface of the surrounding landscape. They would actually have made undermining easier, despite the risk of collapse. Moreover, some ramparts were even constructed of sand. If the rampart is understood as a means of deterring sappers, then the use of the rampart in the highlands of the southern Levant would have only have aided the efforts of sappers by allowing them to dig through earth rather than chisel away at bedrock as they approached the wall. The greatest obstacle to the success of those digging tunnels would not, therefore, have been the defenders on the walls rather than the composition of the rampart, even though this dictated the rate at which tunneling operations could be completed.

In short, the strategic value of the rampart and fosse was that they offered obstacles to advances against the town's fortifications. Devices such as the battering ram, ladders, and siege towers were positioned against the wall by sheer muscle power. Thus, the rampart hindered the approach of soldiers laden with armor, bearing battering rams, and ladders, and pushing siege towers. However, of these three general threats it must be emphasized that the most likely reason for the construction of ramparts was undoubtedly the introduction of the siege tower. While battering rams and ladders could be moved up the slopes with a little more effort, the siege tower was not a mobile feature that was easily advanced once it was built. Likewise, it was not adapted to sloped surfaces. These towers functioned, instead, as fixed firing platforms, as noted above. The fosse, furthermore, enhanced the ramparts' effectiveness by limiting the range within which the siege tower could be brought against the wall and the ease with which it could be withdrawn. Therefore, the fosse and rampart served first and foremost against the siege tower, and only incidentally complicated the efforts of attackers to move both ladders and battering rams against the fortifications.

3. Close-Range Weapons

Following a successful breaching of a town's fortifications soldiers and their kinsmen would have employed almost every means possible to protect themselves and their town from total ruin. The weapons they wielded would have included the mace, club, staff, and even the occasional millstone. Although least effective, it is evident from New Kingdom reliefs that during times of desperation Canaanites even hurled stones from the walls against their Egyptian aggressors (e.g., Wreszinksi 1935:pl. 109).

Equally ineffective would have been the use of the staff or club against armed soldiers as they entered the town. That the club was still used at the end of the third millennium and early second millennium among Asiatics is evident from Middle Kingdom tomb reliefs at Beni Hasan (see esp. Tombs 2 and 3, Newberry 1893:pls. 16 and 30–31). The clubs depicted in these reliefs are wide on the striking end, narrow at the handle, and curved like a bow, similar to those shown in reliefs among contemporary Egyptian soldiers. In light of the introduction of the ax during the Early Bronze Age, it would seem that the club rarely, if ever, functioned as the sole short-range weapon of Levantine soldiers. The three Asiatics depicted on the East Wall (south side) of Tomb 2 at Beni Hasan each bear another weapon in addition to the club; the front warrior bears both the eye-ax and the club, while the two soldiers after him bear clubs and spears (Newberry 1893:pl. 16). It appears that the desired complement of weapons for the average soldier included weapons of varying ranges, such as the spear and the club. This appears also to have been the case in the early second millennium as attested in warrior burials, where individuals were usually interred with both a duckbill ax and a spear (see Garfinkel 2001:Table 8.3).

a. The Mace

Although by the third millennium the ax had replaced the mace as the leading weapon in hand-to-hand combat, the mace continued in use as a ceremonial weapon through the second millennium in the Levant (the same phenomenon is also evident in Mesopotamia, Gibson 1964:8–13). In addition to the Egyptian evidence of this, a Middle Bronze Age (Mardikh IIIB) ceremonial mace made of ivory, silver, and stone was discovered in the "Tombeau des citernes" at Ebla (see Matthiae 1980a:fig. 6). Despite such ceremonial uses, finds like the macehead mold from Tel Dan, which is dated to the late Middle Bronze Age or early Late Bronze Age (Minoff 1992), also suggest that the mace was being produced in more than "one-off" or ceremonial quantities. Eli Minoff argues that the stone macehead mold found at Tel Dan (Area K L.6375.) and dated to ca. 1500 B.C. must have been used for repetitive casting. The very plain shape that this macehead featured when taken in conjunction with Minoff's suggestion makes it possible to speculate they were intended for use by soldiers or garrisoned troops acting as a police force. This interpretation is supported by an additional example of a piriform, limestone macehead found in the remains of Fortress M at Ebla (Mardikh IIIB, Matthiae 1980b:123). The discovery of a number of maceheads in private houses at Nippur during the Late Bronze Age may also indicate a similar purpose (Gibson 1964:14).

b. The Spear

More effective than the mace were the spear, the ax, and the dagger. If a town's defenders were fortunate enough to retain some distance between themselves and their attackers, the first weapon to be used, in addition to the bow, would have been the spear. The spear, when employed in hand-to-hand combat, was intended to give the soldier the advantage of striking the enemy first from a distance before addressing him, if necessary, in hand-to-hand combat.

Despite a lack of evidence for a distinction in Old Babylonian terminology, it is necessary to distinguish between the spear and the javelin. The term spear should be applied to thrusting weapons, which were retained by the soldier, although the spear could, if necessary, be thrown short distances with effect. This is in contrast to the javelin, which was designed specifically to be thrown. Given the two primary applications for what appear to have been nearly identical weapons, it must be realized that the specifications for the construction of these weapons would have varied under optimal conditions due to the stresses related to their use.

The greatest stress was, of course, placed upon the javelin, which was required to survive routine impacts at high velocity against potentially hard targets so that it could be reused. In an ideal situation, therefore, the javelin's head would be constructed of the strongest possible metal with the most secure means of attaching it to the shaft. During the Bronze Age the availability of metals able to meet these requirements obviously limited the options and, therefore, modifications in the design of the head using bronze were crucial to successfully fashion a weapon that could be used as a javelin. Despite the theoretical design requirements, distinguishing between the two as archaeological artifacts is particularly difficult and at times seemingly impossible.

The spear was clearly an important weapon among Mesopotamian armies until the end of the third millennium, as the textual, iconographic, and archaeological sources indicate. The depiction of the troops of Lagaš armed with spears in what appear to have been a close phalanx-type formation illustrates its use in open battle during this period (see Yadin 1963:134f.). Middle Bronze Age reliefs from Ebla demonstrate that armies in the Levant employed similar weapons (Matthiae et al. 1995:figs. 290f., pp. 421f., 435).

Despite the prevalence of weapons identified as spears in Mesopotamia, both Graham Philip and Jonathan Tubb have noted that the spear is seemingly absent from the archaeological assemblage of the Levant during the second half of the Middle Bronze Age and much of the Late Bronze Age. The fundamental problem, however, remains that these weapon types are usually identified by the size of the weapon's head. If this principle were applied uniformly, then the smallest weapon heads would be identified as arrowheads, while larger heads would belong to spears, and javelins would fall somewhere between these two.[29] However, the situation is more

[29]Philip assumed, for example, that even MB IIA points a few centimeters long and about 1 cm wide were socketed spearheads rather than arrowheads (e.g., types 7 and 9 of small socketed spearheads 1989: 94, e.g., fig. 24). Philip's neglect, however, of thirty-one arrowheads from the Early Bronze Age Kefar Monash is problematic (see photo in Bar-Yosef et al. 1993). These items are easily distinguished from four daggers and four spearheads in this assemblage (ibid., 613). It is, therefore, difficult to accept Philip's as-

complex than this, especially when, as noted above, arrowheads of various shapes and different weights were in use. Despite such evidence and a number of other methodological problems, Graham Philip concluded that the spear was absent from the warrior's arsenal from the MB IIB–C through the end of the Late Bronze Age (contra Tubb 1985:193f.; Philip 1989:99).

There is simply insufficient evidence to support the assertion that spearheads entirely disappeared from Palestine during the MB IIB–C. The dearth of evidence for the spear in burial contexts during the MB IIB–C may be simply the result of the declining prestige associated with its use. As warrior burials evolved during the later part of the Middle Bronze Age, the status of the elite professional warrior was seemingly no longer associated with the spear, ax, or dagger, but instead with the composite bow and chariot. The ubiquity of spears probably meant that they were no longer symbols of a privileged warrior class and thus were not included in tomb assemblages. Therefore, like the mace the spear became the mainstay of ordinary garrison troops and foot soldiers.[30]

c. *The Ax and the Dagger*

If a siege had advanced to the point where a town's inhabitants were required to engage in hand-to-hand combat in an attempt to repel their attackers, weapons such as the ax (OB *pašum*; *CAD* P, 267f.) and the dagger (OB *paṭrum*; *AHW* 2, pp. 849–51) would have played a vital role. As J. N. Postgate observes, the weapons carried by the soldiers depicted on the southern Mesopotamian "Stele of the Vultures" (ca. 2450 B.C.) included axes, adzes, and spears (1992:248f.). Given the frequency of the battle ax among warrior burials of the EB IV and the Middle Bronze Age, much could be said concerning its role. However, since neither the ax nor the dagger played a major role in siege warfare per se, an extensive discussion of these weapons is unnecessary.

It is noteworthy that the development of the ax (and the adze) as weapons was probably the result of the adaptation to warfare of commonly used tools (see de Sarzec 1894–1912:pls. 3a, 4c). This underscores how many implements could have been used as weapons, although the frequency with which a given type of object was used is difficult to determine. On the basis of the available skeletal evidence, the ax was identified as the culprit in the maiming and probable death of one notable individual, the Egyptian king Seqenenre Tao II (ca. 1550 B.C.). Because hand-to-hand combat constituted the last phase of a siege, and many sieges probably did not reach this stage before a town capitulated, it is possible that far fewer casualties were caused by the ax, dagger, or sword than by projectiles such as arrows (see pp. 44ff.).

4. The Ruse

> All warfare is based on deception. Therefore, when capable, feign incapacity; when active, inactivity. When near, make it appear that you are far away; when far away, that you are near.
> Offer the enemy a bait to lure him; feign disorder and strike him.
> When he concentrates, prepare against him; where he is strong, avoid him.
> Anger his general and confuse him.
> Pretend inferiority and encourage his arrogance.
> When he is united divide.
> Attack where he is unprepared; sally out when he does not expect you (Sun Tzu 1963:66–69).

The essence of this statement by Sun Tzu was well understood in western Asia during the Bronze Age. While the most familiar vignette concerning the ruse is the tale of the Trojan horse in *The Odyssey*, an Egyptian account called *The Taking of Joppa* (*ANET*, pp. 22f.) preserves an equally impressive use of the ruse. This story, which is ascribed to the reign of Thutmose III in the fifteenth century B.C., describes the taking of the port town of Jaffa on the southern Levantine coast. Egyptian soldiers were hidden in baskets that were thought to contain supplies for the town.

Although there is reason to believe that this piece of literature may not refer to an actual historical event, it does indicate that the ruse was well known as one means of entering a besieged town, even though the chances of success for such an operation must have been slim in many instances (see also *Siegecraft* §29.3–10). Aeneas Tacticus warned of the need for passwords because deceitful tactics were commonly employed to gain entry to a besieged city (*Siegecraft* §24.1ff.). Such tactics were no doubt employed in order to reduce the number of casualties incurred by the attacking force.

---

sertion that the spearheads depicted in the Beni Hasan reliefs are examples of the small spearheads he identifies (Philip 1989:99); there is a lack of reliable scale in these depictions by which to make such determinations.

[30] A bronze spearhead dated to the MB IIB–C and inscribed in cuneiform with an individual's name was found in Fortress M at Ebla (Matthiae 1980b:123; see figs. 159–160 in Matthiae 1995). The findspot supports its use by garrison troops. The inscription may be compared with an inscription on an MB IIB–C dagger found at Lachish.

E. Mortuary Evidence for Siege Warfare

While discussing the categorization of weapon types is useful for understanding Bronze Age warfare, in the absence of detailed historical documentation it is still necessary to distinguish between weapons in the archaeological assemblage that only served symbolic purposes and those that were routinely used. This is best demonstrated through the recurrence of iconographic motifs that were retained for symbolic purposes. The motif of the Egyptian pharaoh smiting enemies with a mace, for example, is an archaizing motif particularly once it is employed after the Middle Kingdom; it reflects an era when the mace more or less dominated hand-to-hand combat. Thus, it is necessary to clarify the precise manner in which particular weapons were actually employed.

Human remains from mortuary contexts can help us on both counts. The physical evidence of wounds and casualties suggest the frequency with which various types of weapons were actually employed to bloody effect, as well as indicating exactly how they were intended to be used. The main obstacle to this approach is the small size of the excavated sample. Mortuary evidence from the Levant and neighboring regions provides, nonetheless, considerable insight into the mechanics of warfare in the Middle and Late Bronze Ages.

One of the best-known examples of military casualties during this period is a collection of corpses from Egypt dated to the end of the Eleventh Dynasty. Fifty-nine mummified Egyptian soldiers found in a communal tomb near Deir el-Bahri were identified by Herbert Winlock (1945) as troops of Nebhepetre Mentuhotep II (ca. 2043–1992 B.C.). Most of these soldiers sustained mortal wounds from arrows, many of which appear to have had ebony heads, although a few of the arrowheads were apparently of the blunt type (ibid., 11, 13). Entry wounds on many of the soldiers indicate that arrows were fired from above, as if during a siege (ibid., 14). Four of the soldiers also sustained head wounds prior to their participation in a final battle, and three of these suffered injuries to the left sides of their heads, indicating that their attackers were right-handed (ibid., 9).

In addition to arrow wounds and head traumas inflicted by blunt objects, Winlock observed frontal wounds on fourteen other soldiers that were probably inflicted by stones (if not sling projectiles) apparently hurled from above (ibid., 14f.). It also appears that a number of the soldiers were not immediately killed by arrow wounds but were finished off with blows to the head by blunt objects such as clubs or even maces (ibid., 18), though it is possible that these final wounds were inflicted by objects thrown or dropped upon them from the town wall. After death the soldiers' bodies had lain exposed for some time since flesh was missing on many of them. This led Winlock to speculate that they were gruesomely "pecked at by carrion birds" (ibid., 11). He concluded that these soldiers were probably involved in a siege on Herakleopolis under Mentuhotep II (ibid., 24).[31]

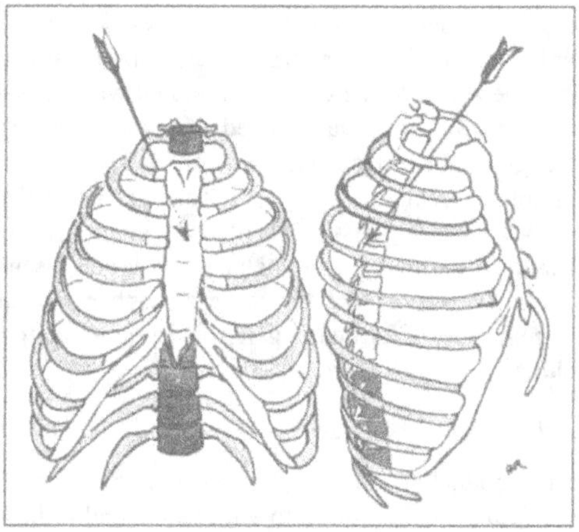

Figure 5. Chest cavity with arrow lodged in vertebrae from Ugarit (after Jarry 1939:fig. 1).

One clear example from the Levant of an individual killed under similar conditions comes from Late Bronze Age Ugarit, ca. 1300 B.C. (Jarry 1939). The remains of a man between eighteen and thirty years of age were recovered (Figure 5). He had been killed by an arrow that pierced his chest from above and in front of him, which became lodged in his spine within his chest cavity. The arrowhead was so deeply embedded that it was impossible to remove it. The angle of the entry wound indicates that the arrow was fired from a sharp angle above the victim, and the depth of penetration suggests that it was shot from close range, perhaps no more than 50 m. The depth of penetration of the arrow also suggests that the powerful bow from which the arrow was fired was of the composite type. Most likely, as with the soldiers en-

[31] In a recent reappraisal of these burials Dorothea Arnold suggests that these soldiers should be redated to the reign of Sesostris II in the Middle Kingdom ("Rethinking the Early Middle Kingdom at Thebes," Second Theban Symposium in Baltimore, September 16, 2000).

tombed a Deir el-Bahri, this individual was killed while participating in a siege and his body was brought back to Ugarit for burial.

Another slain Egyptian from the end of the Middle Bronze Age further demonstrates the effectiveness of the battle ax and the injuries that could be sustained from it. The remains are those of the Theban king Seqenenre Tao II (Dyn. XVII), who was found at Deir el-Bahri and dated to ca. 1550 B.C. (G. E. Smith 1912:1–6, pls. 1f.). This SIP Theban ruler was apparently killed by five blows to the head (no other wounds were found on the body). At least three of the blows are thought to have come from an Asiatic-type of battle ax. He was presumably killed during an early attempt by the Thebans to expel the Hyksos from Lower Egypt. The other wounds to his head were probably caused by a spear or sword, and in one case by a club or the handle of an ax or spear (ibid., 5f.).

To these casualties of war it is possible to add several others known from burials in northern Mesopotamia and the Levant. At Tuttul (Tell Bi'a), for example, the remains of eighty individuals buried in a mass grave were discovered in layers of the central mound dated to the reign of Šamši-Adad, ca. 1719–1688 B.C. (Strommenger 1991:13–15). The corpses were laid haphazardly in a single grave (ibid., fig. 4) and it appears to be possible to distinguish wounded soldiers from civilians. Similar evidence of carnage comes from the siege of Ebla in the northern Levant at the end of the Middle Bronze Age. Here a mass grave dating to ca. 1550 B.C. was recently discovered on the outside slope of the rampart to the east of the Area EE fortress (Matthiae 2000:591, n. 59 and fig. 18). Six soldiers were buried in the grave. Roughly contemporary with these casualties were another seven "armed men" found slain in the guard chamber of the Level VII palace at Alalaḫ (Woolley 1953:80). In the southern Levant the only Middle Bronze Age site that has produced similar evidence of persons slain in combat is Shechem (Tell el-Balatah). Here six skeletons, two of them fully articulated, were found on the inner steps of the East Gate among fallen bricks and the destruction debris dated to the end of the Middle Bronze Age (Campbell 2002:137, 39). At Gezer a woman was found crushed in the debris of the destruction of Stratum XVIII by the Thutmose III, which was dated by Dever to 1468 B.C. (1997b:397).

This array of Bronze Age casualties of war, mostly from the Middle Bronze Age, makes it possible to draw some important conclusions concerning warfare in this period. The first conclusion is that siege warfare was as frequent and as dangerous as has been suggested based on both references among the Mari texts and Egyptian reliefs. Although the sample size is small, the nature of the wounds inflicted on the corpses suggests that sieges and not pitched battles were responsible for many of the casualties incurred by armies in this period. From these casualties we can also determine the types of weapons employed in battle during the Middle Bronze Age. These weapons, not surprisingly, included arrows of various weights and head-types fired from bows, piercing weapons such as daggers and spears, and hacking weapons such as the battle ax. Evidence of blunt trauma on skulls also suggests simple weapons such as the mace or club, which were used to finish off the wounded.

F. The Military Context for Middle Bronze Age Fortification Strategies

From this survey of archaeological, textual, and iconographic evidence for weapons, tactics, and casualties of war during the Early and Middle Bronze Ages, it is possible to offer some general conclusions about the development of warfare over the course of this period. At first glance weapons and tactics might appear to have remained relatively static, centering on siege warfare and a limited arsenal. However, subtle yet noteworthy developments in military techniques did occur, which can be traced by outlining the main developments in offensive and defensive weapons and defensive architecture (see Table 4 below).

As noted above, siege warfare dominated military engagements in Egypt and Mesopotamia during the first half of the third millennium. Although we lack textual sources for the Levant during this period, we can infer that similar conditions prevailed there because strongly fortified walled settlements are amply attested throughout the EB II–III (ca. 3100–2400/2300 B.C.). By all indications siege warfare in this period was executed using early, unrefined versions of many of the same implements that would be gradually but significantly improved between 2300 and 1700 B.C. The most important of these weapons were the composite bow and the battering ram, which coupled with the introduction of the siege tower drastically increased the effectiveness of siege warfare (see Figure 3), just as the chariot improved warfare in the open field. Although simple versions of the bow and the battering ram (or siege pole) were available prior to 2300 B.C., there is no evidence that either weapon provided aggressors with an advantage during siege operations against towns fortified by massive walls and protected by well-armed defenders. The odds of surviving a siege in this period were, therefore, quite favorable.

Although we have no direct evidence, it is plausible that the ability of defenders to resist a besieging army in this period led to the development of more effective offensive weapons by aggressive military powers such as the Akkadian Empire. The gradual replacement of the simple bow with the more powerful composite bow and of the siege pole with the larger battering ram ca. 2300 B.C. meant that towns were faced with a new set of defensive concerns. While defenders could adopt the composite bow in order to match the range and firepower of their attackers, no comparable countermeasures existed against the battering ram. If a battering ram could be deployed and kept in place to do its work, a breach would eventually be made regardless of a wall's width. The battering ram's only weakness was the soldiers who sought to employ it. Yet here the aggressors had another advantage, namely that the defenders' firepower, which was concentrated at close range on those operating the battering ram, was distracted from other assaults. Meanwhile, the attackers could harass the defenders with volleys from a distance, especially from siege towers that served as elevated platforms from which suppressing fire was provided in support of the main assault. Such a scenario must have been replayed many times from ca. 2500 B.C. onward.

Although we cannot be certain that siege towers were used in this way during the Early Bronze Age, the Mari texts indicate that by the Middle Bronze Age they were commonly deployed alongside battering rams. For this reason there was a strong incentive to engineer ever more massive fortifications that would inhibit the coordinated use of battering rams and siege towers. As we shall see in the next chapter, this required major architectural modifications in the design of town defenses, which came to be widely adopted in the Levant by the beginning of the second millennium B.C.

Table 4. Major developments in warfare from the third to mid-second millennium.

Period	Developments
Pre-2300 B.C.	Simple bow Equid-drawn "battle car" Spear-bearing infantryman (Mesopotamia) Siege pole (Egypt and Mesopotamia)
EB IVA ca. 2300–2200 B.C.	Composite bow introduced Battering ram introduced (Sum. GUD.SI.DILI)
EB IVB ca. 2200–1900 B.C.	
MB IIA ca. 1900–1700 B.C.	Light, two-wheeled chariot introduced (N. Levant and Mesopotamia) Siege tower (OB *dimtum*) introduced (Mesopotamia) Rampart fortifications introduced (S. Levant)
MB IIB–C ca. 1700–1530 B.C.	Arrowheads common (S. Levant) More evidence of war chariot (Mesopotamia and Egypt) Layout of rampart fortifications improved (Levant)
LB I ca. 1530 B.C. onward	War chariot for open battles (Egypt, Levant, and Mesopotamia) Siege warfare continues

3. Middle Bronze Age Fortifications

Let him who desires peace, prepare for war.
Vegetius Renatus, *Epitome* Book 3, Preface

Many of the main elements of Middle Bronze Age defenses in the Levant, such as earthen ramparts (Pennells 1983), gates (Gregori 1986; Herzog 1986; G. R. H. Wright 1984), and fosses (Oredsson 2000), have been well documented and studied over the last several decades. Various surveys of Middle Bronze Age fortifications have complemented these reports (G. R. H. Wright 1985a; Kempinski 1992c; Fritz 1995:27ff.). Yet no single work provides a comprehensive description of the features, functions, and nature of the construction of Middle Bronze Age fortifications and how these aspects related to the concerns of Bronze Age warfare. In the absence of such a treatment many basic questions remain. For example, which elements are ubiquitous to fortifications during the Middle Bronze Age, and which are rare? When do the various elements develop and how do they relate to the technological developments in weaponry described in the preceding chapter? Is it possible to suggest that particular brick sizes or site types were confined to certain regions in the Middle Bronze Age? Are particular construction techniques or styles limited to certain regions or polities?

The use of simplistic criteria for site typologies, such as the presence or absence of earthen ramparts, has resulted in misconceptions regarding defensive strategies during this period. As noted in chapter 1 there are inherent problems with existing site typologies, which assume that the use of an earthen rampart around one portion of a settlement justifies its inclusion in a separate category of settlements (see especially comments on Herzog's typology on pp. 11f.). Such an approach obscures that so-called "ramparted" settlements were not completely surrounded by a continuous rampart of one type. While sites such as Qatna, Timnah, Sefinat-Nouh, and others may provide examples of this type of settlement—completely surrounded by a single type of defense—conditions at many settlements prohibited the adoption of a single fortification strategy. A combination of defensive strategies was often necessary to prepare a settlement's defenses in light of local conditions, areas occupied at the time of construction, and uneven terrain. For this reason the presence of ramparts or any single architectural feature (e.g., glacis and fosse) does not serve as an adequate criterion for the establishment of a site typology since this may erroneously imply that these features were continuous around the site. Special attention is paid, instead, to the existence of such features in identifiable recurring combinations at multiple settlements, which are likely to suggest a concerted defensive strategy.

Prior to attempting to create a site typology like the one mentioned above it is necessary to establish a typology of Middle Bronze Age defensive elements that accounts for the historical development of each feature and its regional distribution. In the following discussion of earthen ramparts, for example, it is assumed therefore that a single rampart constitutes a continuous built feature of uniform dimensions (i.e., height, width, and slope) running between gates. Likewise, while a glacis may have covered one slope of a tell or an earthen rampart, it cannot be assumed that it featured a similar construction in other areas around the site. Finally, issues of preservation must also be taken into consideration when determining the existence and extent of certain features.

In this chapter an attempt has been made to advance a typology of the characteristic features of Middle Bronze Age fortifications. The focus is upon components such as walls, gates, towers, bastions, ramparts, glacis, revetment and retaining walls, and fosses, which are frequently identified at archaeological sites. Each feature is examined with specific attention to its construction, function, typology, comparanda, and geographic point of origin. The basic elements of their construction, average dimensions, the period during which they functioned, as well as the specific roles that they filled are discussed, as are the materials used in their construction. Finally, a chronological typology of fortified sites in the Levant is offered for the EB IV through the LB I based upon the occurrence and function of these architectural features. Because the conclusions provided in this chapter are drawn from the catalog of fortified Middle Bronze Age settlements that are provided in this work (see catalogs 1 and 2), citations of reports are not included in this chapter. The reader will find, instead, that exemplars of the features discussed are listed in parentheses. (The Akkadian terms for these features are discussed in chapter 5.)

A. Architectural Elements

1. Earthen Ramparts

Earthen ramparts are recognized as an element almost exclusive to Middle Bronze Age fortifications in the Levant. However, the interchangeable use of the terms *rampart* and *glacis*, which were originally borrowed from French terms for medieval fortifications, has obscured the important distinction between these features. As will be demonstrated below, the earthen rampart and glacis protected against the approach of aggressors and the effects of erosion respectively.

The clearest definition of a rampart thus far is found in the glossary for *The Architecture of Ancient Israel* (Reich and Katzenstein 1992), and for this reason I have adopted it in the following discussion.[1] Ronny Reich and Hannah Katzenstein summarize the consensus that has emerged among Levantine archaeologists regarding the use of the term *rampart*, which they identify as an "earthen mound piled up around a city as a fortification or part of it" (ibid., 319). They define the *glacis* as:

> the outer facing of [an] earthen rampart which serves as a fortification of the lower slope of a mound (tell), or the lower outer-sloping foot of the city-wall. Constructed of different materials such as: beaten earth, lime plaster, bricks, stones, etc. (ibid., 316).[2]

Unfortunately, while their definition of rampart agrees with the data provided below, describing the glacis as "the lower outer-sloping foot of the city-wall" does not adequately distinguish between the morphological and functional aspects of glacis and ramparts. The distinction rests on the fundamental observation that an earthen rampart is simply an embankment built to impede the advance of an enemy approaching the fortification wall by elevating the wall, while a glacis is merely a surface treatment for the slope of the rampart or tell, which was intended primarily to protect it from erosion (for discussion of glacis, see pp. 55f.). If the glacis actually impeded the enemy, this was an unintended result. Additionally, the definition of a rampart should emphasize that earthen ramparts were not limited to the Middle Bronze Age since they were already being erected during the third millennium in the northern Levant and northwestern Mesopotamia, as demonstrated below. Furthermore, based upon their morphological characteristics two distinct types of earthen ramparts can be identified: the freestanding rampart and the supplemental rampart.

a. *Freestanding Ramparts*

A freestanding rampart is defined as an artificial earthen embankment that featured an interior and exterior slope, which was intended to raise the base of the fortification wall above the surrounding plain.[3] Although such ramparts were often built around entire settlements, they were also constructed around portions of settlements that were considered especially vulnerable due to their topography. Freestanding ramparts (as well as supplemental ramparts) were surmounted by defensive walls, although frequently the remains of such walls are poorly preserved if at all (for discussion of fortification walls, see pp. 59ff.).

Freestanding ramparts are the easiest type of rampart to identify because they often contribute to the formation of crater-like tells (Figure 6 below). Settlements located in fairly open areas that lacked any natural defenses were often completely enclosed by freestanding ramparts (e.g., Ebla, Qatna, Tuqan, Dan, Ashkelon, Yavneh-Yam, and Timnah). However, it is also noteworthy that site morphological processes and continuous occupation of settlements have often obscured or even destroyed traces of freestanding ramparts at some sites, particularly in the western Levant where precipitation is much greater than to the east. These "craters" may also have been filled in over time by subsequent occupation, thus erasing the rampart's characteristic topographic signature (e.g., Kabri). This may be the case at sites with saddle-like appearances where gates would have been located in

[1] These definitions find support in G. R. H. Wright's *Ancient Building in South Syria and Palestine* (1985a:183f.), even though he did not explicitly define these terms.

[2] Aharon Kempinski's (1992c) contribution in the same volume ignores the distinction drawn by Reich and Katzenstein between the terms glacis and rampart. Kempinski defines a glacis as "a sloping, external retaining wall" that "protects the base of the city-wall by keeping besiegers away" (ibid., 129).

[3] Kempinski referred to this type of rampart as a "double-sloped rampart" (1992d:175). Peter Parr (1968) and Jacob Kaplan (1975) advanced other terminology to describe earthen ramparts based on the presence or absence of fortification walls. Ramparts with walls within them, which are identified as core walls in this chapter (see p. 54), were called "wall ramparts," while ramparts without walls were identified as "freestanding ramparts" according to Kaplan (1975:1).

the depressions between the two rising ends of the site, where the ramparts were well preserved. Nevertheless, as mentioned in chapter 1, settlements with rampart-like embankments have also been mistaken for Middle Bronze Age ramparts. The Kranzhügeln of Syria have been misidentified as such (see p. 78), as well as Tell el-Yehudiyeh, Heliopolis, and Tell Safut.[4] The morphologies of these settlements are the result of altogether different site formation processes, and therefore caution should be taken in basing the identification of freestanding ramparts solely upon the contours of a site.

The plan of the typical freestanding rampart depended mostly upon the original topography of the site. Where the plain was open and relatively flat and early occupation was limited in extent or had not yet begun (the ideal situation for the construction of ramparts), the rampart could be laid out according to a systematic plan. Layouts were either elliptical (e.g., Achzib, Akko, Alalaḫ, Ashkelon, Byblos, Kabri) or rectilinear (e.g., Qatna, Sefinat-Nouh, Bosra, Deir Khabiye, Dan, Hazor, Yavneh-Yam, Timnah, Lachish, Nagila, Masos, and Haror).

Where the underlying terrain was not sufficiently flat or featured earlier fortifications, modifications were sometimes made to the layout to accommodate one of the two basic plans (e.g., Hazor and Shimron). Adjustments to the layout of freestanding ramparts also appear to have been necessary at settlements with prior occupation (e.g., Tel Dan, Ebla). At Ebla the construction of a series of rectangular fortresses connected by a town wall ca. 1700 B.C., all of which were built on top of the MB IIA and EB IV ramparts led to a straightening of the lines of what were, since the third millennium, probably freestanding ramparts with an overall elliptical plan (see Figure 36). The result was that the plan of Ebla's MB II fortifications, though roughly quadrilateral (perhaps rhomboid) in appearance today, was never actually rectilinear but obtained its present form from an attempt to convert the earlier, third millennium elliptical plan to a new method of fortifying settlements with straight rampart segments. Nevertheless, other continuously occupied sites such as Byblos and Ashkelon maintained their original elliptical layouts despite the addition of successive earthen ramparts.[5]

The difference between the use of elliptical and rectilinear plans for freestanding ramparts appears to correspond to chronological developments in defensive strategies. Freestanding ramparts with elliptical plans appear to have originated in the late third millennium in Mesopotamia (Mari and Terqa), and were introduced to the northern Levant thereafter. They precede the development of freestanding ramparts with rectilinear layouts that are well known in the southern Levant during the MB IIB–C. In addition to examples of sites in Mesopotamia, like Mari and Terqa, which featured ramparts of elliptical plan, sites in the Levant include the EB IV sites of Byblos, Ebla, Tuqan, and Umm el-Marra, and the MB IIA sites of Akko, Burga, Shimron, Zeror, Ashkelon, and Jericho in the southern Levant. While a handful of smaller sites, which were constructed during the second half of the Middle Bronze Age (most located in the highlands of the southern Levant) also featured elliptical plans (e.g., Kabri[?], Gezer, Shechem, and Shiloh), their ramparts were usually of the supplemental type. Furthermore, they were often located atop hills that made the use of rectilinear plans more difficult and costly. In contrast, the evidence for freestanding ramparts of the rectilinear type suggests that they were constructed no earlier than the very end of MB IIA and were built through the end of at least the MB IIB if not through the end of the Middle Bronze Age. The reason for this change may have been due to the realization that straight lengths of fortification were much easier to defend than curving segments. By dividing the defenses into a discrete number of straight lines each archer could protect greater stretches of the town wall and rampart, particularly from towers located at the points where the direction of defenses changed (see Figure 10).

b. *Supplemental Ramparts*

Although freestanding ramparts were commonly constructed at settlements located on a plain where town planning was more or less unrestricted by topography, this type of rampart was not easily employed for settlements built upon tells, nor for those upon natural hills, nor for the fortification of acropoleis. This was most likely because building freestanding ramparts at these sites would have ne-

[4]Safut (MR 228.168), on the northeastern outskirts of Amman has long been identified as a ramparted site. This suggestion, Donald Wimmer (1997:449) notes, was originally made by Farah S. Ma'ayeh (1960) based on a supposed identification of a glacis in a bulldozed section. Unfortunately, this misidentification has been perpetuated in the scholarly literature for the Middle Bronze Age in the works of Roland de Vaux (1960), Rudolph Dornemann (1983:19), James Sauer (1986:6), and most recently by A. Mazar (1990:197).

[5]It is noteworthy, however, that nothing is known of the line of the fortifications atop the ramparts at either of these sites. It is not impossible that similar efforts were made to straighten the line of their fortifications atop existing, though modified ramparts.

cessitated an inordinate expansion of the size of the original settlements to employ freestanding ramparts. Furthermore, the topography of these sites meant that the main advantage gained in the construction of freestanding ramparts, namely elevation, was already present. However, very few sites featured topographical conditions that were so ideal that they required no substantial construction of additional defensive elements. For this reason ramparts appear to have been selectively constructed at key locations around such settlements in order to supplement sufficiently defensible portions of the perimeter. Therefore ramparts at these settlements can be identified as supplemental ramparts (Figure 6).[6] Such ramparts could either supplement a town's fortification wall (e.g., Aphek, Beit Mirsim, Gezer, Lachish, Megiddo, Mevorakh, Poleg, Shechem, and Zurekiyeh) or fortify particularly vulnerable sections of a settlement's perimeter (e.g., Amman, Far'ah North, Shiloh, 'Umayri).

Such ramparts should not only be distinguished from the freestanding variety, but they must also be distinguished from surface treatments such as glacis, which were frequently added to both supplemental and freestanding ramparts. Supplemental ramparts were intended to obtain the advantages of freestanding ramparts but with limited modification of the defenses of an already elevated settlement. Though they were constructed using the same techniques and materials as freestanding ramparts, they were usually much more modest in size and were rarely more than a few meters high. Nevertheless, their slopes were equally steep (see Table 5 below). Because they were constructed atop existing tells, the elevation of the base of the wall above the surrounding landscape was often equal to if not greater than that of the wall atop a freestanding rampart.

[6]Kaplan referred to such ramparts as "wall ramparts" (1975:2). The main problem with his classification, however, is that it confuses the compositional and morphological traits of earthen ramparts. Some ramparts, which Kaplan classified as "freestanding ramparts," could also be classified as "wall ramparts" (based on a compositional analysis) because of the occasional use of core walls against which earth was piled on both sides (e.g., T. Dan, Hazor, Shiloh). Kempinski improved upon Kaplan's analysis by noting that basically only two types of banked fortifications existed, the "double-sloped rampart" (i.e., freestanding type identified here) and the "glacis" type (i.e., supplemental type identified here; 1992: 175f.). Nevertheless, Kempinski's terminology continued to confuse the distinct construction of the glacis (see pp. 55f.) and the earthen rampart whether of symmetrical (freestanding) or asymmetrical (supplemental) design.

c. *General Characteristics*

Although the slopes of both freestanding and supplemental ramparts (and their glacis), appear to be on average approximately 30° (Table 5), many were no doubt steeper prior to erosion following their abandonment, with slopes of perhaps as much as 45°. Only a small percent of ramparts featured slopes of more than 40°, however. The average height of these ramparts above the surrounding plain was usually about 10 m. The width of ramparts depended mostly upon the size of the site but also varied according to whether or not the rampart was freestanding or supplemental. The widest (and longest) ramparts were, of course, the freestanding type as found at Hazor (90 m), Qatna (70 m), and Ashkelon (70 m), while the narrowest were, naturally, of the supplemental type and were often no more than a few meters wide. The dimensions of ramparts will be discussed in greater detail in chapter 5 where the amount of labor required in their construction is discussed.

The terms freestanding and supplemental refer to the main *morphological* characteristics (i.e., their shape as viewed in section, see Figure 6). These observations are distinct from observations regarding their *constructional* characteristics (i.e., the building materials employed). The materials can vary considerably within each of these two rampart types.[7] As noted above similar building materials were used in both types of ramparts. Although based upon the morphological typology offered above one might expect that a relatively limited number of methods of constructing ramparts existed, the archaeological evidence suggests that methods of building both freestanding and supplemental ramparts were as diverse as the construction materials available. In this regard it is impossible to identify an "ideal" landscape where ramparts were most common. Nevertheless, the largest ramparts were built in areas where the greatest quantities of the requisite raw materials, such as loose earth, could be excavated with the least amount of effort (e.g., Qatna, Ebla, Mari, Ashkelon, Hazor). Such sites were often situated in the open plains of Syria and northern Mesopotamia, and in comparable positions along the coastal plain and the Great Rift Valley in the southern Levant.[8]

[7]The lack of distinction between morphology and composition is the chief reason that conflicting typologies of Middle Bronze Age fortifications have been advanced (see Yadin 1955b; Parr 1968; and Kaplan 1975).

[8]The location of sites with the largest ramparts, not surprisingly, also corresponds to regions where the largest tracts of arable farmland were available to support the considerable populations of these settlements.

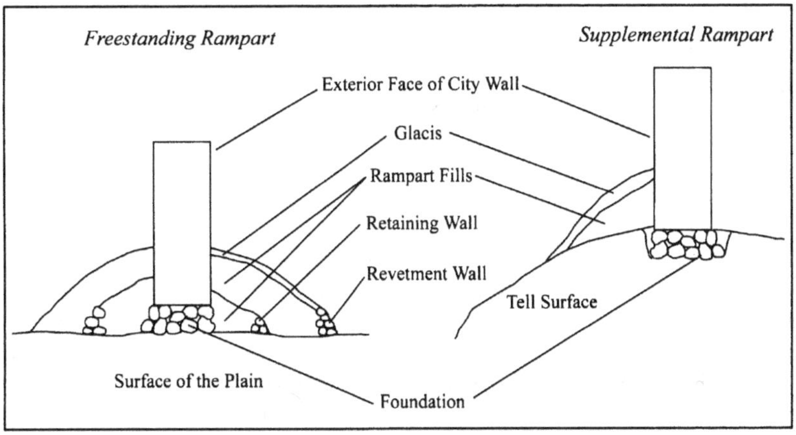

Figure 6. Cross-section of freestanding and supplemental ramparts.

Concerning the composition of ramparts, scholars often mistakenly describe ramparts as composed of *terre pisée* (literally "beaten earth," Yadin 1955; Tufnell 1958:45f.; Kenyon 1979:165; van Seters 1966:27ff.; Stern 1984:49; Mazar 1990:202, etc.). This anglicized French term refers, however, to the construction of a feature using a specific technique, namely the packing of earth into a mold or frame (G. R. H. Wright 1985a:360f.).[9] First, the terminology is misleading, because there is no evidence that the technique was employed for ramparts since they were almost never composed entirely of a single material without the addition of gravel or stone. However, the so-called "sandwich" technique, employing alternating layers of different fills, was frequently employed. The existence of this practice testifies to the unsuitability of a rampart composed solely of earth or soil. Second, there is no substantive evidence for the use of molds or frames in the construction of ramparts in the Levant. Finally and most importantly, there is no evidence for the intentional tamping or beating of a rampart's layers during the construction process. It is more likely that during the process of building the rampart the effect of *terre pisée*, insofar as it is discernible, was the unintended result of the continuous treading of workers and animals up and down the slope of the rampart.

Aside from various types of earth or soil (e.g., alluvium, clay, hamra, loess, marl, sand), rampart fills also included many other materials such as chipped or crushed stone (e.g., huwwar limestone, kurkar sandstone, flint, travertine), pebbles, occupational debris, mudbrick detritus, broken mudbricks, and plaster lenses—all of which were locally obtained materials. Although sand was also used to build ramparts, because it does not serve as a good foundation upon which to build it appears to have been avoided whenever possible. However, where sand was used it was always necessary to provide the rampart with a less permeable surface, which should be identified as the glacis (e.g., Akko and Ashkelon).

Many excavators have suggested that rampart fills probably derived from the excavation of material from the fosse outside the ramparts. These observations are based on the fact that the materials in ramparts correspond, often in reverse order, to the sequence of geological deposits seen in the faces of the fosses outside the ramparts (e.g., Achzib, 'Ajjul, Ashkelon, Timnah, Haror, Hazor). A similar result was noted by Vegetius Renatus concerning the excavation of fosses during the building of a fortified Roman camp (*Epitome*, Bk. I.24, p. 23). This suggestion is also supported because the volumes of fosses associated with ramparts are comparable to the volumes of earth used in these ramparts (for volumes of ramparts, see Table 22). While rampart fills were often sterile, at previously occupied sites fills also consisted of occupational debris, which included quantities of ash, charcoal, pottery, and bones.

[9]The term should be specifically reserved for earthwork constructions like those identified around the cities of Cheng-chou, Yen-shih, P'an-lung-ch'eng, and Tung-hsia-feng in China during the Shang period, ca. 1700–1100 B.C. (Chang 1986:335). The 320 ha site of Cheng-chou in northern Hunan was surrounded by a 7,195 m long beaten earth wall, which survives as much as 9.1 m high and 36 m wide; the wall was built of 8 to 10 cm thick layers of compressed earth (ibid., 205). Apparently, the pisée technique was introduced to Chinese architecture during the Shang period (ibid., 331). It continued to be important during the following Chou Civilization. A ballad from this period preserves details of the construction of the city of Chou using the pisée technique (ibid., 340f.).

Table 5. Slopes of earthen ramparts from EB IV to LB I.

Site	Slope (deg.)	Glacis Composition	Rampart Type
Abou Danne VI	40	—	Freestanding
Achzib	30	Stone cobbles	Freestanding
Akko	30	Mudbricks	Freestanding
Alalaḫ VII	35	—	Freestanding
Amman	40	Stone cobbles	Freestanding
Biruta (MB IIA/LB I)	30/20	Clay/Stone	Supplemental
Byblos VII (ouvrage 3a/3b)	35	Stone cobbles	Supplemental
VIII (ouvrage 4)	15	—	Supplemental
Dan XI (Areas B, K, and Y)	14–43/20	Chipped stone	Freestanding
XI (Area T)	20	—	Supplemental
Ebla IIIA town	30	—	Freestanding
acropolis	9	Mudbrick	Supplemental
Outlying rampart	24	—	Freestanding
Far'ah North (Per. VC)	24	Stone cobbles	Supplemental
Far'ah South (N/W Ramparts)	40/33	—	Freestanding
Gerisa	28	Hamra, Stones, Crushed stone, & Mudbricks	Supplemental
Gezer XVIII (Field I)	45	Chipped stone	Supplemental
Hazor XVI/3 (N Rampart)	25	—	Freestanding
XVI/3 (W Rampart)	18	(Stone inner slope)	Freestanding
XVI/3 (Area G acropolis)	68	Stones	Supplemental
Jericho Period IVa-b	35	Chipped stone	Supplemental
Period IVc	22	—	Freestanding
Kabri	45	—	Freestanding
Kazel	45	Clay	Supplemental
Keisan XV/XVI	25/28	Uncertain material	Supp/Free
Lachish VIII (P–5 and P–4)	29	Lime plaster	Supplemental
Malhata (Per. B)	30	Lime plaster	Freestanding
Masos	17	Pebbles?	Freestanding
Megiddo XIII (Area AA)	16	—	Supplemental
XIII	14	Uncertain material	Supplemental
XI	45	Uncertain material	Freestanding
Nagila VII	35	Chipped stone	Freestanding
Poleg	20	—	Supplemental
Poran	16	Mudbrick	Supplemental
Sefinat-Nouh	19	—	Freestanding?
Shechem XVI/XVII	21/15	—	Supplemental
Shiloh VII	28	Stones & Chipped stone	Supplemental
Timnah XII–XI	15	—	Freestanding
Tuqan (MB I Upper Town)	30	—	Supplemental
Qatna	30	—	Freestanding
Ugarit (LB IA)	45	Carved stone	Supplemental
Yavneh-Yam	40	—	Freestanding
Average	29.43	—	—

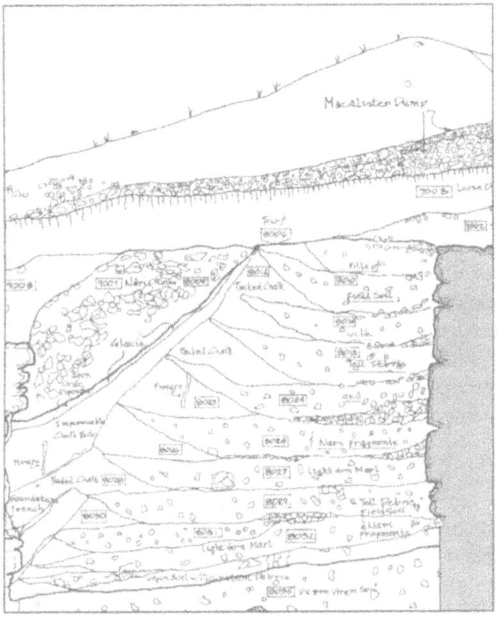

Figure 7. Section of layered fill in rampart at Gezer employing "sandwich" technique (after Dever et al. 1970:plan 2).

The various materials and soil types used in rampart fills have led to considerable speculation concerning why they were used. According to Ernest Pennells, drainage was one of the most important factors involved in the selection of materials for rampart construction (1983). It was not desirable for a rampart to hold water because it would diminish the strength of the rampart and lead to erosion (e.g., mudslides). For this reason Pennells suggests that the "layered fill" technique (elsewhere referred to as the "sandwich" technique) was developed—a technique which would improve drainage of water from within the rampart—since as he noted silts and clays (i.e., materials with the finest particles) absorbed the greatest amount of water (ibid., 57). This technique guaranteed, therefore, that water would drain from earthy layers through rock filled layers below and evaporate through rocky layers above. The best examples of sites where this technique was employed are Gezer (Figure 7), Shechem, Jericho, Dan, and Timnah. While Pennells's observation makes sense of the constructional strategy for the ramparts of most sites, it does not explain why materials such as clay and sand were not completely avoided.

Contrary to Pennells's hypothesis, Ross Voss suggests that materials such as ash were actually preferred because they absorbed "large amounts of rainwater becoming heavier and denser, while resisting the effects of erosion" (Voss 2002:379f.). There is, therefore, some difference of opinion as to how such materials functioned in construction and surface preparation. Voss's suggestion seems to run contrary to the typical observation concerning the use of ash. In fact, ash is usually acknowledged for its water resistant characteristics, such that it was often incorporated into beaten earth floors in domestic contexts and mixed with lime for water resistant plaster (e.g., G. R. H. Wright 1985a:438).

Even if ash were the ideal material for rampart construction, it would never have been available in sufficient quantity for building any large part of a rampart since rampart construction was usually concomitant with the expansion of smaller sites or the foundations of new sites. There is no evidence that ash constituted a great percentage of the mass of any single rampart or that it was a preferred material even when it was readily available. It is possible that ash and occupational debris just like thin lenses of plaster and other crushed layers of stone were actually used in ramparts because of their varied water resistant properties (contra Voss 2002). This explanation would be consistent with Pennells's observation that it was necessary to keep ramparts from becoming saturated because it caused erosion. That other features such as glacis, revetments, and retaining walls were built for the same purpose (see next section) also supports the notion that ramparts were constructed as frequently as possible of materials that repelled moisture. The use of large quantities of earth without chipped stone in entire ramparts appears, in fact, to have occurred most frequently in drier regions (e.g., Qatna and Ebla), whereas mixed fills occur most often in the western Levant (see also discussion of glacis, p. 55).

The evidence thus suggests that a careful process was involved in choosing materials for rampart construction despite, as we shall see, the varied means that were undertaken to build them. For this reason stones larger than gravel size are absent in nearly all fills. This may have been because large stones floating in the rampart fills were considered detrimental to a rampart's integrity. However, this was more likely the result of the process of selecting construction materials. Larger stones of any useful size were routinely employed in the construction of retaining walls, revetments, foundations, and glacis.

The selection process appears to have also taken into consideration the rates at which materials could be excavated. For instance, it is rare to find a single layer of a particular building material with a thickness of more than one meter within a rampart's fill. Structural advantages for building in this manner aside, it would seem that there is another reason why this was the case, namely the speed with which ramparts were constructed (for probable rates, see chapter 6). Speedy construction would have required an efficient management of labor where materials were moved almost as quickly as they were excavated in order to facilitate continued excavation, for example, from a fosse or another source. An excellent example of this can be witnessed in the section of a typical rampart at Achzib, which was built from the bottom up of "red fill . . . dune-sand . . . brown soil . . . grey soil with ashes . . . dune-sand . . . hardened red-brown soil" (Prausnitz 1975:207). That the rampart's center was not built mostly of sand when it is known that it could be used for much of a rampart's composition (e.g., Akko and Ashkelon), suggests that one or more considerations prevailed that prevented this. Considerations regarding the management of laborers dictated a diversification of efforts in obtaining materials. Thus, the excavation of materials proceeded gradually in at least two locations at the same time so that those carrying loads of materials did not have to wait during this process. It is probably also for this reason that it is rarely possible to identify the same sequence of fills within a rampart's construction in different areas excavated around a single site. Instead, fill layers were interspersed and varied from the bottom up in different areas around a site because different work groups were exploiting different sources of material at the same time during the construction process.

In general the evidence for earthen ramparts suggests that they were constructed for the purpose of creating elevated but relatively level embankments, which were intended to significantly improve the defensibility of a settlement's perimeter. At most settlements earthen ramparts appear to have succeeded in elevating the base of a town's walls from 7 to 10 m above the level of the surrounding plain. Although their preserved slopes vary between 20° and 45°, they were also usually covered with a glacis composed of materials that were intended to protect their surfaces against erosion. The materials used in rampart construction varied depending upon the site's location and various techniques were employed in their construction such that it would be incorrect to speak of a single approach to constructing earthen ramparts.

2. Structural Walls

Three types of walls are frequently associated with the construction of ramparts: core, retaining, and revetment walls. The reason for the construction of these walls (and the addition of glacis, discussed below) was at least partially the result of the severe erosion to which earthen ramparts, which ranged between 30° and 45°, were subject. Although some vegetation such as grasses, when not eaten by herds, might have helped to protect the rampart against erosion, it is doubtful that plants were allowed to grow on the rampart's surface because these would have attracted animal herds, which would in turn have contributed to the rampart's erosion. Because the threat of a rampart's layers eroding was more acute with larger ramparts, usually of the freestanding type, two major types of supporting walls were often constructed within freestanding ramparts: small, low retaining walls and core walls built vertically in the center of ramparts.

a. *Retaining Walls*

Retaining walls within a rampart's structure were relatively common (e.g., Dan, Jericho, Keisan, Malhata). They were intended to address potential instabilities at specific points within a rampart, which were probably identified during the course of construction. Such walls, usually built of fieldstones only a few courses high and a few meters long, were buried within the rampart's fills (see W.98 at Jericho in Figure 82).

b. *Core Walls*

If conditions required, a core wall was constructed to provide a foundation for rampart construction during its initial stages. The core wall was a multipurpose feature that, in addition to stabilizing the rampart, also provided a solid foundation for the construction of the town wall that crowned the rampart (see pp. 59ff.). Although they are less common than retaining and revetment walls, the use of core walls within ramparts at larger sites indicates that the planning of ramparts had become increasingly standardized. Although core walls were often built of stone (e.g., Dan, Shiloh), mudbrick walls could also function as core walls (e.g., Hazor). At Dan in Area AB it is evident from the slope of the stone-built core wall that it was erected as the rampart was raised on both sides of it (see Figure 71). It seems that the value of building core walls was realized by the start of the MB IIB–C, whence the main evidence for their construction comes. This may have been as a result of a growing tradition among architects who had acquired experience from previous generations as well as from contact with the northern Levant and Mesopotamia, where ramparts had often been heaped against earlier fortification walls (e.g., Byblos, Ebla[?], Mari, Terqa).

While the building of retaining and core walls represent two different strategies intended to address the same problem, it is important to realize that the use of these techniques was not necessarily identical even across a single site. Rather these techniques, though often used to support freestanding ramparts, represent distinct strategies that were tailored to the varied topography around a settlement just as ramparts themselves were sometimes only constructed at specific points around a settlement. At Dan, for instance, a core wall was used in the construction of the rampart in Area AB, while a core wall and numerous retaining walls were used in Area Y. Likewise, that retaining walls were usually used in freestanding ramparts does not preclude that they were on occasion used to reinforce the construction of supplemental ramparts on existing slopes (e.g., Gezer and Shiloh).

c. *Revetment Walls*

In addition to retaining walls and core walls, revetment walls offered another means of preventing the erosion of ramparts. But unlike retaining and core walls, revetment walls were exclusively built at the foot of earthen ramparts, often with a battered exterior surface. Revetment walls were therefore better constructed than retaining walls, stood between two and ten meters high, and could be between one and four meters wide. While it is possible that some revetment walls may have circumscribed entire sites (e.g., Ebla?), no revetment wall has yet been traced around the entirety of a site. The best known examples appear to have encircled at most one half of a site (e.g., Achzib, Hazor, Jericho, Shechem).

Revetments also often employed what has been traditionally identified as cyclopean masonry. The presence of a steep batter and the use of cyclopean masonry suggest the need to prevent the erosion of the rampart (e.g., Jericho [see Figure 82], Shechem). Although minor erosion would not only have undermined the rampart's structure, it would have resulted in the in-filling of the fosse outside of the revetment wall, obviating the defensive function of both of these features.

That revetments were perhaps not necessarily intended to improve the defenses of sites is evident because many revetment walls were so large that they would have actually provided considerable cover for attackers as they approached the base of the rampart (e.g., Shechem). Nevertheless, such walls did provide an additional barrier to the advancement of siege towers and battering rams, and their presence would have required the use of scaling ladders prior to reaching the town wall, putting the attackers at greater risk. More so perhaps than any of the other types of structural wall, the presence of revetment walls suggests that considerable planning was involved in the construction of defenses.

3. Glacis

As noted in the discussion of earthen ramparts above, glacis construction must be distinguished from rampart construction. The widespread recognition of the distinction between these features has meant that the term glacis has been increasingly used to refer specifically to the intentionally prepared surface of a slope, whether that of a rampart, a mound, or a tell, which prevented its erosion and improved its defensibility.[10] Although the materials used in glacis construction were not particularly different than those available for rampart construction (i.e., earth, plaster, bricks, stone), glacis were always constructed of the most durable and weather resistant materials available, which is apparent when compared with the materials used within the ramparts they protected. While it has been suggested that plaster was the preferred material for glacis construction (see p. 74), it is doubtful that sufficient quantities of any type of plaster were ever produced to cover the entire slopes of settlements (e.g., Jericho, Keisan, Hazor, Lachish, Malhata). For this reason other materials were often substituted, such as: cobbles or larger stones (e.g., Afis citadel, Ashkelon, Biruta, Beit Mirsim, Burga, Byblos, Far'ah North, Haror, Hazor citadel, Hebron); cut blocks (e.g., Ugarit); crushed limestone (huwwar) or other chipped stone such as kurkar, travertine, or flint (e.g., Dan, Gerisa, Gezer, Jaffa, Jericho, Nagila, Qitar, Shiloh, 'Umayri, Yavneh-Yam); pebbles or gravel (e.g., Abou Danne, Hadidi, Malhata, Masos, Megiddo); mudbricks (e.g., Akko, Ashkelon, Ebla citadel, Gerisa, Jericho, Poran); mud plaster (e.g., Haror); "beaten" earth (e.g., Gerisa, Jaffa); or clay (e.g., Beth-El, Deir 'Allā, Kazel, Shiloh, Ta'anach).

While the use of materials in a glacis was often determined by their availability, other environmental factors appear to have played a key role in the type of glacis constructed. The geographic location of sites and the amount of rainfall that these sites received appear to have affected the types of materials selected for use in glacis construction. In this regard it is clear that the greatest efforts were invested in the construction of stone and mudbrick glacis in regions with the

[10] G. R. H. Wright has objected to referring to nondefensive slopes or slicks as glacis since "it is not always a straightforward matter to determine the purpose of this kind of installation" and "such earthworks are not accurately signified by the term *glacis*" (1997:366). Nevertheless, whether originally built for defense or against erosion, glacis served both purposes.

highest precipitation (i.e., the Levantine coast), while areas further to the east featured less substantial glacis when they can even be distinguished from the rampart itself. For instance, although only a few sites featured glacis built of cobbles (see above), all of these sites are situated along the coast (e.g., Achzib, Ashkelon, Biruta, and Byblos [see Figure 31]) where rainfall averaged over 400 to 800 mm per year (see Alex and Wolfner 1984). Ugarit, which featured the only glacis composed of ashlars, was also located in the coastal plain. Although sites with various other types of glacis were also located along the coastal plain and others were further inland (where they were often composed of chipped stone), distinct glacis almost completely disappear east of the Rift Valley (e.g., Ebla, Qatna, Sefinat-Nouh, Tuqan, Umm el-Marra) and in northern Mesopotamia (e.g., Yakaltum, Terqa, Mari).

The geographical distribution of glacis types demonstrates that glacis should be viewed as features intended to protect the time and energy invested by a population in the construction of ramparts against erosion due primarily to rainfall. Confirmation of this is also provided by the construction of drains along glacis (e.g., Ashkelon), through revetment walls (e.g., Aphek, Shechem XVI), and underneath gateways (e.g., Ashkelon XXII Ph. 12, Megiddo XIIIA). At smaller sites where glacis were not constructed of cobbles, the other materials used in glacis construction (e.g., mudbricks, mud plaster, gypsum plaster, and carved stone) were also those typically used to protect fortification walls and gates against the same elements. Perhaps the strongest supporting evidence for this conclusion is the identification of glacis on the inner slopes of the freestanding ramparts at Hazor and Tel Dan. With this in mind it appears that glacis were not constructed to contribute to the defense of a settlement's slopes against attackers, despite the possibility that some glacis types might have impeded efforts to ascend their already steep slopes. Rather, glacis were built primarily to protect the slopes of earthen ramparts against erosion.

4. Fosses

While the terms *fosse* and *moat*, like rampart and glacis, have been used interchangeably by archaeologists in the Levant for many years, a useful distinction can be made in the employment of these terms to describe two separate but related features encountered in archaeological contexts. Although these two terms are, for all intents and purposes, semantically equal as employed in English, in this work the term *fosse* is used to refer to a dry ditch excavated in front of a fortification system. *Moat* on the other hand, which often evokes images of water-filled trenches surrounding medieval castles, is used only to refer to ditches that were probably intended to hold water for defensive purposes, distinguishing them from canals. This distinction may be supported by the occurrence of various terms used in the Mari texts to refer to these ditch-like features (see Table 16). Nevertheless, the results of this survey suggest that there is no clear evidence that any Middle Bronze Age features in the Levant are to be identified as moats. While in certain areas the bottoms of some fosses may be below the water table today, the lack of data for the level of the water table during the Middle Bronze Age at the time when these features were excavated makes it impossible to assert that they were intended to retain water (e.g., Achzib?, Zeror). In fact, evidence from sites such as Lachish suggests that the Middle Bronze Age was actually drier than the preceding Early Bronze Age and that the water table was therefore lower (Rosen 1986:73). While casual water in these fosses during certain seasons may have been advantageous in several respects, as suggested below, the retention of water was never intended to be a primary function of these features. Therefore, all defensive ditches discussed in this work are identified as fosses.

Dag Oredsson has attempted to provide a synthesis of the evidence for fosses in the southern Levant, which he refers to as moats (2000). Wider consideration of Middle Bronze Age fosses throughout the northern and southern Levant is nevertheless possible from the evidence available at a number of sites. In the south these include 'Ajjul, Ashkelon, Achzib, Burga, Ekron(?), Far'ah South, Haror, Hazor, Hebron, Lachish, Malhata, Masos, Nagila, Timnah, 'Umayri, and Zeror, while in the northern Levant fosses were also dug at 'Acharneh(?), Ebla, Hadidi, Qatna, Sefinat-Nouh, Selenkahiye, and Umm el-Marra.

As a defensive feature, the fosse was perhaps the easiest and most primitive type of fortification built. In most regions of the Near East where fosses have been found nothing inhibited the simple excavation of such features around Middle Bronze Age sites (e.g., Hazor, Qatna, Sefinat-Nouh). Even in regions where bedrock was just below the surface fosses were still excavated (e.g., 'Ajjul, Achzib, Ashkelon, Lachish, Nagila, 'Umayri, Ebla, and Hadidi). As Table 6 below illustrates, the dimensions of fosses, like those of ramparts, were extremely varied.

Table 6. Dimensions of Middle Bronze Age fosses and their estimated volumes.

Site	*d.*(m)	*w.*(m)	*l.*(m)	*vol.* (m³)
Achzib	4	?	475	?
'Ajjul	3.7	10.7	1,220	48,300
Ashkelon XXIV (Ph. 14)	5.5	9	?	?
Timnah XI–XII	2	?	800	?
Burga	?	?	1,500	?
Far'ah South	6	44	180	47,520
Haror	4	15	800	48,000
Hazor (W), Str. 4	15	60	600	540,000
Lachish	2	9	140	2,520
Masos	2	3.5	450	3,150
Nagila XI	3	9.75	850	24,863
Qatna	5	70	4,560	1,596,000
Sefinat-Nouh	4.5	20	1,905	171,450
Selenkahiye	3	9	1,430	38,610
'Umayri (Ph. 12)	5	6	?	?
Zeror	4.5	12.5	?	?

Note: Width (*w.*) is an average between *w.* measured at the top and base of the fosse.

It has long been recognized that the most likely source for the earth and stone used in the construction of ramparts was primarily, if not entirely, derived from the excavation of fosses. This led Oredsson to conclude that Middle Bronze Age moats were simply the byproduct of the building of ramparts (2000:52f.). In many respects this phenomenon is comparable to that related to mud extraction pits created during brick production, which have been identified in Mesopotamia (see Wilkinson 2003:109–11). While it is certainly true that the majority of the materials excavated from fosses were used for the construction of ramparts, the lack of evidence for fosses at some sites that featured ramparts suggests that not all ramparts owe their origin to the digging of fosses. This is especially true when supplemental ramparts were employed. At Ebla, for example, materials used in the rampart also consisted of a considerable amount of occupational debris. As the sites listed above demonstrate, the use of the fosse appears to have been restricted to areas around sites where freestanding ramparts were erected (see also discussion of earthen ramparts on pp. 48ff.).

Fosses dating to the Early Bronze Age have been identified in the southern Levant at sites such as Khirbet el-Makhruq, Jericho, er-Rujjum, and Bardawil (Oredsson 2000:40). Although they probably also existed in the north during the Early Bronze Age, evidence from this region is lacking. Interestingly, fosses are not attested at the most familiar of the fortified Early Bronze Age sites in the southern Levant such as Arad, Ai, and Yarmuth, which may be due to their hilltop locations. Nevertheless, it is clear that the fosse was not consistently employed in fortification construction in the Levant prior to the Middle Bronze Age and its use during the Early Bronze Age was, therefore, probably only a local development of a simple feature intended "to overcome topographical disadvantages" (ibid., 46).

There is, however, strong evidence for the use of the fosse in northern Mesopotamia where it appears to have been incorporated into defenses during the early part of the third millennium (e.g., Chuera and Terqa). Unfortunately, limited excavation of the defenses of other Kranzhügeln hamper a more conclusive picture from this region. At least a couple of examples of moats, at Nippur and Aššur, suggest the incorporation of fosses into the defenses of these settlements in the Late Bronze Age (see Oredsson 2000), but the degree to which these features were common prior to this in Mesopotamia remains unclear.

Because the fosse, which is almost always found in conjunction with rampart construction in the Levant, its purpose seems to be first and foremost defensive, though still secondary to walls and ramparts in this capacity. The incorporation of this feature into the defenses at sites that were almost always situated beside wadis or another body of water such as the shoreline of an estuary (e.g., 'Ajjul), a river (e.g., 'Acharneh?), or the Mediterranean Sea (e.g., Achzib Ashkelon, Jaffa), or upon a promontory (e.g., Hebron), also suggests that the fosse was intended to complete the outer line of defenses, which was often already partly provided by natural features. The defensive value of the fosse was emphasized by Vegetius Renatus:

> Fosses before cities should be made as wide and as deep as possible, so that they cannot easily be leveled and filled in by besiegers and, once they begin to fill with water, hardly permit a sap to be continued by the adversary. A subterranean work is prevented from being constructed in two ways, by the depth and by the flooding of fosses (*Epitome*, Book IV.5).

That the characteristics of most fosses were not very conducive to retaining water further supports the conclusion that they were not intended to function as moats. Furthermore, there is no basis for assuming that when these features were excavated during the Middle Bronze Age, which probably occurred during the summer, that the workers continued excavating below the level of the water table in order to intentionally create moats. Instead, the filling of these fosses with water probably only occurred after seasonal rains or gradual, local environmental changes. Constant evaporation and the fact that the bottoms of the fosses were not deeper than the wadis to which they often connected meant that fosses would rarely have held water for any length of time.

Although fosses were probably excavated with defense in mind, settlements did benefit from their creation in various ways during peacetime. First, as noted above, they were a primary source of building materials, not only for other defensive features such as the ramparts, but also potentially as quarries for construction within the settlement. Second, fosses also served to protect ramparts both from rainfall erosion by channeling water away from them but also possibly from mining by enemies (Stager 1991b). Third, although it is not entirely clear that these features ever contained much water, it is possible that for short periods of the year fosses acted as reservoirs for the settlements that they surrounded. Their filling with water in the winter and through the early spring would have also coincided with the period when it was not yet necessary for shepherds to venture great distances from their settlements to find grazing land. These features might be compared to *yarru* (lit., a "pond" or "pool"; *CAD* I, p. 326), which were known from the Old Babylonian period. Tony Wilkinson suggests that such pools may have functioned as nearby mud extraction pits, which often retained water (2003:119). Therefore, herds might frequently have been seen around fosses for a short period after winter or spring rains. Additionally, once the fosses had dried up it would have been easy to reach the water table by sinking wells into their bases—an approach similar to the excavation of wells in the bottoms of wadis during the dry season.

Because the fosse is a defensive feature that is found in many periods throughout the ancient Near East (Oredsson 2000), its origin is difficult to trace. The earliest features that resemble fosses are, of course, the canal systems of southern Mesopotamia. There canals often ran up to, through, and even around settlements. While these features protected settlements from floods, created transportation networks, and enabled the irrigation of crops, they also provided a convenient means of demarcating territory (e.g., *RIME* 3/2.1.1.28). For this reason there is a basis for suggesting that canals probably contributed to the defense of settlements, though their defensive purpose just like that of the fosse in the Levant was a secondary product of their creation.[11]

In light of the evidence from sites such as Tell ed-Der and Sippar (Tell Abu Habbah), where it has been suggested by Paepe et al. that artificial embankments (i.e., ramparts) were intended to limit the effects of floods (1978:33), it is difficult to imagine that the origin of the moat (or fosse) in Mesopotamia was completely unrelated to the creation of earthen ramparts. This would be especially true since ramparts and moats both functioned to keep water away from the foundations of town walls and since the construction of these two types of features involved a single process. The main difference between the use of earthen embankments in Mesopotamia and the Levant was, therefore, that the threat of wall collapse due to erosion was probably the main motivation for their use in southern and even central Mesopotamia. However, in the Levant and northern Mesopotamia such features, particularly ramparts, were of value only for the military advantages they could offer, which were probably already apparent in southern Mesopotamia.

Although the evidence for the spread of the fosse as a "defensive canal" is limited, the preceding reconstruction offers at least one interpretation of its origin. Fosses may have emerged in Mesopotamia, where they developed out of major canal networks into site specific canals or moats. In their second stage, during the early part of the third millennium, they appear to have been incorporated into the defenses of settlements in northwestern Mesopotamia, some of which have been identified as Kranzhügeln (e.g., Chuera, Terqa, and Mari?).[12] While some of these sites, such as Terqa and Mari, were close to perennial rivers, which justified their continued use

[11] This notion parallels J. Kaplan's suggestion that "the builders of *Muriq-Tidnim* did not newly construct this entire wall but probably utilized parts of the existing levees of the main irrigation canals and only added the needed links between them to complete their continuous defence line" (1972b:12).

[12] A fosse from the late fourth millennium identified at T. Hamoukar in the northernmost part of Trench D off the northwest side of the mound (Gibson et al. 2002:47f.) appears to be the earliest example of such a feature attested.

of canals for the management of water and the building of dikes against their walls, most of the other sites were situated near the banks of wadis, which were filled with water only during a brief part of the year. It may have been during this stage of the evolution of the fosse, therefore, that the dry ditch or true fosse came into widespread use. While occasional use of fosses occurred in the southern Levant during the Early Bronze Age, the lack of evidence for fosses at the best fortified sites in this period suggests that use of the fosse in the south at that time was *ad hoc*, and, therefore, unrelated to the gradual and contemporary proliferation of the fosse in northern Mesopotamia from the middle of the third millennium onward.

Despite the defensive benefits of excavating fosses, by the EB IV the fosse had been introduced to the northern Levant as a completely defensive feature. Their excavation for the first time is unquestionably linked to the construction of defensive ramparts (e.g., Ebla[?], Selenkahiye, and Umm el-Marra). Yet the greatest evidence for the continued spread of this feature along with rampart fortifications comes, of course, from the southern Levant where during the MB IIA numerous examples of fosses are known (e.g., Achzib, Ashkelon, Burga, and Zeror). This trend continued and clearly reached its apogee during the MB IIB–C when freestanding rampart construction was at its peak throughout the Levant.

Based upon these data the rampart appears to have served militarily to improve the defensibility of settlements by limiting the approach against the walls with either the battering ram or the siege tower. In certain cases, however, the drawback to the scarps of fosses, like that of revetment walls discussed above, may have been that they also provided cover for enemy soldiers approaching the foot of the rampart before they made their final assault. Even so, the fosse provided the necessary delay that worked to the advantage of the defenders, who it can be imagined may have frequently mounted counter attacks as the attacking force became bogged down in the fosse.

5. Fortification Walls

Although earthen ramparts, glacis, and fosses were an integral part of the defenses of settlements throughout the Levant during the Middle Bronze Age, without fortification walls sloped embankments would not have constituted an improvement on Early Bronze Age fortifications, which did feature fortification walls. Despite the characterization of the evidence for Middle Bronze Age fortification walls built atop ramparts as scant by some scholars (see pp. 12ff.), several lines of evidence demonstrate that fortification walls were built atop earthen ramparts. Solid mudbrick walls, casemate walls, and the exterior walls of buildings all functioned as defensive walls during the Middle Bronze Age. As the following discussion demonstrates, the dimensions of these walls can be determined by means of a judicious examination of various types of evidence.

Figure 8. So-called "Syrian gate" of the mortuary temple of Ramesses III. Photo by A. A. Burke.

The first evidence that supports the existence of Middle Bronze Age defensive walls in the Levant comes from the depictions from Egypt of fortified Middle Bronze Age settlements, which were occupied by Asiatics. All Egyptian depictions of fortified Asiatic settlements, which are thought to have been located in the Egyptian Delta, featured defensive walls, though none of them depict earthen ramparts. The few late First Intermediate Period and Middle Kingdom representations of Asiatics within forts depict walled fortifications similar to those seen in New Kingdom reliefs of fortified towns in the Levant.[13]

[13]For the First Intermediate Period, see plans 1 and 3 of B.

The rounded crenellations shown atop the fortification walls of these towns are the only evidence for this feature dating to the Middle Bronze Age.[14] However, based on the evidence from the so-called "Syrian gate" of the mortuary temple of Ramesses III at Medinet Habu (Hölscher 1951:pls. 19f.), which is dated to the Late Bronze Age, it may be suggested that such crenellations were also employed in Levantine fortifications from the Middle Bronze Age through the Late Bronze Age (Figure 8).

a. Solid Walls

Evidence for the existence of fortification walls also includes the remains of solid mudbrick walls and their stone foundations, which have been discovered at a number of sites throughout the Levant (see Table 7 below). The clearest evidence of this type of wall comes from late third-millennium and early second-millennium sites in northern Mesopotamia (e.g., Mari, Yakaltum, Šubat-Enlil, Terqa, and Tuttul), where mudbrick fortification walls are best preserved due to the arid climate. Conversely, it has been extremely difficult to detect mudbrick walls at sites throughout much of the Levant, especially its western side, and the Egyptian Delta due to erosion (see comments regarding mudbrick erosion by Spencer 1994). There is, therefore, an obvious correlation between the degree of preservation of mudbrick walls and the amount of rainfall and moisture within certain regions. Where the least amount of rainfall is received the walls have been best preserved (see also discussion of the walls of Buhen below).

The remains of solid mudbrick walls and their foundations are preserved to an average height of 2.5 m and an average width of 3 m.[15] It has been generally assumed that such walls stood at least two stories tall based upon the New Kingdom depictions of fortified settlements in the Levant, which often show second story windows. That these walls may have been as much as 10 to 15 m high is suggested by comparisons to the remains of Middle Kingdom fortresses in Lower Nubia, the remains of mudbrick enclosure walls of New Kingdom temples, and references in Naram-Sin's account of the siege of Armānum.

At Buhen, for example, a section of the inner fortification wall, which was about 5 m thick, was preserved to a height of 11 m but was once probably even taller (Emery et al. 1979:4ff.). Further evidence for the height of mudbrick fortification walls can be gleaned from the inner enclosure wall of the temple of Ramesses III at Medinet Habu from the late thirteenth century B.C. The wall, the base of which was 6 m wide gradually tapering to 2 m wide at the top, was preserved to a height of 14.2 m (Hölscher 1941:61). The excavator suggested that the wall was originally 15 m high. The outer enclosure wall or "Great Girdle Wall" of the temple, which was also built of mudbrick, was by comparison 10 to 11 m wide at the base and stood about 18 m high (Hölscher 1951:1f.).

Textual sources confirm that such heights for Bronze Age fortification walls were common. The heights of these walls are, for example, comparable to that of the enclosure which Ramesses III built around the well in the land of Aiyn.

> I made a very great *hnmt*-well in the foreign land of Aiyn, which was enclosed with a wall like a mountain of sandstone consisting of twenty layers (of bricks) as a foundation (and reaching) a height of thirty cubits with the battlements, its portal doorjambs being of cedar, their bolts being of copper with mountings (*Papyrus Harris I*, 77:6–8).

This wall would have stood to a height of approximately 16 m. Naram-Sin's siege of Armānum (see pp. 90f., for discussion) also provides various heights for the walls surrounding that town which were measured in cubits (1 cubit [KÙŠ] ≈ 50 cm). We are informed that the outer *kārum* wall was only 10 m high, while the wall of the lower town (BÀD *danīm*) was 15 m high, and that of the acropolis (BÀD.GAL) 20 m high. The width of these walls is not mentioned, however. Nevertheless, one Old Babylonian text refers to a wall at Larsa as "five reeds [≈ 15 m] high (and) two cubits [≈ 1 m] thick" (see *igartu* in *CAD* I, p. 34). From these various sources it is reasonable to conclude that mudbrick fortification walls of between 10 and 15 m tall were built upon foundations between 5 and 6 m thick. If the wall did not taper, as may have been the case in the Levant and Mesopotamia, then it is possible that narrower walls could have been built to similar heights and would still have supported a walkway along the top of the wall.

With regard to the load-bearing characteristics of mudbricks in such fortification walls, little has been written. It is usually assumed that mudbrick walls less than one meter thick were not intended to support sec-

Jaroš-Deckert (1984). For Middle Kingdom siege scenes, see Alan Schulman (1982).

[14]For evidence of crenellations at the start of the second millennium, see the Egyptian siege of an Asiatic town in the tomb of Intef during the time of Mentuhotep II (see pl. 17 and folding pls. 1 and 3, Jaroš-Deckert 1984): Asiatics are shown defending a fortress against an Egyptian siege.

[15]Compare with a width of 2 m cited by both Herzog (1992:114) and Kempinski (1992c), which was derived primarily from small sites without consideration of the evidence for the width of foundations as evidence of wall width.

ond stories. However, it is more difficult to know to what extent mudbrick fortification walls between 1.5 and 3 m wide often supported a 10 m high mudbrick superstructure. Here it is helpful to examine evidence of modern mudbrick structures built in towns along the Wadi Hadramut in Yemen (Damluji 1992). In a unique architectural style, apartment buildings to this day are built as much as 30 m or about 10 stories high in mudbrick (ibid., 4). Surprisingly, the width of the ground floor walls of these structures is only about one meter (ibid., 128). These structures demonstrate, therefore, that by taking the appropriate precautions, namely the canting of the outer elevation, mudbrick could have been used to build stable structures to a considerable height. This fact has been assumed for some time concerning the so-called Middle Bronze Age *migdal* "tower" temples, which were probably several stories tall (B. Mazar 1968:93f.).

Generally, the mudbrick superstructure of the wall was constructed using the header-stretcher technique built over a foundation of stones of equal width. While the walls could form a continuous straight or curving line, at many sites the presence of towers and buttresses created junctures at which a wall's direction was altered. Where such structures were not present the so-called saw-tooth and offset-inset designs were employed in order to make a thick wall curve as needed around a site (e.g., Shechem, Shiloh, el-Qitar, and Yakaltum), although the base constructional unit was a rectilinear mudbrick. The offset-inset technique seems to have appeared toward the end of the Middle Bronze Age, if the few sites where it has been discovered are reliable indicators of the date of its introduction.

Although mudbrick walls and casemate walls are preserved at many sites, it is often the case that clear evidence for the defensive wall of a settlement is lacking despite targeted excavation of a site's fortifications. While the absence of evidence has led some scholars to conclude that many sites featured no walls at all and others to assert that warfare was not a concern (see pp. 12ff.),[16] it remains possible to extrapolate the width of the town wall from a number of indicators—as is often done for the walls of domestic structures. In such cases it is possible to rely upon the width of the stone foundations, the core wall within the earthen rampart that also served as a foundation wall, as well as the so-called "anchor" walls that flanked the sides of monumental gates. Of these three types it is only necessary to elaborate upon the identification of "anchor" walls.

So-called "anchor" walls, which flanked the sides of gates or terraces as at Akko, Ashdod, Ashkelon, Dan, Far'ah South, Hazor, and Yavneh-Yam, provide substantial evidence regarding the composition of fortification walls. Although the observation that such walls served to "anchor" the gate within the rampart is correct (Figure 9 below), that these walls terminate several meters from the gate in no way implies that they were not a part of a continuous defensive wall (contra Herzog 1997a:134). Instead, as is evident from Hazor (Area P), these short segments of foundations functioned as terraces within the rampart for the purpose of stepping the base of the wall down in order to abut the sides of the gate. The width of the terraces at Hazor also suggests that the Middle Bronze Age wall was as much as 10 m wide and grew to 15 m in width during the Late Bronze Age. If no town walls had been built on the ramparts, terrace walls flanking gates would have been unnecessary. However, the terraces provided each of the wall segments with foundations, which were necessary to keep the wall and rampart from placing lateral stress on the gate's superstructure (see Gregori 1986:83f.).

The function of the breaks between these terraces can be compared with the seams that are placed in the surface of concrete sidewalks in order to keep it from breaking apart. Although sidewalks are often poured as a single concrete slabs, seams are placed at regular intervals along their surfaces so that uneven settling and cracking does not compromise its structure. Likewise, terraces would have enabled the release of varying degrees of stress at the juncture of each terrace over long stretches of the wall's foundations. Thus, any weakening of the gate's superstructure could be avoided as the adjacent sections of the wall settled. The extensive preservation of terraces is due to their being buried during the construction process.

b. *Casemate-Style Walls*

In addition to solid fortifications walls, evidence from a number of sites demonstrates that casemate-style walls were introduced during the Middle Bronze Age. Sites such as Nebi Mend (Kadesh), Kumidi, Abu Kharaz, Dothan, Far'ah South, Hazor, Shechem, Shiloh, and Ta'anach all featured such walls. While it is possible that some of these casemate walls were constructed around the entirety of a settlement, most were employed for only a part of a town's defenses, whether to address the specific defensive concerns of one area or more likely to satisfy space constraints at such within certain quarters (e.g., Far'ah South, Hazor, and Shiloh).

---

[16] This is the case with respect to the question of town walls in the Levant during the Middle Bronze Age (Herzog 1997a:133ff.), as will be discussed below, as well as with New World settlements (e.g., S. A. LeBlanc 2003).

Table 7. Width and composition of solid fortification walls from ca. 1900 to 1500 B.C.

Site	w. (m)	Remains
Abou Danne VI, Ph. 3	2.1, 0.7	Mudbrick
	> 10 citadel wall	Mudbrick
Achzib	6	Mudbrick
Afis citadel	4.3	Mudbrick
Akko	2.5	Mudbrick
Amman	2	Stone foundation
Aphek	2.25–2.5	Stone foundation
ʻArqa	1.7	Mudbrick
Ashdod XXIII, Ph. B	> 2	Mudbrick gate "anchor"
Ashkelon XX, Ph. 10	3.3	Mudbrick
Beit Mirsim G-F	3.2–3.3	Stone foundation
Beth-Shemesh	2.2–4	Stone foundation
Beth-Zur	2.6	Stone foundation
Biruta	0.7	Mudbrick, stone foundation
	1.7	Quarried stone (6 m high)
Bosra	< 6	Stone foundation
Burak	1.2	Mudbrick, stone foundation
Dan XI	7	Mudbrick gate "anchor"
	3.5	Mudbrick
Deir ʻAllā	1	Mudbrick
Dothan	2–2.5	Stone foundation
Ebla	[6, EB IV wall]	Mudbrick
	2.5 citadel wall	Mudbrick
Farʻah North (Per. VC)	2.2–4	Stone foundation
Farʻah South	2.5	Mudbrick
Gerisa	1.7–2.2	MB IIA: Mudbrick
	3	MB IIB: Mudbrick
Gezer XVIII	4	Mudbrick, stone foundation
Hammah	2.8	Mudbrick
Hazor XVII (Area A)	8.8	Mudbrick, citadel wall
(Area P)	15	Stone foundation, gate "anchor"
Hebron	9.5	Stone foundation
Hefer	4	Mudbrick
Jericho (Per. IVa–b)	3.5–5.75	Mudbrick
	1.4–2.3	Mudbrick
Jerusalem	2	Stone foundation
Kabri	> 4	Mudbrick, stone foundation
Kebara	5	Stone foundation
Keisan	1	Stone core wall
Kitan	2	Mudbrick
Kumidi	2	Stone foundation
Lachish	3.35	Mudbrick
Malhata	2.5	Mudbrick?
Megiddo XIIIA	1.5–3	Mudbrick
	5	Mudbrick
Nagila XI	2.3–2.5	Mudbrick
Nahariya II	3.8	Stone foundation
Najjar	4	Stone foundation
Nimrin	3	Stone foundation
Pella XI	2.5–3.5	Mudbrick

Site	w. (m)	Remains
Poleg	2.7	Mudbrick
Poran	?	Mudbrick core
Qashish	1.8	Stone foundation
Qatna	2 (LB?)	Stone foundation
Qitar	2	Stone foundation
Rukais II.2.I	3–3.5	Stone foundation
Selenkahiye	2.25–2.5	Mudbrick, stone foundation
Shechem XX–XVI	2.65–3.75	Mudbrick, stone foundation
XV	3.25–3.75	Mudbrick, stone foundation
Shiloh	2–3.8	Stone foundation and core wall
Timnah XII–XI	2.4	Mudbrick
Tuqan	4.5 citadel wall	Mudbrick
	2.8	Mudbrick
Umm el-Marra IIIC–A	1.5 citadel wall	Mudbrick
	1.5–2	Mudbrick
	7	Mudbrick
Yoqne'am XXIIIB–A	1.5–3	Mudbrick, stone foundation
Zeror	2.5–5	Mudbrick
Zurekiyeh	3.2	Mudbrick, stone foundation

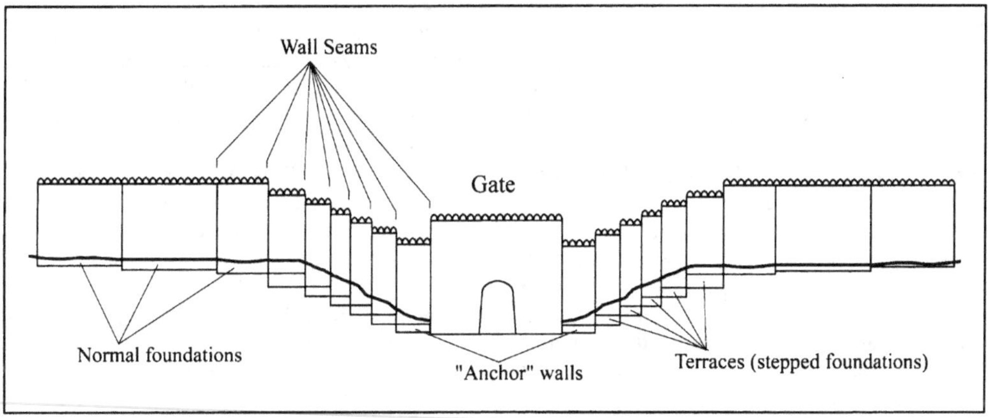

Figure 9. Illustration of the use of terraces and "anchor walls" between fortification walls and gates.

Although the strength of casemate walls by comparison to solid walls cannot be determined with certainty, the average width of these walls (presuming they were filled in when necessary) exceeded the width of regular walls by about two meters. The space between the walls of these casemates probably resulted from attempts to maximize storage space, as is suggested by the number of storage jars found in these types of rooms (e.g., Shiloh, Dothan). Nevertheless, in order to have served as a truly effective barrier it would have been necessary to fill these spaces very compactly with rubble and debris before a siege. To date no such fills have been identified, and thus, whether or not they functioned as true casemates during the Middle Bronze Age remains uncertain.

c. *Unwalled Settlements with Ramparts?*

During the Middle Bronze Age there is no question that unwalled villages and farmsteads existed (see p. 123). The question, therefore, is whether or not larger settlements that possessed earthen ramparts actually lacked fortification walls, or if other explanations exist for why fortification walls have not yet been discovered at these settlements. Despite the widespread evidence for Middle Bronze Age fortification walls of various types, as cited above, a limited number of scholars have maintained that such walls were never built (see pp. 12ff.). Among the sites most frequently identified by these scholars are Akko, ʻAjjul, Beit Mirsim, Dan, Far'ah South, Haror, Hazor, Jemmeh, Jericho, Megiddo X, Mevorakh, Michal, Poran, Shechem, Shiloh, and Yavneh-Yam. However, of these sites it is only the case that no remains of the defensive walls have been identified thus far at ʻAjjul, Haror, Jemmeh, Michal, and Poran. Only MB IIC Tel Michal appears to have lacked a defensive wall. However, this is because the earthen rampart was only intended as a construction platform (see pp. 296ff.).

While there is no question that some sites do lack evidence for fortification walls in this period, such negative evidence only makes it impossible to confirm that they were fortified. It does not constitute proof that they were unfortified. The suggestion that large settlements featuring earthen ramparts actually lacked fortification walls ignores a variety of factors that have led to the conditions discovered at the sites mentioned above. First, at some sites the absence of evidence is entirely the result of limited soundings (and sometimes limited publication) of the fortifications. However, the Middle Bronze Age fortifications of only a handful of sites in the southern Levant have been adequately explored (i.e., with soundings in more than one area: Tel Dan, Hazor). The most notable of the sites lacking sufficient excavation of the fortifications are ʻAjjul and Poran, where only a single section of their fortifications was obtained. However, as mentioned above, intra-site variability in the construction of ramparts and walls to one degree or another appears to have been the norm (e.g., Dan, Hazor, Jericho, Shechem), despite the assumption by most scholars that ramparts with similar profiles were built using nearly identical techniques throughout.

Since the fortification walls were constructed of sun-dried mudbrick (see pp. 73ff.), a second factor contributing to the assumption that walls did not exist is the eradication of the remains of fortifications in certain areas due to erosion (e.g., Jemmeh), mining by farmers for fertile *sebakh* [Ar. soil] (e.g., Haror?) and/or later construction atop the site (e.g., ʻAjjul). Although the fortifications of some settlements were much as four kilometers in length (e.g., Qatna), they are often only explored in a single sounding, which is often no more than five meters wide. The best example of this was the renewed search for the Middle Bronze Age fortifications of Jericho by the Italian expedition, which revealed evidence of a mudbrick wall associated with the rampart immediately adjacent to areas previously excavated by Kenyon and Garstang (see Figure 82). Thus, the remains of the fortification wall were missed by three previous expeditions to the site. Although this wall was preserved only one meter high, it was 5.75 m wide.

A third factor in the mischaracterization of the remains of fortification walls is the result of the limited nature of the remains, which often consist only of a few courses of mudbrick (e.g., Jericho). More frequently only the stone foundations remain (e.g., Tel Dan). A fourth factor contributing to this misinterpretation has been the ignoring of comparative evidence of more substantial mudbrick structures such as gates, towers, and bastions (see discussion below), discovered at the sites in question. (e.g., Hazor). Since in every other period these were accompanied by a fortification wall it is difficult to accept that the ramparts functioned alone to defend these sites in a unique manner during this period. Finally, the assumption that earthen ramparts were the product of cultural rather than defensive concerns of the inhabitants of the southern Levant during the Middle Bronze Age has also been cited as support for this interpretation (for counter argument, see p. 157).

The interpretation of Tel Dan's fortifications demonstrates how nearly every one of these factors has contributed to the assumption that no fortification wall accompanied its earthen ramparts. First, clear evidence of the mudbrick "anchor" wall and gate have been found in Area K (see Figure 73). Second, remains of the lowest courses of the mudbrick wall have been identified in Areas T and Y, though in Area Y the dilapidated remains of the mudbrick wall, which are plainly seen in section, are explained away as an extension of the fieldstone core of the rampart. Third, while no remains of the mudbrick wall were recovered in Area A-B, Biran has previously suggested that the top six meters of the very large earthen rampart were removed in antiquity (see Figure 71).

d. *Exterior Structure Walls*

Another factor may also account for the assumption that earthen ramparts lacked fortification walls. It appears that some sites, whether MB IIB–C unfortified villages or small MB IIC towns, may only have been defended by the exterior walls of buildings within the settlement (e.g., Mevorakh, Michal, Masos, Giv'at Sharett, and Manahat [see Figure 12]). Such defensive practices, though limited in extent, may have been a prelude to Late Bronze Age fortification strategies, as suggested by a lack of evidence for fortification walls at many sites during that period.

6. Towers and Bastions

In addition to the direct evidence of Middle Bronze Age fortification walls discussed above, towers and bastions, which were often built atop Middle Bronze Age ramparts, provide unequivocal proof that earthen ramparts were built for defense. Furthermore, the evidence indicates that almost all towers were rectilinear, while circular and semi-circular towers are only attested at Tuqan and Kannas, respectively. Although the terms *tower* and *bastion* have usually been used interchangeably, the dimensions available for these rectangular structures reveal that these terms can be meaningfully employed to identify two distinct types of defensive buildings incorporated into Middle Bronze Age defenses. Therefore, towers, rectangular structures usually less than 20 m long (length here is measured parallel to the fortification wall) can be distinguished from bastions, which were larger, fortress-like structures (see Table 8 below).

a. Towers

Rectilinear towers were the most frequent type of fortified structure use along town walls. They averaged between 10 and 12 m in length and 5 to 6 m in width. Towers were usually centered within the town wall at evenly spaced intervals ranging from approximately 20 to 35 m,[17] projecting only a few meters from the line of the defenses. The short intervals between most towers probably improved the effectiveness of the archers stationed along the top of the wall by eliminating dead space with overlapping fields of fire (Figure 10), which was similar to the way in which straight lengths of fortification wall also improved the effectiveness of archers. On occasion towers were also attached secondarily to the interior rather than the exterior face of the town wall (e.g., Megiddo).

b. Bastions

Bastions, which like towers were usually rectangular, were also incorporated into fortification walls at regular intervals at some sites. The best examples of these structures have been exposed at Ebla where they are identified as "fortresses."[18] At Ebla bastions were constructed approximately 300 m apart with average dimensions of approximately 27 m long by 13 m wide (Pinnock 2001:22). While it has been suggested that their spacing was optimal for providing suppressing fire between towers (i.e., within 150 m of any tower along the wall), it is doubtful that these bastions functioned without towers between them. This suggestion is supported by towers identified at Gezer (Str. XIX) where a structure that should be identified as a bastion (Tower 5017) was built to the west of the south gate. Contemporary with this bastion were numerous other towers spaced 30 to 35 m apart along the town wall. It is therefore likely that four or five towers were built between every two bastions at Ebla.[19]

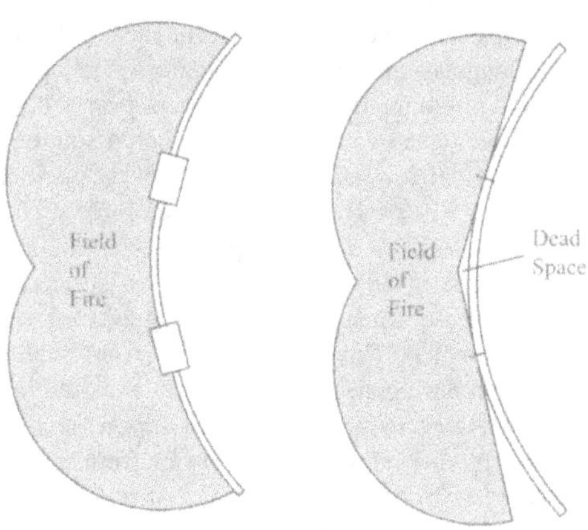

Figure 10. Fields of fire for settlements with and without towers.

While towers and bastions shared many functions as lookout and signal towers, sentry posts, and fixed defensive positions, because of their size bastions are likely to have served a number of more general purposes. They may have served as barracks for soldiers, storage facilities, residences, or administrative buildings. Various finds from the bastions at Ebla, which include weapons, a lexical tablet, and a hearth, provide clues concerning these possible functions.[20]

c. Dimensions, Placement, and Function

Most frequently the ratio of length to width of Middle Bronze Age towers and bastions was approximately 2

[17]Compare the distance attested between the towers along the walls of Late Bronze Age Hattuša and Kuşakli (Sarissa) in Anatolia; geophysical survey of Kuşakli identified the complete circuit of the fortifications including the locations of the gates and towers (see Müller-Karpe 2003, esp. fig. 2), which were spaced ca. 20 to 25 m apart.

[18]It is, however, best to reserve the term fortress for small, fortified settlements of a military nature.

[19]Since 1995 intensive exploration of Ebla's fortifications has focused on areas where large structures could be seen along the surface or were considered most likely to have existed based on the locations of the bastions previously identified. No random soundings have been attempted, which would most likely yield evidence of such towers, nor has geophysical prospection been undertaken.

[20]For discussion of the administrative role of the Western Fortress at Ebla, see Frances Pinnock (2001:31f.).

to 1 (Table 8). This may be related to the observation that the basic building blocks for towers were square mudbricks, which made building rectilinear towers easier than building apsidal or circular towers, which were much more common during the Early Bronze Age. Too great a variation from a rectilinear design necessitated more careful construction and required the cutting of bricks as well as a greater amount of mortar per cubic meter of mudbrick used, thus weakening the structure overall. Solid mudbrick towers with a rectilinear plan (e.g., Poleg) were also easier to construct than circular towers.

Perhaps the most likely reason for the adoption of the rectangular plan was its superiority to apsidal towers in providing the largest platform from which a number of archers could fire down upon a single position. Apsidal towers afforded the fewest positions along the tower's edge from which multiple archers could focus upon a single target (i.e., only a single tangent exists at any one point along a curved surface), especially when the target was against the town wall (Figure 10). This would have been particularly detrimental during siege warfare when it would have been necessary to provide intense suppressing fire against battering rams and siege towers from as many positions as possible. I disagree, therefore, with Kempinski's suggestion concerning the apsidal tower's "superiority in many respects over the rectangular tower" (1992b:72), which is based upon the observation that this tower type was reintroduced in the Hellenistic period and was also used through the medieval period.[21] While it is true that the square corners of towers and bastions would have introduced a vulnerability to prying and battering by the enemy, the advantages to the defenders would have outweighed the disadvantages from the mid-third millennium B.C. onward.

Rectangular towers had become the base unit of construction for all structures appended to fortified walls by the EB III (Kempinski 1992b:68, 72), which in addition to towers included not only bastions but also gates. As will be discussed below, six-pier gates, which were common during the second half of the Middle Bronze Age in the Levant, also employed rectangular bastions parallel to their entryways in order to provide optimal coverage of the gate area by the defenders. Towers (and bastions) were also incorporated into other structures within towns, perhaps in order to defend particular structures or to provide strongholds that could be defended in the event that enemy troops entered the town. A building (L.3245, 3363) incorporated into the Western Palace in Area Q at Ebla featured dimensions (32.5 x 15 m) similar to those of bastions built along Ebla's ramparts, which measured 27 x 13 m (Pinnock 2001:22). A rectangular tower was also built to protect the spring located on the edge of the "City of David" in Jerusalem during the Middle Bronze Age (see pp. 282f.).

Despite their size, the most significant role of both the tower and the bastion was, of course, that their projection from the fortification wall afforded the defenders an advantageous position from which they could more easily protect greater lengths of the fortification system. The most significant difference between settlements with and without towers is, as Figure 10 demonstrates, that towers provide interlocking fields of fire. Without towers dead spaces existed between archers located on opposite sides of a single target.

Although the defensive problems inherent to sites with elliptical layouts were for the most part remedied by the start of the MB IIB (ca. 1700 B.C.) following the introduction of rectilinear plans at most settlements, rectangular towers had already been employed to improve the defensibility of MB IIA sites with elliptical layouts. They are attested during the MB IIA at sites such as Megiddo, Beit Mirsim, Burga, Poleg, and Zeror. They continued in use during MB IIB–C as evidenced at Ashkelon, Dan, Far'ah North, Gezer, and Pella. It is likely that these towers were simply a development of prototypes employed in fortification systems in northwestern Mesopotamia, where rectilinear towers are attested at EB IV sites such as Sweyhat and Selenkahiye.

The ubiquity of rectilinear designs for Middle Bronze Age towers is also evident from the employment of rectangular towers and bastions in Middle Kingdom fortresses in Nubia between the second and third cataracts. At Buhen, Mirgissa, and Semna, for example, rectilinear towers projected from the inner and outer fortification walls at regular intervals and were also accompanied by larger rectilinear bastions. Though the towers were often nearly square, clearly the value of such towers regardless of their origin was already apparent in Egypt by this period. Earlier towers on Egyptian fortifications, like those in the southern Levant during the Early Bronze Age were, however, apsidal. Although it is difficult to be certain why rectangular towers had come to be preferred during the Middle Bronze Age in the southern Levant, it is possible that, as suggested above, the size of the platform that they provided made them preferable to circular and apsidal variants.

[21]Kempinski's suggestion overlooks that from the Hellenistic period onward ballistae were fired against fortifications in order to breach them; employing rounded architecture for stone towers improved their ability to deflect ballistae.

Table 8. Dimensions of Middle Bronze Age towers and bastions.

Site	Area/Locus	*l.* (m)	*w.* (m)	Ratio (*l.*:*w.*)	Spacing (m)
Akko	Building A	15	15	1:1	—
Ashkelon XXI	Grid 2 F.31	14	8	7:4	—
Beit Mirsim G-F	2 towers	10–10.5	5	2:1	23
Beth-Shemesh V	NE tower	9	[6+]	?	—
	W tower	18	11	1.6:1	89
	S tower	11.3	7.3	1.6:1	89
Burak	S and E towers	8	8	1:1	15
Dan XI	Area T L.8931	8	5	2:1	—
Ebla	Area M	27	12.5	2:1	300
	Area V	26	16		
Far'ah North (Per. VC)	L.301–303	12.7	7	2:1	—
Gezer XIX	Tower 1	12	8	3:2	30–35
	Tower 2	16	8	2:1	
	Tower 3	8	12	2:3	
	Tower 4	6	8	3:4	
	Tower 5	13	6	2:1	
	Tower 1005	13	8	1.5:1	
	Bastion 5017	[30]	15.6	2:1	20
Jericho (Per. IVa–b)	Tower E1 (W.268–270)	7.5	3	5:2	—
	"Garstang" tower	16.25	10	8:5	—
Jerusalem	Spring Tower	17	14	8:7	—
Kannas	Bastion 1	7.5	4	2:1	—
Megiddo XIII	4104	13	6.5	2:1	—
XII	Area BB	10	6	5:3	—
Pella XI	Area XXVIII	12	8	3:2	—
Poleg	Area C L.24	15	8	2:1	—
Qashish IXC	Area A	5.5	4.5	1.2:1	—
Selenkahiye	Bastion B	5	5	1:1	—
	Bastion	10	7	3:2	—
Tuqan Phase 1B	Area E	7.5	5.85	5:4	
Phase 1A	Area F	9.4 dia.	NA	14.5–16	—
Zeror	Area D	15	8	2:1	—

Note: Length (*l.*) is measured parallel to fortification wall. Brackets indicate reconstructed dimension.

7. Gates

Middle Bronze Age gates in the Levant have been thoroughly studied on a number of occasions (G. R. H. Wright 1984; Herzog 1986:esp. pp. 37–73; Gregori 1986). The focus of these studies has been the six-pier gate. However, since the present study is focused on a broader assessment of the role of fortifications as a whole, this treatment of gates focuses on the origin and functional characteristics of a variety of different gate types. These include not only the six-pier type but also the four-pier gate and the postern gate, both of which are well attested in the Levant.[22]

The locations of gates can be identified with depressions within the contours of a site's perimeter. These depressions are the combination of the gate itself and erosion along the gate's abandoned passageway (cp. "linear hollows" discussed by T. Wilkinson 1993). At many sites the location of gates was based on the location of earlier gates. The four MB IIB–C gates of Ebla, for example, which were identified with the depressions between the site's embankments, are thought to have been built above the previous gates of the Early Bronze Age mentioned in the Ebla texts (Matthiae 1980b:65). Nevertheless, some gates have been buried in such a manner that the site's contours do not aid in their identification (e.g., Dan).

The location of gates in the Middle Bronze Age, as during other periods, was predominantly determined by the orientation of routes to and from particular destinations. While this is usually true, it is also evident that neither the existence of a gate in a specific location, nor the nature of its construction suggest its relative importance. For example, although the gate in Area A on the southwest side of Ebla is well preserved and certainly leads toward more densely oc-

[22]For gate construction in the Mari letters, see ARM III 10 and 78, and III 11 about the collapse of Saggaratum's gate.

cupied areas to the west of Ebla, it is impossible to assess its significance since there is no way of comparing the amount of use it experienced relative to the other gates (contra Matthiae's suggestion, see pp. 198ff.). Furthermore, there is no way of differentiating between local use and occasional traffic by nonresidents. Social or political motivations may also have restricted the use of a gate for royal purposes, while circumstances such as later blocking of a gate may result in its unusually excellent preservation (e.g., Dan). Functional principles for situating gates persisted during the MB IIB–C.

a. Six-Pier Gates

Because of its first appearance in the Levant during the Middle Bronze Age, the six-pier gate has been considered a hallmark of monumental and defensive architecture of this period. While this type of gate has been variously referred to as an *à tenaille* type gate (Fr. "with pincers," Matthiae 1980b:120), the "three-entrance" gate (Gregori 1986), the four-chamber gate (Ger. *Vierkammertor*; Weippert 1988:222), the "Syrian" gate (Kempinski 1992d:197; 1992c:134ff.), the "fort-gate" (Herzog 1997a:134), and the "triple gate" or "three way gate" (Kaplan 1975:12ff.), all of which emphasize particular aspects of this gate type, the gate is probably most simply described as the six-pier gate. This term avoids restricting the origin of the gate type to a particular region since it is not conclusive that it originated in the northern Levant (i.e., western Syria), and it does not tie its identification to the number of chambers or doorways that the gate featured since these were not necessarily the same at each site, despite the adoption of a generally similar overall plan. The term "three-entrance" gate is also inappropriate since we cannot be certain that six-pier gates were equipped with three sets of doors, as this term suggests; even when all the door sockets are found in situ they seem to indicate that only two sets of doors located inside the inner and outer piers were employed (e.g., Area A gate at Ebla).

The six-pier gate consisted of a direct-axis passage flanked by two rectangular bastions separated by three sets of piers, which were often connected by at least two sets of doors. Such gates are known in the northern (e.g., Alalaḫ, Ebla, Tuqan, and Qatna) and the southern Levant (e.g., Dan, Far'ah South, Gezer, Hazor, Megiddo, Shechem, Shimron, and Yavneh-Yam), as well as in northwestern Mesopotamia (e.g., Yakaltum).[23] Since no contemporary gate structures have been excavated in southern Mesopotamia, it is impossible to determine whether this type of gate was employed there or that it did not originate there, though at present this seems unlikely. There is also no evidence from Anatolia to suggest that this gate type originated there (Naumann 1971:285ff.).

Although it is probable, as Barbara Gregori suggests, that the development of this type of gate was directly related to an increase in the width of fortifications as a result of the use of earthen ramparts (1986:83), it is impossible to demonstrate this especially since even larger gates (with eight piers) were also built in the Iron Age at sites without earthen ramparts. Instead, the appearance of this gate type would seem to correlate with the need to improve the strength of gates against the frequent use of the battering ram (see pp. 39f.). Nevertheless, whatever the reasons for its invention and wide adoption, this gate was especially uniform in size and construction,[24] as Matthiae has previously observed (1980b:204).

Gregori notes that two main types of rectangular towers were incorporated into these structures (1986:85f.): towers with solid lower levels (e.g., Tuqan, Gezer, Hazor) and towers with chambers (e.g., Alalaḫ, Beth-Shemesh, Far'ah South, Hazor, Shechem, and Yavneh-Yam). Some of the narrow chambers of these towers appear to have featured staircases, which were built of wood (e.g., Alalaḫ, Shechem, Yavneh-Yam, Beth-Shemesh) with the lower steps constructed of stone (e.g., Alalaḫ). Other rooms within the gate's bastions served multiple purposes, just like the rooms in most bastions (see p. 65).

Despite the formidable construction of the six-pier gate, it may often have been the case that these gates were supplemented with a smaller outer gate. Such was the case for Ebla's southwest gate where a four-pier gate was constructed outside of the main gate and was separated from it by a large courtyard. At Ashkelon a similar outer gate was added in Phase 12 below the main gate. In both instances, the indirect approaches to these gates were oriented at an angle to the main gate. Though this bent-axis design may have been intended to impede direct entry, it seems more likely that in both cases the location of these gates traced the original approaches to these sites, which were probably established during earlier phases of occupation. In some instances the steepness of the ap-

[23] If the reconstruction of the gatehouse at T. Taya in northern Iraq is correct (Reade 1968), this gate would be the only extant example of a similar gate type in eastern or northern Mesopotamia.

[24] The west gate at Qatna, the "Water Gate" of Kar-Kemiš, and the northeast gate at Yakaltum (Munbaqa) can no longer be included as Middle Bronze Age gates due to the inconclusive evidence for the date of their construction (see discussion on pp. 213ff., 211, and 181ff. respectively).

proach to the gate could be reduced by curling the ramp around the slope (i.e., of the ramparts or side of the tell). This almost always resulted in the creation of a bent-axis approach, which was tailored to the needs of everyday traffic and was not, therefore, intended to deter the approach of enemy soldiers. If bent-axis approaches reduced the effectiveness of enemy attempts to enter gates, then we would expect ample evidence, of which there is none, for the adoption of this strategy in the construction of Mesopotamian gates (see Damerji 1987:esp. 181ff.). Instead, this approach, which is peculiar to the Levant and Anatolia, was dictated by the topography of these sites.

The number of excavated examples of six-pier gates is sufficient to reconstruct a chronological understanding of the development of this gate type. That it first appeared in the northern Levant toward the start of the MB IIA has already been demonstrated by examples from Ebla (Mardikh IIIA) and Tuqan. During the transitional period MB IIA/B this gate type first appeared in the southern Levant (e.g., Dan XI, Hazor Str. 4, and Yavneh-Yam), probably having been adopted gradually throughout the northern Levant, as for example at Qatna. Following this period numerous examples are attested in the southern Levant, particularly during the late MB II (e.g., Shechem XVI, Yavneh-Yam phase 2, Hazor Str. 3, Beth-Shemesh, Far'ah South, Gezer XVIII), and they were constructed through the end of the MB IIC throughout the Levant (e.g., Alalaḫ VII). The latest of these gates appear to have remained in use even at the start of the Late Bronze Age (e.g., Hazor Str. 1). It may even be the case that some examples that have been dated to the Late Bronze Age were actually built in the late Middle Bronze Age (e.g., Megiddo VIII; this may account for the lack of evidence for separate gates belonging to Levels X–IX). Nevertheless, it is clear that when possible these gates were maintained and renovated as necessary.

What remains to be debated, however, are the developments that led to the invention of the six-pier gate. In this regard several gates, which have not previously been addressed in existing typologies of EB IV and Middle Bronze Age gates in the Levant, may provide some understanding of the development of this gate type. These include the EB IV gates of Selenkahiye and Byblos (Level VI), and the gate of Ashkelon XXII (Ph. 14) dated to the start of the MB IIA (see Figure 62). Since the role of the load-bearing, barrel-vaulted arch is thought to have been integral to the construction of the six-pier gate, it is necessary to recognize that such architectural elements were in use in the Levant and northwestern Mesopotamia as much as five hundred years prior to the development of six-pier gates. Among the earliest preserved mudbrick arches is one found in the palace at Beydar that has been dated to ca. 2300 B.C. on the VLC (Bretschneider and van Lerberghe 1997; Oates 1973).

Table 9. Dimensions of six-pier gates (in meters).

Site	Passage *l.*	Passage *w.*	Entry *w.*	Pier *w.*	Bastion *w.*	Figures
Alalaḫ VII (NW Gate)	16	5	2.75	2.6–2.8	9.2	Figure 27
Beth-Shemesh (S Gate)	14.3	4.5	3	2.2–2.5	5.5–7	—
Dan XI (Area K)	13.5	10.6–11	2.5	1.65–2	—	Figure 72
Ebla SW Gate (Area A)	21.3	5.8	2.8	4.35–4.45	—	
NE Gate (Area BB)	20	4.7?	3.2?	4.5–4.75	—	Figure 38
Far'ah S. (S Gate)	18	6.6	3.6	2.2–3.8	7.6	—
Gezer XVIII (S Gate)	12.9	4.6	2.75	2–2.7	8.5–8.7	Figure 76
Hazor Str. 4 (Area K)	9	3(?)	1.8(?)	?	9	—
Str. 3 (Area K)	16.5	7.3(?)	3.1	2–3	6.5	—
Str. 4 (Area P)	20.6	?	?	?	9.5	—
Str. 3 (Area P)	25	?	2.65	3.2–3.6(?)	11	—
Megiddo X?–VIII (Area AA)	18	6.8	2.8	2.6–2.8	—	—
Yakaltum (NE Gate)	17.5	7.2	3.1	2.5	2.5	Figure 24
Qatna (E Gate)	>22	8	3.5	4	?	—
Shechem XVI (NW Gate)	18.3	7.32–7.55	2.8	1.91–2.3	4.9–5.5	Figure 95
Tuqan NE Gate (Area A)	>18	7.3	2.6	3.2–3.75	10.2	Figure 49
SE Gate (Area F)	17.5	7.5	2.7	3.5–3.9	>7.5	Figure 50
Yavneh-Yam (Gate III)	18	5.2	2.8	2.4–2.5	9.5–10	—
Averages	17.9	6.2	2.9	3	8.4	—

Note: Bastion length equals passage length.

Table 10. Dimensions of four-pier gates (in meters).

Site	Passage *l.*	Passage *w.*	Entry *w.*	Pier *w.*	Bastion *w.*	Figures
Akko (Area F)	7	4.4	2	2	—	—
Ashdod (Area G)	6.5	2.9	3?	1.8	—	—
Ashkelon XXIV (Ph. 14)	19	3.6	2.5	1.82–2.8	—	Figure 62
XXIII (Ph. 13)	20	3.6	2.3>2.1	1.82, 2.8	—	Figures 63–64
XXII (Ph. 12)	29	7	5.5	—	—	Figure 65
XXI (Ph. 11)	7.3	3.9	1.5	1.8–2	—	Figure 66
Beit Mirsim E–D	?	6.25	3.25	1.5	—	Figure 68
Ebla (Area A)	10.5	4.75	3	2.7	?	Figure 37
Far'ah North (Per. VC)	7	6	3.2	2	—	Figure 75
Megiddo XIIIA (Area AA)	6	4	1.5	1.8	—	Figure 88
Rukais II.2.I	10	2–3	1.5	2	5–6.5	—
Shechem XV	13.9	8.35	3.3	3.2–3.35	5.45–5.5	Figure 95
Averages	12.4	4.8	2.7	2.1	5.6	—

Evidence that similar arches were incorporated into architecture in the Levant prior to the Middle Bronze Age is attested at Byblos and Selenkahiye. The northeast gate at Byblos (Level VI, ca. 2300 B.C.) bisected the rampart of that period. The interior passage of this gate was 18 m long and 4.8 m wide and was roofed by a barrel-vaulted ceiling that was supported by the canted, stone-built walls that are preserved to a height of more than two meters (see Figure 32). That such a barrel vault, over such a wide passage, which lacked any internal piers was innately unstable is suggested by the spaces left in the walls and floors for the timbers that were required to support the vaulted ceiling. A contemporaneous gate of smaller proportions has also been exposed at Selenkahiye (see pp. 218f.). Its passage was 10 m long and 3 m wide, which is comparable to that of Middle Bronze Age six-pier gates. Although little is known of this gate's superstructure, nor is it certain that it featured a barrel vault, its dimensions and date suggest that it must be considered within the evolution of the six-pier gate type.

The clearest evidence of the evolution from the late third-millennium gates of Byblos and Selenkahiye is, however, the MB IIA gate at Ashkelon (Str. XXII, Ph. 14). This gate featured four arched piers, the outer two of which were constructed in stone with the inner two built in mudbrick, enclosing a barrel-vaulted chamber 9 m long and 3.6 m wide (see Figure 62). The construction of this gate, when compared to the gates of both Byblos and Selenkahiye, suggests that there was a need to reduce the original length and width of the vaulted space for structural reasons. This is further illustrated in the gate at Ashkelon by the addition of buttresses in the center of the chamber's walls in Phase 13 (see Figure 63f.)—a precursor, if not replication, of the function of the central piers of six-pier gates. Though it might be argued that at some point the gate at Ashkelon should have been constructed with a six-pier plan, the narrowness of the original passage of the Phase 14 gate made such a modification impossible. The dating of the original gate at Ashkelon—to the mid-MB IIA (after the EB IV gate at Byblos)—and the evidence that the builders of Ashkelon's gate were unfamiliar with the six-pier gate type supports the suggestion that this gate can be placed in an evolutionary sequence between the EB IV gate type and the MB IIB–C six-pier type. Thus the six-pier gate, which gained widespread use during the MB IIB–C, appears to have evolved from a late third-millennium monumental gate type attested in the northern Levant.

b. *Four-Pier Gates*

The four-pier gate, sometimes referred to as the double-entry gate (Table 10), was the most ubiquitous gate type throughout the Bronze Age. As such it also featured the most variation in orientation. As Gregori observes, this gate type, which she identifies as the "chamber-shaped" type, though present during the Middle Bronze Age, had very clear precursors in the Early Bronze Age (1986:90). It consisted of two main subtypes, which are differentiated as either long, where vaulting may have been used for the passage (see discussion above), or short.

Examples of four-pier gates with a single, long vaulted chamber are attested at Byblos (Level VI) and Ashkelon (XXIV, Ph. 14), as mentioned above. Such large four-pier gates, as suggested above, probably contributed to the emergence of the six-pier gate type. However, smaller four-pier types are attested at Akko, Ashdod, Ashkelon, Beit Mirsim, Far'ah North, and Yavneh-Yam.[25] Most of the exam-

[25] The gate of the Middle Bronze Age palace at Mari was also a four-pier gate (Parrot 1958:pl. 11).

ples of four-pier gates appear to have been sufficiently wide to accommodate wheeled vehicles. Aside from the absence of central piers within the gate, the major distinction between this and the six-pier type was that the four-pier type usually lacked towers (except at Shechem). There also does not appear to be any evidence for the use of vaulted passages in four-pier gates of the Middle Bronze Age like those incorporated into six-pier gates.

c. Postern Gates

Although postern gates are relatively unknown during the Middle Bronze Age, they persisted in use from the Early Bronze Age through the Late Bronze Age—undoubtedly a testament to their important role for pedestrian traffic. This type of gate represented the simplest gate plan, which was essentially composed of two piers. Most postern gates, which because of their simple nature have often escaped identification, were narrow, nondescript passages through the town's wall intended to serve only "foot and hoof" traffic (e.g., Tuttul). Such gates usually constituted an opening in the town wall rather than a fortified structure.

One particular subtype of postern gate built during the Middle Bronze Age featured corbel vaulting. The best example of this type of postern is the corbel vaulted gate on the northwest side of Ugarit dated to the Late Bronze Age (see Figure 54). Such gates appear to have been the only type of gate that was constructed from floor to ceiling entirely of well dressed stone masonry. In addition to the gate at Ugarit, which may in fact be of MB IIC date (see pp. 224ff.), we can also add the lesser known postern gate of identical design at Akko, which Moshe Dothan dated to the MB IIC but identified as a drain (1993a:19f.). This gate did not feature steps within, but simply led below the third rampart into the site. This gate type continued to be built and used in the Late Bronze Age, when examples are attested mostly from Anatolia at sites such as Hattuša, Alaça Höyük, and Alishar Höyük (von der Osten 1937:figs. 27–29).

d. Other Gate Types

Other less common gate types were also built throughout the Levant during the Middle Bronze Age. At Beth-El a U-turn gate was excavated on the northwest side of the site. The complete length of its passage was over 20 m. The NW Gate at Byblos (Level VI) is also unique (see Figure 34). It featured a set of stairs descending the earthen rampart to the northwest and northeast from the northwest side of the site. What could be referred to as pierless direct-axis gates have also been identified at Qitar.

e. Gates and Traffic

Because most travel in the ancient Near East was done on foot, the majority of traffic through Middle Bronze Age gates was pedestrian. In recent years it has been observed, however, that domesticated herd animals were stabled on the first floors of houses inside walled communities during the second and first millennia B.C. (e.g., Stager 1985:13ff.; Schloen 2001:338ff.). Therefore, animal traffic would have been common at town gates as these animals were taken to and from pasturage. What is less clear, however, is the extent to which wheeled traffic was able to pass through certain types of gates.

It has often been assumed that gates were constructed with passageways of sufficient width to permit the entry of chariots, insofar as they were in use at the start of the second millennium (see artist's rendition of chariot passing through Middle Bronze Age gate at Ashkelon in Gore 2001:72). However, this reconstruction has been advanced purely on the basis of the suggestion that gate passages were widened between the Early and Middle Bronze Ages (Herzog 1986:61). Nevertheless, this has not been conclusively demonstrated by means of a comparison of the width of passages and the length of chariot and cart axles, especially since no Middle Bronze Age examples are preserved (Littauer and Crouwel 1979:53f.).[26] Therefore, to adequately answer this question comparative evidence for the length of chariot axles, which can be derived from Late Bronze Age exemplars, must be invoked. These indicate a range from 1.98 to 2.36 m for their axle widths (ibid., 78), suggesting that gates with passages wider than 2.5 m were sufficiently wide to accommodate chariots and carts, provided that no other obstacles such as low ceilings or steps were present. This, therefore, serves to adequately demonstrate that most of the six-pier gates could have accommodated such traffic (see entry widths in Table 10). The gate at Tel Dan with a passage of barely 2.5 m, in contrast, probably did not accommodate such traffic, which is also confirmed by the existence of cobbled stairs leading up to the gate on both sides.

Some gateways indicate that measures were also taken to block them at different times, in order to limit the amount or type of traffic. At Ebla it is evident that at some point during the last phase of the Middle Bronze Age (Mardikh IIIB) access to its southwestern gate (Area A) was limited by the con-

[26]That chariots, vehicles intended for higher speeds and maneuverability, required a wheelbase broader than four-wheeled carts and two-wheeled straddle cars or platform cars is demonstrated by Littauer and Crouwel (1979:53f.).

struction of a 6 m long, 1.5 m wide fieldstone wall across all but three meters of the courtyard between the inner and outer gateways (Matthiae 1970:65 and fig. 2). Although the passage width corresponds to that of the preexisting passages between the inner and outer gates, the sharp turns required to maneuver through this gate reveal that wheeled vehicles were no longer welcomed through this gate or, more importantly, in this quarter of town.

f. Construction

The foundations of gates, in contrast to the foundations of walls, were usually constructed of dressed stone (e.g., Ebla, Tuqan, Qatna, and Hazor). Still the superstructures were probably almost completely built of mudbrick, as illustrated by the nearly completely preserved gates at Ashkelon (Ph. 14), Tel Dan XI, Alalaḫ VII, and Yakaltum. If stone was occasionally required in the superstructure it was used for orthostats, corners, and voussoirs. This is illustrated by the use of stone for voussoirs at Ashkelon in places where mudbricks would have been especially prone to structural failure or where they would have been too easily damaged by traffic through the gate.

In most respects the style of construction of six- and four-pier gates was identical, though orthostats were probably employed more routinely in the construction of the piers and foundations of six-pier gates (e.g., Alalaḫ and Ebla). As Gregori notes (1986:91), these were both aesthetic and functional (i.e., load-bearing), which is demonstrated by their slightly inward canting along the length of the passageway (e.g., Alalaḫ and Ebla). These orthostats probably served to support barrel-vaulted ceilings, like those preserved at Yakaltum and Tel Dan, but attested earlier at Byblos and Ashkelon.[27] The average height of these vaults can be estimated at around 3 m (e.g., Dan).[28]

The use of double doors is indicated by the preservation of door sockets in both four- and six-pier gates (e.g., Ebla). The number of door sockets preserved in six-pier gates clearly indicates that only two sets of doors were in use, though three sets of piers were built. This further supports the observation that the central piers were intended to be load-bearing rather than to serve the installation of a third set of doors.

Although no remains of any original Middle Bronze Age doors have been recovered from these gates, though they were undoubtedly constructed of wood with bronze fittings (see EA 69:24ff.), the original dimensions of these doors can still be derived. These are deduced from the dimensions of preserved portals and the location of in situ door sockets. Two doors were usually paired at each entry point and they were, therefore, probably up to 3 m tall and between 1.25 and 1.5 m wide. If the 30 kg bronze hinge from the thirteenth century B.C. gate at Jaffa is comparable (Kaplan 1956:260; 1972a:81), then the doors were as much as 20 cm thick. Of course, the dimensions and shape of the door would have varied if the door was required to open into a barrel-vaulted passage where it would need to clear the ceiling. The locking and bracing mechanisms for the doors are also almost entirely absent in the archaeological record. However, an example of a socket for a cross bar, which would have been placed behind the inward opening doors of the outer piers, is attested in the Phase 13 gate at Ashkelon.

In general, unlike other aspects of fortifications during the Middle Bronze Age, it is not certain that developments in gate architecture can be clearly tied to the presence or absence of a particular weapon type. That six-pier gates were the most fortified gate type up to that time may in part relate to their association with earthen ramparts. Nevertheless, this gate type may have also been associated with increased use of the battering ram, even though the persistent construction of other gate types such as the four-pier gate and the simple postern gate indicate that defense was not the only concern involved. Unlike other features of Middle Bronze Age fortification systems, the gate experienced nearly continual use, during both peace time and war, when it was probably avoided by attacking forces since it would have been recognized as one of the most strongly defended positions. Just as towers along walls would have been avoided because of the resistance posed by a larger number of soldiers and that entry through these buildings would have been considerably more difficult, fortified gates were likewise avoided (as in the Neo-Assyrian siege of Lachish). Of all of these gate types the one that posed the greatest vulnerability to the defenders would have been the postern gate, which it would have been necessary to block or fill prior to the start of a siege and to conceal its position as best as possible.

8. Water Systems

To ignore the topic of Middle Bronze Age water systems would be shortsighted in any discussion of fortification strategies since without them the inhabitants could not have survived a siege more than a few days. Hence, it may be argued that the prevalence of

[27]This observation supports Gregori's suggestion that the Alalaḫ VII gate featured a barrel vault (1986:93), in contrast to Woolley's reconstruction based on parallels with the Lion Gate at Mycenae and Hittite architecture.

[28]On the use and history of vaulting in mudbrick architecture in Mesopotamia, see works by David Oates (1973; 1990).

siege warfare in this period owed its existence in large part to the creation of elaborate water systems capable of providing especially large settlements with a much needed water supply. Aside from settlements that featured springs within their ramparts (e.g., Tel Dan, Kabri, Beth-El, Ashkelon, Aphek, and Jericho), two basic types of water systems were constructed: open reservoirs and rock-cut tunnels.

The simplest water system for settlements enclosed by earthen ramparts was the open reservoir. These were essentially large pools within the fortification systems that were probably maintained for the purpose of providing a secure water source in the event of a siege, though they may only have held a substantial amount of water in the late spring and early summer. The clearest examples of these are found at Qatna and Bosra. At Qatna the water system appears to have consisted of a depressed area to the north of the palace in the northwest corner of the rampart enclosure that may be identified as a reservoir for water collection (see Figure 44). A comparably large depression, identified as a *citerne*, has been noted by Frank Braemer within the Middle Bronze Age enclosure at Bosra (Braemer 2002:fig. 8). The depression in the southwest enclosure of the lower town at Hazor, which has long been suspected of being a water system, should probably also be identified with this type of water system (see Figure 78).

These depressions may be clay extraction pits (OB *yarrum*) used in the production of mudbricks during the construction of Middle Bronze Age fortifications. Such depressions were especially characteristic of settlements in northern Mesopotamia where they are much easier to identify (see pp. 56f.). However, these features are especially problematic to date.

Probably a later development is the most well-known type of Middle Bronze Age water system, the rock-cut tunnel. Two extensive water systems of this type have been dated with certainty to the Middle Bronze Age in the southern Levant. These are the water systems of Jerusalem and Gezer (see pp. 282ff. and 260ff. for details), although similar systems may have been excavated at Gerisa, Megiddo, and Amman (see refs. in Reich and Shukron 2004:213). Both of these systems enabled the inhabitants of the settlement to access a single water source located deep below ground from within the town's walls by means of an underground, rock-cut system of tunnels. That such enormous subterranean projects were not uncommon during the Middle Bronze Age is evident from the siege tunnels at 'Ajjul (see pp. 230ff.), the tunnels discovered below buildings in Area F of the lower town at Hazor (see Yadin 1972b:43f.; Yadin et al. 1989:135, plan 26), as well as the tunnel in the rampart at Lachish (see pp. 287f. for detailed discussion of these features). Nevertheless, there is no evidence to support the identification of the 'Ajjul, Hazor, or Lachish tunnels as water systems. As evidenced at Jerusalem and Gezer, Middle Bronze Age water systems served their settlements for a very long time, perhaps as late as the Iron Age.

Although almost no systematic effort has been to identify the water systems of Bronze Age sites, there remains little doubt that such efforts would reveal similar systems at other sites. Geophysical survey offers a fairly easy means of identifying these features, though excavation remains necessary to establish the precise date of these systems. Nevertheless, water systems must be considered an essential element of Middle Bronze Age defenses.

B. Building Materials

1. Mudbricks

Sun-dried mudbricks were used in Middle Bronze Age wall construction. There is no evidence for the baking of bricks in the Levant. While the sizes of bricks varied considerably, it does appear that square mudbricks of roughly 35 to 40 cm on a side and 10 to 15 cm thick were most common during the Middle Bronze Age (see Table 11 below). Mudbricks were also trimmed for specific needs after being molded, usually where a regular brick was too large.

The color of bricks used in wall and gate construction, while seeming inconsequential, is more significant than has often been acknowledged. This is because they represent the source of materials available when the defenses were constructed. Reddish or brown bricks, for instance, often reveal the use of clean soil and thus suggest that the materials used in these bricks were taken from outside of the settlement, or that the bricks reveal an early stage of settlement since occupational debris, which is usually gray due to ash content, is absent (Oates 1990:388f.). This is the case, for instance, at Urkeš (Tell Mozan) where red bricks, which were used in early construction, were made from virgin soil taken from the level of the plain because no other material was available (Giorgio Buccellati, personal communication, 2002). Such bricks, however, were of an inferior quality since they contained less temper than bricks made from occupational debris, which were usually gray or black in color. The quantity of ash in occupational debris that was used to make gray bricks may have reduced the permeability of these bricks.[29] While many wall segments and structures within fortification systems were built of bricks using a single source for raw materials, on occasion multiple sources were employed (e.g., Pella).

From these observations it is possible to suggest that newly established settlements, or those that undertook large building projects such as fortifications for the first time, may have been forced to produce bricks from the only material locally available in sufficient quantities, namely virgin soil, which was usually low in clay content. Consequently, this meant that the very buildings into which large quantities of labor were invested and upon which the settlement relied for protection may often have been constructed with inferior materials.

Although Arlene Rosen and Paul Goldberg have thoroughly examined the morphological aspects of mudbricks (1986:75ff.; Goldberg 2004), the data for mudbricks used in Middle Bronze Age fortifications are, unfortunately, inadequate for comparison with their findings. Few such scientific analyses of mudbricks from Middle Bronze Age sites have been conducted (e.g. Jericho, see Cerulli 2000). Future collection of data for mudbricks could, however, yield useful information regarding their use, origin, variation, and adherence to standards of measurement.

2. Mortars

Mudbricks used in construction were usually bonded on all sides with mud mortar, the matrix of which can be primarily, though not always, distinguished from that of bricks by means of color (G. R. H. Wright 1985a:359ff.).[30] This is often the result of the presence of greater quantities of ash within the mortar, which made the mortar grayer than the bricks. Mortar was generally applied in relatively thin sections, about 2 cm thick, but if spacing needed to be maintained in a wall of header-stretcher construction, then it could be applied in thicker quantities to vertical sections between bricks (a process known in contemporary masonry as tuck-pointing). While the use of mortar was important for bonding bricks in walls, the importance of the mortar itself diminished once the structure was completed. At that stage the entire structure also needed to be protected against erosion.

3. Plasters

While mudbricks could be faced with stone blocks, a technique common in Egyptian architecture, in most instances other materials, which were more readily available, were employed to protect mudbrick architecture from the elements. Mud and gypsum plaster were often used on the exterior face of walls and gates, and it is usually assumed that these materials were intended primarily to protect against erosion due to precipitation. However, wind erosion was also responsible for the destruction of mudbrick structures. Deep ruts cut horizontal grooves through the bricks, undermining the strength of exposed structures. One strategy for dealing with the effects of wind and rain erosion was simply to apply mud plaster as often as necessary to the outer surfaces of the bricks. Although mud is no more resistant to wind and rain than are the mudbricks that they cover, the mud coating was easier to replace than an entire wall if the bricks were damaged.

Although the effects of rainfall and wind erosion on mudbrick architecture and exposed slopes have been noted, the long-term and detrimental effects of insects, rodents, and birds are often overlooked (see Rosen 1986:10f.). Insects, such as wasps, bore into mudbrick walls in order to build nests (Voss 2002:381). Termites consume temper materials used in mudbricks and mud mortar such as straw and chaff (e.g., Damluji 1992:148), which sometimes leads to the weakening of the bricks and the breaking away of mud plaster (Kemp 2000). Rodents and birds also burrow into mudbrick walls to build nests that then undermine the base of mudbrick structures. Birds take straw and chaff from the surface of mud structures to build nests and also degrade the surfaces of structures by hunting for insects. While these individual effects upon a mudbrick superstructure may seem minor, prolonged exposure of large mudbrick structures to the combination of these perils in addition to other erosive forces would eventually result in a loss of structural integrity and the need to completely rebuild them.

According to G. R. H. Wright the preferred matrix for mud plaster included as much sand as possible (1985a:360), which was probably the result of the impermeability of silica. Although mud plaster was most frequently used because of its availability, gypsum plaster (and less frequently lime plaster) provided the most effective seal against all of the threats mentioned above. Unfortunately, few studies of plas-

---

[29]Ethnographic evidence suggests that too much ash is considered "cheating" in modern mudbrick production because such bricks are of poor quality and do not last as long (McGuire Gibson, personal communication, 2004).

[30]For an excellent review of mudbrick architecture in Egypt and general practices associated with mudbrick construction, see A. J. Spencer (1979) and Barry Kemp (2000).

ter recovered from archaeological contexts have been done.[31] Yet the results of these studies indicate that the plaster was predominantly a lime compound of calcium oxide obtained usually by burning gypsum but also sometimes lime.[32]

Aside from these reasons for plastering mudbrick structures, an examination of sites throughout the Levant, northwestern Mesopotamia, and the Nile Delta suggests that the selection of plastering materials was primarily dependent upon the availability of materials and the amount of rainfall received in a region. Where drier conditions prevailed (e.g., northwestern Mesopotamia and the interior of the Levant) the greatest concerns were insects and wind erosion, and mud plaster offered sufficient protection. Where greater amounts of rainfall were received (e.g., Aphek, Ashkelon, Beth-Shemesh, Megiddo), particularly along the coast, a less permeable barrier was necessary to protect structures.[33] Gypsum or lime plaster would have provided mudbrick structures and ramparts with greater longevity than mud plaster. Nevertheless, as would have been the case with mud plaster it was also necessary to reapply gypsum and lime plaster to surfaces, albeit less frequently, as is indicated at a number of sites where multiple layers of plaster are preserved.

Furthermore, the use of lime and gypsum plasters in the construction of glacis (e.g., Jericho) indicates that it was also vital to maintaining the integrity of earthen ramparts with slopes between 30° and 45°. An added advantage of plaster along walls and slopes may have been that it made it more difficult for the enemy to hide, especially at night since they could be more easily spotted against the bright background (see suggestion concerning plaster-lined fosse at Buhen by Emery et al. 1979:6). Nevertheless, the use of plaster in glacis construction was infrequent in comparison with mud plaster, probably because it was much more costly and labor intensive to procure.

4. Wood

The use of wood in the construction of fortifications has been attested at a number of sites (e.g., Alalaḫ, Byblos, Akko[?], and Ashkelon). Given its limited availability and the cost of incorporating it into mudbrick and stone fortifications, it would have been primarily intended to cover closed spaces that could not be spanned by other materials, as it was similarly used in palace construction. Otherwise it was used most extensively for the construction of the doors for gates. At Jericho in the excavation of an MB IIA–B tower in Area A, remains of a wood identified as poplar (*Populus*) have been recovered (Marchetti 1998:206f.). This was used in the beams of a tower in the lower town to the south of the mound. Although the tower's length is unknown, the width of the interior space was just slightly less than two meters, meaning that the length of poplar segments were probably only a little more than two meters.

5. Stone

Stone was frequently used in the various types of walls and foundations. No evidence suggests that stone was procured from a great distance during the construction of fortifications. Aside from the effort required to shape large chunks of stone into blocks for the foundations of gates (e.g., Tuqan), the cyclopean blocks used in revetment walls (e.g., Shechem), and delicately carved orthostats (e.g., Alalaḫ, Ebla, Hazor, Shechem), a considerable amount of effort appears to have been expended to obtain crushed or chipped stone, which was used in rampart fills. Still, such materials could be relatively easily procured from exposed sections of stone in wadi beds or from the excavation of fosses through bedrock that accompanied the rampart construction process, particularly of the freestanding type.

6. Soils and Soil Micromorphology

Few of the previous studies of Middle Bronze Age fortifications have investigated the micromorphology of soils used in earthen ramparts and mudbricks. Exceptions include studies by Reuben Bullard (1970) for Gezer, Arlene Rosen (1986), Hanoch Lavee et al. (1993) for Shiloh, and work by Mauro Cremaschi et al. (2002) at Qatna. Geomorphological data reveal, for example, how far it may have been necessary to transport materials for the construction of fortifications. Such analyses could also reveal that elements that are

[31]Plaster from the interior walls of a Middle Bronze Age house at Beth-Shemesh abutting the southern fortification wall were tested; the results indicated that the sample was 82% pure lime, 10% sand, and 8% other materials (Grant and Wright 1939:33f.).

[32]G. R. H. Wright notes that gypsum plaster ($2CaSO_4$ H_2O) was more common than lime plaster (1997:372) because the production of lime plaster ($CaCO_3$) required calcining limestone at temperatures of around 900° C (ibid., 437f.). This, of course, required greater quantities of fuel such as wood, which are not known to have been heavily sought for such uses until the Iron I period. Gypsum plaster could be more easily obtained by calcining gypsum (Gk. *gypsos*, "chalk"), calcium sulphate dihydrate ($CaSO_4$ $2H_2O$), at temperatures of 100° to 130° C (ibid., 373).

[33]This does not, however, appear to have been the reason for using plaster in many places in Egypt. Evidence of plaster has been recovered from the fortifications of Buhen, where it was used on walls, the fosse, and even in the loopholes (Emery et al. 1979:6, 21–23, 27), and at Askut, where it was found on the mudbrick on the surfaces inside the girdle wall and gate (Badawy 1964:50; 1965:131).

no longer visible such as mudbrick walls were present on the surface of stone foundations. Insofar as archaeologists have concerned themselves with the ancient potter's selection of clay for the production of ceramics, they must also begin to consider geomorphological aspects of rampart and mudbrick architecture.

Table 11. Dimensions of mudbricks (in cm) used in EB IV and Middle Bronze Age fortifications.

<table>
<tr><th>Site</th><th>EB IV</th><th>MB IIA</th><th>MB IIB–C</th></tr>
<tr><td>Abou Danne VI</td><td>—</td><td>33 x 33 x 7; 28 x 28 x 7
38 x 38 x 10</td><td>35 x 35 x 8</td></tr>
<tr><td>Afis</td><td>—</td><td>30 x 30 x ?</td><td>—</td></tr>
<tr><td>Ashdod XXIIIB</td><td>—</td><td>—</td><td>50 x 30 x 20</td></tr>
<tr><td>Ashkelon</td><td></td><td></td><td></td></tr>
<tr><td>XXIII, Ph. 13C</td><td>—</td><td>45 x 33–35 x 10–12
54 x 31 x 19</td><td>—</td></tr>
<tr><td>XXIII, Ph. 13B–A</td><td></td><td>51 x 34 x 13
50 x 33 x 11
48 x 33 x 12
32 x 32 x 11</td><td>—</td></tr>
<tr><td>XXII, Ph. 12</td><td>—</td><td>—</td><td>50 x 36 x 10
47–48 x 33–34 x 10–13</td></tr>
<tr><td>XXI, Ph. 11</td><td>—</td><td>—</td><td>35–40 x 35–40 x 10
50 x 36 x 10</td></tr>
<tr><td>Burak</td><td>—</td><td>—</td><td>40 x 40 x 12</td></tr>
<tr><td>Dan XI</td><td>—</td><td>—</td><td>30 x ? x 15</td></tr>
<tr><td>Ebla</td><td>60 x 40 x ?</td><td>—</td><td>32–35 x 30–32 x 10–12</td></tr>
<tr><td>Ferzat</td><td>—</td><td>—</td><td>43–44 x 43–44 x 10–11
36–39 x 36–39 x 10–12
44–57 x 44–57 x 11–15</td></tr>
<tr><td>Gerisa</td><td>—</td><td>35 x 35 x ?
50–55 x 40 x ?</td><td>—</td></tr>
<tr><td>Gezer XVIII</td><td>—</td><td>—</td><td>38 x 29 x 10</td></tr>
<tr><td>Habuba Kabira</td><td>34–38 x 48–50 x 10–12
24–27 x 45–50 x 9–10</td><td>38–40 x 38–40 x 10–12</td><td>—</td></tr>
<tr><td>Hammam et-Turkman</td><td>—</td><td>—</td><td>35 x 35 x 8</td></tr>
<tr><td>Hazor</td><td>—</td><td>—</td><td>40 x 30 x 15
40–50 x 40–50</td></tr>
<tr><td>Jericho (Per. IVa–b)</td><td>—</td><td>42 x 36 x 15
30 x 30 x 10</td><td>—</td></tr>
<tr><td>Kannas</td><td>40 x 40 x 10</td><td>—</td><td>—</td></tr>
<tr><td>Kumidi</td><td>—</td><td colspan="2">36 x 36 x ?</td></tr>
<tr><td>Mari</td><td>—</td><td colspan="2">40 x 40 x 12</td></tr>
<tr><td>Mašin</td><td>—</td><td colspan="2">38–40 x 38–40 x 12–14</td></tr>
<tr><td>Megiddo XIIIA</td><td>—</td><td>35–40 x 35–40 x 10</td><td>—</td></tr>
<tr><td>Mevorakh</td><td>—</td><td>40–45 x ? x 15</td><td></td></tr>
<tr><td>Yakaltum</td><td>—</td><td>—</td><td>—</td></tr>
<tr><td>Pella XI</td><td>—</td><td>—</td><td>40–50 x 40 x 10
70 x 40 x 10</td></tr>
<tr><td>Poleg</td><td>—</td><td>60 x 50 x 12
55 x 35 x 12</td><td>—</td></tr>
<tr><td>Rukais II.2.I</td><td>—</td><td>—</td><td>45–50 x 45–50 x 12–15</td></tr>
<tr><td>Shechem XVI</td><td>—</td><td>—</td><td>38 x 38 x 10–12</td></tr>
<tr><td>XV</td><td>—</td><td>—</td><td>35–43 x 35–43 x 15–16</td></tr>
<tr><td>Sweyhat</td><td>50 x 40 x 10</td><td colspan="2">—</td></tr>
<tr><td>Terqa</td><td>—</td><td colspan="2">34–37 x 34–37 x 8–10
half brick: 34–37 x 17–20 x 8–10
32 x 32 x 10–11</td></tr>
<tr><td>Timnah XII–XI</td><td>—</td><td>—</td><td>50 x 50–75 x ?</td></tr>
<tr><td>Tuqan (Phases 1B–A)</td><td>—</td><td>38 x 38 x 12</td><td>40 x 40 x ?</td></tr>
<tr><td>Tuttul</td><td>50 x 36 x ?
54 x 25 x ?</td><td colspan="2">38–42 x 38–42 x 8–10</td></tr>
<tr><td>Zeror</td><td>—</td><td>30 x ? x 16</td><td>—</td></tr>
</table>

One such study of the rampart at Shiloh has yielded interesting results regarding soil selection for rampart and glacis construction (Lavee et al. 1993). During the MB IIC the fortifications of Shiloh were supplemented by an earthen rampart and a glacis, which probably only existed in certain areas (Areas C and D). The deepest section was excavated in Area D where five main components of the rampart and glacis were revealed. Geomorphological analysis of the distribution of grain sizes in the rampart make it possible to identify the percentages of certain types of soils in each rampart layer. This establishes the degree to which different materials were selected for the construction of the rampart. The differences between the percentages of soils of different particle sizes in each layer revealed that the amount of clay used varied the greatest between each layer (see fig. 13.10 in Lavee et al. 1993). While it is was determined that a high percentage of small particles decreases a layer's permeability, this study suggests that the overall porosity of each layer is directly related to the percentage of particles of the smallest grain size. Since the layers with the highest percentage of silt content belong to the upper layers of the glacis, it is possible to suggest that this was part of a strategy intended to improve the durability of the rampart's surface, by providing it with what constitutes a clay glacis distinct from its rampart. Therefore, according to the excavator, a so-called micromorphological analysis of the matrices of both the rampart and glacis confirms the existence of the glacis around only some parts of the site, which directly correlates with the grade of the slope in that location.

C. Fortifications and the Landscape

While a great deal may be learned about the construction of Middle Bronze Age fortifications by analyzing their individual elements and the materials from which they were built, fortifications as a whole must be studied in relationship to the landscapes in which they functioned. On the one hand, locations were selected in part because of their defensibility. On the other hand, because of the need for improved defenses the construction of fortifications often led to substantial alteration of landscapes. Of foremost concern in this regard are issues relating to site selection.

1. Site Selection

Site selection was the first concern involved in a settlement's defenses in the premodern era. The primary concerns relating to Middle Bronze Age fortified settlements were not substantially different from those of walled settlements throughout the Bronze or Iron Ages. Subsistence concerns aside, three major factors are traditionally thought to account for the location of settlements: the availability of water, defensibility, and proximity to routes. While it is difficult to suggest that any one of these factors was more important than another since to some extent they were integrally related, it is apparent that all things being equal, water was always the most essential element. While the location of a site might lead to the emergence of new routes and likewise fortifications could improve a site's defensibility, it was much more difficult if not often impossible to establish a secure water source if it was not taken into account during site selection. A scarce water supply even limited the ability to build substantial defenses since water was needed for mudbricks. Furthermore, inadequate water meant that a siege could only be withstood for a very short duration.

Many fortified settlements of the Middle Bronze Age also appear to have been located adjacent to wadis. Since water could more easily be accessed during the summer by means of water holes dug into the bottom of wadis it could be argued that an advantage of positioning sites along these features was that water could be accessed more easily. Whether or not this was the case, it seems even more likely, as noted above (see discussion of the fosse above), that the placement of fortified sites along wadis was actually intended to take advantage of the defense provided by this natural feature in what were otherwise rather topographically homogenous landscapes (e.g., Qatna, Jemmeh, Far'ah South, Masos, Malhata).[34] While such a strategy does not on the surface seem especially wise considering the detrimental results of erosion at the base of earthen ramparts, the threat was fairly limited since it was not posed during the critical period of summer when war was waged. Minor repairs and maintenance were no doubt necessary after each winter but were made prior to wartime operations. Thus, it can be argued that the wadi functioned as a ready-made fosse on one side of such settlements.

2. Site Morphological Processes

As with site selection, the processes of site formation and destruction also shed light on the morphology of

[34]The placement of sites along wadis does not suggest their location along trade or transportation routes since land routes did not slavishly trace wadis.

Middle Bronze Age settlements. It is widely recognized that the foundations of many tells in the region date to this period. The construction of earthen ramparts and the digging of fosses during the Middle Bronze Age permanently affected the appearance of the Levantine landscape. The use of earthen ramparts particularly of the freestanding type, which often led to sites resembling craters, only exaggerated the tell-formation process during this period throughout the region. Concomitant with this was the creation of fosses, which in many circumstances probably facilitated the erosion of cultural remains, whether draining into the sea (e.g., Ashkelon), an adjacent wadi (e.g., Qatna), or a river (e.g., 'Acharneh).

To some extent natural processes may also have affected the preservation of Middle Bronze Age settlements along the Levantine coast. David Neev et al., for instance, address the effects of tectonic activity on coastal sites in the eastern Mediterranean (1987). The authors assert that several coastal settlements of the Middle Bronze Age along with their fortifications were "destroyed by the same tectonic movement along the coastline of Israel at the time of the eruption of Santorini" (ibid., 65–67). They argue that Yavneh-Yam, for example, experienced subsidence that resulted in the erosion of the western portion of what was an earthen rampart with a square layout dated to the Middle Bronze Age. Yavneh-Yam, they insist, experienced a "history analogous to that of the first phase of settlement within the Tel Michal site." (ibid.). This hypothesis undoubtedly found its origin in Jacob Kaplan's insistence that the enclosure at Yavneh-Yam once formed a complete square, and therefore that more than fifty percent of the enclosure was destroyed at that time. However, there is no reliable basis for postulating that a rectilinear enclosure surrounded Yavneh-Yam (see Figure 99 and discussion on pp. 315ff.). While it may be true that over the past four thousand years periodic tsunami activity has adversely affected the Middle Bronze Age remains of fortifications along the Levantine coast, it remains impossible to accurately estimate the full extent of this damage.

D. Architectural Evolution

It has been observed that Middle Bronze Age fortifications in the southern Levant bore little resemblance to Early Bronze Age fortifications (Kempinski 1992c; 1992d:175), despite the occasional use of similar defensive features such as the glacis, fosse, and bastion (Parr 1968; Kempinski 1992b:73; Oredsson 2000). Nevertheless, it is possible to trace the evolution of Middle Bronze Age fortifications from northern Mesopotamia through the northern Levant and into the south over the period from the third to the second millennium B.C. In this section the major stages in the development of Middle Bronze Age defenses throughout these periods are described. These consist of two preliminary stages during the Early Bronze Age that led up to the emergence of Middle Bronze Age defenses. These defenses then developed during three stages that were followed by a phase of abandonment during the Late Bronze Age (for overview, see Table 12 below).

1. Kranzhügel Phase (EB II)

During the first phase of this fortification evolution, the Kranzhügel Phase (ca. 3000–2600 B.C.), sites that have been identified as Kranzhügeln were established along the Euphrates River (e.g., Mari, Terqa, Emar)[35] and on the open plain of northwestern Mesopotamia along wadi systems (e.g., Abu Schachat, Bugha, Beydar, Chuera, Glea, Khanzir, Mabtuh Gharbi, Mabtuh Sharqi, Maghr, Mahrum, Muazzar; see Figure 11). Although the debate surrounding these settlements, their origin, and ethnic affiliation has been ongoing since the work of Ursula Moortgat-Correns (1972), the questions that have been raised are gradually being answered by excavations at both Tell Chuera (Orthmann et al. 1995) and Tell Beydar (Ismail et al. 1996; Debruyne 2003; van Lerberghe and Voet 2000), as well as at Tell Muazzar.

As the German term for these settlements suggests, the remains of their *mudbrick fortification walls*, which were built with circular plans, eroded to create "wreath-like" embankments encircling these settlements, which were extensively photographed from the air by R. P. Antoine Poidebard (1934). For this reason such sites can be identified as part of the "signature landscape" of this region during the Early Bronze Age (Wilkinson 2003:7). These sites featured elevated upper towns surrounded on all sides by a lower town. As evident at Beydar (see pp. 168f.), both the upper and lower towns were enclosed by thick mudbrick walls, and the settlements were also often situated adjacent to water courses or wadis.

[35] Although Mari, Terqa, and Emar are not often identified as Kranzhügeln, the evidence for the underlying features, which were responsible for the morphology of these settlements, are identical and contemporaneous with the evidence from Kranzhügeln such as Chuera and Beydar (see catalog 1).

Figure 11. Sites identified as Kranzhügeln.

In this phase, postern gates were used. Although no evidence of ramparts or fosses is known, modest earthen embankments (referred to as dikes by Margueron 2000) and moats are attested among Kranzhügeln along the Euphrates. While these features initially may been intended to preserve walls against being undermined by the Euphrates River (e.g., Mari Ville I and Terqa, Defensive Rings 1–2), they also served as defenses, a function that continued until the mid-second millennium.

2. Northern Levantine Phase (EB III–IV)

The abandonment of most Kranzhügeln in the Ḫabur region during the mid-third millennium created the conditions for the second phase in the evolution of fortification strategies, the Northern Levantine Phase (ca. 2600–1900 B.C.). During this phase a new tradition of settlement defense developed from Terqa along the middle Euphrates to Ebla in the northern Levant. In contrast to the Kranzhügeln of the Jebel Abd-al-Aziz and in the Ḫabur triangle, two sites, Terqa (see pp. 177f.) and Mari (see pp. 171ff.), continued to be occupied. These sites provide the clearest evidence that *freestanding earthen ramparts*, indeed, evolved directly from the tradition of fortification that was characteristic of Kranzhügeln, as Lawrence Stager suggests (1999:237f.).

In addition to settlements along the middle Euphrates like Terqa and Mari, similar fortification practices were adopted from the Euphrates bend to the northern Levant by the end of the EB III and early in the EB IV. The known sites with this type of fortification include Selenkahiye, Sweyhat, Umm el-Marra, Byblos, and slightly later Tuqan. Although less is presently known about the fortifications of Ebla, Afis, and Aleppo in this period, it is likely that their fortifications developed as part of this trend, as suggested by their topographic contours. That the earthen ramparts of these sites were clearly built as defensive features is evident since no river systems or other environmental factors, such as rainfall, threatened to undermine their defensive walls. Like Kranzhügeln these settlements still adhered to a mostly elliptical layout of their defenses (e.g., Byblos, Umm el-Marra, Afis[?], Ebla[?], and Tuqan), though roughly quadrilateral plans also began to be employed for the first time in this region (e.g., Sweyhat, Selenkahiye).[36]

The introduction of the battering ram in the EB IV, as attested in the Ebla archive, may account for the introduction of earthen ramparts during this phase (see summary of weapons developments in Table 4 on p. 46). It is likely, therefore, that the use of this weapon in Mesopotamia and the northern Levant led to the proliferation of freestanding ramparts in this period, an architectural feature that had already been incorporated into the defenses of settlements along the Euphrates River. By contrast the evidence for fosses in this period remains limited (e.g., Umm el-Marra), though this may be due to the depth at which these features are now buried, and it cannot be assumed that they did not exist at all. If, as I have suggested above, fosses were always associated with rampart construction, it may be impossible to identify a specific weapon or technique that they were intended to counter. It may have been the case that under different circumstances the fosse served to counter a variety of siege devices, particularly as attempts by attacking forces were made throughout this period to penetrate increasingly thick fortification walls.

3. Southern Levantine Phase (MB IIA)

The third phase in the development of fortifications is characterized by the gradual adoption of freestanding ramparts in the southern Levant as urbanism was revived at the start of the MB IIA, ca. 1900 B.C. This phase in the evolution of Levantine fortifications, which can be referred to as the Southern Levantine Phase (ca. 1900–1700 B.C.), is characterized less by innovation than by the southward spread of rampart architecture, the refinement of rampart construction techniques, and the adaptation of ramparts to the defenses of previously occupied settlements. There are no architectural or stylistic criteria, either morphological or compositional, by which the elements of freestanding ramparts of the MB IIA in the southern Levant can be distinguished from those of the EB IV in the northern Levant, and for this reason ceramics serve as the only means by which these features can be reliably dated. Nevertheless, the continuity in fortification construction between the northern Levant during the EB IV and the southern Levant during the MB IIA serves as further evidence of the cultural influence that the north exerted on the south at the start of MB IIA.

During this phase, as in the former phase, freestanding ramparts, which featured elliptical (or circular) layouts with mudbrick walls and rectangular towers, continued to be built in the north (e.g., Ebla, Tuqan). They were also gradually constructed for new settlements in the southern Levant, particularly along the main overland trade routes through the Great Rift Valley and the Jezreel Valley to the coastal plain (e.g., Akko, Ashkelon XXIV, Burga, Qana[?], Shimron, and Zeror).

Two innovations are characteristic of fortified sites in the southern Levant during this period. These are supplemental ramparts and the glacis. While all of the largest settlements founded in the southern Levant during the MB IIA were fortified with freestanding ramparts, older sites with previous occupation were fortified during this period with *supplemental ramparts* built against their fortification walls (previously called "wall ramparts"; see Kaplan 1975), which often rested at the edge of the tell or mound (e.g., Aphek [Str. Vd], Jericho [Per. IVa–b], Megiddo XIIIA, Poleg, and Zurekiyeh). These were probably intended to be functionally equivalent to freestanding ramparts, which for topographic reasons could not be built at these mounded settlements. While most sites with supplemental ramparts, like those with freestanding ramparts, were elliptical in plan, rectilinear plans were adopted at a few sites where supplemental ramparts were constructed (e.g., Aphek, Poleg, Zurekiyeh). While this may have been due to the original contours of the landscape, rectilinear plans, which were prevalent in the following phase, may have begun to emerge during this period (see also Qiryat Shemona South).

The second innovation of this period in the southern Levant was the regular adoption of distinct surface treatments for the slopes of fortifications, which should be identified as *glacis*. These were constructed of cobbles, plaster, chipped stone, or brick and were intended to preserve the walls and ramparts of these MB IIA settlements by preventing the erosion of their slopes due to exposure to greater precipitation in the western portions of the Levant than in regions further east. Glacis were used with both freestanding (e.g., Ashkelon) and supplemental ramparts (e.g., Gerisa). While Parr was right to note that similar features were present in Early Bronze Age defenses (1968), the intensive adoption and improvement of this technique for preserving slopes in this period makes its use particularly characteristic of the construction of

[36]The much earlier (ca. 3500–3100 B.C.) rectilinear fortifications of the Uruk colony at Habuba Kabira appear to be unrelated to the evolution described here.

Middle Bronze Age fortifications. Nevertheless, in most instances the glacis appears to have functioned only to preserve fortifications against erosion and thus it probably did not play a defensive role.

In addition to refinements in the construction of ramparts, the introduction of the *six-pier gate* at the end of the Southern Levantine Phase led to a vast improvement in the defensibility of gates. Although Gregori (1986) suggests that this gate type developed as a result of the ever-widening fortifications due to the use of earthen ramparts, it is more likely that its development was multicausal. The development of the six-pier gate no doubt occurred in response to the same defensive concerns as the proliferation of earthen ramparts, namely during the MB IIA when ever-widening hostilities required greater defenses (see chapter 4). Like the introduction of rampart fortifications in the southern Levant, the introduction of this particular gate type appears to have come from the north.

4. Rectilinear Phase (MB IIB)

By the end of the MB IIA new refinements in the construction of ramparts and changes in their layouts suggest that the construction of rampart fortifications had reached its apogee. This period, the Rectilinear Phase (ca. 1700–1600 B.C.), saw the construction of *rectilinear freestanding earthen ramparts* and *fosses* at the largest Middle Bronze Age settlements, once called "fortified camps" by Yadin (1955b). These sites include Qatna, Sefinat-Nouh, Bosra, Deir Khabiye, and perhaps Ugarit, as well as aṣ-Ṣour, in the northern Levant, and Dan XI, Hazor XVII, Jaffa(?), Yavneh-Yam, Timnah XII–XI, Ekron(?), Nagila XI, Lachish, Jemmeh, Haror, ʻAjjul, Malhata (Per. B), Masos, and Irbid in the southern Levant.

Like the Kranzhügeln of the third millennium and the elliptical plans of sites of the Southern Levantine Phase, these sites are responsible for the "signature landscape" of the Levant during the second half of the Middle Bronze Age (Wilkinson 2003:7). While some settlements were newly founded in this period, it is clear that others featured some, though perhaps less extensive, occupation during the MB IIA prior to the construction of their fortifications (e.g., Qatna, Ugarit[?], Dan, Hazor, Ekron[?], Timnah). The plan of these fortified settlements has been recognized as distinct from those of the MB IIA, namely that they were regularly rectilinear rather than elliptical. Therefore, aside from clear improvements made during this period in the know-how of rampart construction, strategic attempts were also made to improve upon the elliptical layout previously employed in freestanding rampart construction. The reason for this may have been the realization, as noted above, that straight wall segments were easier to defend than curvilinear walls. Similar realizations may also have led to the addition of rectangular bastions to walls (e.g., Ebla and Gezer XVIII) along with rectangular towers. These efforts suggest, therefore, a high degree of centralized planning in the process of fortification construction in this period, which can be considered a trademark of the social and political complexity of kingdoms in the Levant during the Middle Bronze Age (see chapter 5).

While sites with massive freestanding ramparts are the main type of monumental architecture attested during this period, supplemental ramparts were probably more common as fortification construction was undertaken at a large number of sites, which included tells. However, because these sites were already substantially raised above their surroundings, much less effort was required to fortify them. During the MB IIB inland sites in the southern Levant were also settled and fortified with supplemental ramparts as the process of settlement infilling continued (e.g., Bira[?], Keisan, and Seraʻ).

5. Highland Phase (MB IIB–C)

While most of the developments of the previous phase coincide with the MB IIB, developments during the following phase that have been traditionally associated with the MB IIC (ca. 1600–1530 B.C.) in several cases appear to have begun during the mid-MB IIB (ca. 1650 B.C.).[37] Thus, during what can be called the Highland Phase (ca. 1650–1530 B.C.) fortification construction of a completely different type was characteristic of highland settlements. However, newly founded settlements elsewhere, as in the coastal plain (which was already intensively settled), may have begun to neglect the construction of earthen ramparts and mudbrick fortification walls during this period (e.g., Michal XVII). Still others continued to feature supplemental earthen ramparts, though the date of their construction is uncertain (e.g., Farʻah South). The only example of freestanding earthen ramparts attributed to this period is found at Sahab.

---

[37]If it could be conclusively demonstrated that no overlap existed between the Rectilinear and Highland phases of fortification construction, then it could be concluded that the Rectilinear phase constituted a quite short-lived phenomenon between 1700 and 1650 B.C. However, the geographic separation between exemplars of these phases prevents greater certainty regarding this question.

During the MB IIB–C larger sites in the hills of the southern Levant were fortified with *supplemental ramparts* in a manner comparable to the fortification of tells that occurred during the previous phase. Among sites that adopted this technique were Dothan(?), Shechem XIX, Far'ah North (Per. VC), Shiloh VII, Beth-El, Hebron, and Beit Mirsim E–D. A similar process occurred in the highlands of Jordan (e.g., 'Umayri [Ph. 12]). In general, earthen rampart construction was refined during this phase to incorporate techniques such as the use of interlaced layers of varied fill types in a systematic fashion (e.g., Gezer XVIII). However, highland sites of this period reveal the first evidence for selective employment of supplemental ramparts, which were used only where the slopes were determined to be insufficient to impede an approach to the wall (e.g., Shiloh). Solid fortification walls were also used along with supplemental ramparts at settlements such as Ta'anach just outside the highlands.

While the intensification of highland settlement during the MB IIB–C resulted in a proliferation of a number of sites fortified by supplemental earthen ramparts, smaller settlements in the highlands and foothills were often only fortified with *solid stone fortification walls* (e.g., Beth-Shemesh[?], Beth-Zur, and Hebron). Other mostly unexcavated examples include Jenin, Ibleam, en-Najjar, Kebara, Kheibar, Qarqaf, Qumei, and el-'Urma. In Transjordan a similar settlement process led to the construction of a few sites with comparable fortifications (e.g., Nimrin and Pella). It was also occasionally necessary in the highlands to construct massive core walls to support the earthen load deposited behind them when supplemental earthen ramparts were constructed (e.g., Shiloh).

One hallmark of fortifications during the MB IIC, was the construction of massive *revetment walls* built of cyclopean masonry. While examples of other revetment walls of cyclopean masonry at Hazor and Ebla, presumably of MB IIB date, were built before those of Shechem XIX, Jericho (Per. IVc), and Byblos (Level VIII), revetment walls were not a regular part of fortification construction during the previous phase. Nevertheless, their widespread use during this phase suggests that there was a recognizable need for this feature during the MB IIC. In many instances the construction of revetment walls may also have been intended to compensate for the absence of a fosse or solid mudbrick fortification wall, especially in the highlands.

During the MB IIC sites that employed massive core and revetment walls in their fortifications appear to have avoided the construction of thick mudbrick walls, which were now replaced by the less effective *casemate-like walls*. At Shechem, for example, this appears to have occurred even though considerable effort was invested in a massive revetment wall, which was built around at least the northern half of the site. Such walls are attested at MB IIC settlements like Far'ah South, Shechem, Shiloh, Beth-Shemesh(?), and Beth-Zur(?), and they appear to have been built in the east as, for example, at Abu Kharaz during the LB I. The population growth experienced by these settlements during the late Middle Bronze Age no doubt necessitated such multipurpose fortification systems, which could provide a modest defense along with the ramparts and the precipitous slopes afforded by their topography. However, this wall type also facilitated the use of space within the area bounded by the defenses.

While on the one hand strategies such those described above were in widespread use during the MB IIC, other strategies were also being pursued. In regions more intensively settled, and thus already well defended by a network of fortified settlements, fortification construction may have been altogether abandoned. A shift toward unfortified settlements, which is considered a hallmark of the southern Levant during the Late Bronze Age (see discussion of "Post-Rampart" Phase below), may therefore have already begun at the end of the MB IIC. Ze'ev Herzog, for instance, argues that the socioeconomic conditions prevalent during the Hyksos period, namely a lively international trade, led to the omission of walls at sites like Tel Michal, Tel Mor, and Ashdod (1989:31–37). The fortifications of Ashdod did, however, include a low earthen rampart, a thin fortification wall, and a simple four-pier gate.

Although Michal XVII was identified as featuring earthen ramparts during the MB IIC and it has been claimed that rampart construction continued into the early Late Bronze Age at newly founded sites (see remarks concerning possible ramparts at Kinneret and Khirbet Rabud by Gonen 1992:218), the evidence from small settlements such as Michal actually suggests that wall construction atop ramparts had essentially ceased during the MB IIC. Michal, like many settlements in the highlands during the same period, appears to have relied upon its steep slopes and a degree of security that was provided by the rear walls of buildings lining the perimeter of the settlement. The proposed shift in defensive strategies attested at some MB IIC settlements may account, therefore, for the cessation of construction of earthen ramparts at Late Bronze Age settlements across most of the southern Levant. However, it is also possible that gradual population growth from the MB IIB–C through the end of the Late Bronze Age placed a premium on space within settlements, which had been limited primarily by the perimeters created by their fortification walls (e.g., Megiddo). The gradual need for more space within these settlements, particularly at smaller sites—a fact also supported by the proliferation of buildings incorporated into defensive perimeters during the late MB II (as discussed above)—and the inability to maintain these fortifications would have led to building on top of and outside of the earlier fortification walls. This may also account for the expansion of the area of some sites by constructing earthen platforms and embankments, which should not to be mistaken for earthen ramparts (e.g., Michal), at the end of the Middle Bronze Age and during the LB I.

6. ABANDONMENT (LB I–II)

Since Rivka Gonen (1984:69) asserts, and most scholars have come to accept, that almost all Late Bronze Age settlements lacked fortification walls, a few observations are in order regarding the nature of Late Bronze Age defenses. Although evidence from Megiddo VIII and later is cited to support this contention (Gonen 1987), the evidence is far more ambiguous than suggested. In the first place, it is generally agreed that a number of settlements retained their Middle Bronze Age fortifications (e.g., Hazor), including walls, ramparts, and gates, during the LB I until they were destroyed, probably as a result of Egyptian New Kingdom campaigns (see p. 101). Gerisa, for example, featured LB I occupation built against the wall of the MB IIC. Part of the continued use of the earlier fortifications also involved the maintenance of *six-pier gates*. Evidence of this practice is found at Far'ah South, Megiddo IX, and Shechem XIV. In other cases new gates may have been built directly over previous gates, such as at Hazor, and integrated within the rebuilt fortifications.

In Jordan a number of settlements with fortifications are attested during the Late Bronze Age, including Tells Abu Kharaz, Fukhar, Irbid, Safut, Sahab, and Sa'idiyya (see Strange 2001:310f.). It is true, however, that their fortifications do not appear to constitute a unified fortification strategy since some fortifications continued the use of Middle Bronze Age walls during the LB I; the fortifications of other settlements were seemingly constructed de novo during the Late Bronze Age, despite having been settled during the Middle Bronze Age. Nevertheless, the evidence is unequivocal that many of the largest Late Bronze Age settlements in Jordan were fortified.

Second, the practice of building earthen platforms and terraces, which began during the MB IIC, appears to have continued into the Late Bronze Age though at few sites (e.g., Kinneret? and Khirbet Rabud). Within such an environment, monumental gates, particularly of the six-pier type, remained in use for some time (e.g., Megiddo VIII), though they were abutted by structures other than a freestanding town wall such as, for example, a casemate wall or the rear walls of buildings—a phenomenon that began during the MB IIC.[38]

In addition to the advent of the incorporation of the rear walls of Late Bronze Age houses to create a defensive perimeter, evidence is available to suggest the later reuse of the foundations of Middle Bronze Age walls during the Late Bronze Age. Among the many sites at which the stone foundations have been observed or suggested to have been reused during the Late Bronze Age (and sometimes later) are Beth-El, Beth-Shemesh (tower), Jerusalem, Shechem, Hebron, and Jericho (Yadin 1982). It is noteworthy that most of these sites are located in the highlands, which was settled and fortified later, and therefore the continued use of the late Middle Bronze Age fortifications at these sites would not be surprising. Furthermore, these sites are not among those named in New Kingdom campaign lists.

Despite the evidence for the retention of fortifications at the start of the Late Bronze Age, it is nevertheless possible that following the destruction of many sites during the LB IA, Egyptian domination led to the widespread abandonment of fortification construction. As Albert Glock suggested, "it was the prerogative of the suzerain alone to fortify cities," and so "without royal permission . . . erecting fortifications, was an act of rebellion (1968:n. 31, pp. 166f.). The same probably applied to Egyptian domination of Levantine states. Nevertheless, since no evidence for such a policy exists it is likely that socioeconomic conditions contributed to the abandonment of earlier strategies. This does not mean that the economic means were insufficient to permit their fortification, simply that circumstances did not warrant or may have actually prevented the mustering of resources necessary for their construction.

Changes in the northern Levant appear to have been more subtle during the first part of the Late Bronze Age, which may have been due to limited imperial presence by Egypt or Hatti. A number of settlements provide evidence of the continuation of earlier fortification strategies, characterized by the maintenance of rampart fortification systems.

---

[38] If Akko featured an LB II "sand" rampart, then it is a unique example of the continuation of this strategy during Egyptian imperial control.

Earthen ramparts of this period are in evidence at Alalaḫ IV, Ugarit, Ras Ibn Hani, Byblos (Level IX), Biruta, Qatna(?), and Kar-Kemiš in the Levant, as well as Ekalte (Munbaqa) and Emar in northwestern Mesopotamia. In the northern Levant Late Bronze Age fortifications are attested at Kumidi, Hadidi, and Qitar.

Middle Bronze Age fortifications in the northern Levant appear to have impacted the tradition of Late Bronze Age fortification construction in Anatolia as well. This process probably began during the Middle Bronze Age as evident at Kaneš. However, by the Late Bronze Age the most convincing evidence for this influence is seen in the fortifications of the Hittite capital, Hattuša (Boğazköy). There the heavy borrowing of Levantine fortification techniques is indisputable. These included the use of massive mudbrick walls with towers, a stone glacis, monumental four-pier gates, and a corbel-vaulted postern gate. A similar adoption of Levantine Middle Bronze Age fortifications can also be seen at nearby Alaça Höyük and Kuşakli/Sarissa (Müller-Karpe 2003).

Table 12. Major developments in defensive architecture from ca. 3000 to 1200 B.C.

Phase	Developments
KRANZHÜGELN EB II–III (ca. 3000–2600 B.C.)	*Walls* serve as the primary type of fortification in Mesopotamia (e.g., Kranzhügeln in northwest); plans are *elliptical.* *Earthen ramparts* and dikes serve as defenses along the Euphrates; *moats* accompanied ramparts to divert water away from mudbrick defenses. In the southern Levant *fortification walls* are used but only occasionally with *glacis* and *fosses*. The only fortifications identified in the northern Levant in this period are at Byblos.
NORTHERN LEVANTINE EB III–EB IVA/B (ca. 2600–1900 B.C.)	*Earthen ramparts* and *fortification walls* are added to Mesopotamian settlements; they are used at northern Levantine settlements for the first time. Settlement plans are still *elliptical* and employ *freestanding ramparts.*
SOUTHERN LEVANTINE MB IIA (ca. 1900–1700 B.C.)	Throughout the Levant fortifications include *freestanding ramparts* built with *elliptical* plans, *fortification walls*, *fosses*, and *glacis*. *Supplemental ramparts* are common at sites with prior occupation. *Rectangular towers* are used with both rampart types. *Rectilinear plans* are gradually employed in the south.
RECTILINEAR MB IIB (ca. 1700–1600 B.C.)	New fortified sites are established in lowland regions. *Freestanding ramparts* are built with *rectilinear* plans and *fosses.* *Six-pier gates* emerge as the most common gate type. Where earlier freestanding ramparts exist their sides are straightened. Large *rectangular bastions* are added at some sites.
HIGHLAND MB IIB–C (ca. 1650–1530 B.C.)	*Supplemental ramparts* are common among highland settlements along with massive *revetment walls.* *Casemate walls* are introduced. Freestanding earthen ramparts are no longer constructed.
ABANDONMENT LB I–II post ca. 1530 B.C.	*Six-pier gates* continue to be used. Wall construction seems to be abandoned. Some features such as *fosses* are co-opted for building space. Features such as *ramparts* and *walls* may serve as a source of building materials during later occupation. Levantine-style fortifications continue to be built in central Anatolia.

PART THREE

Middle Bronze Age Fortifications

& Their Cultural Context

4. A History of Warfare in the Levant (ca. 2500–1460 B.C.)

> Historians refer constantly to war without really knowing or seeming to know its true nature—or natures. We are as ignorant about war as the physicist is of the true nature of matter. We talk about it because we have to: it has never ceased to trouble the lives of men.
>
> Fernand Braudel, *The Mediterranean* (1972:836)

Having reviewed in the preceding two chapters the developments in the art of warfare and defensive architecture between ca. 2500 and 1460 B.C., it is necessary to consider the political, economic, and social conditions in which these fortifications functioned. In this chapter, historical questions relating to periods of military activity, the effects of specific military campaigns, and the frequency of warfare are addressed in an attempt to understand the impetus for the construction of Middle Bronze Age fortifications. The political milieu is further explored in chapter 5, and the social and economic factors related to the construction and use of these fortifications, which are also integral to the historical setting of Middle Bronze Age fortifications, are considered in detail in the chapter 6.

Despite the overwhelming evidence for fortifications during the Middle Bronze Age, there has been considerable debate concerning the need for this particular type of fortification during this period. Against whom, for example, were such fortifications built? Was internecine feuding between rival cities the main threat or did distant powers pose a threat to even the smallest and remotest of Levantine settlements? Alternatively, was this period so politically unstable as to require towns and villages to protect themselves against archaeologically invisible threats such as armed nomads, brigands, or rebels? These are the types of questions concerning the historical setting that must be addressed in order to assess the purposes for which Middle Bronze Age fortifications were constructed.

The size of the geographic region—from the northern border of Egypt to central Anatolia and the middle Euphrates—in which this type of fortification is found over a one thousand-year period is considerable. It would be incorrect to suggest, therefore, that a single set of historical circumstances affected the entire region that would account for the evolution of the fortification types discussed in the previous chapter. Nevertheless, in this region at specific points within this span of time notable events and conditions appear to have encouraged the gradual adoption of a distinctive means of fortifying settlements (see Table 14 below). The geographical and temporal scope of this inquiry can be limited to the major phases of the evolution of Middle Bronze Age warfare and defenses that were identified in the preceding chapters. Therefore, the present chapter investigates the most likely military threats affecting northern Mesopotamia and the Levant during the second half of the third millennium and the first half of the second millennium.

It is noteworthy that military activity during this extended period can be characterized as the result of endogenous and/or exogenous threats. It is usually assumed that the source of military aggression in the late third millennium was exogenous, such as Mesopotamian expansion into the northern Levant and perhaps Egyptian campaigns like that of Weni during the Old Kingdom into the southern Levant. Military aggression in the first half of the second millennium, on the other hand, is usually characterized as primarily the result of internecine warfare. It is necessary to reevaluate whether such generalizations are supported by the data. In order to do so the state of political and military affairs during these periods must be examined through both historical and archaeological sources.

The use of archaeological data for this inquiry is born out of the paucity of historical sources for the Levant that date to the end of the third millennium and the Middle Bronze Age—a condition that is in stark contrast to that prevalent in Mesopotamia. While explanations can be offered for the lack of texts from the Levant for these periods, they are superfluous since, as discoveries at Ebla have demonstrated, our understanding of the region's history can change dramatically, almost overnight. Even so, texts alone cannot answer our historical questions. While texts often help identify the agents of events and their dates, datable archaeological evidence is required to verify the claims made by the authors of these texts, *a fortiori* for claims regarding military victories. In this regard evidence for destruction levels offers a primary means of identifying sites that may have been involved in conflicts throughout this period. Such evidence also makes it possible to assess the degree to which these conflicts affected settlements that are not mentioned in the texts.

For this reason, I will seek in this chapter to bring together all of the available archaeological data for the destruction of sites in the Levant from ca. 2400 to 1460 B.C. Together the historical and archaeological evidence reveal that references to military victories in textual sources were not the product of the imaginations of megalomaniacs; rather, they reflect an important aspect of the milieu of the day, namely the continual military threat posed by large, if distant, political powers. While the lack of Early and Middle Bronze Age sources for the southern Levant has frequently discouraged attempts at historical synthesis of these periods, this synthesis relies on overviews provided by Horst Klengel for the history of and historical sources for the northern Levant (1965–1970; 1992). It also makes use of other sources that to some degree enable a reconstruction of major historical events during this period, if they often lack dynastic intrigue (Astour 1992; 2002; Yuhong 1994; Archi 1998; Oates and Oates 2001).

A. Early Bronze Age Warfare (ca. 2500–1900 B.C.)

The contribution of warfare to the development of centralized power (i.e., the state) during the third millennium has been recognized as a major factor in state formation in this region (Akkermans and Schwartz 2003:275ff.). However, during the third millennium historical developments related to military activity appear to have progressed along separate lines in the northern and southern Levant. While historical references to military activity are essentially nonexistent for the first half of the third millennium, forcing us to rely heavily on archaeological data to reconstruct the evidence for warfare during this period, evidence from Ebla suggests its dominant status in the northern Levant before ca. 2300 B.C. The "Standard of Ebla," for example, portrays bound prisoners and the heads of decapitated enemies held by Eblaite soldiers (ibid., fig. 8.6).

By contrast the major historical and political events during the second half of the third millennium in the northern Levant are usually related to either exogenous military pressure by Mesopotamian kings conducting long-distance campaigns into the northern Levant or endogenous pressure from Ebla, the dominant polity in the region. Unfortunately, the evidence for the political and military history of the southern Levant during this period remains meager. Despite archaeological evidence for the destruction of some settlements in the southern Levant at this time and accounts of military campaigns led by Weni during the first part of the Sixth Dynasty (ca. 2362–2296 B.C.), evidence for Egyptian involvement in the southern Levant is scant. For this reason it is usually assumed that most destructions in this region were the result of internecine feuding between rival settlements or possibly even the result of natural catastrophes such as earthquakes.

1. The Northern Levant

a. Early Mesopotamian Campaigns (pre-2250 B.C.)

The military history of the northern Levant begins with a single reference to a military campaign by the king of Uruk, Lugalzagesi, just after the mid-third millennium (ca. 2350 B.C.). The event in question, which was recorded in Early Dynastic royal inscriptions, is characterized as the subduing of all lands between the "Lower" and "Upper" seas (the Persian Gulf and the Mediterranean Sea respectively) and the placement of "their routes in good order" (*SARI* Um 7.1, p. 94). However, nothing beyond this short statement can be added regarding the frequency or nature of this or any subsequent action in the northern Levant by Mesopotamian kings during the remainder of the Early Dynastic period (EB II–III).

The epigraphic evidence from Mesopotamia in the mid-third millennium provides a picture of the ever-increasing military reach of Mesopotamian polities. Eanatum of Lagaš (ca. 2450 B.C.), for example, had already claimed to have defeated the king of Mari and to have undertaken campaigns well into Elam (*SARI* La 3.5, p. 42). While references to the northern Levant in Mesopotamian texts are also found during the Akkadian period, it is likely that during this same period the kingdom of Ebla filled any power vacuum that might have existed and thus was a likely party to military conflicts that occurred.

b. Ebla and Mari at War (ca. 2250–2200 B.C.)

The history of Ebla's military and political interaction with its vassals as well as its enemies has been summarized by a number of authors (Pettinato 1981; 1991; Klengel 1992; Astour 1992; 2002). However, for this period the Ebla archives provide only a glimpse into the political landscape and military activity in the Levant since, as Klengel notes, "there is little evidence for the political relations of Ebla with other Syrian or Mesopotamian centres" (1992:28) and "as southern Syria is concerned, there are no undisputed . . . identifications of places" (ibid., 30). Still, one Ebla text (TM.75.G.2367), which has been variously described as a "military bulletin" (Pettinato 1991:237–40) and an "introductory letter" (Edzard

1981), provides a record of what appears to have been a series of military actions throughout the northern Levant and northwestern Mesopotamia:[1]

> Thus says Enna-Dagan, ruler of Mari, to the ruler of Ebla: The cities Aburu and Ilgi, the lands of Belan, Anbu the ruler of Mari defeated; tells and ruins in the mountains of Lebanon he left. The cities Tibalat and Ilwi, Sa'umu the ruler of Mari defeated; in the mountain terrain of Anga'i-x he left tells and ruins. The lands of Ra'ak and Irum, Ash'aldu and Badul, Sa'umu the ruler of Mari defeated, in the border of x-an, in Nahal, he left tells and ruins. And the cities Emar and Lalanium and the canebrake (GA.NU.UM), (that) of Ebla, Ištup-Išar, the king of Mari, defeated; in Emar and in Lalanium he left tells and ruins. And the city Gallabi [and x] the liberated canebrake (GA.NU.UM), Iblul-il, the ruler of Mari and Apishal(?), defeated; in Zahiran he left also seven tells and ruins. Iblul-il, the ruler of Mari and Shada and Addali and Arisum, the lands of Burman which (include) the city Sugurum, Iblul-il defeated and left (in) tells and ruins. Also the cities Sharan and Dammium, Iblul-il the king of Mari, defeated; he left two tells and ruins. Against the city Nerad and against the fortress (É.NA$_4$) of Hazuwan Iblul-il, the king of Mari, went forth. And he received the tribute of Ebla in its midst, (in) Nema. Also Emar he defeated(?), he left tells and ruins, did Iblul-il, the king of Mari. And Nahal and Nubat and Shada, the lands of Gasur(?), he defeated in Kanane; he also left seven tells and ruins. Iblul-il, the king of Mari, both Barama—for the second (time)—and Aburu and Tibalat, the lands of Belan, defeated; Enna-Dagan, the ruler of Mari, left tells and ruins. The two nations(?) in oil, the(ir) lands I bound . . . Iblul-il, the king of Mari, . . . (*COS* 3.90).

In this text Enna-Dagan, who may have been the new king at Mari or an Eblaite military commander (see translation by Pettinato 1991:237ff.), recounts to the king of Ebla Mari's history of military victories over rulers from "the mountains of Lebanon" to sites that are probably located in northern Mesopotamia. Although Pettinato suggests that the text refers to a single campaign led by Enna-Dagan (1980), Edzard and others are probably correct to suggest that the letter constitutes a review of military action taken by Mari against its neighbors in the years prior to Eblaite intervention, written in a style comparable to examples from the Neo-Assyrian period (Edzard 1981; Kienast 1980). Unfortunately, Ebla and Emar are the only places mentioned in the text whose locations are certain.[2] Furthermore, it is impossible to know exactly when these towns were defeated since the reigns of many of the kings of Mari identified in this text are otherwise unattested at Mari.

This text serves to illustrate that polities in Mesopotamia posed a threat to the northern Levant in the mid-third millennium. Mari, for example, is approximately 460 km by the overland route via Emar and Ḫalab (Aleppo) from Ebla. If Edzard's interpretation is favored, then the text attests to intervention by Mari in the affairs of the northern Levant during this period and suggests the context for Eblaite military efforts against Mari. It is perhaps in this manner that we can understand why an Akkadian campaign was directed against Ebla within a relatively short time after Ebla's intervention in northern Mesopotamia.

The Enna-Dagan text also demonstrates that Ebla, and possibly yet undiscovered political centers like it in the southern Levant, wielded sufficient military power to enable it to control not only a large area within the Levant but also to defeat powerful polities in Mesopotamia. There are also implications for the southern Levant, if, in fact, Ebla exerted military and political might in the south to the degree that is suggested for regions to the north and east. Despite the threat that Ebla may have posed to settlements throughout the Levant and beyond, much of the northern Levant was still exposed to military incursions by Mesopotamian kings, as was so clearly demonstrated by the destruction of Ebla by Naram-Sin (ca. 2150 B.C.).

Evidence from the upper Ḫabur kingdom of Nagar also provides a glimpse of political and military developments during the mid- to late third millennium B.C. Although neither historical texts nor evidence of the town walls of Nagar in Phase L (ED IIIB), presently identified with Tell Brak, have come to light, the construction of town walls at Šubat-Enlil and Urkeš (see pp. 175f. and 180ff. respectively), as the excavators of Brak note, may have coincided with "the expanding power of the already urban Nagar and/or a threat from Hurrians to the north" (Oates and Oates 2001:380).[3] The fiery destruction of Phase H revealed at Nagar (Tell Brak) in Areas CH (Level 6), ER (Level 5), ST, and DH can be identified either as the result of the arrival of Akkadian power or local conflicts (Oates et al. 2001:382, 91).

---

[1]For bibliographic references concerning this text and its interpretation, see *COS* 3, p. 236.

[2]Michael Astour attempted to locate many of these towns (see map in 1992:27).

[3]There is no historical parallel from this region that supports the suggestion that T. Brak lacked a fortification wall during the mid-third millennium because it was powerful enough that it "needed no such protection" as the excavators have suggested (Oates and Oates 2001:380). It is more likely that the remains of the fortifications have not yet been encountered since as the excavators have noted there have been "only very restricted exposures of this date" (ibid.).

c. The Akkadian Empire (ca. 2237–2060 B.C.)

The inscriptions of Akkadian kings, who recorded their campaigns against towns and regions in the northern Levant, provide the next source for military activity during the third millennium (Table 13). These texts relate Sargon's campaign against Mari, Iarmuti, and Ebla (*RIME* 2.1.1.11–12), as well as Naram-Sin's campaign against Armānum and Ebla (*RIME* 2.1.4.26). It is important to remember that by claiming to have defeated these towns Sargon, who assumed the title of "King of the Land" (*RIME* 2.1.1.1) after he defeated Lugalzagesi, further legitimated his rule by leading an expedition as far as the "Upper Sea."

Table 13. Akkadian kings and the Levantine settlements they claim to have pacified.

King	*RIME*	Settlements
Sargon (2237–2182 B.C.)	2.1.1.11–12	Mari Iarmuti Ebla
Naram-Sin (2157–2121 B.C.)	2.1.4.26	Armānum Ebla Ulišum Amanus

Although the details of Sargon's western campaign, in which he claimed to have conquered Mari, Iarmuti, and Ebla, as well as thirty-four other towns, are not provided elsewhere, the range and effectiveness of such campaigns are certainly plausible given the scope of earlier military activity. In fact, they can be understood to be the projection of Akkadian military might as far as possible to the west (into the northern Levant) as they were also to the east (into Elam). While this assertion does not remove the possibility that Sargon's claim could have been fabricated, it is substantiated by the archaeological record, which consists of destruction levels that are contemporary with the events described. Furthermore, Sargon's admission of a need to put down "rebel cities" in the region (*RIME* 2.1.1.12, lns. 1–10),[4] his display of obeisance to the "god Dagan in Tuttul" (ibid., ln. 6–12), and his engagement in thirty-four specific battles (*RIME* 2.1.1.11, ln. 1ff.)—each of which he claims to have won—are all events that were not previously mentioned by Lugalzagesi. These details emphasize the uniqueness of Sargon's claim to have destroyed these towns, some of which, like Ebla, were not previously mentioned.

The territory that had originally been added to the Akkadian empire during the reign of Sargon was further widened during the reign of Naram-Sin to include new towns such as Armānum and Ulišum, the latter of which was inhabited by "people whom the god Dagan had given to him for the first time" (*RIME* 2.1.4.26, lns. ii 8–19). The text was clearly intended to convey that new territories had been added by Naram-Sin, while also quelling revolts within the empire in towns such as Ebla. Although the exact location of the capital of Akkad has not been established, the Akkadian accounts of conquests in the northern Levant suggest that the range of military action overland against centers in the Levant had in this period nearly doubled with campaigns being made against towns over 750 km away.

Accompanying Naram-Sin's account of this campaign is a lengthy description of the fortifications of Armānum.

> From the fortification wall [BÀD *danīm*] to the great wall [BÀD.GAL]: 130 cubits [≈ 65 m] is the height of the hill (and) 44 cubits [≈ 22 m] is the height of the wall [BÀD]. From the quay wall [BÀD *kārim*] to the fortification wall [BÀD *danīm*]: 180 cubits [≈ 90 m] is the height of the hill (and) 30 cubits [≈ 15 m] is the height of the wall. Total: 404 cubits [≈ 202 m] in height, from ground (level) to the top of the wall. He undermined the city Armānum . . .
> From the river to the quay wall [BÀD *kārim*]; 196 cubits [≈ 98 m] is the height of the hill (and) 20 cubits [≈ 10 m] is the height of the wall. From the quay wall [BÀD *kārim*] to the fortification wall [BÀD *danīm*]: 156 cubits [≈ 78 m] is the height of the hill (and) 30 cubits [≈ 15 m] is the height of the wall (*RIME* 2.1.4.26, lns. iv 20–vi 17)

Although it is impossible to know why the fortifications of Armānum warranted such a detailed discussion, it is likely that the Mesopotamian king's campaign into the Levant had brought him into contact with a completely foreign means of fortifying a settlement. The reconstruction of this inscription offered by Fritz Kraus (1948:fig. 7) and improved upon by Benjamin Foster (1982:35) are both probably very close to what the descriptions were intended to convey. Unfortunately, no consensus has emerged regarding the site's identification with Armānum, although Armi of the Ebla texts, which has been identified with Ḫalab (Aleppo) of the second millennium, has been suggested. However, Wilfred Lambert suggests that Aleppo should be identified with Ḫalam in the Ebla texts (1990), which would leave Armānum unidentified.

[4] The reference to "rebel cities" in this text and Naram-Sin's description of a people given to him for the first time (*RIME* 2.1.4.26) lead one to contemplate whether much of the Akkadian empire in the west was actually built on the military exploits of Lugalzagesi (if not also his predecessors), since no former expeditions by Sargon are mentioned and yet cities are described as rebelling.

That the Amanus region remained open to the Akkadians at least for some time following Naram-Sin's campaign(s) is evident from Šar-kali-šarri's account of his uneventful expedition to obtain cedars from the Amanus Mountains for the construction of a temple for Aštar (*RIME* 2.1.5.5, esp. lns. 53–57). Nevertheless, Akkadian rule over this region was certainly not without challenge, particularly by Amorite groups in the surrounding highlands who were defeated in a number of encounters, according to at least one of the year-names of Šar-kali-šarri (see Frayne 1993:183).

d. The Ur III Dynasty (2018–1911 B.C.)

Although the Akkadian empire's control over Subartu (and possibly the northern Levant) was brought to an end by the Gudeans, the Ur III empire (ca. 2018–1911 B.C.) almost immediately assumed Akkad's role in this region. This may account for the continuation of unrest between the Amorites in this region and the inhabitants of southern Mesopotamia. During the thirty-seventh year of his reign Šulgi conducted raids against the Amorites (Frayne 1997:107). In the same year construction was begun on a wall known as *The Wall of the Land* (also identified as *The Wall Facing the Highland*).[5] As Douglas Frayne notes, that the purpose of the wall was to keep out the Tidnum (i.e., Amorites) is evident from several references that confirm that "troops working on the wall had to contend with attacks by Tidnumite nomads" (Frayne 1997:106). The Mesopotamian military strategy of constructing a defensive line was similar to Hadrian's Wall in Britain.

The strategy continued in force during the reign of Šu-Sin with the construction of another wall of this type twenty-five years later. This wall, called *Muriq-Tidnim* (lit., "It keeps the Tidnum [Amorites] at a distance"),[6] was begun and apparently completed in the fourth year of the reign of Šu-Sin (ca. 1941 B.C.; ibid., 290ff.). Its construction may have been a passive strategy to mitigate the effects of Amorite raids. Nevertheless, Šu-Sin appears to have conducted campaigns into Subartu against Mari, Urkeš, and Tuttul, and in the northern Levant against Ebla, Mukiš, and Lebanon. These place names are attested on a tablet identified as a campaign record most likely attributed to the reign of Šu-Sin (*RIME* 3/2.1.4.2). One text from the reign of Šu-Sin provides an excellent example of the effects of such campaigns on a region.

> The men who had evaded battle and who, like birds, saved their lives (by fleeing) to their cities, did not escape his hand. Against [their] cities he himself screeched like the Anzu bird. He turned their established cities and villages into (ruin) heaps. He destroyed their walls. He blinded the men of those cities, whom he had overtaken, and established them as domestic (servants) in the orchards of the god Enlil, the goddess Ninlil and of the great gods. And the [wom]en of [those cities], whom he had overtaken, he offered as a present to the weaving mills of the god Enlil, the goddess Ninlil and of the great gods (*RIME* 3/2.1.4.3).

While the kings of the Ur III Dynasty had meted out more than their share of destruction upon settlements throughout the Near East, the fate of their glorious capital is very well known. *The Lamentation over the Destruction of Ur* records the ravaging of the town of Ur by its enemies at the start of the second millennium (ca. 1911 B.C.). This literary piece also serves as a contemporary witness to the nature of siege warfare in this period (see p. 25).

e. Archaeological Evidence for Warfare

Aside from the textual sources that discuss military conflicts in this period, there is evidence for the destruction of a number of settlements during the second half of the third millennium that also sheds light on the nature and frequency of warfare in this period. The destructions cluster roughly around two dates, ca. 2200–2150 B.C. and ca. 1950–1900 B.C., and may be the work of the Akkadian and Ur III empires respectively.[7] In addition to the destruction of Nagar (Tell Brak) cited above, destructions are attested at Ebla, Tuqan, Byblos, Ugarit, Hama, Hammam et-Turkman, Tell es-Sweyhat, Tell Hadidi, Selenkahiye, and Bderi.

---

[5]Richard Barnett (1963:20ff.) attempted to identify this wall; his suggestions were subsequently refuted by Julian Reade (1964). For textual references regarding the relative length of these walls, see Barnett (1963:21) and Wilcke (1970).

Although *The Wall of Ur* built by Ur-Nammu is sometimes considered the first of such walls to be built during the Ur III period, the evidence, as Douglas Frayne notes, suggests that the wall belonged to the sacred temenos and not to the city itself (see *RIME* 3/2.1.1.4 and commentary).

[6]See also reference to this wall in *RIME* 3/2.1.4.17.

[7]It is difficult to rely upon the exact dates suggested by excavators for the destruction of sites in order to correlate destructions with the reigns of Mesopotamian kings, since in many cases no specific king was associated with the approximate date offered. Nevertheless, when a historical figure is associated with a destruction, the dates provided here have been adjusted to the very low chronology used in this work. Therefore, since most scholars have relied upon the middle chronology, many of the dates cited in this section could be lowered by at least fifty years. In any event the relative chronology of these events is most significant.

Evidence from Ebla, which is probably best known, attests to its destruction not only ca. 2200 B.C. (Mardikh IIB1), perhaps by Naram-Sin, but also at the end of the EB IV (Mardikh IIB2), ca. 1940 B.C. (Matthiae 1980b:111), perhaps by Šu-Sin (*RIME* 3/2.1.4.2). Although various arguments have been made to suggest that the Akkadians were not responsible for the first destruction of Ebla (see summary in Astour 2002:68–76), the consensus is that the destruction must be attributed either to Sargon (Schwartz and Weiss 1992:238ff.; Akkermans and Schwartz 2003:243f.) or to Naram-Sin (Matthiae 1980b:177). The evidence consists of the destruction of Palace G located on the acropolis. Unfortunately, no carbon-14 dates have been provided and the debate concerning the date of its destruction remains mired in relative ceramic chronologies. Nevertheless, it was probably contemporary with the destruction of Mari's Ville II (Margueron 2004:308–12).

Byblos also experienced two destructions, which were nearly contemporaneous with those at Ebla in the second half of the third millennium B.C. Despite the general difficulties inherent in relying upon archaeological data from Byblos, Maurice Dunand asserted that the first destruction of Byblos could be dated ca. 2200 B.C., attributing this destruction to the arrival of the Amorites.[8] The first phase of the northeast gate was destroyed at that time. However, Muntaha Saghieh's analysis suggests that the second phase of this gate was also destroyed by fire around the end of the third millennium [MC] (1983:131f.), making its second destruction contemporary with that of Mardikh IIB2, ca. 1900 B.C. [VLC].

Hammam et-Turkman, Ugarit, and Hama were also destroyed ca. 2200 B.C. Excavations at Hammam et-Turkman (ancient Zalpah) have revealed two extensive destruction layers dated to the last quarter of the third millennium, ca. 2200 and 2000 B.C. (van Loon 1988:699). These are roughly contemporaneous with those attested at Nagar (Tell Brak), of Phase L, which was discussed above, and of Phase N (Oates and Oates 2001:394). The second destruction of Zalpah actually resulted in an occupation gap at the site during the early second millennium (van Loon 1988:699). Although it is usually suggested that Ugarit experienced a period of abandonment between ca. 2150 and 1950 B.C. (Yon 1997b:258), Claude Schaeffer actually suggested that a conflagration took place at the site (Ancien 2) between about 2350 and 2250 B.C. (1948:35f.). Still, he never intimated who or what may have been responsible for this destruction. Hama J5 also appears to have been destroyed around 2250 B.C. according to Ejnar Fugmann (1958:271f.).

Other destructions at the end of the third millennium (ca. 1950–1900 B.C.) in northwestern Mesopotamia can be found at Sweyhat, Hadidi, and Selenkahiye. Tell es-Sweyhat was supposedly destroyed by fire during the Ur III period (Holland 1977:58). Tell Hadidi is said to have been destroyed ca. 1950 B.C. (Dornemann 1978:24; 1985:50). Selenkahiye was destroyed in a conflagration following Period IV at the end of the EB IVB (ca. 1950 B.C.), which has been confirmed by radiocarbon dates (van Loon 2001:603ff.). Finally, like Ebla, Tuqan also suffered destruction by fire at the end of the EB IVB, ca. 1940 B.C. (Matthiae 1979:8).

In reviewing the evidence for the destruction of sites in northern Mesopotamia and the northern Levant at the end of the third millennium, it is not difficult to understand how Kathleen Kenyon was led to believe that perceptible changes evident in the archaeological record of the southern Levant in this period were in some way related to the volatile events attested in the northern Levant (see Kenyon 1966). Nevertheless, a direct link between these two regions has never been clearly established. The "Amorite Hypothesis" advanced by Kenyon, which posited a violent migration of Amorites out of the northern Levant into the south that would have resulted in the destruction of settlements in the southern Levant at the start of both the EB IV and MB IIA, remains, therefore, only a hypothesis. That subsequent research has not, however, eradicated the notion of an Amorite cultural diffusion from the north into the southern Levant remains the clearest testimony of the continued relevance of the hypothesis to the study of the Middle Bronze Age in the Levant (e.g., Lönnquist 2000). While scholarly perspectives regarding the nature of the movement of groups from northern Mesopotamia into the Levant has certainly changed from one of swift conquering hordes to that of a gradual diffusion of cultural traits by means of various mechanisms, there is consensus among many scholars that, unlike other proposed migrations of ethnic groups in the ancient Near East, some movement or diffusion related to the Amorites cannot be entirely dismissed (Mendenhall 1992a; 1992b; Lönnquist 2000).

2. The Southern Levant

To better understand the contrast between the northern and southern Levant during the EB IV (ca. 2300–1900 B.C.) it is necessary to briefly review the archaeological and historical evidence for conflict in-

[8] Muntaha Saghieh suggests 2150 B.C. (1983:131).

volving southern Levantine settlements up to ca. 2300 B.C.[9] The evidence consists on the one hand of the destruction of sites in the southern Levant and on the other hand of several Egyptian campaigns led by Weni during the Sixth Dynasty. Although Egyptian expeditions to the Sinai were probably commonplace during this period, as they had been during earlier dynasties, it is impossible to infer if or how such economic expeditions influenced the defensive strategies of the inhabitants of the southern Levant.

a. *Early Bronze II–III*

It has been suggested that the onset of the EB III (ca. 2700 B.C.) brought about a new phase in relations between Egypt and the southern Levant that resulted in a bypassing of the southern coast in favor of maritime trade with coastal Lebanon, specifically Byblos (Ben-Tor 1992b:118–20; Stager 1992:41). Despite the apparent disinterest of Egypt in this region at this time, a few sites still flourished as large fortified towns in the southern Levant, occupied from the EB II through the end of the EB III. Destruction levels at some of these sites attest a degree of, perhaps intermittent, internal unrest (A. Mazar 1990:141). The evidence does not present, however, a set of contemporaneous and related events like those in the northern Levant and Mesopotamia. These destructions occur throughout the entire EB II–III period (more than seven hundred years), and the sites were often resettled following their destruction. EB II–III sites destroyed include Tell el-Far'ah North (late EB II, de Miroschedji 1993:436), Khirbet el-Makhruq (end of EB II and EB III, Z. Yeivin 1993:929f.), Tell es-Saidiyeh (late EB II, Tubb 1993:1300), Ai (end of EB III, Callaway 1993:39), and Jericho (end of the EB III, Kenyon 1993:678f.). There is no evidence to suggest that a systematic approach to fortifications in the southern Levant existed during this period, as noted in the previous chapter, which therefore makes it difficult to suggest how these fortification strategies related to potential threats.

In addition to the evidence for the destruction of settlements in the EB II–III, there is the historical evidence of military expeditions led by the Egyptian commander Weni in the Sixth Dynasty during the reigns of Teti I, Pepi I, and Merenre (ca. 2362–2296 B.C.). In his tomb chapel autobiography Weni described his role in several campaigns against the "Sand-dwellers," a general term for the inhabitants of the southern Levant.

> His majesty sent me to lead his army five times, to attack the land of the Sand-dwellers as often as they rebelled, with these troops. I acted so that his majesty praised me [for it beyond anything].
> Told that there were marauders among these foreigners at the nose of the Gazelle's-head, I crossed in ships with these troops. I made a landing in the back of the height of the mountain range, to the north of the land of the Sand-dwellers, while half of this army was on the road. I came and caught them all and slew every marauder among them (*AEL* 1, p. 20).

The location of this final campaign, possibly his sixth (mention of "five times," plus one), is usually identified as north of the Carmel Range in the vicinity of modern Haifa. Unfortunately, beyond this description there is little that can be said and we remain uncertain about the frequency of such Egyptian campaigns.

b. *Early Bronze IV*

The following period, EB IV (ca. 2300–1900 B.C.), is traditionally characterized as a de-urbanized period in the southern Levant, when many EB II–III settlements were abandoned and others shrank considerably in size. The possible reasons for this shift in settlement are numerous, but one of the differences routinely noted in this period is the near complete absence of fortification systems around settlements, whether they were newly established or continued since the EB III. The main exceptions to this are found predominantly in Transjordan (e.g., Palumbo 1990:58–60, p. 132, and fig. 25) and the best example of such a site is Khirbet Iskander (see p. 272). Therefore, a substantial gap appears to have existed in the construction of fortification systems in the southern Levant between the urban phase of the Early Bronze Age and the return of urbanism during the Middle Bronze Age. Nevertheless, this interruption did not occur in the northern Levant. Whatever the reasons for the nature of EB IV settlement in the south, the limited data mean that we are unable to determine the nature of political relations between settlements in this and neighboring regions in this period.

Our only data concerning the relations of the inhabitants of the southern Levant with their neighbors come from Egyptian sources, consisting primarily of several literary sources and tomb reliefs (Redford 1992:66ff.). These illustrate Asiatics in Egyptian battle scenes during the Eleventh Dynasty (ca. 2116–1973 B.C.), which corresponds to a large part of the

[9]Various dates have been proposed for the start of the EB IV: 2350 B.C. (Dever 1995:282), 2300/2240 B.C. (A. Mazar 1990:151ff.), 2300 B.C. (Gophna 1992b:127), and 2250 B.C. (Stager 1992:41). The difficulty, of course, is partly related to the nature of sites with limited periods of settlement, and the possibility that changes in the material culture upon which our periodizations are based were gradual. In the present work the median date of 2300 B.C. is adopted.

EB IV (see discussion of these scenes by A. Schulman 1964). Donald Redford has observed that from the reign of Mentuhotep II (ca. 2043–1992 B.C.) reliefs suggest that Asiatics were once again the target of Egyptian military activity (1992:69f.) However, that these reliefs cannot be so easily interpreted as evidence for attacks against Asiatic settlements, let alone Asiatics themselves, has been noted by Alan Schulman (1982:168, 72–75).

The question that remains for all of these reliefs is: do the fortified centers, which are depicted as defended by Asiatics, represent Asiatics living in settlements in Egypt during the Eleventh Dynasty, or are these settlements located in the southern Levant? Schulman suggested that all of the scenes dated to the Eleventh Dynasty from the reign of Mentuhotep II onward belong to various depictions of the siege of Herakleopolis by this king (ibid., 182f.). However, this remains to be proved. This is particularly true in light of the collection of dead soldiers (see pp. 44f.) who have been traditionally identified with the reign of Mentuhotep II but now have been redated by Dorothea Arnold to the reign of Sesostris II (1878–1872 B.C.). This would undercut the argument of interpreting scenes of warfare from the latter half of the First Intermediate Period and the first part of the Middle Kingdom solely in relation to Mentuhotep II's famous battle (see n. 31, p. 44).

B. Middle Bronze Age Warfare (ca. 1900–1460 B.C.)

Although we still lack substantive historical records for a little more than the first century of the Middle Bronze Age (ca. 1900–1800 B.C.)—the period during which Amorite dynasties were founded (see Execration Texts)—the Mari texts provide us with a relatively detailed understanding of the political and military affairs of the northern Levant for the second half of the Middle Bronze Age. After 1750 B.C. we are informed of numerous Amorite kingdoms not only in the northern Levant but also in northern Mesopotamia (Klengel 1965–1970; 1992:39ff.). These kingdoms include Yamḫad (centered on ancient Ḫalab) and Qatna in the northern Levant as well as Assyria and Mari in Mesopotamia. While limited references are also made to Ugarit, Byblos, and Hazor, only the military affairs of the largest kingdoms, Yamḫad and Qatna, are related in any detail in the Mari texts (see Table 14 below). Thus, the Mari texts illustrate the degree to which hostilities were present and the reasons why it was considered worth investing resources in the construction of defenses.

1. The Northern Levant

a. Assyria Rising (ca. 1750–1719 B.C.)

Although historical sources for the northern Levant are wanting for the period from ca. 1900 to 1800 B.C., events in neighboring regions suggest that the northern Levant was probably not without its share of military concerns during this period. While rivalries between neighboring Babylonian states dominated in southern Mesopotamia, it seems that in the north Assyria was likewise unrestrained and for this reason it was militarily active in distant regions (see Yuhong 1994). Unequivocal archaeological evidence indicates, for example, that Kültepe-Kaneš was destroyed, presumably by the Assyrian king Puzur-Assur during the eighteenth century B.C. Its destruction brought to an end the Level II settlement of the *kārum*, though the evidence for the destruction of the mound is less certain (Özgüç 1997:266). If Assyria is correctly identified as the aggressor, then the campaign covered at least 840 km in one direction to achieve its objective. Unfortunately, the campaign of Puzur-Assur would seem to be the only Assyrian campaign before the reign of Šamši-Adad I that can, with some certainty, be reconstructed, although it was probably not an isolated event during the Old Assyrian period. Nevertheless, after several decades of abandonment Kaneš was resettled (Level Ib) and flourished before it was again destroyed during the second half of the eighteenth century (Bryce 1998:24). It is not impossible that this was the work of Šamši-Adad I of Assyria.

b. Mari, Yamḫad, and Assyria (1719–1680 B.C.)

Our sources resume with the Mari texts, which relate events during the reign of Yaḫdun-Lim king of Mari (ca. 1724–1714 B.C.), to whom Margueron attributes renovation of Mari's fortifications (2004:444). They indicate that Mari was engaged in a struggle for supremacy with the kingdom of Yamḫad, that Mari probably carried out a campaign to the Mediterranean, and that it engaged in battle in the same year with a coalition of rulers from the middle Euphrates Valley led by Yamḫad (Klengel 1992:50):

> In that same year,—La'um, king of Samanum and the land of the Ubrabium, Bahlukullim, king of Tuttul and the land of the Amnanum, Aialum king of Abattum and the land of the Rabbum—these kings rebelled against him. The troops of Sumu'epuh of the land of Yamḫad came as an auxiliary force (to rescue him) and in the

> city of Samanum the tribes gathered together against him, but by means of (his) mighty weapon he defeated these three kings of . . . He vanquished their troops and their auxiliaries and inflicted defeat on them. He heaped up their dead bodies. He tore down their walls and made them into mounds of rubble (*COS* 2.111).

Yaḫdun-Lim succeeded, therefore, in defeating this coalition at Samanum and thus maintained control over this part of the Euphrates. The only archaeological evidence that can be related to these events comes from the second phase of Palace A at Tuttul (Tell Bi'a), which was destroyed during the Middle Bronze Age, possibly by Yaḫdun-Lim (Einwag 2002:146). Associated with this destruction was a mass grave containing eighty individuals. Ultimately, Ilakabkabu, father of Šamši-Adad killed Yaḫdun-Lim and thus initiated the gradual replacement of Mari's regional dominance by Assyria (Klengel 1992:51).

The defeat of Mari by Šamši-Adad in 1685 B.C. probably contributed to efforts by Sumu'epuh, the first attested ruler of Yamḫad, to consolidate power over an area that may have stretched from the Orontes River on the west, to the Euphrates on the east, and to the border with the kingdom of Qatna on the south (Klengel 1992:52). However, expansion to the east by Yamḫad in this period was quickly checked by Šamši-Adad, who installed his son, Yasmaḫ-Adad, on Mari's throne to replace Yaḫdun-Lim's son Sumuyamam.

This inevitably led to hostilities between Šamši-Adad and the kingdom of Yamḫad. Šamši-Adad is recorded thereafter, probably feeling that his rear was secure, to have ventured to the "Great Sea" and the land of "Laban," possibly through Yamḫad's territory, if not that of Qatna:

> At that time I received tribute of the kings of Tukriš and of the king of the Upper Land, within my city, Aššur. I set up my great name and my monumental inscription in the land of Lebanon on the shore of the Great Sea (*RIMA* 1 0.39.1 lns. 73–87).

In a continued effort to cut off Yamḫad from any military support, Šamši-Adad signed a treaty with Qatna and formed a coalition against Yamḫad, some of the members of which had formerly been allied with Yamḫad (Klengel 1992:52ff.). It may have been in direct response to this event that the fortifications of sites such as Ebla and Tuqan (see pp. 198ff. and 221ff. respectively) were renovated ca. 1700 B.C., since this region probably constituted the frontier between Yamḫad and Qatna. The clear buttressing of the wall's defenses at both of these settlements with the addition of towers and bastions would seem to be a clear sign that political and military circumstances necessitated the renovation of their fortifications.

Perhaps because Sumu'epuh conducted raids against settlements in the northern part of the kingdom of Qatna, Šamši-Adad dispatched troops from Mari to Qatna for its defense (Klengel 1992:66). These events eventually precipitated a direct conflict between Šamši-Adad's forces and those of the kingdom of Yamḫad, which probably lasted several years and was continued by Sumu'epuh's son, Yarim-Lim I. The conflict during Yarim-Lim I's rule (1685–1669 B.C.) is a quintessential example of the nature of military alliances in this period, which brought powers into conflicts that did not share borders. Thus, for example, Yamḫad enlisted the assistance of Ešnunna and Babylon against Šamši-Adad, their mutual enemy (ibid., 55). The conflict between the rulers of Yamḫad and Šamši-Adad was ultimately resolved only with Šamši-Adad's death in 1680 B.C., leaving only five major kingdoms: Yamḫad, Qatna, Babylon, Larsa, and Ešnunna (ibid., 57).

It was perhaps during this period, beginning with Yaḫdun-Lim and followed shortly afterward by Šamši-Adad, when non-Levantine powers sought to assert their influence in the northern Levant, that it is appropriate to set the start of the Rectilinear Phase of fortification construction (see p. 81). Although we are certain only concerning the date of the construction of the ramparts of Hazor, Laiš (Tel Dan), and to some extent for Deir Khabiye, it is likely that the ramparts of Qatna, Sefinat-Nouh, and aṣ-Ṣour were also built at this time. In fact, the date proposed for six-pier gates (see pp. 68ff.), two of which have been identified at Qatna, at the start of the second half of the Middle Bronze Age, suggests that construction of the ramparts associated with Qatna's six-pier gates should be dated to this period. Thus we may conclude that the efforts undertaken to fortify sites such as Qatna, aṣ-Ṣour, Tell Sefinat-Nouh, Deir Khabiye (south of Damascus), Laiš (Tel Dan), and Hazor, which employed fosses and rectilinear, freestanding ramparts crowned by thick mudbrick walls, were the direct result of the conflict between these coalitions, which was ignited in the late-eighteenth century B.C. Fortified sites with rectilinear plans in the southern Levant such as at Yavneh-Yam, Timnah, Haror, 'Ajjul, Ekron, Nagila, Jemmeh, and Masos, can also be understood within the context of a common regional strategy, which was adapted to the defensive concerns of the period.

c. *Babylon (1680–1616 B.C.)*

Following the defeat of Assyria and death of Šamši-Adad, Yarim-Lim I of Yamḫad formed an alliance with Zimri-Lim of Mari (Klengel 1992:55) and probably also with Išihi-Adad's successor, Amutpi'el, of Qatna, though this may only have been realized during the reign of Yarim-Lim I's successor, Ḫammurapi I, when Babylon joined this alliance (ibid., 58). For these reasons military activity during the remainder of Yarim-Lim I's reign was probably negligible, although activity within the kingdom of Mari at that time is well attested. Tuttul (Tell Bi'a), for example, appears to have been destroyed by Zimri-Lim of Mari as evidenced by the destruction of the final phase of Palace A (Einwag 2002:147).

Yamḫad appears to have continued peacefully without evident military conflict throughout the reign of its king, Ḫammurapi I (ca. 1669–? B.C.) who was the successor to Yarim-Lim I. Perhaps this was because the king of Yamḫad continued to furnish Babylon with troops (Klengel 1992:59) and was therefore not drawn into the conflict waged by Ḫammurapi of Babylon against Mari. However, Ḫammurapi I of Yamḫad was also able to continue peaceful relations with Qatna to the south, while the kingdom of Yamḫad in this period may have controlled territory as far east as Šubat-Enlil (Tell Leilan), the former capital of Šamši-Adad.

The greatest difficulties faced by Yamḫad during this period, beginning in the reign of Abba'el, son of Ḫammurapi I, were internal, such as the revolt and subsequent destruction of Irrite in northern Mesopotamia. From the reign of Yarim-Lim II, successor of Ḫammurapi I, to the end of the reign of Irkabtum little is known of the foreign relations of the kingdom of Yamḫad. Our information only resumes with a final note regarding a conflict that appears to have revived between Yamḫad and Qatna during the reign of Yarim-Lim III and that now involved Ešnunna (see AT *6 and Klengel 1965–1970:vol. II, p. 106).

While the historical events discussed above are the extent of the textual data specifically for military activity in the northern Levant, the archaeological evidence for the destruction of important Mesopotamian cities during the First Dynasty of Babylon must also be considered. These, too, serve as evidence for the nature of military concerns during the Middle Bronze Age in the Levant since long-distance campaigns were commonplace. The most notorious of these was Ḫammurapi of Babylon's (1696–1654 B.C.) siege of Mari on the middle Euphrates and its ultimate destruction, which is clearly attested in the archaeological record (Margueron 1997b). This destruction was responsible for the preservation of the Mari texts found in the remains of Zimri-Lim's palace.

According to Ḫammurapi's year-names Mari was captured in his thirty-first year (1666 B.C.) and its walls were destroyed in his thirty-third year (1664 B.C.). Although this was the only major foreign conquest by Ḫammurapi, Samsuiluna (1653–1616 B.C.), his successor, was also involved in major siege operations against the large fortified town of Šubat-Enlil (Tell Leilan I) in the Ḫabur triangle around 1632 B.C., and the site was not resettled until modern times (Weiss 1997:347).

Although the exploits of these two Babylonian kings between the years 1680 and 1616 B.C. did not lead them westward into the Levant, the kings of Yamḫad and Qatna, whose territories bordered that of Mari, must have soberly pondered the consequence of the siege of these two massive capitals of their Amorite brethren, the kings of Mari and Assyria. The Amorite kings of Hazor, Akko, and Ashkelon in the southern Levant must also have been aware of these events. Although the probability of such disaster befalling their southern Levantine kingdoms may actually have been low, it is not difficult to understand how the prospect of the complete devastation of their capitals and countryside, and the fragmentation of their kingdoms, would have motivated the construction and maintenance of such massive fortification systems.

d. Pax Amoritica (1616–1554 B.C.)

With the demise of the aspirations of the First Dynasty of Babylon upon the death of Samsuiluna (1616 B.C.), no sustained threat against Levantine lands can be identified. The evidence from the northern Levant from the second half of the Middle Bronze Age (MB IIB–C) suggests, instead, that a large part of this period was characterized by a *Pax Amoritica*. As demonstrated below, a similar period of peace was experienced in the southern Levant at the same time. Despite the existence and maintenance of massive rectilinear enclosures in the south, which can be understood as the perpetuation of a military strategy begun shortly before the subsidence of military incursions, the absence of destruction layers during this period is the clearest confirmation of the existence of a period of peace and prosperity. Nevertheless, this peace is unlikely to have taken effect until after the death of Ḫammurapi of Babylon in 1654 B.C. Thus, it appears that the floruit of MB IIB fortification construction in the southern Levant occurred during the seventeenth century, probably during the reigns of Ḫammurapi and Samsuiluna.

Table 14. Relevant historical events between ca. 2500 and 1460 B.C.

	Date	Event	Phase
ca.	2362–2296	Weni campaigns in the southern Levant	
ca.	2250	Lugalzagesi campaigns by the Upper Sea	
ca.	2250–2200	*Destructions:* Hama (J5), Byblos, Tuttul (palace), H. et-Turkman, and T. Brak (L)	
		Mari defeats Emar, towns in Lebanon, Ebla, etc.	
		Ebla defeats Mari and installs(?) Enna-Dagan as king	
ca.	2237	Sargon (2237–2182) conquers Uruk and defeats Lugalzagesi	AKKADIAN
		Campaigns against Mari (Palace P-1 is *burned*), Iarmuti, and Ebla (*destroyed*)	(2237–2060)
	2157	Naram-Sin (2157–2121) *destroys* Ebla following revolt	
		Attacks Armānum, Ulišum, and the Amanus region; *destroys* Mari	
	2120	Šar-kali-šarri (2120–2099) ventures to Amanus; defeats Amorites at Jebel Bishri	
ca.	2013	Mentuhotep II besieges Herakleopolis(?); Asiatics depicted in siege scenes	UR III
	1964	Šulgi (year 37) raids Amorites and begins building *The Wall of the Land*	(2018–1911)
ca.	1950–1900	*Destructions:* Byblos, Ebla, Tuqan, Hadidi, Selenkahiye, Sweyhat, and Turkman	
	1944	Reign of Šu-Sin (1944–1934) begins	
		Campaigns against Mari, Urkeš, Tuttul, Ebla, Mukiš, and Lebanon	
	1941	Year 4: builds *Muriq-Tidnim*	
	1911	Ur is destroyed	
	1872	Reign of Sesostris III begins (campaigns against *Skmm*)	
ca.	1850–1800	*Destructions/Abandonments:* Nami, Burga, Tel Hefer, Zeror, Poleg, and Gerisa	
ca.	1750(?)	Kārum-Kaneš (Level II) is destroyed by Puzur-Assur(?)	ASSYRIA
	1724	Reign of Yaḫdun-Lim (1724–1714) of Mari begins	(1750–1680)
		Defeats coalition of Yamḫad at Samanum	
		Campaigns(?) to Mediterranean	
ca.	1725–1700	*Destructions:* Kārum-Kaneš (Level Ib, agent unknown); Tuttul (Palace A, ph. 2), Tuqan	
	1719	Reign of Šamši-Adad I (1719–1680) begins	
		Campaigns to the "Great Sea" (Mediterranean)	
	1696	Reign of Ḫammurapi of Babylon (1696–1654) begins	
	1686	Šamši-Adad allies with Qatna against Yamḫad [Ḫammurapi, year 11?]	
	1685	Reign of Yarim-Lim I (1685–1669) of Yamḫad begins [Ḫammurapi, 12/13–28]	
		Šamši-Adad defeats Mari installs Yasmaḫ-Adad	
		Yarim-Lim I allies with Ešnunna and Babylon against Assyria, Qatna, and Mari	
	1680	Šamši-Adad is defeated [Ḫammurapi, year 17]	BABYLON
		Reign of Zimri-Lim of Mari (1680–1666) begins	(1680–1616)
		Yamḫad, Qatna, and Mari make treaty	
		Destruction: Tuttul (Palace A, ph. 1) by Zimri-Lim(?)	
	after 1669	Conflict between Yamḫad and Qatna renewed(?) with the involvement of Ešnunna during the reign of Yarim-Lim III of Yamḫad	
	1666	Ḫammurapi defeats Zimri-Lim and captures Mari [year 31]	
	1664	*Destroys* the walls of Mari [year 33]; Mari texts sealed in destruction	
	1653	Samsuiluna (1653–1616) succeeds Ḫammurapi as king of Babylon	
	1632	Destroys Šubat-Enlil	
	1616–1554	PAX AMORITICA: no destructions attested	PAX AMORITICA
	1554	Ḫattušili (1554–1524) becomes king of Hatti	HATTI & EGYPT
		Besieges Alalaḫ (VII *destroyed*) and Uršu	(1554–1460)
ca.	1554–1494	*Destructions:* Afis, Tuqan (Phase 1A), and Kadesh	
	1540	Reign of Ahmose (1540–1515) begins (start Dyn. XVIII)	
		Siege of Avaris (1540–1538?)	
		Three year siege of Sharuhen (post 1538?)	
ca.	1540–1470	*Destructions* throughout the southern Levant	
	1524	Muršili I (1524–1494) succeeds Ḫattušili	
		Pizikarra of Nineveh(?) *destroys* Ebla (Mardikh IIIB); year unknown	
	1499	Attacks Ḫalab and Babylon	
	1460	Thutmose III (year 23) marches against Canaanite coalition at Megiddo	

Note: Dates for the reigns of Mesopotamian kings follow the VLC outlined in Gasche et al. (1998). *Italics* indicate archeologically attested destructions.

e. *Old Hittite Empire (1554–1499 B.C.)*

Following the end of the Old Babylonian period, the major turn of events for the kingdoms of Yamḫad and Qatna (and Babylon) was the emergence of the Old Hittite empire at the start of the sixteenth century B.C. While this cannot yet be correlated with a single phase in the construction of fortifications in the northern Levant, events related to the rise of Ḫatti serve to underscore the perception of the continual military threat posed by neighboring powers throughout the Middle Bronze Age until the start of the Late Bronze Age. The destruction of Alalaḫ and Ebla are often credited to the military campaigns of Ḫattušili I (ca. 1554–1524 B.C.) or Muršili I (ca. 1524–1494 B.C.).[10]

On the one hand, the fiery destruction of the Level VII palace and temple at Alalaḫ, for example, is usually attributed to the Hittites, most recently to Ḫattušili I (see D. L. Stein 1997:56; Bryce 1998:76, n. 39), although sometimes to Muršili I. The destruction of Ebla (Mardikh IIIB2), on the other hand, which is usually attributed to Muršili I appears however to have fallen to a different foe. Recent reanalysis by Matthiae (in press) of the bilingual Hittite-Hurrian poem *Das Hurritische Epos der Freilassung* published by Erich Neu (1996a:74–97; 1996b) suggests that Ebla was actually destroyed by a hitherto unknown Assyrian king named Pizikarra, an ally of Muršili I. Evidence of the carnage of this battle has been found through the city (Matthiae 2006), including on the exterior slope of the rampart to the east of the Area EE bastion, where six soldiers were buried in a mass grave (Matthiae 2000:591, fig. 18).

Tuqan (Phase 1A) was also destroyed at the same time (Matthiae 1979:9), followed by a renovation of its fortifications with the noteworthy addition of towers to its walls (Matthiae 1982a:315). Likewise Afis, ancient Apsuna, was destroyed toward the end of the MB II (Mazzoni 2002:131). Peter Parr also reported that Kadesh (Tell Nebi Mend) was destroyed at around the same time (1997:115). Each of the sieges that resulted in the destruction of these towns succeeded, therefore, against the largest of settlements despite their massive earthen ramparts, thick mudbrick walls, and strong gate systems.

2. The Southern Levant

Because only a handful of references provide us with any historical data for the southern Levant during the Middle Bronze Age, and most of these concern Hazor in the Mari texts, our understanding of historical events in this region during this period relies heavily upon archaeological evidence despite its inherent ambiguities. The archaeological evidence consists of two sets of data. The first data are dates that can be established for the construction of fortification systems in the southern Levant during the Middle Bronze Age (see pp. 80ff.). Although much more work must be done to examine ceramics (as well as radiocarbon dates) recovered from them in order to establish reliable *terminus post quem* dates for their construction, a tentative chronology for the construction of Middle Bronze Age defensive systems in the southern Levant is possible (see Table 15).

Within the proposed chronological framework for the development of fortification strategies in the southern Levant, the archaeological data for the destruction and abandonment of settlements provide a second line of evidence for assessing events during the Middle Bronze Age. Despite the suggestion that internecine warfare was common in the southern Levant throughout this period (see Dever 1985:73), in general archaeological evidence thus far supports the identification of only two periods during which multiple settlement destructions and abandonments are attested. These events cluster in Phase 3 of the MB IIA and during the transition between the Middle and Late Bronze Ages (i.e., MB IIC/LB IA).

a. *MB IIA, Phase 3 Destructions*

The earliest evidence for the disruption of settlements during the Middle Bronze Age, which probably relates to military conflict, centers on Phase 3 of the MB IIA (see phasing of these sites by Cohen 2002a). The evidence consists of a constellation of destructions and abandonments of settlements including (from north to south) Nami, Burga, Tel Hefer, Zeror, Poleg, and Gerisa, all of which were founded between the Nahal Kishon and the Nahal Yarkon along the coastal plain during Phases 1 and 2 of the MB IIA.

Tel Hefer, which was founded during the first phase of the MB IIA and fortified in the following phase (MB IIA2), experienced three destructions during the MB IIA in Phases B, C (Area A, Phase 15), and E. Although the conflagrations at Hefer during Phases B and C in the early MB IIA1 stand alone, the site joined a number of other sites when it was destroyed again during Phase E (Area A, phase 14), which is equated with the first part of Phase 3 of the MB IIA (see Marcus 2003). Like Hefer, Gerisa and Nami were also burned; Gerisa during Phase 3 of the MB IIA (see p. 259), and Tel Nami, which was first settled during Phase 2 of the MB IIA (Cohen 2002b:89), thereafter during the MB IIA (Artzy and Marcus 1991:13*). Nami was not resettled until the LB II.

[10] Dates after Trevor Bryce (1998:xiii).

Table 15. Initial phases of fortification of Middle Bronze Age settlements in the southern Levant.

Date of Construction	Coastal Plain	Negev	Foothills	Highlands	Jezreel Valley	Rift Valley & Transjordan
MB IIA1	*No fortified sites attested*					
MB IIA2	*North:* Akko *Central:* Burga, Mevorakh, Zeror, Hefer(?), Aphek, *South*: Ashkelon, Poran(?)			Irbid		
MB IIA3	*North*: Kabri *Central*: Poleg, Zurekiyeh, Gerisa *— Destructions —*		B. Mirsim	Jatt, Shush(?)	Megiddo, Yoqne'am	
MB IIA4	Qana			Jerusalem	Qashish	Q. Shemona
MB IIA–B	Yavneh-Yam			Shimron(?), Shechem,		Dan, Hazor, Jericho, Pella, Nimrin(?)
MB IIB	*North*: Achzib, Kabri, Nahariya, Bira, Keisan, Aphek (Kurdane?) *Central*: Jaffa(?) *South*: Nebi Rubin(?), Ekron(?), Timnah, Gath(?), Lachish, Nagila, Haror, Jemmeh, 'Ajjul	Malhata, Masos(?)		Beth-El		
Sites with general attribution to the MB IIB–C	Far'ah South, Me'amer		'Avdon, Esur	Dothan, Hebron, Ibleam, Najjar, Jenin, Kebara, Kheibar, Qarqaf, Qumei, Shuweiket		Abel B.-Ma'akha, Hammah, Amman(?), Rehov(?)
MB IIC	Ashdod, Sera', Mor		Gezer, B.Shemesh	Abu Zarad, Beth-Zur, Far'ah N., Marjama, Shiloh, 'Urma	Ta'anach	Abu Kharaz, Deir 'Allā, Kitan, Rukais, Sahab(?), 'Umayri
LB IA	← *Destructions* →					

Source: Present study and Cohen (2002a:esp. fig. 14).

Several other settlements were abandoned at the end of Phase 3 of the MB IIA. Tel Burga (ca. 20 ha), which was less than 20 km to the south-southeast of Nami, was abandoned during Phase 3 (no Phase 4 settlement is attested, Cohen 2002a:76). Similar abandonments are attested at Zeror (ibid., 96) and Poleg, which was a short-lived settlement during Phase 3 of the MB I (IIA; ibid., 89). The destructions at Hefer, Gerisa, and Nami as well as the abandonments of Burga, Zeror, and Poleg in the late MB I (IIA, Phases 3–4), and the lack of settlement at all of these sites during the MB IIB–C, suggest that they may have experienced the same disastrous plight ca. 1800 B.C. (MB IIA, Phase 3).

Although it may be premature to propose precise dates for the sub-phases of the MB IIA, some of the aforementioned disruptions in the settlement pattern may be related to our only reference to military activity during this period, which concerns an Egyptian military incursion into the Levant (i.e., *Retenu*) during the reign of Sesostris III (1872–1853 B.C.), which is found on the Khu-Sebek stele:

> His majesty proceeded northward to overthrow the Asiatics. His majesty reached a foreign country of which the name was *Skmm*. His majesty took the right direction in proceeding to the Residence of life, prosperity, and health. Then *Skmm* fell, together with the wretched Retenu (*ANET*, p. 230).

Although *Skmm* referred to in this text is traditionally identified with Shechem in the hill country north of Jerusalem (see *ANET*, p. 230, n. 7), the text suggests that the name in this case was not a reference to a specific town but rather to a "foreign country," or the lands of a kingdom by this name. This interpretation is in keeping with the manner in which kingdoms in the Levant were identified, usually as "the lands of GN" where the GN was the name of the leading polity (e.g., Qatna, Hazor). In any case, it may be significant that the reign of Sesostris III, and hence the date of this inscription, fall within the crucial transition between Phases 2 and 3 of the MB IIA, according to Cohen (2002a:130ff., esp. fig. 27), which leaves open the possibility that this particular pharaoh may be to blame for these events.

If correct, the town of Shechem in the hills of the southern Levant may not have been the target of this attack, which would have been unlikely given the meager settlement at the site and in the surrounding area during the MB IIA (see Cohen 2002a:92). Instead, *Skmm* may have referred to a kingdom whose territory included a portion of the coastal plain of the southern Levant, much of the population of which was forced to relocate to the highlands in the east and to the south in newly established MB IIA settlements late in Phase 3 or at the start of Phase 4. While this reconstruction is admittedly speculative, it offers a tangible interpretation regarding the potential target of Sesostris III's campaign and a reason for the abandonment of these settlements. While we can only speculate why military force was required against this kingdom, the strategic location of these settlements in the narrowest part of the coastal plain, at the mouth of the Aruna Pass, and arrayed around the seaport at Nami, would have enabled this polity to effectively hinder Egyptian traffic by sea and along the main overland route to the northern Levant. Similar circumstances may have been at play during the Amarna period when Lab'ayu of Shechem attacked Megiddo (EA 244) and destroyed Shunem (EA 250).

Whatever interpretation might be assigned to the Khu-Sebek inscription, James Weinstein is correct to conclude that there is no evidence for an Egyptian empire in the Levant during the Middle Kingdom (1975). Furthermore, despite Susan Cohen's nuanced analysis of relations between Egypt and the southern Levant in this period (2002a:esp. pp. 128ff.), no new evidence is available to suggest more intensive military or political involvement by Egypt during this period than during earlier periods.

Historical data from Egypt during the Second Intermediate Period (Dynasties XIII–XVII) also provide no new evidence warranting speculation that the construction of fortifications during the remainder of the Middle Bronze Age were motivated by Egyptian incursions. On the whole, no evidence exists, whether textual or archaeological, whereby it can be suggested that this period in the southern Levant was anything but predominantly peaceful until the rise of Egypt's New Kingdom empire. One can only suggest, therefore, that the impetus for fortifying settlements in the southern Levant (MB IIA, Southern Levantine Phase; see pp. 80f.) in a fashion identical to those in the northern Levant from the MB IIA through the MB IIB–C (Rectilinear Phase; see p. 81) was still predominantly related to the obvious and long-standing threat posed by the nearest rival kingdoms such as Qatna, Yamḫad, and most significantly, Hazor, as well as Mesopotamian powers such as Larsa, Ešnunna, and Babylon.

b. *MB IIB–C Pax Amoritica*

While it is certainly significant to note unequivocal evidence for destruction levels throughout the Middle Bronze Age as evidence of political strife, it is equally significant to highlight that some degree of peace or at least stability characterized settlement during the MB IIB–C in the southern Levant. This is suggested, in the first place, by the absence of evidence of any sort for the violent destruction of settlements during this period. It is also supported by the proliferation of small settlements (i.e., of less than 2 ha), some of which were unfortified, that characterize this period (Falconer 1994a:321ff.). These suggest that the population was comfortable with establishing settlements between and beyond major regional centers and their surrounding second-tier settlements as the Middle Bronze Age progressed. While some scholars have suggested that this settlement pattern indicates a "highly fragmented political landscape" (Falconer 2001:282), it can equally be argued that the political environment was sufficiently stable to allow such ventures to be undertaken with frequency across the landscape.

c. MB IIC–LB IA Destructions

The evidence for the destruction of sites in the southern Levant is relegated to the transition between the MB IIC and the LB I (ca. 1530–1460 B.C.). Space does not permit a lengthy reassessment of the archaeological evidence from each of these sites, but these destructions should be understood as a reflection of the events that brought about the eventual demise of the massive Middle Bronze Age fortification systems, regardless of who was responsible for these destructions. James Weinstein compiled a list of the sites with destructions (some of which were followed by abandonments) that have been dated to this period and are usually attributed to Egyptian military activity at the start of the Eighteenth Dynasty (1981). Destruction layers include those at Dan, Hazor (Str. XVI/3), Ta'anach (A. E. Glock 1993:1432), Shechem (Str. XVI), Shiloh (Str. VII), Beth-El, Gibeon 4a, Jericho, Beth-Shemesh (Str. V), Beth-Zur, Malhata (Kochavi 1993:935f.), Beit Mirsim, Lachish (Level P–4, see Ussishkin 2004a:154ff.), Tel Nagila (Str. VII), Tell el-'Ajjul (City III and Palace I), Ashkelon, Tell el-Hesi, Gezer (Str. XVIII, Dever 1993a:501), Gerisa (Herzog 1993a:481), Jaffa (Kaplan 1972a:78), Aphek, Mevorakh (Stern 1984:37), Akko (M. Dothan 1993a:20), Achzib (Prausnitz and Mazar 1993:32), Tel Michal (see Herzog 1989:38), as well as more recent evidence for Kabri (Kempinski 1993a:841), Tel Kitan [Str. IV] (Eisenberg 1993:881), and Timnah (Str. X and IX). From the Egyptian Delta, it is also warranted to add the possible evidence for the fiery destruction of at least part of Avaris (Level D/2 in Area H/I), though certainly also the abandonment of most of the settlement, which is usually related to the historical account of the expulsion of the Hyksos (Dever 1990a:77; Bietak 1996:67). Weinstein was probably correct, however, that no clear evidence has been revealed for destructions at Beth-Shean(?), Tell el-Far'ah North, Tell el-Far'ah South, and Megadim (1981:2, 5).

Although establishing precise dates for the aforementioned destructions is central to the debate concerning the chronology of the period, particularly with respect to the end of the Middle Bronze Age, it is sufficient for the purpose of this chapter to note that this debate is ongoing. It will ultimately be resolved through the discovery of more Egyptian textual sources, which can provide greater details regarding the campaigns of early Eighteenth Dynasty pharaohs, and through the refinement of the means of dating the strata in question (C14 dates for the Middle Bronze Age are nearly nonexistent, see comments by Weinstein 1984:309; and Carmi and Ussishkin 2004).

C. Conclusions

Benjamin Mazar once suggested that the emergence of ramparted tells during the MB IIB–C was to be linked specifically to the reign of Šamši-Adad in Assyria (1968:83), and as this chapter demonstrates this may be a correct assessment. Similarly, Moshe Dothan has suggested that settlements that were established during the last phase of the Middle Bronze Age (i.e., MB IIC) along the southern coastal plain, such as Ashdod and Mor, could be attributed to the "activities of Apophis" (1973). While no scholars since Mazar and Dothan have attempted to identify the particular historical context of the development of fortifications during the Middle Bronze Age, this review of the political and military history of the Levant from 2300 to 1460 B.C. facilitates the identification of specific historical events that are considered to have been pivotal in the evolution of fortification strategies during this period.

A fundamental observation is that from ca. 2300 B.C. the military and political future of the northern Levant was inextricably related to political developments in Mesopotamia. As Table 14 above illustrates, from the first attested intervention by Lugalzagesi of Uruk (ca. 2250 B.C.) through the end of the period documented by the Mari texts, the northern Levant was under a nearly continuous military threat from Mesopotamian powers willing to travel increasingly greater distances to achieve their political and military objectives. Furthermore, insofar as political unrest was not already present between Levantine kingdoms, such as Yamḫad and Qatna, it was surely fomented by Mesopotamian powers, such as Assyria during the reign of Šamši-Adad.

It is also significant that neither historical nor archaeological data appear to support characterizations that assert the prevalence of internecine feuding between small polities or "city-states" in the northern or southern Levant, although we cannot discount that such conditions existed, as they evidently did between Babylonian states. However, it is altogether more likely that the resources of the Levant were sufficient to accommodate the growing population during this period such that a competition for resources between states within the Levant was not a source of conflict to the degree that it may have been in Mesopotamia during the Old Babylonian period (Dalley 1984:413; Stone 1997). Furthermore, the presence of convenient natural boundaries that circumscribed many of the kingdoms of the region, such as Byblos, Ugarit, and Hazor, facilitated a clear recognition of each polity's territory, thus removing another major factor that contributed to conflicts between Mesopotamian kingdoms in this period.

While it is possible that these factors contributed to less warfare in the Levant than in Mesopotamia, it is also likely that a lack of historical sources from the Levant has skewed our perception of military struggles during the Early and Middle Bronze Ages. Certainly Ebla like Mari, if Enna-Dagan's report is any indication, must have leveraged considerable force to establish itself as the dominant polity in the northern Levant during the mid-third millennium. Unfortunately though, the Ebla archives do not preserve these events. We are similarly in the dark concerning the history of the region from the sack of Ebla (ca. 2200 B.C.) to the start of the reign of Yaḫdun-Lim of Mari (1724 B.C.), a period nearly five hundred years long. Nevertheless, archaeological evidence for numerous destructions attested at sites throughout the northern Levant during this period suggests that the political circumstances were no different than during either the preceding or following periods.

It is also particularly clear concerning the political milieu of the Levant between ca. 2300 and 1460 B.C. that there existed few periods when no substantive threat to the security of the region existed. Such a concern, taken together with an understanding of the development of new military technologies such as the introduction of the battering ram and the composite bow, provide a picture of the clear and present danger evident to the Levant's inhabitants throughout this period. Furthermore, understood within the context of the massive building projects of Mesopotamian cities, the construction of large earthwork fortifications, crowned by mudbrick walls, supplemented by towers and bastions, and breached only by well-fortified gates, are understood to have been necessary elements for surviving the Middle Bronze Age arms race (see chapter 3). As we shall see in chapter 6, the toll borne by the population of the Levant for the construction of these fortification systems was not only bearable but was, in light of the historical circumstances highlighted here, entirely warranted.

5. DEFENSIVE STRATEGIES & POLITICAL ORGANIZATION

> The first task is to make the countryside difficult for the enemy to attack, to camp in, and to provision themselves from.
>
> Aeneas Tacticus, *Siegecraft* 8.1
> Greece, mid-fourth century B.C.

STATEMENTS like the one above made throughout Aeneas's work suggest that a settlement's fortifications were regarded as the last line of defense for a region's inhabitants. This sentiment is even conveyed at the outset of Aeneas's treatise:

> When men leave their own territory to meet combat and danger beyond its borders, the survivors of any disaster which strikes them, on land or at sea, still have their native soil and state and fatherland between them and utter extinction (*Siegecraft*, Preface 1).

Given the desperate state in which a siege so often left the citizenry, Aeneas recognized the need to turn every resource, including the landscape, to the advantage of the defenders in order that a siege, if it occurred, would be as short as possible. For this reason he provided practical suggestions for effectively employing the terrain within a regional defensive strategy. Unfortunately, while some of the details of what Aeneas wrote on the subject were included in *Siegecraft*, much of what he wrote in a work very appropriately titled *Preparations*, has been lost.

There is substantial evidence that the fundamental concepts behind Aeneas's suggestions for preparing the landscape were also employed during the Bronze and Iron Ages. Strategic positions were selected for the building of towns, whether atop hills or beside wadis or rivers. When possible, towns were carefully positioned to afford ease of communication, support, and reinforcement. Fortresses were constructed at the intersections of routes or along roads where required, and watchtowers were employed to strategically supplement the viewshed afforded by other settlements.

All of these features are attested in the Levant during the Middle Bronze Age. This chapter addresses therefore how fortified settlements were strategically integrated into the settlement pattern of the Levant, and how this enables the identification of Bronze Age polities. The first part of this chapter suggests that a singular settlement strategy was characteristic of the Levant during the Bronze Age, despite adaptation to different terrains and environments. Seven regional case studies from the Levant and northwestern Mesopotamia are used to illustrate this settlement pattern. In the second section a typology of basic settlement types from the Middle through Late Bronze Ages is presented, as well as the role of each of these settlement types within the defensive networks defined by this settlement pattern. In the light of such evidence an appropriate set of terminology for the political organization of the Levant during the Bronze Age is suggested. Finally, this framework is tested against the identification of the Middle Bronze Age kingdom of Ashkelon and its subsequent development during the Late Bronze and Iron Ages.

A. Bronze Age Settlement Patterns

Because the construction of particular styles of fortifications and the decisions regarding the location of each settlement were integrally related to the physical landscape, seeking to capitalize on its advantages while minimizing its disadvantages, it is necessary to address the functional interrelationship between fortification systems and the landscapes in which they were built. In this section a settlement pattern is identified that is considered characteristic of the Levant during the Bronze Age. While this settlement pattern may appear simple (as perhaps nothing more than a modification of central place theory), its origin must undoubtedly be traced to the Bronze Age and its fullest development attributed specifically to the Middle Bronze Age. The sheer simplicity of the settlement pattern made it highly effective and required only limited augmentation such as the strategic placement of watchtowers and fortresses.

Archaeological and textual sources reveal the relationship between different settlement types and the landscape, and demonstrate how the typical Middle Bronze Age settlement pattern addressed defense among other concerns (e.g., agricultural and hydrological). This is clearly seen when Middle Bronze Age settlement in the Levant is placed within the historical development of Bronze Age settlement patterns (ca. 2500–1200 B.C.) in the arc from the southern Levant to northern Mesopotamia, which

Tony Wilkinson has characterized as "Landscapes of Tells" (2003:100ff.). Within this context it is possible to trace the emergence of both specific settlement types in the Levant during the Middle Bronze Age and the Northwest Semitic terms that described them from the Middle Bronze Age onward. Thus, the underlying settlement pattern throughout much of the Levant during the Middle Bronze Age consisted of archetypal settlements integrated within a hierarchical pattern arrayed around major political centers.

As suggested in Tony Wilkinson's work (2003), Carlo Zaccagnini's study of *The Rural Landscape of the Land of Arrapḫe* (1979) provides a basic outline for the identification of this settlement pattern. Zaccagnini identifies not only the native terminology for settlement types used at Nuzi but also the hierarchical pattern within the kingdom of Arrapḫe into which these settlements were integrated. While his study represents a textually focused interpretation of the settlement types and pattern of northern Mesopotamia during the Late Bronze Age, as Wilkinson notes the major elements of this pattern are reflected in the Bronze Age settlement patterns of northern Mesopotamia and the Levant (2003:esp. pp. 118ff. and fig. 6.16). By reviewing the specific evidence for various Levantine landscapes as well as Mesopotamian landscapes discussed by Wilkinson, the evolution of this settlement pattern can be demonstrated from the third millennium B.C. through the Late Bronze Age. Although Wilkinson emphasized that this settlement pattern, as evident in the landscape of the Jazira (ibid., 109ff.), was primarily motivated by agricultural concerns, the strategic placement of particular settlement types suggests that a defensive strategy was equally significant.

1. The Jazira during the Third Millennium

As a result of increased archaeological research in Syria during the past twenty years, the settlement pattern of the Jazira has received considerable study (e.g., Wilkinson and Tucker 1995; Sallaberger and Ur 2004).[1] The conclusions of these studies regarding settlement patterns have been reviewed by Tony Wilkinson in his discussion of "The Landscapes of Tells" (2003:100ff.). Wilkinson demonstrated that Early Bronze Age settlement in the Jazira represented a "modular landscape" consisting of a hierarchical arrangement of tells each of which was surrounded by its own zone of cultivation. Landscape features such as linear hollows, mud pits (for making mudbricks), and field sherd scatters help delineate the boundaries of settlements within this landscape and their agricultural land.

Although textual sources from sites such as Beydar are at present being studied for details regarding the political relationships between these settlements (see Sallaberger and Ur 2004), as noted above Wilkinson uses Zaccagnini's study of the records for the Late Bronze Age kingdom of Arrapḫe (see pp. 112ff.) as a framework for the settlement pattern described by archaeological survey data for the Jazira during the Early Bronze Age (Wilkinson 2003:118ff.). Wilkinson therefore identifies various features of the Early Bronze Age landscape with the later Akkadian terms (ibid., see fig. 6.16): the central tell settlement (*ālu*), its surrounding fields (*kirû*), its more distant fields (*ugāru*), nearby "mud extraction pits" (*yarru*), hollow ways/roads (*ḫarrānu*), and smaller secondary settlements (*dimtu*). The following sections demonstrate that these elements, along with other features identified by means of textual and archaeological sources, were characteristic of not merely an agricultural settlement pattern in the Levant during the Bronze Age, but that they also reveal a strategy intended to address a regional defensive concern.

An analysis of the settlement pattern of the province of Nabada centered at Tell Beydar reveals the pattern identified by Wilkinson (Sallaberger and Ur 2004). Using textual and archaeological sources the authors of this study have suggested the following hierarchical settlement pattern for Early Jazira III–IV settlements in the province of Nabada belonging to the kingdom of Nagar:

1) A provincial capital (Beydar, 22.5 ha)
2) Smaller tells (ca. 7 to 10 ha)
3) Villages (ca. 2.5 to 4 ha)
4) Hamlets (ca. less than 2.5 ha)

As revealed below, this four-tier hierarchy essentially corresponds to that of the settlement pattern characteristic of the Levant throughout the Bronze Age, although lacking nuance with regard to the smallest settlements (i.e., rural agricultural estates as at Ugarit, see pp. 113ff.). Nevertheless, our understanding of this settlement pattern can be refined by considering the nature of the defenses of the settlements that comprise each tier of this hierarchy.

Because of the widespread abandonment of smaller settlements and the settlement nucleation that occurred in the Jazira during the late third millennium and through the Middle Bronze Age (Wilkinson 2000:235), the nested, hierarchical settlement pattern characteristic of the Early Bronze Age in this region does not appear to have persisted during the Middle Bronze Age. Nevertheless, despite a reduction of the

[1] For a more complete list of surveys, see Tony J. Wilkinson (2000:esp. p. 224, Table I).

population in areas that were distant from the main river basins, major settlements such as Mari, Terqa, and Tuttul were not only occupied during the Middle Bronze Age but remained major political centers. Therefore, there is a basis for viewing this change in the settlement pattern as a response to political and/or environmental factors, particularly since this region was evidently resettled during the Late Bronze Age (ibid.).

2. Early and Middle Bronze Age Ebla

The earliest evidence from the Levant that reveals the settlement pattern discussed above derives from archaeological surveys of the region around Ebla and can probably be attributed to the early to mid-third millennium B.C. Using survey and archaeological data Jan-Waalke Meyer identifies four tiers of settlement within the 800 km^2 territory surrounding Ebla based upon settlement size and fortifications (1996:144ff.). The site hierarchy includes:

1) A single, large fortified town with upper and lower towns (e.g., Ebla, ca. 45 ha)[2]
2) Smaller, similarly organized and fortified towns (e.g., Tuqan and Afis, ca. 15 to 20 ha)
3) Large walled villages lacking acropoleis (ca. 2 to 4 ha)[3]
4) Small unwalled villages (less than 2 ha)[4]

While archaeological soundings and surveys of the landscape around Ebla remain limited, there is no evidence to suggest that the settlement pattern identified by Meyer differed in this region for the period from the late third millennium through the mid-second millennium. This would seem to be supported by the archaeological phases of the main excavated sites in this region such as Ebla (from at least ca. 2350 B.C.), Tuqan (from at least ca. 2150 B.C.), and Afis (from at least ca. 2350 B.C.). This settlement pattern cannot have developed any later than ca. 2350 B.C. in the northern Levant, though it could be older than that.

Evidence for native terminology for settlement types during this period is rather limited. However, in the Eblaite letter of Enna-Dagan several terms for settlements are used (see pp. 88ff.). As during later periods the logogram URU was used to identify a settlement, but unfortunately its use in this context does not provide any information about the nature of such a settlement's defenses. However, two other terms used in this text may have been intended to convey such information. The first term GA.NU.UM (*apum*), which is usually translated "reed thicket" or "canebrake" (*CAD* A/2, pp. 199f.), if it did not also serve as a toponym, may have been an allusion to a particular type of fortified settlement (see *COS* 3.90, lns. 4f., and n. 4), such as the "canebrake of Ebla" or the "canebrake of Gallabi." In this respect the term may parallel Old Babylonian *appārum* (AMBAR), which seems to have referred to a marshy area that surrounded some settlements (see next section). Perhaps slightly more plausible is the reading of the Eblaite logograms E-NA for É.NA$_4$ in line 8 of the Enna-Dagan text as "fortress" (literally "house of stone" as first read by Pettinato; ibid., n. 6). However, beyond these observations there are few textual data for the identification of specific settlement types.

In addition to this evidence it is also possible to identify specific fortified areas within Levantine towns during this period thanks to Akkadian sources. These distinctions appear to have been clear based, for example, on the names used for the different walls that enclosed the town of Armānum during the last quarter of the third millennium as described in Naram-Sin's siege of this town (see pp. 90f.). The walls are referred to as the "great wall" (BÀD.GAL), the "fortification wall" (BÀD *danīm*), and the "quay wall" (BÀD *kārim*), and each wall was progressively taller as one moved from outside the town into the citadel. Because of the descriptions of these walls we can also distinguish three separate parts of the town of Armānum and settlements like it in the Levant during this period. These include the *kārum*, the lower town enclosed by the BÀD *danīm*, and the upper town or citadel enclosed by the BÀD.GAL. While references to upper and lower towns, which are usually a reflection of the different elevations between them (if they do not also imply the presence of a wall between each of these areas), are not commonly identified for Mesopotamian towns, the spaces delimited in Mesopotamian and Levantine towns in this period were essentially the same. Therefore, during the Old Babylonian period references were made to the inner (i.e., upper town in Levant) and outer (i.e., lower town in Levant) towns and the outer town wall. Although a handful of inscriptions, mostly year-names, attest to

[2] Settlement sizes cited for the first two categories here have been reduced to reflect the actual inhabitable area within the fortifications during the EB IV and Middle Bronze Age.

[3] Since none of the settlements in this category, which include Sheikh Mansur, T. el-Agharr, T. Kathri, T. Dadikh, T. el-Hasmi, T. Manabih, T. Ram Handane, T. er-Rabikh, T. Hader, T. Bajer, and T. Mastuma, have been excavated, the sizes provided only reflect their gross size (Meyer 1996:145).

[4] These settlements include T. Abiad, T. Deber, T. el-Mahdi, T. Khursiyyan, T. Suffane, T. Zartine, T. Serji, T. Jalas, T. Qumhane, T. es-Sfine, and T. Denit.

the construction of town walls during the Akkadian and Ur III periods at sites such as Akkad, Ur, and Nippur,[5] these inscriptions do not provide the detailed descriptions that are available in comparable sources during the Old Babylonian period.

3. Old Babylonian Mari and Mesopotamia

The settlement pattern of much of Mesopotamia, particularly along the rivers, presents a distinct set of circumstances, which in many respects was quite different from those that affected settlement in the Levant and the Jazira (e.g., Falconer 1987). Wilkinson notes the contrast between these landscapes, identifying them as "landscapes of irrigation" (e.g., Mesopotamia) and "landscapes of tells" (e.g., the Jazira and the Levant, 2003). By way of comparison, Steven Falconer suggests that the Levant represented "a heartland of villages" (1987), while Mesopotamia was a "heartland of cities" (Adams 1981). Such characterizations, which result from views of the landscape that are focused on agriculture and urbanism respectively, have rightly been criticized (Maeir 2003:63f.).

Although environmental concerns affected settlement patterns and they underlie many of the differences between the economies of these regions, neither of these approaches adequately accounts for a major factor in site selection and the emergence of settlement patterns, namely defensibility. The nature of the archaeological and historical evidence available, especially from Mari, demonstrates that the Old Babylonian settlement pattern of the kingdom of Mari clarifies the nature of Levantine Bronze Age settlement patterns, fortification strategies, and administrative organization. Thus, we may ask to what extent the defensive strategy of most of Mesopotamia compared with that of the Levant during the Bronze Age. Were not Old Babylonian settlements as vulnerable as most settlements in the Levant? The evidence, both archaeological and textual, suggests that settlements in both regions were similarly vulnerable, and that for this reason comparable defensive strategies were pursued in both regions (see pp. 94ff.).

As historical sources suggest and data from both surveys and excavations have revealed (e.g., Adams 1965:esp. p. 49), the first half of the second millennium in Mesopotamia was fraught with warfare between competing states. Despite this evidence very little work has been done to synthesize the data pertaining to Old Babylonian fortifications and fortified settlements and the terminology used to describe them. Nevertheless, Albert Glock addressed some of the Old Babylonian terminology for fortified settlements in the kingdom of Mari (1968).[6] His work provides an initial basis for the creation of a list of fortified settlement types and the fortified complexes within them that are attested for the Old Babylonian period.

It is important to note that the limited occurrence of many of the Old Babylonian terms and our inability to correlate settlement types and defensive features mentioned in texts with excavated remains has often complicated the process of distinguishing the meanings of various terms. For instance, in Glock's discussion of defenses, following in the Assyriological tradition he translated *ālum* (usually lit., "town") as "fortress" in some contexts (ibid., 159; after *CAD* A/1, p. 379.), while in others he noted that *birtum* was used to refer to a "fortress," especially when a garrison, usually of three hundred soldiers (*ṣāb birtum*), occupied these types of settlements (ibid., 110, n. 71). Adding to this confusion is the evidence for the identification of the term *dūrum*, also as "fortress," perhaps specifically a royal fortress, since it often occurs in construct with divine and royal names. Therefore, despite the work on this subject by both Albert Glock and Jack Sasson (1969), further clarification of the Old Babylonian terms for settlements and their defensive features is necessary.

The combined data from archaeological and textual sources for Mesopotamian settlements, like those in other regions, enable the identification of one of four basic settlement types with respect to the nature of their fortifications. These types include large fortified towns, secondary fortified towns, fortresses, and unfortified villages. In the following pages the Old Babylonian terms and the defensive features that accompany them are discussed.

a. Large Fortified Towns

According to the *Chicago Assyrian Dictionary* the Old Babylonian term *ālum* (pl. *ālū*, *ālānū*, or in Mari letters as *ālānû*), which is most often translated simply as "town," had a semantic range that included "fort" or "military strong point" (*CAD* A/1, p. 387).

---

[5]These include the wall of Akkad during Naram-Sin's reign (year-name iii, *RIME* 2.1.4, p. 85), the walls of Ur (*RIME* 3/2.1.1, p. 11) and Nippur (*RIME* 3/2.1.1.38) during Ur-Nammu's reign, and the building of *The Wall of the Land* in Šulgi's sixth year (*RIME* 3/2.1.2, p. 106), as well as the rebuilding of the walls of Ur and Nippur by Ibbi-Sin (*RIME* 3/2.1.5.1).

[6]Unfortunately, Van De Mieroop's recent work on *The Ancient Mesopotamian City* (1997) does not provide an extensive discussion of Akkadian terminology for settlement types, particularly those that were fortified.

However, one may question the translation of the general term *ālum* for town as either "fort" or "station" in Old Babylonian sources, particularly in the Mari texts.[7]

Despite references to activity related to defense within these settlements, such as lighting fire signals (ARM IV 32), the nature of the defenses of the settlements and/or their exclusive function in a particular capacity related to defense cannot be inferred from their identification with the term *ālum*. Simply because textual references may suggest that settlements identified as *ālānû* were fortified does not mean that the unqualified use of the term implied that all such settlements were fortified. As Adams notes, "there is nothing to indicate that the use of the term for city [*ālum*] was tied either to a minimal size of settlement or to the presence of specific urban institutions" (1981:136). Thus, on occasion towns were explicitly described as *ālānū dannūtum* (literally "strong towns") when alluding to their defenses (*CAD* D, pp. 99f.; ARM V 16:15). When they were not "strong" this too was noted (ARM V 16:14f.). Nevertheless, it also cannot be assumed that the adjective *dannum* necessarily meant that a town was fortified. The phrase was probably most often employed to suggest that a town was enclosed by some type of fortified perimeter, though whether this also consisted specifically of fortified gates, a fosse, and/or fortification walls cannot be inferred from the use of this term.

Although it was occasionally necessary to describe in detail the fortifications of Mesopotamian towns,[8] detailed descriptions of the fortifications of the largest Old Babylonian cities relate primarily to their construction, which was probably recorded in a manner that was intended most of all to impress scribes. Detailed descriptions are largely absent, undoubtedly, because the appearances of the fortifications of such large cities were so similar.[9] References to the size and features of the fortifications of the largest Old Babylonian cities are for the most part restricted to inscriptions commemorating their construction. Nevertheless, Old Babylonian inscriptions attest to the building and rebuilding of the main fortification wall (*dūrum*, *CAD* D, pp. 192ff.) of cities such as Isin (*RIME* 4.1.2.3; 4.1.4.5, 4.1.15.1), Larsa (4.2.5.2; 4.2.6.2; 4.2.8.7), Babylon (4.3.9.1), Ur (4.2.13.18–21), Sippar (4.3.6.2; 4.3.6.12; 4.3.7.3), Nippur (4.3.7.2), Kiš (4.3.7.7; 4.8.1–2), Dunnum (4.1.14.1), Me-Turan (4.16.1), Terqa and Mari (4.6.8.1). Although less common than the term *dūrum*, *igārum* (pl. *igartum*) could also be used to refer to fortification walls (see *CAD* I, p. 34). However, the difference between these terms is unclear due to the limited use of *igārum* in the Mari texts (see esp. ARM II 88:17, *igartam ša dūrim*, possibly "the walls of the fortress"[?]; see use in ARM III 71:16, *igaritam ša bitīšu*).

While construction projects at each of the cities mentioned above probably involved work primarily on the town wall, fortifications of these sites also included various other features such as the gate, *abullum* (*CAD* A/1, p. 82ff.; see ARM XV p. 168 for list of refs.) or *bābum* (*CAD* B, pp. 20f.; e.g., ARM II 72:36), and sometimes an outer town wall, *šalḫum* (*CAD* Š/1, pp. 243f.; ARM II 101:11, VI 29:16f., and XIV 24:5). Other infrequent and thus more poorly understood terms were also used to describe fortifications, such as *kilīlum* (*CAD* K, p. 358), usually identified as "battlements," and *sītum* which may have referred to a "battlemented parapet" (*CAD* S, p. 336).

Textual evidence also suggests that some type of trench was often excavated around the perimeter of the largest fortified centers. In some cases the term used was *ḫirītum* (*CAD* H, pp. 198f.), usually identified as a "fosse" or "moat," which was used to refer to trenches around Ur (*RIME* 4.2.13.21), Kiš (where its wall was said to have reached the bank of the Euphrates, 4.3.7.7), Terqa and Mari (4.6.8.1), and towns identified in the Mari texts (ARM I 90:20, 139:6, III 36:19). Another term, though infrequently used, was *amrummum* (*CAD* A/2, pp. 78f.; see also ARM II 30:5–12), which is identified as "moat." It is impossible, however, to be certain whether either term was used to refer specifically to fosses or moats.[10] Similarly, it is unclear whether the occurrence of Sumerian E.KI.SUR.RA (lit., "boundary canal") was intended to refer to a fosse or a moat, which had been excavated at Ur in addition to the *ḫirītum* (*RIME* 4.2.13.21). In other cases, however, the evidence suggests that the trench in question was actually

[7]For a translation of *ālum* as "station" in ARM IV 32:12ff., see A. E. Glock (1968:121; following "lookout station" for the transmission of fire signals from A. L. Oppenheim 1954:143a). For a translation as "fort" or "fortresses" in ARM IV 26:26, see A. E. Glock (1968:159).

[8]The construction of fortifications at what must have been smaller settlements is attested among Old Babylonian year-names (Sigrist 1988; 1990; Horsnell 1999). However, it is difficult to know how large these projects were and if they included some or all of the elements identified as part of the fortifications of the largest cities, though they were certainly of a scale deemed worthy of mention.

[9]The use of the term *city* to describe these settlements is appropriate given the immense size of most of these settlements, as centrally located capitals within their territories, and the intensive effort invested in their fortification.

[10]See also Sasson regarding the use of these terms in lexical lists (1969:n. 6).

filled with water (a moat?) or was connected to a body of water. The term *appārum* (AMBAR), which has often been translated as "swamp" (*CAD* A/2, pp. 179ff.: also "reed marsh, reed bed, lagoon," and "canebrake"), identified defensive features surrounding Sippar (*RIME* 4.3.6.12), Nippur (4.3.7.2), and Kiš (4.3.7.7).

Various Old Babylonian terms suggest the construction of earthen ramparts like those identified in the Levant and northwestern Mesopotamia. Language relating the building of walls and their foundations like "mountains" was probably intended to identify heaped up earthen ramparts. Such terminology was used for the walls of Ur (*RIME* 4.2.13.21) and Sippar (4.3.6.2, 4.3.6.12, and 4.3.7.3). Georges Dossin identified two other terms, *adaššum*[11] and *kirḫum*, which he argued refer respectively to the exterior and interior slopes of earthen ramparts (1972:112f.). From an analysis of the Nuzi texts Carlo Zaccagnini also suggests that the *kerḫu* (OB *kirḫum*), usually translated as "citadel" or "fortified area within a city" (*CAD* K, pp. 404f.), should be identified specifically with the "inner (area of the rampart) of a city" (1979:26f.).[12] This meaning is also suggested by its use to refer to a town's "circumvallation," or in Standard Babylonian to the mountain-like shape of these features (*CAD* K, pp. 404f.). Nevertheless, the former translation "citadel" is not likely for this term, particularly given an abundance of other terms that are identified with "citadel." It is possible, therefore, that the use of the term *kirḫum* in the Mari texts meant that the princesses mentioned in ARM VIII 88:15 took their oaths on the interior slope of the earthen rampart and not "in the citadel" as has been suggested.[13]

As alluded to in the previous section, the largest Mesopotamian settlements could be divided into inner (or upper) and outer (or lower) towns. Although the upper town was sometimes described in the Old Babylonian period as the *ālum elûm* (*CAD* A/1, p. 380), a separate term for the outer town seems to be lacking. The term *birtum*, translated as "citadel" or "castle (as part of a city)" (*CAD* B, p. 261), was often used to identify the acropolis of a large town. Garrisons would be stationed within the *birtum* of a town (e.g., *birātim ša ālānī*; ARM I 20:11'). The citadel or upper town was generally closed off from the surrounding lower town, and the *kārum* was appended to the surrounding settlement and enclosed by its own defenses. If the Naram-Sin text from the second half of third millennium provides a reliable witness (see previous section), then the walls were progressively taller as one moved toward the interior of the town.

b. *Secondary Fortified Towns*

Secondary sites within the territory of the central *ālum* were sometimes identified as *dimātum* (sing. *dimtum*; *CAD* D, pp. 144ff.), perhaps denoting their locations with respect to an *ālum*. According to Rafał Koliński *dimātum* probably functioned in essentially the same manner as those that are described in the kingdom of Arrapḫe (see pp. 112f.), whether the term referred to entire settlements or simply agricultural estates (2001:22ff.). This term, however, does not occur in the Mari texts except in reference to siege towers (see pp. 38f.); most references to *dimātum* are limited to Babylonia, particularly around Larsa, though this might only be the result of the number of Old Babylonian sources and their concern with agricultural production. Sites with a function similar to *dimātum* probably existed around Mari, though clearly a different term was employed. Unfortunately, while we may infer that many of these types of settlements were fortified, whether they always were and how they were cannot be concluded from the use of this term. It is perhaps noteworthy that *dimtum* like *dūrum* ("fortress") was often used in construct with a divine or personal name (e.g., Dimti-Enlil; see *RlA* 2, pp. 226f.).[14] When used in this manner the foundation of the settlement appears to have been of considerable significance, though precisely why is often difficult to discern.

c. *Fortresses*

In addition to fortified cities and towns, various types of fortresses are also attested in Old Babylonian texts. One type of fortress may have been the *birtum* (pl. *birātum*; *CAD* B, p. 262). According to the *CAD* these types of fortresses were usually located "in strategic locations outside of cities and villages, at borders, passes, etc." Yet no single specific named settlement is called a *birtum* in Old Babylonian texts. As a placename, however, the term may be preserved for Late Bronze Age sites in Northwestern Mesopotamia and the Levant. One settlement by the name of *Birtu* is mentioned in a Middle Assyrian text from

[11]This term was not adequately understood at the time of the completion of volume A of the *CAD*.

[12]Zaccagnini's observation that "only houses are located *ina kerḫi*" (1979:27) suggests that gardening and other use of this area may have been considered detrimental to the integrity of the rampart.

[13]It is possible that these two terms refer simply to inner (*kirḫum*) and outer (*adaššum*) towns.

[14]At least one occurrence of the term was used in construct as the name of a *durum*; *Dūr-Dimat-Enlil* was built in seventeenth year of Samsuiluna (*RIME* 4.3.7.5).

Ekalte (Tell Munbaqa) and was located nearby (see *RGTC* 12/2, p. 55).

No ambiguity exists, however, with regard to the use of the term *dūrum* which, rooted in an identification of a settlement with its fortification walls, referred to royal fortresses or military strongholds (*CAD* D, pp. 192ff.). One example of a site so named within the kingdom of Mari was Dūr-Yaḫdun-Lim (*RIME* 4.6.8.1; for a list of references in Mari texts see also ARM XVI/1, p. 10), but the greatest number of these sites, as with *dimātum*, were located in Babylonia (see *RlA* 2, pp. 241ff.), where their names were written in construct with a divine or personal name.

As suggested from textual sources the *dūrum* probably featured the most regular construction of any fortified settlement type since they were intended to garrison troops for the defense of particular regions. One of the clearest episodes of fortress construction occurred during the reign of Samsuiluna when six fortresses were built within two months (*RIME* 4.3.7.5). Another inscription from the reign of Samsuiluna relates that during the foundation of Dūr-Samsuiluna:

> He dug its (surrounding) moat (*ḫirītum*), piled up its earth there [built its rampart?], formed its bricks, (and) built its wall. He raised its head like a mountain (*RIME* 4.3.7.8).

Therefore, the systematic process required in the construction of a royal fortress is very well established by excavations and textual sources.

The remains of various settlements suggest how a *dūrum* was planned and constructed. The only Old Babylonian settlement that is identified as a *dūrum* and has been excavated is Dūr-Samsuiluna, identified with Mound B at Khafajah in the Diyala due to the discovery of a cylinder inscription (Delougaz 1990).[15] Nevertheless, only limited exposure of its north wall was possible in 1937. The dimensions of the fort, the complete plan of which was not exposed, can be estimated based upon the site's contours at approximately 250 by 400 m, or about 10 ha in size (ibid., fig. 28). The only other settlements that might be identified with the term *dūrum* are Haradum (Khirbet el-Diniye) 90 km southeast of Mari down the Euphrates (Kepinski-Lecomte 1992) and Šaduppum (Tell Harmal), though neither has been identified as a *dūrum* in historical sources. Haradum is approximately 135 m square or about 1.56 ha with an orthogonally planned internal layout and a 3 m wide fortification wall (ibid., 90). Despite these data the characterization of the site as a *dūrum* is uncertain. The plan of Tell Harmal is comparable and exhibits very similar regular planning of the settlement within its walls (see fig. 82 in Invernizzi 1992:55), however its precise characterization is also uncertain.

We also learn from textual sources that in addition to fortification walls—which explains the origin of the name for this type of settlement—the defensive features of the *dūrum*, like those attested for large fortified settlements such as Mari, Terqa, Kiš, Larsa, and Sippar, included the use of earthen ramparts for the foundation of the wall as well as the fosse/moat (*ḫirītum*). While the presence of a rampart seems to be implied in the inscription relating to the construction of Dūr-Samsuiluna discussed above, excavation has yet to reveal any feature that can be identified as an earthen rampart. Nevertheless, similar construction processes to those attested for Dūr-Samsuiluna, which also included the excavation of a moat, are described by Yaḫdun-Lim for the foundation of the fortress bearing his name (*RIME* 4.6.8.1). Such fortresses were also often built on the banks of rivers or canals (*nārum*; *RIME* 4.3.9.2).

In addition to the major categories of fortified settlements and fortresses, which were presumably located within the heartland of Old Babylonian polities, other settlements referred to generally as "border towns," *āl pāṭim* (*CAD* A/1, p. 381), are also attested. Like the term *ālum*, this term does not imply any qualification regarding a settlement's defenses. However, Glock suggested that a term did seem to exist for this purpose. The term, *baz(a)ḫātum* (*CAD* B, p. 184), was used in the Mari texts to refer to border forts or outposts along a frontier (A. E. Glock 1968:98). Unfortunately, it is not clear whether all such sites were necessarily fortified (see ARM II 67 r7′, 88:7, 92:21, 102:18, 103:10, 105:6; III 12:7, 17:21, 30:18; VI 64:3). Yet as Glock observed the role of individuals stationed in these fortresses was clearly one related to policing activities.

d. *Unfortified Villages*

The final category of Old Babylonian settlements consisted of unfortified villages, which were usually called *kaprum* (pl. *kapartum*, *CAD* K, pp. 189f.). The *CAD* notes that:

> the semantic range of kaprum extends from "village" in agricultural surroundings, "farm" for producing of barley, "settlement" of shepherds of a more or less permanent nature, to suburban agglomerations around cities. In the plural (kaprātu) the word refers also, in a general way, to out-of-town regions (ibid., 190).

In addition to *kaprum*, *maškanum* (*CAD* M/1, 369ff.) may also refer to unfortified settlements that were associated with the presence of a "threshing floor" if they were not simply "small agricultural settlements."

[15]It should be noted, however, that one cannot be certain that only one fortress was so named.

Table 16. Terminology for settlement types in Old Babylonian, Nuzi, and Ugaritic texts.

Settlement Type	Old Babylonian	Nuzi	Ugaritic
Town	*ālum* (URU)	*ālu* (URU)	*qarītu* (URU)
		dimtu (AN.ZA.GÀR)	*ʿr*
			*pdr**
Fortified town	*ālum dannum*	—	—
Gate	*abullum/bābum*	*abullu/bābu* (KÁ.GAL)	*ṯǵr*
Wall	*dūrum* (BÀD)	*dūru* (BÀD)	*ḥāmītu*
	igārum (É.SIG$_4$/IZ.ZI)	—	—
Rampart (exterior)	*adaššum*	—	—
Rampart (interior)	*kirḫum*	*kerḫu*	—
Moat	*ḫirītum*	—	—
Fosse	*amrummum*	—	—
Swamp	*appārum* (AMBAR)	—	—
Border town	*āl pāṭim*	—	—
Unfortified town	[*ālum*] *ul dannū*	—	—
Divisions of town			
upper town	*ālum elûm*	—	—
citadel	*birat ālim*	—	—
merchant enclave	*kārum* (KAR)	*kāru*	*maḫḫadu* (KAR)
Fortress	*Dūr*-PN/DN (BÀD)	*Dūr*-PN/DN (BÀD)	*ʿamqu* (AN.ZA.GÀR)
			*ḏmr**
			*ḥl**
settlement within	*āl-dūrim*	—	—
on border(?)	*baz(a)ḫātum*	—	—
Village (unfortified?)	*kaprum*	—	*kaparu*
	maškanum (MAŠ.KÁN/KI.UD/É.KI.GÁL)		*maškanu*(?)
			madinatu
Watchtower	*birtum*(?)	*birtu*	*magdalu*
		[MA *madgaltu*]	(*mdgl**)
Agricultural estate	*dimtum* (AN.ZA.GÀR)	*dimtu* (AN.ZA.GÀR)	AN.ZA.GÀR (*gittu/dimtu*)

*Vocalization uncertain.

Although references do not describe how, if at all, such settlements were defended, as mentioned above an *ālum* could also be specifically identified as unfortified (or at least insufficiently fortified) with the phrase *ul dannū* (lit., "not strong"; see ARM V 16:14f.).

The Old Babylonian terminology for fortified settlements and their constituent elements are summarized in Table 16. Based on the discussion above there appear to have been four main categories into which Old Babylonian settlements in Mesopotamia can be classified based upon the nature of their defenses, which as demonstrated below also correspond with archaeological survey data from southern Mesopotamia (Adams 1981:136ff.):

1) Large walled towns between 40.1 ha and 200 ha (50.2% of settlement area; e.g., Nippur, Isin, Uruk, Larsa, and Umma).[16]
2) Medium sized towns between 10.1 and 40 ha, which were probably also walled (20.2% of settlement area).
3) Small towns between 4.1 and 10 ha.
4) Settlements of less than 4.0 ha which may have consisted of a number of different types of settlements both fortified and unfortified.

The last two categories together account for only 29.6% of the area occupied by Old Babylonian settlement according to Adams's survey. Similar percentages pertain for the Ur III-Larsa period that Adams also identifies (see 1981:Table 12, p. 138). Despite similarities in the hierarchical organization of settlements, it cannot be ignored that the settlement pattern of southern Mesopotamia was inextricably linked to watercourses, as Adams emphasizes (ibid., 155ff.).

e. *The mātum and Its Districts*

The Mari texts, which are our best source for reconstructing Old Babylonian political administration,

[16]Umma appears to have been the only settlement to have reached over 200 ha (see Adams 1981:166, fig. 33).

facilitate a reconstruction of the kingdom of Mari and its political dependents during this period. They suggest, therefore, the manner in which settlements in this region related to one another politically and militarily. In this regard we learn that the kingdom (*mātum*) of Mari was administered through a number of districts or provinces (*ḫalṣum*).[17] Daniel Fleming defines the term *mātum*, which is most often associated with the kingdoms mentioned in the Mari texts, and addresses the role of the districts within their territory (2004:104–69):

> The *mātum* is the political unit that can make its own decisions about making peace or waging war. At root, the *mātum* was defined not by place but by population, and in practice, it was adapted to peoples who were defined in both tribal and nontribal terms. By the early second millennium, kings and their kingdoms dominated the Mesopotamian landscape, and most often we see the *mātum* as what the king rules. In fact, however, this royal domain was not a tract of territory but the sum of all the people who accepted the king's leadership. Even aside from the question of nomadic movement or tribal identities, we should not imagine the political map of ancient Mesopotamia to have consisted of states with fully enclosed boundaries of the sort drawn on modern maps. Political control would have been defined in terms of population and routes of communication, with greater and lesser degrees of effective power and certain regions that could not be fit into the framework at all. This definition of a world of *mātum* polities accommodates entirely the possibility of active tribal entities that could straddle the territorial heartlands of two or more *mātum* kingdoms, and such fluid relationship to territory may have been more common in earlier periods. Kings who desired full domination over as large a population as possible would naturally tend to pursue total political control over core territories, and the *mātum* then took on a strong geographical aspect. It must be kept in mind, however, that although the *mātum* is most often translated as "land," the word does not define identity by place and therefore at some level has much in common with tribal categories (ibid., 29–30).

With regard to the function of districts (*ḫalṣum*) in the kingdom of Mari, Fleming notes:

> One substantial feature that could distinguish large "lands" from their smaller neighbors was the administrative subdivision of the larger and more complex kingdoms into "districts" called *ḫalṣums*. These *ḫalṣums* were governed by servants of the king, with the same political power and administrative conception as the kingdom center, as extensions of royal authority into more local affairs. Like the *mātums* under which they functioned, *ḫalṣums* were defined in terms of regional populations that reached beyond any individual town...The *ḫalṣum* was a political unit that was never conceived to be autonomous. It served only the administrative needs of a *mātum* that was ruled by a king in a system of individualized ("exclusionary") power (ibid. 133).

Although we do not know the names of all of these districts, because of the number of texts available, the districts of the kingdom of Mari that can be identified include Mari, Terqa, Dūr- Yasmaḫ-Addu, Saggaratum, Qaṭṭunan, Suhum, Ḫarran(?), Šubat-Šamaš, Tuttul, and Ida-Maraṣ(?), in addition to various other unnamed eastern districts (Villard 2001:62ff.). As it pertains to the archaeological signature of this form of political organization, it is especially noteworthy that like the kingdom to which these districts belonged, districts were usually identified with the largest city within and usually at the center of their territories, and each district was governed by a *šapīṭum* (*CAD* Š/1, 459). The military obligations of these districts within the kingdom of Mari are discussed below (see pp. 124f.).

4. Late Bronze Age Mukiš

While there are insufficient data regarding the overall settlement pattern and organization of the kingdom of Yamḫad (Ḫalab) during the Middle Bronze Age to allow a thorough characterization of its settlement pattern, there is a growing corpus of settlement data from the 'Amuq Valley around Alalaḫ, the capital of the district (Akk. *ḫalṣu*) of Mukiš in the kingdom of Yamḫad. Like the kingdom of Ugarit (see pp. 113ff.), Mukiš administered an area of probably more than 2,000 km^2 (Serangeli 1978:100), but was enclosed on all sides by mountains except where the Orontes and Kara Su rivers entered the valley. The actual agricultural land within the kingdom was, however, probably on the order of around 1,000 km^2 (cp. 30 x 30 km from Casana and Wilkinson 2005b) minus the area of Lake Antioch.[18]

Casana and Wilkinson have described the settlement pattern of the region during the Bronze Age as a "Landscape of Tells" (2005b:37–40), and suggested that the settlement hierarchy consisted of three tiers of settlement:

[17] Old Babylonian *ḫalṣum* is yet another term that has been translated as "fortress" or "fortification" (*CAD* 6, pp. 51f.), although the term only designated a "district" or administrative unit in the Mari texts (e.g., ARM I 16:26ff., 73:20; II 120:3; III 10:6; VI 14:6).

[18] Flavia Serangeli suggests that only 40% of the 2,000 km^2 was cultivated (1978:103).

1) A single, large fortified town with an upper and probably lower town (Alalaḫ [AS 136], ca. 22 ha)[19]
2) Smaller secondary tells evenly spaced across the plain (ca. 5 to 15 ha)[20]
3) Small villages (less than 5 ha)

Although it has been suggested that the nucleated settlement pattern of the 'Amuq lacked a fourth tier of "small, outlying, dispersed farmsteads," as in the case of the lower town settlements around Alalaḫ, these are accounted for by their small size, which has meant that they were easily buried beneath alluvium if they were not destroyed by other natural and cultural processes (Yener et al. 2000:187). Nevertheless, one line of evidence can actually demonstrate that a fourth tier of settlement did exist.

It can be observed, as Flavia Serangeli demonstrates, that in the Alalaḫ IV archives no fewer than fourteen villages are known in Mukiš, which consisted of ten or fewer families, twelve with between ten and twenty families, and eleven with between twenty and thirty families (1978:114). If we assume that these families consisted of joint family households with an average of ten persons each (cf. Schloen 2001:147), then there existed fourteen villages with approximately one hundred inhabitants, twelve with one to two hundred inhabitants, and eleven with two to three hundred inhabitants. If we presume a coefficient of two hundred and fifty inhabitants per hectare, then during the Late Bronze Age in the kingdom of Mukiš there may have existed at least thirty-seven villages of approximately 1 ha or less in size.[21] The Alalaḫ archives suggest, therefore, that a fourth tier of settlement did exist and that it is adequately accounted for in surveys of the 'Amuq, although not separately identified. Thus, the third-tier consists of small villages between 1 and 5 ha, while the fourth tier includes sites less than 1 ha in size.

Certain observations can also be made about the continuity of settlement in the 'Amuq during the Bronze Age. The location of the central site of the Middle and Late Bronze Age (e.g., Alalaḫ) was basically retained from the Early Bronze Age when it was located at Tell Ta'yinat less than a kilometer to the north, and continued even into the Iron Age shifting from Alalaḫ back to Ta'yinat. However, as Casana and Wilkinson demonstrate, this pattern was basically characteristic of settlement in the 'Amuq from the Early Bronze Age through the Iron Age and therefore it only emphasizes the considerable continuity in the settlement pattern of the 'Amuq from the Bronze Age well into the Iron Age (Casana and Wilkinson 2005b:37–40).

The radial arrangement of settlements around central sites was another aspect of the settlement of the 'Amuq that demonstrates continuity in settlement patterns throughout the northern Levant and northern Mesopotamia. Casana and Wilkinson point to the example of Tell Salihiyya (AS 129), which appears to have been surrounded by smaller dependent settlements (AS 123, 131–33, 139–40, 142–3, and 147; ibid., 39 and fig. 2.21). Therefore, the settlement pattern in the 'Amuq during the Middle and Late Bronze Ages (Phases L and M) appears to have been identical with those of Ebla and throughout the Jazira during the third millennium. As we shall see below, it was also comparable to that of the southern Levant during the Middle and Late Bronze Ages.

5. Late Bronze Age Arrapḫe

That the nucleated tell-based settlement pattern persisted through the Late Bronze Age in Mesopotamia is evident around Nuzi in the land of Arrapḫe. There are two main types of sites in Arrapḫe that are attested in the Nuzi texts, which Carlo Zaccagnini identifies (1979). The first of these is the central

[19] Although Alalaḫ is identified as 22 ha in area (Yener et al. 2000:185), this figure does not account for the possibility of a lower town, the existence of which is suggested by sherd scatters to the east of the mound and Woolley's earlier observations (see catalog 2.A). This lower town would have been buried under later alluvium. Therefore doubling this figure may adequately account for the lower town.

[20] These include T. Salihiyyah (AS 129, ca. 4.5 ha), Chatal Höyük (AS 167, ca 9 ha), T. Hasanuşağı/Yurt Höyük (AS 99, ca. 7 ha), T. Uzunarab/Bozhöyük (AS 84, ca. 5.4 ha), Karatepe (AS 86, ca. 11.4 ha), T. al-Judaidah (AS 176, ca. 3 ha), Akpınar Höyük (AS 52, ca. 3.2), Acarköy/Halilağa Höyük (AS 12, ca. 3.3 ha), Esentepe/al-Kanisah (AS 29, ca. 12 ha), and T. Davutpaşa/Daud Paşa (AS 164, ca. 1.4 ha, see Yener et al. 2000:187; Casana 2003:232; Casana and Wilkinson 2005a). Since the size of these settlements accounts only for settlement on their tells the total occupied area may have been substantially greater as suggested for T. Davutpaşa.

[21] Given that it is impossible to adequately distinguish Middle Bronze Age settlement from that of the Late Bronze Age in archaeological surveys of the 'Amuq (see Casana and Wilkinson 2005b:37f.), Bonnie Magness-Gardiner's suggestion that "there probably were not many very small sites" during the period of the Alalaḫ Level VII archive cannot be accepted (1994:40). While it may not be possible to identify "farmsteads" of the type discussed by William Leemans in Babylonia (1982), we may, nonetheless, assume the existence of a number of small villages during both the Middle and Late Bronze Ages. Furthermore, many of the large and medium tells in the 'Amuq Valley were, no doubt, themselves small villages during the Early Bronze Age.

settlement, which as in earlier periods was referred to as an *ālu* and was often situated on a tell. These settlements were surrounded by a network of the secondary settlements, which were identified as *dimātu* (sing. *dimtu*), though they were on occasion also referred to simply as *ālāni* (see Zaccagnini 1979:20, n. 23). These two settlement types and the various terms for elements within both, as well as their parallels with Old Babylonian terms, are listed in Table 16 above.

According to the Nuzi texts, a constellation of *ālāni* and their territory, and the *dimātu* and their territories within the territories of each *ālu* were identified as a *ḫalṣu* or "district" (Zaccagnini 1979:15). But as Zaccagnini noted, although some *dimātu* were acknowledged to be "outside the perimeter (*limītu*)" of an *ālu*'s territory, one or another of the *ālāni* were still considered responsible for the safety of *dimātu* in the "open country" (*ṣēru*; ibid., fig. 1, p. 21). In its abstract form, therefore, this arrangement of settlement as a hierarchically nested settlement pattern was identical to those in earlier periods: a single major *ālu* with its sustaining area of cultivable land was surrounded by other *ālāni* (sometimes referred to as *dimātu*) and their sustaining areas; these *ālāni* were often surrounded in turn by their own cluster of even smaller settlements which were also identified as *dimātu* (see Wilkinson 2003:fig. 6.16).

The Nuzi texts provide two other insights regarding the settlement pattern of the kingdom of Arrapḫe. The first observation is that the territory of Arrapḫe was not devoid of significant landscape features, which required that the idealized settlement pattern presented above be modified as necessary. The landscape was broken up not only by rivers but also forests (ibid., 11ff.). This adaptation of the settlement pattern would have been similar in principle to what was required in areas where hills and mountains more severely interrupted the settlement pattern, as in the Levant. The second observation is that the Nuzi texts illuminate how settlements within the territory functioned as part of a defensive network, which is discussed below.

6. Late Bronze Age Ugarit

Studies of the settlement of the kingdom of Ugarit (e.g., Saade 1979–1980; Heltzer 1982; Liverani 1982) and recent analysis of its topography by Wilfred van Soldt (1996; 1997; 1998; 1999) permit a closer examination of the settlement landscape of the kingdom of Ugarit. The kingdom, which is estimated at approximately 2,000 km^2 in size, was primarily bounded by the Jebel al-Aqra' (Mt. Ṣapanu) to the north, the Jebel el-Ansariye range to the east, and Mt. Siyanu to the south (see van Soldt 1997). Although it was comparable in size to Mukiš, Mario Liverani suggests that only about one-third of this area was cultivated (1982:251). Van Soldt summarizes the characteristics of the settlement of the kingdom of Ugarit based on Ugaritic toponyms:

> . . . in the mountainous northern and northeastern provinces of the city-state of Ugarit only small villages were situated . . . in the center of the state, with its hilly countryside, the towns were generally bigger . . . [and] really large towns have to be sought in the fertile plains to the south and southeast of Ugarit (van Soldt 1999:760).

Van Soldt's studies also assess the evidence for the identification of districts within the kingdom of Ugarit. The precise Ugaritic term for "district" is unknown. The identification of districts within the kingdom of Ugarit is based primarily upon recurring groups among lists of towns and that the "names of towns are sometimes followed by another name" (van Soldt 1999:764). Unfortunately, he concludes that while several such districts must have existed, neither the terms traditionally associated with districts, which include *ʾarr* (the southern district south and southeast of Ugarit), *ǵr* (the "mountain" district east and north of Ugarit), Ṣapunu (for the north) and Ḫuraru, nor the lists of settlements associated with them enable the precise delineation of these districts (ibid., 763ff.). Nevertheless, van Soldt suggests that a metropolitan district including Ugarit and several other towns must be added to these three.

While these observations represent the limits of the consideration given to the settlement pattern of the kingdom of Ugarit to date, Ugarit's significance for the study of Bronze Age settlement patterns is, however, its preservation of a lexicon of Northwest Semitic terms for Levantine settlement types. Although a number of cognates of Ugaritic or Northwest Semitic terms for settlements continued to be employed in the south during the Iron Age, as evident in the Hebrew Bible (e.g., Heb. *qiryat, gath, kefar*, and *migdāl*), these terms seem to have been prevalent throughout the Levant as early as the Middle Bronze Age. Among the terms found in Ugaritic texts are *qarītu, ʿr, kaparu, pdr, ʿamqu, ḥl, ḏmr, magdalu*, and *gittu/dimtu*.[22]

Based on the examples of Bronze Age settlement patterns discussed above, these Ugaritic terms can be correlated with at least six major settlement types that were characteristic of the Levant during the Middle

[22]Vocalizations, primarily from the *UVST*, are provided only for those that have been suggested.

and Late Bronze Ages (Burke 2007). These settlement types arranged in relative size from largest to smallest include: (1) large well-fortified centers (e.g., *qarītu*, *ʿr*, or *pdr*), (2) smaller fortified towns (e.g., *dimtu*), (3) unfortified villages (e.g., *kaparu*), (4) fortresses (e.g., *ʿamqu* or *ḫl*), (5) military watchtowers (e.g., *magdalu*), and (6) rural agricultural estates (e.g., *gittu*). Many of the Ugaritic terms are cognates of Old Babylonian terms (see Table 16 above).

a. Large Fortified Towns

The basic term for town in Ugaritic was *qarītu* (*UVST*, pp. 175, 286), which was semantically equivalent to the Akkadian term *ālu* as demonstrated by the Ugaritic *Syllabary A Vocabulary* (entry 39). The word occurs, however, as both *qrt* (*DULAT*, pp. 712f.) and *qryt* (*DULAT*, p. 715), both of which may be cognates of Akkadian *qarītu* meaning "storeroom" or "granary" from the Old Babylonian period onward (*CAD* Q, pp. 132f.). The restricted occurrence of *qrt* in administrative texts and as a reference to the capital (i.e., Ugarit), when compared to the use of *qryt* in literary texts, suggests the possibility that *qrt* featured a specialized meaning related to storage or some other administrative function. However, the exact implications of these terms are unclear. Nevertheless, based on numerous references to belligerent acts against *qrytm*, it is probable that sites identified by this term were well fortified (see refs. in *DULAT*). One Ugaritic toponym, which featured two spellings, is undoubtedly related to *qrt* (see *Qarātu* and *Qurṭu* in *RGTC* 12/2, pp. 224f., 230).

Continued though limited use of the term during the Iron Age in Phoenician and Hebrew suggests the association of this term with prominent towns. The standard Phoenician term for town and the element of at least the most famous Phoenician settlement was *qart* (e.g., Qart-Ḫadašt > Carthage). The Hebrew cognate *qiryat* was also the name of several important pre-Israelite towns: Qiryat-Yeʿarim (e.g., Josh 9:17), Qiryat-ʿArba (e.g., Gen 23:2), Qiryat-Huzoth (Num 22:39), Qiryat-Sepher/Sanna (e.g., Josh 15:15, 15:49), Qiryat-Baʿal (Josh 15:60), and Qiryataim (Num 32:37). That three of these settlements were renamed during the Iron Age, including Qiryat-Arba (later Hebron, see Gen 23:2), Qiryataim (later Baʿalah, see Josh 15:9), and Qiryat-Sepher/Sanna (later Debir, see Josh 15:15), suggests that *qiryat* toponyms predate the Iron Age, and thus were contemporaneous with Ugaritic *qrt/qryt*. Hebrew occurrences of *qiryat* also connote the meaning of the Ugaritic term since in Hebrew *qīr* means "wall," although usually in reference to the wall of a house. It seems possible, therefore, that the term *qarītu* was often used in Ugaritic and Northwest Semitic to denote large fortified settlements that functioned as administrative centers.

b. Secondary Fortified Towns

Two Ugaritic terms can be identified with secondary fortified towns in the kingdom of Ugarit (i.e., the tier below Ugarit). The most common Ugaritic term for town was *ʿr* (*DULAT*, p. 178), which was used primarily in literary texts. References to attacked and conquered *ʿrm* (pl.) allow the inference that they were fortified (since unfortified towns do not need to be besieged), although their defenses are not described in detail and it is not possible to infer the specific characteristics of their fortifications.

A second and parallel Ugaritic term to *ʿr* was *pdr*. Although it occurs less frequently and its origin is unknown, the term appears to have been used to identify a number of secondary fortified towns; one text refers to seventy such towns (see *DULAT*, pp. 662f.). That such settlements were fortified can be inferred from references to *pdr* being besieged. Two settlements in the kingdom of Ugarit incorporated this term into their toponyms (see *RGTC* 12/2, p. 219).

c. Unfortified Villages

Three terms appear to identify villages that infilled the landscape around fortified towns. The first term is *kaparu* (*DULAT*, p. 452), which is clearly a cognate of the Akkadian word *kaprum* (see pp. 109f.). Based on the characterization of the Old Babylonian *kapartum*, *kaparūma* may be identified as unfortified. No Ugaritic toponyms attested include this term.

A second term, *maškanatu* (*UVST*, p. 181), which is a cognate of Akkadian *maškanum*, may also refer to villages. Nevertheless, its meaning is uncertain since it occurs only once in Ugaritic. The term is thought to refer to either some sort of "habitation(?) field" (*PRU* 6 47:5) or as in Akkadian to a "threshing floor" or other type of "emplacement." A third Ugaritic term for village was *madinatu* (*DULAT*, p. 528), for which there is an Aramaic cognate, *medinah*. Nevertheless, the term is rare in Ugaritic.

d. Fortresses

No fewer than three terms appear to refer to fortresses in the kingdom of Ugarit, and for this reason their nuances are unknown. John Huehnergard suggests that *ʿamqu*, which was usually employed as an adjective (*DULAT*, p. 165), was one term for "fort" or "stronghold" (*UVST*, p. 160). The term probably originated with the Akkadian noun, *emūqu*, meaning "strength." One town in the kingdom of Ugarit, "Kumba, the fort," was identified with this term (see *PRU* 3:152:3).

Another Ugaritic term with the same semantic range was *ḥl* (*DULAT*, p. 359). In addition to the meanings "strength" and "vigor," this term is also translated as "fortress, stronghold" or "tower, farm, or estate" (e.g., RS 1.16 I 7–9). Two Ugaritic toponyms employed the term (see *RGTC* 12/2, pp. 106), and the Hebrew cognate is often translated "rampart" or "fortress" (*BDB*, p. 298; as "rampart" in 2 Sam 20:15 and Ps 26:1 [NRSV]).

Yet another Ugaritic term, *ḏmr*, is also translated "fortress" (meaning no. II in *DULAT* p. 287). Its verbal root means "to protect, guard." No toponyms are attested using this term.

e. Military Watchtowers

The Ugaritic term *magdalu*,[23] which like *dimtu* and *gittu* was identified with the logographic writing of AN.ZA.GÀR in Akkadian texts at Ugarit, is clearly to be identified as a "tower" or "watchtower" (*DULAT*, p. 530). The cognate of *magdalu* is Akkadian *madgaltu*, which is also identified as a "watchtower" and is attested among Middle and Neo-Assyrian sources (see *CAD* M/1, p. 16). At least one place name, URUMadgalte, is attested in the Middle Assyrian period, and the term occurs as an Akkadogram in Hittite. The verbal root of the Akkadian term is *dagālu*, meaning "to look," and its earliest attestation is during the Old Babylonian period (*CAD* D, pp. 21–25). Although an entirely different etymology, related to the root *gdl*, has often been postulated for this word (e.g., van Soldt 2005:177), it seems that the term is actually related to the Akkadian term and that the change in spelling was the result of a metathesis of the consonants *gimel* and *dalet*.

As suggested by the elevation and distribution of these settlements, the purpose of *magdalūma* was almost certainly defensive (Burke 2007). These settlements almost always featured a viewshed over routes to and from major centers such as Ugarit that was provided by the natural topography. At least three *magdalūma* in the kingdom of Ugarit can be identified by means of the Arabic cognate *májdal*. One of these, *Magdalā*, had been previously identified with the modern town of Mejdel near Bdama located to the northeast of Ugarit in the pass between the Jebel Aqra' and the Jebel al-Ansariye (see van Soldt 1998:742). The other two are Májdal-Kikhia and Májdal-Saleh. All three *magdalūma* were located in passes leading into the kingdom of Ugarit through the Jebel al-Ansariye range on the east (see nos. 1, 2, and 3 in Burke 2007).

[23]Vocalization uncertain.

The limited attestation of this site type in Akkadian suggests that the landscape of Mesopotamia did not, by and large, require the construction of towers, which served an almost exclusively military purposes. In Mesopotamia such a function was usually served by the highest building within a tell settlement, which would have provided the best vista of the surrounding countryside, and in many cases similar structures were probably used throughout the Levant. Although more than sixty Arabic *majādīl* toponyms can be identified throughout the Levant, it is important to note that not all *magdalūma* would have been so named and the location of other *magdalūma* are long since lost to us due to the lack of preservation of a related toponym. Only on occasion did *magdalūma* develop into larger towns (e.g., EA 69, 70, 185, 186, 234, and 256). These settlements, by virtue of their nearly continuous occupation, are the ones that are therefore the easiest to identify by means of the preservation of the Arabic toponym *májdal*.

f. Rural Agricultural Estates

The most commonly discussed Ugaritic settlement type is the *gittu* or *gintu* (pl. *gittūma*). Although archaeological evidence for the large number of settlements identified as *gittūma* is still wanting, Michael Heltzer's work on *The Internal Organization of the Kingdom of Ugarit* remains the central study of this settlement type (1982:esp. pp. 49–62). He reports that seventy-eight of about two hundred villages (39%) in the kingdom of Ugarit were identified as *gittūma* (ibid., 52f.).

Heltzer concludes that the Ugaritic term *gittu* is most likely the semantic equivalent of the Akkadian term *dimtu*, which first "appears in an agricultural context" during the Old Babylonian period (1982:49; see also Koliński 2001). Although *dimtu* was used in Akkadian to refer to both a tower and a secondary fortified town (see discussion in previous sections), as it was used in Akkadian texts at Ugarit (for list, see Heltzer 1982:50, n. 10; see also *DULAT*, p. 274), according to Heltzer *dimtu* was employed in Ugaritic as a synonym for *gittu* (1982:52). The term *dimtu* appears to be preserved among several placenames in the kingdom of Ugarit, such as, for example, Dumatu, Dumatu-qidši (RS 17.335+), Dumat(u)-'Agimi (see van Soldt 1998:741). [24]

[24]Dumatu is usually identified with T. ed-Damat located 15 km north-northeast of Ras Shamra (see Yon 1997:fig. 6, p. 21). Gabriel Saade also suggests identifying the location of one of the other *dumātu* with T. al-Qifr 12 km to the east (1979–1980:220). Lending support to these etymologies is

Like the term *dimtu*, the semantic range of *gittu* was broad; it could mean "wine or olive press" but also "farmstead" or "estate" (*DULAT*, p. 310ff.). The context of references to *dimātu* in Akkadian texts at Ugarit, which were identified with the determinative for "house" or "building" (e.g., É.AN.ZA.GÀR), suggests that *gittūma* consisted of a complex of buildings (Heltzer 1982:50–52). They were usually identified by association with a nearby village, an individual, a profession, or even a particular city gate (ibid., 52ff.; see also *RGTC* 12/2, pp. 82–95), and multiple *gittūma* were associated with larger villages (ibid., 56). They could serve a number of purposes including the storage of tools, animal fodder, and agricultural produce, and the stabling of animals (ibid., 56–61).

At least two Hebrew terms, in addition to the Hebrew cognate *gath*, that relate in some way to the various functions identified for Ugaritic *gittūma* further underscore the functions associated with the Ugaritic term, although precisely how these terms related to the Ugaritic term remains to be demonstrated. One term, *gn*, is identified as "garden" (*HALOT* 1, p. 198). Another, the verb *gnn*, meaning "to protect" (*HALOT* 1, p. 199), has an Akkadian cognate in the nominal form, *ganūnum* (*CAD* G, pp. 42f.). This Akkadian term, meaning "storage room or building," occurs as early as the Old Babylonian period and was in use until the Neo-Assyrian Period. Its range of usage and similarity suggest that it is the most likely origin of the Ugaritic term. One of the nuances of the term also identifies buildings so-named with "living quarters" (ibid.).

Parcels of land were also identified with each Ugaritic *gittu* that was under the direct control of an individual or settlement (ibid., 62). Thus, *gittūma* represent a type of settlement, often small and probably unfortified, that was associated with a plot of land within the vicinity of a larger settlement. To what extent *gittūma* were permanently inhabited is uncertain. Nevertheless, during much of the year, if not year round, they were probably inhabited by field hands and their families.

To the extent that confusion existed in Ugaritic between the use of terms *dimtu* and *gittu*, it may have resulted from the identification of these settlements with structures that were identified as towers (i.e., *dimtu* or even *magdalu*). The contexts of references to *gittūma* suggest, therefore, that *gittu* did not refer to a particular building but instead to a rural agricultural estate, hence a distinct settlement type.

the toponym Dimat, a settlement in the province of Mukiš during the Middle Bronze Age (see *RGTC* 12/2, p. 64).

7. The Southern Levant

General overviews of the settlement pattern of the southern Levant during the Middle Bronze Age include those by Amihai Mazar (1990:see pp. 176–79, for the MB IIA, and 97f. for the MB IIB–C), Aharon Kempinski (1992d:see pp. 166–67, for the MB IIA and pp. 82–84 for the MB IIB–C), and Helga Weippert (1988).[25] Demographic studies of Middle and Late Bronze Age settlement have also been undertaken (Broshi and Gophna 1986; Gonen 1984). Nevertheless, the interpretation of the settlement pattern of the southern Levant that has prevailed for nearly two decades is the result of the work of William Dever (1987:152ff.) and his student Wade Kotter (1986).

Based on Kotter's study, the following settlement hierarchy is considered characteristic of MB IIB–C settlement in the southern Levant:

1) Large urban centers (ca. 8 to 70 ha, 5%)
2) Medium-sized settlements (ca. 3 to 8 ha, 10%)
3) Villages and hamlets (ca. 0.2 to 3 ha, 85%)

While this hierarchy most closely resembles the settlement hierarchy identified for the ʿAmuq Valley during the Middle and Late Bronze Ages, it is essentially a scaled-down variant of the Old Babylonian settlement hierarchy that Adams identifies. Dever suggests, this pattern was especially characteristic of "a highly urban culture" during the late Middle Bronze Age (1987:153), which has come to be identified with "city-states" that dominated the Middle Bronze Age landscape. Yet neither site size, Thiessen polygons, nor the presence of fortifications at centrally located sites provide unequivocal evidence for the city-state model that is so frequently invoked (see discussion below). Nevertheless, use of these techniques has led to widespread identification of both Hazor and Ashkelon as the "leading" Middle Bronze Age city-states in the southern Levant and careful avoidance of their identification as kingdoms. Other smaller city-states in the highlands, along the coast, and in Transjordan have also been suggested using similar means (e.g., Sharuhen according to Oren 1997b; and Kabri according to Peilstöcker 2003). The

[25] Although similar studies have yet to be undertaken for the northern Levant, site and survey data from the regions around various Bronze Age settlements in the north are available (see recent bibliography by Gunnar Lehmann, 2002). These include the ʿAmuq Valley (Braidwood 1937; Haines 1971; Magness-Gardiner 1994; Yener et al. 1996; Yener 1998; Yener et al. 2000; Casana and Wilkinson 2005b; 2005a), the Akkar Plain (Thalmann 2002), the Beqa Valley (Kuschke et al. 1976; Marfoe 1998), the Syrian Plain (Calvet 2002), the Orontes Valley (Courtois 1973), the region around Damascus (von der Osten 1956), and the Hauran (Braemer 1984; 1993; Betts et al. 1996).

settlement patterns associated with these polities are evaluated in this section.

a. The Kingdom of Hazor

Since Yadin's excavation of Hazor and in light of the biblical passage regarding its role in Canaan (see Josh 11:10), Hazor has often been mistakenly regarded as the capital of the only true kingdom in the southern Levant during the Middle Bronze Age. This characterization has been further advanced in light of a number of references to Hazor found in the Mari texts (see ARM VI 23 and 78, VII 236, XII 747, XXIII 556, XXV 43 and 103, XXVI 375) and because Hazor (ca. 66 ha) dwarfs all settlements in the southern Levant other than Ashkelon during this period. The characteristics that are usually noted include that Hazor (1) was the largest fortified site in Canaan, (2) is referenced in the Mari texts, (3) has produced a number of cuneiform tablets, and (4) conforms to northern Levantine architectural and artistic traditions. Other archaeological characteristics that would support the identification of the kingdom of Hazor have not, however, been adequately discussed.

While such characteristics definitely support the identification of Hazor as belonging to the cultural sphere of the northern Levant, they fall short of the criteria necessary to identify the site as the capital of a kingdom during the Middle and Late Bronze Ages. A number of questions remain, therefore, to be answered regarding the kingdom of Hazor. What settlements, for example, were included in its territory? If we presume that Tel Dan, Kedesh, or Tel Kinneret, for example, belonged to Hazor, what political basis other than proximity is there for such an identification? Why should Middle Bronze Age Hazor be identified as a kingdom, while during the Late Bronze Age the site is commonly identified with other settlements as a "city-state"? Furthermore, with more than 350 km^2 of suitable agricultural land in the Upper Jordan Valley surrounding Hazor from Tel Dan to the Sea of Galilee, not including the Golan Heights or land along the west side of the Sea of Galilee, what constituted its economic raison d'être? In the final analysis, while there may be little doubt that Hazor was a leading city of Canaan at the head of a territorial kingdom comparable to those in the northern Levant, much more work must be done to address the types of evidence that are expected for such a kingdom.

b. MB IIC Highland Polities

The highlands of the southern Levant present a different example of settlement and defensive strategies during the MB IIC. Although on the surface settlement in the highlands does not appear to have adhered to the basic settlement pattern attested in the plains and valleys of the northern or southern Levant (or Mesopotamia), settlement in this region was essentially an adaptation of the archetypal settlement pattern identified above. Like the adaptations that were made for this settlement pattern in Mesopotamia, where the role that rivers played in communication and transport between settlements were accounted for, settlement in the highlands was similarly forced to account for the location of immovable features in the landscape and the availability of cultivable land (Kotter 1986:490ff.). Such considerations were also factors in the settlement of Bronze Age kingdoms in the northern Levant. Ugarit, for example, which encompassed a total area of approximately 2,000 km^2, probably only had one-third of this area, or about 700 km^2, available for cultivation owing to the topography of its territory (Liverani 1982:251). Similarly, Tunip's actual area of cultivation would have been restricted to about 600 km^2, and that of Mukiš to about 1,000 km^2. These areas can be compared to those calculated for Ashkelon (ca. 1,700 km^2; see pp. 125ff.) and Hazor (see above). There is, therefore, no basis for assuming that insufficient cultivable land was available for the emergence of distinct kingdoms in the highlands, although these polities were smaller.

Settlement in the highlands of the southern Levant during the Middle Bronze Age was faced with a quite different set of circumstances since the terrain did not permit agriculture on a scale comparable with that of other Levantine kingdoms. In an attempt to account for these factors, interpretations of the development of highland settlement during the late Middle Bronze Age have varied considerably. However, few if any of these interpretations account for the role of horticulture and viticulture in the economies of the region before the Iron Age. The only consensus that exists is that the majority of settlement in this region occurred during the MB IIC with the emergence of fortified settlements at Shechem, Shiloh, Jerusalem, and Hebron, where some type of elite class must have resided (see Ilan 1995:304f.).

Less certain, however, are interpretations positing the emergence of autochthonous, polymorphous societies, which undertook the construction of ramparts as part of a process of conspicuous consumption (e.g., Bunimovitz 1992:225ff.; Finkelstein 1992:207; Herzog 1997a:268f.). Nor does Michael Rowton's model of "enclosed nomadism" (1974), adopted by Israel Finkelstein (1993b), provide an improved understanding regarding the Middle Bronze Age populations and subsistence in the highlands. All such hypotheses remain unconvincing because they imply a unique process of settlement in the highlands that bears no relationship to lowland settlement, despite similarities in their material culture and architectural traditions.

Neither artifactual nor architectural evidence suggests the presence of a different culture or ethnic population in the highlands that necessitates an altogether separate explanation for the region's settlement. With regard to fortification systems, as argued in chapter 3, their construction in the highlands must be viewed as the adaptation of the defensive architecture characteristic of the lowlands during the first half of the MB IIB to the local geographic and environmental conditions (see pp. 81ff.). This is also true with respect to the use of stone for construction. This assertion is supported by the quantity of stone used in defensive architecture in the highlands, which appears to have been directly proportional to what was readily available, a phenomenon comparable to the occurrence of stone glacis in the regions with the greatest precipitation (see pp. 55f.). Nevertheless, insofar as we are able to determine, in every other manner the MB IIC fortifications of both regions were identical, employing earthen ramparts (e.g., Shechem and Shiloh), stone revetment walls (e.g., Shechem), rectangular towers (e.g., Jerusalem), and even the six-pier gate (e.g., Shechem). In addition to fortifications, the settlement pattern of highland settlements evinces the influence of lowland settlements, thus undermining theories that imply the development of an indigenous Middle Bronze Age settlement pattern in the highlands. For example, the largest highland settlements were sufficiently distant from each other to sustain their own populations, while smaller, probably politically dependent settlements filled in the landscape around them.

If Jerusalem and Shechem are identified as the largest MB IIC polities in the highlands, then it may be possible to reconstruct the MB IIC settlement pattern of the highlands and to suggest the following settlement hierarchy:

1) Primary fortified towns at Jerusalem (ca. 4 to 5 ha?) and Shechem (ca. 4 ha), which were located roughly 50 km apart and were about 4 or more ha in size. (Possibly also at Hebron [*Qiryat-ʻArba*], ca. 4.9 ha.)
2) Smaller, secondary fortified towns such as Abu Zarad (2 ha), Beth-El (2 ha), Beth-Zur (1.5 ha), Dothan (< 4 ha), en-Najjar (3.2 ha), Far'ah North (3.1 ha), Khirbet Marjama (3 ha), Khirbet Qumei (2.2 ha), Khirbet Kheibar (2 ha), Shiloh (1.7 ha), and Khirbet 'Urma (1.5 ha). These sites, which were less than 4 ha in size, were located along major routes through the highlands within 25 to 30 km (a single day's journey) of the larger centers.
3) Unfortified villages such as Manahat (ca. 3 ha) can also be identified (see small sites [S] in Broshi and Gophna 1986:79–82).
4) Small fortresses of less than 1 ha in size such as Jenin (1 ha), Khirbet Qarqaf (0.8 ha), and el-Kebara (0.2 ha). Kebara featured a 5 m thick wall and Qarqaf was perched 70 m above the main road.
5) Military observation towers strategically located along routes to major settlements (possibly Májdal Bani Fadl [no. 40] from Shechem and Khirbet Májdal-Ba' [no. 43] from Hebron, see Burke 2007).
6) Very small, probably unfortified, rural agricultural/horticultural estates of less than 1 ha in size (see very small sites [VS] in Broshi and Gophna 1986:79–82), that filled in the landscape around larger settlements.

This settlement hierarchy when taken in combination with the observation that the fortifications of Shechem, Shiloh, and Tell el-Far'ah North (Lederman 1985) and those of Jerusalem and Hebron are remarkably similar in construction supports the identification of these settlements as the leading highland settlements during this period. However, the distribution of these sites reveals two clusters of settlements, which are identifiable as the distinct highland polities of Shechem and Jerusalem. Their identification is supported by the fact that these two settlements are the earliest highland settlements mentioned among Middle Bronze Age texts (Jerusalem in the Execration Texts and Shechem as *Skmm* in Sesostris III's inscription) and that both were clearly the leading highland settlements until the Amarna Period. Nevertheless, other aspects of their material culture should be further examined to test this hypothesis.

The settlement pattern of the highlands described above consisted not only of similarly fortified settlements but relied upon the adaptation of the highly-centralized settlement pattern that was characteristic of the lowlands to the terrain of the highlands. On the one hand the smaller size of highland settlements was dictated by the limits of technology to sustain this population. Both water and arable land were in limited supply in close proximity to the largest and most centrally located settlements. On the other hand, at the end of the MB IIB it appears that the military concerns of highland settlements were considered identical to those in the lowlands. For this reason ramparts continued to be built, extensive water systems were added within these towns, and a centralized settlement pattern was maintained.

In light of the lack of direct evidence for the emergence of an autochthonous elite and the recognition of the adaptation of settlement strategies comparable to those attested in the lowlands, how should the settlement pattern of the highlands during the MB IIC be understood? Given the data, the impetus for the settlement pattern appears to have been external to the highlands. The evidence for the growth of the populations of highland settlements, which climaxed in size during the MB IIC, suggests that the timing of this phenomenon relates directly to lowland settlement since intensive settlement in the highlands occurred only *after* most available land in the lowlands had been settled (a process identified as MB II settlement infilling). Although how this process occurred remains to be determined, a number of processes such as colonization, trade and exchange, and the gradual infiltration of lowland settlers may have contributed to this process.

c. *Late Bronze Age "City-States"*

The study of Late Bronze Age settlement and the identification of Canaanite polities are usually viewed as critical for understanding Middle Bronze Age settlement and political organization. This is because the location, character, and extent of Late Bronze Age polities are usually retrojected in order to fill the void created by an absence of textual references that would enable the identification of Middle Bronze Age polities in addition to Hazor in the southern Levant. A number of studies, therefore, have attempted to address the nature of Late Bronze Age polities and their boundaries (e.g., Bunimovitz 1989; 1993; Finkelstein 1993b; 1996b; Na'aman 1997; 1999; Jasmin 2006). Such studies have almost unanimously adopted the term *city-state* for Late Bronze Age polities.

Based on settlement pattern data for the southern Levant for the second half of the Middle Bronze Age (MB IIB–C), discussed above, and the terminology invoked for polities during the Late Bronze Age, Middle Bronze Age polities are also often identified as *city-states*. This view is summarized by William Dever:

> Palestine, by contrast, remained at an intermediate level of political development, which is usually referred to as that of the *city-state*. Although the term *city-state* is frequently used, it is rarely defined. Often the implication seems to be that while the regional urban centers each control their own hinterland, they are in turn united in a larger centralized confederation—that is, they constitute a true state. Yet there is little evidence of that in Palestine of the Middle Bronze Age. Rather, it seems to have been bound together only by what we may call a common southern Canaanite culture. Politically it probably remained divided: each city-state enjoying quasi-independence and dominating the surrounding countryside, most likely rivaling other urban centers (Dever 1987:165).

In this manner MB IIB–C city-states such as Hazor, Sharuhen, and Kabri, mentioned above, are posited. Those dissatisfied with the term *city-state* have adopted terms such as *polity* or *gateway* (Knapp 1989:145, n. 58; Ilan 1995:305). Yet employing such terms, in an effort to avoid the imagery associated with traditional terms such as kingdom or city-state, offer no explanatory value; they address neither the nature of the state nor its relationship to neighboring states.

Although it is necessary to identify an appropriate terminology for southern Levantine polities, all of the above terms exaggerate perceived differences between political organization in the northern and southern Levant during the Middle and Late Bronze Ages (see observations in Bunimovitz 1993). This is an attempt to avoid identifying sites in the south with discernible *kingdoms* comparable to northern Levantine kingdoms of the MB IIB–C, such as Yamḫad, Qatna, Byblos, and Ugarit. Yet none of these northern kingdoms, many of which were made Hittite vassals during the Late Bronze Age, is ever identified by scholars as a *city-state*. Thus, the assumption that few if any large political centers actually existed in the southern Levant during the Middle Bronze Age in contrast to the evidence for political organization in the north is an argument from silence.

The artificial contrast drawn between Middle Bronze Age polities at the two ends of the Levant is dependent upon two methodological errors. The first is an overreliance on textual sources for the identification of Middle Bronze Age polities, whether in the north or the south, in order to establish a hierarchical settlement pattern that can actually only be identified archaeologically. This is most apparent for Hazor, the largest Middle Bronze Age settlement in the southern Levant and the only such settlement aside from Laiš (Tel Dan) that is mentioned in the Mari texts. While on the one hand textual sources for the northern Levant have shed light on major polities like Yamḫad and its vassals such as Alalaḫ, the lack of references to polities in the southern Levant, aside from Hazor, has resulted in the assumption that Hazor was the only southern polity of such size. This interpretation has often been argued on the basis of rank-size analysis of MB IIB–C settlement and a later reference to Hazor at the "head of all those kingdoms" in Joshua 11.

The second reason for the erroneous contrast drawn between political organization in the northern and southern Levant during the Middle Bronze Age relates directly to the prevailing interpretation of the political organization of the southern Levant during the Late Bronze Age. The general understanding of the character of political organization in Canaan during the Late Bronze Age is simply retrojected to account for a dearth of textual evidence for political organization in Canaan during the Middle Bronze Age. However, it can be argued that, in fact, the nature of political organization in Canaan during the Late Bronze Age is itself inadequately understood since it relies upon a single snapshot of circumstances in the middle of what is clearly a dynamic process lasting from the end of the Middle Bronze Age through the Iron I. The snapshot is provided by the Amarna texts, which were written in the mid-fourteenth century B.C. and a large number of which record correspondence between Egypt and its Canaanite vassals.

The Amarna correspondence has served as the basis for suggesting that the main characteristics of Late Bronze Age polities, referred to as *city-states*, were identical to those of Middle Bronze Age polities. In the Amarna texts Canaanite rulers are described as mayors (Akk. sing. *ḫazannu*) ruling over territories or "lands," which were designated by the traditional terminology used to describe Old Babylonian kingdoms and territories (e.g., Akk. *māt-TN* or TN^{KI}). In the Amarna texts the lands of Jerusalem (*Urusalim*), Gezer (*Gazru*), Ashkelon (*Ašqaluna*), and Lachish (*Lakiša*) are all identified (see EA 287), as well as those of Seir[?] (*Šeru*, EA 288), Shechem (*Šakmu*, EA 289), and Jaffa (*Yapu*, EA 138). These are among the territories routinely identified by archaeologists as *city-states*. The same terminology that was used for southern Levantine polities was also employed for northern Levantine polities including Byblos (*Gubla*, EA 98), Ugarit (EA 98), Yarimuta (EA 85), Amurru (EA 101),[26] and Mitanni (*Naḫrima*, EA 288). Thus, it appears that Egyptian New Kingdom officials maintained the terminology familiar to Canaanites in the administration of their Levantine vassals. The collective territorial possessions of the pharaoh were described as "the lands of the king" (e.g., EA 286 passim), consisting of a multiplicity of distinct territorial units.

Despite the forced allegiance of Levantine kingdoms to the Egyptian throne, many of the Late Bronze Age settlements mentioned in these texts are traditionally characterized as *city-states* following Dever (e.g., Leonard 1989; Bunimovitz 1995:326f.). Yet the term *city-state*, if it is to be used consistently, is as inappropriate for the Late Bronze Age as it is for the Middle Bronze Age. Behind the choice to apply this term to Late Bronze Age Canaanite states is the fundamental comparison that is implied between Canaanite polities and small Greek city-states of the first millennium B.C. before Alexander the Great's unification of Greece.

Thomas H. Charlton and Deborah L. Nichols have addressed this phenomenon in the following manner:

> The term *city-state* links the phenomenon of the state and a particular kind of settlement pattern. In general we understand city-states to be small, territorially based, politically independent state systems, characterized by a capital city or town, with an economically and socially integrated adjacent hinterland. The whole unit, city plus hinterlands, is relatively self-sufficient economically and perceived as being ethnically distinct from other similar city-state systems. City states frequently, but not inevitably, occur in groups of fairly evenly spaced units of approximately equivalent size (Charlton and Nichols 1997:1).

The criteria for city-states, therefore, include their being (1) small in size, (2) territorially bounded, (3), ethnically distinct (to some extent), (4) economically self-sufficient, (5) evenly distributed, and (6) politically independent states. Excluding the characteristics shared with kingdoms, city-states were essentially characterized by their *size*, *autonomy*, and perceived *ethnic* identity. One might argue that city-states represent, therefore, the smallest territorial unit possible for a state, namely a single, economically self-sufficient, politically independent unit that perceives itself to possess a separate ethnic character (e.g., Sparta).

During neither the Middle nor Late Bronze Age was the average Levantine settlement (i.e., those below the first tier) politically autonomous or "perceived as being ethnically distinct" from its neighbors to the extent required by this widely accepted definition of the city-state. Textual sources for the Middle Bronze Age do not provide a basis for the identification of city-states. Indeed, the weight of the evidence for political organization in the northern Levant, namely from the Mari texts, suggests the identification of these states as kingdoms. The examples of Hazor, discussed above, and Ashkelon (see discussion below) in the southern Levant reveal the ephemeral nature of the distinction drawn between political organization in the north and south; all of the differences that have been noted are related to differences in the scale (i.e., the size and number of settlements).

---

[26] Amurru is one of the few references that featured the plural designation of land for its territory (see EA 145).

In light of the Amarna texts it also is incorrect to characterize Late Bronze Age polities as city-states since they were neither politically nor militarily autonomous, nor is there evidence that they perceived themselves to be ethnically distinct. Instead, Canaanite polities *attempted* to exert their control over neighboring polities during the Late Bronze Age despite, or perhaps for, the purposes of their Egyptian overlords. It is the assumption of independence in military affairs, which is the result of a reliance on the Amarna texts, that has been mistaken, therefore, for political autonomy during the Late Bronze Age and retrojected in order to facilitate the characterization of MB IIB–C polities as city-states.

When considering the nature of Levantine polities we are well served to remember that once the Greek city-states of the Iron Age were unified, as they ultimately were under Philip of Macedon, the very identity of the seven hundred and fifty Greek city-states—their political and military autonomy and to some extent their ethnic isolation—disappeared (I. Morris 1997:103). Interpretations of Late Bronze Age polities should be shaped, therefore, by our understanding of Middle Bronze Age polities, and *not* vice versa. Furthermore, subtle differences in the cultural assemblage of the northern and southern Levant should not be assumed to reveal substantial differences in political organization (contra Dever 1987:165). Close examination reveals more in common than different in nearly every aspect of their cultural assemblages.

d. The Patrimonial Household Model (PHM)

In the absence of models derived from contemporaneous periods in neighboring regions, such as Mesopotamia, a variety of theoretical models have been suggested to identify the nature of Levantine society and economy during the Middle and Late Bronze Ages. The political and social organization within this region during these periods is, however, most easily understood by means of the patrimonial household model (PHM) advanced by J. David Schloen (2001) based upon native terminology from the Levant. The PHM yields a much more reliable and consistent picture of the development of states within their social context throughout the Levant during the Bronze Age.

As Schloen notes, terminology used in Ugaritic texts reveals "a social hierarchy consisting of the king, his immediate servants, and their servants" (2001:252). He observes that this nested hierarchy is even adapted to account for Ugarit as a vassal of the Hittite empire, whereby the king of Ugarit became the "servant" of the Hittite king:

> . . . thus the patrimonial hierarchy of households extended as far as the imperial court. The social and political organization of vassal kingdom and empire alike was structured on the simplest possible basis as a self-replicating household of households, in which the social actors' understanding of household-based relationships of authority and obedience remained the same at every level. (ibid.)

Fundamentally, as revealed in the Late Bronze Age archives of Ugarit, Levantine society had developed an extremely simple hierarchical model of domination and administration:

> The replication of the household model at various levels of the sociopolitical hierarchy provided a simple but effective means of integrating the entire kingdom as a political unit. Under Bronze Age conditions in which means of communication and transportation were limited, direct supervision and enforcement by the palace would not have been feasible. The hierarchical structure envisioned in the patrimonial household model permits decentralization of authority by means of dyadic relationships, which for any one person would have been few enough to be manageable, while still providing a simple chain of command leading up to the king. (ibid.)

The PHM also provides a framework for understanding socio-political developments in the Levant from the end of the Middle Bronze Age through the end of the Late Bronze Age. These developments center around a process that can be described as the balkanization of large territorial states of the Middle Bronze Age. This process was entirely responsible for the territorial extent of and political relationships between polities that emerged during the Late Bronze Age. In this light and by means of the PHM we can reconsider, for example, Egyptian administration of the Levant during the Late Bronze Age. New Kingdom imperial policies involved the replacement of the rulers of large territorial states of the late Middle Bronze Age (i.e., those ruling from Hazor and similar states) with a selected ruler (Akk. *rābiṣu*), while maintaining a local ruler (Akk. *ḫazannu*) within each district (Akk. *ḫalṣu*) of these kingdoms. Thus, these newly appointed individuals gained prominence by achieving an essentially equal rank with the kings of earlier territorial kingdoms.

In this respect the division of the southern two-thirds of the Levant by Egypt into two or three administrative provinces during the Late Bronze Age (see Moran 1992:xxvi ff.) resulted in the replacement of the titular and functional role formerly played by the kings of several major Levantine kingdoms at the end of the Middle Bronze Age. The administration of these kingdoms, which could previously be described by means of the phrase "the land of TN" (Akk. *māt-TN*), was replaced by several provincial centers established at nearby, smaller and more manageable towns such as Gaza, Kumidi (Kamid el-Loz?), and Sumur. These locations were probably specially selected to facilitate continued control specifically over the former vassals or districts comprising the largest kingdoms, of Ashkelon (discussed below), Hazor, and perhaps Byblos. While they did not use the Middle Bronze Age centers as their bases of operation, the settlements they chose were, in fact, located in strategic proximity to these old capitals. Because Egypt's control over territories further north was of limited duration, the establishment of other administrative centers to replace Qatna and Ugarit, which ultimately fell under Hittite rule (Klengel 1992:95), was never realized. Nevertheless, by the time this part of the northern Levant was overrun by the Hittites in the fourteenth century similar practices were employed to maintain control there. Šuppiluliuma, for example, installed his sons as administrators at Ḫalab (Yamḫad), Kar-Kemiš, and Emar (Bryce 1998:190f., 94, 203f.), and accepted oaths of loyalty from his vassals at Ugarit and Nuḫḫašše (ibid., 189). For a number of reasons, therefore, it is becoming increasingly clear that previous assumptions regarding the political organization of the Levant in the Middle Bronze Age cannot be maintained. As demonstrated by the evidence for the Middle Bronze Age kingdoms in the southern Levant, questions regarding large urban centers comparable to Qatna must be addressed by means of archaeological evidence since only a handful of textual sources exist and these are inadequate to suggest the nature of political organization in this region. Furthermore, given the extent of the textual evidence from Ugarit, the burden of proof rests on those who would assert that a model other than the PHM was in operation for any polity in the Levant during the Bronze Age.

B. Defensive Settlement Networks

The preceding discussion illustrates that settlement patterns in the Levant during the Bronze Age conformed to a nested hierarchical model composed of a limited number of settlement types each of which are described using specific terms. While the agricultural basis for this pattern has been the primary focus of studies of settlement patterns (e.g., Wilkinson 2003), the pattern can also be considered in light of the defensive role that individual settlements played within settlement networks. In order to understand how such defensive networks were formed the classification of settlement types must rely on a native understanding of different settlement types as described, for example, in Ugaritic, and not simply on criteria such as settlement size and fortification type (e.g., Herzog 1997a:268ff.).

This section demonstrates that settlement data combined with an emic understanding of site types can be employed to identify defensive networks that centered around clusters of politically affiliated Bronze Age settlements. Such networks suggest that they were organized primarily around districts, as evident in Mesopotamia. The function of such networks is corroborated by textual references to the strategic role played by settlements within their district. In this section it is suggested that defensive networks can be identified in the Levant by employing the terminology for Bronze Age settlement types discussed above, and settlement pattern data that allow for the identification of administrative or quasi-military districts.

1. Settlement Types

Despite parallels with Old Babylonian settlement patterns and site types, the evidence reviewed above for settlement types suggests that Ugaritic terms provide an appropriate framework for the classification of Levantine Bronze Age settlements (see pp. 113ff.). This is supported, on the one hand, by the frequency of the use of certain Ugaritic terms and, on the other hand, by the persistence of many of these terms in Northwest Semitic Languages such as Hebrew during the Iron Age. Thus, six primary settlement types are identified with examples of each identified among excavated Bronze Age settlements.

It is noteworthy, however, that the role of settlements, particularly with respect to defense, could change significantly over time, and the site's morphology with it. It is possible, for example, that as the political landscape changed, settlements were required to modify their defensive strategies and thus may have changed names at that time. In other instances their names did not change to accommodate their changing role. Magdala in the Amarna texts (EA 185), for example, became a substantial settlement and served as more than a military watchtower; nevertheless it retained its original toponym. Such changes were due to shifting borders and the vicissitudes of regional politics and military confrontations.[27]

[27] Koliński observes that Dimti-Dada and Dimti-Enlil were later transformed into fortresses (*dūrum*) based on a change of their names (2001:27; similarly, see Leemans 1982:247).

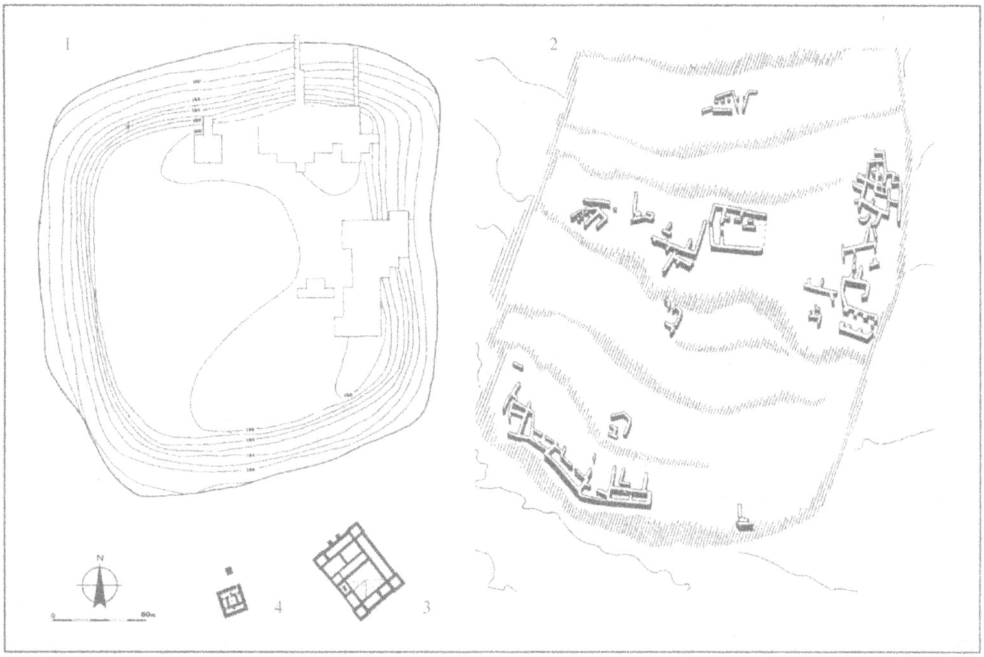

Figure 12. Examples of main MB–LB settlement types: (1) secondary fortified town (Timnah, after A. Mazar 1997b:fig. 2), (2) unfortified village (Manahat, after Edelstein 1993:1281), (3) fortress (Tell el-Burak, after Kamlah and Sader 2004:Abb. 5), and (4) military observation tower (Amman airport structure, Burke 2007:fig. 5).

a. *Large Fortified Towns*

The largest towns in the Levant, as in the kingdom of Ugarit, were always fortified. Freestanding ramparts (and their accompanying features) were characteristic of towns such as Ḫalab(?), Ugarit, Ebla, Qatna, Byblos, Hazor, Kabri, Akko, and Ashkelon (cp. Emar, Yakaltum, Mari, and Šubat-Enlil in Mesopotamia). However, large primary settlements fortified by supplemental ramparts also include, in particular highland sites, such as Shechem, Jerusalem, and Hebron. That a specific size is a relative criterion for the identification of sites in this category is, however, demonstrated by Byblos, which was only 5 ha in size (cp. Qatna at 90 ha). In light of the use of *qarītu* as a generic reference to Ugarit, the royal capital (*DULAT*, p. 713), it is the most appropriate term for large fortified Middle and Late Bronze Age settlements (cp. Phoenician *qart*). Such "royal cities" often featured upper and lower towns in addition to fortifications, as already noted by Jan-Walke Meyer (1996:144f.).

b. *Secondary Fortified Towns*

Smaller fortified settlements, which were usually satellites of large centers, comprise the second type of settlement. These can be identified using the Ugaritic term *ʿr* (cp. Heb. *ʿīr*). While these towns often possessed similar layouts and fortifications, as the largest fortified settlements, *ʿrm* were usually located within one day's journey (ca. 30 km) of *qarītu*. Examples of *ʿrm* include Afis and Tuqan, which were 12 to 15 km from Ebla, Tell aṣ-Ṣour 27 km east of Qatna, Deir Khabiye 20 km south of Damascus, Biruta 28 km south of Byblos, Tel Dan 27 km north of Hazor, and Yavneh-Yam, Ekron, Haror, Lachish, and ʿAjjul, which were about 30 km from Ashkelon (see pp. 128ff. and Figure 13 below). Secondary fortified settlements varied in size, but never exceed the size of the royal capital (see Figure 12, no. 1). *ʿrm* of different sizes might also be located around other *ʿrm* (cp. *dimātum* in Mesopotamia).

c. *Unfortified Villages*

The third category of settlements, which can be described using the Ugaritic term *kaparu* (cp. Hebrew *kefar*) were unfortified, although an enclosure was often formed by the rear walls of the settlement's buildings (see p. 64). The main function of such an enclosure was probably not defensive, but rather as an animal pen or sheepfold (see Figure 12, no. 2). It was probably the case that the inhabitants of such estates would have sought refuge within the walls of a nearby *ʿr* in the event of a military crisis.

As evident from Magen Broshi and Ram Gophna's survey (1986), the vast majority of Middle Bronze Age sites in the southern Levant were unfortified, though most of the sites that have been identified belonged to the MB IIB–C and were located in hill country (A. Mazar 1990:198). Examples in this category include Giv'at Sharett (2 ha) near Beth-Shemesh (Bahat 1993), Manahat (3 ha) in the Valley of Rephaim less than 5 km from Jerusalem (Edelstein 1993), Tell el-Hayyat (0.5 ha, Falconer and Magness-Gardiner 1993), Nahal Rephaim (5 ha, Eisenberg and Edelstein 1993), as well as Tel Michal.

d. Fortresses

The archaeological evidence for Middle Bronze Age fortresses in the Levant (see Figure 12, no. 3), which may be described using the Ugaritic term *ḥl*, is restricted to a handful of examples, which are noted for their strategic positions. Probable examples of such sites include Tel Mevorakh, Tel Poleg, ʿAin Zurekiyeh, Tel Malhata, Tell el-Burak, and Qiryat Shemona, which are between 0.1 and 1.2 ha in size and are located along the coastal plain. Mevorakh's position on the narrow coastal plain between the Mediterranean and the Carmel range afforded it an excellent view of approaching traffic and a position from which to relay information to nearby Tel Burga. The locations of Poleg and Zurekiyeh were functionally equivalent since they too were located along the main north-south road through the coastal plain. Similarly, Tel Malhata (1 ha) served as a road fortress through the Negev. Tell el-Burak, located along the narrow coastal plain south of Sidon, provides perhaps the best example of the layout of fortresses of about 0.1 ha in size (see Figure 29).

e. Military Watchtowers

Military watchtowers, which were identified using Ugaritic *magdalu* (cp. Heb. *migdāl*) and were probably smaller than fortresses (perhaps less than 0.1 ha in size), constitute yet another type of settlement within Bronze Age defensive networks in the Levant (Burke 2007). Survey data suggest that *magdalūma*, while small, were strategically positioned on ridges along major routes. Despite a lack of excavation data for *magdalūma*, which makes it difficult to assert the nature of their layout or defenses, the Amman airport structure, albeit dated to the Late Bronze Age, probably functioned as a watchtower (see Figure 12, no. 4).

f. Rural Agricultural Estates

The Ugaritic term *gittu* (EA *gintu*), as discussed above, designated small rural agricultural estates (cp. Heb. *gath*), which were probably unfortified simply because of their size. Like the inhabitants of villages, those of *gittūma* would also have sought refuge within the walls of nearby *dimātu* during a military crisis. However, the primary distinction between a *gittu* and a *kaparu*, both of which probably served agricultural or horticultural functions, would have been size. *Gittūma* were probably between 0.1 and 0.5 ha in size. Nevertheless, like *magdalūma*, *gittūma* also graduated in size to become villages if not eventually fortified towns (cp. Gath [Tell es-Safi]). A number of settlements may have served as *gittūma* including Kefar Ruppin (0.4 ha, Gophna 1979), Hamadiya (Maeir 2000), Tel Par (Gal 2000), and Abu Snesleh near Irbid in Jordan (Lehmann et al. 1991).

2. Administration

a. Districts

Within a given region the settlement patterns that emerged were a product of a careful balance between the agricultural, hydrological, and defensive needs of the inhabitants. It is no accident, therefore, that the Old Babylonian term for district *ḫalṣum* sometimes also meant "fortress," as noted above. In light of the archaeological and textual evidence for the types and patterns associated with fortified settlements in the Levant and northern Mesopotamia during the Bronze Age, it can be proposed that fortified settlements comprised what might be identified as the *strategic defenses* of the kingdoms to which they belonged. This term is employed by military historian John Keegan to refer to "deliberately conceived" systems of fortified settlements (1993:142). He suggests that Middle Kingdom cataract forts, for example, were among the earliest examples of the employment of such a strategy. However, the arrangement of Bronze Age settlements in the Levant and Mesopotamia also reveal the existence of the strategic defenses of kingdoms in these regions. In addition to what has been said above several sources suggest how Bronze Age strategic defenses actually functioned. Texts emphasize two main aspects related to defense, the military obligations of a region's settlements and their role within communication networks.

b. Military Obligations

The military obligations of the district governor (OB *šapīṭum*) and town mayor (OB *ḫazannum*) are revealed in the royal edicts from the Old Babylonian period through the Late Bronze Age. For the Old Babylonian period the specific duties required of the *šapīṭum* and the inhabitants of settlements within these districts are recorded in the Mari letters (Marzal-Gonzalez 1969:48–91, esp. 90–1). The district governor is described as protecting (ARM VI 76:28–9), inspecting (IV 44:10), and administering (I 62:10 passim) the district as well as arresting deserters and fugitives (II 35:30–1; XIII 118:23).

A major duty of the *ḫazannum* or mayor was also defense (Marzal-Gonzalez 1969:121–23). A settlement could be required to take refugees from the surrounding lands and nearby settlements (ARM III 30; V 37, XIII 117). Inhabitants of settlements could be made to fortify or prepare the fortifications of particular settlements against an approaching threat (ARM II 3). Rulers could also be forced to conscript troops to garrison settlements (ARM I 42, 20; II 97) as well as to provision troops located outside their district (ARM III 27). These and other specific duties are discussed in greater detail in the works of Albert Glock (1968) and Jack Sasson (1969).

A royal edict from the Late Bronze Age that was issued by Mušteya the king of Arrapḫe further elucidates the military obligations of settlements, especially the mayor (*ḫazannu*), in contributing to the strategic defense of the kingdom:

> The king (Mušteya) has established the following order (for) the mayor of Taššuḫewe. Any mayor must watch over the territory of his city, in its (total) perimeter. If there is a *dimtu* in the open country of his city, which is left abandoned, the mayor should (likewise) watch over it. In the territory of his city no robberies must be committed, nor must be (encountered) enemies who kill or plunder . . . If there is a *dimtu* (with-)in the territory of this city, which is left abandoned, the mayor will bear the responsibility for it. The governor of the district shall send tablets to each of the *dimtu*-officials, to give them the following orders: 'In case there are (people) who went out from this *dimtu* in order to commit robberies, or there are enemies who kill and plunder, the official of this *dimtu* has committed a sin, and I shall take the *dimtu* away (from him) . . . '(HSS 15, 1, Zaccagnini 1979:19)

Here we are provided with a glimpse of the administration of the defenses of a region within the Late Bronze Age kingdom of Arrapḫe and the role that each *dimtu* and the mayor (*ḫazannu* or *bēl dimti*) of each settlement within the district (*ḫalṣu*) played within this network. Explicit attention was paid to the responsibility that governors bore to guarantee that settlements, which were not only within their district, but also lacked oversight by a mayor, were looked after, particularly if they were abandoned. This may be interpreted to mean that there existed a general concern that abandoned settlements in peripheral areas could be used as bases of operation by enemies or brigands. Such a concern may have also been expressed in the Egyptian literary masterpiece *The Wisdom of Merikare*: "As a man's name is not made small by his actions, so a settled town is not harmed" (*COS* 1.35). The implication is clearly that an unsettled town runs the risk of being occupied by groups either disloyal to the crown or openly hostile to it. Additionally, Mušteya's edict underscores that the defense of a kingdom was actually the duty of each of its constituent districts and the various settlements within these districts, thus forming a fractal model (for further details, see Kendall 1975:52–55). This conforms with the PHM that was characteristic in the administration of Bronze Age kingdoms, as discussed above (see pp. 121f.).

c. *Communication Networks*

A main advantage of adherence to the settlement pattern characteristic of the open plains in the Levant and Mesopotamia, which is conspicuous for the level of central organization it exhibits, is especially evident with respect to inter-site communication, a vital component of strategic defenses. The communication and intelligence systems of the kingdom of Mari in particular have been discussed by Albert E. Glock (1968:115ff.). In the settlement pattern discussed above, outlying settlements (e.g., *magdalūma*) and fortresses, which were among the smallest settlements, probably often served primarily as observation posts, providing advanced warning of approaching threats as part of a rapid communication network. Fire signals appear to have been the main means of rapid communication during periods of heightened tensions (see ARM IV 31, 32, and discussion by Dossin 1938; Oppenheim 1954:143; and A. E. Glock 1968:119ff.). Most of the time messages were probably carried by couriers between these fortresses on foot or by donkey (A. E. Glock 1968:118f.).

C. A Case Study in Bronze Age Polities: The Kingdom of Ashkelon

In light of the potential for identifying polities by identifying defensive networks of fortified settlements, it is necessary to consider whether any previously unknown polities can be identified in the southern Levant. The obvious candidate for such a regional analysis is the heavily settled southern coastal plain, where a combination of survey (e.g., Gophna and Ayalon 1980; Gophna and Beck 1981; Broshi and Gophna

1986; Gophna and Portugali 1988; Cohen 2002a:esp. 123ff.) and excavation data enable the identification of the Middle Bronze Age kingdom of Ashkelon. As noted earlier, because of Ashkelon's size (ca. 50 ha), it is considered by many to have been of comparable stature with Hazor to the north. Furthermore, recent studies have also emphasized Ashkelon's role in the southern Levant as a trade entrepôt between Egypt and Canaan. Thus, various lines of evidence suggest that a closer examination of the settlement history of this polity is warranted.

1. Early Bronze Age Settlement

Lawrence Stager has recently noted that Ashkelon's role as a Canaanite entrepôt during the Middle Bronze Age was, in fact, simply the resumption of its role during the Early Bronze Age (Stager 2001; 2002). Ashkelon benefited primarily from maritime traffic between Byblos (ancient *Kbn*) and Egypt during the EB I–III period (Stager 2001). However, its importance as a port was furthered by the emergence of dendritic trade networks that brought a number of inland products, such as olive oil, wine, and copper, to the coast. Unfortunately, limited archaeological evidence (except for pottery sherds) is available from Ashkelon and its immediate environs, since EB I–III occupation levels have not yet been reached on the tell, from which to flesh out the details of its Early Bronze Age settlement (ibid., 630). Nevertheless, at this early stage there is no basis for suggesting that Ashkelon politically or militarily appropriated resources in its hinterland (ibid., 629). Thus, the "port power" model is one of an economic system capable of penetrating "diverse ethnic, cultural, and political boundaries . . . to interconnect with them with a minimal use of force" (ibid.).

2. MB IIA Settlement

a. *The Amorite Kings of Ashkelon*

Although we lack data regarding Ashkelon's political identity during the Early Bronze Age (and there is no evidence for its occupation during the EB IV), during the MB IIA its identity as an independent polity, and perhaps the largest Canaanite kingdom, is certain. Attested in the Egyptian Execration Texts (both the Berlin and later Brussels groups) alongside reference to the city as *ʾIsqʾnw* (cp. later cuneiform *Ašqaluna*) are the names of three of its rulers, dated to the MB IIA: *Ḫʾykm* [e23], *Ḫktnw*(?) [e24], and *Mwri* [E2] (see Ritner 1997:51). That these rulers should be identified as kings—and not chiefs—is supported by the size of Ashkelon at this time, which was comparable in size to the capitals of MB II kingdoms in the northern Levant, which were ruled by Amorite kings.

Starting in Phase 2 of the MB IIA Ashkelon was enclosed by a fosse and ramparts of the freestanding type following an elliptical layout (see Figure 61). The best preserved remains of its fortifications and, in fact, of MB IIA Ashkelon, are those of the northern rampart, fosse, and gate belonging to Phases 14 through 12. The mudbrick, barrel-vaulted gate provides a glimpse into the massive amount of effort invested in the construction of these fortifications during the MB IIA2 (see Figure 62). However, its defenses may have been considered more than was necessary since the fosse was filled in during the following phase (MB IIA3) when the earthen rampart was expanded (see Figure 63). This may be interpreted, therefore, as the first indication of the degree of political stability and security during this period in the southern coastal plain around Ashkelon.

The area inside the 1,900 m of Ashkelon's earthen ramparts was around 50 ha, which may have been inhabited by as many as 12,500 people (cp. 250 per ha). As a result of this study, the disparity in actual size between the largest fortified settlements in the southern Levant at the start of the MB IIB–C, inside the earthen ramparts of Hazor (ca. 63) and Ashkelon (ca. 50 ha), has been considerably reduced (see pp. 265ff. and pp. 237ff. respectively).

While we possess only limited evidence for MB IIA settlement around Ashkelon, archaeological and toponymic data connected with a cemetery located 5 km east-northeast of Ashkelon suggest the identification of an MB IIA *magdalu*. The site is known by its Arabic name as Májdal (MR 1118.1198). Following, Koucky (n.d.), it may be identified as a *magdalu* that served to provide advanced warning of threats approaching Ashkelon, which was situated at sea level and was, thus, unable to see along the coastal road (see no. 41 in Burke 2007). The MB IIA4 burials support its Middle Bronze Age date (Gershuny 2002), although no architecture belonging to the settlement has yet been exposed.

b. *Settlement of the Southern Coastal Plain*

Recent study of the settlement of the southern Levant during the MB IIA by Susan Cohen (2002b), which was based on the seriation of MB IIA ceramics from Aphek, provides the first phasing of settlements in the southern coastal plain during this period (see Table 17 below). From Cohen's work we observe that during Phase 1 settlements such as Aphek, Hefer, and probably Akko, as well as even Ashkelon[28] were

[28] The earliest MB IIA ceramics from Ashkelon to date are contemporaneous with Phase 2 at Aphek (Stager 2002).

founded along the coastal plain, probably as trading ports and way-stations (Stager 2001; 2002). These sites were also fortified at this time (2002b:107ff.), thus suggesting a degree of insecurity in the region, although the reason for this insecurity remains unknown (see discussion of Phase 3 below).

During Phase 2, a number of new settlements along the coast, predominantly north of the Yarkon River, such as Nami, Burga, Zeror, 'Ain Zurekiyeh, and Gerisa were founded and fortified, filling in the landscape around the earlier settlements. Settlement infilling continued during Phase 3 with foundations in the north at Kabri, Bira, and Mevorakh, which were fortified, and in the south at a number of sites mostly between the Yarkon River and the Nahal Sorek, such as Yavneh-Yam, Nebi Rubin, and Gezer. Poleg also appears to have been first settled in Phase 3.[29] A degree of instability is, however, apparent along the coastal plain during perhaps Phase 3 when a number of these sites were destroyed (see pp. 98f.). The progression of settlement continued southward during Phase 4 with the founding of Lachish and Tell el-Hesi, both as unfortified villages.

The picture that emerges, therefore, is one of an almost ink-blot like progression down the coast of the southern Levant from Phase 2 of the MB IIA onward, gradually spreading inland along wadi basins from coastal centers, but with the majority of settlement still in the coastal plain. Cohen identifies the pattern associated with contacts between coastal centers and inland settlements as indicative of the existence of "dendritic networks" (Cohen 2002a:126–28). While the evidence for such networks is strongest in the north, such a system can also be inferred for Ashkelon. In sum Cohen's analysis bridges the gap between the identification of Ashkelon as an important Amorite center during the MB IIA and its recognition as a "large, comparatively integrated urban system—a united polity" during the MB IIB–C (Bunimovitz 1995:323).

3. MB IIB Settlement

Although the Mari texts preserve references to Qatna, Hazor, and Laiš (Tel Dan), there are no references to Ashkelon in this archive.[30] In order to account for this gap in our knowledge many scholars have retrojected understandings of the political organization of Canaan during the Late Bronze Age. Yet, as noted above, this has led to incorrect perceptions regarding the nature of MB IIB–C political organization in the southern Levant (see p. 119ff.). Nevertheless, archaeological and survey data from both the city of Ashkelon and settlement in its hinterland facilitate an understanding of the region's political organization during this period.

a. *Ashkelon and Its Hinterland*

During the MB IIB–C Ashkelon continued to be fortified as its rampart fortifications and gate were rebuilt. In Phase 12 at the end of the MB IIA the earlier gate, which was growing increasingly unstable, was intentionally destroyed to accommodate a wider gateway providing easier access in and out of the city (see Figure 65). Later, during Phase 11 in the MB IIB, the new gate was intentionally filled in and a pedestrian gate was built over it (see Figure 66). Outside of this gate, at the foot of the rampart, a building was found that included a "calf-shrine." The expansion of the gate at the start of the MB IIB suggests a thriving economic center that needed to accommodate increased traffic. Furthermore, the construction of buildings outside of the rampart enclosure suggests that Ashkelon was experiencing a space shortage and that its inhabitants viewed the period as one of sufficient political stability to warrant construction outside of its walls.

We possess some limited evidence to indicate the type of settlements that occupied the hinterland within a radius of about 15 km of Ashkelon (see Figure 13). Within less than 5 km of Ashkelon in addition to Májdal, which may have been abandoned during the MB IIB, were probably a number of satellite villages. One example, identified as Ashkelon (b), was ca. 0.2 ha in size and located only 2 km south of Ashkelon. Approximately 10 km to the south was another village at Erez (0.2 ha) along the Nahal

[29] Cohen notes that the distinction "between the Phase 2 and Phase 3 ceramic assemblage is hardly well defined" and that, therefore, "it often proved impossible to determine whether a site belonged to Phase 2 or to Phase 3" (2002a:109).

[30] ARM VI 23 may, however, include a reference to Ashkelon as the final destination in an itinerary. A portion of this text reads: "Further, a group of travelers in transit from Babylon, Ešnunna, Ekallatum, Karana, Qatna and Arrapḫa on their way to Yamḫad, Qatna, Hazor, and [. . .] have arrived here" (as trans. by Yadin 1972b:4). In Jean Kupper's transcription of the text he suggested that four signs belonged in the broken portion, though only the second sign could be suggested (as Borger's no. 354 1981), thus reading X-šu(?)-X-X(ki). However, neither Alt's reading of "Egypt" (1954) nor Yadin's of "Laiš" (1972b:n. 4, pp. 4f.) employed this sign in their readings, and no alternative reading has been proposed. Given the deteriorated state of the sign it is possible that the Old Babylonian sign, no. 15, which has the value *ka*, is the correct reading. If correct, the new reading would enable a reconstruction of the name as [Aš]-ka-[lu-na(ki)]. The itinerary of this text would, therefore, include stops at all of the capitals of major kingdoms on the road between Yamḫad and Ashkelon.

Shiqma. The largest settlement around Ashkelon was Tel Poran (ca. 10 ha), which features a poorly dated earthen rampart enclosure. The land around Poran also supported its own satellite villages as suggested by occupation at 0.2 ha Tel Poran (b) and a cemetery to the east of Tel Poran. To the northeast of Poran lay Ashdod, which was founded and fortified during the MB IIC on the road leading north. Ashdod, like Tel Mor and Tel Michal, appears to have been the product of the increasing settlement along the coastal road between Egypt and the southern coastal plain. The hinterland within 15 km of Ashkelon consisted, therefore, of an area of approximately 350 km^2.

b. Ashkelon's Hinterland Centers

An MB IIB–C ceramic sequence comparable to Aphek's MB IIA sequence does not yet exist, thus complicating our attempts to establish a clearer understanding of settlement in the coastal plain during the second half of the Middle Bronze Age. Perhaps for this reason the settlement pattern identified of the coastal plain (e.g., Gophna and Ayalon 1980; Gophna and Beck 1981; Broshi and Gophna 1986; Bunimovitz 1989) has been characterized simply as "settlement infilling" between the earlier MB IIA sites. However, using the criteria for settlement patterns and settlement types established in this chapter, and archaeological data pertaining to the fortification of settlements in the region, a more nuanced understanding of the settlement pattern of the southern coastal plain during the MB IIB–C is possible.

The settlement process that began during the MB IIA, discussed above, continued unabated during the MB IIB. The focus of settlement was, however, one of so-called infilling, and the greatest number of new settlements in the coastal plain was established *south* of the Yarkon River. In addition to Ekron, to the southeast of Ashkelon new settlements were founded at Nagila, Haror, and Jemmeh and fortified shortly thereafter. Meanwhile, sites around Ashkelon such as Yavneh-Yam, Timnah (Batash), Lachish, and 'Ajjul, which were settled during the MB IIA were fortified for the first time during the MB IIB. It is the nature and contemporaneity of their fortifications, however, which employed massive freestanding earthen ramparts with rectilinear plans (see p. 81), that suggest the activities of a single, central polity during a relatively short period. By the end of the MB IIC no fewer than five secondary fortified settlements accompanied by smaller secondary fortified settlements were dispersed to the east of Ashkelon from the north to the south at an approximate distance of 30 km. The following discussion highlights, therefore, the evidence for the dates and construction of the MB IIB–C fortifications at each of these settlements beginning in the north (for a detailed discussion of their fortifications, see catalog 2.B).

Table 17. Correlations for stratigraphy of Middle Bronze Age settlements in the southern coastal plain.

Phase	Yavneh	Ashkelon	Timnah	Lachish	Nagila	'Ajjul	Gezer
MB IIA1	—	—	—	—	—	—	—
MB IIA2	Str. 9	Ph. 14 (Gate 1)				Courtyard Cemetery	XXII (Tombs)
MB IIA3		Ph. 13 (Gate 2)	*Sherds*	*Sherds*			XXI
MB IIA4	Gate III	Ph. 12 (Gate 3)					
MB IIB	Gate II	Ph. 11 (Gate 4)	Rampart XII	Rampart Palace	XI	City I Fosse	XX
MB IIC	Gate I	Ph. 10	XI		X–IX	City II	XIX (Rampart)
LB IA	—	Ph. 9†	X† IX†	P-4†	VII† VI	City III† Palace I	XVIII†

†Destructions of these settlements are attributed to Egyptian campaigns.

Table 18. Primary settlements in the kingdom of Ashkelon at the end of the MB IIC.

Settlement	Area (ha)	MB Phases	Coordinates (MR)	Settlement Type
Ashkelon	50	IIA2–B/C	107.119	*qarītu*
Ekron	22.2	IIB–C	1356.1315	*ʿr*
Yavneh-Yam	16.9	IIA2/3–B/C	1212.1479	*ʿr*
Haror (Yurza?)	16.2	IIB–C	11257.08795	*ʿr*
T. el-ʿAjjul (Sharuhen?)	10	IIA2–B/C	0934.0976	*ʿr*
T. Poran	10	IIA-?	1137.1242	*ʿr*
Nebi Rubin	10	IIA2/3–C	1245.1485	*ʿr*
T. es-Safi (Gath)	8	IIA?–B/C	1359.1237	*ʿr*
Gezer	7	IIA3–B/C	1425.1407	*ʿr*
Ashdod	5.3	IIC	1178.1295	*ʿr*
Lachish	5	IIA3/4–B/C	1357.1083	*ʿr*
T. el-Farʿah South	3.1	IIB–C	100.076	*ʿr*
T. Zeitah	3	IIB–C?	134.115	*ʿr*
Jemmeh	2.9	IIB–C	097.088	*ʿr*
Nagila	2.6	IIB/C	127.101	*ʿr*
Beth-Shemesh	2.5	IIA–B/C	148.129	*ʿr*
Timnah (Tel Batash)	2.25	IIA–B/C	141.132	*ʿr*
T. Seraʿ	1.5	IIC	119.088	*ʿr*
T. el-Hesi	4.4(?)	IIA4–C	124.106	*kaparu*
T. Harasim	4	IIC	1337.1272	*kaparu*
T. Malat	0.7	IIB–C	1374.1404	*kaparu*
T. Mor	0.6	IIC	117.136	*ḥl* (?)
T. Milha	0.2	?	1288.0968	*ḥl* (?)
Májdal	0.1	IIA4–C?	1118.1198	*magdalu*
Givʿat Ha-Parsa	0.1	IIB–C	1188.1403	*gittu*
Ayanot (school)	0.2	?	1284.1468	*gittu*
Ashkelon (b)	0.2	IIB–C	1007.1175	*gittu*
T. Poran (b)	0.2	IIA–C	1139.1227	*gittu*
Aseret	0.2	IIB–C	1259.1374	*gittu*
Erez	0.2	IIB–C	1086.1075	*gittu*
Duheisha, Khirbet ed-	0.2	IIB–C	1286.1432	*gittu*
Gat-Galon Road	0.2	IIA3/4–B/C	1327.1154	*gittu*
Gedera	0.2	IIA–C	1290.1361	*gittu*
Mazkeret Batya, east of	0.2	IIA4–C	1359.1404	*gittu*
Kefar Menahem	0.2	IIB–C	1351.1282	*gittu*
T. el-Haraz	0.2	IIB–C	1193.1426	*gittu*
Revadim quarry	0.2	IIA2/3–B/C	1330.1327	*gittu*
Yavneh (Shikun Holot)	0.2	IIB–C	1255.1428	*gittu*
Yavneh, coast	0.2	IIA–C	1200.1443	*gittu*
T. Ridan	0.2	IIB–C	0821.0881	*gittu*
Telluliot Batash	0.1	IIB–C	1415.1325	*gittu*
T. Burnah	?	IIB–C	1380.1154	*gittu*
TOTAL SETTLED AREA	191.65			

Note: Settlement sizes have been adjusted so as to exclude uninhabited area occupied by fortifications. MB IIA subphases are based on Cohen (2002).

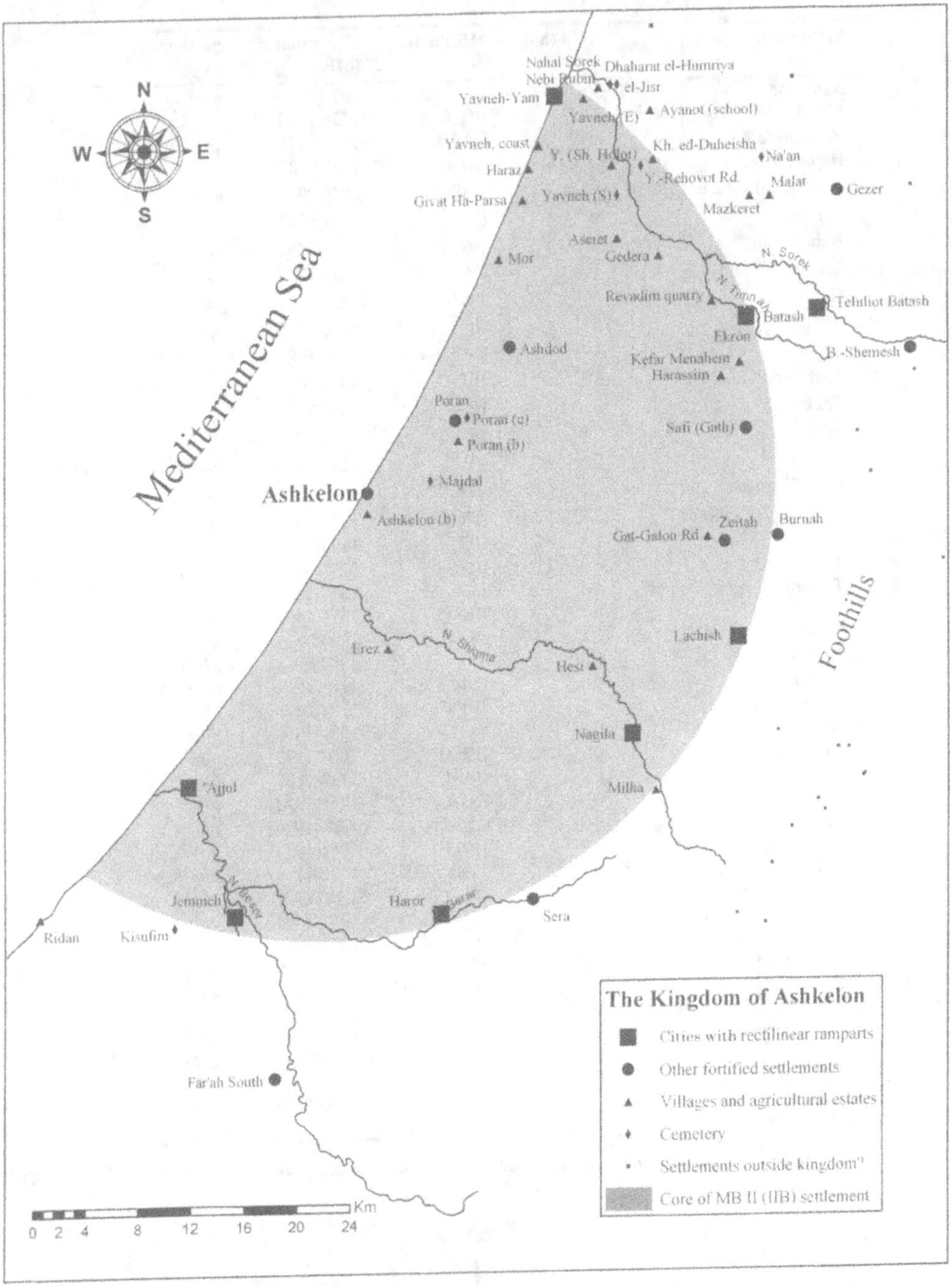

Figure 13. The MB IIB–C kingdom of Ashkelon.

Yavneh-Yam. Yavneh-Yam is located 32 km up the coast from Ashkelon and was first occupied during Phase 2–3 of the MB IIA (Cohen 2002a:94). It was surrounded by a freestanding earthen rampart of rectilinear plan, which Jacob Kaplan suggested was constructed during the MB IIA (see Figure 99). However, the construction of the rampart and Gate III probably date to the start of the MB IIB. Thus Gate III, a six-pier gate, was contemporaneous with Ashkelon's widened Phase 12 gate, and Gate II, a four-pier gate, with the four-pier pedestrian gate of Phase 11 at Ashkelon. If Gate I predated the LB I as Kaplan suggested, then it might have been contemporaneous with Ashkelon's Phase 10–9 gate. Yavneh-Yam's inhabitable area was probably only around 16.9 ha (contra Kaplan's 64 ha, 1993) during this period. At one day's distance from Ashkelon, Yavneh-Yam probably served as a stopping point for caravans along the coastal road leading to and from Ashkelon.

The settlement pattern around Yavneh-Yam enables the identification of a number of satellites within a 13 km radius. The largest of these was Nebi Rubin (ca. 10 ha), which may have been fortified with an earthen rampart. Yavneh (E), Khirbet ed-Duheisha, Ayanot (school), Yavneh (Shikun Holot), Aseret, the Yavneh coast site, Tell el-Haraz, and Giv'at Ha-Parsa were probably *gittūma* of about 0.2 ha around Yavneh-Yam. MB IIB–C cemeteries suggest still other smaller settlements nearby. During the final phase of the MB IIC Tel Mor (0.6 ha) was settled.

Ekron. Thirty-one kilometers northeast of Ashkelon and 22 km southeast of Yavneh-Yam in the coastal plain lay Ekron (Tell Miqne). According to Trude Dothan and Seymour Gitin, the site's overall shape was the result of the "fortifications that encompassed both the upper and lower cities in the Middle Bronze Age" (1997:30). The straight lines formed by the ramparts and the fosse can be discerned in aerial photographs of the site (see T. Dothan and Gitin 1993:1051). If these lines are, in fact, those of the Middle Bronze Age fortifications (see Figure 74), then these defenses were approximately 2,200 m long and enclosed a settlement of about 22.2 ha.[31] Ekron's orientation between Ashkelon and the mouth of the Sorek Valley suggest its vital role as an entrepôt between Ashkelon and settlements in the highlands up this valley.[32] While there does not appear to be evidence for Ekron's settlement during the MB IIA, there is limited evidence for its occupation during the MB IIB–C (Dothan and Gitin 1997:30).

Like Yavneh-Yam, Ekron was a secondary fortified settlement that was itself surrounded by a number of satellites. Foremost of these is Timnah (Batash), which is located 5.4 km east of Ekron at the mouth of the Sorek Valley. Recent excavations by Amihai Mazar have confirmed the hypothesis that the site took its shape from the initial, rectilinear plan of the rampart fortifications that were erected during MB IIB on the south bank of the Nahal Sorek (1997b). The ramparts and wall, which belong to Strata XII–XI, were excavated in areas A and B on the north side of the mound (see Figure 74). The ramparts, which are about 600 m long, enclosed a settlement of only 2.25 ha. The planned character of the settlement and the amount of work required in building its ramparts suggests that it functioned primarily as a fortress-settlement along the route leading through the Nahal Sorek into the highlands from Ekron.

A number of other settlements have also been identified within Ekron's hinterland. In addition to Timnah and its satellite, Telluliot Batash, these include what were probably unfortified villages or *gittūma* at Kefar Menahem, Gedera, the Revadim quarry (0.2 ha), and the larger Tel Malat settlement (0.7 ha) and its 0.2 ha satellite east of Mazkeret Batya. The fortification of Beth-Shemesh (2.5 ha) during the MB IIC was probably the result of settlement encroachment up the Sorek Valley by this period, though who was responsible for its fortification remains difficult to assert. The fortification of Gezer (7 ha) during the MB IIC was probably also the result of Ekron's influence in the region.

Gath. Directly to south of Ekron 8 km and 30 km to the east-northeast of Ashkelon was Gath, which is identified with Tell es-Safi. Although of all of Ashkelon's secondary centers it is the least known, it is likely that it too emerged as a secondary fortified center around Ashkelon during the second half of the Middle Bronze Age. The July 2006 excavations revealed that the 8 ha town of the MB IIB–C(?) was fortified by an earthen rampart featuring stone towers (Aren Maier, personal communication, 2006). Although further exploration will be necessary to identify the extent and nature of its fortifications, these preliminary results and Gath's position with respect to Ashkelon and other fortified centers suggests that the settlement's early history can be understood within this regional framework.

Lachish. Although the site featured MB IIA occupation (Ussishkin 2004a:144ff.) belonging to Phase

---

[31] Barry Gittlen speculates that it may have been as much as 32 ha in size (1992:52*). However, no basis for this figure is provided.

[32] That Gezer did not emerge in a similar capacity during the MB IIA or MB IIB suggests that no coastal polity located north of the Nahal Sorek was prepared to exploit this part of the highlands.

3–4 (Cohen 2002a:86), the construction of the rampart at Lachish during the MB IIB was most likely responsible for the shape of the mound today (see Ussishkin 2004b:55). Thus, the site is roughly rectilinear—almost square (335 m x 235 m)—with its four corners oriented to the cardinal points of the compass (see Figure 86). With a rampart 30 m wide and approximately 1,020 m long, the site was only 5 ha in size. In addition to the fortifications the site also featured an impressive palace during the MB IIB–C built atop the remains of the MB IIA settlement in Area P (Ussishkin 2004a:146ff.). MB IIB–C sherds recovered from the rampart's layers and the date of the founding of its MB IIB palace are suggestive of an MB IIB date for the fortification of the settlement. Lachish's location in the western foothills of the hill country approximately 30 km east-southeast of Ashkelon and along the Nahal Lachish (Wadi Ghafr) was probably crucial to its economic development and its gradual emergence as its own polity during the Late Bronze Age. As suggested by the etymology of its name, Lachish may have been connected with exploitation of local pine forests for wood and resin (Zadok 2004:121).

Only a few settlements are known within the immediate area of Lachish. Tel Nagila is located along the Nahal Shiqma approximately 27 km southeast of Ashkelon and less than 10 km southwest of Lachish. The Stratum XI settlement (MB IIB) was founded after an occupational gap. Area C on the northwest provided the most information about the site's fortifications, which consisted of a rampart, glacis, town wall, and a fosse at the foot of the rampart. Although the excavators were unable to determine during which part of the MB IIB–C (Strata XI–VII) the fortifications were constructed, the settlement's comparable fortification style suggest a foundation roughly contemporaneous with Lachish. Elsewhere along the Nahal Shiqma Milha (0.2 ha) and Tell el-Hesi (4.4 ha?), both of which were villages, were occupied. Other villages were identified on the Gat-Galon road and at Tel Burnah, as well as a potentially fortified settlement at Tell Zeitah.

Tel Haror. The enclosure of Haror, biblical *Gerar*, is located along the Nahal Gerar 31.6 km to the south-southeast of Ashkelon. It was first settled at the start of the MB IIB and grew to approximately 16.2 ha in size during this period when earthen ramparts were constructed around the lower town (Oren et al. 1991:73). These rectilinear freestanding earthen ramparts, formed a trapezoidal enclosure against the base of the upper town (see Figure 77), which can be clearly seen in aerial photographs (Oren 1996:22; 1997a:474).

The only site that can be identified as a subsidiary settlement within Haror's territory is Tel Sera' (1.5 ha), which probably functioned as a way-station on the road leading toward the highlands from Haror. The negligible amount of settlement within a 15 km radius of Haror suggests (in contrast to the settlements discussed above) that its role may have been primarily as an entrepôt between pastoralists and the kingdom of Ashkelon which served to exploit the western Negev. It is possible that settlement at Tell el-Far'ah, which was only fortified during the MB IIC was a satellite settlement of Haror.

Tell el-'Ajjul. Located 25.5 km to the southwest of Ashkelon along the coast, 'Ajjul is the southernmost of Ashkelon's secondary fortified settlements. In addition to the reasons given by Kempinski for identifying this site as Sharuhen (1974), it is the first major Canaanite settlement encountered on the road to Ashkelon from Egypt. This may explain why it was the first Canaanite town besieged by Egypt following the expulsion of the Hyksos (see discussion below). Petrie identified the fosse that enclosed the settlement on three sides, which is probably dated to the MB IIB–C (see Figure 58). The excavated limestone from the fosse was used to build up the rectangular freestanding rampart enclosure of about 10 ha. Aharon Kempinski also observed that Tell el-'Ajjul's layout had all of the earmarks of a "planned settlement" lacking an organic phase of growth (1992e:125f.). This suggests that, like the plans of the settlements discussed above, 'Ajjul's orthogonal layout was established before large-scale settlement occurred at the site. As at Lachish, 'Ajjul also featured remains of a late MB IIB–C palace (see discussion of Palace I in Herzog 1997a:127) that was likewise destroyed during Egyptian campaigns.

Although archaeological data for Gaza and its environs are limited, several smaller settlements around 'Ajjul support its hegemonic role in the local settlement hierarchy. Tel Ridan (0.2 ha) on the coast less than 15 km to the south of 'Ajjul may have served as a way-station on the road to Egypt. The cemetery at Kissufim about 5 km to the west of Jemmeh suggests the existence of another still unidentified village. However, Tell Jemmeh was the most important settlement near 'Ajjul.

Jemmeh, is located 32.7 km south-southwest of Ashkelon and up the Nahal Besor from Tell el-'Ajjul, was first occupied during the MB IIB (Van Beek 1993:668). Both Petrie and Van Beek excavated portions of the site's poorly preserved MB IIB freestanding earthen rampart. Originally the site's northern defenses were built along the Nahal Besor, but the wadi has now greatly eroded much of this side of the mound (see Figure 80). That the site's plan was probably rectilinear is suggested by three lines of evidence: (1) the roughly square shape of the mound in Petrie's original plan (1928:pl. 4), (2) the straight lines of the rampart and fosse that are evident in black and white aerial photographs of the site (Van Beek 1993:667), and (3) the orientation of the Eighteenth Dynasty buildings excavated by Petrie on the west side of the mound—the west wall of this complex was found to be parallel to the line of the west rampart and fosse (see Petrie 1928:pl. 6). If this reconstruction is correct then Jemmeh was roughly square in shape with sides 170 m long and approximately 2.9 ha in size.

c. *Ashkelon's Settlement Types*

If the fortified settlements discussed above are to be identified as second- and even third-tier fortified towns (i.e., *ʿrm*) within the kingdom of Ashkelon, the many small settlements distributed around Ashkelon and its outlying centers belonged to the full range of settlement types identified earlier in this chapter (see Table 18). A number probably represent unfortified villages (i.e., *kaparūma*), such as Tell el-Hesi. In addition to these there are numerous small MB IIB–C settlements, possibly rural agricultural estates (i.e., *gittūma*) averaging 2 dunams in size, such as those near Poran, around Yavneh-Yam, and at Aseret, Erez, Khirbet ed-Duheisha, the Gat-Galon Road site, Gedera, Giv'at Ha-Parsa, Kefar Menahem, Tell el-Haraz, Tel Milha, Revadim, and Yavneh (Shikun Holot). Furthermore, Middle Bronze Age cemeteries at Májdal to the northeast of Ashkelon, at Kisufim, and in the vicinities of Poran and Yavneh-Yam also suggest the presence of other very small, unidentified villages and *magdalūma* within this region during this period. Later settlement at Tel Mor provides the clearest example of a what was probably a Middle Bronze Age fortress (i.e., *ḥl*)

d. *The Land of Ashkelon*

Based on the above data MB IIB–C settlements at Yavneh-Yam, Ekron, Lachish, Haror, 'Ajjul, and possibly Gath, all of which were between 5 and 22.2 ha in size, may be identified as secondary fortified centers around Ashkelon (see Figure 13 above). Around each of these settlements existed a third tier of similarly fortified settlements such as Nebi Rubin(?), Timnah, Nagila, Sera', and Jemmeh, which ranged in size from 2.5 to 10 ha and were located at nearly the same distance from Ashkelon as the secondary fortified centers—about 30 km away or roughly one day's journey. While these third-tier settlements were under the control of second-tier settlements 30 km away from Ashkelon, Tel Poran and Ashdod are identified as third-tier settlements despite their size because of their proximity to Ashkelon (less than 15 km away).

Within the framework of the settlement pattern and administrative systems discussed in the previous section, a number of criteria suggest that the settlements identified as secondary settlements around Ashkelon served as district capitals of the kingdom of Ashkelon. First, they were located about one day's travel from Ashkelon (ca. 25 to 33 km), an ideal distance for administrative purposes, but sufficient to warrant major settlements with their own sustaining areas. Second, they were strategically positioned to guard and administer major routes that led to the port of Ashkelon, whether up and down the coast (e.g., Yavneh-Yam and 'Ajjul), down the many wadi systems from the highlands (e.g., Ekron, Gath[?], and Lachish), or into the northern Negev (e.g., Haror). Third, like Ashkelon they were surrounded by their own network of settlements that consisted, notably, of similarly fortified satellite settlements, which were likewise oriented to provide maximum protection and observation of routes leading from the highlands toward Ashkelon. If sites such as Ekron had been fortified in response to a threat posed by Ashkelon, as might be suggested in another model, then settlements such as Timnah would be expected between Ashkelon and these settlements.

The fourth criterion for the identification of such a state, the massive investment of labor in the construction of the fortifications of secondary settlements and their satellites, suggests the central role that Ashkelon played in organizing the kingdom's defenses. The dependence of other settlements on Ashkelon for their security is revealed, in particular, through analysis of the labor required for the construction of the fortifications of the secondary (e.g., Haror) and tertiary (e.g., Timnah) fortified settlements (see Table 27). Large projects such as these at such small settlements could not have been completed sufficiently quickly by their inhabitants alone or even all of the inhabitants within their district (see Table 26). These projects clearly required the organization, labor, and resources of an entire kingdom in order to be completed within a reasonable timeframe (see discussion in chapter 6).

If this reconstruction is correct, it would affect the interpretation of several other artifacts within the Middle Bronze Age landscape of Ashkelon. First, the MB IIB–C palaces at Lachish and 'Ajjul might be identified not as palaces of local kings but of district governors, Old Babylonian *šapīṭum*, who did the bidding of the king of Ashkelon. Second, it suggests that the absence of references to settlements here identified as Ashkelon's district capitals (or its other smaller settlements) in foreign correspondence of the Middle Bronze Age, such as the Mari texts, since international correspondence was the domain of the capital. Thus, correspondence with Ashkelon is what would be most anticipated among Old Babylonian sources from the northern Levant and Mesopotamia, alongside the possible reference to Ashkelon in an itinerary, as I have suggested earlier (see n. 30, p. 127). Consequently, if textual references to Ashkelon's district capitals are preserved, they are most likely to occur at Ashkelon or its district capitals. However, given the manner in which most Akkadian sources have come down to us—through the destruction of settlements—the lack of destruction layers during the floruit of Ashkelon's rule suggest that such sources may not have been preserved. The recovery of a fragment of a Late Bronze Age cuneiform lexical text from Ashkelon (Huehnergard and van Soldt 1999), which was found by chance in a later wall, illustrates the nature of this problem. Destruction layers dated to the late MB IIB–C/LB IA probably came too late to preserve earlier documents.

Although a few of the settlements discussed above, particularly those in the south, were occupied for the first time during the MB IIB, most of the northern settlements were already settled by the end of the MB IIA. When the identical MB IIB–C fortification strategies employed at these settlements are viewed within the framework of the settlement pattern characteristic of the Levant during the Middle and Late Bronze Ages, the historical and rank-size primacy of Ashkelon reinforces the observation that it played a central role in this process. Collectively, therefore, these settlements form the backbone of Ashkelon's settlement network at the start of the MB IIB, and in this light a "united polity"—or more precisely within the milieu of the Middle Bronze Age—a kingdom, can be identified (see Table 18). The kingdom is, therefore, best understood by means of a network model (e.g., M. L. Smith 2005) that emphasizes not only rank-size relationships among its settlements but also the nature of its rule, which employed a nested hierarchy of dependency that reflected its underlying patrimonial character (i.e., PHM).

While natural features did not contain settlement and expansion related to the kingdom, several features do circumscribe Ashkelon's territory, thus providing it with what might be identified as conceptual boundaries. The importance of these are borne out by their identification as such in later sources such as the Hebrew Bible. Such conceptual boundaries included the Nahal Sorek on the north, the foothills leading into the highlands on the east, the Nahals Besor and Gerar on the south, and of course, the Mediterranean Sea on the west.

As the study of Iron Age roads in ancient Israel also indicates, the main sites in the kingdom were connected to their nearest neighbors by a network of roads, which also established a ring-road around Ashkelon between its secondary centers (see Dorsey 1991). The nucleus bounded by this belt of settlements around Ashkelon and defined by the borders identified above would have been at least 1,700 km^2, which is comparable to the size postulated for Early State Modules (ESM). These polities were on average 1,500 km^2 in area and were usually situated about 40 km apart (Renfrew 1975:12–18, fig. 6). The size of the kingdom of Ashkelon in the MB IIB–C also falls within the attested range of other Bronze Age kingdoms in the northern Levant, most notably the kingdoms of Mukiš and Ugarit (see discussions above). The considerable size of the polity is supported, however, by the observation that most of this land was cultivable. As suggested by Stager's "port power" model, discussed above, agriculture and horticulture were probably the mainstay of the economy of this region.

While Ashkelon itself comprised approximately 25% of the settled area within its territory, its two tiers of secondary fortified centers, which ranged in size from 2 to 22 ha (see Table 18), comprised another 40% of its settled area. Together with Ashkelon they constituted more than 130 ha of settled land by the end of the Middle Bronze Age and sheltered approximately 70% of the inhabitants of the territory. Thus, as major population centers within the kingdom these nine settlements facilitated Ashkelon's domination of its hinterland during the second half of the Middle Bronze Age.

4. MB IIC Settlement

a. *Dendritic Growth*

Sites that were just beyond the belt formed by secondary fortified settlements during the MB IIB probably should also be identified with the growth of the kingdom of Ashkelon and the solidification of its control over trade networks during the late Middle Bronze Age. These settlements include Tell el-Far'ah South (3.1 ha), Gezer (7 ha), and Beth-Shemesh (2.5 ha), which were all fortified during the MB IIC, though they also appear to have been occupied to one extent or another during the MB IIB. (If Gezer's MB IIA occupation was fortified, the site likely served as a way-station or fortress on the road to the highlands.) These settlements were all approximately 42 km from Ashkelon as the crow flies and reflect the continued growth of dendritic trade networks from Ashkelon. Each of these settlements possessed similarly planned fortifications including the construction of six-pier gateways. Perhaps because their fortifications, including their gates, were built during the MB IIC(?) they no longer conformed to the rectilinear plan that had been characteristic of Ashkelon's secondary fortified settlements and typical of major settlements during the MB IIB.

The date of the fortification of these settlements suggests that they were the result of a later stage of economic growth of the kingdom of Ashkelon. For the same reasons noted above regarding Ashkelon's secondary fortified centers, there is no basis for identifying these settlements as competitors or enemies of Ashkelon during this period. Thus, at the end of MB IIC the greatest growth of settlement around Ashkelon occurred beyond the ring formed by the original land cultivated by Ashkelon's settlements, which had been fortified during the first half of the MB IIB (see Figure 13). The growth of Ashkelon's trading activity may also, therefore, account for a surge in settlement and fortification construction in the highlands to the east of Ashkelon's heartland during the MB IIC. Thus, highland settlements such as Jerusalem may have grown and strengthened substantially as a direct result of the growing demand for horticultural and viticultural products, namely olive oil and wine, which were shipped out of Ashkelon's port to Egypt, as Stager suggests (2001).

b. *Population Estimates*

With a total settled area of approximately 190 ha by the end of the MB IIC (see Table 18), the kingdom of Ashkelon may have been inhabited by as many as 47,500 persons (if 250 person/ha). To what extent settlements north of the Nahal Sorek must also be included within the kingdom of Ashkelon during the Middle Bronze Age remains to be determined. Yet, if this reconstruction is correct, then roughly one-third of the population of the southern Levant west of the Jordan Valley (cf. Broshi and Gophna 1986) would have been under the immediate control of Ashkelon by the end of the Middle Bronze Age.

c. *A Kingdom Destroyed*

The important role played by Ashkelon's largest outlier settlements within the Canaanite landscape is revealed by what is suggestive of a process involving their targeted destruction during the transition between the Middle Bronze Age and the Late Bronze Age. In addition to the destruction of Ashkelon, all of its secondary fortified centers except those in the vicinity of Haror (*Yurza*?) were destroyed or abandoned (see Table 17). Yavneh-Yam appears to have been abandoned, but destruction levels have been noted at Timnah (Str. X–IX), Gezer (Str. XVIII), Lachish (Level P-4), Nagila (Str. VII), and 'Ajjul (City III and Palace I), as well as Beth-Shemesh (Str. V). 'Ajjul (Sharuhen?) endured the three-year siege recorded in the Egyptian account (*AEL* 2, pp. 12–14), and it may be that the Upper and Lower Tunnels identified by Petrie relate to this siege (see pp. 230ff.). In addition to fortified centers, villages such as Tell el-Hesi were also destroyed, while other towns such as Tell el-Far'ah South appear to have been simply abandoned. Haror and Jemmeh appear somehow to have escaped destruction at the start of the New Kingdom. The destruction of these settlements constitute nearly a third of the destructions attributed to the Egyptians by means of archaeological data (see p. 101). The events surrounding these destructions further suggest shared political and strategic relationships between these cities and, thus, may expose alliances that existed between them.

5. Late Bronze Age Settlement

The proposed reconstruction of the kingdom of Ashkelon during the Middle Bronze Age is supported by historical and archaeological data for the political organization of settlements in the coastal plain during the Late Bronze Age. These data reveal the fragmentation or balkanization of Ashkelon's Middle Bronze Age kingdom along the lines of what may be identified as its districts, which included Ashkelon and those associated with its secondary fortified centers during the MB IIB–C, as defined in the previous section. Of principal importance in this reconstruction are the Amarna letters from the mid-fourteenth century B.C., which record correspondence between the Egyptian pharaoh and the rulers of these territories who are identified using the Akkadian term *ḫazannu*.

a. *"City-States" of the Southern Coastal Plain*

By means of the Amarna letters a number of Canaanite polities including Akko, Ashkelon, Gezer, Gath, Lachish, and Jerusalem have been identified. Since the work of Nadav Na'aman (1975) these texts have served as the primary basis for the identification of so-called Canaanite "city-states" (Bunimovitz 1989; 1995; Finkelstein 1996b), as discussed above (pp. 119–121), though with some notable reservations regarding the size of their territories (Na'aman 1997). More recently Na'aman has emphasized the significance associated with the identification of the very individuals who ruled these "city-states" and who corresponded with the Egyptian pharaoh as mayors using the Akkadian term *ḫazannu* (1997:601f.).[33] However, in contrast to Finkelstein (1996b:603) he notes the degree to which the Amarna archive is incomplete, and therefore can only be used to produce an incomplete list of Canaanite city-states during this period.[34] Nevertheless, Na'aman observes that within the entire Amarna corpus the southern coastal plain is the best-represented region of the southern Levant (1997:607ff.), a view echoed in recent work by Michaël Jasmin (2006).

A number of towns mentioned in the Amarna correspondence, which are often identified as city-states, held sway over territory that formerly belonged to the Middle Bronze Age kingdom of Ashkelon according to the reconstruction discussed above. These include Ashkelon (*Ašqaluna*), Gezer (*Gazru*), Gath (*Gimtu*), Lachish (*Lakiša*), *Yurza* (Haror?), and Tell el-Hesi (ancient name uncertain). Still other polities mentioned in the Amarna letters, the identifications of which remain uncertain, include *Zuḫra* (EA 334), *Aḫtiashna* (EA 319), *Na-x-ḫa-x* (EA 272), *Ṣab/puma* (EA 273–274), *Tianna* (EA 284, 298, and 306), and possibly *Zilu* (EA 288 and 335), which Na'aman notes was probably a neighbor of Lachish (Na'aman 1997:608; contra Albright 1924). Nevertheless, it is uncertain whether or not these remaining "polities" were located in the southern coastal plain, and therefore to what extent they are relevant to this discussion.

Correspondence between the *ḫazannu* of Ašqaluna, Yidya, and the Egyptian pharaoh comprises eight letters (EA 320–326, 327[?], and 370), with another reference to Ašqaluna found in EA 287:14. This correspondence indicates that Ashkelon: (1) possessed an Egyptian royal residence[?] (EA 321:24–26; 322:15ff.; 370), (2) served as a way-station and supply depot for Egyptian troops (EA 324:10ff.; 325:15ff.), and (3) produced glass for export (EA 323:13–16, also 327[?]; see also Yurza in EA 314). Despite a considerable reduction in the overall size of its territory following the Middle Bronze Age, that Ashkelon continued to be a center of Canaanite power on the coastal plain even by the end of the Late Bronze Age may explain why Merneptah was forced to march against the Canaanite town around 1207 B.C., as recorded on the Merneptah Stele (*AEL* 2, p. 77) and at Karnak (Yurco 1986).

Five letters are preserved between the *ḫazannu* of Gazru and the pharaoh (EA 298–300, 369, and 378). Gezer is also mentioned in EA 253:22, 254:22; 287:14, 290:8; and 292:43. According to EA 273 Ayyaluna (biblical *Aijalon*) belonged to Gezer, and EA 299 notes Gezer's problem with the 'Apiru.

The correspondence of Lakiša includes five letters (EA 328–332). The slaying of Zimredda, one of its *ḫazannu*, is mentioned (EA 288:43), as is Lachish's seizure of Muḫraštu (EA 335), though from which mayor is unclear. Lachish is also mentioned in EA 287:15. A letter found at Tell el-Hesi (EA 333) that was sent from one Paapu indicates that Lachish was threatening Tell el-Hesi, suggesting that the site was not considered part of Lachish's territory.

Three more letters belonged to the *ḫazannu* of Yurza (EA 314–316). Although Yurza is identified traditionally with Tell Jemmeh (Goren et al. 2004:299–301), it is not impossible, as Na'aman suggests that Tel Haror should be identified as Yurza (Na'aman 1997:607), which is accepted here. Like Ashkelon, Yurza supplied Egypt with glass (EA 314).

That 'Ajjul, possibly ancient *Sharuhen*, is not mentioned among the Amarna texts is not particularly surprising given that it suffered the brunt of the blow from Egypt during the late sixteenth century, which is amply recorded in *The Autobiography of Ahmose son of Abana* (*AEL* 2, pp. 12–14). Furthermore, a new political center, Gaza (*Ḫazzatu*), located within Sharuhen's territory, had emerged and was located only a few kilometers north of 'Ajjul along the coastal road. While 'Ajjul clearly featured Late Bronze Age occupation, no evidence remained of its Middle Bronze Age fortification wall (although the fosse remained functional), suggesting an especially deliberate effort by Egypt to guarantee that the city remained no threat to Gaza, its new administrative headquarters for the southern Levant.

Finally, it may be the case that Tianna, which Na'aman identifies with Ashdod (Na'aman 1997:612ff.), is to be identified with Late Bronze Age Ekron. This is a more plausible identification in light of Na'aman's argument, which correctly recognizes Tianna's conflict with its neighbors, Gazru (EA 298), Gimtu (EA 284), and Ašqaluna (EA 306), as the primary basis for identifying this town. First, Gezer as a neighbor of Ashdod overlooks Ekron's position between these two towns. Second, Ekron (ca. 22 ha), which is not otherwise identified in the Amarna letters, would have posed a greater and more formidable threat than Ashdod (ca. 5.3 ha) had it rebelled. Ekron by its sheer size would have allowed it to contend with all three of its neighbors at the same time, as might be suggested by these letters. Finally, if Muḫḫazu is to be

---

[33]For a brief discussion of terms and their meanings see William Moran (1992:xxvii and n. 73).

[34]Finkelstein asserts that "the material at hand enables a full or almost full reconstruction of the territorial map of Late Bronze Age Canaan" (1996b:603).

identified as a border town of Tianna (Na'aman 1997:613), then this is likely to be the name of Timnah (Tel Batash) during the Late Bronze Age, given the proximity of Ekron and Timnah and their position along the southern border of Gazru.

b. "City-States" or Administrative Districts?

Despite the general acceptance of the conclusions discussed above, the potential significance of previous kingdoms such as Ashkelon in the south and Hazor in the north, and their districts has been overlooked in studies of Late Bronze Age political organization in the southern Levant. Although it is true that little has been known with any certainty regarding political organization during the Middle Bronze Age, the equation proposed between Late Bronze Age polities in the southern coastal plain and earlier Middle Bronze Age districts in the kingdom of Ashkelon is well grounded in political realities of the northern Levant and northwestern Mesopotamia during the Old Babylonian period and their subsequent trajectories. Thus, it is reasonable to conclude that the so-called Late Bronze Age "city-states" of the southern coastal plain, which were governed by *ḫazannu*, most likely constituted the remnant districts (OB *ḫalṣum*), each of which was also ruled by a *ḫazannu*, that belonged to polities of the late MB IIB–C.

The identification of administrative continuity between the Middle and Late Bronze Ages in the southern coastal plain is supported by several lines of evidence. First, it explains the continuity of settlement shared by almost all of the largest fortified population centers of the MB IIB–C in the kingdom of Ashkelon. Thus, the former Middle Bronze Age towns of Ashkelon, Ekron (Tianna?), Gezer, Gath, Lachish, Haror (Yurza?), and 'Ajjul (Sharuhen?) continued to be inhabited throughout the Late Bronze Age. Only Yavneh-Yam appears to have been abandoned at the end of the Middle Bronze Age.

Second, all of these major towns are identified as administrative centers (i.e., polities, or "city-states") during the Amarna period. According to the Amarna texts each of these towns was governed by an individual identified as a mayor (Akk. *ḫazannu*), but who ruled at the local level the "land" around his town (Akk. *māt*-TN) like a king and was, therefore, identified locally using the Akkadian term *šarru* (see Na'aman 1988:183f., n. 18). It is the identification of these individuals as *ḫazannu* in the Amarna letters, which may also have been the title of their MB IIB–C predecessors during Ashkelon's hegemony, that the identification of their towns as the administrative centers of districts (Old Babylonian *ḫalṣum*) within the Middle Bronze Age kingdom of Ashkelon is made possible. This is further supported by observation that these districts were overseen by an Egyptian commissioner who ruled from Gaza (Ḫazzatu), which had usurped the role of Ashkelon on a regional level and 'Ajjul's (Sharuhen?) role at the district level.

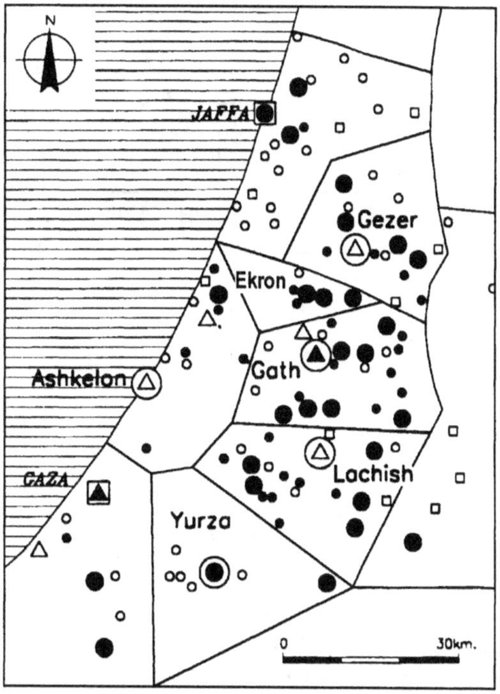

Figure 14. Late Bronze Age "City-States" of the southern coastal plain represented as polygons with adjustments (after Finkelstein 1996b:map 2, p. 255).

Third, the territories associated with the Late Bronze Age polities discussed above provide a reasonable starting point for the identification of the territory administered by these towns during the Middle Bronze Age. The most frequently cited reconstruction of the Late Bronze Age territories associated with coastal Canaanite "city-states" is Israel Finkelstein's, which uses Thiessen polygons (see 1996b:esp. map 2, p. 255; and 1996a:fig. 1). Na'aman notes that a more nuanced approach is necessary since Finkelstein overlooks the importance of what were probably several other "city-states" (1997). Such considerations are taken into account by Michaël Jasmin in his reconstruction of the political organization of the southern coastal plain during the LB II (see esp. 2006: fig. 6).[35]

[35]Jasmin's approach is an improvement upon Finkelstein's in two ways; first, he avoids the use of Thiessen polygons to demarcate political territories and, second, he accepts that wadis served as "natural and political," if permeable, borders.

Table 19. Historical evolution of Middle Bronze Age districts around the district of Ashkelon with the names of known rulers through the Late Bronze Age.

MB IIB	MB IIC–LB I	LB II	Iron I	Iron II
ʾIsq ʾnw §*Ḫ ʾykm* §*Ḫktnw* §*Mwri*	Ashkelon	*Ašqaluna* §*Yidya*	→	*Ashkelon* (P)
T. Miqne†	T. Miqne†	*Tianna* (T. Miqne)†	→	*Ekron* (P)
Yavneh-Yam†	Yavneh-Yam†	Yavneh-Yam†	→ *Ashdod* (P)	*Ashdod* (P)
Haror†	Haror†	Haror (*Yurza*?)† §*Pu-Baḫla*	?	?
ʿAjjul	ʿAjjul (*Sharuhen*?)	» *Ḫazzatu* (Gaza)	→	*Gaza* (P)
?	?	*Gimtu* (Tell es-Safi)	→	*Gath* (P)
Lachish	Lachish	*Lakiša*†† §*Zimr-Edda* §*Šipṭi-Baʿla* §*Yabni-Ilu*	→	*Lachish* (I)
Gezer	Gezer	= *Gazru*†† §*Adda-danu* §*Milk-Ilu* §*Yapaḫu*	→	*Gezer* (I)

†Ancient name uncertain.
††Towns listed as conquered by Israel in biblical tradition (see Josh 12).
»Denotes shift of center to a nearby settlement. P=Philistine pentapolis town. I=Israelite administrative center.

With minor modifications Finkelstein's map serves as an adequate visual aid for the above proposal (see Figure 14). Based on the historical reconstruction offered the necessary modifications include the identification of: (1) Yurza's with Haror (rather than Jemmeh), (2) Ḫazzatu as the capital of its own district (i.e., polygon), (3) Ekron as a separate polity (probably Tianna), and (4) an adjustment of the territories associated with the remaining polities. All the while it should be remembered, of course, that polygons are purely of a heuristic value for identifying administrative jurisdictions.

Since the territory associated with the Late Bronze Age polities discussed above essentially accounts for the entire landscape surrounding the kingdom of Ashkelon during the MB IIB–C, this model suggests that Late Bronze Age settlement in the southern coastal plain should not be viewed to be in decline following the Middle Bronze Age. Rather, Late Bronze Age settlement was a product of the realignment of earlier political districts within larger polities such as Ashkelon and Hazor. This also explains why there is a dearth of references to independent kingdoms during the late Middle Bronze Age comparable to the references to polities in the Amarna letters; this is because they did not exist as such. Furthermore, within this framework the so-called "city-states" of the Levant can hardly be identified as such since they were neither politically independent, which as noted earlier in the chapter is a quintessential hallmark of a city-state, nor ethnically distinct.

6. Iron Age Settlement in Philistia

Further support for the present reconstruction of the Middle Bronze Age kingdom of Ashkelon and its Late Bronze Age fragmentation comes from the organization of the Philistine pentapolis during the early Iron Age. The five major Philistine settlements, although they do not all correspond with Late Bronze Age "city-states," formed a settlement pattern around Ashkelon not dissimilar to that identified for the Late Bronze Age (Finkelstein 1996a). As Lawrence Stager demonstrates, the core area dominated by the Philistine pentapolis in the Iron I can be identified with an area associated with the presence of Philistine monochrome ware that was bounded by Egyptian fortresses (1995).

Four of the five settlements of the pentapolis—Ekron, Ashdod, Gaza, and probably Gath—that effectively delineated the borders of the Philistine heartland during the earliest phase of their occupation, were located along the belt of second-tier Middle Bronze Age settlements around Ashkelon, with Ashkelon's position retained. Only in the second phase of Philistine settlement, during the Philistine bichrome phase, did the Philistines begin to settle beyond the heartland of the pentapolis, a phenomenon that repeated that of the expansion of the kingdom of Ashkelon at the end of the Middle Bronze Age. That the initial phase of Philistine settlement corresponded with the political landscape of the Middle Bronze Age kingdom of Ashkelon and its districts is not surprising however. In fact, it suggests that the Philistines were aware of and exploited administrative hierarchies that belonged to the earlier Canaanite states. As in so many other cases involving political change in the ancient Near East, the most expedient means of ruling a region with preexisting socio-political networks was to replace the seat of power, which in this case was Ashkelon, but also included second-tier settlements inland from it, such as Ekron, Gath, Gaza, and Ashdod. Furthermore, it is notable that the remaining administrative centers, such as Gezer and Lachish, eventually fell under Israelite control (see Table 19).

7. Conclusions

Ashkelon and its hinterland during the MB IIB–C serves as an excellent example of how in the absence of historical records archaeological data enable the identification of a polity, its territory, districts, and a defensive network of settlements. In the case of Ashkelon during the MB IIB–C the primary data include: (1) the dominant position of Ashkelon based on rank-size analysis, (2) the distinct pattern of settlement associated with second- and third-tier settlements around Ashkelon, (3) the nearly identical and contemporaneous fortifications of its second- and third-tier settlements, and (4) the identification of likely geographical limits for the proposed polity and its districts. When these aspects are viewed in light of the historical developments of the political landscape in the southern coastal plain during the Late Bronze and Iron I periods, this reconstruction gains greater probability.

6. The Socioeconomic Impact of Fortification Construction

Where the army is, prices are high;
when prices rise the wealth of the people is exhausted.
When wealth is exhausted the peasantry will be afflicted with urgent exactions.

Sun Tzu, *Art of War* (p. 74.)
China (fourth century B.C.)

During the past decade and a half much of the research on Middle Bronze Age fortifications in the Levant has concerned itself with assessing the impact of their construction on the society and economy of the region (see pp. 12ff.). Whether or not one accepts that earthen ramparts during this period were intended to serve primarily as fortifications and not merely as a "cheap means for delimitation of the city's borders" (Herzog 1992:1035), questions related to the costs and value associated with the construction of fortification systems during the Middle Bronze Age must be addressed. These questions include: how long did these defenses take to construct? How can the number of people involved in the construction of these fortifications be reliably estimated? Of whom did the labor force consist and where did it originate? Was this labor force drawn from corvée or *ilkum* service, the army, hired hands, or slaves? This chapter, therefore, explores a number of issues relating specifically to the impact of fortification construction on the society and economy of the Levant during the Middle Bronze Age.

The first section of the chapter addresses the issue central to the construction of Middle Bronze Age fortifications in the Levant, namely the organization of labor. In the second section an attempt is made to establish reliable figures for estimating the rate of construction. These figures are employed in the third section in order to estimate the total labor consumed in the construction of the Middle Bronze Age fortifications at a number of sites. The fourth section sets forth various scenarios for the construction of fortifications involving the three most likely sources of labor during this period. Finally, the understanding afforded by these considerations is used to evaluate the role of fortification construction in the development of social stratification in the Levant during the Middle Bronze Age.

A. The Organization of Labor in the Levant

Understanding the organization of labor is essential to determining the labor and resource requirements for building fortifications, yet no sources exist from the Levant during the Middle Bronze Age that enable the identification of the labor force in question. Nevertheless, in light of the political organization invoked for the Levant during this period (see chapter 5) and its parallels to Old Babylonian kingdoms, data for understanding the organization of labor in Mesopotamia during the Old Babylonian period are especially helpful (e.g., Powell 1987). Old Babylonian sources provide a framework for evaluating the nature of the organization of labor in the Levant during this period.

1. Old Babylonian Labor Sources

a. *Army Labor*

Old Babylonian sources reveal that the construction and maintenance of fortification systems was often undertaken by the army. This appears to have been the nature of the military service for many of Ḫammurapi's soldiers who labored on the wall of Sippar for three years, from his twenty-third to his twenty-fifth year:

> . . . at that time I, Ḫammurapi, . . . by the supreme might which the God Šamaš gave to me, with the levy of the army of my land, I raised the top of the foundation of the wall of Sippar with earth (until it was) like a great mountain. I built (that) high wall (*RIME* 4.3.6.2).

Similarly, Samsuiluna reports that the army was employed during his reign in the rebuilding of Kiš and its fortifications (*RIME* 4.3.7.7). Given the nature of army service, which was provided for the duration demanded, we may assume that these inscriptions imply continuous labor over the course of the years required to build these fortifications.

a. Corvée Labor

A considerable body of literature is available on the topic of labor during the Old Babylonian period (Walters 1970; see also bibliography in Klengel 1987). From these sources we learn that "there was a regular obligation to perform a certain number of days' labour per month or per year" in what Horst Klengel has referred to as "public labour service" (1987:161), usually referred to as corvée labor. This work often consisted of "digging and cleaning irrigation canals and strengthening and repairing dikes," as well as working the fields of the palace, or contributing labor during crises. A number of other infrequent tasks of this group, which for this reason may not have been mentioned, may have included the construction, modification, and maintenance of a settlement's defenses.[1] However, corvée laborers mentioned in Late Bronze Age texts, such as EA 365, seem to be involved in agricultural activity and not in construction projects. For corvée labor it is generally assumed that as much as three months per year could have been required, but not more, because individuals needed to tend to domestic labor, which included agricultural work.

b. Hired Labor

Laborers could also be hired for the building of fortifications. From at least two inscriptions we are informed of the wages paid to the laborers working on a city wall (*RIME* 4.2.8.7 and 4.2.13.21—both record the same wage), which indicates that labor could be hired by the crown when necessary. For example, Nur-Adad (1769–1754 B.C.) of Larsa reports that:

> At that time I built the great wall of Larsa like a mountain in a pure place. The wages of each worker were 3 *ban* of barley, 2 *sila* of bread, 2 *sila* of beer, 2 shekels of oil; thus they received this in one day (*RIME* 4.2.8.7).

Texts from Mari suggest that laborers could be sent to a settlement when it was urgent to rebuild a collapsed section of wall, as at Saggaratum where two hundred laborers from Terqa were requested to complete the work in ten days (ARM XIV 24). On other occasions, however, there appears to have been no immediate hurry to complete repairs (ARM II 101). We may assume that the primary distinction between hired labor and the other forms of labor discussed above was simply that, if paid sufficiently, the wages of the laborers compensated for the other labor obligations these individuals may have had. Thus, hired laborers could be put to work for extended periods of time, certainly longer than corvée laborers.

c. Other Sources of Labor

Obligatory service, referred to as *ilkum*, in which a kingdom could obtain public service from its citizens, may have provided yet a fourth source of labor. During the Old Babylonian period *ilkum* was identified as "service performed for a higher authority in return for land held" (*CAD* I, p. 74). It can be imagined, therefore, that during certain short periods in the life of each settlement members may have fulfilled their *ilkum* service by participating in fortification construction. Because of the nature of this labor, which on the one hand was integrally related to military activity but on the other was in the public's interest, it is possible that those occasionally fulfilling *ilkum* in this manner labored beside corvée laborers, soldiers, or hired hands.

During actual crises women, slaves, and even children may have labored alongside these groups, as was the case in the rebuilding of the wall of Jerusalem (see discussion below). It appears, therefore, that various sources of labor could be sought for the completion of fortification construction project when necessary. Nevertheless, the epigraphic evidence does not unequivocally suggest a single labor source that was employed for fortification construction.

2. LABOR IN PERSIAN PERIOD JUDAH

The most detailed account of the organization of a labor force during the construction of fortifications in the Levant dates to the Persian Period and is preserved in the book of Nehemiah. The account records that the inhabitants of the town of Jerusalem ca. 445 B.C. rebuilt the town's fortifications within a remarkable fifty-two days (Neh 6:15). Their feat was greatly aided by the involvement of additional workgroups from the villages of Jericho, Tekoa, Gibeon, Mizpah, and Zanoah (see Neh 3). This project provides evidence of several socioeconomic factors involved in the construction of fortifications. First, it demonstrates that workforces were drawn from surrounding towns and villages when it was necessary to build the fortifications of a settlement, particularly a political center rapidly. Second, it illustrates how a large fortification project could be undertaken through the delegation of responsibility for particular sections of the defenses to residents, households, and local dis-

[1] Seth Richardson observes that building or maintenance of the fortifications of Larsa occurred about every thirty-five years, or more or less once a generation (personal communication, 2004). This basically coincides with the archaeological evidence from Middle Bronze Age fortification systems in the Levant. Thus, we may imagine that in Mesopotamia at most individuals would have contributed service on such projects only one time during their lifetimes.

trict rulers. Finally, it demonstrates that fortification construction or reconstruction did not require a specialized labor force, but was probably often accomplished by a labor force, which with only a limited supervision and a few specialized craftsmen, was simply drawn from a region's population and could include women (Neh 3:12).

3. Specialized and Unspecialized Labor

If any of the situations described above are indicative of the organization of labor in the Levant during the Middle Bronze Age, it appears that unspecialized labor was predominantly used in the construction of fortifications. There is no basis for assuming that any of the groups mentioned above, other than the army, would have had prior experience in fortification construction. The absence of textual evidence indicating the presence of specialized laborers among them or reference to any such titles supports this conclusion.

Archaeological evidence also confirms that very little specialized labor was required in the construction of Middle Bronze Age fortifications. As noted in the previous chapter, only the largest political centers featured regular layouts, which resulted from their being constructed in the most systematic fashion. However, a closer examination of the means employed in the construction of the fortifications of individual sites suggests that separate labor teams worked to build the fortifications. These teams may have been derived from each quarter of the town or levied through extended kinship groups. At Tel Dan, for example, where multiple sections of the ramparts have been excavated, it appears that a limited set of specifications were adhered to during the building of the earthen ramparts and core wall (see p. 250). At the most these seem to include the desired elevation of the top or "head" of the rampart, the width of the base of the rampart, and some guidelines concerning the materials to be used in the construction of the core wall and the fill deposited within the rampart. Work teams were probably allowed to monitor their own progress in order to meet a deadline for the completion of each phase of the project (i.e., rampart building, wall construction).

If specialized laborers were called upon, their job would have been to oversee the integration of structures such as gates, towers, and bastions into the overall plan. Such individuals were probably identified by the Old Babylonian term *itinnum* (*CAD* I, pp. 296f.), which often referred to "house builders." These individuals were familiar with the mudbrick construction required in complicated features such as barrel-vaults that were used in gate construction. Such masons were therefore dispatched to oversee work on a town's wall as necessary. This appears to have occurred at Saggaratum where a particular mason (*itinnum*) was called upon to assist with the repair of a breach in the wall that may have been caused by the Ḫabur River (see ARM II 101, and Coleson 1982:108ff.). In the absence of greater textual evidence it does not, however, seem necessary to assume that very much specialized labor was actually required in the construction of Middle Bronze Age fortifications.

B. Construction Rates and Resource Requirements

Setting aside exceptional circumstances when women, children, and slaves participated in the construction of fortifications, in light of the evidence cited above three groups were probably employed in the construction of Middle Bronze Age fortifications: (1) the army, (2) corvée laborers, and (3) hired laborers. While each of these groups would have required oversight to one degree or another by specialized craftsmen, as noted above, the timeframe involved in the construction of fortifications would have differed greatly when undertaken by each of these groups. Thus, the likeliest labor source can be identified if the amount of labor required in construction of the fortifications can be determined reasonably well. This is because the amount of time that could be contributed by each of these groups was directly related to the duration associated with their service. The durations range from all the time necessary (i.e., hired laborers or the army) to only two or three months at a time (i.e., corvée workers).

In this section an attempt is made to assess the labor and resources required to build Middle Bronze Age fortifications. This begins by identifying the individual processes involved in these projects. Ethnographic data are then applied to determine the rates of construction that were likely to have been typical for each of the major phases of the construction process.

1. The Construction Process

With sufficient evidence estimates of the labor and resources can be broken down into separate estimates for at least four distinct stages common in the construction processes. These include (1) the excavation of fosses and the building of earthen ramparts (related processes), (2) the laying of stone walls (e.g., core, retaining, or revetment walls), (3) the raising of mudbrick walls (e.g., core or fortification walls), and (4) the construction of gates, bastions, and towers. This process in almost exactly this order is attested in various inscriptions including two of Samsuiluna:

Table 20. Previous estimates of the labor required in the construction of earthen ramparts.

Site	*Vol.* (m^3)	No. of workers	Duration in days	Excavation rate (m^3/person-day)	Person-days
Dan (Biran 1990:65; 1994:71)	1,000,000	1,000	1,000	1	1,000,000
Dan (Finkelstein 1992:208ff.)	1,000,000	2,000	900	0.56	1,800,000
Haror (Oren 1997b:257)	150,000	400	180	2.08	72,000
Hazor (Finkelstein 1992:208ff.)	1,000,000	2,000	900	0.56	1,800,000
Mevorakh (Finkelstein 1992:208ff.)	40–45,000	200	480	0.47	96,000
Michal (Herzog 1989:32)	31,500	200	120	1.31	24,000
Michal (Finkelstein 1992:208ff.)	31,500	200	240	0.69	48,000
Shiloh (Finkelstein 1992:208ff.)	40–45,000[†]	600	469	?	281,400
Sweyhat (Zettler 1997a:170f., n. 42)	170,000	?	?	3	?
Timnah (A. Mazar 1997b:250)[††]	140,000	300	467	1	140,000

[†]According to Finkelstein 20,000 tons of stone, which has been accounted for in the estimated duration presented here, should be added to the figure for rampart volume.

[††]Mazar's figure has been adjusted from the 180,000–199,800 total person-days of labor (or 600–666 days of labor for 300 workers), which accounts for the building of the ramparts, fortification walls, and the citadel, to reflect only the labor required for the construction of the earthen ramparts.

> At that time, Samsuiluna, the mighty, by the means of the force of his army built the city of Kiš. He dug its canal, surrounded it with a moat, (and) a great deal of earth made its foundations firm as a mountain. He formed its bricks and built its wall. In the course of one year he made its head rise up more than it had before (*RIME* 4.3.7.7; see also *RIME* 4.3.7.8 cited on p. 108).

Nevertheless, of these four phases it is only possible to attempt reliable estimates for the first and third stages of construction since the types of walls constructed in the second stage were often buried within the rampart and, therefore, their dimensions and construction requirements cannot be reliably estimated.[2] The duration and resources required in the fourth stage are also essentially accounted for by assuming that the area occupied by gates was instead occupied by a section of solid wall and rampart (i.e., the amount of brick and labor required for the section of wall passing through the gate area is reasonably close to the amount required for building the gate itself). Lack of information regarding the number and location of gates, towers, and bastions for most sites also complicates the estimate of labor required in the construction of these features. Furthermore, no sources provide reliable estimates for the labor requirements for gate construction in this period. Therefore, by estimating the labor required in the construction of the largest elements of these fortifications, their ramparts and mudbrick fortification walls, we can establish fairly reliable estimates for the amount of labor and time required in the construction of fortifications.

2. Construction Rates

As suggested above, various sources—most of which have never been consulted for this purpose—exist from which reliable estimates can be established for the labor required during fortification construction. These include Mesopotamian texts attesting to the labor accomplished by a single individual excavating earth, the results of experimental archaeology, and ethnographic evidence for the rates of mudbrick production and mudbrick wall construction from the modern Middle East. In this section rates are proposed for the two major processes that can be calculated: earthen rampart and mudbrick wall construction.

a. Earthen Rampart Construction Rate

Mesopotamian textual sources provide verifiable data concerning the quantity of earth that a single laborer was expected to excavate within one day of work. This figure, which was 3 m^3 per day (10 GÍN), was discovered by Albrecht Goetze in texts related to the labor of individuals who dug canals during the Old Babylonian period (1962; see also Powell 1990:490). Although this figure has been often cited by scholars in relation to canal excavation in Mesopotamia, it is veritably unknown to Levantine archaeologists. It is also necessary to note that this is a median and not maximal figure, meaning that greater and lesser rates are attested among Old Babylonian texts concerning canal excavation. Because the process of fosse excavation and rampart construction bore considerable resemblance to the excavation of canals, if there was not, in fact, a direct relationship between the two, 3 m^3 per day provides a

[2]It is also not certain that it is worth estimating the requirements of constructing these walls since it is quite possible that procurement of stones, whether from old structures or the countryside, required less labor than filling the same volume with excavated earth.

reasonably reliable rate of fosse excavation and rampart construction in this period.

Although assertions have been made by several scholars that less earth would have been moved by each laborer per day in the Levant during the construction of earthen ramparts due to the presence of different soil types and bedrock, we possess no reliable basis for suggesting that the expectation regarding the quantity excavated was lower in the Levant than in Mesopotamia. The basic rate of excavation employed by Finkelstein (1992) and others in their estimates (see Table 20 above), which was around 1.42 m^3 per worker per day, is based on poor comparative data. Closer scrutiny reveals that the latter rate of excavation was drawn from an English experiment whereby archaeologists sought to determine the excavation rate for a Neolithic ditch (e.g., Atkinson 1961; Ashbee and Cornwall 1961; Cotterell and Kamminga 1990). In doing so they established that by using "antler picks, shoulder-blade shovels and wicker carrying-baskets" (1961:129) about 1.42 m^3 of "solid chalk" could be excavated, transported, and deposited in a ten hour day (ibid., 131f.). However, they also observed that "with modern tools" the rate of excavation—of chipped stone—was basically doubled or 2.84 m^3. Therefore, with metal tools like those used in Mesopotamia, the rate of excavation is almost identical to the figure attested in Old Babylonian sources, as cited above.[3]

Given the similarities between the landscape of Upper Mesopotamia and much of the northern Levant we may conclude, therefore, that 3 m^3 (10 GÍN) is a reasonable average for estimating the quantity of earth that was usually excavated by a single worker in one ten hour day.[4] It must also be remembered that when such quantities of earth were excavated, the procurement of these materials served to produce not only a rampart (or bank), but also a fosse (or canal). While it is probably true that in the highlands earthen debris was more difficult to procure, it was probably for this reason that the scale of rampart construction projects in the valleys and plains of the Levant dwarfed those at the few highland settlements that featured earthen ramparts such as Shechem and Shiloh.[5]

b. *Construction Rate for Fortification Walls*

In addition to the rate of earthen rampart construction it is possible to establish a reliable rate for the construction of mudbrick fortification walls that improves estimates for the labor requirements and duration of Middle Bronze Age fortifications.[6] The following discussion, therefore, provides a summary of the results of various ethnographic and architectural studies from the Near East that furnish useful data for establishing construction rates for mudbrick walls. The results of this review are presented as the average number of bricks and their total volume that were produced by a single person per day. The results presented in Table 21 below are also sorted by increasing volume of bricks produced per day, since the rate of brick construction in Mesopotamian tradition, which began in the Akkadian period in Mesopotamia (see Powell 1990:490f.), was usually measured in volume perhaps because the dimensions of bricks varied (see Table 11).

The figure calculated from George Reisner's statistic for brick production in early twentieth century Egypt, of about 7.68 m^3 of bricks per laborer per day, is well outside the normal range for the quantity of bricks produced per day as revealed in other ethnographic sources (see Table 21). Excluding Reisner's data, a range between 1.28 and 2.67 m^3 of mudbrick produced per person per day is revealed. An average of the six remaining volume totals for mudbrick production derives an average of about 1.8 m^3 of bricks produced per laborer per day. If Mallowan's results are compared, his bricks being similar in size to Middle Bronze Age bricks, and two and a half days are assumed to produce and install a quantity of bricks (i.e., to procure all the materials, mix the mud for bricks and mortar, mold the bricks, dry and lay tem), then about 0.7 m^3 of mudbrick were added on average to a wall per laborer per day. This figure does not account, however, for the volume of mortar used with these bricks.

---

[3] Various efforts to account for different soil types have only exaggerated estimates to the point that in many instances the actual excavation rates used in calculations are below 1 m^3 per person per day (see Table 20).

[4] McGuire Gibson observes that this rate of excavation is the same to this day in Syria and Iraq (personal communication, 2004). Workers are hired with the expectation of excavating 3 m^3 of earth per day. Laborers will often work rapidly to accomplish this in order to be hired for other jobs.

[5] The difficulties of procuring materials and the late emergence of fortified Middle Bronze Age settlements in the highlands also accounts for the small size of these sites and a reduction in the materials and labor used in their construction. It is also possible that some of the workforce involved in these projects may have been drawn, as suggested by Finkelstein, from other sites, perhaps even from the coastal plain. It is also noteworthy regarding settlement in the highlands that the construction of fortification systems during the MB IIC, like the introduction of terracing during the Iron I, required the availability of tools made of new and harder metals: bronze during the MB IIC and Iron during the Iron I (see Stager 1985:5ff.).

[6] This was not attempted in most earlier studies since it was assumed that most, if not all, ramparted settlements lacked defensive walls.

Table 21. Mudbrick production rates per person per day according to various sources.

Region of study	Brick dimensions (cm)	No. made	Produced (*vol.* m³)	No. laid	Rate of wall constr. (m³/day)
1. Iran	20 x 20 x 4	800	1.28	?	?
2. Egypt	23 x 11 x 7	750	1.33	?	?
3. Turkey	38 x 25 x 12	120	1.37	?	?
4. Turkey	30 x 15/30 x 10	265	1.8	?	?
5. Iraq	35 x 35 x 13	150	2.39	100	0.96
6. Yemen	46 x 31 x 5	375	2.67	125	0.67
7. Egypt	40 x 20 x 12	800	7.68	?	?
Average	40 x 40 x 15	75	1.8	50	0.7

Notes for each region:

1. The figure for the number of bricks made is a reduction by about one-third of the number of bricks that, theoretically, could be produced in ten hours by one individual molding bricks at a rate of 250 per hour (Wulff 1966:109f.). We may conclude that bricks were not being produced at this rate by a single individual, if that individual were also required to obtain and mix his own materials. Based on the evidence from other estimates I assume, therefore, that the workgroup consisted of at least three individuals who were able together to produce 2,500 bricks in a ten hour day.
2. Hassan Fathy (1969:255) notes that 660 bricks measuring 23 x 11 x 7 cm (or 1.17 m^3 of bricks) required 1 m^3 of earth, 1/3 m^3 of sand, and about 45 lbs. of straw (ibid., 252).
3. See Y. Dalokay (1969:119).
4. Number of bricks made is the total for 66.25 sets of bricks, each of which consisted of four bricks: two measuring 30 x 30 x 10 cm and two measuring 30 x 15 x 10 cm (G. Hall et al. 1973).
5. Dimensions for these bricks are an average figure cited by Max Mallowan for Assyrian bricks at Nimrud (1966:53, n. 1)—which were identical in size to those used in the construction of his dig house—thus ranging in size from 35 to 48 cm^2 (ibid., 407, 416, 463f., esp. 464). The figure for bricks made is the median between 139 and 167 bricks per day, depending on whether Mallowan's reference was to five or six days within a single week for a crew of six workers responsible for making 5,000 bricks during this period of time.
6. See Salma Damluji (1992:130ff.).
7. George Reisner noted that in modern Egypt a "pair of brick makers and a mud mixer," three laborers, could produce a maximum of between "4,000 to 6,000 bricks a day, but the usual production is less owing to the conditions under which they work" (1931:73). Therefore, his figure of 2,000 to 3,000 produced by a gang of three laborers would have been more normative. Thus, one can estimate that around 800 bricks were being produced per person (i.e., ca. 2500 divided by 3). It is difficult to imagine that a single worker ever produced over 15 m^3 of bricks per day as Reisner reported.

If we account for mortar used in the wall, assuming for instance that it was only two centimeters thick between bricks and covered half of every mudbrick (while the other half of the mudbrick was covered by the mortar used on adjacent mudbricks), then about 20% of any mudbrick superstructure consisted of mortar:

Brick (40 x 40 x 15 cm) = 24,000 cm^3 (80%)
Mortar (2 cm thick) = 6,216 cm^3 (20%)

Each brick (for average dimensions of mudbricks see p. 73) was therefore covered with mortar equaling more than 25% its own volume. Thus the ratio of brick to mortar was four to one. Using this figure we can then determine how much more volume was added to walls when a mortar of about 2 cm thick was applied to bond 0.7 m^3 of mudbricks. The actual volume of wall constructed per person per day would be about 25% greater, which equates to about 0.9 m^3 of wall constructed per laborer per day. Since the figures in Table 21 have been used to establish an average rate for a laborer who was functioning as part of a wall construction team, we may reasonably assume that the procurement of material for brick production and mortar are both accounted for. Therefore, it is possible to conclude that *each* laborer working as part of a team of workers to procure materials, mix mud for bricks and mortar, mold mudbricks, and lay them within a wall, contributed approximately 0.9 m^3 in bricks and mortar to a wall per day under average working conditions.

C. Estimating Labor Consumption

The construction rates established in the previous section can be used to evaluate the reliability of previous estimates of the labor required in the construction of earthen ramparts at a number of Middle Bronze Age settlements. Among the most common figures cited are those calculated for the ramparts at Tel Dan, Hazor, Michal, Shiloh, and Timnah (Tel Batash) in the southern Levant (Table 20 above). This section demonstrates that all prior estimates are grossly in excess of the figures calculated using the rates of construction established in the preceding section. However, by recalculating them and also accounting for the labor required in the construction of the mudbrick fortification walls of these settlements, it is possible to provide detailed estimates for the construction of entire fortification systems.

1. Labor for Earthen Ramparts

Each of the estimates provided in Table 20 includes (1) the total volume of earth used in the ramparts, (2) an estimate of the number of workers involved, (3) a calculation of the duration of the entire project in days, (4) the excavation rate, and (5) the total number of person-days required to complete the construction of the ramparts. Each of these figures is discussed separately in this section.

In general, estimates of the ramparts' total volume, which were usually obtained by simply multiplying the dimensions of the ramparts, represent the most consistently derived figure. However, volumes were almost always determined by assuming that the dimensions of the largest section of the ramparts were representative of the dimensions of each section of the ramparts, which the present study has shown not to be the case. Estimates of the number of workers involved in these projects also vary considerably, which is in part a result of the excavation rates assumed and the estimated volume of construction materials. As demonstrated below, this is also a result of differing assumptions among scholars concerning the organization of labor during this period. Rates of excavation were adopted from comparative or ethnographic sources and adjusted selectively, as noted in the previous section. Each of these three figures contributed to the duration established for each project, which has been converted in Table 20 into a figure representing the total number of person-days required for the purpose of comparing them to estimates presented here (see Table 22 below).

Since the estimates provided in Table 20 for the ramparts of Timnah (Tel Batash), Tel Dan, Shiloh, Tel Haror, Tel Michal, and Tell es-Sweyhat were collected from various types of archaeological reports and were not part of a systematic study of fortification construction, they must be regarded as provisional. However, in light of the observations made above concerning the variability in construction rates employed in the calculation of these estimates, each must be individually addressed in order to demonstrate how such figures were derived and how they can be improved. This is made possible by comparing them to revised estimates using the rates of construction established in the previous section, and the specific data collected for the sites in question (Table 22). In what follows the estimates for these sites are addressed in the order in which they were published.

Table 22. Revised estimates of the labor and earth required for earthen ramparts.

Site	*Vol.* (m^3)	Person-days
Shiloh	28,955	9,650
Timnah	90,000	30,000
Dan	236,250	78,750
Haror	91,275	30,425
Sweyhat	170,000	56,670
Hazor	961,670	320,560

a. *Tel Michal*

Although a reappraisal of the evidence from Tel Michal suggests that no Middle Bronze Age ramparts existed at the site (see p. 296), since Ze'ev Herzog first attempted such an estimate (see Table 20) and since aspects of his approach have been adopted in subsequent estimates, it is necessary to carefully review his methods. Nevertheless, for the reasons cited below the site has been omitted from further calculations and therefore excluded from Table 22 above. The first observation regarding Herzog's conclusions is that he assumes that two different rates of excavation were required for the different materials used in the rampart, 1 m^3 per person per day for hamra soil and 2 m^3 for sand (1989:32), although he does not provide a source for either of these rates. The average rate of excavation was, therefore, about 1.31 m^3 per person per day, which is less than half the figure derived from Old Babylonian texts. With a total estimated volume of 31,500 m^3, according to Herzog the project required about 120 days of labor from about 200 men, or 24,000 person-days. As surprising as Herzog's figures are for such a small site, more surprising still is Finkelstein's suggestion that Herzog's estimate should be doubled to 48,000 workdays, particularly since the materials excavated, primarily sand and hamra, were not as difficult to excavate as "solid chalk" that formed the basis for the use 1.42 m^3 per day as a rate (Finkelstein 1992:210).

The starting point for any critique of Herzog's (and likewise Finkelstein's) method is the realization that it is impossible to reliably establish any ratio for the quantity of hamra to sand for features identified by Herzog as part of the Middle Bronze Age rampart, based on the very small section published. Therefore, the rate of 1.31 m^3 is essentially arbitrary. There is also insufficient evidence to suggest that the interior of the site (i.e., the "crater" formed by the Stratum

XVII ramparts) was filled in to create a platform, and therefore the fill for this platform must be subtracted from Herzog's figure. Thus, the dimensions of the remaining ramparts (if they existed) could have been at most 20 m wide and 8 m tall over, at most, a length of about 120 m. Quite possibly, however, they would only have existed on the north, east, and south, thus being only 90 m long. In any event, these figures would reduce Herzog's estimate for the volume of Tel Michal's earthen fills under reasonable circumstances to only 9,600 m^3 or to about 30% of his original estimate. This is also prior to the recognition that topographical and geological conditions similar to those attested at Tel Mevorakh (see discussion below), which was of roughly the same size and was likewise "situated on a kurkar cliff" above the coast (Herzog 1993b:1036), also prevailed at Tel Michal. To whatever degree this was the case, this would have even further reduced the quantity of material required for the construction of ramparts at this site.

As observed for Tel Mevorakh, there is insufficient evidence from which to accurately estimate the volume, let alone the labor, that was required to build Tel Michal's "ramparts." This should have been apparent when it was estimated that the 24,000 person-days would have been required to build the "ramparts" of a settlement approximately 0.1 ha in size. Since the site was not of any military significance when it was founded (it does not sit on the coastal road, which was further inland) and no evidence for a military presence was found at the site, it is unrealistic to posit that a workforce would have been brought to build sand ramparts for a site that was not strategically significant or in any perceptible danger.

b. Tel Dan

Similar rates of construction have been employed by Avraham Biran to estimate the time required for the construction of the earthen ramparts of Tel Dan (see Table 20). In addition to an excessive estimate for the size of Dan's ramparts, Biran has chosen an excessively slow rate of 1 m^3 per person per day for its building and he was forced, therefore, to assume an unnecessarily large workforce, conscripted for three months each year over no fewer than ten years (Biran 1990:65, n. 17; 1994:71). Using similar means and Biran's figures for the volume of Dan's ramparts, Finkelstein also concludes that the ramparts of Dan took about ten years to build but required 2,000 laborers.

My revised estimates for Tel Dan demonstrate that previous estimates for the construction of its earthen ramparts are grossly inflated. Specific calculations of the volume of each of the distinct sections of Tel Dan's ramparts are possible as a result of the many soundings of the defenses around the site (see Table 29). These produce an estimate of only 236,250 m^3 of earth used in the construction of its ramparts, which would have required as few as 78,750 person-days to excavate and to collect fieldstones for the core walls in the ramparts.[7]

c. Shiloh

The next group of estimates are provided in Israel Finkelstein's article, *Middle Bronze Age 'Fortifications': A Reflection of Social Organization and Political Formations* (1992; reprinted in Finkelstein et al. 1993:378ff.).[8] Finkelstein concludes that the construction of Shiloh's ramparts took about 470 work days for a workforce of 600 over the course of five years (i.e., no more than 3 months of labor per year for these individuals). Despite the usefulness of attempting such calculations, his conclusions still prove difficult to accept because he arrives at his figures by using maximal dimensions for the defenses and the lowest figures available for the rate of excavation. First and foremost, he presupposes that the fortifications of each settlement were predominantly homogenous throughout, which I have demonstrated was rarely the case (see chapter 3) and therefore he uses the largest dimensions cited for these fortifica-

[7]It is assumed that the fieldstone core of much of Dan's ramparts, the volume of which was 65,536 m^3, was constructed at the same rate since fieldstones could be collected from around the settlement without considerable difficulty. Thus, this volume is added to that of the earthen fills of the ramparts, which alone is estimated to have been about 170,714 m^3.

[8]There is a discernible inflation in each successive estimate provided for the labor required in building ramparts at the sites discussed by Finkelstein in this article. His assertion that it is "more reasonable to assume that much of the work at each site was accomplished in a relatively short time" is nowhere demonstrated in his estimates. Although I agree with this statement—if for different reasons, as is later discussed— Finkelstein's further assertion that "there are only a few sites where chronological phasing [of the fortifications] can be proven" and "in most cases the 'stages' are no more than constructional elements" (1992:210) is unfounded. It is largely influenced by the archaeological evidence from the late Middle Bronze Age settlement at Shiloh. While this may be true for the calculations of ramparts at sites such as T. Dan and Hazor, which were erected within a short period of time, the fortifications of many settlements feature a heterogeneous composition that makes reliable estimates almost impossible in the absence of multiple soundings of the ramparts. Furthermore, it is necessary to acknowledge that the fortifications of many sites attained their present dimensions only after the latest phase of their construction.

tions to derive these estimates. At Shiloh, for example, the dimensions of the fortifications on the northeast side of the site are used (1993:208)—the only place where a rampart was identified—although everywhere else the site was essentially defended by a modest casemate-type wall. Furthermore, taking a cue from Herzog's work at Tel Michal he ventures to estimate, without verifiable evidence, the size and volume of the earthen fills used inside the fortifications during the construction process at Shiloh.

Other factors also affect the reliability of Finkelstein's conclusions. As discussed above he employs an extremely low rate for excavation, ca. 1.42 m^3 per person per day, and does not account for the excavation of materials such as sand, earth, or gravel, which were more easily excavated and did in fact provide the bulk of the composition of the ramparts Finkelstein discusses. Finally, Finkelstein's method results in further inflated figures by supposing that materials were transported from some distance, at least as much as one kilometer to Shiloh (1992:208), for example, an assertion which finds no support in the present study (see also Kotter 1986:490).[9] As a result of this Finkelstein suggests that five years were required for the construction of Shiloh's fortifications, presupposing that only twenty percent of the population was involved in the construction project during at most three months each year (ibid., 208f.), following the conclusions of K. Mendelssohn (1977:esp. pp. 143, 92) and Colin Renfrew (1984:esp. p. 238). This, however, is a total of 469 days of labor by 600 individuals moving 20,000 tons of stone (in 416 days) and 45,000 m^3 of rampart fill (in 53 days) according to Finkelstein (1992:208f.).

According to the calculations offered here, which are based on the dimensions of Shiloh's rampart (see pp. 310ff.), Finkelstein's estimate of the eastern rampart must be reduced to about 12,250 m^3. It is also difficult to accept that another 45,000 m^3 of fill, a figure that cannot be verified due to the limited extent of the excavations, were used inside the stone wall on much of the east side of the site. It is reasonable to suggest that an amount at least equivalent to the rampart itself was required against the inside of the core wall of the rampart, or another 12,250 m^3. In order to account for the estimated 4,455 m^3 of stone used in the core wall of the rampart, this figure was added to the rampart's volume. It is assumed that materials for this feature could have been obtained at the same rate as the earth excavated, since much of it may have been taken from exposed outcrops of stone, perhaps during the leveling of the hill.

Even if the rampart on the east had consumed most of the labor during the construction of the fortifications, the project was still within the reasonable limits of the work required to build most Middle Bronze Age fortification systems. However, it is not unreasonable, as Finkelstein suggests, that the construction of Middle Bronze Age fortifications in the hill country such as those at Shechem, Shiloh, Hebron, and possibly Jerusalem, did require a disproportional amount of work when compared with fortifications constructed in the coastal plain, for example. They were smaller settlements to begin with and required considerable modification of surrounding topography prior to beginning the construction of their fortifications. While the source of the extra labor will remain a topic of debate, it is clear that labor like that which was required for most rampart projects has been overestimated.

d. *Hazor*

Although revised calculations suggest that Hazor's ramparts were almost as voluminous as Finkelstein suggests (see Table 35), he does not reveal how he arrived at this figure (1992:210). He only notes that the same quantity of materials were used in Hazor's ramparts as had been used for Tel Dan's, though Hazor was, of course, a much larger site.[10] Furthermore, despite the precise information available from which to estimate Hazor's population and thus its available workforce, he does not note that the labor requirements for the construction of Hazor's ramparts were considerably lower per capita than for all of the other sites he includes in his study. However, with the detailed information available for the dimensions of Hazor's earthen ramparts and the rates of excavation identified earlier, as few as 320,560 person-days were probably required to build Hazor's ramparts, compared with Finkelstein's estimate of 1,800,000 person-days.

e. *Tel Mevorakh*

Finkelstein's suggestion that the ramparts of Tel Mevorakh required even twice the amount of labor as those of Tel Michal, at around 96,000 workdays, is unfounded since it was based on the supposition that

---

[9]Finkelstein does not account for the fact that Shiloh was built on a hill, which was shaped differently than it is today, and that leveling before construction of the fortifications provided much of the material used in the infilling of the settlement's "platform" and rampart.

[10]With respect to the labor required to build Hazor's rampart, Finkelstein seems to have ignored the site's size. This may be because Hazor's population was sufficiently large to complete the project in a far shorter time period than would have been possible at T. Dan with its smaller population.

the earth was "more difficult to dig . . . than at Tel Michal" (1992:210). Despite Finkelstein's attempt to estimate their volume and the labor required in their construction (see Table 20), since insufficient sections are available from which to establish the dimensions of the ramparts of Tel Mevorakh in Stratum XIII, it does not seem possible to accurately estimate the quantity of materials used in the ramparts. Ephraim Stern also notes that this rampart "was erected above a natural hill in the form of a rectangle of which the north, west, and south sides were almost completely aligned with the slopes of the natural hill" (1984:50). Furthermore, Stern's remarks concerning the source of the building materials for the rampart during this period demonstrate that the project was probably never a very large one:

> During the course of the construction of the rampart, the buildings in the centre of the site and those projecting above it, were almost totally razed, though all the walls of both stratum XIV and of the fortress of XV were left standing and were buried under the rampart. These walls served to add strength to the rampart itself and as an economical supply of building material (1984:68).

In light of these observations it is impossible to attempt a revised estimate of the labor consumed in the construction of Mevorakh's ramparts.

f. Timnah (Tel Batash)

One of the more recent estimates of labor used in the construction of Middle Bronze Age ramparts was produced by Amihai Mazar for Timnah (1997b:250). In many respects Mazar's estimate was well constructed, accounting for the transportation of construction materials and the effort required to build the mudbrick wall and citadel. For this reason, however, his calculations are adjusted in order to be directly compared with those provided by Finkelstein and others in Table 20. Nevertheless, Mazar surmises that a workforce of 300 individuals living in "temporary camps" could have built the fortifications (i.e., ramparts, wall, and citadel) of Timnah in two years (ibid.). This would be about 180,000 to 199,800 person-days of labor for the entire project or about 140,000 person-days of labor for the earthen ramparts alone and about 40,000 to 50,000 person-days of labor for the mudbrick walls.

Although Mazar observes that previous estimates did not account for the distance from which materials were brought (1997b:250), it has been demonstrated in this work that such a factor was basically insignificant in the calculation of the labor required since most of the necessary building materials were readily available within meters of the construction site. Furthermore, they were often extracted from the fosse that was excavated outside the ramparts. Thus, it is probably unnecessary to speculate about transportation distances, which unnecessarily inflate the figures for required labor.

Mazar's estimate is inflated for a number of other reasons. First, he assumes that the complete length of the fortifications were 800 m when actually the ramparts' lengths must be measured along their crest, which reveals them to be about 600 m long. This provides an initial reduction of Mazar's estimate by about 25%. Second, he relies upon an exceptionally low rate of around 1 m^3 per person per day for the excavation of earth. Finally, since no rate for the construction of mudbricks and mudbrick walls is provided, simply increasing the existing estimate by 30% to 50% as Mazar does negates the effort invested in attempting to accurately calculate the labor requirements. The net effect of each of these assumptions is, again, an inflated figure for the length of time required to build the ramparts of Timnah. However, employing the new figures, it is possible to suggest that only 30,000 person-days were required to build the ramparts of Timnah. The construction requirements of its walls are discussed in the following section.

g. Tel Haror

For Tel Haror Eliezer Oren concludes that a workforce of 400 men were involved in the construction of its ramparts (Oren 1997b:257), providing 72,000 person-days worth of labor (see Table 20). His figures suggest the acceptance of a much faster rate for the excavation of earth for the ramparts, of around 2 m^3 per person per day, though still below the figure of 3 m^3 suggested in this work. Thus, reconsideration of the labor required for Tel Haror using this new rate and revised figures for the ramparts' dimensions (see pp. 264ff.) reveal that the earthen ramparts could have been built with about 30,425 person-days of labor (see Table 22).

h. Tell es-Sweyhat

Sweyhat provides the most reasonable basis for estimating the rate of construction for earthen ramparts, although no figures for the labor force or the total number of days required were provided by Richard Zettler (1997a:170f., n. 42). Nevertheless, the rate of excavation which he assumes to be appropriate, approximately 3 m^3 per day, comes from the same Mesopotamian sources used to establish the rate discussed above. Hence, Zettler's estimate, though not fully fleshed out in his own work, would result in 56,670 person-days worth of labor for the construction of Sweyhat's ramparts (see Table 22).

i. Jerusalem

Another set of estimates has been provided for the construction of the Middle Bronze Age defensive system of Jerusalem (Boas-Vedder 2001). However, since these fortifications did not feature a rampart these figures have not been included in Table 20 above. Diny Boas-Vedder concludes that the building of the Middle Bronze Age wall of Jerusalem was "executed within 2.5 years, without inflicting considerable stress on the local community" (ibid., 151). While the article's content is at times thought provoking for the specific issues it raises with respect to fortification construction, the study and resulting estimates are exceedingly speculative and to a degree imaginary. This is due to the paucity of evidence for the size of Jerusalem in this period and the specific dimensions of the fortifications being considered.[11] As in the case of Tel Michal and Mevorakh, there is insufficient evidence by which to accurately establish the dimensions of Jerusalem's fortifications and their course. Therefore, along with these two sites it is not worthwhile to include Jerusalem in further estimates of the labor and resources required in wall construction, which are provided in the next section.

j. General Observations

As demonstrated above, earlier studies that have sought to consider the labor requirements for building Middle Bronze Age ramparts do not provide accurate estimates of the amount of labor required in their construction. This is largely due to efforts to treat the fortifications of each site individually, establishing different rates for their construction whenever possible. However, such rates are often subjectively derived. Too often the excavation rate has dropped to an unrealistically low 0.47 to 0.69 m^3 per person per day, which is less than one-quarter the figure suggested by Mesopotamian sources. For this reason most estimated excavation rates must first of all be adjusted upward and, second, the labor required in the construction of the mudbrick fortification walls built atop the ramparts must be accurately established.

2. Labor for Fortification Walls

Estimates of the quantity of mudbricks used in wall construction are only relevant for solid mudbrick walls. There is no value in estimating the labor required for building casemate-like walls such as, for example, those at Shiloh, Kumidi, and Shechem. Since these structures were multipurpose and were often used as storerooms (as attested by the frequent finds of storage jars inside them), we may safely assume that buildings of similar dimensions would have been built even if they had not been incorporated into the site's defenses.

By applying the average rate identified earlier for the construction of solid mudbrick walls (0.9 m^3 per person per day) to the walls for the sites that possess evidence of solid mudbrick walls, we may further refine our labor estimates. Thus, estimates are provided for the labor required to build the fortification walls of Shiloh, Timnah, Dan, Haror, Sweyhat, and Hazor (see Table 23). These figures are based upon the data for estimated wall height presented on pp. 60ff. and for the width and length of each site's walls (see Table 7). These dimensions suggest that mudbrick walls were probably about 10 m high (a median between the low suggestion of 5 m and an upper limit of 15 m).

3. Total Labor Requirements

Of particular interest with regard to the figures for rampart and wall construction is the total amount of labor required for both of these tasks (see Table 24). The height and width of walls and ramparts were, of course, the most significant factors in determining which task took longer. Sometimes these two features required nearly the same amount of time to build. In other cases, however, where the ramparts were particularly pronounced, as at Timnah, Dan, and Hazor, the ramparts took considerably longer to build than the walls. In still other cases, particularly where walls were thicker and the ramparts less pronounced, wall construction appears to have consumed more labor, as at Haror and Sweyhat. Such varied evidence contradicts the suggestion that rampart construction consumed more resources and labor than wall construction, undermining the assertion that ramparts were intended as a form of conspicuous consumption by ruling elites during the Middle Bronze Age.

[11] Although Boas-Vedder has marked each estimated figure in his study with **, the reliability of the estimated figures varies considerably. Particularly significant assumptions in these calculations include that the wall: (1) was only 5 m high and was built entirely of stone (Boas-Vedder 2001:137); (2) was 1,000 m long; (3) featured indentures along its entire length; (4) required the construction of support ramps to build (p. 138). Other assumptions include that (5) the foundation stones should be accounted for separately (p. 138); (6) only two gates existed; and (7) materials were also consumed in the creation of a path for access to the wall (p. 139).

Table 23. Estimated labor required in the construction of fortification walls.

Site	Avg. wall *w.* (m)	Est. wall *L* (m)	Wall *vol.* (m^3)	Person-days required
Shiloh	3	565	16,950	18,835
Timnah	3	600	18,000	20,000
Dan	4	1,200	48,000	53,335
Haror	4	1,300	52,000	57,780
Sweyhat	9	1,675	150,750	167,500
Hazor	5	2,980	149,000	165,560

Note: Since much of the wall of Shiloh consisted of the rear walls of domestic structures, the new estimate may still be as much as 50% in excess of the labor actually required. Also, the width used for the wall at T. Dan is an average of attested widths from 3 to 7 m in different areas.

Table 24. Labor requirements for Middle Bronze Age fortification construction.

Site	Person-days for rampart constr. (see Table 22)	Labor ramparts (%)	Person-days for wall constr. (see Table 23)	Labor wall (%)	Total person-days	Total person-mos.
Shiloh	9,650	34	18,835	66	28,485	950
Timnah	30,000	60	20,000	40	50,000	1,667
Dan	78,750	60	53,335	40	132,085	4,403
Haror	30,425	35	57,780	65	88,205	2,940
Sweyhat	56,670	25	167,500	75	224,170	7,472
Hazor	320,560	66	165,560	34	486,120	16,204
Average	87,676	52	80,501	48	168,178	5,606

Table 25. Pertinent data for estimating the available workforce for sites listed in Table 24.

Site	Site size (ha)	Est. pop.	Est. pop. of adult males
Shiloh	1.7	425	85
Timnah	2.25	560	112
Dan	11.1	2,775	555
Haror	16.2	4,050	810
Sweyhat	30	7,500	1,500
Hazor	63.1	15,775	3,155

Table 26. Duration of fortification construction employing corvée or hired laborers.

Site	Total person-mos.	Male labor force	Corvée labor (*years* required at 1 mo./year)	Corvée labor (*years* required at 3 mos./year)	Hired labor (*mos.* required)
Shiloh	950	85	11.2	3.75	11.2
Timnah	1,667	112	14.9	5	14.9
Dan	4,403	555	7.9	2.6	7.9
Haror	2,940	810	3.6	1.2	3.6
Sweyhat	7,472	1,500	5	1.65	5
Hazor	16,204	3,155	5	1.65	5

Table 27. Numbers of soldiers required to build fortifications within two to eight months.

Site	Total person-mos.	2 mos.	3 mos.	4 mos.	5 mos.	6 mos.	7 mos.	8 mos.
Shiloh	950	475	320	240	190	160	140	120
Timnah	1,667	834	556	417	334	278	238	208
Dan	4,403	2,200	1,470	1,100	880	735	630	550
Haror	2,940	1,470	980	735	590	490	420	368
Sweyhat	7,472	3,740	2,490	1,870	1,500	1,250	1,070	934
Hazor	16,204	8,100	5,400	4,050	3,240	2,700	2,315	2025

D. Scenarios for the Construction of Fortifications

Combining the number of person-days required for the construction of earthen ramparts and the mudbrick fortification walls yields the total number of days of labor required to fortify these sites (Table 24). Yet these figures alone do not identify the labor force involved. When presented as a number of days it remains difficult to compare the scale of the labor requirements to projects described in textual sources, which are described in terms of the months required to complete them. Furthermore, corvée and *ilkum* service are often thought of as labor contributions in terms of months contributed per person per year. For various reasons, therefore, converting this figure into the number of person-months of labor that were required to complete the construction of these fortification systems (last column of Table 24) facilitates their comparison with the manner in which labor is usually represented in Near Eastern sources.

From here it is possible to proceed to identify the labor force employed in the construction of Middle Bronze Age fortifications. As demonstrated in section A.1, Old Babylonian sources suggest that one of three sources of unspecialized labor were routinely involved in fortification construction projects: (1) corvée laborers, (2) hired laborers, or (3) the army. In this section each of these sources of labor will be considered as candidates for the workforce most frequently involved in fortification construction. By considering the characteristics of the labor usually contributed by these three groups (i.e., the number of months they could be employed per year) it is possible to increase the likelihood of identifying which of these groups were most routinely involved in fortification construction.

We begin by considering the scope of these projects with respect to the maximum adult male workforce available from each site. The data pertinent for determining the size of this labor force for the six sites discussed above are presented in Table 25. The population of each of these settlements can be reasonably estimated assuming a density of approximately 250 persons per settled hectare having, of course, subtracted the area occupied by the fortifications themselves (cf. figures in Broshi and Gophna 1986). The number of adult males, who usually represented the bulk of any construction force, however it may have been organized (see scenarios below), can be estimated, assuming that they usually comprised about 20% of the population (see Mendelssohn 1977; also Renfrew 1984). Such figures reveal that the male labor force from Shiloh would have been quite small, while that of Hazor was more than sufficient for a massive undertaking.

a. *Corvée Labor*

The service available from corvée laborers—the type of workforce advocated by Finkelstein and others—can for our purposes be estimated to have been no more than one month per year per person from a given settlement (see discussion below). Since in this scenario projects would have lasted multiple years, the significant figure is not the total number of months of labor contributed, since this would have been consistent per annum regardless of the type of project undertaken. Rather, the figure of importance is the duration of the projects, which would have lasted years (see Table 26). Under these conditions the male inhabitants of a settlement would have been expected to labor one month each year between a minimum of 3.6 years (e.g., Haror) and a maximum of about 15 years (e.g., Timnah) depending upon the size of the settlement. Even if corvée laborers were compelled to contribute three months of service, the fortifications of even the smallest settlements would still have required more than a year and a half to complete.

Despite the appeal of suggesting that corvée labor was employed in the construction of these fortifications, several problems are encountered in this scenario. First, it assumes that a complement of adult males equal to that of the settlement during its peak was available when construction began or that laborers would be routinely drawn from the countryside for this task. Second, it is only realistic to assume that corvée laborers could contribute about one or perhaps two months of labor per year, beyond which their normal subsistence activities would have been adversely affected. Third, the amount of time required of corvée laborers, which is comparable to those estimated by other scholars (see discussion below), supports the conclusion that obtaining labor in this manner would have been extremely inefficient if security was of any concern. (It cannot be assumed that any of the Old Babylonian towns, for which we have descriptions of the construction of their fortifications, possessed no prior fortifications.) Furthermore, archaeological evidence does not support the suggestion that the construction of the defenses of Middle Bronze Age settlements occurred over such prolonged periods of time.

b. Hired Labor

If corvée laborers could not supply a labor force capable of completing the work within a reasonable timeframe, then perhaps hired laborers might have completed these projects at a faster rate, which would be in keeping with the archaeological and historical evidence suggesting that the largest projects lasted at most a couple of years. Still assuming the limitations of a locally hired labor force—of equal size to the male populations of the settlements being fortified—we can estimate the total number of months that would have been required to complete their construction (see last column in Table 26 above). Thus, projects would have lasted anywhere from 3.6 to 15 months with continuous construction.

A number of factors mitigate against the use of hired laborers, however. First, there is little basis for identifying a large pool of available workers in the Levant that would have been available for all of the projects being undertaken. Second, it is unlikely that hired laborers, most of whom would have held other occupations, could have offered their services full-time (ten hours per day) for the required number of months until these projects were completed. Larger labor pools than those suggested in Table 26 would have been necessary to maintain the desired pace. Therefore, if hired labor was to be employed, a larger workforce would be needed from across a larger area—a scenario already discounted above—or the duration of these projects would need to have been extended, which as already noted is not supported by archaeological evidence.

c. Army Labor

Scenarios involving corvée and hired labor, if drawn from the settlement that was to be fortified, demonstrate that freestanding rampart fortifications of the largest sites, such as Hazor, would have required the least work per capita in their construction. Therefore, the larger sites were, the less inordinate the work that was required by their inhabitants in the construction of their fortifications. Such conditions would, no doubt, have allowed for greater social stratification at these sites, as there was certainly room for choices to be made about who would participate (or be conscripted) in the menial labor of wall and rampart construction, with plenty of openings for those who would not. Yet there is little evidence to suggest large-scale settlement at these sites prior to the construction of their fortifications, and thus we cannot presume that such large and diverse populations provided the reservoir from which laborers were drawn. This poses a considerable problem for the scenarios posited above, which is not easily resolved: if we cannot reliably determine the size of the population of these and surrounding settlements prior to the erection of their fortifications, then by what criteria are we to determine how long such projects took?

In light of the challenges posed by the use of both corvée and hired labor, the employment of army labor seems an altogether more likely source for the labor required in the construction of fortifications during the Middle Bronze Age. As suggested by Old Babylonian sources, the army provided not only the massive labor force that was required, but the ability to organize such work groups. Army labor could be conscripted from across a broad geographic region for building the fortifications of individual settlements as necessary, an observation that is also fundamental to previous hypotheses regarding the labor force involved. Furthermore, the use of army labor is in line with the recognition that the Levant during the Middle Bronze Age was primarily divided among large territorial kingdoms such as Hazor and Ashkelon, as advocated in chapter 5.

If, for example, a detachment of two thousand soldiers was regularly conscripted to build fortifications, the task could be completed at sites of different sizes within fewer than eight months (see last column in Table 27 above). The fortifications of sites up to about 16 ha in size (e.g., Haror) could have been built in less than three months. Those of larger settlements such as Sweyhat (ca. 30 ha) would have required almost four months, while the fortifications of the largest order of settlements like Hazor would have required a little more than eight months.

Alternatively, the fortifications of these settlements could have been completed with army labor by the end of two consecutive summers, if construction had been limited to between two and three and one-half months of service each summer. Nevertheless, such a scenario presupposes that security concerns did not warrant the prompt completion of these defenses and that insufficient labor was available from which to conscript a force of two thousand soldiers for the entire duration necessary (i.e., up to eight months), even if individuals were routinely replaced by fresh recruits. This scenario presupposes, of course, that during the early stage of a settlement's urban development when fortifications were being built, the army would not have been engaged in campaigning and thus was free to be engaged in building campaigns.

Historical texts describing the construction of fortification systems of major cities and the building of fortresses throughout Mesopotamia provide one of the most important sources for determining the duration, cost, and scale of these projects. They also make it possible to assess the accuracy of the assertion that the army was involved in these construction efforts. The textual evidence for the durations of these fortification projects is particularly useful for comparative purposes. An inscription attributed to year twenty-one of Gungunum of Larsa (1836–1810 B.C.) reports that:

> in the course of one year [Gungunum] made the bricks and built the great wall of Larsa named Utu-kibale-sadi ('the god Utu overtakes the rebellious land') [*RIME* 4.2.5.3].

The construction (or reconstruction) of the wall of Sippar during the reign of Ḫammurapi appears, however, to have required at least two but possibly as many as three years (i.e., between years 23 and 25, 1674 to 1672 B.C.). The raising of "the top of the foundation of the wall of Sippar with earth (until it was) like a great mountain" and Ḫammurapi's note that "I built (that) high wall" appears to suggest that this construction project may have resembled those of rampart-fortified settlements in the Levant (*RIME* 4.3.6.2). Ḫammurapi's project, as noted earlier, was also accomplished by employing the army. Samsuiluna also records completing the fortifications of Kiš, which included a canal, a moat, and heaped up earth in addition to raising its wall within one year (ca. 1630 B.C., *RIME* 4.3.7.7).

The evidence also indicates that smaller fortresses (i.e., a *dūrum*) could be built in as little as two months. Samsuiluna reports for his seventeenth year (1637 B.C.) that he produced the bricks for and constructed six fortresses within two months (*RIME* 4.3.7.5). Elsewhere Samsuiluna records the construction of Dūr-Samsuiluna and again confirms that two months were required for its construction (*RIME* 4.3.7.8). The dimensions of the fort, which does not appear to have featured a rampart, can be estimated based upon the contours of the site with which it is identified, Mound B at Khafajah. It was probably about 250 by 400 m or about 10 ha in size (Delougaz 1990:fig. 28). Since the majority of the work of building this settlement probably consisted of building its wall, we can estimate the quantity of labor required in its construction. The above dimensions of the fortress (which are maximal) yield a perimeter of about 1,300 m. The wall is known to have been about 4.7 m thick. If it was 10 m high, then about 61,100 m^3 of mudbrick were required for its construction. If it was completed within two months (60 days), then an average of approximately 1,020 m^3 of mudbrick were deposited per day. The number of laborers required for this task would, therefore, have been about 1,130, if we assume that they built the wall at a rate of roughly 0.9 m^3 per person-day as suggested above. The completion of this project with a labor force of this size is comparable to our estimates for the construction of the fortifications of Haror within two months (see Table 26).

In sum, the evidence from these historical sources suggests that smaller fortresses without ramparts (*Dūr-PN/DN*) on the order of 10 ha in size could be constructed in as little as two months. This figure is in keeping with the five to six months calculated above for Levantine settlements of comparable size, though these also included the construction of earthen ramparts. Furthermore, the construction or reconstruction of the fortifications of much larger cities in Babylonia also appears to have required one year but could also require as many as three years.

The results of the three scenarios are revealing. It can be safely concluded that there were no fortified settlements in the southern Levant that could not have been fortified in less than about eight months by employing a workforce of just two thousand soldiers. Any increase in this number would have improved the rate of construction or the number of projects that could be undertaken throughout a region at the same time. In light of the political organization suggested for the Levant during the Middle Bronze Age in chapter 5, there is little doubt that these fortifications were most often, if not primarily, built using conscripted troops during typical periods of enlistment (i.e., summer months) as described in Old Babylonian sources.

E. Social Complexity and the Construction of Fortifications

Evidence for the integration of fortified settlements into polities such as the kingdom of Ashkelon and the labor force required to fortify them during the Middle Bronze Age suggest a high degree of social complexity. Despite studies of Middle Bronze Age fortifications and their relationship to the settlement patterns and political centers of the southern Levant (see pp. 12ff.), only one of which has addressed fortifications as a "social phenomenon" (Bunimovitz 1992), very little work has been done to relate these developments to social complexity (e.g., Dever 1993b; Falconer 1994b; Maeir 2003). Therefore, it is worthwhile to consider how the socioeconomic impact of fortification construction during the Middle Bronze Age relates to increasing social complexity in the Levant.

1. Increasing Social Complexity

In addition to references to Amorite kingdoms in the northern Levant in the Mari texts and the listing of kings and well-known cities of the southern Levant in the Execration Texts, the clearest lines of evidence for increasing social complexity in the Levant during the Middle Bronze Age include the emergence of urban centers, the construction of massive fortification systems, and the labor invested in them, craft specialization, and burial customs. Of these, only urbanism (e.g., Kotter 1986; Falconer 1987; Kempinski 1992e; Falconer 1994b; Falconer et al. 2003) and burial customs (e.g., Stiebing 1970; Hallote 1994; Garfinkel 2001; Hallote 2002; Baker 2003) have received close study. Therefore, the labor requirements for constructing Middle Bronze Age fortification remain to be considered within the framework of increasing social complexity. Likewise, settlement patterns, especially when viewed in relation to defensive concerns, serve as a barometer of political and social stratification.

The patterns characteristic of both settlement and fortification construction throughout the Levant during the Middle Bronze Age suggest an advanced state of political and social complexity not unlike that of Mesopotamia. As suggested by the evidence for a hierarchical settlement pattern in the southern Levant during the Bronze Age (see pp. 103ff.), it is incorrect to suggest that the evidence for settlement patterns "invalidate any hypothesis of urban nucleation, or urbanization of any kind" in the southern Levant (see Falconer 1987:190f.). This perspective is based, first, on the observation that there existed far fewer "cities" in the southern Levant, which only ranged in size from 10 to 60 ha, than in Babylonia where they ranged from 10 to 200 ha. The scale of Mesopotamian society, which was obviously the result of irrigation agriculture, should not obscure the recognition of the fundamentally identical agricultural focus of both cultures. Thus, the scalar difference between these societies is irrelevant for evaluating the degree of relative stratification, both political and social, in the Levant.

Falconer's suggestion also relies upon the assertion that the economy of the Levant, which is usually identified as "rural," was fundamentally different from that of Mesopotamia, which is often identified as "urban." As David Schloen demonstrates, the traditional invocation of the urban-rural dichotomy is misplaced in the analysis of Bronze Age cultures in the Near East (2001:101–04, also esp. at Ugarit, pp. 334–42). It is clear, for example, that political integration of the type characteristically associated with Mesopotamian society, which resulted in the emergence of an integrated settlement pattern, existed albeit on a smaller scale (see pp. 116ff.). In the Levant this pattern exploited not only the agricultural potential of each settlement's territory as noted by Wade Kotter (1986), but it also led to the selection and fortification of strategic sites in the defense of political districts within each kingdom.

Despite the limits of historical sources available for the southern Levant during the Middle Bronze Age, there is no basis for doubting that the development of social and political stratification paralleled that of the northern Levant and to some extent Mesopotamia. Discovery of the sepulchers of Amorite rulers at Ebla, Byblos, and most recently, Qatna, reveals the customs associated with the Amorite ruling class who resided at capitals and administrative centers, which were also furnished with palaces during this period (e.g., Ebla and Qatna). The absence of evidence thus far for royal burials at identical settlements, such as Hazor and Ashkelon, is no more than a matter of chance.

2. Middle Bronze Age Kingdoms and Their Labor Resources

Previous discussions regarding the political organization of the southern Levant during the Middle Bronze Age by both William Dever (1987) and Israel Finkelstein (1992), while differing in their details, find considerable agreement with respect to the identification of particular sites and their territories as Middle Bronze Age polities. In light of the historical and socioeconomic context presented in this study, such analyses are now granted greater significance. The traditional view that resulted in the identification of a number of independent Middle Bronze Age polities defined by discrete territorial units must be altered. These polities are identifiable as districts within what must have been five or six large kingdoms that dominated the southern Levant in a manner not unlike the kingdoms of Yamḫad and Qatna in the northern Levant. Therefore, most of the second-tier settlements including Lachish, es-Safi, and Gezer, which were once identified as "city-states" by Finkelstein and others, are better identified as administrative centers or district capitals subordinate to larger kingdoms like Ashkelon (cp. Old Babylonian organization, see pp. 106ff.), Hazor, Akko, Shechem, and Jerusalem during the Middle Bronze Age.

Given that the evidence used in this study to identify political organization in the Levant has been available for some time, Finkelstein's assertion of the existence of "polymorphous chiefdoms" in the highlands of the southern Levant during the MB II is unnecessary (1992:207). While it is true that these fledgling highland settlements were smaller than those along the coast, as with Falconer's comparison of Mesopotamian and Levantine societies, differences in scale do not necessitate differences in political (i.e., administrative) or social organization. Unfortunately,

Finkelstein's interpretation is fueled by exaggerated calculations for the labor required for earthen ramparts which, in his opinion, reveal that they were "more sophisticated and complicated to build than simple walls" (1992:213). However, this conclusion is supported neither by technical assessments of their construction (see pp. 48ff.) nor by reassessments of their labor requirements.

Likewise, Finkelstein's argument that massive resources were invested by a mostly rural population in a limited number of urban political centers, as he suggests for the highlands, is incorrect. It is evident from the territory around Ashkelon that the greatest quantity of labor was invested in the construction of the defenses of all of the secondary fortified settlements within these kingdoms, such as Yavneh-Yam, 'Ajjul, Haror, Ekron, and Lachish, though they were integral to Ashkelon's defense. The investment of labor resources was not, therefore, restricted to the capital of these polities. If such an interpretation of the highlands is advanced, it is solely the result of limited data for fortified settlements around Jerusalem and Shechem.

Based upon the available evidence there is no basis for positing a unique social structure or political organization in the highlands during the Middle Bronze Age, despite that, because of ecological differences, the subsistence economy may have differed in character and scale from that of the coastal plain or Jordan Valley. Similarly, even though we are still unable to account for the chronological development of the interactions between the indigenous population of the highlands and what may have been new settlers arriving from the coastal plain, the evidence is insufficient to suggest that the inhabitants of the highlands viewed Middle Bronze Age fortifications at, for example, Shechem and Shiloh, as distinctive reservoirs of ideology and propaganda (e.g., Bunimovitz 1992; Ilan 1995:esp. pp. 316f.). As demonstrated in chapter 3, the evidence suggests that in the highlands these structures functioned as fortifications as they did throughout the Levant, and thus they do not require the identification of a different socio-political organization.

3. Conspicuous Consumption or Strategic Investment?

In light of these observations, suggestions by Israel Finkelstein and other scholars that the construction of earthen ramparts (if not also fortification walls) consumed astronomical quantities of labor and thus only constituted conspicuous consumption must be dismissed. The labor postulated by these scholars, which required groups of between three hundred and two thousand corvée laborers working between 24,000 and 1.8 million person-days (see Table 20), is unnecessary since as demonstrated above on average between 20% and 40% of this labor was required for the construction of earthen ramparts (cp. Tables 20 and 24). While the previous scenarios required imagining decades of ceaseless labor by large numbers of the inhabitants of the southern Levant in the construction of fortifications as they moved from site to site laboring only a few months each year, this study has demonstrated that such a scenario is unwarranted. To judge from the archaeological evidence, fortifications were actually erected within relatively short periods of time to a limited set of specifications (see pp. 48ff.), and although they were built using unspecialized labor, the army would have provided the most likely source for such labor on the occasions that it was necessary.

While the estimates provided in the preceding section illustrate that when necessary the construction of freestanding earthen rampart fortification systems were well within the capacity of Levantine society, it should also be remembered that such fortifications were much less common than supplemental ramparts, which required far less labor while achieving the same level of defense for a settlement's inhabitants. Furthermore, the potential employment of a combination of different workforces would only have increased the speed with which fortifications of either the freestanding or supplemental type could be erected. If the involvement of the army is discounted or presumed to be less than the size suggested here, the remainder of the required labor force could always be provided by employing corvée or hired laborers.

4. Craft Specialization

While it is clear that the construction of fortifications required only a limited amount of specialized labor, the regularity of the layouts and constructional styles of gates, towers, and bastions does suggest the employment a group of specialists capable of planning for and integrating these structures into each site's fortifications. In this respect, such craft specialists were comparable to those thought to have produced the frescoes attested at sites such as Avaris, Kabri, and Alalaḫ during the later stages of the Middle Bronze Age. If the Old Babylonian term *itinnum* was used to refer to such individuals, they were probably hired when necessary to provide the planning required in fortification construction (see pp. 143).

In such a context it is possible to make better sense of the notion of "competitive emulation" among Levantine polities as has been postulated by Shlomo Bunimovitz (1992:228), though it is unnecessary to suggest that it occurred between "neighbouring tribal groups." Instead, it would appear that these fortification construction projects reflected the social and political stratification of Middle Bronze Age kingdoms in the Levant. Thus, Middle Bronze Age fortifications were the product of, rather than the driving force behind, "social development in Palestine toward the re-emergence of a stratified, urban society" as Bunimovitz suggests (ibid).

5. Conclusions

Despite this critique of former attempts to assess the impact of fortification construction projects on the society and economy of the Levant during Middle Bronze Age, we may still accept that the construction of fortifications with and without ramparts involved important social and economic implications, some of which can be assessed and others cannot. To a certain extent there is no debating that they involved a degree of conspicuous consumption of labor and resources, as does the construction of any fortification system. Given the political context in which these sites were settled the consumption of labor and materials, however conspicuous, appears to have been primarily intended for the defense of their inhabitants against threats posed by the weapons available at the time wielded by ambitious, though often distant, powers.

7. CONCLUSIONS

> I am aware, after all, that I have told you many things you must have known before and, perhaps, may have considered as well as myself; but this I did, as I told you I should, so that you might perfectly comprehend the nature of true military discipline and the art of war, and for the instruction of others who may not have had the same opportunity of learning them that you have.
>
> Niccolò Machiavelli, *The Art of War*, Book VII (1965)

The preceding chapters constitute an initial attempt at a synthesis of issues relating to the construction of fortification systems in the Levant during the Middle Bronze Age. Of particular interest is the enclosure of so many settlements with massive freestanding earthen ramparts crowned by thick mudbrick fortification walls. Nevertheless, other strategies, such as the use of supplemental earthen ramparts at previously occupied sites and the construction of fortresses, also provide the data necessary for evaluating the context of Middle Bronze Age fortification strategies. In light of the archaeological data collected in this work, answers to the many questions raised in chapter 1 have been sought by considering technological developments in warfare (chapter 2), the evolution of defensive architecture (chapter 3), the significance of military events (chapter 4), the identification of defensive settlement networks (chapter 5), and data enabling the identification of sources of labor (chapter 6).

Of first importance this study presents unequivocal evidence for the identification of earthen ramparts, fosses, gates, and their walls as features employed in a distinctive approach to defensive architecture that was characteristic of the Levant and Upper Mesopotamia from the late third millennium through the first half of the second millennium B.C. The origin of this defensive strategy can be traced back to a group of sites located in Upper Mesopotamia that are identified as Kranzhügeln, which date to the first half of the third millennium. While these sites lacked ramparts, their characteristic encirclement by a thick mudbrick fortification wall served as the base plan for the evolution of fortifications to follow. Included among these sites are Mari and Terqa, where the earliest evidence for most of the elements that are characteristic of Middle Bronze Age fortifications were identified. These were determined to be part of a multifaceted approach to defensive architecture, which modified the Kranzhügel plan with the addition of a rampart to both sides of the fortification wall. Despite the abandonment of all Kranzhügeln in the Khabur region, the uninterrupted evolution of the fortifications of Mari and Terqa can be traced through the mid-second millennium by which time their fortifications had been replicated throughout the Levant at settlements such as Ebla, Qatna, Hazor, and Ashkelon.

While archaeological evidence for fortification strategies in the southern Levant during the third millennium do not appear to bear any relationship to the spread of fortification strategies from upper Mesopotamia to the northern Levant, the adoption of Mesopotamian-style fortifications in the northern Levant by the mid-third millennium was probably the direct result of military intervention by Mesopotamia. The accounts of long-distance campaigns against prominent political centers in the northern Levant such as Armānum (Aleppo?) and Ebla by Lugalzagesi, Sargon, and Naram-Sin support that a palpable threat of Mesopotamian intervention existed by at least ca. 2300 B.C. Likewise, Naram-Sin's detailed account of the fortifications of Armānum and archaeological excavations at Byblos and Ebla suggest that such defensive strategies had already been implemented at major towns.

Although specific environmental or topographical circumstances may also have contributed to the development of the fortifications characteristic of Mari and Terqa during the third millennium, the standard employment of this type of defensive architecture during the thousand years that followed occurred almost entirely out of a functional concern to improve the effectiveness of these defenses against improvements made in siege tactics and devices. Among the most notable of these developments in siege warfare was the introduction of the battering ram (Sum. GUD.SI.DILI; OB *yašibum*), which is believed to have occurred around 2300 B.C. Other siege techniques that were also being improved during this period included the siege tower (OB *dimtum*) and tunneling (OB *pilšum*), both of which are attested in the Mari texts, as well as the composite bow, which facilitated the use of the flame arrow in siege warfare. Because the gradual modification and general enlargement of the individual architectural features of ramparted fortifications had by the start of the second millennium already resulted in the development of the type of defenses known at the start of the Middle Bronze Age (ca. 1900 B.C.), we may conclude that siege techniques mentioned in the Mari texts two hundred years later had been around since the last quarter of the third millennium.

At the start of the second millennium, therefore, settlements in the northern Levant were enclosed by freestanding earthen ramparts often nearly fifteen meters high, which were crowned by thick mudbrick fortification walls with rectangular towers and bastions, and well-fortified gates. The justification for the continued adherence to such a regimented approach to the defense of settlements at the start of the second millennium is clear in light of the destruction of a number of Levantine settlements ca. 1900 B.C. These include Byblos, Ebla, Tuqan, and Hadidi, not to mention Selenkahiye, Sweyhat, and Hammam et-Turkman in northern Mesopotamia. These destructions, which were basically contemporaneous with the destruction of Ur dated to 1911 B.C., suggest a basis for continued concerns regarding security at the start of the Amorite period. Thus, the construction of fortifications during the first half of the Middle Bronze Age was not a simple matter of competitive emulation, but the product of legitimate security concerns.

While in the past efforts have been made by scholars such as Kathleen Kenyon to connect destructions during the last quarter of the third millennium and start of the second millennium with Amorite raiders, the archaeological and historical evidence do not support this conclusion. Efforts by Šulgi (1962 B.C.) and Šu-Sin (1941 B.C.) to build defensive walls in southern Mesopotamia against Amorite raiders do not provide an adequate basis for suggesting that early second-millennium destructions of settlements in the Levant must be attributed to the Amorites. In fact, no destruction of a major center is attributed to Amorites by Mesopotamian historical sources. The destruction of Ur, for example, which occurred only thirty years later, is attributed to an Elamite invasion and not an Amorite group. Thus the threat posed by Amorites, insofar as it existed between ca. 2200 and 1900 B.C., must have consisted primarily of raids against small, mostly unfortified rural settlements. The walls built by Ur III rulers would, therefore, have been most effective for this purpose and not against those armies that were well equipped to wage siege warfare.

Based on events during the last quarter of the third millennium and indications in the Mari texts, it is possible to infer that during the first half of the Middle Bronze Age (ca. 1900 to 1700 B.C.) sporadic warfare began between kingdoms in the northern Levant as well as with Mesopotamia powers. Despite the historical lacuna, the Mari texts illustrate that the resumption of long-distance military campaigns from ca. 1750 to 1632 B.C. by allied armies with the intent of besieging enemy capitals had become commonplace. Quite often these events brought Mesopotamian powers into the Levantine sphere (a pattern that continued from the late third millennium) through alliances with, for example, either Qatna or Yamḫad against the other. Similar campaigning brought about the well-known destructions of Kārum-Kaneš (1750? B.C.), Tuttul (ca. 1750 B.C.), Mari (1664 B.C.), and Šubat-Enlil (1632 B.C.).

Despite a lack of textual evidence in the southern Levant for intervention by Mesopotamian powers, not to mention Qatna or Yamḫad, similar concerns regarding security were expressed through the widespread construction of freestanding earthen ramparts. These were built following elliptical plans during the first half of the Middle Bronze Age and they are attested at a number of sites including Akko, Burga, Shimron, and Ashkelon. That these concerns were not unwarranted is revealed by a cluster of destructions and abandonments of a number of settlements along the central coast of the southern Levant by about 1800 B.C. (MB IIA Phase 3). Among the sites affected were Nami, Burga, Tel Hefer, Zeror, Poleg, and Gerisa.

It is as a result of the temporal connection between the meticulous employment of earthen rampart fortifications during the early Middle Bronze Age by a ruling class of Amorite ethnicity (see Execration Texts) that this type of fortification can be identified as one element of an Amorite cultural *koiné* that began to take shape at the start the Middle Bronze Age. This shared material culture, which reflected a common ethnic, cultic, religious, and social, if not also political, identity also included the so-called *migdāl*-style temple, burial customs, as well as cultic and royal iconography. Room remains, however, for further consideration of the degree to which together these elements of the material culture of the Levant reveal Amorite identity.

Although there is an absence of evidence for the destruction of settlements in the southern Levant that would indicate the region experienced siege warfare throughout the second half of the Middle Bronze Age (ca. 1700–1530 B.C.), the construction of freestanding earthen rampart fortifications continued unabated. During the first half of this period (i.e., MB IIB) news of long-distance campaigns by Ḫammurapi of Babylon against Mari (1666 B.C.) and Samsuiluna against Šubat-Enlil (1632 B.C.) undoubtedly reached not only the northern Levant but also the south, warranting a continued policy of fortification construction as is attested at Hazor and throughout the southern coastal plain.

The systematic employment of rectilinear plans in the construction of freestanding earthen rampart fortification systems during the MB IIB also permits the identification of what by about 1700 B.C. constituted a well-defended kingdom centered on Ashkelon. Ubiquitous features of fortification systems at this time included the incorporation of massive bastions, rectangular towers, and six-pier gates. Although the identification of many other Middle Bronze Age kingdoms in the northern and southern Levant by means of the same criteria may be possible, preliminary studies of the settlement patterns and fortification history within these kingdoms do suggest that the political organization of the northern Levant is revealed through a similar organization of fortified settlements. The landscape around Ebla in the north provides one example, and that around Akko in the south another.

Although historical and archaeological evidence point to a period of peace and prosperity throughout the Levant during most of the MB IIB–C, which may be identified as the *Pax Amoritica*, the construction of rampart fortifications continued. The focus of expansion and fortification during this period occurred primarily beyond the bounds of intensively settled regions, such as the coastal plain, and thus fortified settlements emerged in areas such as the highlands and the Negev. While the use of six-pier gates continued at many sites, fortifications in these regions were adapted to local geological and topographical conditions and, therefore, large quantities of quarried stone were increasingly employed in the construction of fortifications atop well-defended promontories. Nevertheless, the fortifications of Shechem, Shiloh, Jerusalem, and Hebron strategically addressed similar security concerns.

Despite the limitations of textual sources, in light of the evidence for similar hierarchical settlement patterns, the building of palaces and *migdāl*-style temples, and the employment of fortification strategies, the southern Levant clearly exhibits a level of political complexity identical to that of the northern Levant during the Middle Bronze Age. By identifying relationships between fortified settlements in the Levantine landscape the areal extent of individual kingdoms can be reconstructed. This is possible through the recognition that these kingdoms consisted of the individual territories associated with each of their subordinate (second-tier and lower) settlements that were located around the major fortified centers. Using the terminology found in Old Babylonian sources and retained in the Amarna letters, the territories associated with secondary fortified centers or district capitals, which were administered by mayors (OB *ḫazannum*), can be identified as districts (OB *ḫalṣum*) within these kingdoms (OB *māt*-TN). Thus, smaller Middle and Late Bronze Age politico-territorial units of the southern Levant, about which there has been considerable scholarly consensus, represent districts within a number of large Middle Bronze Age kingdoms, rather than a large number of very small but independent "city-states" during the Middle and Late Bronze Ages.

This reconstruction of the political organization of the Levant is also supported by likely scenarios by which the labor and resources required in the construction of Middle Bronze Age fortifications could be obtained. Renewed estimates using excavation rates drawn from Old Babylonian sources indicate that previous estimates for rampart construction are on average 20% to 40% in excess of the labor actually required. The shortcomings of previous estimates are due, on the one hand, to the use of inaccurate and exaggerated dimensions for the earthen ramparts and, on the other hand, to the adoption of extremely slow rates for construction, particularly for the excavation of materials for the ramparts. Nevertheless, even after adding the labor required for wall construction using an average construction rate derived from ethnographic data, the resources required are often still lower than earlier estimates.

Consideration of construction scenarios, intended to facilitate the identification of the labor force involved in the construction of these features, suggests that the use of army labor as attested in contemporaneous Mesopotamian sources was more likely than the use of corvée or hired laborers. Furthermore, there is no evidence to suggest that slaves or wider segments of the populations of the settlements in question were conscripted for fortification construction on a regular basis. While the evidence does suggest that rampart fortification construction projects of the Middle Bronze Age were the largest and most conspicuous fortification projects undertaken in the Levant during the premodern era, given their clear military function, the labor that was invested in their construction was worthwhile.

If the construction of fortifications appears to reflect the conspicuous consumption of labor resources during the *Pax Amoritica*, during the sixteenth century military intervention by the Hittite and Egyptian empires tested the efforts invested in the defenses of Levantine towns. Beginning with the reign of the Hittite king Ḫattušili I (1554–1524 B.C.), the northern Levant and Mesopotamia came under the new threat posed by the rising Hittite empire. The destruction of Alalaḫ VII is, for example, usually attributed to Ḫattušili, while his successor, Muršili I (1524–1494 B.C.), may have been responsible for the archaeologically attested destructions at Ebla (Mardikh IIIB), Afis, Tuqan, and Kadesh. Muršili is, however, best known for his infamous and seemingly unprovoked attacks against Ḫalab and Babylon in 1499 B.C., which revived a tradition of long-distance campaigns against enemy capitals that had not been seen since the days of Samsuiluna.

During the second half of the sixteenth century the southern Levant experienced a similar fate, albeit at the hands of the pharaohs of the Eighteenth Dynasty. Following the siege of Avaris and the expulsion of the Hyksos, Egypt pursued a territorial empire in the Levant as it sacked countless Canaanite towns. Almost all of the settlements in the kingdom of Ashkelon were destroyed as were Hazor and numerous settlements in the north. Although the precise dates for the destruction of these towns are difficult to establish, between ca. 1530 and 1460 B.C. the fortifications of many, if not most, fortified Levantine towns, which had been erected during the Middle Bronze Age, were razed by Egyptian sieges. All that remained of most fortifications were the monumental gateways, which in most instances continued to function as the settlement's gate.

A number of questions remain unanswered regarding the demise of Middle Bronze Age defenses and their relationship to the fortification strategies of the succeeding period. For example, to what extent were the destroyed MB IIC/LB IA settlements that remained inhabited during the Late Bronze Age fortified? If they were, in what manner were they fortified, and how did these fortifications relate to those of the Middle Bronze Age? How are Egyptian iconographic representations, which represent Levantine towns as fortified throughout the Late Bronze Age, to be understood in light of archaeological evidence for a lack of defenses at so many settlements? Furthermore, is there evidence for an explicit Egyptian policy against the re-erection of fortifications?

The present work has addressed, and hopefully answered, many of the questions that have been posed thus far regarding Middle Bronze Age fortifications in the Levant. However, further excavation, particularly in the northern Levant, will make it possible to address the questions that remain and the reconstructions advanced in this work. Nevertheless, until such data are presented it is hoped that the catalog of fortified settlements provided in this study will encourage further consideration of how Middle Bronze Age fortification systems relate to the society and economy of the Levant and its neighbors.

PART FOUR

A Catalog of Middle Bronze Age Fortified Settlements in the Levant and Mesopotamia, and Their Forerunners

General Notes

1. Site Selection

The bases for the inclusion of sites in this catalog are articulated in chapter 3. Catalog 1 consists of all sites in northwestern Mesopotamia fortified in the Middle Bronze Age, as well as a majority of sites that were fortified during the Early Bronze Age. Catalog 2, which is divided into the northern (2.A) and southern (2.B) Levant for the sake of convenience, consists of fortified sites dated from the Early Bronze IV to the Middle Bronze Age. Catalog 2.C summarizes data from Tell ed-Dab'a in Egypt. In certain instances sites or specific details of the development of a settlement's fortifications during earlier (i.e., pre-EB IV) and later periods (i.e., Late Bronze Age) have also been included, when they are considered relevant to the history of a settlement's fortification strategy.

If there is any question about whether a site was fortified during the Middle Bronze Age an asterisk follows the entry [e.g., Michal, Tel (Dhahrat Makmish)*].

2. Site Names

The following conventions are used insofar as possible for the site names. Priority has been given to known Bronze Age place names for sites (e.g., Ebla instead of Tell Mardikh). However, when a historical name is unknown or uncertain, the most common names for sites are employed [e.g., 'Ajjul, Tell el- (Sharuhen?)]. In all other cases sites are simply identified by their tell names.

3. Site and Fortification Data

Each entry features the following information. (1) The location of the site in UTM coordinates, or the old Palestine grid references (MR), which is in addition to a textual description of their proximity to other sites discussed in the catalog (see Figures 15, 25, and 56). (2) A complete listing of relevant references, plans, sections, and photos or illustrations; for plans and sections preference is given to those presented in final reports, but when such items are cited from preliminary reports or articles the listed items include only the most recent version (i.e., an earlier plan is not listed if a revised plan was published later). In the event that a discrepancy exists more than one set of plans is cited. (3) The most complete textual descriptions of the key elements (which are described in chapter 3.A) of each site's fortifications are also provided for the defensive features. When an identification of a defensive feature is definite or reasonably certain, the terms used for these features are *italicized.* If the identification of a feature appears in doubt or appeared uncertain to the excavators the term for the feature remains in normal font. When possible, a summary of the major parts of the site known to have been settled during the period in question, an assessment of topographical features as they pertain to defense, and an estimate of site size in hectares (minus the area occupied by ramparts) are provided. Although a few of the illustrations provided for site plans or sections of their fortifications are published for the first time in this work (e.g., Byblos and Jaffa), the majority of the illustrations have been reproduced, sometimes with modifications (e.g., Michal), in order to orient the reader or to clarify misunderstandings. Nevertheless, the original plans should be consulted for any questions requiring close scrutiny of the section or plan in question.

Inter-site comparisons of information found in this catalog are made in tables in chapter 3. Measurements are metric and in certain cases have been converted from British imperial measurements. For this reason discrepancies may exist between the figures provided here and those suggested in other *secondary* sources for the same features. Compass directions are abbreviated and occur in small caps (e.g., N, NNW, NW) except when used to refer to a specific feature (e.g., N Gate).

Conventions regarding chronology follow guidelines outlined in chapter 1 (see Table 1).

4. Dates and Archaeological Phases

In most instances archaeological phases are cited in the catalog are those provided by the excavators. However, when it is clear that dates used by an excavator rely upon the middle chronology, these dates have been adjusted to coincide with VLC dates used in this work (see pp. 18f.). However, no dates before the mid-third millennium B.C. have been lowered.

1. Northwestern Mesopotamia (ca. 3000–1500 b.c.)

The sites included in this catalog are located in eastern Syria between the bend in the Euphrates River to the west, the borders of modern Syria to the north and east, and the Euphrates River Valley to the south.

Figure 15. Sites in northwestern Mesopotamia discussed in catalog 1.

BDERI, TELL

UTM: 37 N 662959.4028410

Refs.: Pfälzner (1987a; 1987b; 1989–1990; 1990).

Plans: Site (Pfälzner 1989–1990; 1990:fig. 1); Gate (1989–1990:fig. 19).

Sects.: NA.

Photos: S slope (Südhang, Pfälzner 1989–1990:fig. 20); Gate (Pfälzner 1987b:figs. 3f.; 1987a:fig. 5; 1989–1990:fig. 21).

Tell Bderi is located on the E bank of the lower Ḫabur River. Although the site was abandoned during the Middle Bronze Age, the third-millennium settlement has yielded information concerning the site's Early Dynastic defenses (Pfälzner 1990:66f.). The site during this period was approximately 5 to 6 ha in size (ibid., 67). The fortifications of this period were exposed in a N-S step trench through the site's S side or Südhang (ibid., see fig. 1). Here remains of the town wall, a glacis, a tower, and gate, all of which belong to Level 25 (ca. 2800 to 2200 B.C.) were exposed.

The roughly 2.8 m wide *fortification wall* was built of mudbricks, which measured 20 x 20 cm (Pfälzner 1989–1990:216ff.). The accompanying 1.7 to 2 m wide *glacis* was constructed of pisée and was renewed and possibly widened once or twice (1987b:294). In addition to the town wall a *tower* measuring 1.8 m square was later added to the wall and was constructed of "brick debris and mud with a framing of upright standing mudbricks" (ibid., 294). The remains of a *gate* were exposed within the line of the town's defenses also on its S side.[1] The gate was dated by Pfälzner to the ED I (ca. 2800 B.C.) and suggested to be the predecessor to Middle Bronze Age orthostat-type gates (1990:67f.). Its passage was 3 m wide and was lined with stone orthostats, one of which measured 125 cm tall by 95 cm wide (1989–1990:218, and fig. 21).

Although the site was abandoned around 2200 B.C. and throughout the Middle Bronze Age, it was reoccupied during the Late Bronze Age (Pfälzner 1990:67).

BEYDAR, TELL (NABADA?)

UTM: 37 N 641697.4066975

Refs.: Quenet (1997); Bluard (1997); Bretschneider (1997); Suleiman (2003).

Plans: Site (Lebeau and Suleiman 1997:fig. 5); Gate (Bluard 1997:fig. 1); Area G (Quenet 1997:fig. 2; Suleiman 2003:plans 29f.); Area H (Bretschneider 1997:fig. 3).

Sects.: Area G (Quenet 1997:fig. 1; Suleiman 2003:fig. 307).

Photos: Site aerial (Poidebard 1934; Lebeau 1997b: fig. 3); tell profile (Lebeau 1997b:figs. 8f.); Area G (Quenet 1997:fig. 3; Suleiman 2003:photos 1–3); Area H (Bluard 1997:figs. 2f.; Bretschneider 1997: figs. 2, 4).

Tell Beydar (Early Dynastic *Nabada*) is located in the Ḫabur triangle. It is among a group of third-millennium sites identified as Kranzhügeln because of their topographic appearance. The site was occupied during the Early Dynastic I–III (i.e., EB II–III). During this period the site consisted of a 7 ha central mound, probably the upper town, situated 27.5 m above the plain, with a citadel perched 7.5 m above the upper town (Lebeau 1997b:9f.). The entire upper town was surrounded by a 20.9 ha lower town which was itself enclosed by a fortification wall (ibid.; see Figure 16).

The embankments that now surround the site have only been explored in Area H on the N side of the site (Bluard 1997; Bretschneider 1997).[2] The embankments are in fact the remains of a mudbrick *fortification wall* built during the ED I (ca. 2950–2750 [EB II], Lebeau 1997b:9). The earth used in the construction of this wall was probably taken from an area roughly 80 m wide around the foot of the upper town (i.e., the material was taken from within the settlement). This is inferred from the half-meter difference in elevation between the area inside the wall (359.5 m ASL), which is not thought to have been built upon during the Early Bronze Age, and the area outside the wall (360 m ASL). The remains of the embankment today are approximately 1,885 m long, 90 m wide, and 7 m high (ibid.), allowing us to determine at least the original length and minimum height of the wall.

The dating of the embankment containing the ED I wall was possible as a result of materials obtained from the tombs and houses built upon its dilapidated remains which were excavated in Area H (Bluard 1997). Here the embankments consisted of debris and settlement built over the eroded remains of a defensive wall, outside (i.e., N) of which tombs were located (ibid., 180). It appears that a 4.5 m wide mudbrick *fortification wall* was probably built during the earliest occupation of the site in ED I, or by at least early ED II, on the level of the plain (ibid., 181).[3] During ED II (ca. 2700–2600) houses were built against and over the wall, and a necropolis was opened outside of it. Burials at the necropolis contin-

[1] A life size reconstruction of this gate can be seen at the regional archaeology museum in Deir ez-Zor, Syria.

[2] Although Lebeau identifies this embankment, or at least its lowest portion, as a "rampart" (1997a:111), there was no rampart below the wall. As noted above, in fact, the wall itself is only slightly above the surface of the plain.

[3] Judging from fig. 3 in the summary of Area H (Bretschneider 1997) the S elevation of the wall could be traced down to about 361 m ASL, only 1 m above the plain.

ued through the ED III (ca. 2600–2350), by which time the wall had already fallen out of use (ibid.).

The outer wall was interrupted by at least seven distinct depressions probably corresponding to the site's *gates*, identified by Bluard as leading to the major ED destinations such as Mardin to the N, Urkeš to the ENE, Ninua to the SE, Nagar to the SSE, the Ḫabur River to the SSW, the Abd al-Aziz to the SW, and Chuera to the WNW (1997:fig. 1). Seven depressions in the upper town also align with these breaches. These gates have not yet been excavated, though the divine names given to them are attested in ED tablets found at Beydar (see texts 1, 5, 28f., 52 in Ismail et al. 1996).

The defenses of the upper town dating to between 2900 and 2500 B.C. were also exposed in Area G (Quenet 1997; Suleiman 2003). Based on the evidence from this area the fortifications of the upper town appear to have featured three phases of separate construction each consisting of a mudbrick *fortification wall* and gray mudbrick *glacis* (Suleiman 2003:303). The earliest mudbrick wall (7917) was only 1.85 m wide and was accompanied by a gray mudbrick glacis; the second mudbrick wall (7904), which was less than 1.5 m wide, also featured a similar glacis, as did the latest fortifications, which consisted of 4.5 m thick mudbrick wall (7662) at about 377.66 m ASL. Although the base of the latest wall was never exposed, it was probably located around the 371 m contour line, which if traced around the acropolis would have been at least 785 m long. While Suleiman has asserted the identification of a moat at the base of the fortification system based upon soundings there in 1998 (ibid., 305), this identification cannot be supported by the published sections since they only reached 360.28 and as noted above the depression around the citadel is now as low as 359.5 m ASL (see pls. III and IV in Suleiman 2003).

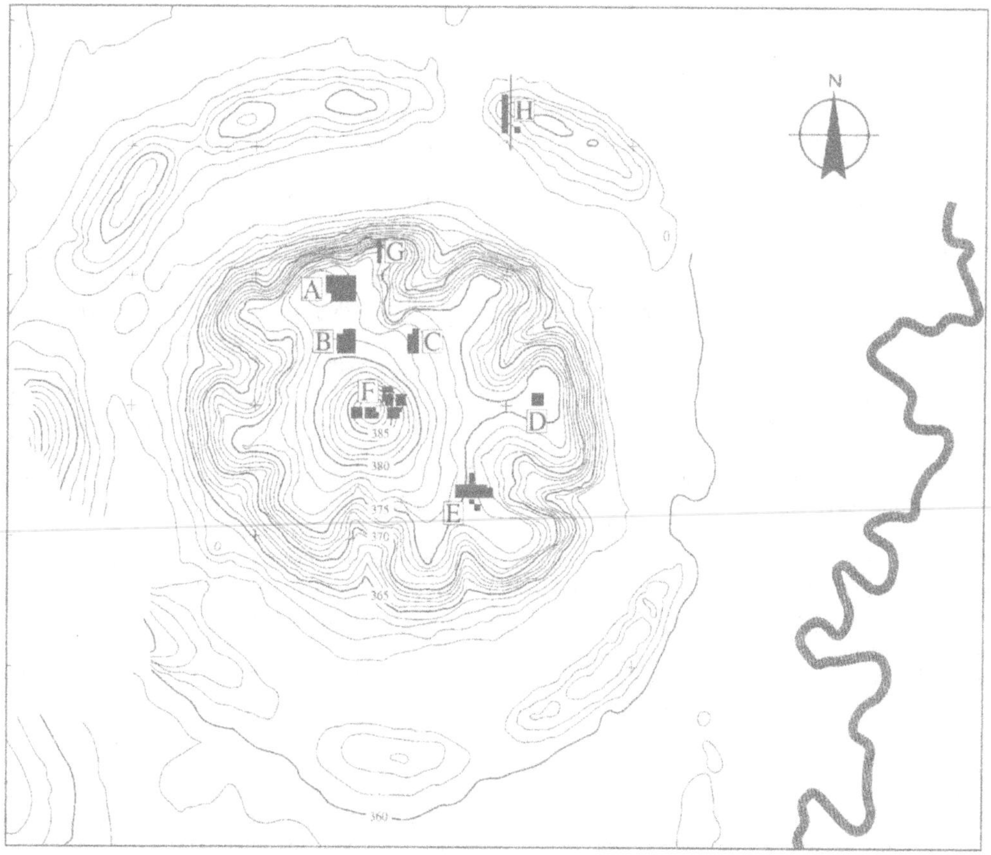

Figure 16. Plan of Tell Beydar with excavation areas (after Lebeau and Suleiman 1997:fig. 5).

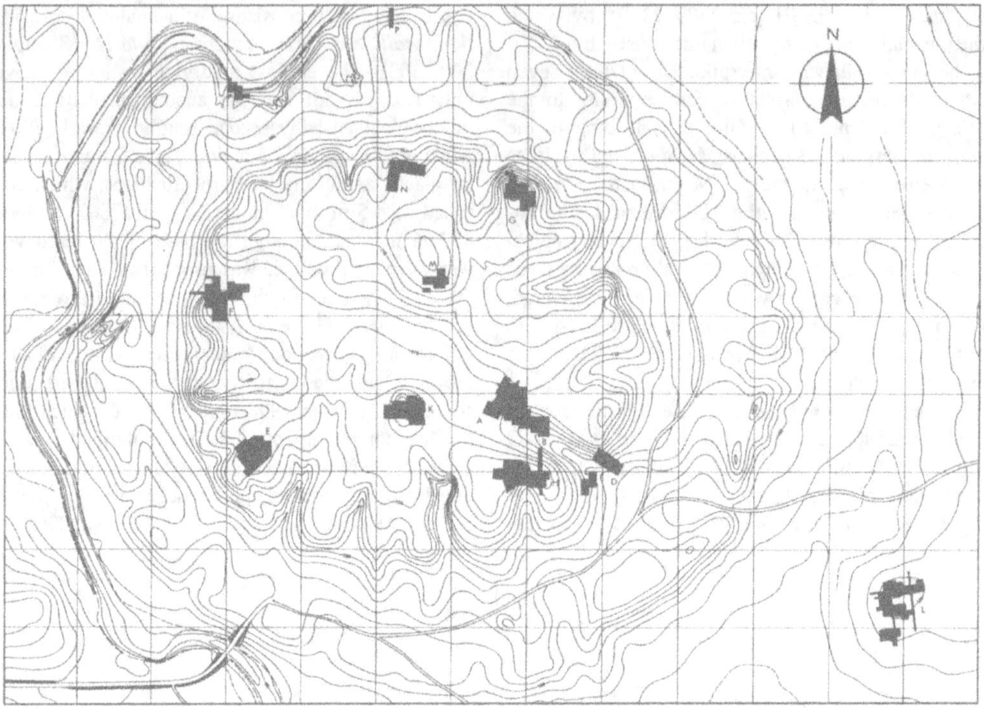

Figure 17. Plan of Tell Chuera (after Orthmann et al. 1995:supp. 2).

CHUERA, TELL

UTM: 37 N 0544295.4055805

Refs.: Novák (1995); Orthmann et al. (1995); Orthmann (1997).

Plans: Site (Orthmann et al. 1995:supps. 1f.); Area P (Novák 1995:fig. 87).

Sects.: Area P (Orthmann et al. 1995:supp. 17); Area P schematic (ibid., fig. 109).

Photos: Area P (ibid., pls. 21f.).

Tell Chuera, located between the Baliḫ and Ḫabur Rivers, covers an area of 65 ha and is therefore the largest of the early third-millennium sites identified as Kranzhügeln (Orthmann 1997:491). Although Chuera does not appear to have been occupied from ca. 2350/2300 to 1600 B.C., its development as a settlement during the EB III (Per. I, ca. 2800–2300 B.C.), sheds light on the type of fortifications known in the region during this period and may also relate to the subsequent development of fortifications at sites further W. The site consists of an upper town of roughly 43 ha surrounded by a lower town of 22 ha. It was flanked on the W by Wadi Chuera which added to its defensibility.[4]

[4]For phases of settlement at Chuera, see Table 2, p. 15 in Orthmann (1995).

ED (EB III): PERIOD IC

During Period IC when the site reached its peak size it was surrounded by a mudbrick *fortification wall* and 15 m wide *fosse* (Oredsson 2000:45). The wall around the lower town has been investigated on the N side of the site in Area P (Orthmann et al. 1995:supp. 2). Period IC is the earliest period to which the wall may be attributed with any certainty, though it may have functioned during Period IB (Novák 1995:175). The wall itself was constructed of mudbrick more than 5.5 m wide, probably before any settlement began in this part of the site. It was preserved to a height of at least 4.4 m (ibid., supp. 17) and its circuit was probably about 2,810 m long. At that time a 1.2 m wide mudbrick *revetment wall* ("Vormauer") was also built about 6 m in front of the town wall. A "glacis," or more properly speaking an *earthen rampart*, of alternating layers of complete and fragmentary bricks and grayish-black ash with an approximate slope of 19° was constructed between the early town wall and the revetment wall (ibid.). No surface treatment (i.e., glacis) was noted. Excavations in 2002 revealed the same series of walls on the W side of the site in a long and wide trench cutting across the site's defenses. Period IC closed with a fiery destruction that was accompanied by the skeletal remains of a number of victims (Akkermans and Schwartz 2003:256).

LATE ED–AKKADIAN: PERIOD ID–IE

During Period ID the *revetment wall* was renovated and a stone foundation for a new mudbrick superstructure was built upon it (Novák 1995:175). Between this new "Vormauer" and the town wall a new *earthen rampart* was built up with a preserved slope of 35°. Prior to the completion of this rampart (visible in the excavated section) a supportive buttress or *tower* was added to the outside of the town wall (ibid.). This second phase of the town's fortifications appears to have remained in use until Period IE (Late ED/Akkadian).

The upper town appears to have also been at least partially enclosed by a separate *fortification wall* (Orthmann 1997:491).

HAMMAM ET-TURKMAN (ZALPAH)

UTM: 37 N 0505144.4037521
Refs.: van Loon and Meijer (1987); van Loon (1988); Meijer (1988; 1989; 1997).
Plans: Site (Meijer 1989:fig. 1); K 22 (ibid., fig. 4); J23 (Meijer 1988:pl. 39b), O18 (van Loon 1988:pl. 34).
Sects.: H 23 (Meijer 1989:fig. 2), O16–O18 (van Loon 1988:pl. 32).
Photos: O18 (van Loon 1988:pl. 33b).
Hammam et-Turkman, identified by some with ancient *Zalpah* from the Mari texts, is situated approximately 58.5 km N of Tuttul. The site appears to have experienced near continuous occupation from the late fourth millennium through the Iron Age (Per. V–IX).

EB IV: PERIOD VI

Thus far excavations have been limited to the mound, which is thought to have been approximately 250 x 150 m, or 3.75 ha during the EB IV—an area marked out by the 335 m contour line (van Loon and Meijer 1987:1) about 25 m above the plain (Meijer 1989:1). That the EB IV *fortification wall* (ca. 2350 B.C.) was constructed along this contour is suggested by a portion revealed in 1986 in the W trench on the acropolis (van Loon and Meijer 1987:1). The remains of the town wall or the W elevation of a tower along the wall dating to this period were found there. Excavation also revealed that the wall was destroyed by the first of two fires ca. 2200 B.C. (van Loon 1988:699). Upon the ashy remains of the first EB IV wall a second phase of EB IV defenses were built, which featured an unimpressive wall with suites of storerooms built along its interior (see H 23 in fig. 2, Meijer 1989). These were also subsequently destroyed at the end of the third millennium in a fiery conflagration that resulted in the deposition of a layer of gray and orange ash (van Loon 1988:699). This was then followed by an occupation gap before occupation resumed in the Middle Bronze Age (ibid.).

MIDDLE BRONZE AGE: PERIOD VII

Middle Bronze Age occupation of Hammam resumed (perhaps in late MB IIA or IIB) with the construction of a mudbrick *fortification wall* exposed in square O18 on the N side of the mound. Here a 5.5 m wide wall (Wall V) was built of more than sixteen rows of gray mudbricks measuring 35 x 35 x 8 cm (van Loon 1988:80). The wall's base was at approximately 340.5 m ASL, about 30 m above the plain (see E section of N trench, van Loon 1988:pl. 32). Despite this evidence, because the wall was only preserved to a height of a little more than 1 m van Loon concluded that the structure might in fact have been something other than the town wall (ibid., 699ff.). But the evidence of erosion at the site and in other similar cases mitigates against the need for any other explanation. In square J23 on the W side of the mound a feature probably to be identified as the remains of a *tower* (W) were excavated (Meijer 1988:84). This was accompanied by a 5 cm thick mud plaster *glacis* with a slope of 18° (Meijer 1989:2).

LATE BRONZE AGE

Hammam et-Turkman is thought to have been unfortified during the Late Bronze Age (Akkermans and Schwartz 2003:352).

MARI (TELL HARIRI)

UTM: 37 N 0673319.3824909
Refs.: Parrot (1958); Margueron (1983; 1984; 1987; 2000; 2004).
Plans: Site (Margueron 2004:564); Trench J_1 (ibid., fig. 49); Trench J_3 (ibid., figs. 51, 108, 323, and 325); western fortifications? (ibid., p. 344); SW quarter (ibid., fig. 1); OB Palace enclosure (Parrot 1958: foldout).
Sects.: Trench C_5 (Margueron 2004:figs. 46, 106, and 320), Trench J_3 (ibid., fig. 109); schematics of the evolution of the outer fortifications (ibid., figs. 48, 107, 321, and 420).
Photos: Trench C2 (Margueron 2004:fig. 419); Trench J1 (ibid., figs. 47 and 50); Trench J3 (ibid., figs. 75 and 323f.); OB Palace enclosure (Parrot 1958:pls. 60f., figs. 405, 12f.); isometric drawings of defenses (Margueron 2004:figs. 112, 328, and 421–23).

Mari is located on the S bank of the middle Euphrates just to the W of the modern border between Syria and Iraq and about 60 km SE of Terqa. An examination of the topographical plan of Tell Hariri and recent studies of the Euphrates Valley in the vicinity of the site have led to the conclusion that the site during the third and second millennia consisted of a centrally located upper town accompanied by a lower town, both of which were enclosed by a ring of rampart-style fortifications (Margueron 1987; 2004). Aside from indications that the exterior rampart served as the site's primary defensive line, a similar defensive structure is thought to have been built around the upper town and excavated in Trench J_1 (the "rempart intérieur," Margueron 2000:fig. 1) comparable in character to the wall found around the palace (see Parrot 1958:329ff.). Today the SW quadrant of the site is all that remains of this overall plan (see Figure 18 below), while the remainder of the site is presumed to have been washed away by the meandering of the Euphrates River.

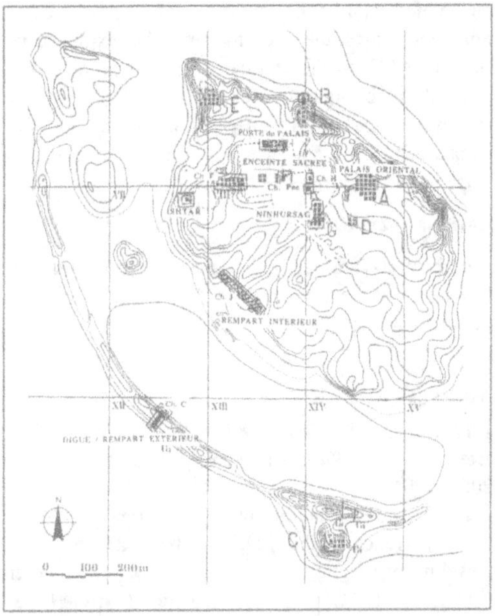

Figure 18. Plan of Mari (after Margueron 2000:fig. 1).

The so-called outer *earthen rampart* (Fr. rempart extérieur) of Mari in the third and second millennia has been explored in a series of trenches in Area C (C_1–C_4) located on the SW side of the tell. The first of these trenches was opened in 1979 on a small projection S of the rampart in squares F–J/12–13 in Grid XIV, but no built structures were exposed (Margueron 1982:29f.). Exploration of the ramparts resumed in 1980 with the opening of Trench C_2 in Grid XII more than 500 m to the WNW from the previous trench (Margueron 1983). Its excavation was completed in 1982 and Trenches C_3 and C_4 were then opened, both of which were directly N of Trench C_1 (Margueron 1984:26–30). The results obtained in Trenches C_3 and C_4 served only to confirm the conclusions drawn from the excavation of Trench C_2. Information regarding the fortifications of the inner town at Mari (Fr. rempart intérieur) has been provided by excavations in Trenches J_1 and J_3 from 1997 to 2000 (Margueron 2004:86f.).

Based on recent conclusions concerning the construction and phasing of the exterior rampart presented by Margueron (2000; 2004), it appears that the excavations in Areas C and J have yielded sufficiently detailed information to enable a detailed reconstruction of Mari's fortifications. Margueron's reconstruction of Mari's fortifications consists, therefore, of three main phases associated with each of Mari's main phases of settlement, designated Ville I to Ville III, which span from ca. 2900 B.C. to 1664 B.C. (see Figure 19 below).

ED I–II: VILLE I (CA. 2900–2600 B.C.)

Mari's earliest settlement, so-called Ville I, appears to have been surrounded by what Margueron calls as an earthen dike (Fr. digue) that was excavated in Trench C_5 (Margueron 2004:85f.). It consisted of the remains of a more than 1 m high stone foundation that was buried by the heaping up of earth against its surfaces, thus creating the foundation for a mudbrick wall and an earthen rampart (ibid.:fig. 48). The width of the *earthen rampart*, which was constructed of yellow clay and chipped limestone, was about 8 m. Only a few courses of a 1.9 m wide mudbrick *fortification wall* were preserved atop the foundation (for section, see Margueron 2004:fig. 46).

Margueron had once suggested that the original purpose of the dike, which featured a diameter of 1,900 m and would have been as much as 5,970 m long, was probably not defensive but served as an outer dike, 300 m outside Mari's main fortifications, which was intended to protect the settlement against inundation by the Euphrates (Margueron 2000:910; also 2004:86). This notion, however, seems unnecessary since Margueron, as noted above, has also identified the mudbrick wall belonging to Ville I, which rested directly atop this dike. Therefore, these features appear to constitute a single phase of fortification construction associated with Ville I, which consisted of a mudbrick fortification wall built on a stone foundation and accompanied by an earthen rampart (Figure 19).

The interpretation of these features as elements of Mari's fortification system during the early third millennium is corroborated by the excavation of the inner town's fortifications in J_1 and J_3. In both of these trenches the remains of a mudbrick *fortification wall* 6 m wide were found atop a 1.5 m high stone foundation (Margueron 2004:86f.). In Trench J_1 the wall was only preserved 0.6 m high, though according to Margueron the wall attained a height of about 8 m. Projecting 2 m from the wall in Trench J_1 was a 11.55 m long *tower* built on a stone foundation.

In Trench J_3, which was 80 m west of Trench J_1, half of a *four-pier gate* of the inner town was revealed in 1998. The gate, like the tower mentioned above, projected outward from the line of fortifications of the inner town. The interior chamber is estimated to be 7 x 4 m, and its walls were 4 m thick (Margueron 2004:87). The location of the gate continued through the Ville III period (see discussion below).

Ville I was followed by a short period of abandonment (see stratigraphic chart in Margueron 2004:9).

ED III: VILLE II (CA. 2550–2150 B.C.)[5]

Evidence for the Mari's Ville II defenses is less clear than for the defenses of either the preceding or following periods. The evidence of Mari's Ville II outer fortifications consists of a rebuilding of the outer *earthen rampart* and its mudbrick *fortification wall* excavated in Trench C_5 (Margueron 2004:139, fig. 138). The new wall, which was 3 m wide and was preserved to a height of 2 m, was built directly atop the dilapidated remains of the Ville I rampart and wall (see Figure 19 below).

In Trenches J_1 and J_3 only a portion of Ville II's inner fortification wall was recovered and its former width cannot be determined (Margueron 2004:139, fig. 108). Margueron notes that this may have been because the fortifications of Ville I were sufficient or that subsequent construction during Ville III left no evidence of the earlier fortifications (ibid., 139f.). Ville II was destroyed in a massive conflagration ascribed to Naram-Sin (see stratigraphic chart in Margueron 2004:9).

EB IVB–MB: VILLE III (CA. 2150–1664 B.C.)

The Ville III settlement provides the greatest evidence of Mari's fortifications (see Figure 19 below). Each of the phases of the construction of the fortifications throughout this period evinces a town's fortifications under continual renovation and improvement (Margueron 2000; 2004:343–46, 443–46). In the first phase of the outer fortifications, which is identified as the *first earthen rampart* and is attributed to the period of the Šakkanaku, a 3 m wide mudbrick *fortification wall* was added above the Ville II fortification wall along with new earth to both sides of the rampart (Margueron 2000:921).

After this structure had fallen into disrepair and was abandoned it was replaced by a new 4 m wide mudbrick *fortification wall* with expansion of the rampart's surface on both sides of the new wall. This was identified as Phase A of the *second rampart*. This was followed by Phase B when the wall's width was doubled by the addition of mudbrick walls along its interior and exterior sides, resulting in a total width of 9 m. The final phase of the second rampart, Phase C, was characterized by massive expansion of the slope on the interior side of the rampart, which Margueron refers to as the "glacis intérieur," which was composed of gravel and supposedly intended to counter efforts to undermine the wall. While the destruction of this final fortification system was clearly the work of Ḫammurapi of Babylon and thus the wall can be attributed to Amorite hegemony at Mari,[6] historical events associated with the preceding phases are unknown.

An exact correlation between the various phases of the Ville III outer fortifications and the inner fortifications has not been proposed. As with the inner fortifications, Margueron suggests that Yaḫdun-Lim was responsible for improvements to the outer fortifications (2004:444). Evidence for the fortifications of the inner/upper town where the four-pier gate of Ville I, which was rebuilt writ large in the Ville III period, was recovered from Trench J_3 (Margueron 2004:345f.). The Ville III *four-pier gate* featured walls estimated to be over 10 m thick with a passageway of about the same width (ibid., fig. 325). The *fortification wall* was approximately 10 m thick where it joined with the gate, but appears to have been reduced to about 6 m in width at about 17 m from the gate. Margueron suggests that if the entire upper town were enclosed by the same type of fortification wall as was found near the gate, then 1,570,000,000 bricks measuring 40 x 40 x 12 cm would have been required in its construction. In addition to the recent evidence for Mari's fortifications, André Parrot excavated a portion of the palace *fortification wall* (Parrot 1958:329–39). Its average width was 3.5 m (ibid., 328).

---

[5]Because LC dates are employed throughout this work Margueron's date for the end of Ville II and the start of Ville III has been adjusted from 2240 to 2190 B.C.

[6]Margueron suggests specifically Yaḫdun-Lim (2004:444), noting his inscription which reads "I built the wall of Mari and I dug its fosse" (for inscription, see Sollberger and Kupper 1971:244, fig. 415).

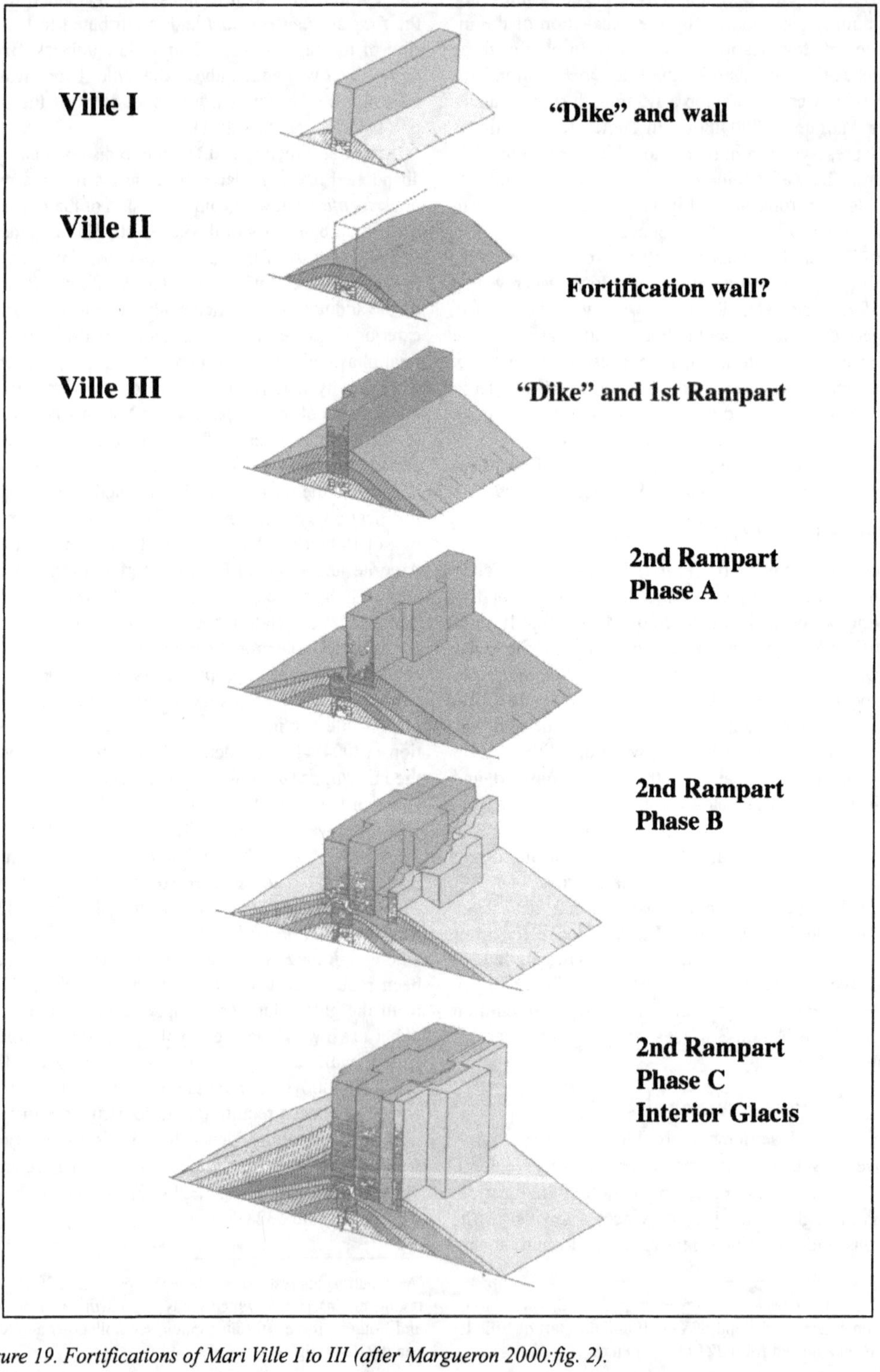

Figure 19. Fortifications of Mari Ville I to III (after Margueron 2000:fig. 2).

The entire extent of Mari's fortifications is difficult to ascertain, as mentioned above, since much of the settlement has been erased by the meander of the Euphrates River. Nevertheless, it is worth attempting to determine the total extent of Mari's fortifications. Margueron notes that the diameter of the existing settlement's outer rampart is approximately 1,900 m (Margueron 2000:910). The extrapolation of these dimensions, presuming that the present site only constitutes approximately one-third of the ancient settlement, would yield a site of approximately 283.5 ha with a circumference of around 5,970 m. If the acropolis were not much bigger than it is today, then it would have been around 45 ha leaving 238.5 ha for the lower town. Such a settlement size would be comparable, therefore, to settlements in lower Mesopotamia: ED I Uruk at ca. 400 ha (Postgate 1992:75), Old Babylonian Isin at ca. 225 ha (Hrouda 1997) and Larsa at ca. 360 ha (Margueron 1997a).

With regard to the volume of the rampart Margueron suggests the following figures (2004:444):

Rampart *h.* x *w.* ≈ 7 x 15 m
Vol. of 1 m across fosse ≈ 55 m^2
Rampart *l.* ≈ 5,600 m
Outer rampart *vol.* ≈ 308,000 m^3

While the calculations are basically correct, presuming that the site was once as large as Margueron suggests, it is unlikely that the quarries from which the rubble was taken were as far from Mari as Margueron states (ibid.). It is also improbable that three million trips by donkeys each carrying 100 kg of debris from a distant quarry were necessary. Thus his conclusion that more than eight years were required with one thousand donkeys making one trip to the quarry per day is baseless. It is important to observe that of the six trenches that have been excavated through Mari's ramparts, none has reached the level of the plain and thus it has been impossible to confirm the existence of a fosse that is most likely to be found at the foot of the rampart. This would have been the main source for materials used in the ramparts' construction.

NAGAR (TELL BRAK)

UTM: 37 N 683931.4060098
Refs.: D. Oates, J. Oates, and McDonald (1997).
Plans: Site (Oates et al. 2001:fig. 2); Area TW (Oates et al. 1997:fig. 166).
Sects.: NA.
Photos: Area TW (Oates et al. 1997:figs. 165f.).
Tell Brak, located in the Ḫabur triangle 45 km SE of Beydar, has been identified as the capital of the third and second-millennium kingdom of *Nagar* (Oates and Oates 2001:379ff.). The "violent destruction and fire" revealed at Tell Brak at the end of Phase L in Area CH (Level 6), Area ER (Level 5), and in Areas ST and DH is considered to be the work of either the Akkadians or local wars (Oates et al. 2001:382, 91). While no remains of the fortifications of the third-millennium site, which was between 40 and 65 ha (ibid., 380), have yet been exposed, the fortifications of the early second millennium have been encountered in Area TW and may be evident near Trench D and Area FS (Oates et al. 1997:142f., figs. 65f.). However, little else can be added regarding its Old Babylonian fortifications.

ŠUBAT-ENLIL (TELL LEILAN)

UTM: 37 N 0723171.4092880
Refs.: Weiss (1985; 1990; 1997); Schwartz (1988); Stein (1991); Weiss et al. (1991).
Plans: Site (Ristvet and Weiss 2004).
Sects.: Operation 1 (Schwartz 1988:figs. 3–5); Operation 4 (Weiss et al. 1991:fig. 22).
Photos: NA.
Tell Leilan, which is probably to be identified with the third-millennium town of *Šekhna* and second-millennium *Šubat-Enlil*, is located in the Ḫabur triangle in northern Mesopotamia.

LEILAN IIID (CA. 2600–2350 B.C.)

The first fortifications at Tell Leilan were constructed as part of the Leilan IIId settlement ca. 2600 B.C. During the Leilan IIId period, when the site was first fortified, the site grew from 15 ha to approximately 90 ha in size. It remained this large through the Leilan IIa period (ca. 2350–2250 B.C.). The site consisted of an upper town of less than 15 ha and a lower town of around 75 ha (Weiss 1997:342). Based on the size of the settlement and its topography it is possible to infer that the mudbrick *fortification wall* around the acropolis was about 1,415 m long while the one around the lower town was about 2,250 m long.

LEILAN IIA (CA. 2350–2250 B.C.)

During the early part of the EB IV Leilan was again fortified. These fortifications were encountered on the NW corner (Operation 1) of the upper town where the remains of a 2.5 m wide mudbrick *fortification wall* were found in a phase of occupation that preceded Akkadian control (Weiss 1997:343). Presumably these fortifications are to be identified with walls 2 and 3 in Operation 1 published by Glenn Schwartz (1988:26f.). Weiss has also suggested that the lower town was unfortified during the Leilan IIa phase since the period IIa town wall around the lower town was not identified in Operation 4 on the NE side of the site (1997:343). However, this may only be because the excavation trench did not reach the highest point of the embankment there (G. J. Stein 1991:555).

LEILAN IIB (CA. 2250–2150 B.C.)

The later Akkadian dominated settlement was surrounded by a massive mudbrick *fortification wall* measuring more than 16 m wide. It consisted of two sets of 8 m wide walls with a "walkway" between (Akkermans and Schwartz 2003:262). An *earthen rampart* has also been identified along the northern slope of the tell.

LEILAN I (CA. 1850–1632 B.C.)

After Akkadian hegemony and following a hiatus in occupation between ca. 2150–1850 B.C., the site was resettled and became the capital of the Old Assyrian kingdom of Šamši-Adad called Šubat-Enlil. The Old Assyrian settlement appears to have been approximately the same size as that of Leilan II and during this period also included settlement in a lobe to the S of the tell that has been suggested to be the location of a *kārum* (ca. 10 ha) appended to the original settlement (Weiss et al. 1991:534f.). This additional settlement probably necessitated the construction of another 750 m of town wall and rampart. During this time it appears that the entire site was fortified and that in the lower town houses were built "against the early second-millennium BCE city wall" (Weiss 1997:345). Weiss suggests that the wall of the Old Assyrian town was approximately 3,700 m long, tracing the 100 m contour line around the lower town and rising 5 to 15 m above the surrounding plain (Weiss et al. 1991:534). This mudbrick *fortification wall* was explored in Operations 2 and 4 situated on the E and NE ramparts, respectively. In Operation 4 the Leilan I town wall appears to have been built toward the end of Period I occupation in this area (G. J. Stein 1991:554). The bricks used in this 5 m wide wall were "identical in appearance to the bricks used in the Period I temple construction on the Acropolis" (ibid.). The wall appears not to have been maintained throughout the period and it already began to collapse before the end of Period I (ibid., 555).

SWEYHAT, TELL ES-

UTM: 37 N 0432998.4014657
Refs.: Holland (1976; 1977); Zettler (1997a; 1997b); Peregrine et al. (1997).
Plans: Site (Holland 1976:figs. 1, 17; Zettler 1997a:fig. 1.4); Area IV (Holland 1976:fig. 3; 1977:fig. 1); Operation 25 (Zettler 1997b:fig. 3.13).
Sects.: NA.
Photos: Site aerial (Zettler 1997a:fig. 1.5); geomagnetic maps (Peregrine et al. 1997:figs. 4.1, 4.3); Area IV (Holland 1976:pl. 5a); Operation 25 (Zettler 1997b:fig. 3.14).

Tell es-Sweyhat includes a tell, which sat 15 m above the Euphrates Valley measuring approximately 300 m N-S by 250 m E-W, and a roughly rectangular enclosure of the lower town with gross dimensions of 700 m N-S by 600 m E-W (Holland 1976:36).[7] Recent exploration has led to the conclusion that only the upper town was settled during the early third millennium, though by the third quarter of the third-millennium settlement had expanded to the N and E into the lower town (Zettler 1997a:169). Although the site's gross size is cited as 42 ha, its net size within the wall was actually only about 30 ha (Figure 20).

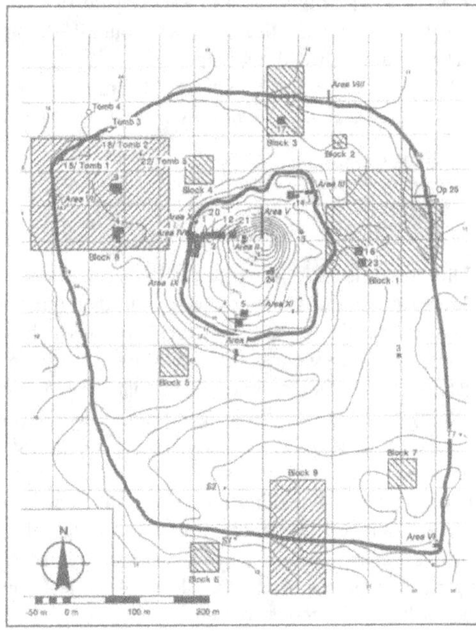

Figure 20. Plan of Tell es-Sweyhat showing course of EB IV fortifications (after Zettler 1997a:fig. 1.4).

The fortifications of the upper town were only excavated on its W side in Areas IV and X (Holland 1976:49, fig. 3). Here a 2.5 m wide mudbrick *fortification wall*[8] preserved to a height of 1.17 m built on

[7] Michael Astour identified Sweyhat with the Early Bronze Age settlement of Burman (2002:167).

[8] This wall's width was corrected in the second preliminary report, as a result of excavation in Area IV Trench D (Holland 1977:37), from the previously reported width of 7 m (1976:49).

stone foundations was exposed. The mudbricks used in this wall measured 50 x 40 x 10 cm in size. Mudbrick detritus found in Trench IIIJ on the NE side of the site located along the same contour line may also be related to this wall. Although it is not certain whether the wall enclosed the entire upper town, it is doubtful that this structure should be otherwise interpreted since clay sling bullets discovered in a room on the interior of the wall probably suggest a defensive purpose for this wall (Holland 1976:49; Stout 1977). The 7 m thick portion of the wall in Area IV is now interpreted as a *tower* 7 m wide though only 5 m long (Holland 1977:37), an interpretation that seems appropriate in light of the finds in adjacent structures. To the S of these excavations, a gap in the W fortifications of the upper town in Trench IVS accompanied by a stone pavement may be evidence of a gate to the upper town (Holland 1977:41f.). Evidence from the rooms in Area IV inside the wall suggests that this wall was in use at the end of the third millennium (Holland 1976:49; 1977:37). If the suggested line of the acropolis defenses is traced, then its wall perimeter would have been about 720 m long (Zettler 1997a:fig. 1.4).

Soundings were made of the ramparts surrounding the lower town in three main locations by the first expedition (Areas VI–VIII, see Holland 1976:fig. 17). In a slit trench across the N fortifications in Area VIII the fortifications appear to have consisted of "alternating layers of wadi gravels and *libn* packing" (Holland 1976:62). Subsequent exploration of the lower town's defenses were completed by the second expedition to Tell es-Sweyhat (Zettler 1997b). Excavation in Operation 25 exposed an outer fortification line assigned to a second phase of occupation in this area (ibid., 49ff.).

This was part of an *earthen rampart* 18.5 m wide whose outer face featured a stone *revetment* with a 38° slope and the inner face of which was supported by a "1.15 m wide *retaining wall* [emphasis added] that consisted of mud bricks set on substantial stone footings three to four courses high." The later excavators have compared the construction of the site's fortifications with those of well-known Middle Bronze Age sites in the Levant such as Shechem, although they have preliminarily dated the structure to the third quarter of the third millennium. The structures identified in Operation 25 would, according to the excavators, have supported a separate wall on a leveled surface which was 8 to 10 m wide and about 6 m high. The ramparts are estimated to be 17 m wide on average (Zettler 1997a:170). If the projected line of the ramparts is correct (see Zettler 1997a:fig. 1.4), then the fortifications were approximately 1,675 m long. Zettler estimates the volume of the rampart as approximately 170,000 m^3 (ibid., 170 and n. 42).

A *gate* was identified on the E side of the site in a magnetometry survey but has not yet been excavated (Zettler 1997a:170). Unfortunately, this gate has been plowed under by local farmers since 2001, according to Richard Zettler (personal communication, 2004). The late third-millennium fortifications of the upper town were destroyed some time during the Ur III period in a conflagration (Holland 1977:58).

TERQA (TELL ASHARA)

UTM: 37 N 0643326.3865678

Refs.: Buccellati (1979:42–83); Buccellati and Kelly-Buccellati (1978; 1983; 1997b).

Plans: Site (Buccellati 1979:fig. 2; 1983:fig. 1); Area D (ibid., fig. 18); .MP 13+ (ibid., fig. 13); SG 17 (ibid., fig. 19).

Sects.: Schematics (Buccellati 1979:44, figs. 16, 19); isometric reconstruction (1983:fig. 2); Area B (1979:fig. 9); MP11 (ibid., fig. 11); MP 13+ (ibid., fig. 12); Defensive Rings (ibid., fig. 16).

Photos: Site aerial (Buccellati and Kelly-Buccellati 1983:photo 1); Area B (Buccellati 1979:photo 63); Area D (ibid., photos 46–62, 66–75).

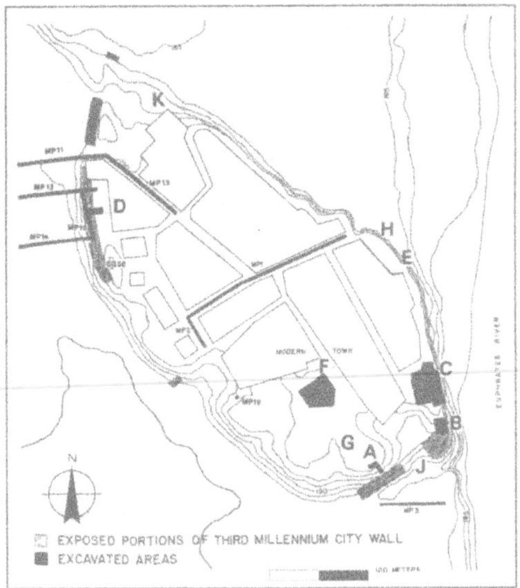

Figure 21. Plan of Terqa and excavation areas (after Buccellati and Kelly-Buccellati 1983:fig. 1).

Terqa, 60 km upstream from Mari on the Euphrates River, has yielded considerable evidence of its fortifications during the third and second millennium (Buccellati and Kelly-Buccellati 1997b:189). It is thought that Terqa, like Mari, was originally founded with a circular plan, despite its 10 ha size, and that it too experienced massive erosion of its N side as a result of the meander of the Euphrates River from the second millennium onward (1983:47). If indeed half of the site has been lost, then the site may have been as much as 20 ha in size. The evidence for the fortifications of Terqa is derived from excavations in Areas MP 11 and 13+, D, and SG 17 on the W side of the site and from Area B on the S side (Figure 21).

Evidence from Terqa consists of three distinct phases of construction of the town wall during the third millennium, each dated by C14 samples and diagnostic pottery (Buccellati 1979:70ff.). Each of the three walls (Defensive Rings 1–3) was built upon an *earthen rampart* that today only survives on the W side of the main mound (ibid., 47f.). The *fosse*, which was excavated outside these walls, is also thought to have functioned with each of these three third-millennium walls. It is believed to have been less than 15 m wide (ibid., 53).

Defensive Ring 1 (ca. 2900 B.C.)

The earliest defenses of Terqa, referred to as Defensive Ring 1, have been identified with the original construction of the *inner fortification wall* (see "Early Wall" in Figure 22; see City Wall 1) upon an *earthen rampart* ca. 2900 B.C. (Buccellati 1979:78f.). Only about a century later (ca. 2900 B.C.) this wall was rebuilt 5 to 6 m wide (City Wall 1A) with projecting *towers* (ibid., 79). The bricks used in this wall's construction measured 45–46 x 20–22 x 10 cm (Buccellati and Kelly-Buccellati 1978:21). Evidence of a "sally port" (*postern gate*) was also found in SG 50 (1983:47). Localized rebuilding of this wall appears to have occurred in some places around the site (Buccellati 1979:78).

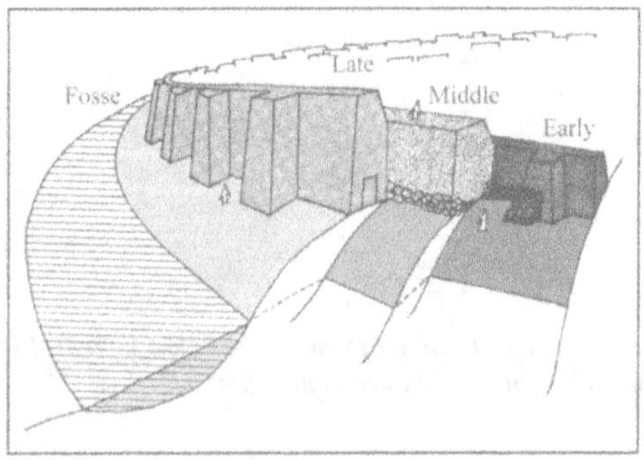

Figure 22. Reconstruction of Terqa's third-millennium defensive system (after Buccellati and Kelly-Buccellati 1983:fig. 2).

Defensive Ring 2B (ca. 2800 B.C.)

During the second phase, which is dated to ca. 2800 B.C., a 9 to 10 m wide mudbrick *fortification wall* (see "Middle Wall" in Figure 22), was added to the exterior face of the original wall and identified as Defensive Ring 2B (Buccellati and Kelly-Buccellati 1983:47). It is thought that the boulder foundation, which was revealed against the foot of the inner wall, belonged to the "middle wall" (1978:20). Although Giorgio Buccellati identifies it as a glacis (Buccellati 1979:79), it appears to have functioned as a footing that protected the base of the second wall.

Defensive Ring 2C (ca. 2700 B.C.)

In the third and final phase, identified as Defensive Ring 2C (ca. 2700 B.C.), a 4 to 6.5 m wide *fortification wall* (see "Late Wall" in Figure 22) was added to the second wall leaving a 2 m wide passage between the outer and middle walls (Buccellati and Kelly-Buccellati 1983:47). It is possible, however, that the space between these walls was the result of erosion or cutting away by the builders of the third phase (1978:21). The outer wall, like the inner wall, also featured projecting *towers* (1983:47).

The cumulative width of these defensive rings at the end of the final phase of use was roughly 20 m. If the site was indeed circular in plan and only fifty percent has been lost, then the projected length of the ramparts, wall, and fosse would be approximately 1,800 m.[9]

Defensive Ring 3 (ca. 2500 B.C. onward)

Based on the evidence from the excavations, Giorgio Buccellati has concluded that following the final phase of construction during the mid-third millennium (Defensive Ring 3) the wall continued to serve as the town's primary defensive line at least until the end of the third millennium, and quite possibly into the second (Buccellati 1979:42, chart 11 and pp. 82f.). Unfortunately, it has been difficult to date most of these modifications because of a lack of pottery within the structure and the impossibility of associating any floors with these renovations (G. Buccellati, personal communication, 2002). Some localized rebuilding of the wall took place during both the late third millennium and the early part of the second millennium, the latter of which is evident based on brick sizes associated with Middle Bronze Age construction at Terqa (Buccellati 1979:82f.). These bricks are usually square

[9]Only 640 m of the defenses probably remain along the W and S sides of the tell, half of which is visible on the surface as mudbricks and detritus (see Buccellati 1979:fig. 2).

in contrast with the earlier rectangular bricks. It is also clear that the reconstruction of the third-millennium defenses during the Middle Bronze Age led to the defenses being increasingly elevated above the plain. A final *fortification wall* during the late Middle Bronze Age appears to have been built after the previous fortifications had been out of use for some time (ibid., 83). Although evidence is scant it has been suggested that a *fosse* also accompanied this structure.

TUTTUL (TELL BI'A)

UTM: 37 N 05504252.3979224

Refs.: Strommenger (1981–1982); Herbordt et al. (1982); Arns et al. (1984); Czichon and Werner (1998); Miglus and Strommenger (2002).

Plans: Site (Miglus and Strommenger 2002:pl. 5); Hügel A-C (Miglus and Strommenger 2002:pls. 16–18, 26, supp. 2); Hügel C (Miglus and Strommenger 2002:pls. 23, 25f., 65, supps. 3, 15f.; Herbordt et al. 1982:supps. 7f.; Arns et al. 1984:supp. 3); Hügel M (Miglus and Strommenger 2002:pls. 9–12).

Sects.: Hügel C (Miglus and Strommenger 2002:pls. 24f., supp. 4; Herbordt et al. 1982:fig. 13, supps. 7–8); Hügel K (ibid., supp. 5); Hügel M (Miglus and Strommenger 2002:supp. 1).

Photos: Site aerial (Miglus and Strommenger 2002:pl. 5); Hügel A-C (Miglus and Strommenger 2002:pls. 14f., 19); Hügel C (Herbordt et al. 1982:fig. 15; Arns et al. 1984:figs. 19f., 26; Miglus and Strommenger 2002:pl. 14); Hügel K (Miglus and Strommenger 2002:pl. 27); Hügel M (ibid., pl. 12).

Tell Bi'a, on the N bank of the Euphrates upstream from the confluence of the Baliḫ and Euphrates Rivers, was settled during the third millennium and first half of the second, when it was known as *Tuttul*. Extensive excavations of the site from 1980 to 1995 have provided a considerable overview of the site's settlement during these periods in the following areas: Hügel M on the S, Hügel C on the W, and Hügel K on the N. Within the confines of the Early Dynastic and Middle Bronze Age walls the site appears to have been approximately 31 ha in size. In the final report on the town's fortification walls the Early Dynastic wall is referred to as the "ältere Mauer," while the Middle Bronze Age wall is referred to as the "jüngere Mauer" (Miglus and Strommenger 2002:9–22).

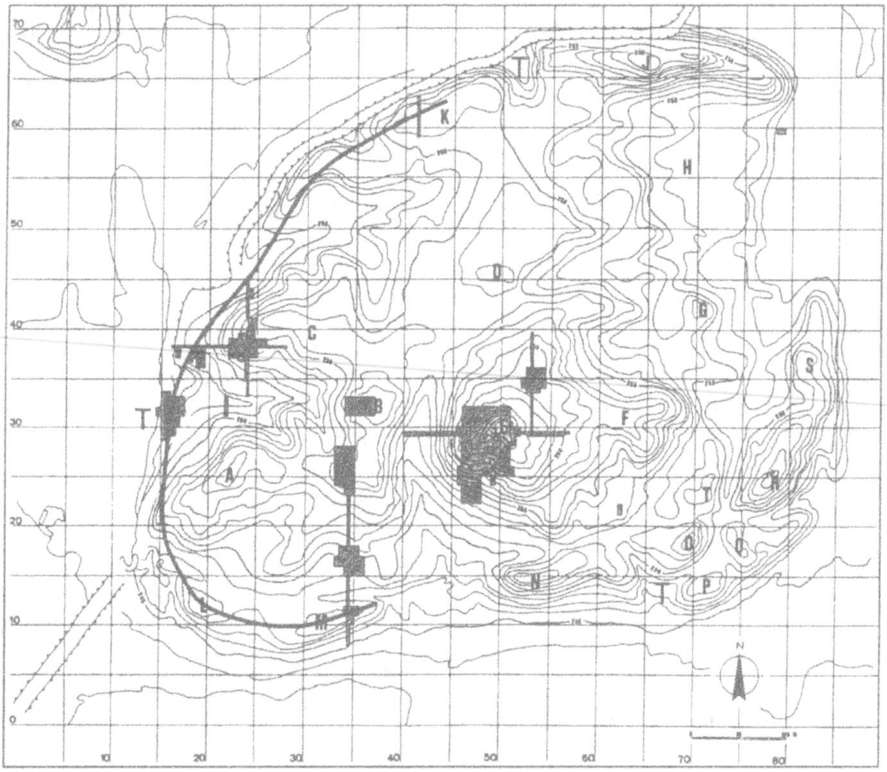

Figure 23. Plan of Tell Bi'a (after Miglus and Strommenger 2002:pl. 5).

Early Dynastic Fortifications

The town's Early dynastic *fortification wall* has been explored in trenches on the S (Hügel M) and W (Hügel C) sides of the site. Although the exact date of this wall was not ascertained (Miglus and Strommenger 2002:21), it is certain that the wall was erected prior to the Akkadian period based on an Akkadian tomb dug into the "glacis" in Hügel M. In the 1.5 m wide trench in Hügel M this wall was 6 to 6.3 m wide and was built of mudbrick (Miglus and Strommenger 2002:9f.). Bricks used in this wall varied in size and included both rectangular bricks (38 x 34, 42 x 36, 43 x 30, 47 x 32, 48 x 38, 49 x 31 cm and often 50 x 36 cm), and square bricks (30, 36, 43, 46, or 48 cm on a side), and their half brick variants.

When the wall was later rebuilt a *glacis* of hard mudbrick material was added against its exterior face. Later a 1.8 m wide outer wall was added parallel to this wall but about 16 m S of it on the remains of an earlier wall that measured 1.2 m wide (ibid., 10). The outer wall's bricks measured 42 x 42 x 10 cm. During the phase associated with the outer wall child burials were added in front of and behind the outer wall.

The complete length of the Middle Bronze Age fortifications may have been as much as 2,120 m, while that of the Early Dynastic period was slightly shorter. At no time does it appear that the walls were founded more than 2 m above the level of the surrounding plain and to date there is no evidence for the existence of an earthen rampart during the Middle Bronze Age.

The Early Dynastic town wall was also revealed between Hügeln A and C where a 25 m stretch of the wall was exposed to a height of 2 m and up to 6.3 m wide (Miglus and Strommenger 2002:9). On the W side of Hügel C the wall measured about 6.4 m wide and was built of rectangular mudbricks measuring 54 x 35 cm (ibid., 17). Buttresses measuring roughly 1.5 m wide were added to the interior face of the wall and spaced about 3 m apart. A 1.8 m projection on the exterior of this wall may be part of a *tower* (see Miglus and Strommenger 2002:pl. 23). Rooms built against the interior side of the wall in this area have been dated to the mid-Early Bronze Age based on ceramics recovered from them (Arns et al. 1984:47). The exterior, W walls, of the structures built against the interior of the town wall are 2.3 m wide (ibid., 48) and added to the width of the wall, which averaged 8.7 m wide in this period. On the N extension of the excavations in Hügel C the wall was also revealed to be 4.7 m wide (Miglus and Strommenger 2002:17f.).

Remains of the W *gate* of the Early Dynastic town have also been exposed between Hügeln A and C (Miglus and Strommenger 2002:13f.). Unfortunately, the structure was poorly preserved and only part of the stone foundations in addition to three of the gate's door sockets remain. Nevertheless, the plan of the gate has been reconstructed as belonging to the four-pier type, and it clearly featured two sets of doors, which were located at both ends of the chamber. The width of the passage was about 3 m. In the vicinity of the gate the town wall appears to have been about 8 to 8.3 m wide (ibid., 20). The location of two other gates have been suggested on the N and S sides of the site (ibid., pl. 5:1).

Middle Bronze Age Fortifications

Excavation to the W of the Early Dynastic wall in Hügel C (Sq. 38/17 and 37/16) also revealed the line of the Middle Bronze Age *fortification wall*, which measured 4.7 m wide and was preserved to a height of 3.8 m (Miglus and Strommenger 2002:18). There was also evidence of a bend or buttress on the inside of the S end of the wall in Square 37/16. The wall was built on occupational debris without any special foundation and was covered with white plaster 0.2 to 0.3 cm thick. Further to the SW between Hügeln A and C remains of this wall were also preserved to a width of 3.6 m built of mudbricks measuring 38–40 x 31–34 x 8–9 cm (ibid., 14).

This same fortification wall up to 4.8 m wide was also revealed on the N side of the site in Hügel K (Miglus and Strommenger 2002:18f.) Here it was constructed of mudbricks measuring 42 x 42 x 9–10 cm.[10] A passage through the wall found in the W side of this trench is identified as a *postern gate*. Mud plaster on the original wall was well preserved. During a second phase of the fortifications, the wall was prepared inside and out for a new superstructure to be built over the remains of the former wall. The wall was then reinforced on its exterior face with a 1.9 to 2.0 m wide exterior wall.

Urkeš (Tell Mozan)

UTM: 37 N 0677510.4103143

Refs.: Bunnens and Roobaert (1988); Kelly-Buccellati (1990); Buccellati and Kelly-Buccellati (1997a); Buccellati (1998); Dohmann-Pfälzner and Pfälzner (2002:181–83).

Plans: Site (Buccellati 1998:figs. 2f.); Area K (fig. 13).

---

[10]Similar size bricks were found during the excavation of the Middle Bronze Age palace in Area E (Strommenger et al. 1986:13).

Sects.: Area K (Buccellati and Kelly-Buccellati 1988:figs. 12, 14–16).
Photos: Site (Buccellati and Kelly-Buccellati 1988:ill. 2–4).
Tell Mozan, which has been positively identified as ancient *Urkeš*, is located in the upper Ḫabur triangle. The site appears to have consisted of an 18 ha upper town (High Mound) rising 25 m above the plain and a lower town (Outer City) of approximately 135 ha. Although the lower town is surrounded today by an earthen embankment only a couple of meters high and about 3,870 m in length (see ill. 5 in Buccellati and Kelly-Buccellati 1988), which was identified as the remains of the wall of the lower town (Buccellati and Kelly-Buccellati 1997a:60), the site's fortifications have only been excavated in Area K on the E side of the High Mound (Kelly-Buccellati 1990:122ff.).

By the early third millennium the upper town was fortified with a *fortification wall* of perhaps as much as 1,610 m in length. It has been explored in Areas KW, on the E side, and in S1. This wall was located at the base of the mound's steep slope, but no exact date can be suggested for its construction since the base of the wall has not been reached (Buccellati 1998:12). Although the wall was preserved to at least 6 m in width and 6 m high, it may have been as much as 8 m wide (Buccellati and Kelly-Buccellati 1997c:79); it was built of bricks 32–33 x 8.5–9 cm (Bunnens and Roobaert 1988:fig. 12, p. 62). Its outer face appears to have been plastered and against this face an *earthen rampart* (referred to as a "glacis") 3.5 m thick and 10.5 m wide was added (Kelly-Buccellati 1990:123).

Excavation of Area KW also indicated that this wall, its rampart, and fosse(?) fell out of use by the mid-third millennium (ED III or later), with the fosse having been filled in (Buccellati 1998:13). Although a single, though limited sounding in Area OH 1 did not reveal any town wall (ibid., 18), recent magnetometry has corroborated the existence of the lower town's *fortification wall* (Dohmann-Pfälzner and Pfälzner 2002:181ff., fig. 19). In addition to revealing a 100 m stretch of this outer wall, a *gate*, a roadway leading to the southeast quarter of the lower town, and an outer "ring" road located 35 m outside the town's wall were identified.

YAKALTUM/EKALTE (TELL MUNBAQA)

UTM: 37 N 0426016.4001620
Refs.: Kühne and Steuerwald (1979); Machule and Rentschaler (1971); Orthmann and Kühne (1974); Orthmann (1976); Machule et al (1992; 1993; 1996); Werner (1998).
Plans: Site (Czichon and Werner 1998:plan 1); Kuppe (Machule et al. 1986:figs. 5f.); N Gate (Machule et al. 1993:fig. 1); NE Gate (Orthmann 1976:fig. 4; Kühne and Steuerwald 1979:plans 1–3); E Gate (Machule et al. 1993:fig. 12); S Gate 1 (Machule et al. 1993:fig. 18).
Sects.: Kuppe (Eichler et al. 1984:fig. 9; Machule et al. 1986:fig. 7); rampart section (Machule and Rentschaler 1971:supp. 9); NE Gate (Kühne and Steuerwald 1979:plan 4).
Photos: Kuppe (Eichler et al. 1984:figs. 10–11; Machule et al. 1986:figs. 8–9; Werner 1998:fig. 17); rampart section (Machule and Rentschaler 1971:fig. 33); N wall (Machule et al. 1993:fig. 3; Werner 1998:fig. 35); N rampart (Werner 1998:fig. 41); N Gate (Machule et al. 1993:fig. 10); NE Gate (Orthmann and Kühne 1974:figs. 27–32; Orthmann 1976:fig. 5; Kühne and Steuerwald 1979:figs. 1–12); computer-reconstructions (Werner 1998:fig. 45).
Tell Munbaqa, OB Yakaltum/MBab Ekalte, lies on the E bank of the Euphrates River about 75 km E of Ḫalab. It was situated only 5 to 10 m above the floor of the river valley (Werner 1998:158f.). The first two periods of occupation of the site during the late EB III to EB IV and during the second half of the Middle Bronze Age were limited to settlement on the upper town proper, referred to as the Kuppe (lit., hilltop).

EB III–IV (PHASE IV-KU-1/2)[11]

From 2500 to 2050 B.C. the settlement was fortified by a 5 m wide *casemate wall* built of mudbricks (ibid., 38ff.). The inner wall of this structure (as evident in squares 30/30–31/30) was as much as 2.8 m thick with a mudbrick superstructure built of mudbricks measuring 50 x 30 x 9 cm (Eichler et al. 1984:73). The rectangular rooms of the casemate wall were about 1.2 m wide with an exterior wall less than 1 m wide.

Domestic structures were later built against the exterior wall of the casemate when the settlement expanded to the E in this area during EB IV (Machule et al. 1986:68ff.). This settlement, which cannot date later than EB IVA (ibid., 85), was enclosed by a more than 5 m wide stone wall preserved to a height of 4.25 m (ibid., figs. 5–6). The line of these fortifications enclosed the Kuppe, an area 170 m by 100 m, approximately 1.7 ha in area with a perimeter of about 540 m. Gates for this settlement leading inland were identified on the S and NE sides, while another gate let out from the center of the site's W side to the Euphrates River. As the settlement grew structures were eventually constructed against the wall and outside of it.

[11]Phasing follows Werner (1998:37).

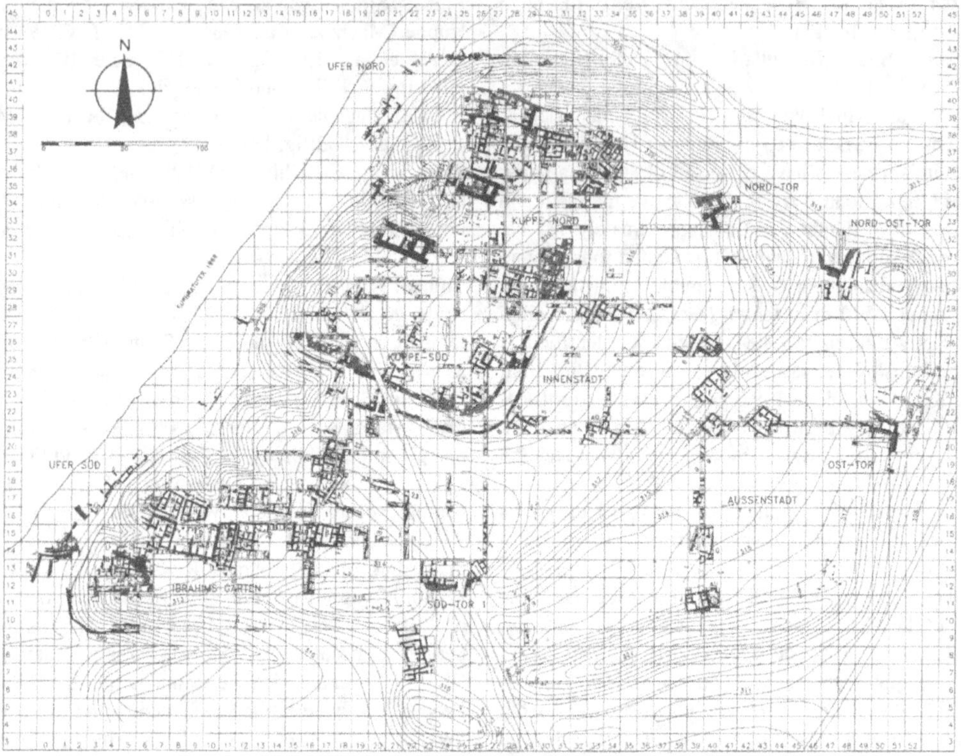

Figure 24. Plan of Tell Munbaqa (after Czichon and Werner 1998:plan 1).

MB II (PHASE III-KU)

Following an abandonment of the site during the MB I the site was resettled and fortified during the MB II (Werner 1998:45ff.). Evidence for the line of the Middle Bronze Age fortifications from the W end of the N side of the site (Sq. 28/40–29/40) suggests that the settlement was reestablished with a rectilinear plan parallel to the Euphrates River similar to that of the EB III settlement. Remains of the defenses in this area consisted of a 64 m stretch of mudbrick *fortification wall* with saw-tooth design preserved over 1.5 m high (Machule et al. 1993:76f.).

One section was made through ramparts on the N side of the Kuppe (Machule and Rentschaler 1971:supp. 9). The section, cut along a NW-SE line (ibid., 53–55, supp. 9 and photo in fig. 33), revealed an *earthen rampart* of alternating layers 10 to 20 cm thick of fine and coarse pebbles (ibid., 54). The core of the rampart and many of its subsequent layers consisted of gravel. On the exterior of the rampart two stone walls (nos. 5 and 6), which may be *revetment walls*, running parallel to the rampart appear to have served to support the rampart. Mudbrick detritus overlay this, which may be the remains of an original mudbrick wall built atop the rampart (ibid., 55). The rampart's section reveals that it was approximately 10 m high. The exact date of this rampart is uncertain but an MB II date is likely.

LB I FORTIFICATIONS (PHASES II-IG 4–6)

The Late Bronze Age settlement, for which Munbaqa is best known (i.e., Ekalte), expanded over the ramparts of the original Kuppe to include several new quarters (Werner 1998:49). At the start of the LB I settlement on the Kuppe and over its ramparts totaled 2.1 ha when the Innenstadt ("Inner City") to the E along with the area of Ibrahims Garten ("Abraham's Garden") to the SW were added (4.4 ha) and enclosed by *earthen ramparts*. As was the case with earlier settlements at Munbaqa, despite an expansion of the settlement during the first part of the Late Bronze Age to the E and S followed by the circumvallation of these areas with a rampart, separate attention was again given to fortifying the Kuppe. The Kuppe, therefore, can be identified as the upper town, which was surrounded by a 4 m wide stone *fortification wall*. This area included several temples and administrative buildings, but no palace according to the excavators, since none was excavated or detected during a magnetometry survey. Two gates gave access to the upper town having been built in the same positions as the Early Bronze Age gates: one on the NE and another on the S.

As mentioned above the lower town settlement in the Innenstadt and Ibrahims Garten were also surrounded by an *earthen rampart* crowned by a mudbrick *fortification wall* (Werner 1998:52). The wall featured *six-pier gates* on the N (Nord-Tor) and S (Süd-Tor 1). The N Gate (Nord-Tor) found in squares 39–41/33–34 was 16 m deep with piers 3.5, 2.5, and 3 m wide, all of which protruded 1 m (Machule et al. 1993:86). The minimum width of the passage was 4.2 m (ibid.). Bricks used in the wall's construction measured 45 x 45 x 12 cm, while those used in the arch were rectangular (45 x 55 x 12 cm). The excavators have observed that the white mortar used in the construction of the mudbrick arch was actually harder than the mud in the bricks themselves. The floor of the passageway of the gate was sealed with gravel and the wall abutting the gate was 3.2 m wide (ibid., 85).

Limited information for the S Gate (Süd-Tor 1) is available, though it was determined that the gate was 3.5 m wide (Machule et al. 1993:89). Although the Süd-Tor was not completely excavated, the *fortification wall* around the Innenstadt here was constructed with cyclopean masonry (Machule et al. 1992:29–30, fig. 12). It measured 4 m wide and was preserved to a height of 2 m. It has also been suggested that another gate led from the lower town to the river bank, although it was never excavated (Werner 1998:52).

LB II FORTIFICATIONS (PHASES II-IG AND 1–3)

During the LB II the Außenstadt ("Outer City," 2.9 ha) further to the E, the Uferzone ("Riverbank," 0.5 ha), and the area atop the LB I ramparts (3.4 ha) were inhabited. The Außenstadt was then enclosed by *earthen ramparts* and a wall. In all, this was an increase of 11.6 ha in settlement from the end of the Middle Bronze Age to the end of the Late Bronze Age.

With the later addition of the Außenstadt and its ramparts, more *gates* were constructed of which at least two have been identified: the NE Gate (Nord-Ost Tor) and the E Gate (Ost-Tor). The NE Gate is the most familiar of these structures, and its dating has been the subject of considerable debate. According to Kühne and Steuerwald (1979) the earliest phase of the gate (Bauschicht 5) appears to date to the late Middle Bronze Age. Twenty meters of the *fortification wall* (no. 18/29) approaching the gate's E side were also exposed (ibid., 204). This wall was built with a mudbrick (brick dimensions 40 x 40 x 10 cm) superstructure 2.5 m high and 1.6 m wide on a stone foundation 3 m wide and 4.5 m high, both of which were plastered with mud on their exterior surface. (The N Gate, whose wall was 2.2 m wide, lies at a right angle to this wall (18/29); ibid., 205).

The NE Gate was preserved as a result of the later addition of a stone wall parallel to the gate's N elevation and located to its N (ibid., 206); 1.2 m of the original 2.2 m wide mudbrick arch gateway was preserved (ibid., 205f.). The passageway was 3.1 m wide and must have originally been 3.5 to 5 m high. The overall dimensions of the gate are 12.5 m wide by 17.5 m deep (ibid., plan 3). Although not much is known of the gate chamber to the S due to its poor preservation, the gate is thought to be of the *six-pier type*. It is only evident that a stone wall (15) was added to the W side of the chamber for later support (ibid., 207). The gate was later destroyed by fire and functioned as a pedestrian gate for a short while later (ibid., 212). In the LB II (Bauschicht 4) the fortifications in this area were renovated and reoriented (ibid., 213). In its final phase the gate was blocked by a massive stone wall (no. 7). The complex has been variously dated by Cypriote pottery found in association with burials (ibid., 213f.).[12]

The E Gate (Ost-Tor) is 22 m long with a 3.25 m wide passageway (Machule et al. 1993:99). It does not appear to have featured any piers. Its first interior space is 10 m long and 4 m wide. Its second short space constricts to 3.25 m wide where two tower-like constructions protrude into the gate. The length of its final space has not been determined owing to incomplete excavation. The gate is flanked by the *fortification walls*, which were over 3 m wide. Analysis of the gate's position suggests that it was built during or after the construction of the ramparts.

[12]In the original preliminary report H. Kühne observed that Schichten 5 and 4, to which the gate belongs, dated to "die Mitte und zweite Hälfte des zweiten Jahrtausends v. Chr." (Orthmann and Kühne 1974:79). This date was later reasserted (Kühne and Steuerwald 1979:211, and n. 17) and H. Kühne still maintains that it is possible that it was built at the end of the Middle Bronze Age (personal communication, 2002). Nevertheless, Peter Werner has advanced a Late Bronze Age date for both Schichten as stated in the latest reference concerning this gate (Werner 1998:52–54).

2. The Levant (ca. 2400–1500 B.C.)

This catalog of sites includes evidence pertaining to the fortifications of sites in the Levant during the Middle Bronze Age. It was compiled from published data as well as personal communications with excavators. For ease of use it has been divided into catalogs of fortified settlements in the northern Levant (2.A) and southern Levant (2.B).

Due to the nature of the evidence for earthen rampart fortifications in the Levant, which have long been considered to be of northern Levantine origin (see chapter 1), a catalog of sites in the northern Levant cannot be conveniently restricted to Middle Bronze Age sites. In fact, northern sites of Middle Bronze Age date (ca. 1900–1550 B.C.) would prove insufficient for identifying the tradition after which Middle Bronze Age sites in the southern Levant were fortified. Therefore, in order to demonstrate the priority of fortified settlements in the northern Levant within the development outlined in this work it has been necessary to include sites that predate the fortified Middle Bronze Age sites of the southern Levant (e.g., Afis).

A. The Northern Levant: Western Syria, Lebanon, and the Hatay

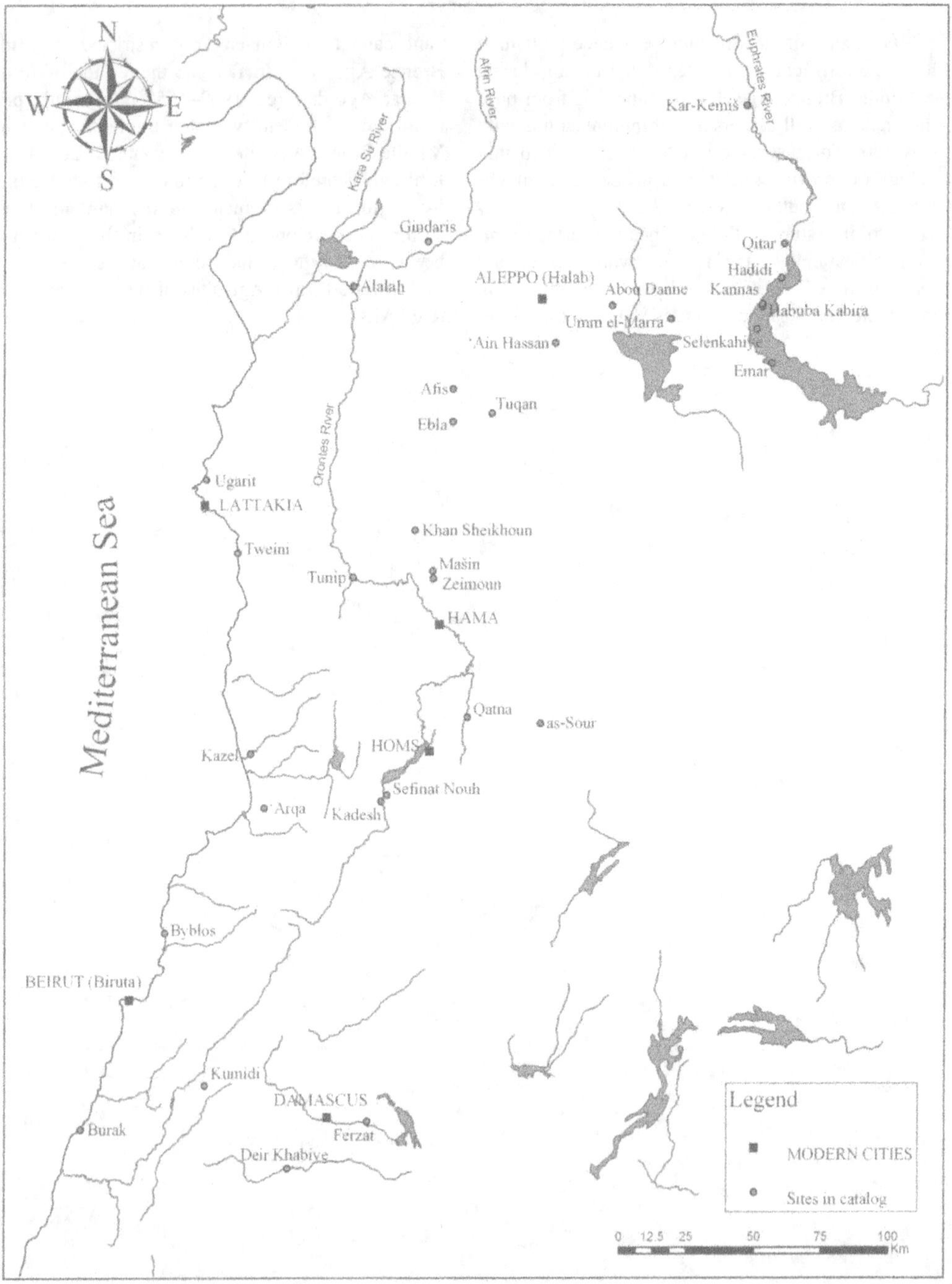

Figure 25. Sites in the northern Levant discussed in catalog 2.A.

ABOU DANNE, TELL

UTM: 37 N 036739.4004915
Refs.: Tefnin (1978–1979; 1979a; 1979b; 1979c; 1981–1982; 1983).
Plans: Site (1979b:fig. 1); Area B (fig. 2); Sondage I level VI (1979a:fig. 12).
Sects.: Sondage I-East (1979a:pl. 1); West (fig. 14).
Photos: NA.

Excavations at Tell Abou Danne 25 km E of Ḫalab have exposed the site's Early and Middle Bronze Age fortifications (Tefnin 1979a). Although the acropolis was fortified during the EB I and II, there is no evidence that it was fortified during the EB IV even though the site itself appears to have been occupied (see discussion of Areas S IV and XXII in Tefnin 1981–1982:201).

EB I–II (LEVEL VII) & EB IV

The E section of Sondage I on the acropolis revealed that the N elevation of the EB I–II mudbrick wall (Level VII) was preserved over 7.5 m high (1979a:pl. 1), its upper portion being slightly canted backward (ibid., 197). Some of the bricks used in the construction of the Early Bronze Age wall (18–20 x 6–7 cm) are described as Riemchen bricks (1978–1979), of the type known among late fourth millennium Uruk colony sites. It is uncertain whether or not the Early Bronze Age wall was built upon a berm or upon bedrock, but the depth of the wall seems to indicate that it was built at ground level when the site was founded and not on an artificial mound. As mentioned above there is no evidence of EB IV defenses on the acropolis, though an EB IV structure was built on the W side of the mound directly on top of the remains of the EB I–II fortification wall (1981–1982:201).

MIDDLE BRONZE AGE: LEVEL VI, PHASE 1

During the Middle Bronze Age three phases of construction are evident for the enclosure wall of the acropolis (1979a:192ff.). In the first stage (Level VI) a 2.1 m wide *fortification wall* built of mudbricks averaging 33 x 33 x 7 cm was built on a foundation of small stones above the preserved remains of the EB I–II wall. The construction of this wall probably followed the addition of an embankment in front of the Early Bronze Age wall, which perhaps should be identified as an *earthen rampart*; it was approximately 9.3 m wide and 6 m high with a slope of 40° (ibid., pl. 1, p. 193). While a narrow wall of 0.85 m was found parallel but outside of this wall in Sondage I (ibid., 192), it is unlikely that this wall was present around the entire site and its function remains unclear. According to Tefnin, in the same phase of construction or slightly later, two parallel mudbrick walls (36 x 36 x 8 cm bricks) were built on the interior of the original enclosure wall. Evidence of a *glacis* composed of loose, limestone gravel, and compact earth belonging to the first stage was also found (ibid., 193). The end of the first phase of the Middle Bronze Age defenses was marked by a site-wide destruction.

MIDDLE BRONZE AGE: LEVEL VI, PHASE 2

After the rebuilding of the town wall, during the second stage, structures identified as *towers* built of gray mudbricks (38 x 38 x 10 cm) were added to the wall (Tefnin 1979a:193). In addition to these structures, the rebuilding during the second stage included the addition of a new *glacis* over top of the first glacis. The town wall of the second stage appears to have eroded in the area of Sondage I following its violent destruction.

MIDDLE BRONZE AGE: LEVEL VI, PHASE 3

Despite two considerable destructions of the site during the Middle Bronze Age, the inhabitants engaged in a final phase of fortification construction. In the area of Sondage I they built a modest *fortification wall* only 0.7 m wide (35 x 35 x 8 cm bricks, 1979a:193f.). However, on the acropolis they appear to have undertaken the construction of a more substantial *fortification wall*, which was over thirty bricks wide (1981–1982:202) or more than 10 m wide. Although it has not been confirmed, Tefnin suggests that an earthquake contributed to the final destruction of the site at the end of the Middle Bronze Age (ibid., 197).

LATE BRONZE AGE

The site is thought to have been unfortified during the Late Bronze Age (Akkermans and Schwartz 2003:352).

AFIS, TELL (APSUNA)

UTM: 37 N 0301287.3975662
Refs.: Mazzoni (1994); Gianessi (1998); Cecchini and Mazzoni (1998).
Plans: Site (Cecchini and Mazzoni 1998:5f.); Area E_1 (Gianessi 1998:fig. 13).
Sects.: Area E_1 south (Cecchini and Mazzoni 1998:4f.).
Photos: Area E_1 (Gianessi 1998:figs. 14.1–15.2).

Tell Afis (18.5 ha), which is located 12 km N of Ebla, was settled from at least EB II through the Late Bronze Age (Cecchini and Mazzoni 1998). It is probably to be identified as *Apsuna*.

MB IIA

To date, what is known of the fortifications of Tell Afis is limited to the defenses of the W side of the citadel (Area E_1), which is roughly 1 ha in size (Figure 26). Although the citadel was settled during the EB IV, the earliest evidence for the fortifications of the citadel are actually the remains of a 4.3 m wide *fortification wall* (M.1115) in Level 16 (Middle Bronze Age) built of mudbricks (30 x 30 cm) on foundation stones roughly 40 x 50 cm and 30 x 35 cm (ibid., 104). The area had been leveled for the construction of this wall and a stone *glacis* was added to the slope of the citadel; in Area E storage rooms were added to the interior of the wall (Mazzoni 1994:148).

MB IIB–C

In the following phase of settlement on the acropolis (Level 15), probably in the later half of the Middle Bronze Age, another *fortification wall* (M.1116) was built against the interior of the earlier wall (M.1115; Gianessi 1998:104); red and yellow mudbricks were used in its construction. It is not possible to estimate its dimensions (ibid., fig. 13).

Future excavations in Area F, N of the citadel, might provide an understanding of what type of fortifications enclosed the lower settlement (Cecchini and Mazzoni 1998:5).

‘AIN HASSAN, TELL*

UTM: NA [DZ 288.456].
Refs.: Nigro (2002a:315, n. 99).
Plans: NA.
Sects.: NA.
Photos: NA.

Tell Hassan is about 17 km SSE of Ḫalab. The site, though unexcavated, was examined in 1994 by Lorenzo Nigro and Nicolò Marchetti who observed that a modern cut on the W side of the site revealed the Middle Bronze Age mudbrick *fortification wall* (L. Nigro 2002:315, n. 99). This wall they suggest is dated to the early MB I based on ceramics that were associated with it. Furthermore, it could be traced for a distance of more than 50 m.

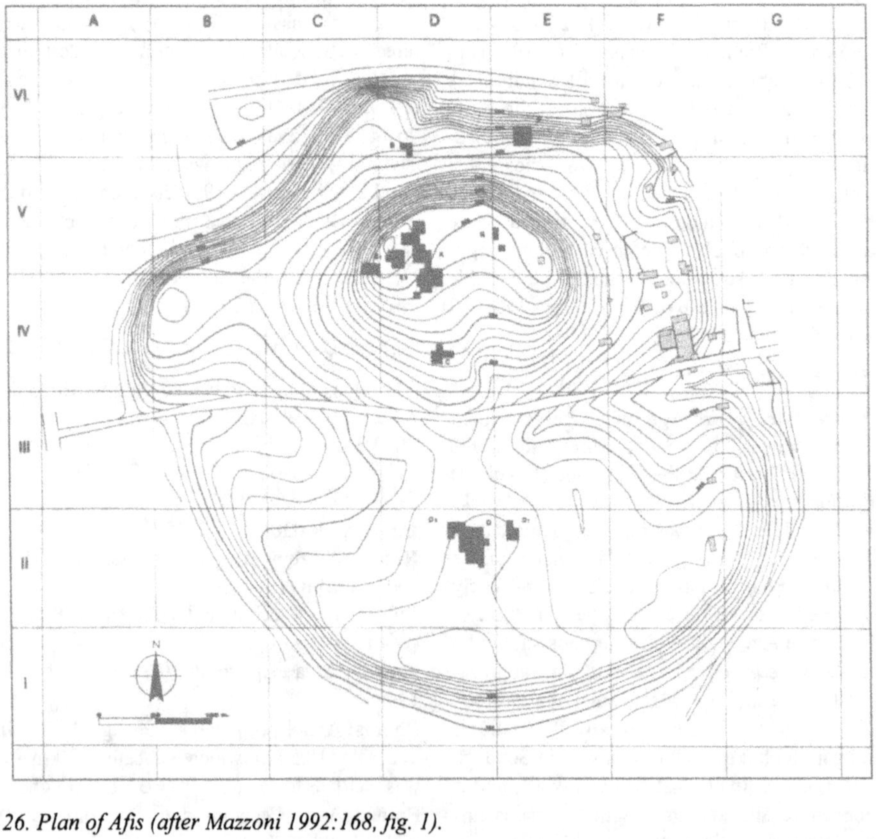

Figure 26. Plan of Afis (after Mazzoni 1992:168, fig. 1).

ALALAḪ (TELL ATCHANA)
UTM: 37 N 264917.4013507
Refs.: Woolley (1955); Yener et al. (2000); Yener (2005).
Plans: Site (Woolley 1955:pl. 14); Level VII N Gate (ibid., figs. 55f.); Level VII fortifications (Yener 2005:fig. 4.27).
Sects.: Trench F N section (Woolley 1955:fig. 52).
Photos: Level VII N Gate (Woolley 1955:pls. 29–30a).

Alalaḫ (Tell Atchana) is located on the southern side of the 'Amuq Valley just E of the Orontes River. Although the site was extensively excavated by Sir Leonard Woolley (1955), excavations of the Middle Bronze Age fortifications were limited to a sounding of the rampart on the NE side of the site in Trench F and exposure of the N Gate (Figure 27). The Middle Bronze Age features in these areas belong to Level VII, which is usually identified as the last phase of Middle Bronze Age occupation of Tell Atchana, which ended ca. 1550 B.C. The site's gross size in this period is estimated to have been at about 22 ha (Yener et al. 2000:185) and its inhabitable area within the fortifications was approximately 19.5 ha (see below). Woolley acknowledged that there was probably a lower town with its own defenses, but he was not able to identify its limits or any fortifications that might have been associated with it (Woolley 1955:132f.).

The evidence for the site's Middle Bronze Age fortifications consisted primarily of an early *earthen rampart*, which was excavated in Trench F on the NE side of the site (ibid., fig. 52a; see pl. 22 for location of Trench F). It was built before Level VII and then widened to feature a slope of at least 35° (ibid., 137). This sounding revealed that it was composed of trash, ash, ceramics, and "debris of all sorts" heaped to a height of more than 4 m above the modern level of the plain (ibid., 135), perhaps as much as 9 m above the level of the plain at the time of its construction (ibid., n. 2, p. 135). If the lines of the rampart in the section are projected, assuming that the rampart was of the freestanding type, then the rampart was about 30 m wide. However, the fortifications of the upper town may have featured only a supplemental rampart.

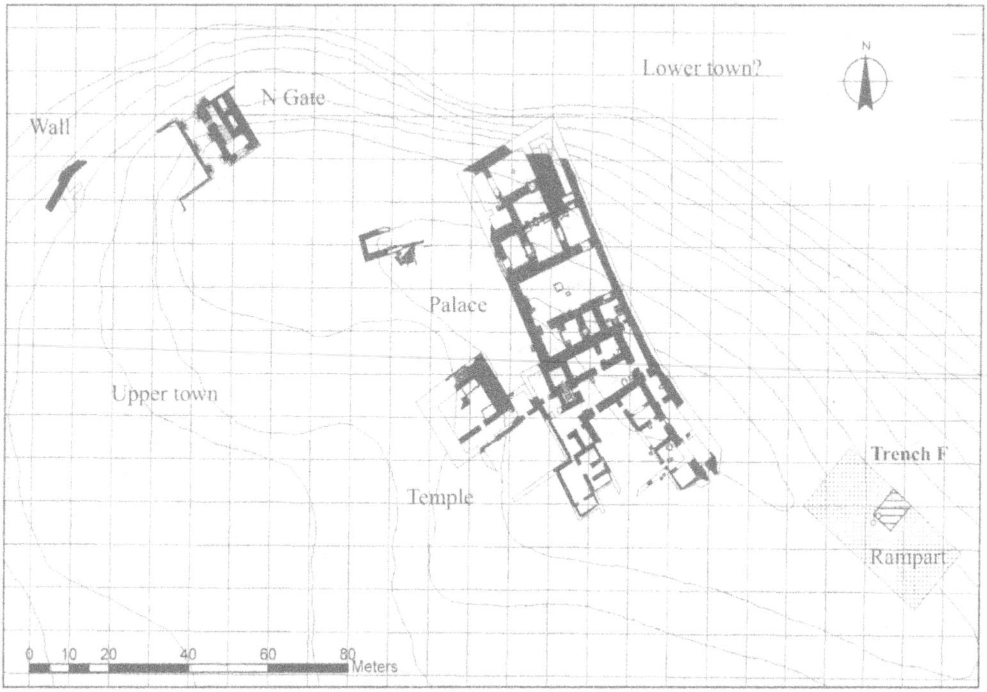

Figure 27. Plan of Alalaḫ Level VII showing location of earthen rampart (after Yener 2005:fig. 4.27).

No evidence of either the town's mudbrick fortification wall that crowned the ramparts nor the fosse that might have existed outside of the rampart have been encountered, although Woolley suggested that part of the E wall of the Level VII palace (in room 29—although its wall was reconstructed here, see fig. 35 in Woolley 1955) continued in use as a defensive wall in Levels V and VI (ibid., 137, 139; see also section in fig. 2). He also noted that the site's fortifications probably varied in character around the town (ibid., 139). However, Woolley's suggestion of a wooden palisade located at the top of the early rampart (ibid., 137) is without any parallel in Near Eastern archaeological excavations at Middle Bronze Age sites.[1]

The circuit of the mound is about 1,650 m and thus the rampart's volume, if it enclosed the entire mound, was approximately 44,550 m^3. This rampart, the earliest of two ramparts which were completely exposed—the other is dated to Level IV—was attributed by Woolley to Level VII because it contained a considerable quantity of pottery from Levels XII through VIII (ibid., 135f.).

The Level VII gate, a *six-pier gate* centered in Square W7 (not V7) is the earliest gate exposed at the site (Woolley 1955:145ff.). It led from the mound to the NW. The structure's façade was 23 m wide and its passage was 17 m long. It was built of mudbricks on stone foundation and was preserved about 1.5 m high. The faces of the piers and corners of the entrance were lined with limestone orthostats. The top of the orthostats still lean into the gate's passage indicating that it probably featured a barrel-vaulted corridor (cp. Woolley's reconstruction of a "relieving-arch" built of wood and mudbrick, p. 148; see 1955, fig. 55). The space between the piers averaged 2.7 m wide. There was also evidence of wood timbers used in the construction of the gate, above the orthostats and along the floor below the mudbrick superstructure. The E half of the gate was exposed to reveal an entrance leading to what Woolley identified as a "sentry-chamber and guard-room" where a set of stairs led to a second story (ibid., 150). The W half was constructed of a mass of mudbrick (ibid., 148). It is presumed that the gate's plan changed considerably during later periods and gradually shifted E (ibid., 151).

There is considerable evidence for a fiery destruction of the Level VII town and its gate, which is attributed to Ḫattušili I (D. L. Stein 1997:56). Vitrified bricks had fallen into its passage (Woolley 1955:148).

[1]Woolley's speculation regarding the existence of a wooden palisade can be traced back to the identification of the remains of wooden palisades atop the earthen ramparts of Iron Age hill forts in the British Isles (see Cunliffe 1983).

'ARQA, TELL (IRQATA)

UTM: 37 N 0228906.3824971
Refs.: Thalmann (1978; 1991).
Plans: Sq. AK21 (1978:fig. 3; 1991:fig. 2).
Sects.: NA.
Photos: Site aerial (1991:fig. 1).

Tell 'Arqa (ancient *Irqata*)[2] is located on the Akkar plain in N Lebanon inland about 5 km from the Mediterranean Sea and about 20 km SSE of Tell Kazel (see below). Preliminary reports of the excavations of Tell 'Arqa indicate that the settlement was first fortified in Level 13, during the MB II (Thalmann 1978:62–67; 1991:32–34). The fortified settlement of the mature Middle Bronze Age replaced the unfortified EB IV (Levels 16–15) and MB IIA (Level 14) settlements (1991:25–32). Remains of the Middle Bronze Age fortifications were found in square AK21. Although only a portion of a Level 13 *tower* has survived (see Thalmann 1978:fig. 3), the continuation of the mudbrick *fortification wall* from the tower, which had completely eroded in the excavated area, is presumed to have been of a similar composition measuring 1.7 m thick and was built on a stone foundation (1991:32).

BIRUTA (BEIRUT)

UTM: 36 S 0731896.3753736
Refs.: Marquis (1995); Badre and Thalmann (1996); Finkbeiner (1997); Finkbeiner and Sader (1997); Badre (1997a; 2002).
Plans: Site (Badre 1997a:fig. 1a); Area BEY 003 (ibid., 1b); Area BEY 020 (Finkbeiner and Sader 1997:fig. 4).
Sects.: Schematic (Badre and Thalmann 1996:fig. 3); Area BEY 003 (Badre 1997a:fig. 10b).
Photos: Area BEY 003 (Badre and Thalmann 1996:figs. 4f.; Badre 1997a:fig. 10a; 2002:pl. 3); Area BEY 020 (Finkbeiner and Sader 1997:pls. 1f.).

Tell Beirut is located 28 km SW of Byblos. The elliptically shaped tell of Beirut measures approximately 250 x 120 m (Badre and Thalmann 1996:91), covering a gross area of 3 ha. Fortifications, including a gate dated to the Middle Bronze Age, have been recently exposed in areas BEY 003 and 020 on the SW side of the site (Figure 28).

MB IIA: COMPLEX I

The earliest fortifications of the site are dated to the MB IIA and are referred to as part of Complex I (Badre 1997a:26). They consisted of a 0.7 m wide mudbrick *fortification wall* that rested on a stone

[2]See EA 62, 72, 100, 139, 140.

foundation (W.353), which in turn was built directly on bedrock. Although only 15 m of the length of its foundations were exposed in BEY 003, which were preserved to an average height of 1.7 m, the existence of a mudbrick superstructure is supported by the amount of mudbrick detritus found atop the foundations. Badre has observed that the wall (or at least its foundation) appears to have been built in sections with the addition of each length after one was complete. Once completed the wall was coated with clay plaster and only later was a *glacis* of "rammed clay" built against this wall directly upon bedrock with a slope of 30° (Badre 1997a:26f., fig. 8b; 2002:3). Badre estimates the dimensions of the earliest settlement to have been around 115 m by 70 m or roughly 0.8 ha (2002:3).

MB IIB–C: Complex II

During a later phase of the Middle Bronze Age (Complex II) another *fortification wall*, preserved to an average height of 6 m in BEY 003, was built 13 m outside the original wall but parallel to it (Badre 1997a:28). The wall (W.329/398), which was about 1.7 m wide, was constructed of large quarried stones of up to 2.85 m in length and 0.2 m in width, which were laid in alternating rows of long and short blocks. Additionally, pillars protruded slightly from the wall and were incorporated into the wall at 5 m intervals (ibid., fig. 11a). A stone wall excavated in area BEY 020 in 1995 was probably the extension of this wall from BEY 003 to the W (Finkbeiner 1997:124ff.).

Wall 329/398 in BEY 003 was the first for which evidence of a *gate* was recovered, though it is possible that the gate was opened after the construction of this wall (Badre 1997a:28). This is indicated by the pillars in the wall, which are not spaced in relation to the gate's opening, and because the town wall (W.329/398) is continuous below the gate's threshold. The gate's passage, which is 2.1 m wide, is a simple direct-axis gate featuring high steps (obviating the use of this gate by wheeled vehicles). Evidently, the gate experienced another modification when an L-shaped wall (W.320), which Badre referred to as a "chicane," was added outside the gate (1997a:30). Its addition created a bent-axis approach, which meant that the town would be more difficult to enter through this gate.

The wall of Complex 2 continued to the E into BEY 020 where 12.5 m were traced, and E of that into BEY 013 for 35 m where it was preserved to a height of up to 7 m (1996:4; Karam 1997:107–09). Badre suggests that the settlement associated with Complex 2 was about 225 m long by 80 m wide (2002:5), or roughly 1.8 ha. Perhaps at the same time that the gate was modified if not a little later, W.329/398 was reinforced by another wall against its interior measuring 3.75 m wide (Badre 1997a:30).

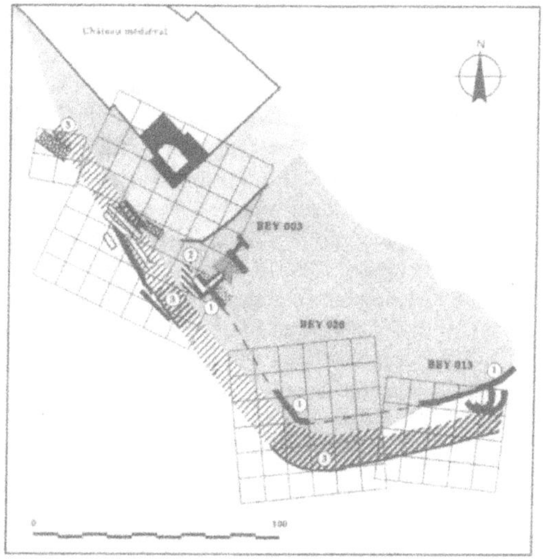

Figure 28. Plan of fortifications excavated at Biruta (after Badre and Thalmann 1996:fig. 2): (1) MB IIA gate and ramparts; (2) MB IIB–C rampart; (3) Late Bronze to Early Iron Age rampart.

MB IIC/LB I Modifications

This fortification system was used throughout the remainder of the Middle Bronze Age until it was overhauled, either at the end of the Middle Bronze Age or at the start of the Late Bronze Age with the addition of a stone faced *glacis* (Glacis I) to the exterior face of the earlier fortification system (Badre and Thalmann 1996:92, fig. 5). Glacis I was built of large limestone boulders with a slope of 20° and stood more than 2.6 m high (Badre 1997a:48). Its irregular curve blocked the earlier bent-axis gate and led to its falling out of use. Badre suggests that Glacis I may have been built in two stages: during the first stage a glacis was built as part of the defenses along the line of W.329 and only later was the gate blocked by the extension of the glacis in an arc toward the S (Badre 1997a:50). To date it has not been possible to test this hypothesis because a modern concrete pillar has disturbed the junction between these two sections of glacis.

At the end of the Late Bronze Age a second *glacis*, the most impressive exposed at ancient Beirut, was built against the earlier one (Badre 1997a:60ff.). It appears to have functioned into the Iron Age.

BURAK, TELL EL-

UTM 36 S 743795.3778586

Refs.: Kamlah and Sader (2004).

Plans: Site (Abb. 2, Taf. 8B); Area 1 (Abb. 4–5).

Sects.: Area 2 (Taf. 9B).

Photos: Area 1 (Taf. 12); Area 2 (Taf. 9A).

The 1.3 ha mound of Tell el-Burak lies approximately 9 km S of Sidon on the Lebanese coast. Recent excavations of the small mound have revealed the southern half of a small MB IIB–C fortress built in Area 1 as well as a revetment wall in Area 2 (Kamlah and Sader 2004:131–38). It was constructed at about 10 m ASL on top of what was probably an artificial mound built during the Middle Bronze Age (see Kamlah and Sader 2004:131, Taf. 9B). The mound featured a stone-built *revetment wall*, identified as a "retaining wall," that was preserved more than 4 m high. Fills behind the wall were deposited during the Middle Bronze Age and thus the revetment wall is so dated.

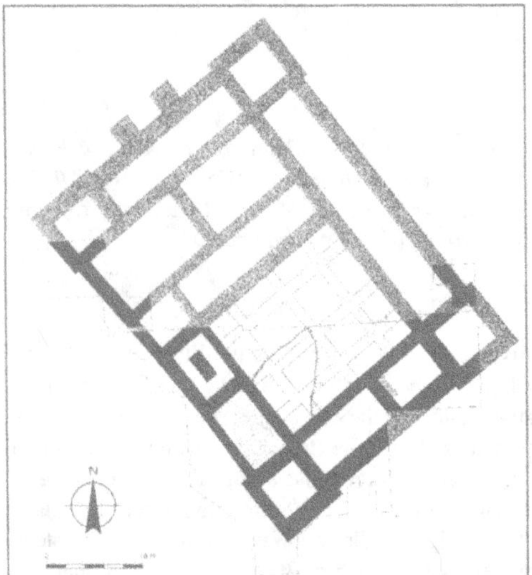

Figure 29. Tell el-Burak fortress (after Kamlah and Sader 2004:Abb. 5).

The fortress in Area 1 is estimated to be more than 40 m long (full length not preserved) and about 31 m wide (Figure 29).[3] It was constructed with an outer *fortification wall* approximately 1.2 m thick, which was built of mudbricks measuring 40 x 40 x 12 cm (ibid., 134). Square *towers* measuring approximately 8 m on a side were exposed on the S and E corners of the fortress. The edge of a third tower on the W side of the fort was only partially revealed. A gate is surmised to have existed on the N end of the fortress, although it has not been exposed. A courtyard of more than 15 x 15 m was part of the original design of the fortress.

The closest parallel for the fortress at Tell el-Burak is found at Tel Mevorakh (see p. 295). Both were 0.1 ha in size and sat approximately 10 m ASL above the coastal road that they guarded. The MB IIB–C fortress continued in use during the Late Bronze Age, as suggested by the presence of LB ceramics (Kamlah and Sader 2004:134).

BYBLOS (GEBAL)

UTM: 36 S 744051.3778594

Refs.: Dunand (1949–1950; 1954:124f., 270; 1955a; 1955b; 1955c; 1956a; 1956b; 1961; 1964; 1968); Saghieh (1983).

Plans: Site (Dunand 1939:pls. 202–04; 1954:pl. 212); NE Gate in Sqs. 26–28/10 (1954, Levels 28–27).

Sects.: Schematic (Dunand 1967:fig. 2).

Photos: Site aerial (Dunand 1954:pl. 212); N ramparts (1949–1950:pl. 7; 1954:pl. 47); NE Gate (1954:pls. 48–50); Hyksos rampart near NE Gate (1954:pl. 50; 1956a:pl. 2; 1956b:pl. 4; 1961:pls. 1–3).

Byblos (*Gebal*), located 28 km NE of Biruta on the coast of Lebanon, is among the most important Middle Bronze Age sites in the northern Levant. Unfortunately, working with the site's publications continues to prove challenging.[4] Many of these difficulties were discussed by Muntaha Saghieh who also adds, especially with regard to the fortifications, that Dunand employed different terminology at different times in the phasing of the same features (1983:65f.). Nevertheless, today plans and photos from the excavation's archives are posted on-site at each major excavation area. These provide the most coherent summary of Dunand's findings that has been made available to date, and they have facilitated the interpretation that follows.

[3]The proportions and size of the fortress are comparable to those of the Iron Age fortress at Kadesh Barnea (see Meshel 1992:298).

[4]The first reason for this is that the site was almost completely excavated by Dunand, making it nearly impossible to verify many of the results that were obtained between 1926 and 1965. The second reason is that Dunand worked in arbitrary meter intervals rather than in coherent and contemporaneous stratigraphic phases. Finally, the absence of complete publication of his final seasons makes it difficult to fully assess the results of his excavations between 1938 and 1965. For an overview of Dunand's methodology, see Saghieh (1983:x–xi).

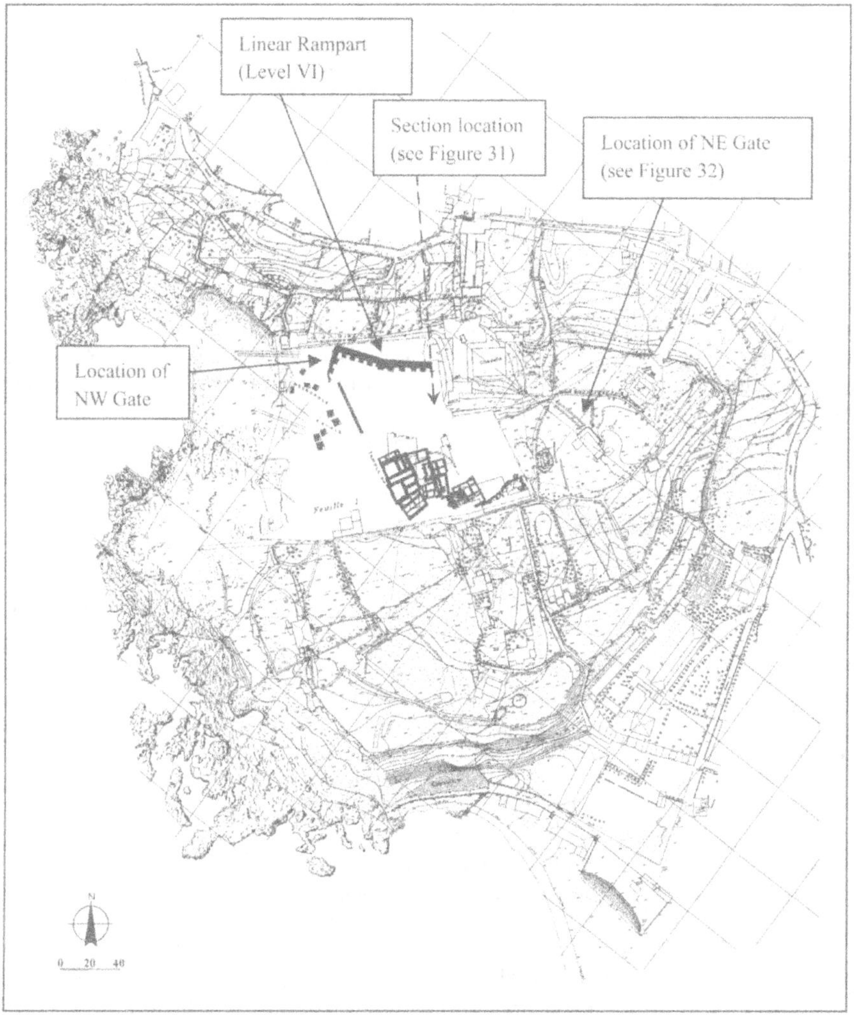

Figure 30. Plan of Byblos (composite after Dunand 1939:pls. 202f.).

The clearest presentation of Dunand's results concerning the third-and second-millennium fortifications of the site are to be found in one of his later preliminary reports (1955b). Therein Dunand summarized what he discerned to be seven *ouvrages* (construction phases) in the development of the fortifications of the site (ibid., 18ff.), the main elements of each being a wall and rampart, both of which were identified with French *rempart*. Each of these seven phases is recognizable in the unpublished section of the N fortifications which is displayed at the site today (Figure 31 below). Of these various phases only the first four directly concern this study, while the fifth phase includes a Late Bronze Age rampart. The last two remaining phases belong to the Iron Age ramparts. Byblos, like many other sites in the Levant, demonstrates that a site's earliest fortifications (i.e., those from the third millennium) often had a direct bearing upon the development of later fortifications (i.e., during the Middle Bronze Age).

During the Early Bronze Age the line of the fortifications appears to have limited the settlement to an area of approximately 5 ha (Dunand 1968:21). Although this line moved outward slightly, it is probable that the site was not much larger during the Middle Bronze Age. Whether or not the area outside of the main settlement was occupied in these periods has not been confirmed. In the summary that follows the major phases of occupation at Byblos that have been identified by Dunand are referred to as "Levels" based upon the summary of the occupational sequence that Dunand published in his overall plan of the site in *Fouilles de Byblos II* (see Dunand 1954:pl. 212).

EB III: LEVEL V (OUVRAGE 1)

The history of the fortifications of Byblos begins during the EB III, Dunand's Level V or Saghieh's Period KIII (1983:65f., 84, 130f.). At that time the site received its first fortifications (ouvrage 1), which consisted of two large stone walls (Dunand 1954:pl. 212, no. 23).[5] The S fortification wall was located about 2 m behind the N fortification wall (Dunand 1955b:20), which is largely conjectural though it probably corresponds with a single course of foundation stones of a 2.5 m wide wall preserved in the W section to the S of the later Early Bronze Age buttressed wall (Figure 31). Dunand suggested that the space between these walls was filled with earth taken from the construction of a sacred lake (1954:pl. 212), though this cannot be confirmed. Although these walls are collectively identified as the "linear rampart,"[6] only the interior wall has been traced from the N side of the site to the E of the castle where it turns to the SE and abuts the NE Gate; from there it continues on the S side of the NE Gate for a short distance. Although Saghieh was not particularly concerned with a study of the fortifications, he suggests identifying the earliest phase of the NE Gate (see Figure 32 below) and the first ramp within the gate with this set of fortifications (1983:66), though it is not possible to verify his hypothesis. However, Dunand dated the earliest phase of the NE Gate to the following phase.

EB III–IVA: LEVEL VI (OUVRAGE 2)

The site's original fortifications were supplanted in the next period, Dunand's Level VI (so-called "Pre-Amorite" phase) or Saghieh's Period KIV (1983:67, 84, 131), which corresponds roughly with the period of EB III–IVA (Dunand 1954:pl. 212, no. 30). These fortifications (ouvrage 2) remain perhaps the most distinctive discovered at Byblos because the stone *fortification wall* on the N side of the site featured large interior buttresses measuring between 3 and 4 m wide (Figure 30), which were spaced approximately 2.5 m apart, while the exterior surface was built of dressed stones. The wall was 4 m wide and was preserved up to 6 m high (Dunand 1939:421ff.; 1955b:19f.).

On the NE side of the site Dunand suggested identifying the continuation of the fortifications of this period with an *earthen rampart* about 24 m wide into which a gate was then built (Dunand 1954:pl. 212). It is uncertain, however, that this rampart should be associated with the "linear rampart" on the N side of the site. Nevertheless, if these fortifications encircled the entire mound at this time then their length would have been approximately 850 m. The existence of a *fosse* associated with ouvrage 2 can be asserted based upon a carved depression in the bedrock below ouvrage 3.

Two gates were added during the expansion and renovation of the fortifications in this period. A simple *direct-axis gate* was added on the NW. Another much larger *pierless gate* was also added on the NE side of the site, which would have led directly inland (Figure 32). This gate's passage was 4.8 m wide and 18 m long and was built into the rampart on the NE side of the site (Dunand 1961:84). A wooden frame built into the floor and along the walls supported the sloping sides and roof of the gate's passage, which judging from the canting of the stone walls of the passage probably featured a mudbrick barrel vault. If so, this gate may very well have been the largest vaulted stone structure in this region during the late third millennium, which would suggest a high degree of architectural proficiency.[7]

Based on archaeological evidence from the gate and around the site it seems that the beams used in this gate were burned in a conflagration toward the end of the third millennium that ultimately caused the gate to fall out of use. This destruction, which also claimed many other structures throughout the site, was identified by Dunand with an Amorite conquest. However, given the known involvement of the Akkadian kings in this region during this period they would seem to be a more likely culprit. This conclusion also fits Saghieh's suggestion that the end of Period KIV roughly coincided with the end of the Sixth Dynasty in Egypt, ca. 2176 B.C. (1983:131).[8]

---

[5]Because the final report of the excavations of Byblos from 1933 to 1938 was only published in 1954, Dunand was able to include a preliminary presentation of the major phases of the fortifications of Byblos (see Dunand 1954:pl. 212). However, a clearer textual description was offered in a short article the following year (Dunand 1955b).

[6]This term is drawn from the on-site posters provided by the Antiquities Department of Lebanon, which derive from the Dunand archive.

[7]Because Saghieh identifies the earthen rampart of this period as the second one, he suggests identifying the NE Gate in this period as the second gate, which was also accompanied by a second phase of the ramp leading to the gate from within the site (1983:66).

[8]Although Dunand suggested ca. 2100 B.C. for the date of this destruction, it is notoriously difficult to date the stratigraphy of Byblos. It is best, therefore, to place this destruction between 2200 and 2000 B.C. until greater clarity is possible. A sign posted at the gate (A.D. 2002 citing the Dunand archive) suggests a date between 2300 and 2100 B.C.

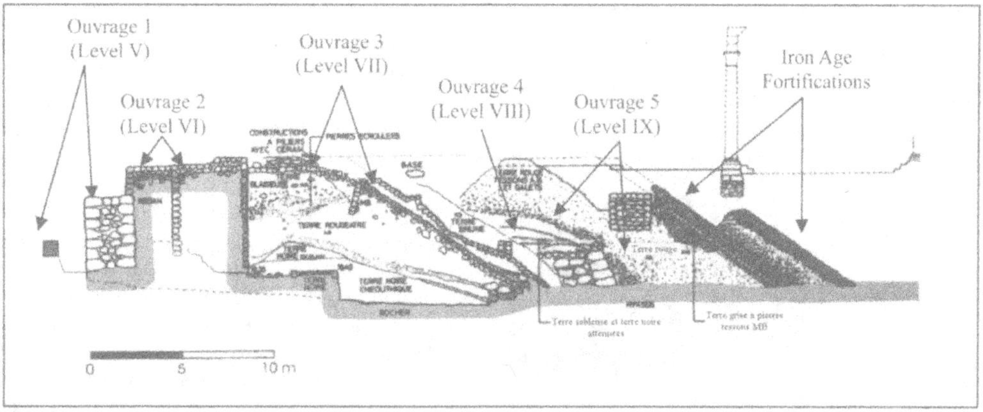

Figure 31. West section through northern fortifications of Byblos. After on-site display.

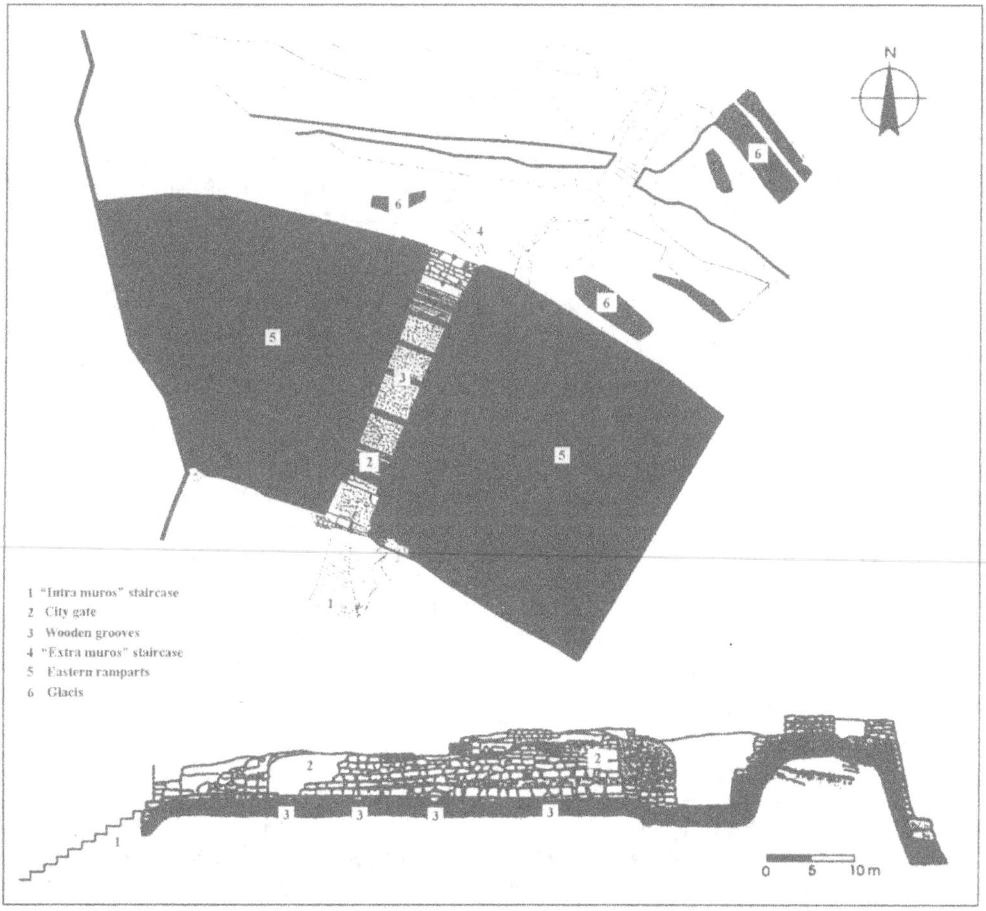

Figure 32. Plan and section of NE Gate and earthen rampart. After on-site display.

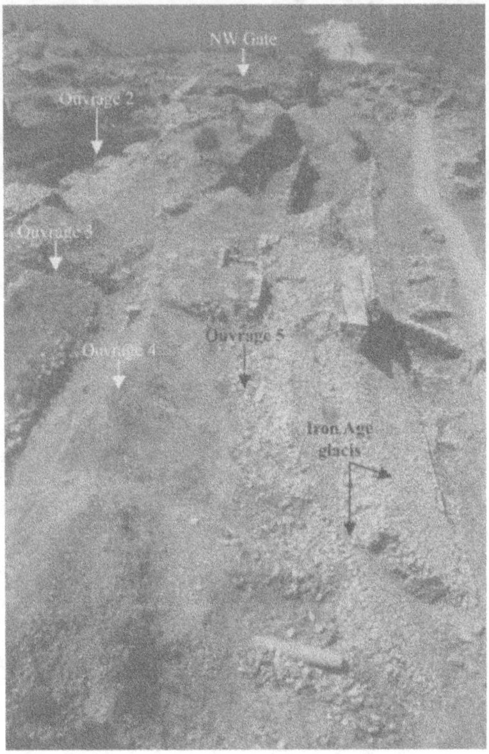

Figure 33. Northern ramparts at Byblos. View to the west. Photo by A. A. Burke.

EB IVB: SAGHIEH'S PERIOD JI/II

Although Dunand's presentation of the stratigraphy and remains of the fortifications of Byblos do not provide a separate reconstruction of the site's defenses during the last phase of occupation in the third millennium (late Level VI), Saghieh suggests that another phase of the fortifications can be detected on the NE side of the site, which he identifies as JI/II. According to him this phase represents the third and final phase of the NE Gate and ramp leading to the gate from within the site (ibid., 66). This phase of the gate was also destroyed in a fire (1983:131f.).

MB IIA: LEVEL VII (OUVRAGE 3)

The next major phase in the development of the fortifications of Byblos belongs to Level VII, which consisted of Middle Bronze Age occupation traditionally associated with the Amorites (Dunand 1954:pl. 212). The distinct phases of the fortifications of this phase (ouvrage 3), which include at least two earthen ramparts and their cobble glacis, are best preserved just to the N of the Early Bronze Age fortifications of Level VI (Figure 31). These ramparts were also identified to the SE of the Medieval castle outside of the line of the Early Bronze Age rampart (Dunand 1961:64f.). However, here the ramparts were clearly damaged by late expansion of the settlement (see Dunand 1954:pl. 212).

The first *earthen rampart* was built against the exterior face of the Early Bronze Age buttressed wall (ouvrage 2). It consisted of several stages of building; its lower half was composed of red earth filled with Chalcolithic sherds, while above this it was composed of black earth filled with Early Bronze Age sherds along with a few Middle Bronze Age sherds found in the highest part of the rampart (Dunand 1955b:19). Though the combined width of the various phases of the Middle Bronze Age rampart was given as approximately 19 m (Dunand 1949–1950:62), based on the section it appears to have been only about 16 m wide and 8 m high. The *glacis* of this rampart was very crudely constructed with limestone cobbles (cp. Ashkelon).

The second *earthen rampart*, which was added slightly later, consisted of less than one meter of earthen fill above the first glacis, which was then covered by a second *glacis* made of cobbles identical to those used in the first glacis. A small *retaining wall* was also identified in the section a few meters from the bottom of the slope of the second glacis. The slope of both ramparts with their glacis was about 35°. The 2 m thick layer of brown earth above the second rampart (see *terre brune* in Figure 31 above) is either mudbrick detritus associated with the erosion of the fortification wall atop the second earthen rampart or an intentional fill deposited before the construction of the so-called "Hyksos" rampart during the next phase.

One *gate* also has been identified with these fortifications on the W end of the N rampart. It appears to preserve the location of the Early Bronze Age (Level VI) "Sea Gate" (see Dunand 1954:pl. 212, no. 38). It featured two stone-built staircases just outside the gate, one descending toward the port to the NW and another set descending to the NE, possibly to the location of a lower town (ibid., pl. 212, nos. 39 and 40). Evidently no wheeled vehicles were intended to use this gate. This staircase can still be traversed today (Figure 34 below).

MB IIB–C: LEVEL VIII (OUVRAGE 4)

Toward the end of the Middle Bronze Age (Level VIII) a new *glacis* and *earthen rampart* of interspersed layers of sand and black earth (ouvrage 4) were constructed against the earlier ramparts (no. 42 on pl. 212, Dunand 1954; 1955b:19). A *revetment wall* was built on bedrock of large roughly hewn sandstone blocks at the foot of the rampart (1961:84).

Figure 34. NW Gate with double staircase at Byblos. Photo by A. A. Burke.

This is the most precisely dated rampart due to the late Middle Bronze Age sherds found in its construction (Dunand 1955b:19). These allowed Dunand to suggest that it should be dated to the eighteenth century B.C. (probably MB IIB–C) and led to its identification as the "Hyksos" rampart. Despite being able to date the rampart, it is much more difficult to define its dimensions; it was roughly 5.6 m high and 8 m wide with a preserved slope of about 15° (see Figure 31). Although less is known of the *gate*, which was newly constructed in this period, it was located to the NE of the Early Bronze Age NE Gate where the Middle Bronze Age rampart departed from the line of the Early Bronze Age rampart. The rampart was moved further to the NE in this area probably in order to expand the town's overall size.

That no evidence of a Middle Bronze Age wall was found above any of the Middle Bronze Age ramparts should not be surprising. Considerable building activity through modern times makes it remarkable that any part of the site's early defenses has remained intact. Furthermore the site's location on the coast guaranteed its exposure to intense precipitation over many centuries.

Late Bronze Age: Level IX (ouvrage 5)

The final Bronze Age rampart (ouvrage 5) is dated to the Late Bronze Age (Level IX). It featured an *earthen rampart* of red earth, which was laid against the preceding Middle Bronze Age rampart, and a stone *glacis* built of cobbles larger than those used in the two later Iron Age ramparts (Dunand 1955b:19). Dunand observed that the surface of the glacis was gently parabolic, which probably refers to its curve at the top (Figure 31).

Deir Khabiye

UTM: 37 N 0235920.3694830
Refs.: von der Osten (1956:13f.).
Plans: Site (fig. 3).
Sects.: NA.
Photos: NA.

The small MB–LB site of Deir Khabiye is located about 20 km SSW from Damascus on the S outskirts of the modern settlement of Deir Khabiye. In addition to being transected by the modern road leading S from the town, the site today is occupied by a military base that has badly pitted the site's surface with trenches and tank emplacements. The site's contours and general plan reveal, however, the Middle Bronze Age ramparts that enclosed it (Figure 35).

Although one cannot survey the site thoroughly, the section cut by the road through the N rampart reveals mudbrick architecture and provides clear evidence of the rampart's construction. The site's plan is rectangular, as evident in plan; it measures 310 m E-W by 220 m N-S with its summit 12 to 14 m above the surrounding plain (von der Osten 1956:13). Its inhabitable area (2.2 ha) is easily defined by the ramparts that enclose the site on four sides. The defenses were examined in Area A, a NNW-SSE road-cut section through the SE corner, where they were found to consist of a mudbrick *fortification wall* built on an *earthen rampart*. The presence of a cylinder seal dated to the First Dynasty of Babylon has been used to date the rampart; ca. 1798–1499 B.C. (ibid.).

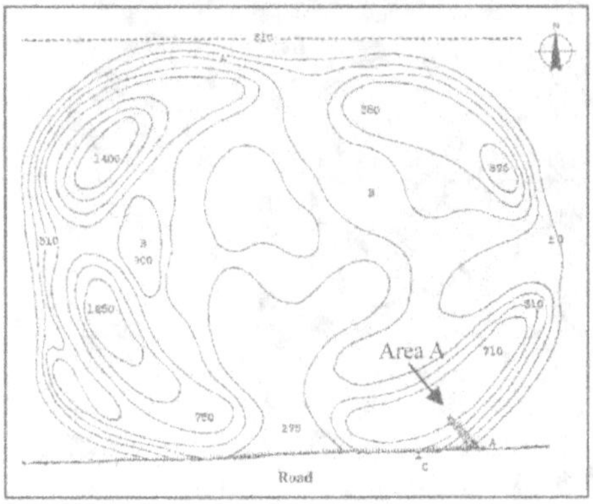

Figure 35. Plan of Deir Khabiye (after von der Osten 1956:fig. 3).

The rampart also features openings, no doubt the location of its four main gates, oriented to the cardinal points (ibid., 14); the basalt orthostats(?) of one of these gates appear to have been visible on the surface during the early 1950s (ibid.). The difference in elevation of the W area within the ramparts indicates that this would have been the location of the most important public buildings (see fig. 3; ibid.).

EBLA (TELL MARDIKH)

UTM: 37 N 03031051.3963882

Refs.: Matthiae (1967; 1968; 1970; 1980b; 1996; 1997a; 1997b; 1997c; 1998; 2000; 2002); Matthiae et al. (1995); Guardata (2000); Peyronel (2000); Pinnock (2001).

Plans: Site (Matthiae 2000:fig. 1; Pinnock 2001:fig. 1); Area A (Matthiae 1970:figs. 2f., 16; 1997b:fig. 14.2); Area M (Pinnock 2001:fig. 15); Area V (Matthiae 2000:fig. 10; Peyronel 2000:figs. 1–3); Area AA (Matthiae 1997c:fig. 23; 1998:figs. 16, 18; 2000:fig. 12); Area BB (Matthiae 1998:fig. 19); Area DD (Matthiae 2000:figs. 21, 24); Area EE (ibid., fig. 16).

Sects.: S section Area R (Pinnock 2001:fig. 4); Area Z (Matthiae 1998:fig. 10).

Photos: Area A (Matthiae 1995:fig. 75, 77–83); Area G (Matthiae 2000:fig. 4); Area H (ibid., fig. 74); E rampart (ibid., fig. 3); Fortress M (ibid., figs. 115f.); Area R glacis (Matthiae 1997b:fig. 14.3); Area V (Matthiae 1998:figs. 13–15; 2000:figs. 2, 11; Peyronel 2000:figs. 4f.; Nigro 2002a:fig. 29); Area Z (Matthiae 1998:fig. 11; 2000:fig. 5); Area AA (Matthiae 1998:fig. 17; 2000:figs. 9, 13f.); Area BB (Matthiae 1998:figs. 20f.); Area DD (Matthiae 2000:fig. 3); Area EE (ibid., figs. 17–19).

Ebla (Tell Mardikh) is located 55 km SW of Ḫalab (Figure 36).[9] Although a great deal has been written on the archaeology of Ebla since excavations began in 1964, no final reports of the architectural remains have been published. The description that follows has been drawn, therefore, from summaries of the expedition's findings (Matthiae 1980b:114–24; 1997b:380–84; Matthiae et al. 1995; Pinnock 2001), as well as preliminary excavation reports published in *CRAIBL* in order to provide an overview of Ebla's defenses from the beginning of the EB IV through the end of the Middle Bronze Age. These reports have also been supplemented by a recently published summary of the Middle Bronze Age fortification system (Matthiae 2002).[10]

EB IVA–B: MARDIKH IIB1–IIB2

Because Ebla's Middle Bronze Age ramparts overlay its EB IV defenses, it has until recently been impossible to do more than speculate concerning the fortifications and size of Ebla during the late third millennium. However, evidence of the EB IV mudbrick *fortification wall* has recently come to light beneath the Middle Bronze Age forts in Areas AA and V, on the NW and W sides of the site respectively. These discoveries suggest that Ebla's size during the EB IV was approximately the same as that during the Middle Bronze Age (Pinnock 2001:21, n. 6) confirming Matthiae's original hypothesis (1980b:114). In Area AA the wall (M.7357) was at least 6 m wide and was preserved to a height of 3 m (Matthiae 2000:580 and n. 33). The size of the mudbricks used in the wall (60 x 40 cm), which were identical in size to those used in Palace G, suggests that the wall probably dates to the EB IVA and not EB IVB (ibid.). Due to the impossibility of safely reaching the earliest levels of the rampart it has not been possible to provide evidence of an EB IV rampart below the EB IV wall.

With this information it is possible to estimate the size of the EB IV and Middle Bronze Age settlements

[9] I would like to thank Nicolò Marchetti for his suggestions regarding improvements to this entry.

[10] Matthiae considers the recent excavations to be part of the sixth phase of archaeological exploration of T. Mardikh (1997c:3). It includes exploration of Area V begun in 1995 (Matthiae 1996), AA in 1996 (Matthiae 1997c:12), BB in 1997, DD in 1999, and EE (Pinnock 2001:fig. 1).

at Ebla.[11] Using the published plans to trace the top of the rampart's contours, the length of the rampart (and by extension the length of the EB IV wall and Middle Bronze Age rampart) can be estimated as ca. 2,790 m (Table 28). By the same means the gross size of Ebla including its ramparts is revealed to have been ca. 56 ha, the size suggested by Frances Pinnock (2001:13) for the Middle Bronze Age settlement. If the EB IV settlement was approximately the same size as that of the Middle Bronze Age, then the area occupied by the exterior slopes of the ramparts where no occupation was found in either period would have totaled roughly 8.6 ha or approximately 15% of the site's gross size.[12] This leaves 47.4 ha for occupation within the ramparts. After subtracting the gross area of the acropolis (3 ha), the net area of which was 1.5 ha, the lower town would have been 44.4 ha.[13] These figures are, therefore, the most reliable estimates of the size of both the EB IV and Middle Bronze Age towns.

Table 28. Estimated length of ramparts according to sections identified by Matthiae (1995:32ff.).

Section (Orientation)	Est. *l.* (m)
1. W (SSE-NNW)	250
2. W (S-N)	280
3. W (SSW-NNE)	300
4. N (W-E)	220
5. N (WNW-ESE)	340
6. E (NNW-SSE)	260
7. E (N-S)	250
8. E (NNE-SSW)	280
9. S (E-W)	310
10. E (SE-NW)	300
TOTAL	2,790

At the end of the EB IVA Ebla experienced a massive destruction, the best evidence of which comes from the remains of Palace G. This destruction is attributed to the Akkadian ruler Sargon, although Naram-Sin also claims to have conquered Ebla. Another destruction is presumed at the end of the EB IVB (Matthiae 1997a:181), although the evidence is not as clear as that for the earlier destruction due to the near complete eradication of EB IVB remains from the lower town, presumably as a result of the building of the MB I ramparts (2002:33).

MB IIA: MARDIKH IIIA

In contrast to the EB IV remains, the excavation of Ebla since 1964 has resulted in a thorough understanding of the Middle Bronze Age town (Mardikh IIIA–B) and its network of fortifications. The defenses of Ebla, built at the start of the MB IIA, were constructed directly atop those dated to the EB IVB with no gap after the destruction of Mardikh IIB2.[14] In their completed state the elements of the Middle Bronze Age fortifications included earthen ramparts, a number of gates, a fortification wall, fortresses along the rampart, and the defenses of the acropolis.

Matthiae suggests that both the location of the Middle Bronze Age gates of Ebla and the general layout of its ramparts were dictated by the layout of the EB IV town (1997b:382), having been built "at the beginning of Mardikh IIIA phase" (1980b:119). This has been confirmed in recent years with the discovery of the remains of the EB IVA town wall under the Middle Bronze Age rampart (see discussion above). The *earthen rampart* rises 20 to 22 m above the level of the surrounding plain, but only 12 m above the lower town, and its base was 40 to 50 m wide (Matthiae 1997b:382).[15] It is not possible to accurately determine the original slope of the rampart from published plans, but the proposed figure of 22° for the slope of the W rampart in Area Z is probably not steep enough (see S section of rampart in Pinnock 2001:fig. 11).[16] In contrast, Matthiae had earlier suggested that the rampart in Area A probably originally featured a 45° slope, although at most only 30° of this now remains (1970:61).

[11] Matthiae has previously suggested that the EB IV settlement was almost 50 ha (1997a:180) while that of the Middle Bronze Age was 60 ha (ibid., 181).

[12] Although Pinnock (2001:18) suggests 40 m as the rampart's original width, Matthiae's original suggestion that it was 50 m wide (1980b:118) seems to better fit the data (see south section of rampart in Area Z, Pinnock 2001:fig. 11). Recently Matthiae reiterated this dimension (1997b:382). The figure for a larger rampart yields a more conservative estimate of the size of the site, which is appropriate given the degradation of the rampart over time and the realization that it would have been far larger originally.

[13] These figures were arrived at from observations made by the author along the crest of the rampart in August 2002.

[14] The lack of a gap is demonstrated by the absence of MB IIA pottery in the rampart fills (Matthiae 1980b:118). The reasoning for Pinnock's suggestion that a "gap not exceeding fifty years" might have existed is unclear (2001:n. 11).

[15] It is difficult to provide exact dimensions for the rampart since much of the top is assumed to have eroded (Pinnock 2001:fig. 11) and excavations have only reached a limited depth so far.

[16] The reconstruction offered for Area Z is entirely speculative. An examination of the position of the Area Z trench (Pinnock 2001:fig. 1) indicates that it is located in a depression in the W rampart and thus does not provide a true profile of the rampart.

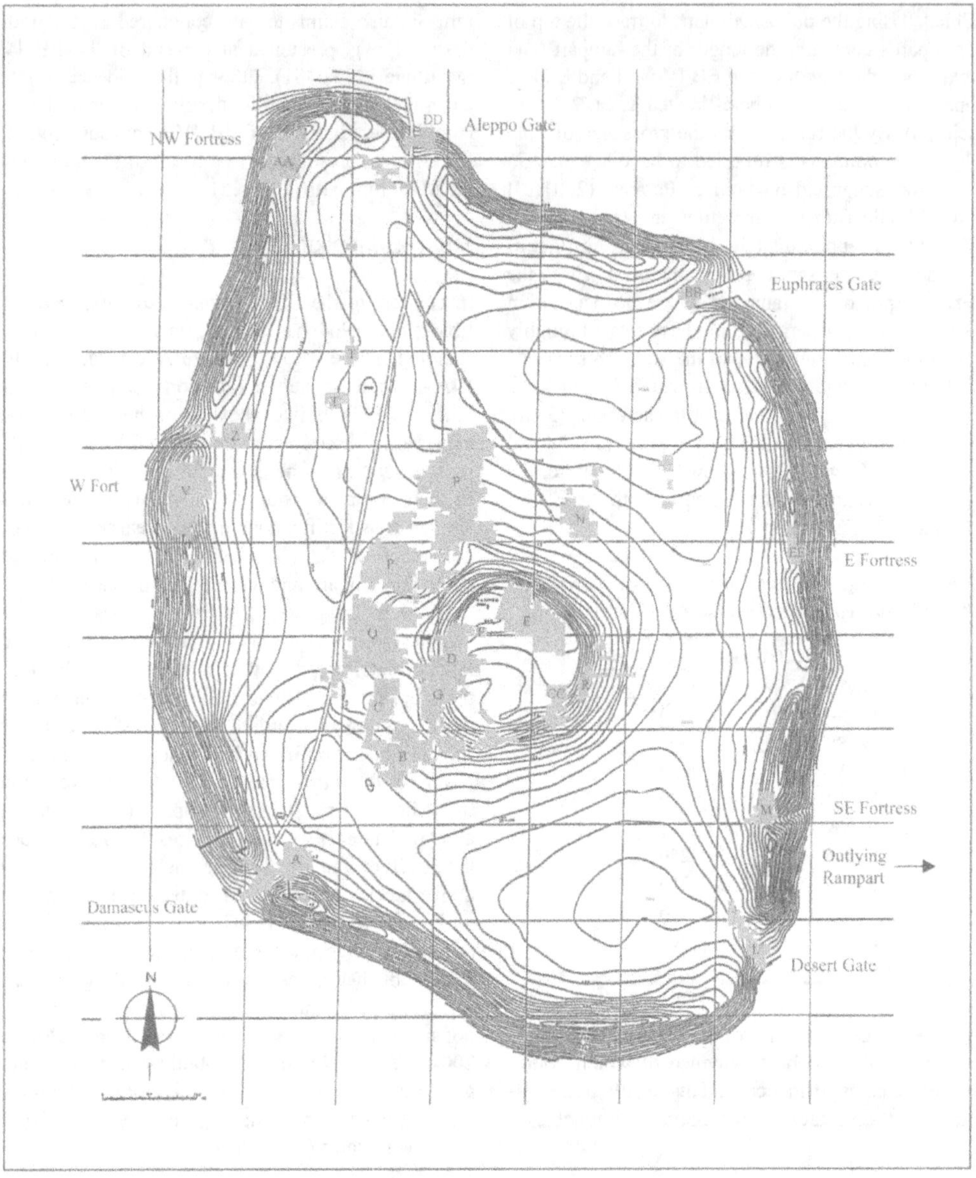

Figure 36. Plan of Ebla showing excavation areas (after Matthiae 2002:31).

The rampart fills consisted of white, chipped limestone, red clay-like soil, and occupational debris. Most of the limestone and clay material was probably obtained from a *fosse* that was dug on the outside of the base of the rampart during this period, as is evident on the SW and S sides of the site, though it dimensions have not been ascertained. The remainder of the rampart's fills consisted of ashy EB IVA (Mardikh IIB1) and IVB (Mardikh IIB2) occupational debris obtained from within the original EB IV enclosure (ibid.). The date of this material along with the evidence from MB IIA burials in the rampart and below many of the fortresses along the rampart (Pinnock 2001:n. 12) provide conclusive evidence supporting an early MB IIA date for the construction of the earthen ramparts. If the dimensions cited above are correct then the total volume of the ramparts can be estimated to have been more than 1,318,275 m^3.

As recently revealed, the core of the Middle Bronze Age rampart consisted of the remains of Ebla's EB IV defenses, which included the EB IVA mudbrick fortification wall (Pinnock 2001:n. 4) that served as the *core wall* for the construction of the Middle Bronze Age rampart. The base of the exterior of the rampart was reinforced by a low *revetment wall* of dressed stone (M.325 in Area A, Matthiae 1970) preserved to a height of about 1.5 m that Matthiae speculates was originally between 3 and 5 m high (1997b:382; 2002:32). Remains of the foundations of the fortification wall along a 180 m stretch were identified east of the Damascus Gate on the southern rampart (Matthiae 1997a:181; 2002:41). Elsewhere nothing remains of this wall, which was probably destroyed by the fortresses constructed on the rampart during the MB IIB.

In four areas at Ebla excavation or examination of the site's contours have also enabled identification of the major gates dated to the MB I, which continued to function through the MB IIB–C. These are all of the *six-pier gate* type: the Aleppo Gate (Area DD) on the N, the Euphrates Gate (Area BB) on the NE, the Steppe or Desert Gate (Area L) on the SE, and the Damascus Gate (Area A) on the SW (for the names of the gates see Pinnock 2001:fig. 1, p. 21).[17] Of the gates thus far identified only those in Areas A (Damascus), L (Qatna), and BB (Euphrates) have been completely excavated (ibid., 21). The gate in Area DD may lie outside the line of the ramparts or if it was situated like the others within the ramparts it has been completely destroyed (Matthiae 2000:n. 75).

The gate in Area A, known as the Damascus Gate (Figure 37), was the best preserved of the gates at Ebla (Matthiae 1980b:119–23). Making a gradual left turn upon entry from an ENE approach, it consisted of three parts: an outer four-pier gateway, a central courtyard, and an inner, six-pier gateway. The outer gateway lying on a WSW-ENE axis was a *four-pier gate* with two outer piers (M.341 and 358) and two inner piers faced with basalt orthostats (M.326 and 356). The outer gateway possessed only one set of doors, which based on the location of the two door-sockets were located on the inside of the first pier when approaching the town. These doors opened inward and would have measured slightly more than 1.8 m wide. The outer gate was 10.5 m deep with 3 m wide entries. Between the inner and outer gates was a trapezoid-shaped court on a SW-NE axis. On the S side of the courtyard stood a massive *retaining wall* (M.365) of rough-hewn stones 2.75 m high with a stone stairway that led to a bastion and sentry walk, according to Matthiae (1980b:122).

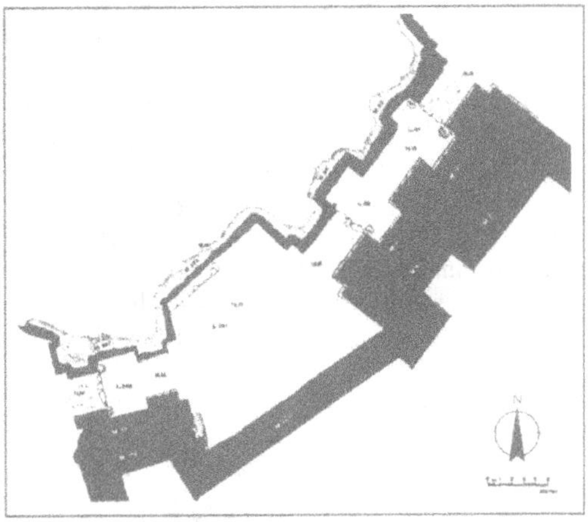

Figure 37. Damascus Gate in Area A at Ebla (after Pinnock 2001:fig. 14).

The inner, *six-pier gateway* was constructed of rough-hewn stones that protruded 1.32 m from the gate's sides (Matthiae 1968:16). Only the E half of the gate was faced with orthostats, to a height of 1.8 m, the piers with basalt orthostats (M.22, 18, and 14), and the two intervening spaces with limestone orthostats. All the orthostats had smoothed exterior surfaces, though the interior surfaces of the basalt orthostats were not smoothed, and they featured circular holes spaced regularly to accommodate a masonry superstructure. The basalt orthostats also were trapezoidal in section, supposedly in order to function as foundations for the voussoirs, which are thought to have supported a barrel vault constructed over the passageway. The orthostats, therefore, should not be understood as simply ornamental since they were also integral to the structure of the gate. Although Matthiae (1980b:121) suggests that the weight of the masonry structure was to blame for the cracks in the orthostats, this seems unlikely since they may have cracked as a result of the final collapse of the gate and its fiery destruction. The inner gate was roughly 21.3 m deep.

The construction of the gate complex is dated to the MB IIA (Mardikh IIIA) based on ceramics found on the surface associated with its initial construction. It experienced only modest modifications during the following period (Mardikh IIIB), which included a new *retaining wall* (M.41) that resulted in the trapezoidal courtyard becoming smaller in area. Although

[17]It is noteworthy that the modern designations of these gates follow the typical Near Eastern convention of naming them according to the major settlement or region to which one could travel directly from the gate.

a building excavated in Area A just inside the gate was identified as the guards' quarters (Matthiae 1970:66, fig. 3), it is not yet possible to confirm this designation.

The Steppe Gate in Area L on Ebla's SE side, though poorly preserved, was probably a *six-pier gate* identical to the one in Area A. Although the Area BB Euphrates Gate on the NE was also excavated (Matthiae 1998:584–87, fig. 19), indications are that it too was rather poorly preserved (Figure 38). Only two of the six piers of this *six-pier gate* on its N side survived the effects of erosion on the NE side of the site. The dimensions of the gate passage (ca. 3.2? x 20 m) are nevertheless comparable to those of other Middle Bronze Age gates (ibid., 586 and n. 55).

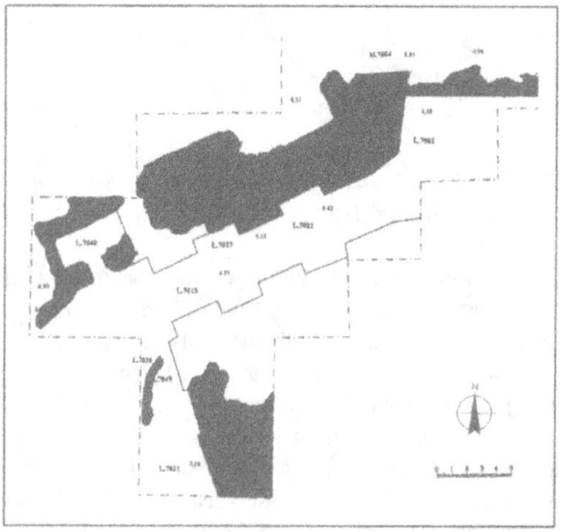

Figure 38. Euphrates Gate in Area BB at Ebla (after Matthiae 1998:fig. 19).

MB IIB–C: Mardikh IIIB

As further evidence of the site's defenses during the Middle Bronze Age, a number of *bastions* (so-called "fortresses") have been identified atop the earthen ramparts. These have been dated to the renovation of the site's defenses during the transition between the MB IIA and MB IIB. These include the fortifications of the SE Fortress (Area M), the W Fort (Area V), the NW Fort (Area AA), the NE Fortress (Area EE), and the remains of a bastion detected on the surface in Area A (see Pinnock 2001:fig. 1 for their locations) and in grid BV to the S of Area V (Matthiae 1998:n. 46 and pp. 587f.). As Matthiae notes the placement of the bastions was probably determined by the juncture between separate stretches of the fortifications (ibid.). The construction of these bastions must be dated to some period after the initial phase of the MB IIB (Matthiae 1998:582, 88) since they overlay MB IIA burials that were dug into the rampart before the bastions were built (Guardata 2000). While excavations of many of these structures have only recently been conducted, their existence had already been posited based on surface remains (Matthiae 1980b:118f.). These bastions are thought to have functioned as armories during peacetime, for the defense of the rampart in the event of siege (ibid., 123), and even as barracks for troops (Peyronel 2000:1358). As Matthiae notes, they conform to a rectangular plan of roughly 12 m wide by 25 m long, are spaced an average of 250 to 300 m apart, feature a staircase on one side of a court, and have an internal plan consisting of two rows with three rooms to a side (Matthiae 1998:574f., 87). Pinnock also notes that during their construction the bastions were adapted as necessary to the lines of the original EB IV rampart (2001:24).

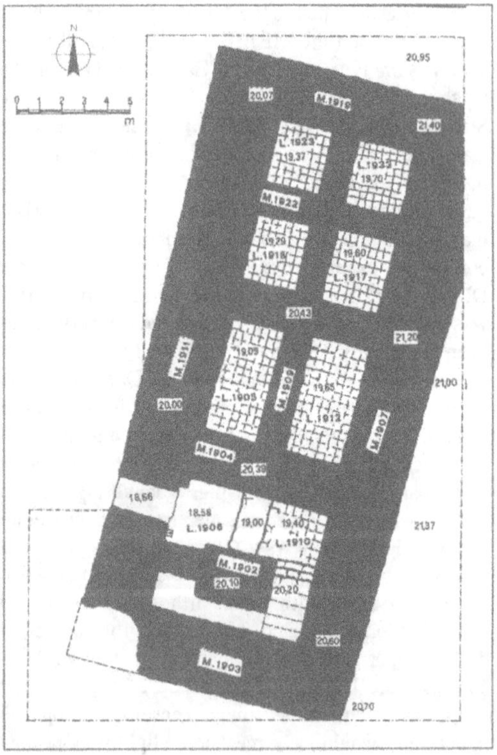

Figure 39. Bastion in Area M at Ebla (after Pinnock 2001:fig. 15).

Bastion M or the SE Fortress (Figure 39), for which the most has been published, is located on the interior slope of the E rampart N of the SE Gate. Like the other bastions it too was rectangular in plan (12.5 x 27 m). Its long side was aligned with the rampart on a N-S line and its exterior walls, built of mudbrick, measured approximately 3 m wide. The structure, which consisted of six small rooms paved with mudbrick, also featured a second story as attested by the staircase at its S end where a ground floor entrance on the W side of the building was also located. Having been destroyed at the end of the Middle Bronze Age (Mardikh IIIB) several remains were found in situ amidst the ash. These included a piriform limestone macehead and two spearheads found beneath a staircase, one of which was inscribed with a name in cuneiform (Matthiae 1980b:123).

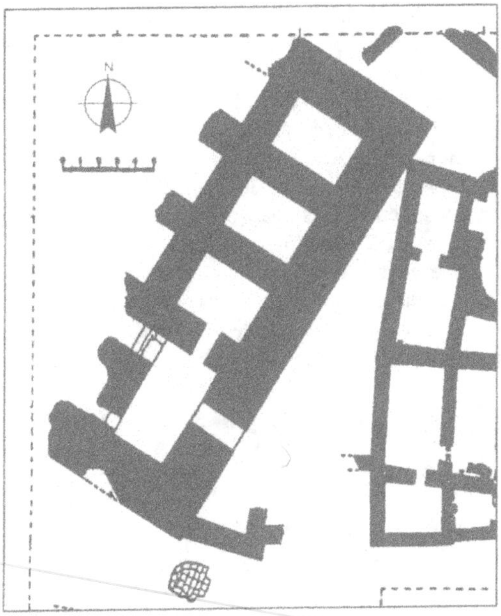

Figure 40. Bastion in Area V at Ebla (after Pinnock 2001:fig. 13).

In Area V another fortress, referred to as the W Fortress (Figure 40), was identified (Peyronel 2000). It appears that this fort was constructed about a century into the late MB IIA (IB) when use of the MB IIA (IA) rampart as a cemetery had come to an end (ibid., 1354). In that period both the N and W ramparts were used as cemeteries. The bastion, which measured 26 m N-S by about 16 m E-W, was in continual use until it was destroyed at the end of the MB IIB–C (ca. 1550 B.C.). Its walls were slightly less than 3 m thick and built with bricks of similar dimensions to those of other Middle Bronze Age structures on the site and measured 30–32 x 32–35 x 10–12 cm (ibid., n. 9). Of particular interest is the use of orthostats and that those between rooms L.6516 and L.6522 were angled so as to support an arched doorway (see Peyronel 2000:fig. 5; and Nigro 2002a:fig. 29).

In 1996 part of the bastion in Area AA known as the N Fortress was excavated (Figure 41). Although it is less well-preserved than the others, featuring only its W peripheral wall (M.6958), its E supporting wall (M.7330), and a small part of the rectangular fortress (Matthiae 2000:fig. 12, pp. 584f.), its plan is comparable to that of fortresses M and V. Its exterior walls were more than 2 m thick. The N continuation of the E supporting wall (M.7330), M.7369, which was built of stone and mudbrick was found preserved to a height of more than 3.5 m (ibid., 587). Matthiae suggests based on the discovery of a hearth with three or four vertical openings in room L.6906 of the N Fortress that it was used for sending smoke signals (1997c:12; 1998:580). He asserts that the absence of evidence for cooking or metallurgy around the hearth support this suggestion. Additionally, an Old Babylonian lexical tablet was found in this fort (ibid.).

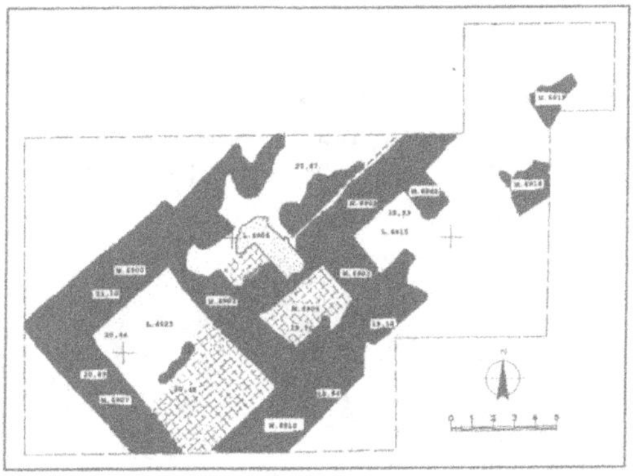

Figure 41. Bastion in Area AA at Ebla (after Matthiae 1997c:fig. 23).

The NE Fortress in Area EE was neither very well preserved nor completely excavated, but appears to have been of a comparable plan to the other rectangular bastions known (Matthiae 2000:fig. 16, pp. 588f.). The bastion in Area EE provides potential insight into the development of the rampart defenses of Ebla during the Middle Bronze Age in three phases. According to the most recent publication of the findings, during the earliest part of the MB I just after the construction of the rampart a fosse seems to have been dug into the rampart, perhaps along its crest, which was then later filled with refuse (Matthiae 2000:593). Later in the MB IIA some structures, apparently of smaller size than the MB IIB–C fortresses, appear to have lined the crest of the rampart providing a defensive line composed of the back walls of structures (ibid., fig. 16). Only during the MB IIB–C, as also evident from the other fortresses, was the Area EE fortress constructed, again on the interior slope of the rampart (ibid., 591). While almost no data have been provided concerning the mudbrick fortification wall that crowned the rampart, it may have been at least 3.5 m thick during the MB IIB–C if the plans of the exterior walls of the fortresses spaced along its length are any indication (Pinnock 2001:figs. 12f., 15).

Middle Bronze Age Acropolis Defenses

In addition to all of these features fortifications also surrounded the acropolis in the center of the lower town during the MB I and II (see Pinnock 2001:fig. 4). From bottom to top these consisted of a stone *revetment wall* (M.4573) preserved 1 m high and 2 m wide, which was originally about 4.15 m high, a 31 m wide *glacis* with a slope of 9°, a citadel *rampart* ("mudbrick revetment" in Matthiae 1997b) atop the glacis at the foot of the citadel's wall 5.7 m high and 7 m wide covered with mudbricks, and a mudbrick citadel *fortification wall* (M.4500) over 2.5 m wide the original height of which is unknown (Matthiae 1997b:382f., fig. 14.3). The combined height of these defenses was 14 to 16 m (ibid., 383). *Towers* constructed of large stone blocks protected the approach to the citadel (ibid., 384).

Outlying Eastern Rampart

Another structure that is not discussed among Ebla's fortifications bears mentioning, specifically as part of the site's Middle Bronze Age settlement and defenses. The feature consist of a massive outlying rampart to the east of the ringed settlement (Matthiae 2002:49), plans of which have never been published. It can be seen to the E from the top of Ebla's E rampart (see Figure 42). This *earthen rampart* lies in a N-S alignment between 275 and 290 m E of the site's E rampart and is approximately 515 m long, 60 m wide, and over 12 m high with a slope between 23° and 26°. It occupies 3.1 ha with a volume of approximately 185,400 m^3.

Although it is nearly three-quarters of the length of the E rampart, it is not stratigraphically linked to the settlement itself. The date of its construction, its purpose, and its relationship to the settlement remain, therefore, enigmatic. The Italian expedition has, nevertheless, considered several possibilities regarding its construction (Matthiae et al. 1995:49). Among them are the possibilities that it: (1) was intended to provide greater defense of the E rampart, (2) represents a phase of aborted expansion of the settlement to the E, or (3) was intended to protect a besieging army encamped to the east of Ebla.

Figure 42. Ebla's outlying rampart as seen from eastern rampart. View east. Photo by A. A. Burke.

EMAR (TELL MESKENE)

UTM: 37 N 0419764.3982911
Refs.: Margueron (1995); Finkbeiner and Leisten (1999–2000); Finkbeiner (2001); Finkbeiner and Sakal (2003).
Plans: Lower town (Finkbeiner and Leisten 1999–2000:fig. 11); Squares 082–085/048–049 (fig. 20); Upper and lower towns (Finkbeiner 2001:plan 1); Tempelbezirk (Finkbeiner and Sakal 2003:Beilage 1).
Sects.: Schematic (Finkbeiner and Leisten 1999–2000:fig. 22); Sq. 059/054 (Finkbeiner 2001:Abb. 4).
Photos: Wall (Finkbeiner and Leisten 1999–2000:pl. 40).
Tell Meskene (Emar) is located 88 km ESE of Aleppo along the Euphrates (now on the shore of Lake Assad). Although little is known of Early and Middle Bronze Age Emar, J.-C. Margueron suggests that its layout as early as the mid-third millennium resembled that of Mari and Terqa (1995), namely a tell encircled by ramparts. Although excavations have revealed a considerable portion of the Late Bronze Age settlement, excavation on the S edge of the upper town has revealed the remains of a fortification wall and glacis that date to the Middle Bronze Age (Finkbeiner and Leisten 1999–2000:29ff.).

Portions of the fortification wall were detected along the S side of the upper and lower town. As seen in the schematic section (Finkbeiner and Leisten 1999–2000:fig. 22), the defenses, excavated in square 085/048 in the lower town (Unterstadt), consisted of a mudbrick *fortification wall* over 3 m thick built on a stone foundation with a *glacis* (ibid., 29). The same wall with virtually identical construction was identified in a number of locations as it continued W into the upper town (Oberstadt) and then along the W edge of the upper town (Sq. 059/054) adjacent to the Late Bronze Age temple to Astarte (see Finkbeiner 2001:plan 1). In Square 060/054 a mudbrick *tower* measuring slightly less than 7 m long by 3 m wide was also exposed (Finkbeiner and Sakal 2003:Beilage 1). The Late Bronze Age layout of Emar, which conforms to a rectilinear plan (Margueron 1995:128), may suggest the nature of the enclosure during the Middle Bronze Age, despite Margueron's hypothesis regarding the plan of the Early Bronze Age settlement.

FERZAT, TELL (TELL ES-SALIHIYEH)

UTM: 37 N 0265512.3711020
Refs.: von der Osten (1956).
Plans: Site (fig. 19); Stratum XII (fig. 18).
Sects.: E section (fig. 98).
Photos: E section (fig. 55); Walls (figs. 53f.).
Salihiyeh, known today as Tell Ferzat and occupied by a military base, lies about 17 km E of the center of Damascus. Founded no later than the start of the MB II, its strategic importance is evident as Wayne Pitard notes since "it is possible to see all the major entry points into the Damascene basin" from the site (Pitard 1987:38). Although its topographic signature appears irregular in plan today, von der Osten suggested that the Middle Bronze Age site (Str. XII–XI) measured about 250 m N-S by 300 m E-W (1956:15). However, a closer examination of the contours which overlay the MB IIB–C rampart and walls in 1951 (ca. 113 m contour line, von der Osten 1956:fig. 19) suggests that the citadel was probably limited in size by a 400 m long wall built around it. Thus it would only have been about 1.4 ha in size; whether or not the surrounding area within the dimensions provided by von der Osten was inhabited during the Middle Bronze Age cannot be verified. While it is not certain that the earliest occupation of Salihiyeh has been identified, preliminary examination by von der Osten suggests that the MB IIB–C fortifications of the citadel were the earliest defenses constructed for the site shortly after two short-lived phases of occupation, only a small part of which has been exposed (ibid., 79). There is no evidence of an enclosure around the site of the type known at nearby Deir Khabiye (see p. 197).

Stratum XII began around the end of the nineteenth century B.C. and constitutes the first phase of fortification construction on the site. A 3 m high *earthen rampart* was raised around the citadel in Stratum XIIb1a (von der Osten 1956:38).[18] It is uncertain whether or not this rampart featured a wall at that time. Unfortunately, the interior slope of the rampart was not preserved in the area excavated since it had been carved away in antiquity.

In a second phase of construction, following the removal of much of the interior of the rampart's slope, the area was leveled with a fill 2.8 m deep (XIIα). Upon this fill a 1.2 m wide mudbrick wall (XIIb2) was built. This wall was then subsequently leveled in order to create a large flat area (XIIβ), effectively extending, or renovating the earliest rampart. During this third phase of construction, directly above the extant remains of the earliest rampart (i.e., on the exterior edge of this large embankment with a flat top), a 1.9 m wide *fortification wall* (XIIa1) was built of gray and yellow mudbricks (43–44 x 10–11 cm). The front of the wall was faced with two rows of mudbrick (XIIA); today five courses of this structure are preserved. This wall identified as a *revetment wall* (Vormauer) stood 2.1 m in front of a mudbrick wall (XIIδ) of identical width (1.9 m); it was only preserved to a height of 0.9 m (bricks 36–39 x 10–12 cm).

---

[18]The trench in which these results were obtained on the N side of the site has been backfilled.

It is possible that both of these walls belonged to a bastion or casemate construction as part of the defenses of the citadel. Both were destroyed at the same time (ibid., 39) and then built over in Stratum XI with a single, thicker *fortification wall* (XI1), 2.5 m thick, which was built again of mudbrick (44–57 x 11–15 cm) as the fourth and final phase of the site's Middle Bronze Age fortification (ibid., 35). All fortification walls on the site lacked any evidence of stone foundations.

GINDARIS, TELL

UTM: 37 N 0292824.4029372
Refs.: Khadour et al. (1997); Sürenhagen (1999).
Plans: Site (Sürenhagen 1999:fig. 2).
Sects.: NA.
Photos: NA.
Preliminary evidence is gradually surfacing regarding this Middle and Late Bronze Age settlement approximately 5 km W of the Afrin River in NW Syria (Khadour et al. 1997). It is possible that the extent of the 12.4 ha mound, which is almost square in plan, represents an original MB IIA settlement in the Afrin valley.[19] However, much of the mound's 20 m elevation above the plain is certainly the result of settlement during later periods.

Examination of the S, W, and NW slopes of the site reveal a very clear line of large stones undoubtedly the foundations of a *fortification wall* as Khadour notes (ibid., 119), resting 15 m above the elevation of the plain to the S. If this wall surrounded the entire mound, then its length would have been approximately 1,200 m. These stone foundations are about 7 m lower than the Middle and Late Bronze Age occupation in Area I. The fortification wall is suggested to have been 8 m wide and may have rested at the top of an earthen rampart, which gave the mound its rectilinear plan. At least two gates may have existed: one in the SW corner in the area of a depression and another in the center of the N side in the area of a second depression evident on the contour plan.

HABUBA KABIRA, TELL

UTM: 37 N 416484.404787
Refs.: von Schuler et al. (1969); Ludwig, Strommenger, and Opificius (1970); Schmid et al. (1971; 1973); Strommenger (1976; 1979; 1980); Heusch (1979).
Plans: Site (von Schuler et al. 1969:fig. 8); Level 5 (Heusch 1979:165; Strommenger 1976:fig. 3), Level 6 (Heusch 1979:167; Strommenger 1979:fig. 17; 1976:figs. 1f.), Level 7 (Heusch 1979:169), Level 10 (Heusch 1979:170; von Schuler et al. 1969:fig. 9; Ludwig et al. 1970:supps. 2, 4; Schmid et al. 1971:supp. 3), Level 11 (Heusch 1979:173), Level 15 (Heusch 1979:175; Schmid et al. 1971:supp. 4), Levels 17–20 (Heusch 1979:177).
Sects.: Level 10 (Ludwig et al. 1970:supp. 3).
Photos: Level 6 (Strommenger 1979:18), Level 10 (von Schuler et al. 1969:figs. 10–12; Ludwig et al. 1970:figs. 3f.); Level 15 (Schmid et al. 1971:fig. 5).
The 0.4 ha site of Tell Habuba Kabira, which is located about 1.2 km N of Tell Kannas on the W bank of the Euphrates River, sits 14 m above the floor of the valley at 303 m ASL (von Schuler et al. 1969:41). Today, like Tell Kannas, it has been inundated by the waters of Lake Assad. Excavations revealed twenty strata (*Schichten*) of occupation on the SE side of the tell that span the period from the mid-fourth to the mid-second millennium (Strommenger 1979:73).[20]

EARLY DYNASTIC: STRATA 2–10

Stratum 1 is dated to the Late Uruk Period (Heusch 1979:161). The next major period was dated to the Early Dynastic, Strata 2–3. The N building located on SE side of tell in this phase was identified as a work area (ibid., 163). The exterior surface of the outer walls of this structure as well as its interior surface were renovated in this period and appear to have been plastered or resurfaced with a chalky mud (ibid., 164). The site does not appear to have been enclosed by a freestanding wall in this period; rather the rear walls of the various structures served as a perimeter or casemate wall for the settlement. In this wall a 1.5 m wide gate was identified on the S side of the settlement. This passage remained open and in use through Stratum 10.

The next major alterations in the site's plan are visible in Stratum 6. In this phase the outer wall of the settlement, which had until Stratum 5 consisted only of the exterior wall of the settlement's E structures, was widened with the addition of a 1 to 1.2 m wide *fortification wall* outside of it (Heusch 1979:166). The space between the original wall and the new wall was filled with clay. During this phase a 2.2 m wide breach was added to the E side to permit access to the river through this part of the settlement. Few modifications of the interior structures seem to have taken place in this phase. Stratum 7 featured the improvement of the settlement's exterior wall after

[19] Khadour's figure of 20 ha (1997) is too large based upon my reconnaissance of the site.

[20] The data from the preliminary reports are insufficient to provide more precise dates for these archaeological phases.

modification of the earlier perimeter wall led to it being made taller (ibid., 168).

Stratum 10 represents the next significant change in the settlement's plan in this area (Strommenger 1979:168). The major renovation in this phase included the replacement of the E opening with a formal gate built on a stone foundation (Heusch 1979:168; von Schuler et al. 1969:44). The new gate featured a steep pavestone ramp leading toward it with a 1.6 m wide entrance (Heusch 1979:168). The direct-axis gate, which is E-W aligned and was 5 m long, features side rooms 3.8 m deep (Ludwig et al. 1970:35). The bricks used in its construction measured 24–27 x 45–50 x 9–10 cm (von Schuler et al. 1969:44). The walls added from the S side of the gate to the E wall, which enclosed a newly dug well, also featured interior buttresses for additional support (ibid., 171). The E wall was 1.2 m wide and built on a stone foundation, which was 1.5 m wide (Heusch 1979:171) of bricks measuring 34–38 x 48–50 x 10–12 cm standing to a height of 1.5 m (Ludwig et al. 1970:38). The buttresses averaged 1.45 m deep by 1.25 to 1.30 m wide. Another 3.5 m wide entrance was added in the S half of the excavated area (Heusch 1979:171).

AKKADIAN PERIOD: STRATA 11–13

In the following phase, Stratum 11, the E gate, which was opened in the previous phase, was renovated and converted into an *indirect axis gate* (Heusch 1979:172). This was accomplished with the addition of a massive *fortification wall* more than 3 m wide on its E and N sides. The gate's entrance during this phase emptied to the S and an opening was added between the junction of the town wall and the buttressed wall that had been added in the previous phase. This phase appears to date to the Akkadian period (Schmid et al. 1971:18). Strata 12–13 saw improvement of the fortifications constructed in Stratum 11 and eventual modification of the E gate again (Heusch 1979:174).

UR III–OLD BABYLONIAN: STRATA 14–20

Major modifications of the settlement's fortifications also took place in Stratum 14. The *fortification wall* of this period, which was added as an extension S from the corner of the buttressed wall in square S13, was built 2.8 m wide and at least 5 to 6 m high (Heusch 1979:174). It was constructed at an average level of 295.44 m ASL of fired bricks approximately 38–40 x 38–40 x 10–12 cm without a stone foundation (Schmid et al. 1971:18). The structure is dated by special finds to the period between Ur III and the First Dynasty of Babylon (ibid., 18).

Stratum 15, which immediately followed, was predominantly of the same plan as Stratum 14. Although the entrance during this phase, which was located in the S side of the settlement, was not well preserved, it appears to have been a simple *direct-axis gate*.

By the final phases of occupation at Habuba Kabira, Strata 17–20, the earlier walls appear to have fallen out of use and almost no evidence is available from which to formulate conclusions regarding the settlement's defensive strategy. Despite the limited data that are available in preliminary reports for the dating of the architectural phases at Tell Habuba Kabira, the site offers a unique perspective on the defensive strategies of settlements that were almost continuously occupied from the Late Uruk through the Middle Bronze Age in northern Mesopotamia.

HADIDI, TELL (AZU)

UTM: 37 N 423723.4013702
Refs.: Dornemann (1979a; 1979b).
Plans: Site (Dornemann 1979b:fig. 1); Area B, Level D (1979a:fig. 25).
Sects.: NA.
Photos: Areas A (Dornemann 1979a:fig. 29; 1979b:fig. 12); Area B—LB (1979b:fig. 13).

Tell Hadidi (ancient *Azu*) was located on the W bank of the Euphrates (now under the waters of Lake Assad). The EB IV settlement of Tell Hadidi included occupation of both the upper town on the W [1.8 ha] and the lower town to the E [2.2 ha] (Dornemann 1979b:217). By the completion of excavations at the site scant evidence existed for any fortification of the site during this period (1979a:116).[21]

MIDDLE BRONZE AGE

During the Middle Bronze Age, when the settlement retracted to the 1.8 ha upper town, it was surrounded by the first fortification system built at Hadidi. The site's Middle Bronze Age fortifications were excavated in Area A on the W, in a long step trench in Area B on the N, and on the S in Areas G, J, and P (1979a). In Area A 17.5 m of the more than 4 m wide *fortification wall* was exposed (1979b:225), while in Area B the remains of the stone fortification wall were revealed to be only 3 m wide (1979a:132). Although the phasing is not entirely clear from the limited publications that exist, a

[21]This statement cannot be taken as an indication that the site was unfortified during the EB IV. The excavations, particularly of the lower town, were extremely limited in both duration and extent as Dornemann notes and the fact that EB IV fortifications were not revealed in Area B, where the best evidence for the Middle Bronze Age fortifications was encountered, does not conclusively demonstrate that the site was unfortified in this period.

gravel *glacis* appears to have complemented this fortification wall during the first part of the Middle Bronze Age. The function of this glacis was replaced by a *revetment wall* (Level D), which was built on the interior surface of the Middle Bronze Age *fosse* toward the end of the period (1979b:225).

In Area G the fortification wall was found to be 7 m thick, although Dornemann admits that this may indicate the location of the foundation of a tower (1979a:141). While no gates were excavated, the location of a gate is suggested by depressions located between Areas G and P on the S side of the upper town and in the center of the W side (ibid.). An enclosure wall around the entirety of the upper town was estimated to be approximately 1,425 m long, and would have contributed to the roughly rectangular plan of the upper town.

LATE BRONZE AGE

During the Late Bronze Age the site returned to its earlier grandeur (4+ ha) with occupation of both the upper and lower towns (1979b:217f.). Expansion of the site during this period is evident as the settlement's LB I *fortification wall* was moved some 20 m beyond the line of the earlier Middle Bronze Age fortifications (ibid., 225). In Area B two phases of the Late Bronze Age defenses, using river gravel as fill behind them, were revealed. During the first phase (Level C in Area B) a *fosse* about 4.25 m wide also appears to have been cut into the bedrock to replace the Level D (MB II) fosse, which had by now been filled in (1979b:225; 1979a:141). In the following phase (Level B) a new 20 m wide fosse was dug in Area B and a *revetment wall* of cyclopean masonry was added to the interior slope of the fosse with a batter of 30° (1979a:141). In Level A gravel was added behind the rebuilt revetment wall in Area B (ibid.).

ḪALAB (ALEPPO)*

UTM: 37 N 0334787.4007617
Refs.: Sauvaget (1929); Suleiman (1983; 1985); Nigro (1997–1999).
Plans: Reconstructed Middle Bronze Age site plan (Nigro 1997–1999:fig. 8).
Sects.: NA.
Photos: NA.

The mound of Aleppo (ancient *Ḫalab*) is located 55 km NE of Ebla. To date no excavation of the lower town of Aleppo has been possible on a wide scale and the defenses of the acropolis also remain unknown. Despite this lacuna in the study of ancient Aleppo's fortifications, recent excavation of the acropolis since 1997 by the Syro-German expedition in addition to minor examinations of the lower town enable speculation regarding its Middle Bronze Age layout.

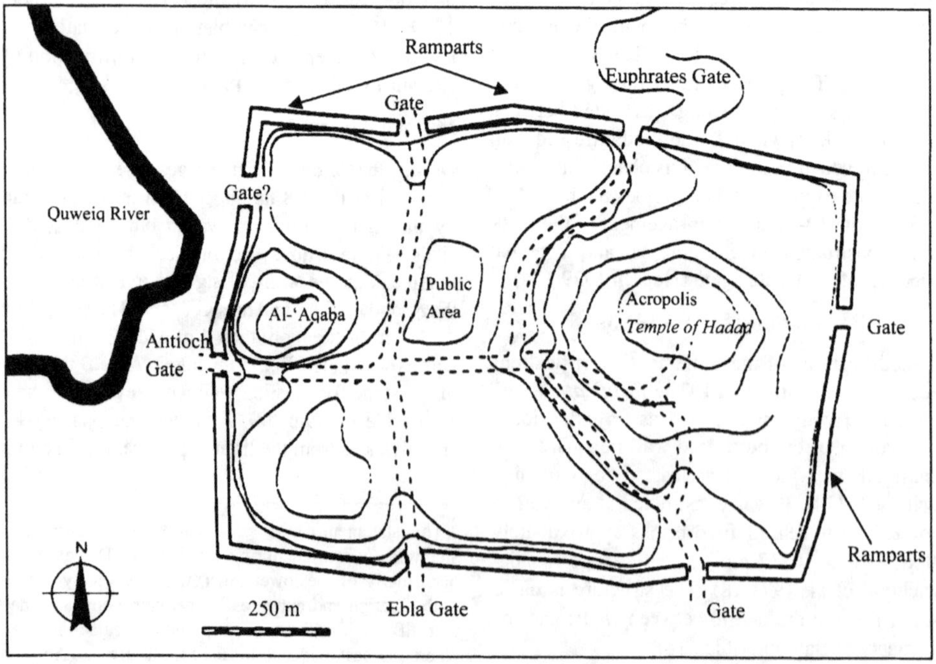

Figure 43. Reconstructed plan of Middle Bronze Age Ḫalab according to L. Nigro (after 1997–1999:fig. 8).

The most recent attempt to reconstruct the settlement's layout is that of Lorenzo Nigro (1997–1999) who proposed that Middle Bronze Age Aleppo featured a rectangular plan, supposedly measuring 1200 m E-W by 900 m N-S and thus forming a settlement of approximately 108 ha (ibid., 49, fig. 8). He bases his conclusion on the position of five important urban features that he identifies around the mound of Aleppo (see Figure 43 above). These include: (1) an occupied acropolis beginning in the third millennium, (2) the remains of what may have been a Middle Bronze Age rampart, (3) the hypothetical location of six gates, (4) a central terrace, which Nigro speculates accommodated public structures in the Middle Bronze Age as it did in later periods (ibid., 53f.), and (5) the mound of al-ʿAqaba on the W side of the site just inside the fortifications (ibid., 49).

Since the acropolis at Aleppo is presently under excavation, nothing of its defenses is yet known and, as stated above, nothing can be definitively stated regarding the Middle Bronze Age fortifications of the lower town. For this reason Nigro draws what little evidence there is for the ramparts around the lower town from a variety of sources, though the excavators of these features were not primarily concerned with the third- or second-millennium fortifications (ibid., 51ff.). Nevertheless, Nigro reports that the *earthen rampart*, which he refers to as a revetment, was composed of crushed limestone, fieldstones, and clay (ibid., 53). The rampart's height was between 15 and 25 m, composed of "broken mudbricks and limestone blocks," and it was built at the start of the second millennium (ibid.); but the date of the rampart's construction is, unfortunately, not confirmed by ceramic evidence. Furthermore, the general composition of the rampart including questions regarding the composition of its core such as the presence or absence of a core wall, the existence of a stone revetment wall along its exterior foot, or the presence of a fortification wall on top of the rampart cannot be addressed. Although Nigro does not discuss it, there is also no evidence for a fosse outside the rampart.

In his proposed reconstruction, Nigro suggests identifying "six gates of Old Syrian Aleppo" (ibid., 53), as well as the possible location of a seventh gate (ibid., fig. 8). The seven gates include, clockwise (Figure 43 above), (1) the Antioch Gate on the W side, (2) the NW Gate leading to the Quweiq River, (3) the N Gate, (4) the NE or Euphrates Gate, (5) the SE Gate, (6) the Ebla Gate, and (7) the E gate (ibid., 53). While Nigro argues that the Ayyubid period Antioch Gate preserves the plan of "second-millennium bent axis city gates" with even a "triple-arched" passage (ibid.), it is doubtful that one can conclude that the inspiration for this particular gate was at all derived from what lay hidden under the soil more than two thousand years.[22] Nevertheless, while we cannot yet confirm the existence of each of these gates at these locations, Nigro's model serves as a basis for future exploration of the second-millennium plan of Aleppo.

The validity of his reconstruction can be called into question on several grounds. First, as Nigro himself notes, the N border of this layout is not clear since "modern buildings have damaged its ancient boundaries" (ibid., 51). Second, also unexplained by this reconstruction is the unusual placement of the acropolis on the leeward side of the site downwind from most of the town, which has no known parallels in Middle Bronze Age town planning in the Levant. At Ebla, for example, the placement of the acropolis, containing the palace, in the center of the site (a third-millennium layout) is compensated for because nonindustrial, public structures such as temples and administrative buildings occupied nearly the entire area to the W between the acropolis and the W rampart. Nigro's reconstruction leaves, however, the focal point of Aleppo downwind from everything in the town. The third but perhaps strongest objection to Nigro's reconstruction is that it relies too heavily upon the plan of Medieval Aleppo, one which likely has no bearing upon its second-millennium plan (see n. 22). Finally, Nigro's reconstruction of Middle Bronze Age Aleppo presumes a situation wherein the plan of the third-millennium town, which is presently unknown, did not have any bearing upon its Middle Bronze Age layout. This is very significant since, on the one hand archaeological evidence confirms that Aleppo was occupied in the third millennium, and on the other hand Aleppo has been identified by some with Armānum whose impressive fortifications were mentioned in Naram-Sin's inscription of his siege of the town (see p. 90ff.).

Due to these objections regarding Nigro's reconstruction of Middle Bronze Age Aleppo I believe it is warranted to set forth the most basic conclusions concerning the layout of the site during the third and second millennia. This is possible without attempting to identify all of the town's major defensive features since they are mostly obscured by modern construction. A visitor today who stands outside the gateway of the medieval citadel can observe that the land around the citadel slopes downward considerably from E to W. The slope is substantial enough that it is indicative of the original underlying topography of the area.

[22]In fact, if this were the case, Aleppo's continuity of settlement and urban planning would be unique in the region of Syria-Palestine, if not the entire Near East, since to my knowledge no other equally late phase of occupation has preserved elements of the Bronze Age city plan.

As noted above, excavations on the acropolis, which have yielded the plan of a third-millennium temple (ca. 432 m ASL), make it absolutely certain that the medieval citadel overlies the site's Bronze Age acropolis. Because the citadel's medieval glacis has encased the tell it is not possible to be certain of the location of the entrance to the citadel during the Bronze Age, a fact which probably bears upon the location of much of the settlement below. Nevertheless, given that the area to the W of the citadel slopes away from the mound and that even in medieval times the gate was not located on this side, but rather to the SSW, we may presume that the ramp from the citadel led most likely from a position on the S to E sides of the tell. It is possible that the ramp led in a counterclockwise fashion, the most defensible type of ascent, beginning at the present location of the citadel gate and emptying to the E, making the ascent the most gradual possible since here the difference in elevation between the upper and lower towns was the least of anywhere around the citadel. Such a location for the gateway might also explain the outcrop upon which the medieval gateway was built.

These observations are significant because they are difficult to rectify with Nigro's reconstruction of the Middle Bronze Age town since the ramp from the citadel would lead directly out of the town (or into the town for attackers) to the E. Given that points of access to acropoleis do not usually lead directly to an exit from the settlement without first passing through the lower town (presumably more for the purpose of increasing accessibility to the lower town rather than preventing an easy exit of the settlement from the acropolis) and that the acropolis is usually located on the windward side of the settlement, it is reasonable to suggest that the settlement was predominantly located to the E of the tell itself. Furthermore, the topographical contours provided by Nigro when considered in this interpretation lend support to the conclusion that the continuous depression leading from the Euphrates Gate, NNW of the citadel, to the proposed gate S of the citadel may constitute the W limits of the settlement during the Bronze Age. This would mean that all the structures built to the W of the line marked by this depression were first built only after the Middle Bronze Age (as the evidence seems to support) and that it is highly doubtful that any early structures exist below them. At the present time, however, it is not possible to speculate further concerning the settlement of the third and second millennia.

KADESH (TELL NEBI MEND)

UTM: 37 N 0272389.3826751
Refs.: Pézard (1931); Parr (1983; 1991; 1993; 1997); Bourke (1989); Marfoe (1995:159).
Plans: Site (Parr 1983:figs. 1f.); Wall 1 [Mur X] (Parr 1983:fig. 3; Pézard 1931:pl. C).
Sects.: Trench I (Pézard 1931:pl. C).
Photos: Trench I (Pézard 1931:pl. 13, fig. 3).

The mound of Tell Nebi Mend is situated between the Orontes River, which provides a natural boundary on its E side, and the Orontes tributary known as the al-Mukadiyah to its W. According to Peter Parr (1983:101f.), the site today consists of three main components: the main mound, the lower mound(s) extending to the S, and an enclosure to the W. The main mound or acropolis measures approximately 450 m NNE-SSW x 250 m ESE-WNW (ibid., 101), or less than 11.3 ha assuming that its steep slopes were not inhabited. The highest part of the acropolis today is about 30 m above the level of the plain (Bourke 1989:155). The lower mound, or more properly mounds since there are two, project SSE from the acropolis about 450 m and are about 300 m wide (Parr 1983:102), comprising approximately another 13.5 ha. The N of these two mounds, adjacent to the acropolis, is approximately 7 m high, while the S one, separated from it by a depression to its SSE, is slightly lower.

To the W of the mound is a rectilinear "enclosure" (Parr 1983:102), which measures approximately 750 m WNW-ESE by more than 900 m NNE-SSW (ibid., fig. 1). The lines of the enclosure only appear to be preserved at its SW corner and along 800 m of its W side. In these places its remains consist of a ditch about 30 m wide with an embankment, at its highest reaching 5 m, on its interior (ibid., 102). Parr suggests that a "bay" or depression just N of the SW corner of the enclosure may be the location of a gate in the enclosure at this point (ibid.). Various dates for the enclosure have been suggested including a Middle to Late Bronze Age date, identifying the structure with the moat depicted in Egyptian reliefs of the Battle of Kadesh. However, according to Parr, excavation has demonstrated that this enclosure cannot predate the fourth century A.D., thus making it Byzantine in date (ibid., 102, 108). This dating is borne out by stone architecture, which can be seen along the surface of this embankment today.

Middle Bronze Age fortifications have only been identified, therefore, in Trench I on the E side of the acropolis (Parr 1983:106, fig. 3). There a 2.5 m wide mudbrick *fortification wall* (Wall 1) was built on a foundation of fieldstones supported on the inside by buttresses. Maurice Pézard (1931:pl. C) had identified this wall (Mur X) exposing it to a height of 3.67 m (twenty-four courses?); its exterior sloped back 11° and the wall was encased on both sides by mudbricks that were identified as different than those used in the core, two rows on the exterior and one on the interior. Although it no doubt served as a defensive wall as indicated by its location, it may also have served as the wall of a palace built on the acropolis. Therefore, Parr identifies this wall and its interior rooms as part of a "double casemate fortification" system (1983:106). This wall is identified as Middle Bronze Age in date by its stratigraphic association with floors on its interior, which yielded pottery identified as belonging to Phase H at Hama (ibid., 106), and more specifically it has been dated to the mid- to late seventeenth century B.C. (Bourke's phase H–G, 1989:162; see also Parr 1991:83). According to Bourke's (1989:163) phasing of Middle Bronze Age occupation, which was based on ceramics from Trench I, the wall was built in Phase H dating to the mid-seventeenth century. The wall was destroyed in a conflagration at the end of the seventeenth century, which is the end of Phase G (ibid., 164).

Parr also identifies an artificial *earthen rampart* dated to the Middle Bronze Age in one part of the site, but did not specify where this was located (Parr 1997:114). The Middle Bronze Age occupation of the site was brought to a close "with the destruction of the fortifications and associated buildings" (ca. 1550 B.C.) after which the site may have been temporarily abandoned (1997:115). The only other relevant data concerning the fortifications of Kadesh are reliefs of the Battle of Kadesh in Egypt (see Yadin 1963:238). Despite suggestions that these reliefs provide a pictorial representation of an earthen rampart around Kadesh, it is far more likely that the branches of the Orontes River are simply shown encircling Kadesh as they did.

KANNAS, TELL

UTM: 37 N 416484.4003535
Refs.: Strommenger (1979).
Plans: Tower (Strommenger 1979:fig. 9).
Sects.: NA.
Photos: Site (Strommenger 1979:fig. 1); Tower (figs. 7f.).
Tell Kannas is located along the W bank of the Euphrates River (now submerged below Lake Assad). The data for the EB IV and Middle Bronze Age fortifications of Tell Kannas are limited to a single report (Strommenger 1979). Traces of the *fortification wall* of Kannas, probably EB IV in date (possibly Middle Bronze Age), were detected from the SW to the NE corners; bricks along this wall measured 40 x 40 x 10 cm and were laid upon a stone foundation (ibid., 83). This late Early Bronze Age wall featured mudbrick *semi-circular towers* with stone foundations (ibid., 84). These appear to have been replaced during the Middle Bronze Age by stone-built, *rectangular towers*; the dimensions of one of these towers (Bastion 1) is 7.5 x 4 m (ibid., fig. 9).

KAR-KEMIŠ (JERABLUS)*

UTM: 37 N 411474.4076803
Refs.: Woolley (1952:71); Parr (1968).
Plans: Site (Woolley 1921:pl. 3); W Gate "inner" town (pl. 10); S Gate "inner" town (pl. 12); Water-Gate (pl. 16).
Sects.: NA.
Photos: "Inner" town W Rampart (Woolley 1921:fig. 7); W Gate "inner" town (pl. 11a); S Gate "inner" town (pl. 11b); Water-Gate (pl. 17a).
Kar-Kemiš is located on the Euphrates along the border between Turkey and Syria. Despite the site's importance, the main observations regarding its fortifications are limited to the initial findings of the British Museum expeditions between 1911 and 1914 (Hawkins 1997:423). While Woolley remarked that the site consisted of an acropolis and an inner town during the Middle Bronze Age—the "outer" town was only occupied after ca. 1200 B.C. according to Woolley (1921:48)—the certainty with which its remains can be ascribed to the Middle Bronze Age remains uncertain. The features that have frequently been identified as Middle Bronze Age consist of the Water-Gate at the base of the citadel on the E side (e.g., Gregori 1986; Herzog 1986) and the rampart around the inner town with its W and S Gates (Woolley 1921:44f.). The earthen rampart running from the S side of the inner town to the N around the W, stood about 20 m high and was supplemented by a fosse, which is said to have been as much as 5 m deep (Woolley 1921:44). While identifying some of these features with those of the Middle Bronze Age is tempting, particularly by way of comparison with Munbaqa to the south, caution is certainly warranted in assuming that these features of Kar-Kemiš's defenses should be identified as its Middle Bronze Age fortifications. This is all the more the case in light of Parr's suggestion that these features belonged to the Iron Age (1968:30ff.), which has been supported by Irene Winter (1973:167).

KAZEL, TELL (SUMUR?)

UTM: 36 N 0773534.3844798
Refs.: Badre et al. (1990); Bounni (1997).
Plans: Site (Badre and Gubel 1999–2000:fig. 1).
Sects.: NA.
Photos: NA.
Tell Kazel (ancient *Sumur*?) is located a few kilometers from the Mediterranean coast in the Akkar plain about 20 km NNW of Tell 'Arqa. The site's contours

reveal that Tell Kazel's plan was roughly circular (Badre and Gubel 1999–2000:fig. 1), measuring 310 x 280 m or about 8.7 ha (Bounni 1997:275). The site was settled during the Middle Bronze Age, though the exact phases of this early occupation have not yet been established (Leila Badre, personal communication, 2002).

At that time the site stood only about 10 m above the plain (Bounni 1997), the underlying layers including, it is thought, some Early Bronze Age occupation (Leila Badre, personal communication, 2002). During the Middle Bronze Age a stone-built fortification wall and glacis encircled the site, as was revealed during excavations in Area III (Badre et al. 1990:87). Excavations on it NE side exposed the remains of a stone *fortification wall* more than 0.8 m wide against which a *glacis* of hard, yellow beaten clay was laid at a 45° angle (ibid.). Unfortunately, excavation of the fortifications is incomplete and the exact date of their construction is unclear.

KHAN SHEIKHOUN, TELL

UTM: 37 N 0286362.3924781
Refs.: du Mesnil du Buisson (1932).
Plans: Site (pl. 32).
Sects.: Schematic (fig. 2).
Photos: Site aerial (pl. 31).
Khan Sheikhoun, 42 km SSW of Ebla on the road between Hama and Ḫalab, was settled during the EB IV and Middle Bronze Age. Du Mesnil du Buisson suggested that the mudbrick *fortification wall* that he exposed was built during the Middle Bronze Age (see Zone B, 1932). The site's circular layout averages 175 m in diameter with an area of approximately 2.4 ha and a perimeter of about 550 m (ibid., pl. 32). The site consisted of an area on its N side, which resembles the remains of an *earthen rampart*, elevated about 3 m above its central area, in addition to the remains of what might have been a citadel on its SW side. This small rise was only one meter above the settlement surrounding it on the N and E sides, but much more elevated above the terrain to the S and E, where a depression is suggestive of the position of a gate. As a small site, it may have had one other gate on its W side.

KUMIDI (KAMID EL-LOZ)

UTM: 36 N 0761703.3723986
Refs.: Bertemes (1986); Hachmann (1991); Marfoe (1995); Badre (1997b).
Plans: Site (Marfoe 1995:fig. 71); ID 15 (ibid., fig. 58); IIE6-IIG7 (Bertemes 1986:supp. 1).
Sects.: S1–S6 (Bertemes 1986:supp. 1).
Photos: NA.
Kamid el-Loz, which is identified as ancient *Kumidi*, is located in the Beqa Valley. Its Middle Bronze Age fortifications (presumably MB IIB–C in date) have been excavated on the N and E sides of the site. On the N the *fortification wall* (Wall 3) ran from square ID 15 to IIC1 and IID1 (Marfoe 1995:104). Its fieldstone foundation was more than 2 m wide and preserved to a height of more than 2 m "with bastion-like *towers* and projecting walls" (ibid.). The superstructure was built of mudbrick as "substantial collapse levels in the east sections of IC 16 and IC 17 testify to [its] once extraordinary height" (ibid.). These fortifications were probably contemporary with those revealed on the E side of the site since they were similarly planned with rooms 2.3 to 3.5 m wide built against its interior side (ibid., fig. 58, p. 107). The mudbrick superstructure of the wall appears to have averaged between 0.7 and 1 m wide between the N and E sides of the site (ibid., 107). Although it has been determined that this fortification wall belonged to Phase 7 (Bauperiode 7), which included building levels 17–20 for the intramural structures, it has not been possible to distinguish any stages in the construction of the wall (ibid.). This wall belonged neither to the earliest nor latest phase of Middle Bronze Age settlement, as there are building levels below this (e.g., Phase 21) that also extend N of the wall's position (p. 104) and Middle Bronze Age occupation after it during Phase 6 (ibid., 110).

The Middle Bronze Age fortifications of Kamid el-Loz were also excavated on the E side of the tell during rescue operations in 1978 after they were accidentally exposed by local workers (Bertemes 1986). Excavation in squares E6, F6, F7, and G7 (salvage operation S1 to S6) revealed an offset-inset style wall that has been interpreted as the outer face of a Middle Bronze Age *casemate-type wall* (ibid., fig. 6). The evidence of its construction over a length of 32 m consisted of fieldstone foundation walls averaging 1.3 m wide with, at certain points along its length, a mudbrick superstructure (ibid., 80–82, supp. 1). The mudbricks used were 36 cm square. At least three phases are distinguishable from the excavations, as reconstructed by Bertemes (ibid., fig. 6). In the first phase the exterior walls appear to belong to the E side of two rectangular structures, the N of which is slightly to the W of the S room (fig. 6.1). During the next phase it appears that a larger rectangular structure was built to the N of these two preexisting structures, but further out from the tell to the E. The walls of this structure were notably wider and the mudbricks on the wall were still preserved (Walls 1 and 3; pp. 80f.; fig. 6.2). The final phase of this wall is less clear. It appears that the original casemate structure, around which the other structures to the N and the S had been added, had by this time fallen out of use and it was probably replaced by another similar, rectangular structure to the W (fig. 6:3).

The site was also fortified during the Late Bronze Age (Badre 1997b:265).

MAŠIN, TELL

UTM: 37 N 292771.3909822
Refs.: du Mesnil du Buisson (1935b).
Plans: Site (pl. 47).
Sects.: NA.
Photos: Site aerial (pl. 46); Tell (pl. 48:1).
Tell Mašin is located more than 20 km N of Hama and just to the W of the road to Ḫalab. The site was occupied during the EB IV and Middle Bronze Age as is evident from ceramics recovered during excavation (du Mesnil du Buisson 1935b:pls. 49f.). The site was roughly square with dimensions of approximately 125 m N-S by 140 m E-W for a total of 1.8 ha (ibid., pl. 47). These can be compared with du Mesnil du Buisson's figures of 185 m x 140 m (ibid., 123). The site also appears to have featured a small citadel in the S quarter of the site rising only 2 to 3 m above the settlement.

Only limited evidence for the fortifications is available. This includes excavation in Trench III on the W side of the site where a Middle Bronze Age wall was encountered (ibid., 131). The *fortification wall* was built on stone foundations 3 m wide and was preserved to a height of 1.9 m. It was built of square mudbricks: 38–40 x 38–40 x 12–14 cm.[23] The length of the glacis, wall, and rampart can be estimated at around 530 m. Although additional details of the features identified on the site in the preliminary report are limited, it does appear from the plan that a *rampart*(?) was identified in trenches III and IV on the W and NW sides of the site (ibid., pl. 47). While these features are probably to be dated to the Middle Bronze Age based on the dimensions given for the bricks and the report of the stratigraphy given by du Mesnil du Buisson, the nature of the settlement with regard to fortifications during the EB IV remains uncertain.

QATNA (TELL MISHRIFE)

UTM: 37 N 0304745.3856779
Refs.: Ronzevalle (1914); du Mesnil du Buisson (1926; 1935a); al-Maqdissi (2001); Cremaschi et al. (2002).
Plans: Site (du Mesnil du Buisson 1935a:pls. 1–1a; Novák and Pfälzner 2001:fig. 1); E Gate (du Mesnil du Buisson 1927:pl. 60); W Gate (du Mesnil du Buisson 1926:figs. 1–5, 6–7).
Sects.: N Gate (du Mesnil du Buisson 1926:fig. 11).
Photos: Site aerials (du Mesnil du Buisson 1926:pl. 50; 1935a:pl. 2; al-Maqdissi et al. 2002:figs. 3–4); Ramparts (1926:pls. 52f., 57); N Gate (ibid., fig. 10); E Gate (1927:pl. 63); E interior gate[?] (al-Maqdissi 2001:fig. 9); W Gate (du Mesnil du Buisson 1926:pls. 55f.); E wall (du Mesnil du Buisson 1927:pl. 62).
Qatna (Tell Mishrife) is located to the NE of Homs and SE of Hama. The site was first excavated by R. P. S. Ronzevalle in 1906 and 1912.[24] However, it was only as a result of the excavations of du Mesnil du Buisson (1926–1929) that the site was first identified as a traditional fortified enclosure of the Middle Bronze Age type.[25] Archaeological work on the site's defenses by the DGMA under the direction of Michel al-Maqdissi have been conducted since 1994 when excavations were renewed in Areas E and F in the vicinity of the W and E gates (al-Maqdissi 2001).

Qatna, as alluded to above, is best known for its distinctive and well-preserved, rectilinear *earthen ramparts* that enclosed the site on all four sides during the Middle and Late Bronze Ages (Figure 44). They formed a square enclosure nearly one kilometer on a side (after du Mesnil du Buisson 1935a:40) and about 3,980 m long. However, the large gaps, which indicate the locations of its Middle Bronze Age gates, account for more than 400 m of this perimeter, thus reducing the total length of the earthen ramparts and fortifications to about 3,520 m. The ramparts appear to have been erected to a height between 15 and 20 m above the surrounding plain (al-Maqdissi et al. 2002:9), with an exterior slope that du Mesnil du Buisson observed to be 30° (ibid.); the slope was no doubt slightly steeper before more than three thousand years of erosion. The average width of the ramparts was about 70 m. With these dimensions the ramparts' total volume is estimated to be approximately 1,848,000 m^3. Subtracting the uninhabited area on the exterior slopes of the ramparts yields a net size of approximately 78.4 ha.

Despite the prominence of the site's ramparts and its identification as a Middle Bronze Age fortification system, no soundings have been made through the rampart itself by any archaeological expedition. For

[23]Similarly sized bricks measuring 40 x 40 x 6–9 cm were used in the construction of what was probably the palace located on the citadel (du Mesnil du Buisson 1935b:124).

[24]Interestingly, Ronzevalle identified the enclosure as the remains of a Late Bronze Age camp established by the Sea Peoples who had landed in Amurru (1914).

[25]Although he published his excavations in a single volume (du Mesnil du Buisson 1935a), these must be supplemented by numerous preliminary reports published in *Syria* (vols. 7–9, 11 from 1926 to 1930).

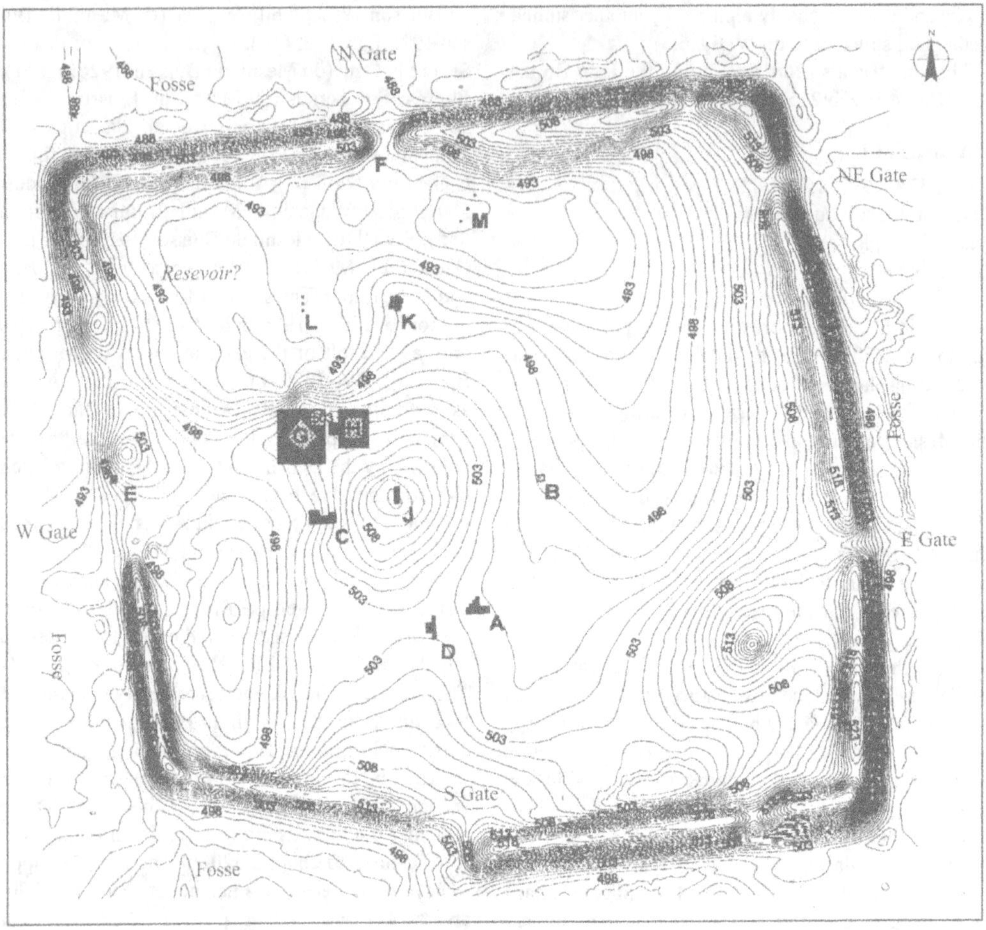

Figure 44. Plan of Qatna (after Novák and Pfälzner 2001:fig. 1).

this reason it is necessary to rely on du Mesnil du Buisson's observations, which were based on his excavations in the area of each of the gates. According to him the rampart was composed of chipped limestone and earth (ibid., 41). He also suggested that the materials for the ramparts were probably taken from the fosse, which is visible outside the ramparts in several locations (1935a:pl. 1). While du Mesnil du Buisson never identified a mudbrick wall crowning the rampart, he was convinced of its existence on the basis of a 2 m wide stone foundation wall, presumably belonging to the *fortification wall,* which was exposed along the top of the ramparts to the S of the E Gate (du Mesnil du Buisson 1927:281, pl. 62). Although its height was not provided, based upon these photographs it appears to have been at least one meter tall. While the date of this foundation remains uncertain, it may belong to the Late Bronze Age. However, an Iron Age date is not out of the question. Still, this wall also provides some evidence for the relationship between the wall and rampart, suggesting that it was built behind the present crest of the rampart. No traces of this foundation wall can be seen along the tops of the ramparts, which may be the result of stone robbing for construction in the village of Mishrife, which was built within the ramparts of Qatna or possibly because it simply lies atop the rampart but below mudbrick detritus from the wall's superstructure.

Recent work has been done on the soil micromorphology of the site by the Italian expedition in an attempt to answer questions concerning the construction of each rampart section (Cremaschi et al. 2002). Based on soil colors and micromorphological characteristics three distinct constructional units were identified: the white, the pink, and the brown ramparts (ibid., 19).

Figure 45. Fosse and southern ramparts at Qatna. View west. Photo by A. A. Burke.

The E rampart and the E ends of the N and S ramparts, not including the SE corner, are identified as the white rampart, which was probably constructed first. The small portion of the SE corner not included with the white rampart is to be identified as the pink rampart. The entire W rampart is identified as the brown rampart. Stratigraphic data suggest that most of the N, S, and E ramparts, which were composed of the white material, were constructed first, followed by the W rampart (brown rampart) and at the same time or later the SE corner (pink rampart). This study also brought to light the first evidence that the rampart was originally crowned by a mudbrick *fortification wall*, even though the date of this wall has not yet been established archaeologically (ibid., 20).

The *fosse* outside of the ramparts on all four sides was noted by du Mesnil du Buisson to have been as much as 100 m wide in places (1935a:41), though on plans it appears to average no more than about 70 m wide. Since it has never been explored its original depth remains unknown, through it was at least 5 m deep. If the fosse encircled the entire site it would have been approximately 4,560 m long and the volume of earth removed from it would have been about 1,596,000 m^3 (cp. with rampart's estimated volume of 1,848,000 m^3).

Based on the topography of the site du Mesnil du Buisson identified four primary and five secondary gates to the town (1927:293f.).[26] While the primary entrances to which du Mesnil du Buisson refers are apparent on the plan of the town (Figure 44), only two of the so-called secondary passages, which were located in the NE corner of the enclosure, are visible. The main gates, on the other hand, are evident as gaps of up to 70 m in the ramparts: the N Gate in the center of the N rampart; the W Gate in the center of the W rampart; the S Gate in the center of the S rampart and directly in line with the N Gate; and the E Gate in the E rampart approximately 370 m N of the SE corner of the enclosure. Given the representation of these openings in the rampart, it is not entirely clear why du Mesnil du Buisson considered the depression (or the NE Gate) approximately 100 m to the S of the NE corner of the enclosure to be a secondary gate since the size of the gap is comparable in size to that associated with the E Gate.[27] It seems most rea-

[26]It is not possible to be certain, however, that he meant for all of these gates to be identified as Middle or Late Bronze Age gates. See, for example, discussion of W Gate below.

[27]It is reasonably certain that du Mesnil du Buisson's conception of the city plan conformed to an ideal resulting from his

sonable, therefore, to identify five rather than four main gates, which included the N Gate, the NE Gate, the E Gate, the W Gate, and S Gate.

Although du Mesnil du Buisson supposedly identified the plans of the W and E Gates within the depressions located in the ramparts, it is clear that only one of these gates dates to the Middle Bronze Age. The E Gate (or "Desert Gate"), excavated in 1927, is clearly of the *six-pier type* with foundations constructed of roughly hewn limestone blocks. Its plan, perpendicular to the ramparts, was perfectly integrated into the overall plan of the site in a manner familiar at other Middle Bronze Age settlements. Its interior passage measured approximately 22 m long by 3.5 m wide. Recent reexamination of the publication of Clermont-Ganneau's soundings inside the E Gate (du Mesnil du Buisson 1935a:42ff.) have led al-Maqdissi to identify an interior gate just inside the main gate (2001:148, figs. 8f.). He suggests that the structural remains on the S end of Clermont-Ganneau's trench were the remains of the S half of an inner gate featuring a single pilaster and a set of interior steps (ibid., fig. 9). While his reconstruction is interesting, insufficient material has been published to test this proposal. Furthermore, parallels for an inner gate within the main gate are lacking, though outer gates are known (e.g., Damascus Gate at Ebla). The topography lines published on the original plan also appear to counter the proposed reconstruction.

Although the E Gate should probably be identified as one of Qatna's original Middle Bronze Age gates, the W Gate, which was excavated the year before (in 1926), was only identified as a Middle Bronze Age gate after the excavation of the E Gate in 1927. In the initial preliminary report for the excavations of the W Gate du Mesnil du Buisson provided no clear basis for the dating of the structure (1926:294–301). Since this dating was first advanced by the excavator other scholars have assumed that his dating of the structure was correct without considering reasons against the suggested date (Kaplan 1975:fig. 14; Herzog 1986:60; Gregori 1986:fig. 12). The most convincing evidence against a Middle Bronze Age date for the W Gate includes that the structure was neither of comparable dimensions, similarly oriented, nor built with comparable masonry to any of the other Middle Bronze Age gates at Qatna. The orientation of the W Gate was also notably different than that of the E Gate. The W Gate's axis is parallel rather than perpendicular to the line of the ramparts like both the gates on the N and E sides. This is particularly problematic since it is clear that the E elevation of the gate was also exposed as is evident from the boss on the E face of the stones in Wall XI (du Mesnil du Buisson 1926:figs. 1, 6). The W Gate was also of considerably smaller dimensions (cp. figs. 11 and 12 in Gregori 1986). Additionally, the masonry of the W Gate is very different from that attested in the other Middle Bronze Age gates. An examination of the stone-by-stone plans and photos of the W Gate reveals the use of a technique unfamiliar to any of the so-called six-pier gates in Syria during the Middle Bronze Age (du Mesnil du Buisson 1926:figs. 1–7, pls. 55f.). The stones used in the gate feature a distinguished, wide and flat perimeter around their rough-hewn faces and they were fitted together very carefully with sides that were chiseled flat (ibid., pl. 56). Furthermore, none of the worked stones featured holes in their upper surfaces for fixing the superstructure to its foundations as was common in the Middle and Late Bronze Ages (e.g., du Mesnil du Buisson 1926:pl. 55:3). This argues against either a Middle or Late Bronze Age date for the gate. In fact, the type of stone construction used in the W Gate closely resembles the stone masonry of the Hellenistic period in the southern Levant (see Samaria Sebaste). If such evidence were not sufficient, the discovery of what appears to have been infant jar burials under Wall XI on the E side of the gate suggests at least that this was not the original Middle Bronze Age gate and that no Middle Bronze Age gate has yet been discovered to account for the gap in the W ramparts (ibid., 301).[28]

Du Mesnil du Buisson also explored the N Gate in the gap between the two parts of the N rampart (1926:301–03). Unfortunately, little remained of the gate except traces of two parallel lines of fieldstones that suggest the placement of a direct-axis gate (ibid., fig. 10). The masonry here is again comparable to that of the E Gate, but entirely different than that of the W Gate, adding further weight to the redating of the W Gate suggested above. Although du Mesnil du Buisson explored the vicinity of the S Gate (1927:283), his

perception of the very regular layout of T. Mishrife's ramparts. Thus, he assumed that each of the four ramparts would have required only one primary gate. C. Clermont-Ganneau had already explored the area of the NE Gate, and it is therefore possible that the lack of impressive finds dissuaded du Mesnil from suggesting that a major gate was responsible for the unimpressive gap in the rampart there.

[28]Given the criteria set forth here for dating the W Gate, and that it appears to be neither Middle Bronze Age nor Late Bronze Age in date, I would suggest that at the earliest an Iron Age date is possible for the gate. Unfortunately, the pottery recovered from the gate was meager and is not helpful for establishing its date (see du Mesnil du Buisson 1926:300f., figs. 8f.).

findings do not provide any evidence concerning the Middle Bronze Age fortifications in this area.

Despite considerable exploration of Qatna's gates, important data concerning the dating of its defenses are still lacking. Based on the plan alone, Levantine archaeologists have suggested dating the enclosure to the MB IIB (ca. seventeenth century B.C.), comparing its plan with sites such as Tel Dan and Hazor in northern Israel. Similar dates could be proposed on the basis of historical references to the kingdom of Qatna in the Mari texts (Klengel 1992:65ff.), as well as the dates proposed for the beginning of the Middle Bronze Age palace at Qatna (Novák and Pfälzner 2002:111); these suggest that the construction of the defenses were probably undertaken during the settlement's Middle Bronze Age floruit. Nevertheless, in the absence of excavation of the ramparts themselves, the date of the construction of Qatna's defenses remains uncertain.

QITAR, TELL EL-

UTM: 37 N 424938.4026392
Refs.: Culican (1983); McClellan (1984–85; 1986).
Plans: Site (McClellan 1984–85:fig. 1); Area X (ibid., fig. 2); Area Y (ibid., fig. 3; 1986:fig. 3).
Sects.: NA.
Photos: NA.
El-Qitar was a small but prominent site on the Upper Euphrates 60 km S of Kar-Kemiš (now submerged under Lake Assad). While the site was certainly fortified during the Late Bronze Age, it also appears to have also been occupied at least in the latter half of the Middle Bronze Age. Excavations from 1982 to 1984 revealed the settlement's defenses in Areas X and Y, which consisted of two gates and a curtain wall with towers, though only the stone foundations of these features remain. Only one course of stones, probably the foundations of the *fortification wall*, were found around the site and in most places these were laid directly on bedrock. In Area Y the wall was preserved 2 m wide and appears to have been built in an offset-inset manner (Culican and McClellan 1983:33). It is possible to estimate that the wall was about 1 km in length and enclosed an area of about 5.6 ha. As is obvious from the topographic plan, it does not appear to trace the contours of the tell (McClellan 1984–85:fig. 1).

The River Gate located on the E side of the site (Area X), was a *direct axis gate* aligned N-S with 4 m wide walls with a 4.5 to 5 m wide passageway about 10 m long. The gate was flanked by towers on the E (Tower 6) and W (Tower 7); one of these towers was preserved to a height of 4 m (ibid., 37).

Another *gate complex* was excavated on the W side of the site (Area Y) and thus identified as the W Gate (ibid., 37f.). This complex consisted of two gates: the Lower and Upper W Gates. The Lower W Gate, less well preserved, was located about 19.5 m W of the W line of the curtain wall and featured flanking orthostats and two sets of piers along its 4.5 m wide by 10 m long passage (McClellan 1986:fig. 3). The Upper W Gate was probably a second element of the defenses of the W Gate complex, though its stratigraphic relationship to the latter is uncertain (Culican and McClellan 1983:38). Although no orthostats or piers were identified, it appears that the gate's passage was approximately 5.5 m wide and 10 m long.

Two *towers* identified as the N (Tower 5) and S (Tower 4) towers were revealed in Area Y at the N and S ends of the upper settlement. Tower 5 was the best preserved at the site (ibid., 35). A crushed limestone *glacis* 1 to 2 m thick composed of "coarse yellowish-white limestone chunks" and "darker layers of stony fills and patches of fine dark grey ashy soil" was found in the vicinity of Tower 4 around the SW corner of the site (McClellan 1986:88f.). The glacis appears to have been held in place with a *revetment wall* (W.747/749) built at its foot (ibid., 90). The revetment wall and glacis run parallel here to the *fortification wall* (W.741/745) behind it, which may date to the Middle Bronze Age (ibid., 105). The combined width of the revetment wall, glacis and fortification wall averages 17 to 18 m (ibid., 90). This defensive system may have been made obsolete during the Late Bronze Age if, as it appears, these features were built over. Although a fosse was never located during excavations, it may have been detected as an anomaly by magnetometry conducted in the final season (ibid.).

SEFINAT-NOUH, TELL

UTM: 36 N 0274639.3828956
Refs.: Mousli (1989/1990); G. R. H. Wright (1969:163).
Plans: Site (Mousli 1989/1990:fig. 110).
Sects.: NA.
Photos: Site aerial (Tallon 1956:pl. 10 from R. P. A. Poidebard).
Sefinat-Nouh lies about 3.2 km NE of Kadesh (Tell Nebi Mend). Majed Mousli identifies this rectangular site, which sits 7 m above its surroundings, as a fortified enclosure of the Middle Bronze Age (1989/1990:300) and compared with Qatna and Khan Sheikhoun since the work of P. L. Ronzevalle (1914). Although soundings within the enclosure have demonstrated Middle Bronze Age occupation (Mousli 1989/1990), no sections of its fortifications have yet been made.

The site is surrounded on all four sides by a 20 m wide and 4 to 5 m deep *fosse* (Tallon 1956:60). It was 1,905 m long and covered an area of approximately 3.8 ha (Figure 46). The volume of earth removed for the construction of the fosse can be estimated to have been around 171,450 m^3. *Earthen ramparts* (and the wall that crowned it) about 1,585 m long were built from the inner scarp of the fosse to a height of about 7 m. The width of the rampart from its base to its crest is approximately 20 m, though if it was originally constructed as a freestanding rampart its cross-section may have been trapezoidal with a base about 40 m wide. Its volume can, therefore, be estimated to have been approximately 221,900 m^3. This is only slightly more than the volume of the fosse itself.

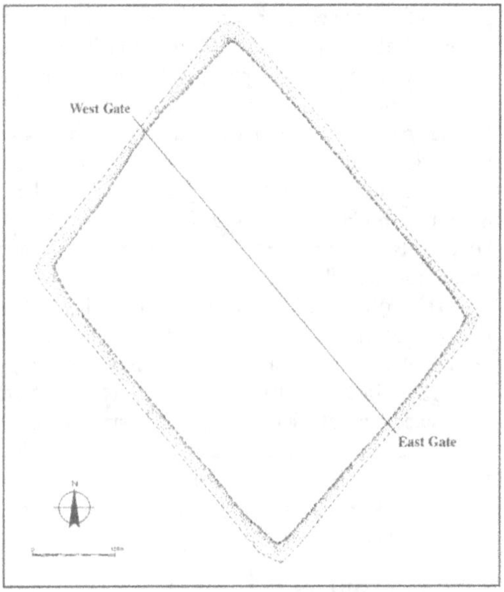

Figure 46. Plan of Tell Sefinat-Nouh (after Mousli 1989/1990:fig. 110).

The uninhabitable area covered by the ramparts would have been approximately 3.2 ha. The site's inhabitable area was therefore about 15.2 ha, or an area with average dimensions of 435 m by 350 m. Although no gates have been excavated and their location is not entirely evident from the site's contours, Mousli suggests that the site featured gates at least on the NW and SE sides of the site (1989/1990:301). But there can be little doubt that the site also featured gates on its other two sides since the route to the SW led to Kadesh and the route to the NE led to Qatna. Unfortunately, until further excavations are conducted details regarding Sefinat-Nouh's defenses will remain limited.

SELENKAHIYE, TELL ES-

UTM: 37 N 414292.3995707

Refs.: van Loon (2001); Meijer (2001).

Plans: Site (Meijer 2001:fig. 3.2); Area B (fig. 3.18); Area B, O/P26 (fig. 3.14), Bastion B in Q21 (fig. 3.19); Area C, T06/S04 (fig. 3.30); Gate in Q3 (fig. 3.32).

Sects.: N section of Area B, P26 (Meijer 2001:fig. 3.15); W section of Area SSS07 (fig. 3.31); O26 (fig. 3.36).

Photos: Area B: O26–P26 (Meijer 2001:pl. 3.10), Wall IV (pl. 11), Bastion B (pl. 3.23).

Selenkahiye (now submerged under Lake Assad) was established along the W bank of the Euphrates River around 2350 B.C. and occupied until 1850 B.C. (EB IV to MB IIA [IA]). In addition to benefiting from the defensibility afforded by the river on its E side, the site was also surrounded by a series of fortification walls during this period. The settlement of Selenkahiye measures approximately 11.5 ha inside the town walls that stretched around the settlement tracing approximately the 305 m contour line for about 1,430 m and forming a roughly rectangular plan.

Although a sequence of four fortification walls has been identified for the five hundred years of occupation mentioned above, all four walls were not identified in all areas and the town wall was not excavated on the E side of the site by the river. The clearest sequence of the fortification walls was identified in Squares O26 through Q26 in Area B (Meijer 2001:51). Here it appears that the first mudbrick *fortification wall* (Wall I), which only resembles rubble 40 cm high, was built upon a boulder foundation 1.5 m high and 2.5 m wide that had been laid upon a pebble surface (see Meijer 2001:pl. 3.10). After the wall had fallen out of use a second mudbrick *fortification wall* (Wall II) was built upon the ash layer overlying Wall I. Pebbles were again used as a foundation for Wall II, which measured 2.5 m wide (ibid.). A third mudbrick *fortification wall* (Wall III) was later built inside the line of Wall II after much of the earlier wall had been removed.

Meijer identifies pebbles laid against Wall III and covering the remains of Wall II as a 4.5 m wide *glacis* (Meijer 2001:53, fig. 3.15), though much of its upper portion has supposedly eroded away. At some point after its construction it appeared necessary to double the width of Wall III. In some areas this was accomplished by adding what has been identified as Wall IV inside Wall III, while in other areas the rooms of houses adjacent to Wall III were filled with laid bricks (ibid., 53). After its construction the interior face (E side in Area B) of Wall IV was re-plastered. Although it has been suggested that the site contracted in size after the abandonment of Wall IV, this suggestion is based upon speculation that a later wall may have been constructed further back upon the tell. However, this is mainly an attempt to explain the mudbrick detritus found above the four walls discussed above (ibid.). Portions of a *tower* (Bastion B) built onto the town wall were also identified in Area B; it measured 5 m long (N-S) and 5 m wide [E-W] (ibid., 53 and pl. 3.23).

The town's defenses were also excavated in other areas at Selenkahiye. While these areas have proved helpful in the reconstruction of the nature of the town's defenses during these periods, it has not been possible to correlate each of the features from these areas with the sequence of walls outlined above (ibid., 103). This is in part due to the various methods of construction and repair that were employed in different areas around the site and the different interpretations of various excavators.[29] In Squares T06–04 the NW corner of the *fortification wall* was also identified (ibid., 83). Against the interior of the corner (Sq. S05–06) a mudbrick structure was built diagonally, apparently with the intention of reinforcing the corner of this town wall (ibid., 87). In Area D on the N side of the site a red brick *fortification wall* built upon a pebble foundation (as in Area B) and measuring 2.25 m in width should probably be identified with the eroded remains of one of the town's early walls. The remains of a second tower or a portion of a second phase of the town wall also appear to have been identified in Square TTT07 (ibid., 87). Another *tower* that was built against the exterior of the town wall and measured 10 m long and 7 m wide was excavated in Square Z07 (ibid.).

Although evidence of a rock-cut *fosse* surrounding the site was brought to light during the excavations the exact correlation of these remains with the walls of various periods is uncertain. Nevertheless, a 3 m deep by 9 m wide fosse was excavated in Squares N25–26 W of the town wall in Area B (ibid., 93f.). Its volume can be estimated at about 38,610 m^3. Despite the ambiguity of Meijer's conclusions, the discovery of this fosse at the foot of the mound would seem to support the suggestion that the extramural settlement W of the tell was later than the construction of this fosse. It does certainly appear to belong to an earlier defensive system than Wall A and Bastion B in this area (ibid., 89).

One *gate* was identified in Square Q3 (ibid.). Its 3 m wide and 10 m long passage was flanked by two towers, each 10 m long along the gate's axis and 8 m wide, that only projected beyond the town wall on its interior side. This gate served the NW quarter of the town and provided access to the W and possibly to the lower town on that side. It is also possible that a second gate may have existed in the vicinity of Squares P26–Q26 based on the depression evident in the site's topography (ibid., 105). This gate would have served the S end of the town. A River Gate may have been located on the opposite side of the town from the gate in Q3 and another may have been located on the W side around Square Q26–27.

Ṣour, Tell aṣ- (Ṣourane)*

UTM: 37 N 0331974.3854119
Refs.: Tallon (1956).
Plans: Site (Tallon 1956:fig. 1).
Sects.: NA.
Photos: From various locations around the site (Tallon 1956:pls. 1–9).[30]

Tell aṣ-Ṣour is located 28 km ESE of Qatna.[31] It was first identified as a ramparted enclosure of the type comparable to Qatna and Sefinat-Nouh by Maurice Tallon (1956). Aṣ-Ṣour was transected in antiquity by the road from Qatna leading E into the desert (Figure 47). The site consisted of both an upper and lower town, both of which were enclosed by separate sets of ramparts.

The upper town, lying to the N of the E-W road is less than 9.2 ha in size and was roughly square in plan, while the portion appended to the site on the S, which was of similar layout, measured less than 13.2 ha in area. The entire site as represented by these two halves was, therefore, rectangular in plan measuring 545 m long (N-S) by 485 m wide (E-W).[32] The site sits around 6 m above the surrounding plain that slopes gently from N to S away from the mountain to the NW. The modern road following the course of the wadi now divides the upper and lower towns. This line also

[29] It should also be noted that no consensus was reached between the director (M. van Loon) and the main supervisor (D. J. W. Meijer) concerning the interpretation of the fortification walls in addition to other aspects of the settlement (see Meijer 1988:104).

[30] While some of the locations of the photos are discernible from the texts and their captions, due to the quality and poor textual description of some of the photos they cannot all be related to the site's plan.

[31] It is noteworthy that the use of the Arabic word *ṣour* in the name of this site refers to the visibility of its walls or ramparts. Similarly, the ruins of Qatna are referred to as Ṣour-Mishrife and a site to the north of Homs is called Ṣourane.

[32] The measurements provided here, which were taken onsite in August 2002, are more reliable than those of 662 m by 420 m provided by Tallon.

demarcates the line of drainage within the town's ramparts: to the N the upper town drains S, and to the S the lower town drains N. It is perhaps on this basis that one may observe that the W and E gates at both ends of this line served as the main gates of the town during the site's floruit.

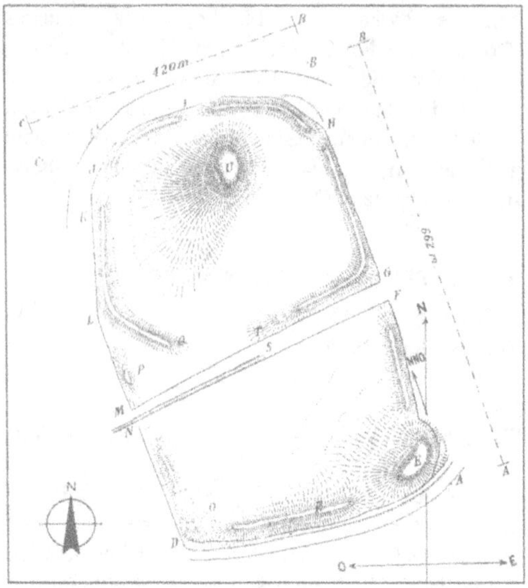

Figure 47. Plan of aṣ-Ṣour (after Tallon 1956:fig. 1).

The upper town was surrounded on all sides by *earthen ramparts* except where four depressions suggest the locations of gates leading to the N (point I), to the E (point H), to the S (between points Q and T) into the lower town, and to the W (point J), which probably led toward Qatna (Figure 47). Considerable remains of the stone foundations of the fortifications are preserved one course high for the E half of the N Gate at point I. Here the wall's foundation was constructed of roughly hewn stones 5 m wide. The stones flanking the W side of the gate near point Q are also still preserved in situ. The so-called upper town also appears to have featured an acropolis in its center, located at point U on the plan, which was defended by similarly constructed walls of unknown thickness; the lines of the stone foundations of these walls are still traceable today.

The lower town which lies directly to the S was separated from the upper town by the E-W wadi running between them. It too was enclosed by *earthen ramparts* on its three remaining sides. Two main depressions in its ramparts just N of hill E and E of the SW corner indicate that in addition to the W and E gates adjacent to the upper town, two more gates were located at the S end of its E side and on its SW corner. As part of its defenses aṣ-Ṣour appears to have featured walls (and perhaps bastions or towers) built upon its ramparts (ibid., 56, 59; pl. VII:3). It may even have featured a double wall around the citadel above the upper town (ibid., 55, pl. III:1). Evidence of these walls is present today on the site and the lines formed by the foundation stones are extremely clear. In addition to the ramparts and walls, the site was also surrounded by a *fosse* that measured anywhere from 8 to 30 m wide; it was detected outside the S rampart of the lower town and in the vicinity of the gate at point I on the NE corner of the upper town (ibid., 54–57). The fosse is 3 to 4 m deep on average. Tallon was probably correct to suggest that the fosse provided most of the material for the building of the ramparts, which surround both the upper and lower towns (ibid., 51), given the bedrock protruding on both the exterior and interior slopes of the fosse on the N and S ends of the site.

As Tallon noted, while a date comparable to the other known sites with rectilinear enclosures seems appropriate, evidence for later settlement at aṣ-Ṣour such as during the Roman period precluded his suggesting beyond doubt that the defenses described above were Middle Bronze Age in date. Nevertheless, no such late ceramics were identified during the author's visit that would confirm a late date for the fortifications of the settlement. Furthermore, the earthen ramparts, fosse, and evidence of the stone foundations belonging to a fortification wall along the rampart suggest that the site was contemporaneous with Qatna and probably dates to the Middle Bronze Age.

TUNIP (TELL 'ACHARNEH)*

UTM: 37 N 0263182.3908077
Refs.: Fortin (2006a; 2006c:125–33).
Plans: NA.
Sects.: NA.
Photos: Site aerial (Fortin 2006b:figs. 6 and 9).

Tell 'Acharneh, the site most likely to be identified as ancient *Tunip* (Klengel 1995; Frayne 2006), is located toward the S end of the Ghab Valley E of the Jebel al-Ansariye and NW of Hama. Current exploration of the site is being undertaken by Michel Fortin of Laval University and Lisa Cooper of the University of British Columbia. Although the site was clearly occupied during the Iron Age, as evident from the fortifications on the N side, which have been dated to the eighth century B.C. (2006c:125–33), the tell no doubt attained much of its size (gross size 70 ha) and height during its Bronze Age settlement (Fortin 2006a:214). The size of the tell (2 ha) and its accompanying lower town (52.8 ha) make it one of the most impressive settlements of the Bronze and Iron Ages in the Ghab Valley, with an inhabitable area inside of its ramparts of about 55 ha. The question concerning 'Acharneh's Middle Bronze Age settlement is, therefore, to what extent the earthen ramparts around the site are Bronze Age in date despite Iron Age fortifications that have been encountered above them.

TUQAN, TELL

UTM: 37 N 0315441.3966571
Refs.: Matthiae (1979; 1982a; 1983; 1989–1990; 1997b); Guardata (1990).
Plans: Site (Guardata 1990:fig. 1); Area A (ibid., fig. 2b); Area E (Matthiae 1982a:fig. 11); Area F (Guardata 1994:15); Area G (1990:fig. 4).
Sects.: Area E schematic (Matthiae 1982a:figs. 14f.).
Photos: Area A (Guardata 1990:fig. 2a); Area E (Matthiae 1982a:figs. 12f.); Area F (ibid., figs. 18f.); Area G (Guardata 1990:fig. 4a).

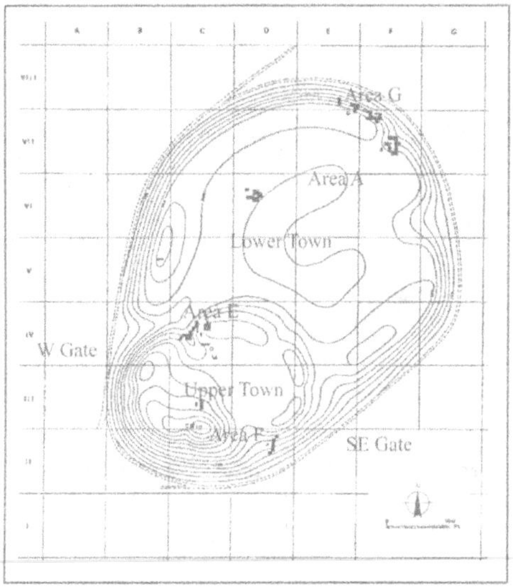

Figure 48. Plan of Tuqan (after Guardata 1990:fig. 1).

Tell Tuqan is situated on the open plain 14.6 km to the ENE of Ebla and 45 km SSW of Ḫalab. The site was settled from the start of EB IVB through the end of the Middle Bronze Age. Tuqan's settlement peaked at approximately 16.5 ha during the Middle Bronze Age (Figure 48),[33] at the end of which it was destroyed and the lower settlement was abandoned (Matthiae 1979:7–9). The upper town continued to be occupied during the Late Bronze and Iron Ages. The lower town itself was approximately 12.5 ha in area within the line of its *earthen ramparts*, which are around 1,240 m long and rise about 10 m above the plain (Matthiae 1979:fig. 1). The upper town of 4 ha, which sits 6 to 8 m above the lower town, was surrounded by another 240 m of fortification wall.

EB IV: PHASES 2A–C, LOWER TOWN

According to Matthiae, Tuqan's settlement in the EB IVB was the direct result of the destruction of Ebla (Mardikh IIB1). Unfortunately, the nature of the site's fortifications during the EB IV remains unclear and it is only during the Middle Bronze Age that we first have any information concerning their construction. Nevertheless, the lower town was settled in this period (Phases 2A–C).

MB IIA: PHASE 1B, LOWER TOWN

At the start of the MB IIA the entire settlement was probably enclosed by a fortification wall built atop the earliest ramparts, which featured strongly fortified six-pier gates at four key locations. The lower town featured at least three gates that may be identified with the major depressions in the site's surrounding embankment (Figure 48). These are the W Gate located on the W side adjacent to the NW side of the upper town (Grid B IV), the SE Gate also located adjacent to the upper town on its SE corner (Grid E III), and the NE Gate located across from the upper town (Grid F VII/Area A).

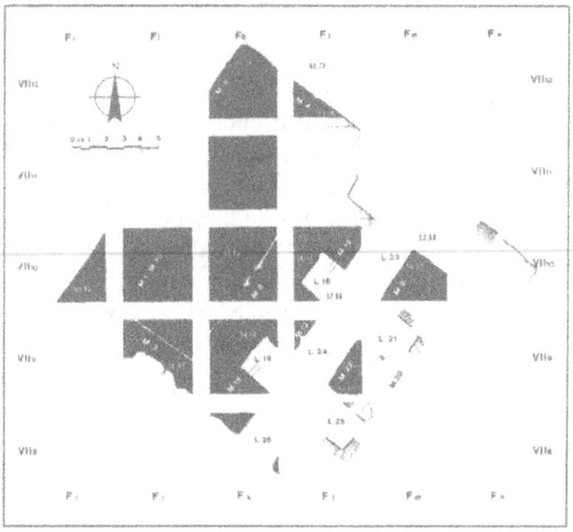

Figure 49. Gate in Area A (after Guardata 1990:fig. 3).

Of the lower town's gates, only the NE Gate in Area A (Figure 49) has been excavated (in 1978 and 1981). This gate was a *six-pier gate* with a 14 to 15 m long and 2.6 m wide passage presumably giving ac-

[33]Compare with the gross figure of 27 ha reported by Matthiae (1979) or 30 ha reported by Guardata (1990:64).

cess to a source of water outside the site (Matthiae 1979:8; Guardata 1990:65). The gate structure was over 18 m deep. The three piers, which are preserved at the ends of the gate (M.18, 9, and 15), measured 3.25 m wide, while the center piers (M.13 and 22) measured 3.75 m wide (Matthiae 1979:8). The gate featured a mudbrick superstructure preserved to a height of 4.5 m that was built with bricks measuring 38 x 38 x 12 cm, and faced with limestone blocks (Matthiae 1979:8).

The upper town during this period was enclosed by a second fortification wall that included an earthen rampart, though only its final stage has been clearly defined. The upper town's fortifications and related features have been excavated in Areas E (Grid C IV) and F (Grid D II). The final size of the rampart around the upper town at the end of the MB II was approximately 19.5 m wide in Area E, though it is believed that it grew to this width over several phases of building (Matthiae 1979:66). In Area E excavations also demonstrated that the surrounding embankment featured a 4.5 m wide mudbrick *fortification wall* (M.130) built on stone foundations which was preserved to a height of 5.4 m (Matthiae 1982a:316). An *earthen rampart* 4.9 m high appears to have been built against the wall in at least two building phases with a slope of about 30° (ibid., 318). For whatever reason it appears that towers were not thought necessary during the MB IIA when the wall was first built.

MB IIB–C: PHASE 1A, LOWER TOWN

At the start of the MB II Tuqan appears to have experienced a destruction that may have precipitated a renovation of the site's fortification system (Matthiae 1982a:315). These changes appear to have included the renovation of the upper town's rampart, the addition of towers to the exterior of the upper town's wall, the construction of at least one additional gate in the upper town, and the construction of a new fortification wall with circular towers around the lower town.

The MB IIA fortification wall was destroyed but was subsequently rebuilt during the MB IIB with towers (Matthiae 1982a:318). A rectangular *tower* (7.5 x 5.85 m) without a stone foundation was built on the exterior of a stretch of this wall located in square C IV13 during the MB IIB–C (ibid., 323).

In Area F (Figure 50) on the SE side of the upper town a so-called *six-pier gate* was revealed with a 17.5 m deep and 14 m long passage (Matthiae 1982a:figs. 16f.; Guardata 1990:65). Examination of the structure revealed massive rough-hewn limestone masonry similar to that employed in the NE Gate in Area A along with a decorated flagstone paving (ibid., figs. 20f.). The passageway of the SE Gate was 2.7 m wide and the niches on the W between the piers were smaller than those on the E (Guardata 1994:14). Two of the door sockets for the two leaves of the inner door on the N end, and one of the two sockets for the S outer doors were found just inside the two sets of piers. Although the gate was partially blocked at the end of the Late Bronze Age by fallen stones, it continued in use until the end of the Iron II (ibid.). Another gate to the upper town may have been located in Grid D IV (Matthiae 1979:6).

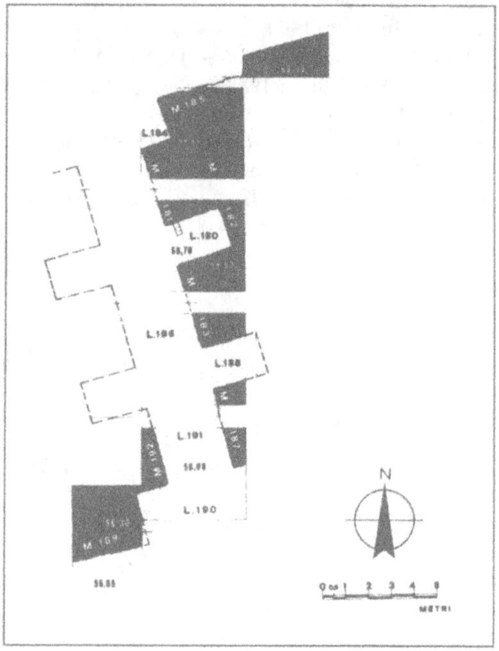

Figure 50. Gate in Area F (after Guardata 1990:fig. 3).

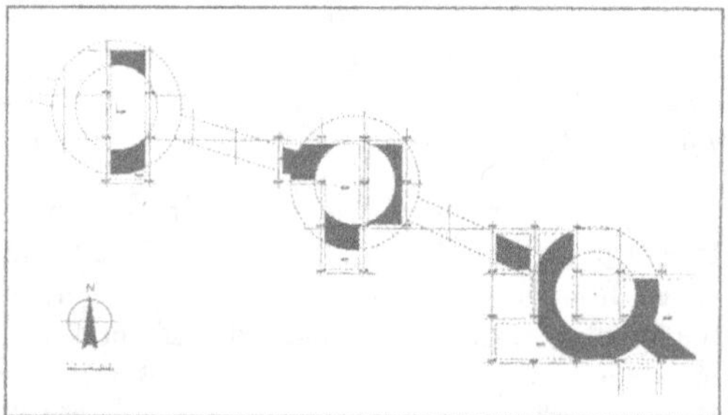

Figure 51. Towers in Area A (after Guardata 1990:fig. 4b).

Circular towers were added to the fortification wall of the lower town at even intervals along its length (Figure 51). The *fortification wall* (M.371/330/301) around the lower settlement was excavated in Area G to the NW of the gate in Area A. Along the 2 m wide wall three *circular towers* were identified with diameters of 9.4 m each spaced 14.5 to 16 m apart (Guardata 1990:65f.). These towers were constructed of mudbrick (40 x 40 cm) with walls 2.8 m wide which lacked stone foundations. The town wall and towers have been dated to the MB IIB–C. That the site was destroyed at the end of the MB IIB–C was evident in the excavations of Areas A and H (Matthiae 1989–1990:337f.) and also Area D (Matthiae 1982a:315).

TWEINI, TELL

UTM: 36 N 0765202.3917389
Refs.: Bretschneider et al. (2004).
Plans: Site (fig. 2).
Sects: Wall (figs. 3, 15–16).
Photos: Wall (fig. 17).
Tell Tweini, identified with ancient Gabala, is located 1.7 km inland from the Mediterranean along the Rumailiah River about 30 km S of Lattakia (Bretschneider et al. 2004:215). Fortifications were exposed in 2001 and 2002 in Area C on the E side of the site. These are thought to date to the Middle Bronze Age, presumably MB IIB–C, and may have continued to function during the Late Bronze Age. The remains consist of a 100 m stretch of cyclopean masonry preserved to a height of more than 5 m (ibid., 226). This feature is probably to be identified as a *revetment wall.*

UGARIT (RAS SHAMRA)

UTM: 36 N 0752182.3943409
Refs.: Schaeffer (1939; 1951); Lagarce (1984); Callot (1986); Yon (1992; 1997a); Calvet (2006).
Plans: NW fortifications (Yon 1997a:fig. 15); Postern and bent-axis gates (Yon 1997a:fig. 18a); Gate (Yon 1997a:fig. 18b).
Sects: NA.
Photos: NW fortifications (Schaeffer 1939:pls. 42f.; Lagarce 1984:figs. 5, 8–17; Yon 1997a:fig. 16); Postern interior detail (Schaeffer 1939:fig. 13; 1951:pl. 4).
Ugarit (Ras Shamra) is located on the coastal plain of the northern Levant 140 km SW of Ḫalab and about 165 km N of Byblos. In the absence of definitive evidence concerning the Middle Bronze Age fortifications at Ras Shamra and in light of the possibility that the earliest phase of its fortifications may date to the MB IIB–C (Yon 1997a:41), it is worth examining the evidence for the Late Bronze Age fortifications of Ugarit as known from excavations since 1938. Exploration of the fortifications from 1938 to 1939 was limited to the NW corner of the site around the royal palace. There Claude Schaeffer exposed the well-preserved remains of a monumental gate as well as a postern gate. With the conclusions of a later study by Jacques Lagarce (1984), it is now possible to better understand the results of the original excavation, although many questions must remain unanswered until further excavation of the defenses in other areas can be conducted.

Figure 52. Glacis at Ugarit. Photo by A. A. Burke.

MB IIB–C/LB IA

The construction of Ugarit's fortifications resembles many of the Middle Bronze Age settlements along the Levantine coast particularly with respect to the use of stones in the construction of its glacis (Figure 52; cp. Byblos, Biruta, Ashkelon). This construction is probably the combined result of the availability of these materials in the region,[34] the importance of Ugarit as the capital of its kingdom, and the need to keep the earthen rampart from eroding. That the underlying *earthen rampart* is MB IIB–C in date has been recently confirmed based upon the ceramics recovered by Claude Schaeffer from the sounding of the northern rampart (Calvet 2006:191).

Probably also dated to the late Middle Bronze Age if not LB IA (Lagarce 1984:173) is a *tower* built of dressed stones (ibid., 162ff.), which stood directly overtop of the path leading into the monumental gate. It is square and measures 14 m on a side (Schaeffer 1939:289).

[34] A quarry, which probably served as the source for most of the stone used in construction at Ugarit, has been located to the north of the site (Yon 1992:23).

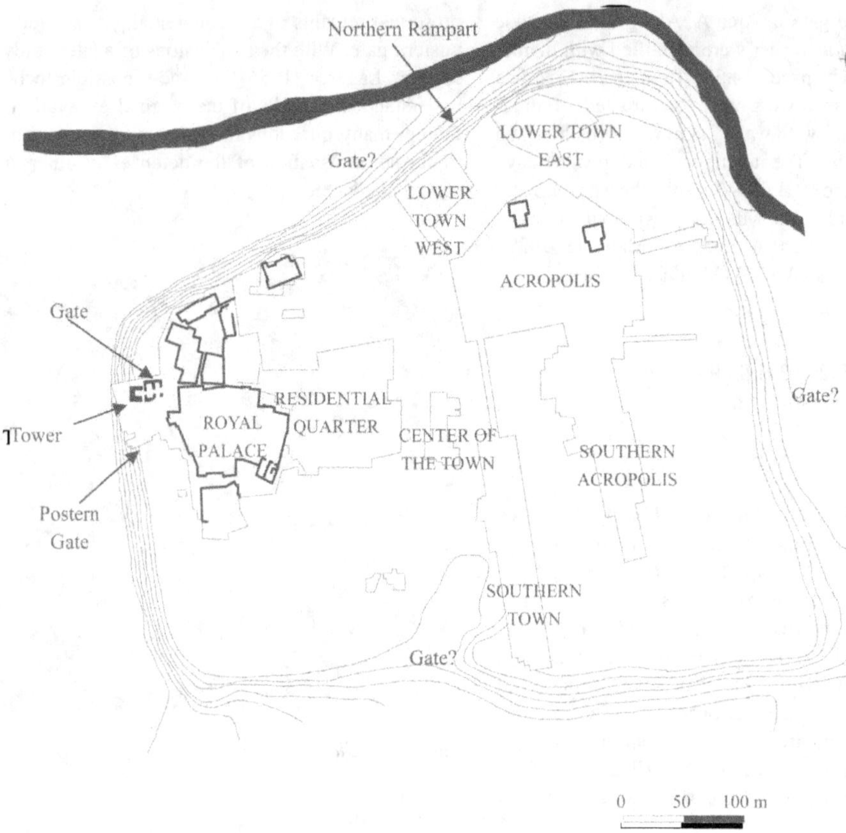

Figure 53. Plan of Ugarit (Ras Shamra). Reprinted by permission (Schloen 2001:fig. 18).

LB I (15TH–14TH CENT. B.C.)

The effort invested in Ugarit's Late Bronze Age fortifications is nowhere more evident than in a second phase of its fortifications which included a stone-built *glacis* (see Figure 52), presumably constructed after the tower and vaulted postern gate (Lagarce 1984:162f., 73). An examination of the face of the glacis reveals, as Lagarce notes, that larger blocks, often more than 1.4 m long, were used in the lower part of its construction while smaller blocks were used in the upper courses (1984:156f.). The slope of the glacis was 45° (Schaeffer 1939:289).

Because the *postern gate* (see Figure 54) was built into the face of the glacis it was probably contemporary with the glacis, and can therefore also be tentatively dated to the fifteenth to fourteenth centuries B.C. (Yon 1997a:41; after Lagarce 1984:173). This structure was corbelled (e.g., Akko) like the roofs of several tombs found below houses at Ugarit, and its ceiling was about 5 m high (Schaeffer 1939:290).

During the final phase in the development of the fortifications around the royal palace, dated to the thirteenth or early twelfth century, the postern was closed (see pls. V:1–2 in Schaeffer 1951) and buried (ibid., pl. IV:4), and the *gate* leading into the NE quarter was modified (Lagarce 1984:167–73). The modifications included a change in approach with a ramp built to lead up from the S, passing through a small tower before turning right into the main gate—if the reconstruction presented by Marguerite Yon is correct (1997a:fig. 18b). The plan of the monumental gate of this period included three, or possibly four, sets of piers (Lagarce 1984:171), but it was not regular in plan like typical six-pier gates. The gate was about 3 m wide at its narrowest point (ibid., fig. 7). Aside from the W gate, Ugarit probably also featured at least one other gate on the S located in the depression evident on contour plans of the site (Yon 1992:26). This is presumably where most merchandise was brought and access toward the sea was obtained. The wall in this period was a *casemate* structure (Lagarce 1984:171).

Figure 54. Postern gate at Ugarit. Photo by A. A. Burke.

LB II (13TH–12TH CENT. B.C.)

Though estimates of the site's size suggest that Ugarit was about 20 ha during its final phase of occupation at the end of the Late Bronze Age, measurements from the top plans suggest that the site was actually around 22.11 ha above its slopes (see Yon 1992:23). However, as Yon observes, much of the N part of the site has been destroyed, and it is not possible to accurately determine how much of the area of the site was occupied by the wall during the Middle or Late Bronze Age. Although the perimeter of the tell at the top is currently 2,225 m long, the wall and rampart around the site must have been longer because some of site has eroded since 1200 B.C. on the N by the Nahr Shbayyeb (ibid., 20). The site was also flanked on the S by the Nahr ed-Delbeh. These two rivers no doubt aided the defense of Ugarit. By the end of the Late Bronze Age the site stood 20 m above the surrounding plain (ibid., 23).

UMM EL-MARRA

UTM: 37 N 0382483.3999608

Refs.: Tefnin (1979b); Warmenbol (1980); Curvers and Schwartz (1997); Schwartz et al. (2000; 2003; 2006).

Plans: Site (Schwartz et al. 2000:fig. 2); NW Area (Schwartz et al. 2003:fig. 27)

Sects.: W Area A (Schwartz et al. 2000:fig. 9).

Photos: Area F (Tefnin 1983:pls. 4f.); W Area A (Schwartz et al. 2000:fig. 10); NE Area glacis (Schwartz et al. 2003:fig. 25); Acropolis N gate (ibid., fig. 26).

Umm el-Marra (ancient *Tuba*?) is located on the road to the Euphrates Valley 45 km ESE of Ḫalab. During the Middle Bronze Age (Per. IIID–IIIA) the site was approximately 25 ha in size (20 ha inside the ramparts) and it appears to have been the largest settlement in the Jabbul between Ḫalab to the W and the Euphrates River (Schwartz et al. 2006:603, n. 2).

EB III

During the late EB III (ca. 2500 B.C.), after the site's initial settlement, an *earthen rampart* was constructed around the site (Schwartz et al. 2000:426). This may have given Umm el-Marra its original circumvallated shape. Although the exact date of the rampart remains uncertain (G. Schwartz, personal communication, 2004), it was composed of brown soil and alternating layers of ash deposited at a 45° angle. The only useful material for establishing the rampart's date comes from a pottery kiln dug into the rampart that contained Early Bronze Age wasters (Schwartz et al. 2000:426).

EB IVA–B: PERIODS V–IV

Following the rampart's construction, a "red" *glacis* of reddish-brown earth, probably taken from outside the site creating the *fosse* found there, was built against its exterior surface (i.e., to the W) during the EB IV or early Middle Bronze Age (Schwartz et al. 2000:427). The front edge of this glacis featured "tongues" of clay protruding into the glacis (ibid., fig. 6). The glacis was held in place by a *revetment wall* of boulders 1.4 m high and 1.6 m wide (ibid., 427). Not much later the "white" *glacis* composed of limestone chips and pebbles was added outside the earlier "red" glacis. Although Schwartz and Curvers suggest that the stone structure under the base of the "white" glacis is to be identified as another revetment wall (ibid.), its dimensions and position are different than the revetment wall of the "red" glacis, suggesting a different function. By contrast this wall measured 2.3 m high and 5.3 m wide and featured a flat surface on top (ibid., fig. 9), which was overlaid by the W end of the "white" glacis.

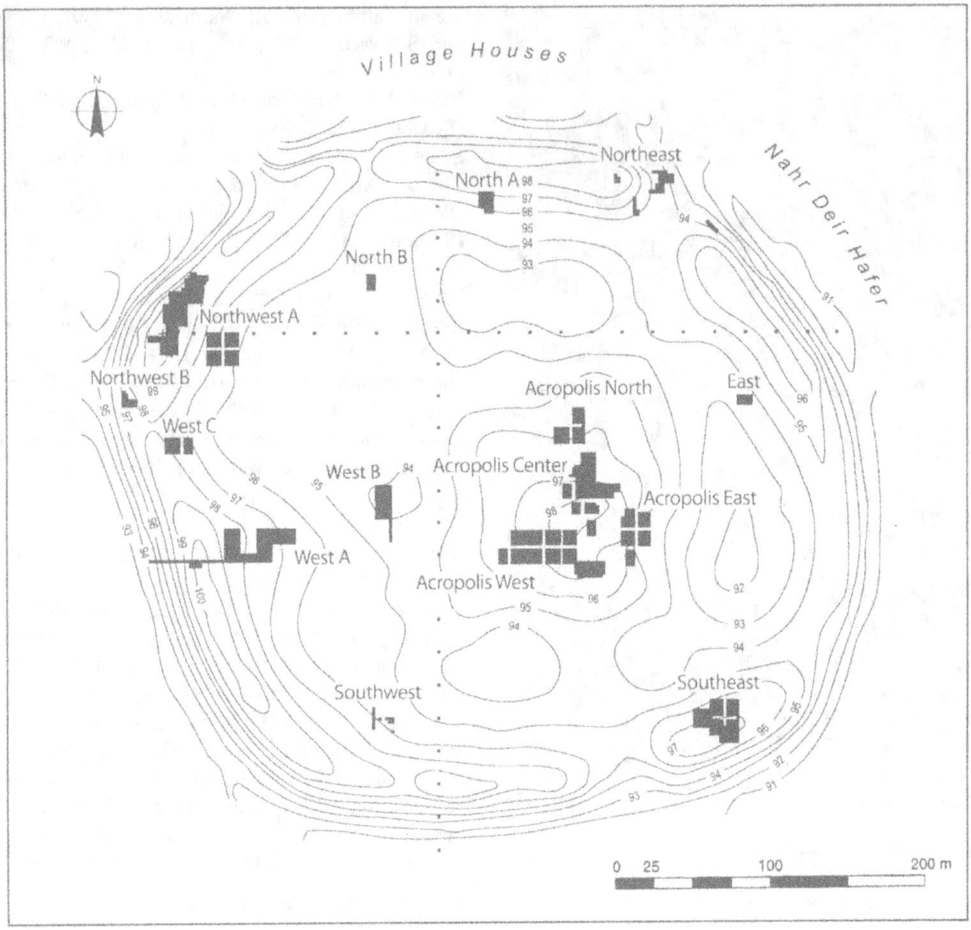

Figure 55. Plan of Umm el-Marra (Schwartz et al. 2006:fig. 2). Reprinted by permission.

While it only remains possible to date the sequence of fortifications after the Early Bronze Age rampart to the period from EB IV to the start of the Middle Bronze Age (ibid., 427), further excavation will help to clear up details concerning the phasing of Umm el-Marra's fortifications. Nevertheless, given the absence of EB IVB sherds in either of these glacis it seems probable that their construction should be dated to the late EB IVA.

MB IIB–C: Period IIIC–A

Following the MB IIA (Per. IIID), during which no new fortifications were built, the site's fortifications underwent construction again during the MB IIB–C. On the highest point in the NW Area just below the surface the remains of a mudbrick *fortification wall* dated to the late MB IIB–C standing 1.4 meters high and about 1.5 to 2 m wide was exposed during the 1999–2000 season (Schwartz et al. 2003:342). This wall was built on top of the so-called "red glacis" of the EB IV with 4.5 to 6 mudbricks wide above a stone foundation. It also featured thick buttresses along its exterior face.

On the W side of Area A the remains of a 7 m wide MB IIB–C *fortification wall* were also exposed, featuring alternating courses of flat lying gray bricks and vertically standing red bricks (Schwartz et al. 2000:426, fig. 9). The 7 m of red mudbrick standing on end and overlying the gray bricks below were originally thought to have been the collapsed remains of the mudbrick wall (Curvers and Schwartz 1997:215). It appears that this mudbrick wall may have been preceded by a stone wall since excavations in W Area A revealed that the foundation of this mudbrick wall cut into an earlier stone wall (G. Schwartz, personal communication). If the MB IIB–C wall traced the top of the embankment around the settlement, then it would have been approximately 1,350 m long.

Tefnin had previously speculated that at least three of the town's gates could be identified based on the topography of the site (1979b:72): (1) the NW Gate to Ḫalab, (2) the NE Gate to the Euphrates, and (3) the S Gate to the Jabbul. The NE Gate, referred to as the "Euphrates gate," was explored by Tefnin (1983) and remains the only gate excavated thus far. The gate's structure was preserved to about 2 m high with a passageway about 3 m wide and 7 m long (ibid., 143). More evidence of the EB IV and Middle Bronze Age fortifications and one of the Middle Bronze Age gates is also coming to light in the NE area of the present excavations (Schwartz et al. 2000:429f.).

In addition to the settlement's outer defensive wall and gates, the acropolis in the center of the site was also surrounded by a separate set of fortifications in this period (Schwartz et al. 2003:341–42). Here a wall and gate were excavated on the E and N sides of the Acropolis. The *citadel fortification wall* appears to have been built of four rows of mudbricks to about 1.5 m wide with buttresses on its interior face, which were about four bricks wide. At least two phases can be distinguished in this wall's construction on the E side of the acropolis. On the N side of the acropolis the wall featured a stone foundation, evidence of a *tower*(?), and a clay *glacis* outside the wall. Schwartz suggests that the acropolis wall was short-lived and functioned only during the MB IIB and was afterward built over.

A direct axis two-pier *gate* was built on the N side of the acropolis into the Period IIIC–A wall, which was also revealed to its W (Schwartz et al. 2003:342). The gate was flanked by two piers, which were about 3.8 m wide, which were protected with a stone facing. A stone-slab threshold lined the passage between these towers, which was between 1.2 and 1.4 m wide.

The site appears to have lacked a defensive wall during the Late Bronze Age (Schwartz et al. 2000:426).

B. The Southern Levant: Israel, Jordan, and the Palestinian Territories

Figure 56. Sites in the southern Levant discussed in catalog 2.B

ABEL BETH-MA'AKHA* (T. ABIL AL-QAMH)

MR 2046.2960
Refs.: Dayan [Nevo] (1963); Dever (1986a).
Plans: NA.
Sects.: NA.
Photos: NA.

Abel Beth-Ma'akha is located on a rise 80 m above the N end of the Huleh Valley about a half kilometer E of the Nahal 'Ijon (Kotter 1986:112ff.). Although Dayan conducted some initial soundings on the mound (1963), the excavated remains have not been published. The site was, however, surveyed by William Dever (1986a), who notes that it was nearly 14.2 ha in size. It consists of an acropolis that rises approximately 10 m above the oval shaped lower town. Dever noted that a bulldozer cut on the E side of the mound revealed the Middle Bronze Age *earthen rampart* (referred to as a "glacis," ibid., 217). He suggests dating the "lower city, glacis, and major city walls" to the MB IIA, though he suggests that there was probably "considerable buildup of fortifications" during the MB IIB (ibid., 220).

ABU KHARAZ, TELL

MR 200623.206196
Refs.: Fischer (1998; 2000).
Plans: Site (Fischer 1997:fig. 1).
Sects.: NA.
Photos: Area 2 (ibid., fig. 6).

Tell Abu Kharaz, located 4 km to the E of the Jordan River and just W of Pella, was less than 12 ha in size during the MB IIC (Fischer 2000). During this period (Str. 2a–c) the site appears to have been fortified by a 4.2 m wide *fortification wall*, of which the stone foundation survives on the W side of the site in Area 2 (Fischer 1998:215). The fortification wall itself was constructed on the foundations of the EB IB/II (ca. 3250–3000 B.C.) town wall as witnessed in Area 1. The Early Bronze Age wall, which measured 3 to 5 m wide and featured stone foundations 3 to 4 m high (ibid., 452), was a part of the earliest settlement on the site. It was built directly on bedrock and continued in use through the end of the EB II, ca. 2700 B.C. (ibid., 448). On the inside of this wall casemate rooms were constructed (Fischer 2000:455), as well as perhaps a square *tower* with 1 m thick walls (1998:215).

The MB II wall was later replaced by an early LB I casemate wall (Fischer 2000:457).

ABU ZARAD, TELL (SHEIKH)*

MR 17195.16790
Refs.: Finkelstein (1988; 1993a).
Plans: NA.
Sects.: NA.
Photos: Site (Finkelstein 1988:fig. 45).

Tell Abu Zarad (biblical *Tappuah)* is located on a hilltop in southern Samaria about 13 km SW of Shechem (Finkelstein 1988:fig. 54). It was identified as a fortified MB IIC settlement (1993a:1313), approximately 2 ha in size (1988:152).

ACHZIB

MR 1598.2727
Refs.: Prausnitz (1973; 1975; 1993); Oren (1975).
Plans: Site (Prausnitz 1975:fig. 2).
Sects.: Area D (ibid., fig. 3).
Photos: Area D (ibid., pl. 24a–b).

Achzib is located on the Mediterranean coast about 5 km N of Nahariya. The Middle Bronze Age site was approximately 7 ha in size with an additional 5 ha area "consisting of anchorages" (Prausnitz 1975:202). While the site was occupied during the MB IIA as indicated by ceramic remains, the fortifications date to the MB IIB (Oren 1975). According to Prausnitz's assessment the site was very favorable because it was surrounded by vital defensive features (Prausnitz 1975:202). On the W it was bounded by the Mediterranean, on the N by the Nahal Keziv, and on the E by a fosse which ran from the Nahal Keziv S to a bay of the Shahal River (Figure 57). This, therefore, in one sense made Achzib an island-like settlement.

The Middle Bronze Age fortifications were explored in Areas D, E, and X (Prausnitz 1975:207ff.). They were best preserved in Area D where Prausnitz was able to identify all the major elements of Achzib's defensive system including the earthen rampart, fortification wall, glacis, revetment, and fosse. The *earthen rampart* (cp. "glacis," ibid., fig. 3:B), which formed the core of this defensive system was described by Prausnitz as a "sloping, tamped and smoothed earth wall" (ibid., 208). Although it is difficult to discern the slope of this earthen core from the top portion that was excavated, it does appear to have been about 30° rising to a point more than 9 m ASL, though it was no more than 18 m wide in Area D. The fill of the rampart was varied but consisted of several identifiable sources of material, which Prausnitz suggested were taken from the excavation of the fosse. From earliest to latest deposition these included "red fill . . . dune-sand . . . brown soil . . . grey soil with ashes . . . dune-sand . . . hardened red-brown soil." The surface of the rampart appears to have been protected by a *glacis* (identified as "stone packing," see 1975:fig. 3:C) made of a couple of courses of cobbles. The glacis transitioned into the *revetment wall* (ibid., fig. 3:F), which after 7 m assumed a nearly vertical slope and was 1 m thick on the inner slope of the fosse. The revetment wall and glacis appear to maintain the 30° slope of the underlying rampart.

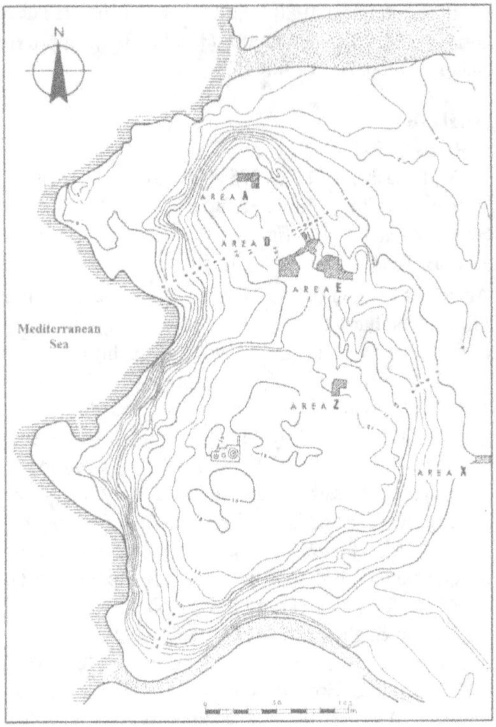

Figure 57. Plan of Achzib (after Prausnitz 1975:fig. 2).

The *fosse*, the bottom of which was never reached in Area D, may have been about 4 m deep and only about 2 to 3 m ASL at its base (Prausnitz 1975:207). It was also identified in Area X where it was found filled with pottery dating to the late Iron Age and later periods. Its width remains unknown though it was probably 475 m long from N to S. While Prausnitz has speculated that springs in the vicinity of the site, which may have fed into the fosse, turned the fosse into a moat and the site effectively into an island, without the excavation of the bottom of the fosse this suggestion cannot be substantiated. Furthermore, this argument is undermined because the fosse was at least a couple of meters above sea level and would not, therefore, have collected any significant amount of water.

Atop the rampart Prausnitz (1975:207) identified the remains of the town's mudbrick *fortification wall* ("brick core," ibid., fig. 3:A). Although he suggested that the earthen rampart was "heaped against" this wall, the published section betrays that this is not the case and that the wall was built on the interior slope of the lower layers of the original rampart with subsequent layers being added after the lower courses of the wall were constructed. The distance between the edge of the revetment and the wall was about 12 m. Because Prausnitz did not excavate to a depth which revealed the lowest courses of the mudbrick wall, his suggestion that "the 'brick core' could not have been part of a very large wall" because its base was "too wide" and the foundations were set "too low" (ibid., 208) cannot be accepted. This 6 m wide mudbrick core, which should definitely be identified as the settlement's fortification wall, is of comparable width to the wall identified at Jericho and is situated in a similar manner within the rampart as were the walls of other sites. The entire width of the fortification system (minus that of the fosse) was approximately 20 m rising at least 10 m above the landscape (ibid., 209). The curve of the fortifications around the E side of the site would have been approximately 475 m in length, and if the W side of the site featured at least a rampart and wall, this would have meant another 400 m of defenses.

Soundings in Area E, where the location of a gate was suspected, confirmed that the destruction of the settlement and its fortifications dated to the end of the MB IIC (Prausnitz 1975:207). According to Prausnitz excavations in Area E confirmed that the construction of the fortifications was a "quick execution" with the "building of the core and the erection of the earthen glacis" as a "simultaneous effort" (ibid., 209). The ceramics support this conclusion, suggesting that the fortification was completed during the first half of the MB II [or IIB] (Oren 1975).

'AJJUL, TELL EL- (SHARUHEN?)

MR 0934.0976
Refs.: Petrie (1931; 1932).
Plans: Site (Petrie 1931:pl. 45).
Sects.: Fosse (Petrie 1932:pl. 45).
Photos: Fosse (Petrie 1931:pl. 12).

Tell el-'Ajjul (ancient *Sharuhen*?) is located on the Mediterranean coast about 25.5 km SW of Ashkelon. Early excavation of Tell el-'Ajjul by W. F. Petrie (1930 to 1934) make it important in the discussion of Middle Bronze Age fortifications. Unfortunately, although 'Ajjul clearly possessed a rampart, fosse, and fortification walls, the site is of lesser comparative value due to the excavation methodology employed (see also comments in Parr 1968). Albright's analysis of the stratigraphy and remains (1938a) and the posthumously published work of James R. Stewart (1974) on the stratigraphy of 'Ajjul provide the most complete synthesis of the Middle Bronze Age defenses.

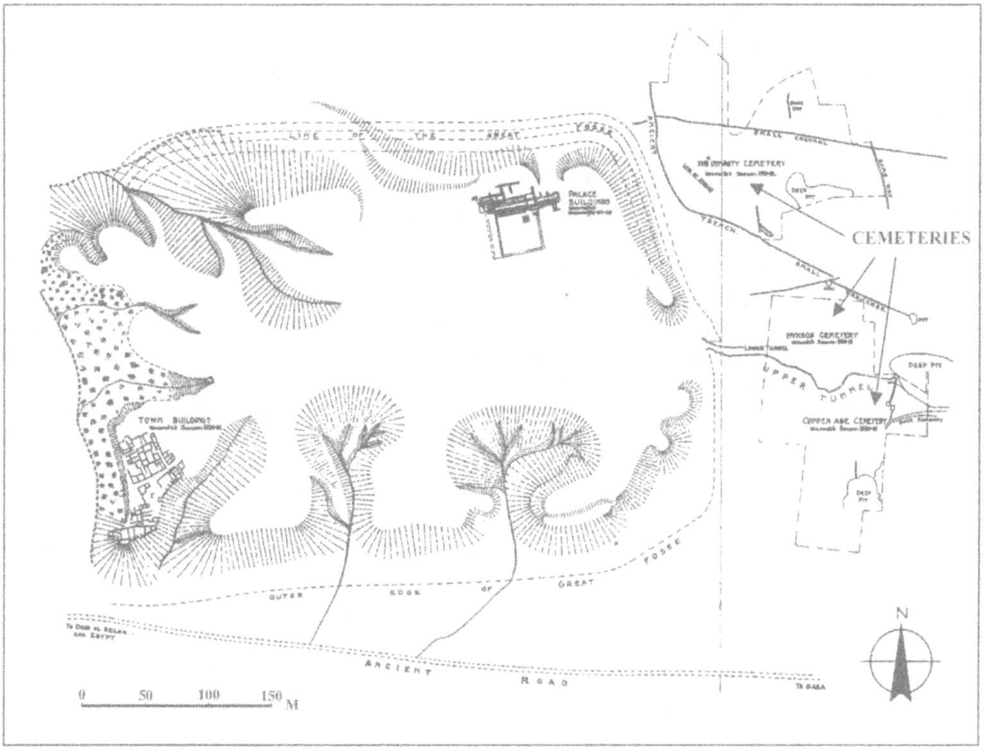

Figure 58. Plan of Tell el-'Ajjul (after Petrie 1931:pl. 45).

Recent work by Peter Fischer and Moain Sadeq has also sought to improve our understanding of the site, but thus far renewed excavation of the fortifications has not been undertaken (Fischer and Sadeq 2000). While Petrie had originally identified 'Ajjul as ancient Gaza, Kempinski suggested that the site should be identified with Sharuhen mentioned in *The Autobiography of Ahmose son of Abana* (*AEL* 2, pp. 12–14) and Josh 19:6. His identification, which seems considerably more plausible than the identification of Tell el-Far'ah South with Sharuhen, is based on several lines of evidence (see Kempinski 1974).

FORTIFICATIONS

From Petrie's excavation reports it is clear that the Middle Bronze Age settlement was enclosed on three sides by a *fosse* 18 m wide and 6 m deep that had been dug into the limestone (Figure 58 above).[35] The inclusion of Tomb 1166 in the E section drawing reveals that excavation of the fosse took place off of the NE corner of the site. The section is, therefore, a view of the SSE side of the cut. When the fosse was cut in antiquity it left a slope in the bedrock down the side of the mound, a flat bottom, and a more or less vertical face on the distant side of the fosse. LB I tombs were later cut into this outer, vertical face of the fosse.

Although Petrie does not provide the exact length of the fosse, according to the plan of the site it would appear to have been about 1,220 m long (Petrie 1932:pl. 41). Petrie originally noted that the fosse was of the "North Syrian type," similar to that at Homs, and that its quarrying provided most of the stone necessary for the foundations of buildings on the site (Petrie 1931:1ff.). The great length of the fosse is interrupted only on the NE side by a 7.62 m wide strip of bedrock that probably served as a causeway "edged with a balustrade of round-topped stones" leading out the gate toward Gaza. Although

[35]There is a lack of consensus regarding the average width and depth of the fosse at 'Ajjul. Measurements range from 10.7 to 15 m wide and 3.7 to 10 m deep (cp. Yadin 1963:67, 15 m wide x 10 m deep; cp. Tufnell and Kempinski 1993:50, only 6 m deep). These measurements may reflect the varied dimensions of the fosse throughout its length. Nevertheless, Petrie's plans support a depth of 6 to 7.5 m, a width at its base of 4.5 m, and a width of 18 m at the top (1932:pl. 45).

Kempinski suggested that the fosse was filled with seawater (1974:150), it is unlikely that it held standing water. Furthermore, it would have complicated exiting the site during peacetime leaving only the Gaza gate to serve the population of the 10 ha town. To this we might also add that LB I tombs would not have been added to the fosse if water was routinely present in it. It is impossible to establish a precise date for the digging of the fosse, but it was probably dug during MB IIB–C since there is no extensive MB IIA occupation of the site, and it occurred prior to the placement of the LB I tombs. Where the SW side of the mound was unprotected by the fosse, an estuary is thought to have abutted the lowest part of the rampart.

Although it is possible to state that an *earthen rampart* was built as part of the fortifications of Tell el-'Ajjul, the only part of it exposed was the *glacis* during excavation of the fosse (Petrie 1931:11). The settlement on the mound, due to the layout of the ramparts and the fosse, featured a rectangular plan (see Figure 58).

Although only limited traces of a *fortification wall* of black brick were found by Petrie in the vicinity of the Palaces, Stewart suggested that a mudbrick wall built on a stone foundation most likely rested on the inside of the perimeter created by the fosse (1974:11). This wall enclosed Palace I on the NE corner of the site during the Middle Bronze Age, but was built over—indicating that it had fallen out of use by the time of Palace III, identified by Albright as an Egyptian fort dated to the Eighteenth Dynasty. Unfortunately, neither brick sizes nor the thickness of the wall are mentioned in any of the reports and it is, therefore, difficult to propose a height for the remains of this wall.[36] Peter Fischer's assertion, based on his current reexamination of the site, is that the rampart and fosse were constructed during the MB IIA. However, within the typology of Middle Bronze Age fortifications proposed in this work and based on the limited evidence for the settlement's occupation during the MB IIA, the earthen ramparts were probably constructed at the start of the MB IIB.

No gates have been excavated on the site, aside from the one suggested on the NE side leading toward Gaza. However, another gate on the N side of the mound at the sunken area near the location of the palaces has been inferred by the presence of stone weights. These were suggested by Petrie to have been related to maritime trade that would have used a gate near this area (1934:13).

Siege Tunnels

Two interesting features present themselves as elements that may illuminate the nature of the defense of the site. These consist of two tunnels, referred to as the Upper and Lower tunnels (see Figure 58), which begin on the E face of the fosse on the NE side of the mound around the causeway that interrupts the fosse (Petrie 1931:11f.). Both tunnels run to the NE underneath the E cemetery. The so-called Upper Tunnel measures over 150 m in length ranging from 1.35 to 1.85 m high and 0.84 to 1.35 m wide, while the Lower Tunnel, substantially shorter, was only about 30.5 m in length.

Petrie had suggested that the tunnels were only used for LB I burials based on the pottery from Tombs 168 and 194, recovered from the Upper Tunnel (Petrie 1931:pl. 40). However, since only two tombs were added, and they were not located at one or the other ends of this tunnel—which might have indicated a progression of use—their excavation for this purpose seems unlikely. In both tombs the only published, datable pottery included two Egyptian drop-shaped vessels of Late Bronze Age date (cp. Amiran 1970:pl. 58). Without firsthand examination of the vessels a more precise date is not possible, but the Upper Tunnel seems to have gone out of use by the LB II (Stewart 1974:11).

Although Petrie had also suggested that the tunnels were dug from both ends, this cannot be verified and we can only add that the Lower Tunnel was clearly begun from the fosse on the N side of the causeway. The main argument against the use of the tunnels as part of a mortuary complex is that only two tombs were dug into the walls of the Upper Tunnel during the entire Late Bronze Age, which would not account for the considerable effort undertaken to excavate them. In an effort to understand the function of these tunnels, Tufnell has stated that they were "irrigation systems," comparing them to the Iranian *qanats*, suggesting that they predated the fosse (1993:50). This claim does not, however, make sense of the location of the entrances of the tunnels or the course of the Upper Tunnel. It also does not explain where the water was to be conducted. Since one can neither determine that the fosse was filled with water nor that one end of these tunnels ever connected to a source of water, it seems impossible to assert that these tunnels were dug for this purpose.

In light of the lack of consensus regarding the function of these tunnels, it is possible that these tun-

---

[36]Herzog suggests that there is no evidence for a wall at the site, disputing the identification of the wall near Palace I as a city wall since it does not continue to Area T on the S side of the mound (1997a:125). However, expecting the wall to be equally preserved in both areas is unwarranted.

nels represent an altogether different phenomenon. Several lines of evidence lead to the suggestion that these tunnels may date to the end of the Middle Bronze Age and thus relate to the Egyptian siege attested in *The Autobiography of Ahmose son of Abana* (*AEL* 2, pp. 12ff.).[37] This argument begins with the almost certain identification of 'Ajjul as Sharuhen (Kempinski 1974), which has been previously asserted due to its size, geographical position, and material culture. The length of the siege, three years, suggests that no serious effort was made to enter the town during that time (see Shea 1979). Instead, it is most probable that the Egyptians hoped to starve the inhabitants out. These tunnels also do not represent an Egyptian effort to enter the town. Further evidence of this comes from the Lower Tunnel, which was begun from the fosse and dug eastward terminating after only 30.5 m. Also, the attackers would most certainly have faced a slaughter in the fosse from archers on the wall above if they were forced into the fosse from this tight passage.

Therefore, considering that on the other end of this siege some effort was surely expended during the course of three years to secure supplies and assistance in the struggle against the Egyptians, these tunnels may be identified as the primary means of ingress and egress at that time. Such an effort would have been aided because the surface of the area to the NE of the site was pocked with Middle Bronze Age tombs that would probably have kept the Egyptian encampments outside this area, as well as the obvious need to stay outside the range of archers on the town wall (approx. 200 m). Both tunnels led out the NE side of the site from the causeway toward the nearest major settlement, Gaza, which was only 6 km away on the road leading N along the coast. Although it would have seemed prudent to have begun such an effort from inside the settlement, tunneling below the fosse, it must be remembered that the water table at 'Ajjul was rather high, as attested even during Petrie's excavations, and this would have hampered any effort to dig from within the town below the fosse to the other side. The next favorable location for a tunnel would have been outside the town walls but as close as possible to a gate, which corresponds to the location of the Lower Tunnel. Work here would most likely have been done at night where several workers could slip out the gate and down to the outer face of the fosse to work on its excavation.

This scenario suggests that the Lower Tunnel was begun first. While its location in the fosse suggests that it was the easiest and most clandestine of locations for the original tunneling attempt, this was complicated by the realization that from this location the "Hyksos" cemetery could not be avoided. While this does not on the surface seem problematic, when it is realized that many entrances existed to these tomb chambers and that these entrances would have allowed multiple points of access to the tunnel from the surface along its course, this would have increased the likelihood of the tunnel being discovered by the Egyptians. For this reason a new tunnel, the Upper Tunnel, would have been undertaken in an attempt to thread between the "Hyksos" cemetery to the N and the EB IV Cemetery 100–200 ("Copper Age" cemetery) to the S.

The conspicuous location of the entrance of the Upper Tunnel, directly off the causeway located outside the gate leading NE toward Gaza suggests that its construction was not concerned with continuous, daily traffic through this gate because there would have been no regular traffic during the siege. Evidence from the Upper Tunnel also attests to multiple places along its course where, evidently, "arrangements for blocking" were put in place during its use and ventilation shafts were added (Stewart 1974:11). The shafts shown in the illustrations seem to have been on the W end of the tunnel (Petrie 1931:12 and pl. 12), which would make sense since these would have kept the source of air progressing along with the diggers excavating eastward. These two pieces of evidence not only argue against the use of this tunnel for water management since air is not needed to conduct water, but also against its use as a catacomb. Such efforts, which are not attested in other subterranean construction projects, must surely be an indication of prolonged periods of intense labor by the inhabitants of Tell el-'Ajjul within this tunnel. And it is very likely that the blocking doors, if correctly identified, were intended to hamper efforts by the Egyptians to trace the tunnel had it been discovered. This interpretation accounts, therefore, for the entirety of the evidence available for these tunnels, the *terminus ante quem* date for their construction, and the historical events that precipitated the end of the Middle Bronze Age at Tell el-'Ajjul.

AKKO, TEL (TELL EL-FUKHAR)

MR 1585.2585

Refs.: M. Dothan (1976; 1993a); M. Dothan and Raban (1980; 1984); Raban (1991).

Plans: Site (Raban 1991:fig. 10); Area F (M. Dothan and Raban 1980:38; Raban 1991:fig. 7).

Sects.: Area B (M. Dothan 1976:fig. 5).

Photos: Aerial of Areas A and B (M. Dothan 1976:fig. 4); Area B (ibid., fig. 6; Raban 1991:fig. 3).

---

[37] According to Starkey, Petrie also suggested that these tunnels represented "a secret means of exit" (1934:169).

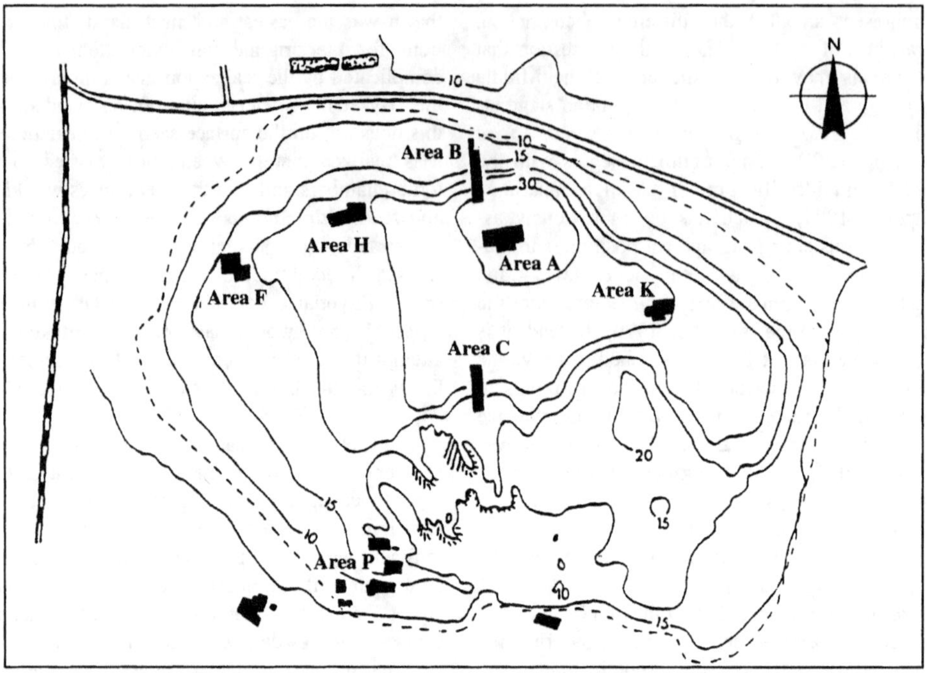

Figure 59. Plan of Tel Akko (after Raban 1991:fig. 1).

Tel Akko is located on the coast of Israel N of Haifa, just east of medieval Akko, and about 10 km S of Nahariya. Remains of the fortifications were found in Areas A, B, F, H, K, and P (Figure 59). While final publication of the fortifications remains forthcoming, several preliminary reports enable a cautious analysis of its defenses. These reveal that Akko was at no time unfortified during the Middle Bronze Age.

MB IIA

The first two ramparts, excavated in Area B, date to the MB IIA (M. Dothan 1993:18; Phase 2, see Cohen 2002a:70). The first *earthen rampart* was constructed of hamra, described as "beaten," which was covered by a *glacis*. The rampart was crowned by a 3.5 m thick "cyclopean" stone foundation preserved 2.5 m high and capped with a 2.5 m thick mudbrick *fortification wall* that has been exposed over a length of 20 m. The exterior face of the mudbrick wall was plastered. This wall is described as "built in segments with recesses between them" and featured a *tower* on its E side. The interior of the rampart in Area AB was reinforced by at least one *retaining wall*. The earliest rampart exposed in Area B may correspond with the earliest rampart found in Area F upon which the Sea Gate was later constructed, though no evidence of this same stone or mudbrick wall was found in Area F.

A second *earthen rampart* discovered in Area B was constructed of hamra, clay, kurkar gravel, and sand with a stone staircase running from the top of this rampart to the top of a mudbrick wall, which may be the wall of the acropolis or a fortress built on the summit (Raban 1991:20*f.). Based upon photographs of the ramparts in Area B the slope of this structure is estimated to be between 30° and 35° (ibid., fig. 3). The plan of Akko suggests that the original layout of the ramparts must have been elliptical and about 950 m in length (Figure 59).

The so-called Sea Gate in Area F on the NW side of the site was probably constructed upon the remains of the first rampart during the building of the second rampart (M. Dothan and Raban 1980; 1984). It is a direct access *four-pier gate* that was built of mudbrick and reinforced on its exterior by two stone walls projecting over 8 m along the route from the gate. The chambers of this mudbrick gateway, preserved to a height of 3 m, are identified as part of a two-story guard house. Two phases of use are suggested for this complex. The stone walls projecting outward from the mudbrick gate narrow from 4 m where they join the gate to 3.6 m at their NW end where with a small buttress they form an entrance less than 2 m wide. Four stone steps lead out of the town at this point, thus restricting its use to pedestrian traffic. Plastered mudbrick benches were found on the inside of the W stone wall indicating that people loitered in this space (3 x 6 m). There is no conclusive evidence that these stone walls were built later than the town wall and mudbrick gate since mud plaster also smoothed over the seams between the structures. Mud plaster was also found on a vertical projection of the W side of the mudbrick gate, which is thought to have served as a point of attachment for a wooden door jamb. Additionally, a 40 x 15 cm space in the floor of the inner entrance of the mudbrick gate is interpreted as the location of a wooden threshold. Although the argument is superfluous, the use of mud plaster is cited as the reason for identifying both spaces of the gate as covered.

Although both Raban and Dothan have identified the mudbrick structure adjoining the gate on the SW as a tower (4 x 4 m), nothing about its construction distinguishes it from that of a mudbrick *fortification wall*. Unfortunately, the width of the town wall adjoining the gate at this point is not known, as it has been damaged by later pitting, but it may have been about 4 m wide. All of these structures are dated to the MB IIA since no MB IIB–C sherds were found in their construction.[38]

MB IIB

At the end of the MB IIA the gate in Area F was filled with sand in advance of the construction of a third *earthen rampart*. This rampart has also been identified to the N of Building A in Area B and in Area P on the S side of the site (Raban 1991:29*). In Area B this rampart appears to have been faced with black mudbricks (see section, M. Dothan 1976:fig. 5), which should probably be identified as a separate *glacis*. Dothan suggests that the site's proximity to swamps and the Na'aman River may have obviated a rampart on the S prior to the building of this rampart. However, this assertion cannot be tested. Evidence from Area P indicates that this area was enclosed through at least the final phase of the Middle Bronze Age. Here a stone-built corbel-vaulted *postern gate* with stone slab floor was discovered. It is dated to the MB IIC by sherds collected from its floor. Dothan identifies this structure as being "apparently used for drainage," a suggestion probably based on its interior dimensions (1.6 m high and 0.6 m wide) (M. Dothan 1993a:19f.). Nevertheless, these dimensions are sufficient for passage of people (cp. postern gate at Ugarit).

[38]W. G. Dever dates the construction of the Sea Gate to the beginning of the MB IIB (1997a:54). However, given the presence of two ramparts dated to the MB IIA and the fact that the gate was constructed atop the first rampart, it seems unlikely that the Sea Gate dated to the start of the MB IIB.

In Area B inside the fortification line of this sand rampart a mudbrick *bastion* identified as Building A was constructed (M. Dothan 1993a:18). This building, the E end of which was not excavated, was preserved to a height of 3 m, and is thought to have been originally two stories high. Its original dimensions are estimated to have been 15 x 15 m (Raban 1991:21*).

LATE BRONZE AGE

Despite Akko's impressive Middle Bronze Age defenses the last rampart and wall seem to have fallen into disrepair during the LB I when they were leveled for residential construction (Raban 1991:29*–31*). However, during the LB II another *earthen rampart* was constructed of sand but without a wall (M. Dothan 1993a:21).

AMMAN (RABBATH-AMMON)

MR 237.151
Refs.: Dornemann (1983); Northedge (1983); Zayadine et al. (1989); Greene and 'Amr (1992).
Plans: Site (Northedge 1983:438); Area III (fig. 5); Area A (Zayadine et al. 1989:fig. 3).
Sects.: Area A (Zayadine et al. 1989:fig. 4).
Photos: Area III (Dornemann 1983:figs. 8).

Amman (biblical *Rabbath-Ammon*) is located on the Transjordanian plateau up the Wadi Sir from the Jordan Valley. The Middle Bronze Age ramparts belonging to its fortifications were discovered below the Roman wall in Area A on the SE side of the Lower Terrace in 1968 by the Department of Antiquities. Rudolph Dornemann also made soundings of the *earthen rampart* in Area III in 1969 (Dornemann 1983:19). He suggests that the structure was a fort comparing it with Ta'anach. However, verification of these observations must await final publication of these excavations. Further exploration of the rampart in 1988 by Zayadine, Humbert, and Najjar also led to the identification of the rampart in Area A of the Lower Terrace to the E of the Hercules Temple (Zayadine et al. 1989:359). The rampart here consisted of an earthen fill about 1.5 m thick covered with a 0.3 m thick layer of chipped flint. This fill contained Chalcolithic and EB II–III sherds, indicating that the structure must be of EB III or later date.

The rampart is retained by a *revetment wall* (2015) at its foot, 1.6 m wide with a preserved height of 1.4 m. The rampart appears to have been topped by a *fortification wall* with a stone foundation (2005) averaging 2 m in thickness and preserved to a height of 1.85 m. Between these walls and covering the rampart's surface a cobble (huwwar) *glacis* was added, which contributed to its 40° slope. Ceramics taken from both walls dated to the MB IIB–C. The full height and width of the rampart were not established.

The Middle Bronze Age fortification wall may also have been encountered in Deep Sounding M 26–27 on the N side of the Lower Terrace (Greene and ‘Amr 1992:116). However, with little information regarding its dimensions it is not possible to verify the suggestion that this feature was “part of a tower or gate” (ibid., 117).

No evidence is available for any Late Bronze Age fortifications at Amman.

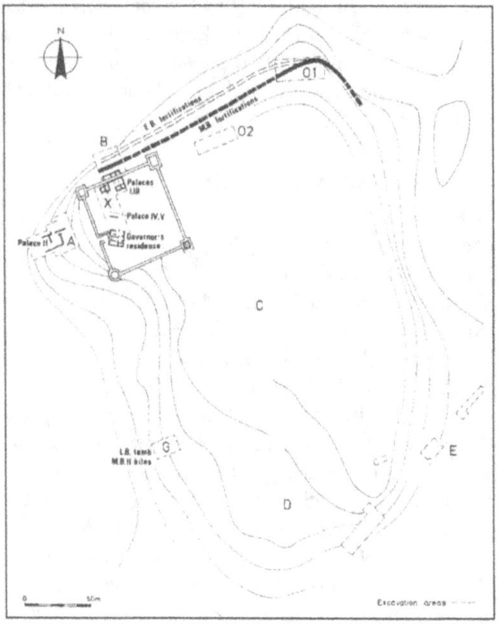

Figure 60. Plan of Aphek (after Kochavi et al. 2000:fig. 1.5).

APHEK, TEL (TELL RAS EL-‘AIN)

MR 1435.1678
Refs.: Kochavi et al. (2000); Gal and Kochavi (2000).
Plans: Site (Kochavi et al. 2000:fig. 1.5); Area B (Gal and Kochavi 2000:figs. 7.15, 7.26, 7.30, 7.39, 7.47).
Sects.: Area B (Gal and Kochavi 2000:fig. 7.47).
Photos: Area B (Gal and Kochavi 2000:fig. 7.2, 16–18, 21).
Tel Aphek is situated in the Plain of Sharon W of the hill country on the N bank of the Yarkon River about 17 km NE of Jaffa (Figure 60). In the first stage of fortification (Area B Stratum Vd) a 2.25 to 2.5 m wide *fortification wall* (C250) was constructed with buttresses (346 and 381) located 15 m apart along its exterior face (Gal and Kochavi 2000:71ff.). A stone foundation of varying height laid in a trench was necessary to provide a level surface for the building of this wall. The wall was also supported on its exterior elevation by a small *earthen rampart*, a part of which was exposed around the W buttress, with layers divided by thin plaster surfaces (ibid., 71f.). A drain (L.1405) was put through the wall “no later than Phase Vb” (ibid., 73). The erosion of wall C250 led to the deposition of mudbrick detritus (L.372) down the slope. According to Gal and Kochavi “the wall suffered repeatedly from landslides and other erosive factors and was regularly repaired” (ibid., 85). “Localized repair or rebuilding of the city wall” represents a later phase of the wall (C261), though its construction was identical to the original wall (ibid., 85).

After an initial period of unfortified occupation associated with the Pre-Palace phase in Area A, a fortification wall and *earthen rampart* were constructed (for relative chronology of occupational phases, see Table 3). While the absence of the town wall in Area A makes it difficult to directly link the MB IIA palace phases with the wall remains in Area B, it seems that the earliest fortification wall (C250) was built in the second phase of MB IIA, which is the Palace I phase in Area A. A later rebuilding of the wall (C261) should be assigned, therefore, to the third or Post-Palace phase (Area A).

The circumference of Aphek’s MB IIA fortifications was approximately 1,330 m enclosing a site of about 10 ha (Broshi and Gophna 1986). The wall fell out of use during the MB IIB–C.

ASHDOD, TEL

MR 1178.1295
Refs.: M. Dothan (1982; 1993b).
Plans: Site (M. Dothan 1982:plan 1); Area G (M. Dothan 1993b:plan 1), Schematic of Gate (ibid., plan 2).
Sects.: Area G (ibid., plan 1).
Photos: Area G (ibid., pl. 1).
Ashdod is located just inland from the Mediterranean Sea about 15 km NNE from Ashkelon. Dothan noted that the Middle Bronze Age fortifications of Ashdod are significant because they provide an example of a late foundation of a fortified town at the end of the MB IIC (M. Dothan 1973). The foundation of Ashdod, Stratum XXIII, represents the earliest architecture and the first fortified settlement of the site. Two successive phases of Middle Bronze Age occupation have also been identified, Strata XXII and XXI; however, neither featured any evidence of fortifications (1993b).

The fortifications of Ashdod consist of evidence of a gate on its N side in Area G along with the remains of a mudbrick fortification wall and earthen rampart associated with this gate. A *four-pier gate* was revealed in

Area G. It was constructed during Phase B of Stratum XXIII on the N side of the site (1993b:19). The structure, which measures 6.5 m along its passage and 8.2 m wide, was positioned on a NW line. Within the gate the width of the chambers opposite the passageway measured 4.7 m, although between the piers the passage was only 2.9 m wide. The four piers were on average 1.65 m wide (along the passage). The gate and its piers were constructed directly on bedrock with mudbricks measuring 50 x 30 x 20 cm. The 80 cm projection of a mudbrick wall from the E tower of the gate (W.4322) may be identified as the means by which the wall along the rampart was anchored to the gate. If this was the case, then the *fortification wall* can be estimated at over 2 m wide. During Stratum XXIIIB the gate may have fallen into disrepair but was later restored, as Dothan suggests (ibid.), though at most the evidence seems to indicate that the floor level simply rose with architectural changes to the gate.

Associated with the gate were the badly eroded and robbed-out remains of an *earthen rampart*. This rampart (L.4385, L.4405, and L.4414) was composed of soil and like the gate was built directly on bedrock. Because of the rampart's condition it is impossible to determine the slope of the rampart's surface, its full height, or its width. Nevertheless, its preserved height, which was only 0.4 to 0.9 m, suggests that it was never an impressive feature.

Since the remains of the gate area are the only part of the MB IIC fortifications thus far identified it is difficult to extrapolate either the course of Ashdod's defenses or the size of the settlement in this period, but it is possible that the site encompassed only the western of the two hills at Ashdod (see plan 1 in M. Dothan 1982). Based upon this suggestion and the observation that the fortifications are positioned along the 45 m contour line, the site was at most around 5.3 ha and no more than approximately 230 m square (cf. Broshi and Gophna's estimate of 8 ha, 1986); thus it had a circumference of approximately 920 m in length.

Dothan suggests that the Late Bronze Age gates (Str. XIX to XIV) on the N side of the site were completely eroded (1993b:9).

ASHKELON, TEL

MR 107.119

Refs.: Stager (1991a; 1993); Wolff (1991:507); Schloen (1997); Voss (2002).

Plans: Ph. 14 (Voss 2002:fig. 1); Ph. 13C (ibid., fig. 2); Ph. 13B–A (ibid., fig. 3); Ph. 12 (ibid., fig. 4); Ph. 11 (ibid., fig. 5).

Sects.: NA.

Photos: Site aerial (Stager 1991a:5; 1993:104); Gate (Schloen 1997:fig. 1); Rampart (Stager 1991a:2).

Ashkelon sits on the coast of the Mediterranean 16 km N of Gaza.[39] The site today consists of a N mound identified as the North Tell, the site's 1,900 m long semi-circular earthen ramparts, and a mound along the shore that is nearly centered within these ramparts that is identified as the South Tell. Although the nature of third-millennium occupation of Ashkelon remains uncertain, substantial evidence of its fortifications from the beginning of the second millennium (Per. XXIV–XX) through the Fatimid period is available as a result of excavation of the Middle Bronze Age gate in Grid 2 on the NE side of the North Tell (Figure 61 below). Grid 2 consists of squares covering an area of more than 2,500 m^2, which is delimited by the SW corner of square 84, the NW corner of square 46, the NE corner of square 50, and the SE corner of square 88.[40]

While multiple phases of the Middle Bronze Age gate were exposed during excavations from 1990 to 1998, a section was also cut through the rampart and its associated glacis N of the gate. A considerable length of the fosse to the N of the gate was also revealed. The phases of the gate, rampart, and fosse were assigned to Phases 14 through 10 of the cultural sequence in Grid 2, which represent continuous Middle Bronze Age occupation of this area from MB IIA through the MB IIC (see Figure 61).

EARLY MB IIA: PERIOD XXIV, PHASE 14

Phase 14 (late Dyn. XII) represents the earliest phase of Ashkelon's fortifications (Voss 2002:379ff.). Although our picture of these fortifications is incomplete and the construction date of this gate remains uncertain (owing to the preservation of the superstructure of the gate of Phase 13C), it is still possible to remark upon three aspects of the fortifications of this period: the fosse (referred to as the "moat"), the rampart, and the gate (Figure 62 below).

A *fosse* and a ramp across the fosse were excavated outside the gate in square 56 of Grid 2. The fosse (F.21), of which only 13 m of its 70 m length were excavated, was cut into the kurkar bedrock as a specific modification of a 20 m wide artificial gap that had been cut into the kurkar ridge that parallels the coast. The fosse drains toward the NW, toward the sea, and was about 5.5 m deep and almost 9 m wide.

---

[39] I would like to thank Lawrence Stager for allowing me to use the records of the Leon Levy Expedition to Ashkelon for compiling this entry.

[40] The description of the fortifications here uses a true north orientation, although the discussion by Voss (2002) refers to north in relation to the Ashkelon grid system, approx. 35° off of true north (see Figure 61).

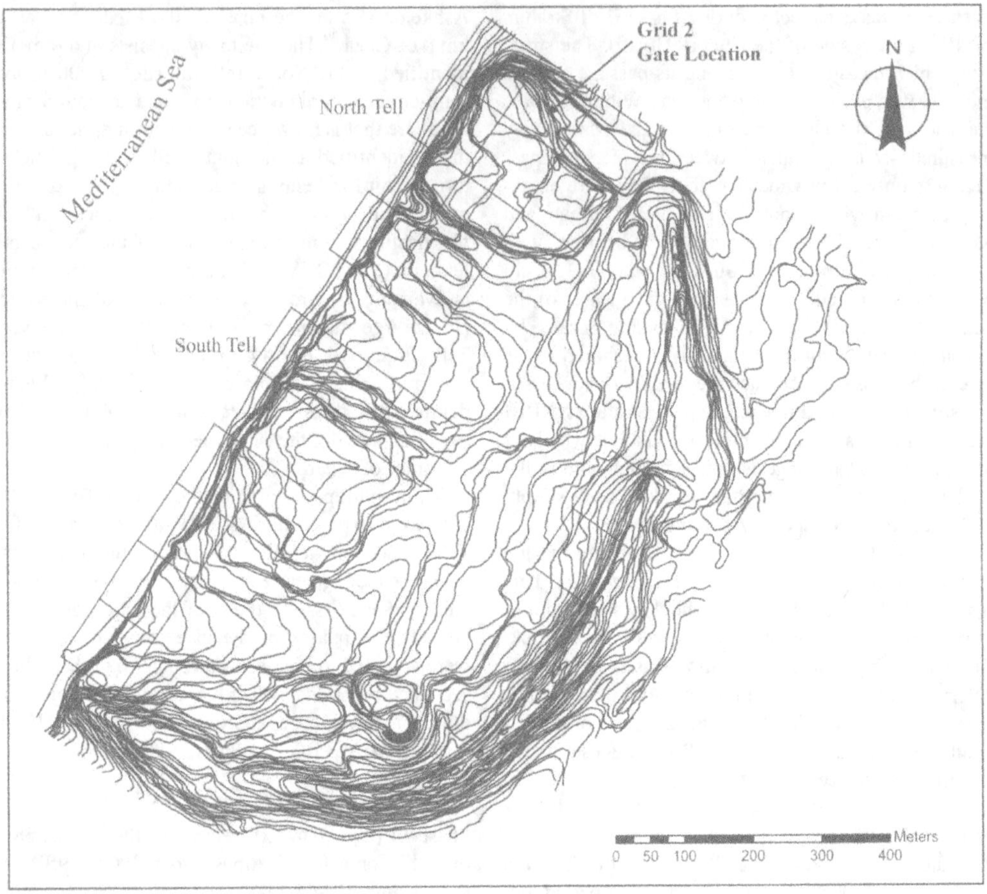

Figure 61. Plan of Tel Ashkelon. Courtesy of the Leon Levy Expedition.

The fosse terminates in a rounded end to the N of the approach road, formed by a ramp of residual bedrock that was aligned with the gate, thus enabling a direct axis approach. There is no evidence of any drainage under this ramp and the continuation of the fosse to the S of the approach ramp has not been exposed, though it is thought to have continued in this direction. As at other sites the fosse is presumed to have been the source of much of the kurkar limestone used in the construction of the "terraces, banked fills, and glacis capping of the rampart" (Voss 2002:379).

The *gate* (ibid., 380f.), which was aligned to the E (ca. 81° off N) throughout its many phases, was 20 m long and 12 m wide in Phase 14. In its original state it featured a total of four mudbrick piers with stone foundations located at both ends; the exterior (E) piers (F.187 and 186), of which only the lowest stone course was preserved, were approximately 1.82 m wide protruding 0.65 m, while the interior (W) piers (F.113 and 124) were slightly wider, 2.8 m, but protruded less than 0.5 m. The narrowest point, between the E piers, was 2.3 m wide (ibid., 381), while the space between the W piers was about 2.5 m (ibid., 380).[41]

The gate's interior entrance opened onto a square space (7.1 x 7.1 m) with a beaten earth surface (F/L.59) before leading into the town. This space was flanked by large mudbrick walls F.71 on the N and F.70 on the S, both of which were of comparable width to the W piers. Wall F.198 and Wall F.195 on the N and S sides, respectively, formed the walls of the gate's passage. Although the passage's ceiling was not preserved, it was presumably covered by a barrel vault of kurkar cobbles coated with mud, the lowest courses of which were preserved (ibid., 381). The passage measured 3.6 m wide and was 9 m long. Its floor (F/L.192)

[41]The width of the passage would seem to suggest that the gate did not accommodate chariot traffic (see pp. 71f.).

and the gate's foundation walls were built upon a fill of ash and silt (L.197, L.196, and L.193). Mudbricks used in this phase of the structure were consistent in size and measured 40 x 35 x 10 cm (ibid., 380).

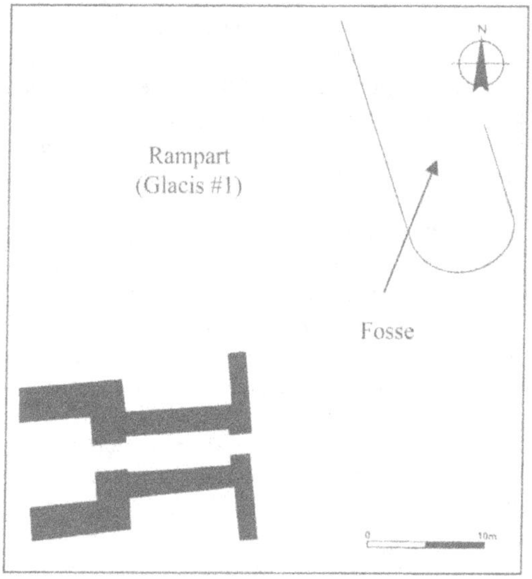

Figure 62. Phase 14 gate (after Voss 2002:fig. 1).

The *freestanding earthen rampart* associated with the Phase 14 gate was identified as Glacis #1 (Voss 2002:380). It was composed of sloped layers of kurkar rubble and sand and may have covered the vaulted passageway of the gate E of the mudbrick gateway. The *glacis* that sealed the rampart was composed of mudbricks. If the rampart's center was aligned with the center of the gate's passage and its foot rested at the edge of the fosse, then it was approximately 40 m wide.

MB IIA: Period XXIII, Phase 13

The fortifications of Phase 13 (early Dyn. XIII) are divided into three sub-phases (C, B, and A), each of which is distinguished primarily by the construction of a new rampart (Glacis #2 to #4) and minor alterations to the gate itself (Voss 2002:381ff.). The fosse of Phase 14 fell out of use when it was filled as a result of the enlargement of the *earthen rampart* (Glacis #2) as much as 15 m from the base of the original glacis. It was intentionally filled with sand and debris at the start of Phase 13 (ibid., 380). This "moat deposit" featured a number of items including Egyptian seals that have enabled precise phasing of the fortifications associated with Phases 14 and 13C (Stager 2002). No fosse has been discovered for any of the subsequent phases of the fortifications.

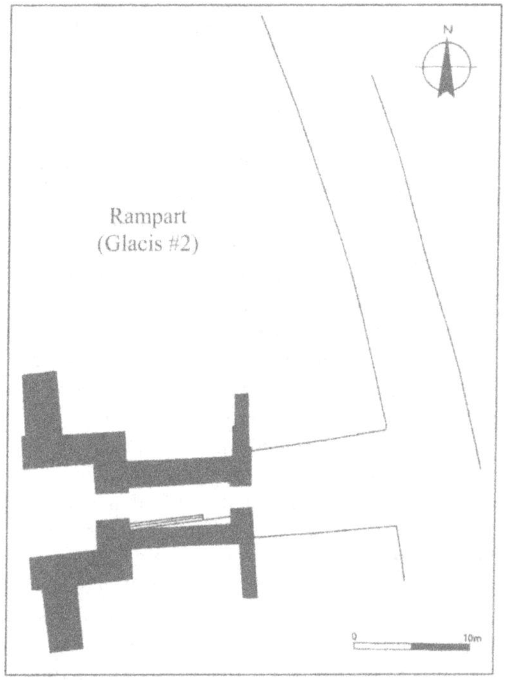

Figure 63. Phase 13C gate (after Voss 2002:fig. 2).

Phase 13C

During Phase 13C the *gate* appears to have been renovated, though this did not drastically affect the overall layout of the gate or its dimensions (Figure 63 above). The most apparent alteration involved the use of ashlar masonry in the gate's bay walls (Voss 2002:381). Two walls (F.42 and F.85), which ran parallel to the rampart and abutted the interior projecting walls, were also added. These mudbrick walls were approximately 3.1 m wide and were constructed on worked stone foundations abutting the interior wings of the gate (F.41 and 40), which were built atop the wings of the previous phase (F.71 and 70). Walls F.42 and F.85 projected about 5.3 m to the N and the S of the gate.

The façade of the gate was composed of N and S piers (F.119 and 120). These piers, which were 1.86 m wide, were built of mudbricks (54 x 31 x 19 cm) on ashlar foundations and they projected about 65 cm from the sides of the gate to form a 2.3 m wide entrance to the passage. Gypsum plaster was found adhering to the surface of the ashlars along the edges of the piers, which were built of stone so as to endure wear and tear. The bay walls (F.112 and 154) constructed of stone were also well preserved above the Phase 14 gate, which originally formed a barrel vault over the passage (ibid., 381). Gypsum plaster was also used on the ashlars of the bay walls to protect the mud mortar between the stones from insect infestation and limit the decomposition of the kurkar due to moisture. Stepped mudbrick buttresses reinforced the N and S bay walls of the Phase 14 gate, which served as the foundation and lowest courses for the Phase 13 gate. During Phase 13 these buttresses also functioned as curbs against carts (ibid., 382) since the reduction of the width of the passage from 2.7 m to 2.1 m would have ruled out the passage of many wheeled carts and most chariots.

The street surface (LF.127) associated with this gate ran continuously from inside the rampart, through the passage, to the exterior of the gate at a slope of about 14°. It consisted in most places of beaten earth surfaces (L.64 and L.156), but within the gate's passage itself the street was lined with mudbricks, many of which were damaged in the center of the passage when a massive drain was added during Phase 12 (F.48). The construction of this drain almost directly on top of this floor, cutting through some of it and the occupational debris that had accumulated on the floor, demonstrate that this floor remained in use throughout Phase 13. This observation supports the conclusion that despite the identification of three architectural sub-phases of the Phase 13 gate, it was probably in use for only a short time. This is further confirmed by the excellent state of preservation of the mudbrick floor, which would be expected to show signs of extensive wear.

Just inside the E piers small niches were cut into both bay walls. These are thought to have been used to brace the gate's doors with a crossbar when they were secured. Evidence of the crossbar consisted of charred wood found in these carved niches. To facilitate the opening of the doors the vault's underside was set back about 1.4 m from the inner face of the piers.

Atop the W piers of the Phase 14 gate new piers were added during Phase 13C (F.108 and F.109). These walls were preserved up to forty courses in some places and were built with mudbricks measuring 45 x 33 x 12 cm (Voss 2002:382). It was against these piers that new walls (F.42 and F.85), mentioned above, were added.

This phase also featured the earliest evidence of a mudbrick *casemate fortification wall* that ran along the top of the rampart. The segment that has been preserved (F.185) was only exposed on the N side of the gate and unfortunately it was not possible to obtain a broad exposure of this structure because it remained buried below later ramparts.

The construction of the Phase 13C rampart above that of Phase 14 led to the abandonment of the earlier fosse (Voss 2002:380). The new *earthen rampart*, composed of ash, earth, and crushed kurkar, provided the core of the rampart throughout Phase 13. The *glacis* of Phase 13C was identified to the N of the gate in square 63 (02.101.F.142=02.63.F.11), where it was found to be composed of kurkar cobbles sealed with a 5 cm thick layer of clay. A plastered drain made of cobbles (02.63.F.12=02.101.F.68) was also identified running along the surface of the glacis. This feature indicates that care was taken to guarantee that water draining along the face of the glacis did not undermine the rampart's construction. The revetted, outer face of this glacis to the N of the gate was detected in square 76 (F.164). Here the revetment turned to the SW and formed the N corridor wall leading into the gate. The stone revetment wall of Phase 13C (F.164) was only preserved one row wide and two courses high. A corresponding corridor wall on the S side of the outer area (F.160) was also built upon the same beaten earth surface (LF.156), which led into the gate in this phase and continued to function in phases B and A. In Phase 13B both of these stone revetment walls were replaced with mudbrick walls.

PHASES 13B & 13A

Phase 13B and 13A constituted drastic changes to the *gate*, which had almost entirely collapsed (Figure 64). The collapsed portions included the stone barrel vault, the W mudbrick arch, and the E stone arch (Voss 2002:382f.). Although the W mudbrick arch was rebuilt, the barrel vault was not. The reason for this collapse is unknown.

The stones of the Phase 13C vault were used to reinforce the mudbrick buttresses of Phase 13C. On the N side the buttress (F.128) was constructed of mudbrick and mud mortar and preserved thirteen courses high. On the S side of the passage a similar mudbrick buttress was added (F.146). Neither of these buttresses ran the entire length of the passage. The stone buttresses were raised to the level of the spring line of the former arch, while the stone bay walls of the former gate were replaced with thinner mudbrick walls. The front arch, which was previously constructed of ashlars, was replaced by a mudbrick arch built on the surviving stone voussoirs of the Phase 13C gate. It is thought that the gate featured a flat ceiling that spanned the width of the gate from the stone ledge on the S bay wall to the N wall.

Outside the gate along the N side of the corridor approach a mudbrick *revetment wall* (F.157), preserved eighteen courses high and three rows wide, abutted the N end of the gate's N pier (F.119). Its E end featured a "rabbeted" corner where the wall turned N into a mudbrick *revetment wall* (F.9) that ran along the base of the glacis. The bricks of this revetment wall measured 51 x 34 x 13 cm and the wall was preserved up to nine courses high and more than 1.3 m wide. The revetment and circumference wall continued in use through Phase 12 (MB IIB). A similar revetment wall was probably built on the S side of the gate's corridor, though it appears to have been replaced by a *revetment wall* of slightly different construction (F.115). This new revetment was built against the gate's S pier (F.120) and was preserved up to twenty-seven courses high of mudbricks measuring 48 x 33 x 12 cm (Voss 2002:383). These revetment walls were probably intended to hold back the newest expansion of the *earthen rampart* (Glacis #3).

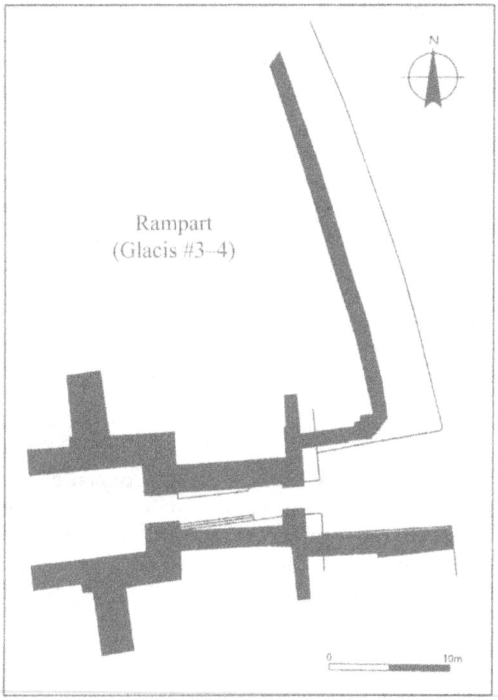

Figure 64. Phase 13B–A gate (after Voss 2002:fig. 3).

Following the addition of these revetment walls sloping mudbrick buttresses were also added to the exterior façade of the gate on both sides of the entryway (F.151 and F.152). These structures were built eighteen to nineteen courses high of mudbricks of comparable size (50 x 33 x 11 cm) as the N mudbrick revetment wall (F.157) with which they were contemporaneous. However, that buttresses F.151 and F.152 were built against the gate and against the new revetment walls (F.114 and F.157) to their S and N, respectively, and were not built into them suggests that these buttresses must have been constructed after the addition of the revetment walls on both sides of the outer corridor, perhaps when structural weaknesses were identified.

Inside the rampart to the W of the gate *revetment walls* were also built to retain the expanding rampart. Extending from the N pier (F.42), which was built in Phase 13C, a mudbrick wall (F.46) was built with bricks measuring 32 x 32 x 11 cm projecting W about 5 m. A similar wall (F.75) projected about 5 m from the S pier (F.70). These revetment walls probably served like those outside the gate (F.157 and 115) to hold back the additional layers of the rampart's interior slope.

Following extensive additions and renovations in Phase 13B it appears that little was done to alter the overall plan of the gate during Phase 13A. Yet as in the previous phase the addition of a new *glacis* and *earthen rampart* was roughly contemporary, if not responsible, for a set of modifications to the gate at that time. Glacis #4 associated with this phase (F.4) grew to the point of extending beyond the revetment wall (F.9), which had been built at the outside foot of the glacis of the previous phase. The new glacis and rampart, as in earlier periods, were again constructed of a single course of kurkar cobbles laid above an earthen fill. Though the lower portion of this glacis was covered by the extension of structures outside the rampart in the following phases, the rampart and glacis continued in use from Phase 13A through Phase 11 at the end of the Middle Bronze Age.

MB IIA/B: Period XXII, Phase 12

The fortifications of Phase 12 (late Dyn. XIII) represent the first significant departure from the original plan of the gate, which had been established in the early part of the second millennium (Figure 65). The nature of the renovations suggests that the Phase 13A gate was intentionally destroyed because the aged superstructure had weakened substantially. The process of designing the new *gate* for Phase 12 was well planned and intended to make the gate significantly larger (Voss 2002:383). Only the bay walls of the Phase 13A gate served as the foundation for the walls of the new gate. The remainder of the gate was buried except for some revetment walls. The new gate was built entirely of mudbricks averaging 50 x 36 x 10 cm in size (ibid., 384).

The first step in the renovation of the gate involved the laying of a corbelled, mudbrick-built drain (F.48), which stood six courses high. It ran along the sloping floor of the passageway of the Phase 13A gate from inside the settlement to the outer base of the rampart of the previous phase (see Figures 64 and 65 65). It was about 34 m long and the bricks used in its construction appear to have averaged about 47–48 x 33–34 x 10–13 cm in size. After the laying of the drain more than 2 m of crushed sandstone fill (L.6, L.7, L.8, L.149, and L.150) was added over the drain and over the floor of the Phase 13A gate in order to raise the foundation level to the tops of the preserved walls of the previous gate (ibid., 383).

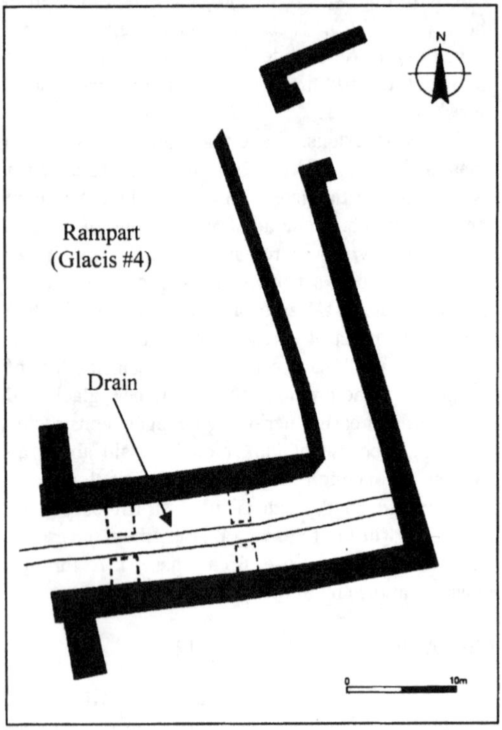

Figure 65. Phase 12 gate (after Voss 2002:fig. 4).

Once this was complete the new walls of the gate were laid on both sides of the passageway. On the N side of the passageway a wall (F.41) was laid over the exterior revetment wall (F.157), N bay wall (F.112), and the N wing wall (F.41) of the previous phase. Similarly, on the S side of the passageway a mudbrick wall (F.40) was laid over the exterior revetment wall (F.115), S bay wall (F.154), and the S wing wall (F.40). The corridor walls were more than 29 m long on both sides of the gate. The new passageway, which was about 6.5 to 7 m wide, was much wider than the original. No evidence was found of any piers associated with this gate that would have constricted it, though it is possible that the later pedestrian gate of Phase 11 destroyed the remains of these piers when it was constructed. Furthermore, no evidence was revealed of any of the arches or roof associated with this gate.

While this gate featured considerable modification of the plan of the previous gate, some features were rebuilt along the lines of the earlier plan. *Revetment walls* inside the gate such as those on the interior end of the gate (e.g., F.71), which ran parallel to the rampart from the exterior faces of the gate's corridor walls, were rebuilt along the lines of the earlier walls (F.42 and F.168). The *revetment wall* on the N side of the gate's exterior corridor with the rabbeted corner (F.157) also remained in use during this phase, though the corresponding revetment wall on the S side of the corridor was built over by the E extension the gate's S corridor wall (F.40).

The gate's plan during Phase 12 was also modified, apparently in order to complicate an approach to the gate. The direct axis approach of Phase 13 was adapted to include a 90° turn to the N upon leaving the gate, which then ran for approximately 30 m and was flanked by walls on both sides (F.9 and 97). Another 90° turn to the NE was necessary before exiting through a lower gate. (This approach as at other sites made it difficult for right-handed attackers trying to enter the town to use their shields.) The lower gate featured an opening only in the outer wall of the corridor. The Phase 12 gate was contemporaneous with the rampart and glacis of the previous phase (i.e., Glacis #4). No remains of the wall associated with the fortifications in this period have been identified.

MB IIB: Period XXI, Phase 11

The Phase 11 gate (early Dyn. XV) comprised the most radical modification of the gate's plan during the Middle Bronze Age (Figure 66). In essence the gate appears to have shifted in function from one of the settlement's all-purpose gateways to what is identified as a pedestrian gate, suggesting that the gate would not have been accessible to wheeled vehicles (Voss 2002:384). This shift must have been preceded by a massive filling project that required several meters of fill to bury the Phase 12 gate. The new *four-pier gate* was built with the same orientation as the earlier gate, but the structure now featured a passage only 1.5 m wide. The gate structure was almost perfectly square, measuring about 7.3 m on a side with mudbrick walls averaging 1.8 m thick, which were built on stone foundations. Bricks used in its construction measured 35 to 40 cm square and were on average about 10 cm thick (cp. 50 x 36 x 10 cm in Voss 2002:384). The floor (02.85.LF.103) of the gate consisted of a layer of mudbricks as well as beaten earth.

The approach to the gate functioned like a funnel leading into the gate. It was formed by cutting away parts of the earlier gate and the surface associated with it. A street made of ash (LF.90 and LF.117) was laid atop the former approach ramp. The revetment walls to the N and S (F.30 and 97) of the gate were built of small fieldstones. It also appears that the earthen rampart of Phase 13A (Glacis #4) continued in use through this phase. However, a mudbrick *tower* (F.31) was added atop the rampart. Its base was roughly 14 m long by 8 m wide. The tower continued in use through Phases 10 and 9.

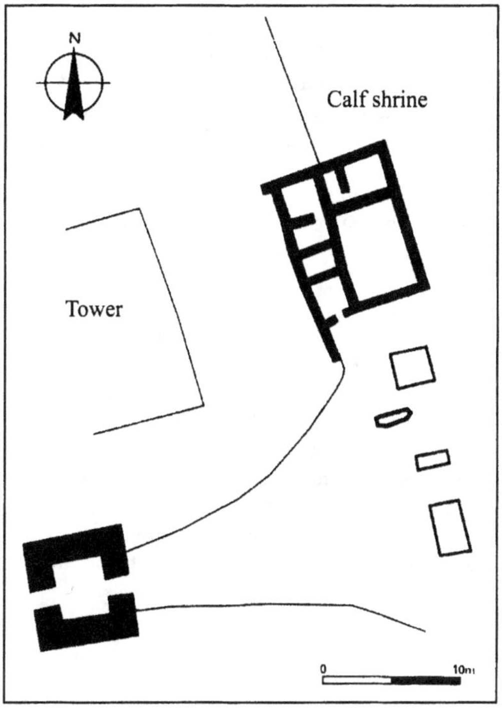

Figure 66. Phase 11 gate (after Voss 2002:fig. 5).

The so-called calf shrine, to the left of the gate's approach, and built tombs were discovered outside the rampart of this phase (Stager 2006). The building, which was built with its rear wall against the mudbrick wall at the base of the rampart (F.9), consisted of a modest plan. In one of the rooms of this structure a ceramic shrine accompanied by a small silver calf was discovered.

MB IIC: PERIOD XX, PHASE 10

During Phase 10 (Dyn. XV) a final *earthen rampart* was constructed, which was 15 m high and 70 m wide (Stager 2006:403). This rampart buried the gate and the calf sanctuary outside the gate that belonged to the previous phase (Voss 2002:384). The fortifications of Phase 10, which are not well preserved, represent the final phase of the Middle Bronze Age fortifications in Grid 2. The *tower*, which had been built in Phase 11, continued in use during this phase. It was supplemented by a mudbrick *fortification wall* (F.46) about 3.3 m wide that was exposed to its S above the remains of the Phase 11 gate. About 12.3 m of this wall were excavated.

'AVDON*

MR 1655.2725
Refs.: Prausnitz (1973; 1975).
Plans: NA.
Sects.: NA.
Photos: NA.
The fortified settlement of 'Avdon sits on top of a mountain 80 m above the surrounding wadis inland from Achzib in northern Israel. It was identified as a fortified settlement by Moshe Prausnitz following his 1964 survey, though it was never excavated (1973). Nevertheless, two fortification systems appear to have surrounded the top of the site and its slopes. These included a *fortification wall*, remains of a *tower*, and an *earthen rampart*(?). Though ceramics were collected on the site, none were collected from the tower and the rampart, both of which have been identified as MB IIB structures. The area between the walls is 200 m long E-W and 130 m wide N-S enclosing about 2.5 ha (ibid., 222). The site's relationship to Achzib has been compared by Prausnitz to that of Tel Bira, an outlying settlement E of the larger settlement of Akko (1975:202).

BAR, TEL (TELL EL-AGHBARIYEH)

MR 1648.2229
Refs.: Raban (1999).
Plans: NA.
Sects.: NA.
Photos: NA.
Tel Bar is a 3 ha Middle Bronze Age settlement located on a hill overlooking the Jezreel Valley, approximately 1.5 km NW of Megiddo and 3 km SE of Yoqne'am (Raban 1999:74*f.). Archaeological survey of the site revealed an *earthen rampart* ("earthen glacis") approximately 10 to 15 m wide and 5 m high that is thought to be responsible for the pentagonal shape of the mound. Entrances to the settlement that may relate to the Middle Bronze Age defenses were traced on the E, S, and W sides of the tell.

BEIT MIRSIM, TELL

MR 141.096
Refs.: Albright (1938b); Eitan (1972); Yadin (1973).
Plans: Site (Albright 1938b:pl. 46), SE (ibid., pls. 49, 54).

Sects.: SE (Albright 1938b:pl. 53).
Photos: SE (Albright 1938b:pls. 6–8, 16).
Tell Beit Mirsim is a small site of approximately 3 ha located about 40 km SE of Ashkelon and about 20 km SW of Hebron in the western foothills of the hill country. Archaeological excavations by William F. Albright first brought to light the remains of Tell Beit Mirsim's fortifications from 1926 to 1934. Because this site was one of the first in the southern Levant to be extensively excavated, Albright's conclusions have been variously interpreted over the years (Yadin 1973; Bienkowski 1989). Any interpretation of its development must be based, therefore, upon subsequent refinements in chronology and understanding of the archaeological sequence. Nevertheless, the site offers considerable data concerning its fortification during the Middle Bronze Age.

MB IIA, PHASES 3–4: STRATA G–F

Following abandonment of the site during the later half of the EB IV it was resettled around Phase 3 or 4 of the MB IIA (Cohen 2002a:75), Strata G–F. Albright's original observation regarding these strata was that this phase represented the earliest Middle Bronze Age fortifications of the town (Albright 1938b:17). His relative phasing of the development of the fortifications was aided by destruction levels that separated each stratum between H and F.

Over 80 m of the foundations of the town's *fortification wall*, which was between 3.2 and 3.3 m wide and built on the ash remains of Stratum H, was exposed in the SE quarter of the site (Figure 67). Even though the tell itself is oval in plan the wall appears to have been constructed of straight-line segments. Buttresses 30 to 50 cm thick were attributed to Stratum F. Albright also identified two *towers* located twenty-three meters apart, which measured 10 to 10.5 m long by 6 m wide; they probably projected about 1.5 m on both sides of the wall (ibid., 18). If the wall's fieldstone foundation roughly traced the 490 m contour line, then it would have been approximately 620 m in length, 13% of which was excavated. Albright also remarked that the masonry of the domestic architecture was basically identical to that used in the fortifications (ibid., 23). No earthen rampart or glacis was found to be in association with the wall in this period. Although according to Albright it was "not possible to reconstruct the gateway" on the E side of the site in this period (ibid., 20), Herzog has done so (1986:40, fig. 37).

Abraham Eitan (1972) argued that the town wall Albright had identified as that of Stratum G, in fact, belonged to the later part of the MB IIA at Beit Mirsim, namely Stratum F (late MB IIA). Eitan's first basis for this was that walls of the houses of Stratum G did not align with the wall, which would have been necessary had the wall been constructed in the same period (ibid., 19ff.). Furthermore, he underscored the difficulties Albright admittedly faced in understanding how the town wall related to Stratum G (ibid., 22).

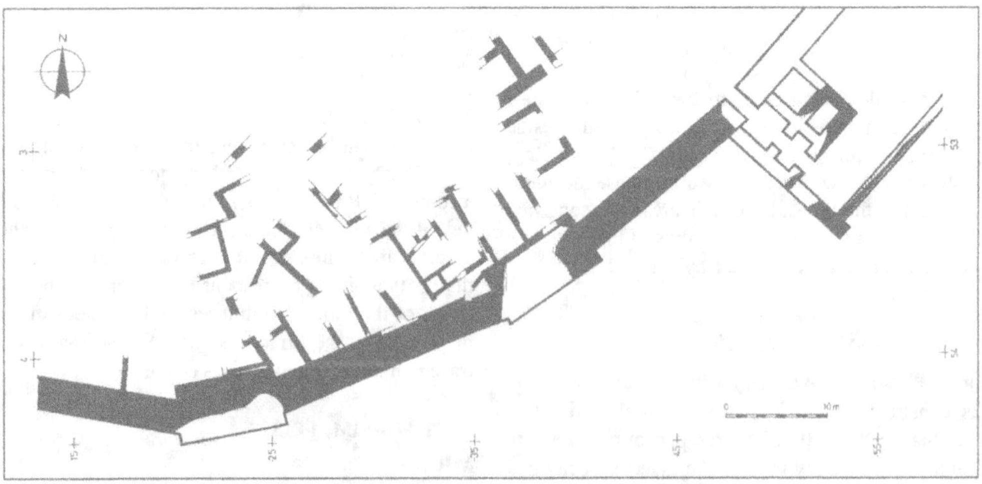

Figure 67. Tell Beit Mirsim Stratum F fortifications (after Herzog 1997a:fig. 4.3).

Figure 68. Tell Beit Mirsim Stratum E fortifications (after Herzog 1997a:fig. 4.15).

Yigael Yadin's (1973) observations concerning the site were the final word on Tell Beit Mirsim's stratigraphy. His first observation was that the buildings ascribed to Stratum G featured foundations considerably lower than those of the town wall also ascribed to Stratum G. As Yadin noted, based upon evidence from other sites it was unlikely that buildings so close to the town wall would have been built below the level of the foundations of the town's wall. Second, he also contended that Eitan's argument about the "radial arrangement" of the houses with respect to the wall was correct, though this applied, as Yadin argued, mostly to Strata E and D (MB IIB–C). Lending further strength to Yadin's argument, though not noted by him, was that sherds of Strata G–F wares were found in the "talus of ash filled earth," which Eitan (1972:21) had argued was an indication of the construction of the wall in Stratum F (instead of G), but which can also be used to argue that the wall was constructed at the start of Stratum E, as Yadin had understood it.

MB IIB–C: Strata E–D

The first *earthen rampart* at Tell Beit Mirsim belonged to Stratum E (MB IIB) and was built against the town wall of Stratum G–F (Albright 1938b:27ff.). This rampart was described as constructed of chipped stone and earth using the "sandwich" technique and was probably later furnished with a stone *glacis* during Stratum D (ibid., 29f., pl. 16). It is not possible to determine the slope, height, or width of the rampart since its section was not drawn. In addition to these elements, half of a *gate*, which may have continued in use into the LB I, was also excavated (ibid., 30f.). The piers of the gate each measured 1.5 m long by 1.5 m wide with a space between them thought to have been approximately 3.25 m; the interior of the passage would have been about 6.25 m wide (Figure 68).

Beth-El (Beitin)

MR 17280.14815

Refs.: Kelso (1968; 1993).

Plans: Site (Kelso 1968:pl. 1); N wall (pl. 91a); NW Gate (pl. 101); W wall (pl. 102a); Abu Tabar W wall (pl. 94a); W wall in "Farmer's house" (pl. 102b); Area III (pl. 11); S wall (pl. 103c).

Sects.: N wall (pls. 93a, 93c); Abu Tabar W wall (pl. 94a); W wall in "Farmer's house" (pl. 102b); Area III (pl. 11); S wall (pl. 103c).

Photos: W wall (pls. 13, 110, 112a); N wall (pl. 96a, 97); NW Gate (pls. 104b–106a, 107b–108, 109b); S wall (pl. 112); Abu Tabar W wall (pl. 99b); Area III (pl. 13); S wall (pl. 112).

Beth-El (biblical *Luz*) is located in the hill country approximately 17 km N of Jerusalem. The site's only geographic advantage was the abundance of water from local springs (Kelso 1968:10). Although Beth-El was occupied during the MB IIA following an abandonment in the late third millennium, the only definitive evidence for the site's fortification dates to the start of the MB IIB.

MB IIB

A 3.5 m wide *fortification wall* was constructed in the MB IIB (Kelso 1993:193). The wall was identified in soundings on the N, NW, W, and S sides of the site. On the N side of the site a 20.5 m stretch of the foundation of the town wall was built upon bedrock with "semi-dressed stones" and preserved to a height of 2.8 m (eight courses) measuring 3.38 m wide (Kelso 1968:10). The limestone blocks used in the construction of this wall were as much as 1.4 m long and averaged around 0.75 m in length. The wall's construction is of the typical boulder and chink style. Although James Kelso suggested that the wall featured a slight batter (approximately 86°, though bowed) this does not confirm that the mudbrick superstructure also had such a batter. At the W end of this wall the exterior face featured a protrusion that might be the remains of a square *tower* projecting from the wall, or a segment of the wall to the W offset from this segment.

The N wall was also accompanied by a clay *earthen rampart* built on virgin soil (referred to as a glacis in Kelso 1968:11). The fill of the rampart contained sherds from all periods of occupation prior to the MB IIB suggesting its contemporaneity with the wall. Unfortunately, the lack of detail in the section (and no photo of it, Kelso 1968:pl. 93c) makes it difficult to discern the exact function of each of the components of the fortifications N of the earthen rampart though they have been dated to the MB IIB–C.

The NW corner of the town *fortification wall* was also identified just adjacent to the NW gate (see below). It is interesting to note that this corner was a nearly perpendicular turn in the wall and therefore this point in the site's defenses would have been particularly vulnerable. To the S of it an 11.5 m stretch of the W wall was identified (Kelso 1968:14f.). Limited exposure of the wall makes it impossible to provide its exact width at this point or to establish if any modifications were made inside the wall to adjust for its vulnerability here. At the S end of this stretch the wall turned to the SW just slightly; the complete excavation of this 3.5 m long section of the wall revealed that the wall was 3.5 m wide, was preserved to a height of 1.75 m, and was built in a similar fashion as the N wall (ibid., 15). The exact temporal and spatial relationship between these two wall segments was uncertain, but the S section of the wall was dated to the MB IIB. More of the settlement's W fortification wall which dated to the MB IIB was identified in two areas to the W of the site.

In the Abu Tabar area "on the W edge of the village and some distance down the hill from the west wall" in an 8 m long trench a part of this wall and a tower were identified (Kelso 1968:15).[42] The *tower* was 5.75 m wide, though its length cannot be established, and its walls were preserved to a height of 3.5 m. Although it is not possible to establish the thickness of the fortification wall here, the tower's outer wall was 1.6 m wide and projected 1 m from the wall. Kelso suggested that this tower was built in the MB IIC and served to fill a breach that was made earlier in the wall. The basis for his argument rests on the recognition that the "breached area not covered by the tower had been carelessly plugged," and "the stones . . . were not interlocked," and "the patch was not dovetailed" (ibid., 15).

Kelso also identified a "glacis," or what should be identified as an *earthen rampart*, consisting of "small loose stones." The clay facing described for this feature should, in fact, be identified as the *glacis*. Ceramics identify the tower as MB IIC in date (ibid., 16). More of the town's W wall was identified in the so-called "farmer's house" (ibid.). The wall here also measured 3.5 m wide, was preserved to a height of 2.8 m, and featured a similar construction to the town wall near the NW gate and N wall.

The S part of the town *fortification wall* of the MB IIB–C settlement was also identified (Kelso 1968:17f.). This wall was preserved to a height of 4.3 m and was comparable in composition and construction to the walls on the N and W sides of the site. Evidence for a wall, which made a right angle to the S town wall, suggests that this was the location of either a tower or gate. The thickness of the wall at this point could not be established, though Kelso speculated that it was similar to that of the W and N sections of the wall (ibid., 18). Evidence of a breach in the wall to the W of the excavated portion of the S wall was also revealed.

Evidence of up to four gates for the site was also revealed by the Kyle Memorial Excavations (Kelso 1968:12f.). These included two *gates* that were excavated on the NE and NW sides of the site as well as supposed traces of gates through the settlement's W and S walls. The excavation of the NE Gate, which does not appear in plans, only consisted of a limited sounding of the W jamb, though the lowest course of the town wall was revealed to be abutting the gate (ibid., 12).

The *NW Gate* was also located and its dimensions more thoroughly explored (ibid., 13f.). The NW Gate is of a unique plan in the Levant. It was clearly built as a separate structure against the N elevation at the W end of the N town wall. Its plan has been described as "U-shaped;" it was entered from a set of steps rising 0.7 m on the E side of the gate, after passing through two

[42]Unfortunately, it seems impossible to precisely locate this excavation area since it does not occur on the site plan (Kelso 1968:pl. 1) and the text description is inadequate.

piers the 3.5 m wide passageway proceeds W for 9.5 m before turning S, where it then ascends several steps rising 0.86 m and turns again to the E. The point of entry through the N wall of the town was never located. Its N elevation is 14.6 m across, the E elevation 9.7 m and the W elevation 9.2 m; its overall plan was described as "rhomboid." Although never exposed, its S elevation abutting the N wall must have been about 15.3 m long. The walls of the structure vary from 1.5 to 1.75 m in width and were preserved as much as 2 m high. Its entrance beyond the first three steps and between the two piers was 2.5 m wide. The floor was composed of 10 to 15 cm of huwwar limestone in places, and only a few flagstones remained at the W end of the N corridor. The gate featured no guard rooms (ibid., 14). Based on the ceramic evidence the gate has been dated to the MB IIB–C, and Kelso suggested that it was contemporaneous with the construction of the N wall. However, given that the gate could have been better integrated into the wall, if it was built at the same time, it may have been later than the building of the town wall. Nevertheless, the gate like the rest of the town's fortifications appears to have been destroyed at the end of the MB IIC.

LATE MB II (MB IIC)

Although certain segments of the town wall clearly dated to the MB IIB while continuing in use during the MB IIC, other parts of the wall appear to have been of entirely later construction. One such stretch excavated in Area III dated to the MB IIC was found preserved to a height of 4 m (Kelso 1968:16). A 15 m section of the wall revealed that "the outer face had here been quarried away in Byzantine times" and "the wall continued in use after the town's fall to the Israelites" (ibid.). The construction of the wall was considered "superior" by Kelso as the squared 0.6 to 0.8 m blocks were well fitted together. Altogether the town wall appears to have averaged about 3.5 m thick during the MB IIB–C.

The MB II walls of Beth-El appear to have functioned at least as the foundation of the settlement's wall in many later periods including the Late Bronze Age, the late Iron Age (Babylonian period), and the Hellenistic period (Kelso 1968:11, 16). Broshi and Gophna (1986) estimated the site to have been about 2 ha in size during the Middle Bronze Age. No new estimate of the site's size is possible since published maps do not provide the site's contours and because most of the site is occupied by a modern village.

BETH-SHEMESH (TELL RUMEILEH)

MR 1477.1286

Refs.: Mackenzie (1911; 1913); Grant (1934:11 ff.); Grant and Wright (1938; 1939); Bunimovitz (2003).
Plans: Site (Mackenzie 1913:pl. 1; Bunimovitz and Lederman 2003:fig. 1); Strata V–IV (Grant 1931:pl. 30; Grant and Wright 1939:fig. 1); S Gate (Mackenzie 1911:pl. 14; 1913:pls. 2f.).
Sects.: S Gate (Mackenzie 1911:pl. 14; 1913:pl. 2); Wall on E side (Mackenzie 1911:pl. 9).
Photos: Site aerial (Grant 1931:pl. 2); Stratum V W wall and towers (Grant and Wright 1938:pls. 2); Wall (Mackenzie 1911:fig. 27; 1913:figs. 5f.); NE Tower (Mackenzie 1911:fig. 26); W Tower (Grant and Wright 1938:pl. 2); SW Tower (ibid., pl. 2); S Gate (Mackenzie 1911:fig. 28; 1913:pl. 15); Area R2D2 (Bunimovitz and Lederman 2003:fig. 2).

Beth-Shemesh is located 40 km ENE of Ashkelon and about 25 km WSW of Jerusalem. Sherds found in excavations at Beth-Shemesh suggest that the site was at least occupied during the EB IV and MB IIA, Stratum VI (Bunimovitz and Lederman 1993:250), though the fortifications of the site were first constructed during the MB IIB–C, Stratum V.[43] The evidence of the fortifications included mostly the remains of the town wall, the NE and W Towers—all of which were revealed by Mackenzie—and the SW Tower and the S Gate both excavated by Grant. Unfortunately because of Mackenzie's method of tracing the wall around the site (see Mackenzie 1911:85f.), it has until recently been impossible to confirm that an earthen rampart had not been built against the wall since the remains of the wall do not appear to have featured a batter (see below). However, recent excavations by Shlomo Bunimovitz and Zvi Lederman in Area R2D2, where brick debris has been found resting just above the bedrock upon which the wall had been constructed, appear to confirm that no rampart existed at this site (2003:234).

The town *fortification wall* exposed around the entire site by Mackenzie was between 2.2 and 2.4 m wide (Grant and Wright 1939:20), though in places it appears to have been as much as 4 m wide (ibid., fig. 2). It was built of large stones in the chink and boulder style directly upon bedrock. Although none of the superstructure was found intact, Grant surmised that it was originally built in mudbrick as suggested by the "masses of fallen brick" found below Strata III and II within the town wall (ibid.). Similar "burnt and fallen bricks" have been identified by Bunimovitz

[43]Mackenzie's original identification of the Bronze Age settlement, refered to as the "Second City," had lumped what Grant later identified as Strata V (MB II) and IV (Late Bronze Age) into a single architectural plan (see Grant and Wright 1939:27).

and Lederman (2003:234), thus confirming that it did carry a mudbrick superstructure. Grant also observed that the lower portions of the wall featured no apparent batter. On the SE corner of the site below the E wall of the Byzantine convent the E fortification wall was 2.25 m wide (Mackenzie 1911:85). This wall could be traced clockwise around this part of the site below the Byzantine building. Based on Mackenzie's plan of the wall the entire length of the fortifications would have been about 610 m and would have enclosed an area of about 2.5 ha.

At least three towers (called bastions by Mackenzie) were identified along the wall (Grant and Wright 1939:20f.). One on the NE side of the site was traced by Mackenzie, another on the SW was excavated by Grant, and a *W Tower* (18 x 11 m) was mentioned by Mackenzie but only identified and drawn by Grant's expedition (Grant 1931:21f.; Grant and Wright 1939:21f.). Unfortunately, it is difficult to be certain of the dates of these towers, though they are likely to have been contemporary with the MB IIB–C town wall. Nevertheless, the *SW Tower* (11.3 x 7.3 m) may date to the early Late Bronze Age (Bunimovitz and Lederman 1993:250). The *NE Tower* was found to be 9 m long projecting 5.95 m from the wall on its E end and 4 m on the W (Mackenzie 1911:86).

The S Gate, which was considered to be contemporary with the settlement's earliest defenses, was identified by Mackenzie (1911:89ff.; 1913:23ff.). If the dating of the gate is accepted, then it appears that this was a *six-pier gate* like those known during the second half of the Middle Bronze Age. The piers projected about 60 cm into the passage and were 2.2 to 2.5 m wide, like the walls of the gate. Between the rectangular towers of both sides of the gate the passage appears to have been about 3 m wide and 14.3 m long. The S chamber on the W side of the gate also featured a guard room with an entrance off of the passage to the W.

Evidence that Beth-Shemesh was destroyed at the end of the Middle Bronze Age (Str. V) or early in the Late Bronze Age (Str. IVa) was found in a number of different areas (Grant and Wright 1939:22ff., 29; Mackenzie 1911:91f.). This evidence consisted of breaches and patches identified along stretches of the town wall. Patches were traditionally narrower than the original wall, about 1.4 m wide in places.

BETH-ZUR (KHIRBET ET-TUBEIQA)

MR 1590.1108

Refs.: Sellers (1933); Sellers et al. (1968); Funk (1968; 1993).

Plans: Site (Sellers 1933:pl. 1); Excavated areas (Sellers et al. 1968:plan 1); SW (Sellers 1933:pl. 3); Field I (Sellers et al. 1968:plan 2); Field III (ibid., plan 5).

Sects.: SW (Sellers 1933:pl. 4); Field I (Sellers et al. 1968:plans 3f.).

Photos: Field I (Sellers et al. 1968:pls. 5a, 6).

Beth-Zur is located 6 km N of Hebron and 26 km SW of Jerusalem in the hill country. The site sits atop a defensible hill established 100 m above the surrounding valley to its N and 22 m above the same valley to its S (Funk 1993:259). While Beth-Zur appears to have been reoccupied during the late MB IIB, its fortifications were only first erected during the sixteenth century B.C. (1993:260).

The *fortification wall* appears to have averaged 2.6 to 2.7 m wide with a foundation built from large stones on bedrock with smaller stones serving as chinking material between the larger ones. The construction of the wall identified on the SW side of the site (Sellers 1933:26) serves as the only reliable identification of the original Middle Bronze Age defensive wall.

While Funk suggested that Wall A in Field III could be identified as the continuation of the wall on the E side of the site, where it measured almost 3 m wide (see sections in plans 3f., Sellers et al. 1968), the almost perfectly straight stretch of the wall between Fields I and III does not in the least resemble the character of the curving wall with towers excavated by Sellers. Although intra-site variability of the construction techniques employed in fortification systems is attested elsewhere, there are no parallels for the use of both straight and curvilinear defensive strategies. Funk and his colleagues had held the same view prior to their final publication of the results of excavation in Fields I and III (1968:5). A Middle Bronze Age date for Wall A can, therefore, only serve as a *terminus post quem* and it is possible that this orthogonal wall belongs to the Iron Age or the Hellenistic citadel. No brick remains were found atop the stone foundations (ibid., n. 3).

The identification of the casemate-style construction in Fields I and III is moot (Funk 1968:6), if, in fact, the outer wall was not itself the defensive wall, as I have suggested above. Nevertheless, a *tower*, with dimensions of 5 x 10 m was exposed in conjunction with the Middle Bronze Age wall on the SE side of the site (Sellers 1933:26).[44]

Although it is nearly impossible to be certain, the fortification wall was probably no more than 500 m long, forming an oval around the top of the hill. The site's size within the Middle Bronze Age wall was

[44]Note that in pl. 3 the outer face of the tower is identified as a Hellenistic buttress (Sellers 1933:26).

between 0.8 and 1.5 ha during the Middle Bronze Age (Funk 1968:4). However, an estimate of the space inside the oval created by the enclosure would favor a figure of 1.5 ha in settled area. The final demise of the fortifications in a fiery destruction has been attributed to post-Hyksos military activity during the LB I (Sellers 1933:9). Nevertheless, these fortifications were "repaired and reused down into Hellenistic times" (Funk 1968:4).

BIRA, TEL

MR 166.256
Refs.: Prausnitz (1993b).
Plans: NA.
Sects.: NA.
Photos: NA.

The 19.8 ha site of Tel Bira, which lies about 10 km ESE of Akko, included both an upper acropolis [6.1 ha] and lower town [13.7 ha] (Prausnitz 1993b:262). Although the date of the settlement's fortifications within the Middle Bronze Age is unclear due to the lack of publication of these remains, the lower town appears to have been fortified at some point before the MB IIC with an outer *fortification wall* and *glacis*. However, apparently during the seventeenth(?) century B.C. (MB IIB?) the lower town, which may be identified as a commercial quarter, was abandoned and construction was begun on a "steep glacis" and accompanying fortification wall around the acropolis.

BOSRA

MR 288.250
Refs.: Braemer (2002).
Plans: S fortifications (fig. 3).
Sects.: S fortifications (figs. 3 & 4)
Photos: Site aerial (fig. 2).

Bosra is located approximately 140 km to the S of Damascus. Although the site may be mentioned among the MB IIA Execration Texts (E.27), Helga Seeden suggests that Middle Bronze Age occupation of Bosra was confined to an 8 ha unfortified village atop the NW tell 8 m above its environs (1986:12, 23). However, in a recent sounding Frank Braemer identified the Middle Bronze Age fortifications that delineate the settlement and determined that it was, in fact, about 21 ha in size (700 x 300 m, see Braemer 2002:72), the outline of which can be seen in an aerial photograph of the site (ibid., fig. 2).

By means of ceramics the Middle Bronze Age fortifications excavated on the southern side of the site have been dated to the MB IIB, ca. eighteenth century B.C. according to Braemer (2002:70).[45] The fortifications consisted of an earthen rampart, fortification wall, and possibly a revetment wall. The best preserved element of the fortifications is the foundation of the *fortification wall*, which was nearly 6 m wide and built of undressed stone blocks averaging 0.8 x 0.6 m in size (ibid., 69). Although a breach was detected in the wall, which was later filled when the wall was reused during the Nabatean period, it is impossible to determine when this breach was made.

Only limited exposure of the top layers of the *earthen rampart*, consisting of alternating layers of loose earth and compact clay, was achieved outside (to the south) of the fortification wall (Braemer 2002:69). The rampart's southern edge may have been contained by a *revetment wall*, though only the top portion of it was encountered (ibid., fig. 3, lowest part of left side of section). Neither a glacis nor fosse was encountered. Based on the surviving remains and that the tell sits 8 m above its environs, it is possible to suggest that only the uppermost layers of the earthen rampart have thus far been encountered. Several more meters of the rampart probably remain unexcavated. Nevertheless, Braemer observes that as an MB II fortification system the basic plan of the site was rectangular (ibid., 73). He has also considered the possibility the massive "citerne," a large open reservoir in the center of the site, may have been included as part of the site's plan.

As mentioned above, the Middle Bronze Age fortification wall was used as a foundation for the narrower fortification walls of the Nabatean and Byzantine periods. No Late Bronze Age fortifications were detected in this area.

BURGA, TEL

MR 1472.2143
Refs.: Kochavi et al. (1979).
Plans: Site (fig. 9).
Sects.: NA.
Photos: NA.

Tel Burga is located just inland from the Mediterranean coast only a couple of kilometers to the SE of Tel Mevorakh. As a result of agricultural plowing, the visitor to Burga today can still see the remains of a large, dark mudbrick *fortification wall* that once crowned the edge of the mound above its slopes. This wall has been dated to the MB IIA by pottery found during the course of a salvage excavation in 1966 in Area C on the W side of the mound, which was not completed due to the start of the farming season. From this limited excavation it was established that

[45] Although Braemer suggests that this date is also supported by the type of Middle Bronze Age fortification system (2002:70), this reasoning is circular.

the *earthen rampart* was "protected from erosion by mudbrick *retaining walls* [emphasis added] and a mantle [i.e., *glacis*] of rubble stones" under which were found EB IV sherds (Kochavi et al. 1979:143). The length of this rampart around the site appears to have been about 1,500 m enclosing an inhabitable area of roughly 17.3 ha; the site has usually been identified as 25 ha in size (Figure 69 below).

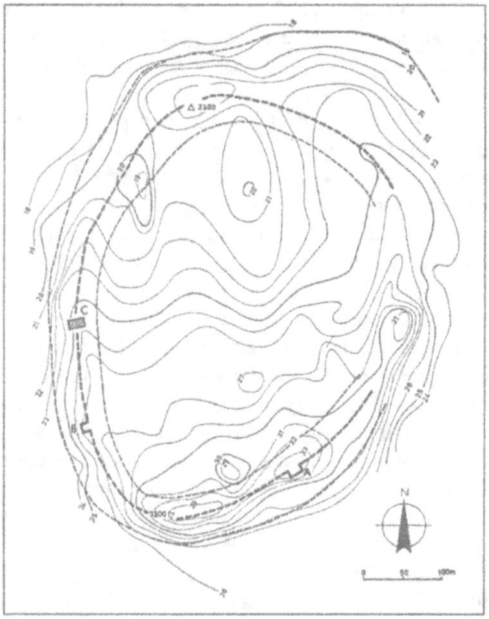

Figure 69. Plan of Tel Burga (after Kochavi et al. 1979:fig. 9).

The site rests an average of 7 m above the plain and was surrounded by a *fosse* the remains of which are now only a shallow ditch. Occupation within the enclosed mound was heaviest in the SW corner covering approximately 4 to 5 ha (no MB IIB–C sherds have been found). The remains of two mudbrick *towers*, Towers A and B, associated with the MB IIA wall were also identified on the S and W sides of the site respectively, though neither was excavated. With regard to the MB IIA defenses it is also important to note that Nahal Tanninim probably abutted the N side of the site and thus contributed to its defense as suggested by Kochavi (1979:n. 33). It seems probable that the location of at least two gates can be identified: between the walls on the E and W sides of the site, leading directly to the coastal plain and the fortress of Mevorakh to the W as well as the inland route E along the Nahal Ada.

DAN, TEL (LAIŠ)

MR 211.294

Refs.: Biran (1980; 1981; 1984; 1990; 1993; 2000); Biran et al. (1996).

Plans: Site (Biran et al. 1996:plan 1), Area K (Biran 1984:fig. 1; Biran et al. 1996:plan 10); Area T3 (Biran et al. 1996:plan 7).

Sects.: Area A-B (Biran et al. 1996:plan 4), Area K (ibid., plan 10), Area K gate (Biran 1984:figs. 2–7); Area Y (Biran et al. 1996:plan 9); Area T3 (plan 8); Area H (fig. 1.30).

Photos: Site aerial (Biran et al. 1996); Area A (fig. 1.4); Area B (fig. 1.19); Area T (fig. 1.46); Area T3 (fig. 1.52); Area Y (fig. 1.54, 1.56); Area K (figs. 1.57–59, 1.61).

Tel Dan, which was known as Laiš during the Middle Bronze Age, is located just to the SW of Mt. Hermon in the upper Huleh Valley (Figure 70). While archaeological remains from burials at Tel Dan confirm that it was settled during the MB I (IIA), the various remains of the Middle Bronze Age fortifications that have been revealed in different areas were only constructed during the transitional MB IIA–B, Stratum XI. Evidence for the rampart fortifications built during this phase of occupation was encountered in Areas A-B, K, Y, and T, making Tel Dan one of the most thoroughly studied fortified Middle Bronze Age settlements.

AREA A-B

Although Biran himself notes that the construction of the defenses was different in each area excavated, in each area the defenses do seem to have employed some type of core wall built of medium and large fieldstones against which earth was piled to form the rampart (Figure 71 below). On the S side of the site, a deep trench in Area A-B revealed that a stone wall served as the *core wall* for the fortifications, which seem to have been built higher as each layer of the rampart was heaped against it (Biran 1993:324; 1994:59–63; Biran et al. 1996:18f.). The stone-built core wall (Wall 1) measured 6.5 m thick and 10 m high. Although Biran suggests that the core and rampart projected at a slope of approximately 42° to 45°, indicating that possibly an additional 6 m of the upper part of the core and rampart are now missing, this height for the rampart and core wall cannot be verified. In section (Figure 71 below) the core wall is shown leaning backward, which suggests that the core was never freestanding but was built simultaneously. In 1996 more of the core wall on the S side was exposed and revealed to be 5 m wide (Biran 2000).

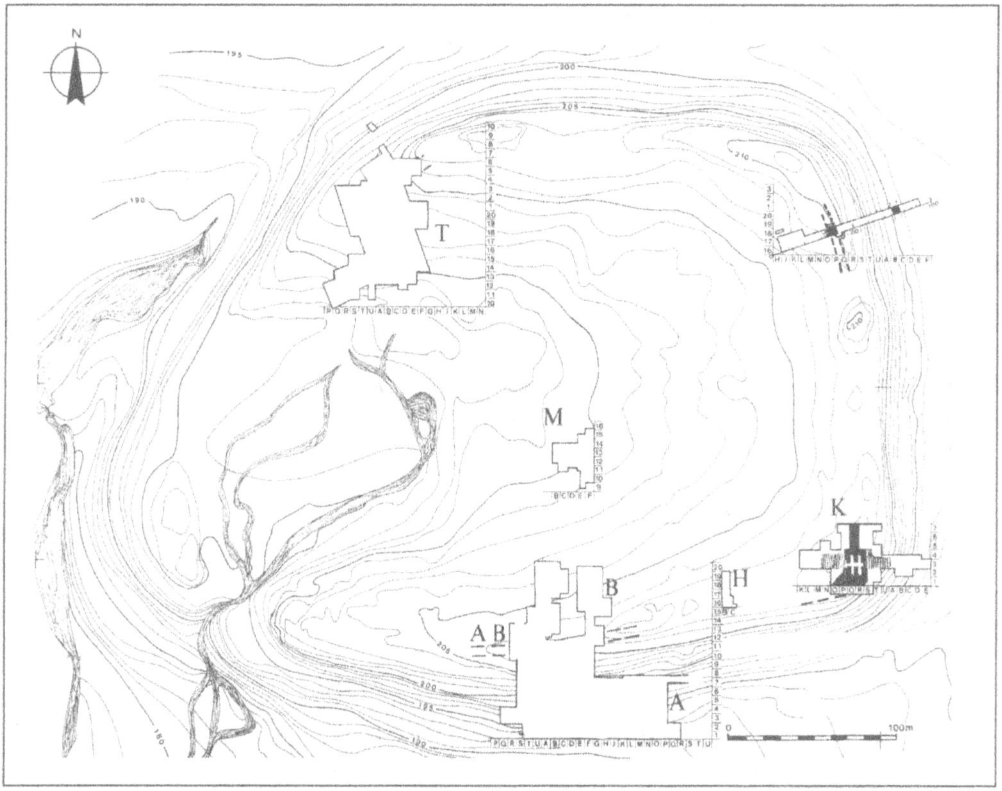

Figure 70. Tel Dan and excavation areas showing MB I–II features (after Biran et al. 1996:plan 1).

The S core wall is projected to extend more than 220 m from Area A-B to Area K on the SE corner of the enclosure.

The *earthen rampart* was composed of hamra and settlement debris including mudbricks, bones, Early Bronze Age and MB IIA pottery, and masonry (Biran et al. 1996:19). The base of the rampart measured 60 m wide in Area A-B. A *glacis* of crushed yellow travertine ($CaCO_3$) sealed both the interior and exterior surfaces of the rampart (Biran 1994:59–63). The date of the rampart was arrived at by the latest sherds found in its layers, which were of MB IIA date, and because the rampart overlays MB IIA burials (e.g., Tomb 23).

AREA T

In Area T on the NW side of the mound a 23 m stretch of an Early Bronze Age mudbrick fortification wall built on stone foundations with buttresses was exposed. These predate the MB IIA–B construction of the rampart (Biran 1993:324; 1994:68–70; Biran et al. 1996:51–53). The earthen rampart was built against these structures at the end of MB IIA. The date of the construction of the 3.5 m wide fieldstone wall (W.8897) and its buttresses (W.8894, 8895, 8898, 8937, 8933, and 9218), two of which along with the wall were topped by seven courses of mudbrick (W.7917)—presumably the superstructure of a mudbrick wall, is thought to be EB III. This date is based on pottery recovered from the fill of the rampart and a quadruped figurine bearing sacks, which was found on the surface of L.9348 upon which the rampart was laid. The buttresses, each 3 m wide and spaced 2.3 m apart, and the wall were preserved to a height of 4 m.

The excavation of the rampart in Area T revealed that measures were taken to shore up the older wall before the fill of the rampart was added, including the addition of stone fills (L.9315) against the wall. When the rampart was constructed its layers reached the mudbrick courses on top of W.8897 creating a slope of 20°, 13.75 m wide, and 5 m high. The construction in this area is that of a supplemental earthen rampart since it does not seem that there was an interior slope

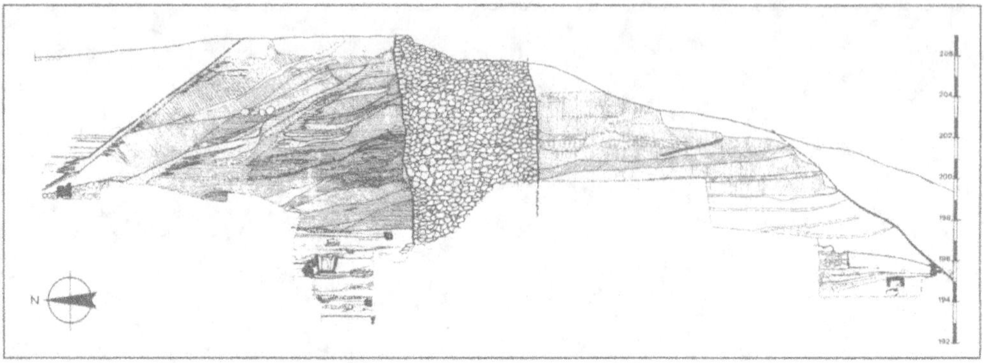

Figure 71. East section of rampart at Tel Dan in Area A-B (after Biran et al. 1996:plan 4).

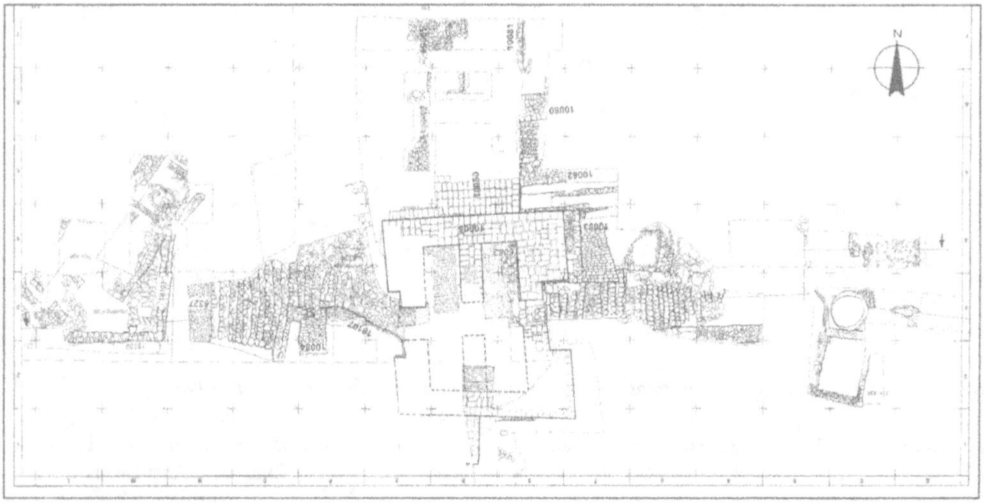

Figure 72. Plan of six-pier gate in Area K at Tel Dan (after Biran et al. 1996:plan 10).

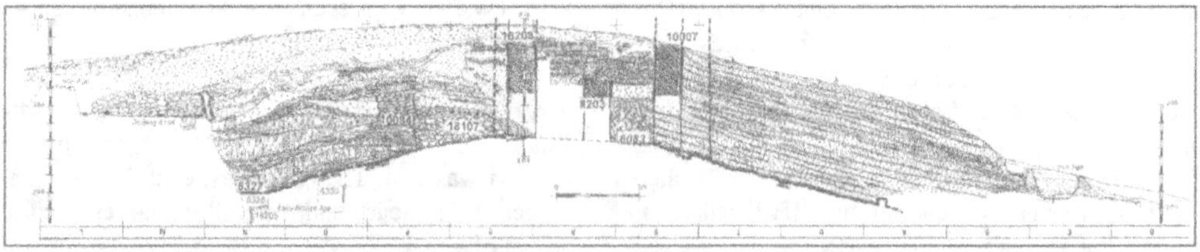

Figure 73. Area K section (after Biran et al. 1996:plan 10). View to the north. Note that image has been reversed to align with north oriented plan (see Figure 72).

to the rampart here, but only a stepped slope on the interior of W.8897 designated W.7901. Also, no remains of a mudbrick fortification wall datable to the Middle Bronze Age were preserved on top of the rampart or earlier wall, which now served as the core of the rampart. Although from plan 7 ("Post-Rampart" Middle Bronze Age "Platform" phase) it seems that a fieldstone wall (700) may have been built for use atop the rampart here, a date for its use has not been published and we can only surmise from the excavator's interpretation that it postdates the rampart. The only other Middle Bronze Age structure in this area was a mudbrick platform (8931) faced with fieldstones measuring approximately 8 x 5 m, which was found to the E of the rampart excavations. It is considered part of a "post-rampart" phase during the Middle Bronze Age and it may have been the base of a tower.

AREA Y

In Area Y a previously existing but somewhat dilapidated fieldstone structure (W.9001) dated to the MB IIA or Early Bronze Age served as the core of the *earthen rampart* on the site's NE side (Biran 1993:324; 1994:63–67; Biran et al. 1996:53–57). Its interior face was sloped, almost terraced, while its exterior face was nearly vertical. The core measures 8.5 m high by 13 m wide. After its construction a series of small fieldstone walls (W.9408, 9400A, 9400B, 9403, and 9407) averaging about 2.5 m apart were built parallel to the exterior of the rampart as *retaining walls* for debris that had been heaped between each of these walls and the rampart's expanding surface. A new retaining wall was added each time the slope became too steep against which to deposit earth. The composition of the debris used in the rampart here appears to have been the same as that used in Area A-B.

Although Biran suggests that the foundation W.9020 and its ten-course mudbrick wall were built above the fieldstone core for the purpose of extending the height of the rampart, several arguments contradict this. First, the easiest way to extend the height of the rampart would have been to collect more fieldstones, of which there is no shortage in this region. One would also expect that if it were an extension of the core wall that it would have rested directly upon it. Second, in Area K (see below) a 3 m wide mudbrick wall clearly accompanies the fieldstone core and is related to the construction of the triple arched gate. It seems, therefore, that the remains of the mudbrick wall measuring 3.5 m wide and only preserved to a height of 1.8 m were actually the remains of the *fortification wall.* This is in contrast to Area A-B, where more than 6 m of the top of the rampart, probably including its mudbrick wall is thought by Biran to have eroded or to have been robbed for materials. It seems, therefore, that these mudbrick remains cannot be dismissed as merely the upper portion of the rampart's core wall in this area. What is observable, however, is that the wall did later subside because its foundation rested above the level of the core wall (W.9001) on fill like that used in the rampart. We cannot, though, determine whether this subsidence occurred during the life of the wall or only after its abandonment.

AREA K

The construction of the *earthen rampart* in Area K (Biran 1984:9; 1993:325f.; 1994:67f.; Biran et al. 1996:58–61), was slightly different than in other areas, though it too used existing structures, probably of Early Bronze Age date, to attain an elevation of about 12 m above the surrounding plain. However, the rampart here is only partially understood since discovery of the famous "triple arched gate" limited exposure of the rampart, which revealed only how the gate was fitted into the rampart (see Figure 72). What appears to be the original stone core (W.10081) for the rampart, which was not extensively excavated, terminates about 9 m N of the gate.

This core of the rampart was crowned by a mudbrick *fortification wall* the remains of which are preserved eight courses high. In actuality this part of the fortification wall was well preserved because it was buried since it served as a terrace (see pp. 60ff.). Another fieldstone wall (W.10054) was built parallel to it and against its W or interior face. Before the gate was connected to the rampart, a 2.8 m long by 7 m wide mudbrick extension wall (W.10050) was built onto the N side of the gate on a stone foundation. To the N of this two parallel fieldstone *retaining walls* (W.10062 and W.10080) were built to support a packed earth fill between the stone core (W.10081) and the mudbrick extension of the gate (W.10050).

As indicated above, the most amazing discovery in Area K was what has become known as the "triple arched gate." It is so-called because three mudbrick "basket" arches (also known as barrel vaults) were incorporated into the construction of the length of this pedestrian gate (Biran 1980; 1981; 1984). This mudbrick structure, which should be designated simply as a *six-pier gate*, measures 15.45 m across its façade, 7 m high (forty-seven courses of mudbrick), and 10.5 m deep. Each arch consisted of three courses of mudbrick which formed a half circle with a radius of 1.25 m. Although its construction did not feature the use of a mudbrick "keystone," on the E elevation seventeen courses were preserved above the arch, and fourteen on the W elevation. The passageway through the gate measured 13.5 m long, 2.5 m wide, and 3 m high. Plaster was used on the surface of the exterior of the gate, its floor, and interior. The E wall of the gate complex was estimated to be 3.5 m thick while the W, interior, wall of the gate was estimated to be only 2.8 m thick. Its entrance lies on an E-W line, and though it was not aligned with the center of the E rampart, it was probably the only gate on that side of the town. Although Biran asserts that the gate is preserved to its full height, he acknowledges that its superstructure may originally have been two stories tall (1984:12). No evidence for the gate's doors has been found. A "recess at the base of the mud-brick wall inside the eastern entrance of the gate" may have been used for a doorway, but this is uncertain.

Table 29. Calculated volume of ramparts at Tel Dan based on dimensions of stone core and earthen fills.

Area	*h.* (m)	*w.* (m)	*l.* (m)	Core *w.* (m)	Stone core *vol.* (m^3)	Earth *vol.* (m^3)
A-B[†]	10	45	300	6.5	19,500	48,000
K[††]	8	81	200	6.25	10,000	54,800
Y[‡]	11	60	315	16	36,036	67,914
TOTAL			815		65,536	170,714

[†]Biran's original height of 16 m in Area A-B (see Figure 71) suggests too much erosion and robbing of the rampart. Likewise, 60 m for the width of the rampart is a maximum based on Area A-B.
[††]It is difficult to estimate the rampart's width in Area K; this is certainly a maximum width based on its 14° slope.
[‡]Because the stone core is not square in section but was terraced on its inner side it is necessary to reduce the estimate of its volume by about 35%.

The gate was set into the ramparts with mudbrick *retaining walls* covered by basalt fieldstone walls (such as W.10083, eighteen courses high) intended to retain the rampart and protect the gate. Ceramic evidence indicates that the gate was built at the same time as the MB IIA–B rampart. The gate seems to have fallen out of use sometime shortly after its construction during MB IIB when it was intentionally filled in to prevent its collapse, hence its excellent preservation. That it may have already experienced structural problems before its use is evident from its E elevation, which leans outward, and that a fieldstone revetment wall (W.18107) had been built against its SW pier inside the town (see Figure 72). This late *revetment wall* pointing to the NW would have constricted passage through this already narrow space, at least causing the gate area to be used only by the occasional passersby. That the gate was structurally unstable can also be concluded because the gate did not feature towers and, therefore, the unusually narrow and very long piers on both sides of the gate's deep chambers were expected to bear the load of the gate's superstructure. Because basalt steps as high as 27 cm and as much as 40 cm wide led up to the gate on both sides, and the gate's width was approximately 2.5 m wide, it is unlikely that it was ever intended for wheeled vehicles to use. On the interior twenty such steps led to a pebble street (L.6327) 3.25 m below the floor of the gate and twenty more steps were found on the exterior leading down 3.5 m to the plain.

Labor Estimates

The excavation of the defenses in each of the areas discussed above reveals that different strategies were employed for the completion of the rampart, which contrasts with Biran's estimate that one million cubic meters of earth were required in what was assumed to be a fortification construction project employing homogenous techniques (Biran 1994:71). However, this figure was the product of the adoption of the dimensions of the rampart as revealed only in Area A-B as a basis for determining the size of the ramparts around the site. When Biran's own observation about the differences in rampart construction are used to revise the calculations, it is clear that the previous figure are inflated (Table 29).

By using the topographic plan of the site and the section drawings from each excavation area to determine the dimensions of each length of the rampart it is possible to provide a more precise calculation of the overall volume of the earthen ramparts and stone core at Tel Dan. The total length of the ramparts was 815 m divided among three dimensions.[46] As Table 29 illustrates, these figures reveal that, in fact, only 24% (236,250 m^3) of the total volume suggested by Biran (1990:65, n. 17) was necessary for the construction of the entire feature. Furthermore, 28% of the rampart actually consisted of core walls constructed of fieldstones and boulders that would have been easily gathered from the surrounding fields. The remainder of the structure (72%) consisted of 170,714 m^3 of earth, much of which consisted of occupational debris filled with Early Bronze Age and MB IIA sherds.

In light of more precise figures for the circuit of Dan's fortifications, the site's size should be reduced to 11.1 ha, especially once the area occupied by the ramparts (4.9 ha) is subtracted from its gross size of 16 ha (cf. Broshi and Gophna 1986:86).

[46]Area T has been excluded from these calculations for two reasons. First, no rampart was suggested by Biran to have been present on the W side of the site and an examination of the topography does not allow for one, though we may guess that a wall was necessary on the W. Second, Area T cannot serve as a basis for calculating the rampart's composition since its core was probably built during the Early Bronze Age with very little effort expended during the MB IIA–B to add an earthen rampart.

DEBBEH, TELL*

MR 2272.98
Refs.: Braemer (1984); Weiss (1991).
Plans: Site (Braemer 1984:fig. 17).
Sects.: NA.
Photos: Site (ibid., figs. 35f.).
Tell Debbeh is located a little less than 80 km to the SSE of Damascus and approximately 50 km E of Tell Ashtara in the Hauran of southern Syria along the Wadi A. Dahab (see Braemer 1984:fig. 2). Survey of the region by Frank Braemer confirms that the oval site, which is about 4 ha in size, featured a *fortification wall* (ibid., 242ff.). This feature was identified in the early twentieth century along with a ramp and *gate* (four-pier?), which had a passage 4 m wide and 14 m long. It led into the site from its NE side. Both of these features should probably be dated to the Middle Bronze Age (ibid., 244). Basalt was used in the construction of these features. Bruce McLaren has confirmed the similarity of the plans of the gates of Debbeh and Rukais (2001:103).

DEIR ʿALLĀ, TELL*

MR 209.178
Refs.: Franken (1961); Franken and Ibrahim (1978); Van der Kooij and Ibrahim (1989).
Plans: Site (Van der Kooij and Ibrahim 1989:fig. 17); Area E (fig. 96).
Sects.: NA.
Photos: Site aerial (Van der Kooij and Ibrahim 1989:fig. 14).
Tell Deir ʿAllā is located 29 km to S of Pella in the Jordan Valley. A mudbrick wall at the base of the SE part of the site in Area E was identified as the 1 m wide town *fortification wall*; it was preserved to a height of 2.5 m (Van der Kooij and Ibrahim 1989:75f.). This wall was built of mudbricks measuring 60 x 40 x 11 cm on a foundation of a single course of fieldstones (ibid., 76). It was assigned to Phase IV and dated to the MB IIC. If it roughly followed the –220 m contour line the fortifications would have been about 530 m long (1 ha?). Although the excavators have suggested that some heaped clay against the S side of this wall may have constituted a *glacis*, further excavation is needed to corroborate this identification.

DOTHAN, TELL

MR 1727.2021
Refs.: Free (1958); Cooley and Pratico (1997); Master et al. (2005).
Plans: Site (Master et al. 2005:fig. 4.5); Area D (ibid., fig. 7.5); Area K (Free 1958:fig. 3).
Sects.: Area D (Master et al. 2005:fig. 6.3); Area K schematic (ibid., fig. 6.22).
Photos: Area K (Master et al. 2005:fig. 7.9).
Tell Dothan, a site of approximately 4 ha in size, is located 22 km N of Shechem on the edge of a small valley within the highlands of Samaria (Cooley and Pratico 1997; Master et al. 2005). Remains of the settlement's Middle Bronze Age fortifications were revealed in two excavation areas, D (to the S) and K (on the W).

AREA D

The MB IIB fortifications of Dothan were partly exposed in Area D and assigned to its phase 7. The section reveals that the Middle Bronze Age fortifications of Dothan were built well after the Early Bronze Age fortification walls (W.445 and 494) had fallen out of use. The stone foundations of the Middle Bronze Age *fortification wall* (W.444) were preserved over a length of 25 m to a height of 1.1 m, and 2.0 to 2.5 m in width (Master et al. 2005:50). A smaller wall (W.467) that was also dated to the Middle Bronze Age and that ran parallel to W.444 but was located 7 m down slope may have served as a *revetment wall* for the earth that had built up against the outer face of the remains of the Early Bronze Age wall, thus creating a *supplemental earthen rampart*. The 18.5 m stretch of W.467 was preserved to a height of 1.5 m and to 0.9 m in width.

AREA K

The Middle Bronze Age fortifications exposed in Area K were assigned to phase 2 (Master et al. 2005:52ff.) and included a *fortification wall* and according to Robert Cooley an *earthen rampart* (1997). Joseph P. Free, the site's excavator, reported that "following the Middle Bronze Age Wall . . . we discovered that it developed into a complex of rooms, dating originally from Middle Bronze with possible reuse and rebuilding in Late Bronze times" (1958:17). Approximately, 5 m of the wall that Free describes consisted of two faces with earth between them (W.351) with a total width of 1.97 m (Master et al. 2005:53). This wall was preserved to a height of approximately 2 m and Middle Bronze Age storage jars were discovered within rooms just inside the wall (Free 1958:14; Master et al. 2005:fig. 7.10). Its construction is, however, enigmatic since another 5 m stretch of this wall that continues to the N appears to consist only of the inner face of this wall according to Master et al. (2005:53). Furthermore, another wall (W.355) ran nearly parallel to W.351 but less than one meter outside of it; its function remains uncertain (ibid., fig. 7.8). A third wall, 13 m of which were exposed, ran at roughly 6 m outside of these walls, and probably served as a *revetment wall*, comparable to W.467 in Area D; it was preserved to a height of 85 cm and a width of 1.2 m.

Perhaps the most interesting feature of the fortifications in Area K was a tower, which was identified as either a defensive tower or a part of a gate (Master et al. 2005:54). As it was oriented perpendicularly to the line of the fortifications, it seems unlikely that the structure served as a defensive tower, which would have been built at regular intervals in the site's defenses (though admittedly a limited stretch of the fortifications was exposed). Instead, its alignment suggests that it should be identified as the S half of the Dothan's W *gate*. The walls of the structure (W.510, 511, 512, and 513), which were constructed in two phases, were between 1.5 and 2 m thick. The tower featured an entrance on its N side (i.e., along the passage of the tower). Further supporting the identification of this tower as part of a gate is the "angle wall" (W.389) that abutted the gate's S elevation and the N end of W.351 and extended to the SW at a 45° angle to the gate and fortifications. This wall also ran at 45° down the slope from the tower to the level of the revetment wall, suggesting that it had been constructed to retain the earthen slope below W.351 and above W.357. A depression in the contour lines at the foot of Area K may also support this identification (see Master et al. 2005:fig 4.5).

While the most recent publication of the excavations of Tell Dothan assists in our understanding of the site's Middle Bronze Age defenses, many questions remain and, therefore, the reconstruction offered here must be considered provisional.

Ekron (Tel Miqne)*

MR 1356.1315
Refs.: Gittlen (1992); T. Dothan and Gitin (1993).
Plans: Site (Gittlen 1992:51*).
Sects.: NA.
Photos: Site aerial (T. Dothan and Gitin 1993:1051).
Tel Miqne, identified as biblical Ekron, is located 22 km SE of Yavneh-Yam and 31 km to the NE of Ashkelon in the coastal plain. Although the publications of the remains of Ekron's MB IIB–C levels (Str. XI) remain forthcoming, the site's basic shape, according to Trude Dothan and Seymour Gitin, was the result of "the fortifications that encompassed both the upper and lower cities in the Middle Bronze Age" (1997:30). Middle Bronze Age sherds were recovered from every field excavated. From an aerial photograph of Ekron it is possible to discern the very straight line of the E rampart and fosse in the fields just to the E of the site and just beyond the E extent the trenches in Field I (T. Dothan and Gitin 1993:1051). If these lines are the lines of the Middle Bronze Age *earthen rampart* and the *fosse*, then the fortifications were approximately 2,200 m long and enclosed a space of about 22.2 ha (not including the area occupied by the ramparts).[47]

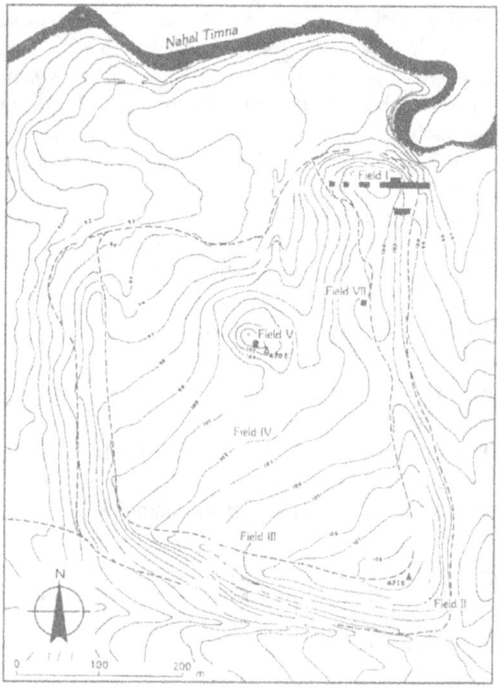

Figure 74. Plan of Ekron (after T. Dothan and Gitin 1993:1052).

The settlement contracted during the Late Bronze Age to occupy only the 4 ha acropolis on the NE side of the site nearest to the water source (Gittlen 1992:51*).

Esur, Tel (Tell Asawir)*

MR 1521.2098
Refs.: Dothan (1993c).
Plans: NA.
Sects.: NA.
Photos: NA.

[47] This estimate is conservative for the size of Middle Bronze Age Ekron since B. Gittlen suggests that the Middle Bronze Age settlement "covered most of the 80 acre tell" (1992:52). The variable here is the lower town which during Iron I was 16.2 ha but which increased to 26.3 ha during Iron II (T. Dothan and Gitin 1997:30).

Tel Esur is located at the S end of the Megiddo Pass (Nahal 'Iron) and was approximately 3 ha in size during the Middle Bronze Age (IIA–IIC) rising 7 to 11 m above the plain. Gophna and Beck identify the site as fortified with an *earthen rampart* (1981:47).

FAR'AH NORTH, TELL EL-

MR 1823.1822
Refs.: de Vaux (1962:237 ff.); Mallet (1973; 1988); Dever (1990b).
Plans: Site (Mallet 1988:plan 2); Area II (ibid., plan 27); Area III (1973:pls. 8, 10); Area IV (1988:plans 3–5, 9, 10f., 14, 27).
Sects.: Area II (Mallet 1988:plans 37, 39f.); Area IV: Trench 7 (plan 18), Trench 8 (plans 20–22), Trench 143 (plan 19), Trench 778 (plans 23f.).
Photos: Site aerial with fortifications (Mallet 1988:pls. 1); Area II (ibid., pls. 6–9); Area IV (ibid., pl. 9); W wall (1973:pl. II); W glacis (ibid., pl. III).
Tell el-Far'ah North is located about 9 km NE of Shechem. Following an abandonment during the second half of the third millennium, the site was reoccupied during the MB IIA–B as an "open village" (Per. VA) in the hill country (de Miroschedji 1993:437). This settlement was followed by another unfortified phase (Per. VB) dated to the MB IIB, which was finally followed by a fortified settlement during the MB IIC, designated Period VC [Str. B3] (ibid., 438). The area enclosed by the MB IIC fortifications of Far'ah North was smaller than the area enclosed by its EB II fortifications, approximately 3.9 ha (approximately 260 x 150 m).

The MB IIC fortifications of the site consisted of a revetment wall, fortification wall, gate, towers, and an earthen rampart. The *fortification wall* (mur 1) revealed in Areas II, III, and IV (on the W side of the site) was 2.2 m wide with large stone foundations and was preserved to a maximum height of 2.3 m (Mallet 1988:43f.). The N part of the MB II wall (Area II), of which a stretch of about 32 m was traced, was built more than 4 m wide on the ruins of the second EB II wall, which had been built of stone. From its E end the N wall then turned toward the ESE, which meant that much of the NE area of the Early Bronze Age town was not included within the Middle Bronze Age settlement's wall. At the W end of the N wall it made an abrupt almost 90° turn to the S. Here the W wall to the N of the gate was built 2 to 3 m wide atop the mudbrick ruins of the EB II wall (mur III). The W wall continued S of the gate where after 29 m a number of stone buttresses were discovered along 23 m of its interior face N of the tower (see below) spaced 2 to 2.5 m apart (ibid., 46f.). These buttresses measured 0.5 to 0.8 m deep and were 1 to 1.4 m wide. More of these buttresses were also found along the 39 m long interior face of the wall (here 2 m wide) S of the tower with spaces between them of up to 2.8 m. In all about 150 m of perhaps a total of 820 m of the settlement's perimeter wall were exposed.

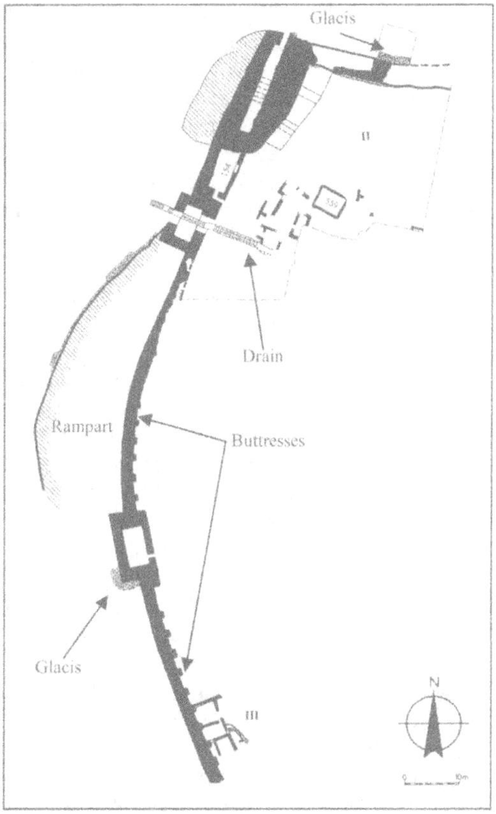

Figure 75. Fortifications of Tell el-Far'ah North (after Herzog 1997a:fig. 4.19).

A *four-pier gate* (locus 152) was built just N of the earlier EB II gate upon the remains of an Early Bronze Age tower, which served as its foundation (Mallet 1973:29f.; 1988:44). The gate's stone foundation was seamlessly bonded to the wall running S from it with this wall also forming the SE pier of the gate. The connection between the gate and the wall to the N has been lost. The gate is a direct axis, four-pier gate measuring 9.5 m wide and 7.5 m deep with an interior chamber measuring 6 m long and 3.5 m wide. The structure protruded 5 m from the line of the wall and featured walls and piers 1.5 to 1.8 m thick. The narrowest point was between the gate's outer piers and measured 2.5 m wide. Below the floor of the passage was a large drainage channel built of stone and exposed for about 22 m (cp. drain under Ashkelon's Phase 12 gate, Figure 65).

A rectangular *tower* (loci 301–303) was identified 50 m to the S of the gate along the SW stretch of the town wall (Mallet 1988:47ff.). Its dimensions were 12.7 m by 7 m and it was divided into two rooms both of which featured entrances. Its walls were about 2 m thick, comparable to the thickness of the town wall in this part of the town. A cobble *glacis* was constructed for the protection of the S corner of the tower, but its extent is limited and it may have dated to the Late Bronze Age.

Although Mallet suggests that the two ramparts, which he referred to as "glacis," constituted separate phases (Ph. B2–B3/4) in the construction of the site's defenses (1988:50ff.), there is little basis for accepting that these elements constitute separate ramparts. The addition of the lowest layers of an *earthen rampart* (Mallet's "glacis de terre noire") composed of black earth on the W slopes of the site appears to have been only the first constructional phase of this feature (ibid., 38). As seen in the section of Trench 8 (ibid., plan 22) this 8 m wide and half-meter thick deposit began about 4.5 m out from the base of the wall where it was retained by two stepped walls opposite the town wall. This was overlaid by a layer of cobbles and the addition of red earth, Mallet's "glacis de terre rouge" of Phase B3–4, the exterior slope of which was retained by a stone *revetment wall* (mur 4) up to 2.8 m high and more than 60 m long. This upper portion of the rampart also rested at the base of the town wall, though in some areas this connection has been clearly lost (ibid., 51ff.).[48] The average slope of the surfaces of the preserved layers of the rampart was about 24°, though the original rampart surface may have been as much as 30°.

The steep slope up from the Wadi el-Daleib on the S side of the site appears to have been a sufficiently difficult ascent so as not to necessitate any defensive additions or modifications other than the town wall. Outside of the wall N of the gate a part of a 6 m section of a stone cobble *glacis* (locus 122) at least 3 m high was exposed (Mallet 1988:50f., 77ff.). This glacis rested against the foot of the wall and must have run the entire length of the N wall.

FAR'AH SOUTH, TELL EL-

MR 100.076

Refs.: Petrie and Tufnell (1930); MacDonald (1932); Gophna (1993a).

Plans: Site (Petrie and Tufnell 1930:pl. 51); Recent site plan (Lehmann and Schneider 2000:fig. 1); S Gate (MacDonald et al. 1932:pl. 77).

Sects.: W side, N section (Petrie and Tufnell 1930:pl. 13 top); N side, W section (pl. 13 bottom).

Photos: Site aerial (Petrie and Tufnell 1930:pl. 3); gully on N side (pl. 52:2).

Tell Far'ah South's position 22 km S of Gaza along the W bank of the Nahal Besor afforded it considerable defense on its E side (Petrie and Tufnell 1930:15). Although Tell el-Far'ah South featured Middle Bronze Age fortifications, like Tell el-'Ajjul the site suffers from the early date of its excavation (1928–1929), which therefore limits the conclusions that can be reached. Nevertheless, several aspects of the MB IIB–C defenses of Far'ah South can be remarked upon. Ravines leading into the wadi adjacent to the N and S sides of the site probably served as a natural defense for the settlement and may, as Petrie suggested, have been excavated in antiquity to enhance this defense (Petrie and Tufnell 1930:16). It is difficult, though, to determine to what extent the meandering of the wadi removed what might have been a considerable part of the E side of the site. The remainder of the site was clearly defended by typical elements of Middle Bronze Age fortifications including a fosse, rampart, glacis, fortification wall, and gate.

The *fosse*, excavated on the W side of the site, measured 24.4 m wide between its upper edges (Petrie and Tufnell 1930:16). The outer bank of the fosse featured a slope of about 40°. Based on Petrie's plan the fosse does not appear to have been more than 180 m long. At the bottom of the fosse was a 1.2 m wide and 6 m deep ditch with a vertical face 2.3 m high serving as a natural revetment at the foot of the rampart (ibid.). The *earthen rampart*, which was heaped up behind this revetment and above the original ground level was 6.3 m high and 11.6 m wide with a slope of 33°; the distance from the top of the rampart to the edge of the revetment along its surface was 18.9 m. The top of the rampart appears to have been leveled for the town wall, though it has not been preserved in the excavated area. On the N side of the site the rampart featured a slightly different construction; here it was 8 m high having been built upon basal sand out of marl in at least two distinct phases (ibid., 17, pl. 13 bot.). The inner slope of the rampart is about 30° while that of the outer slope, which is over 30 m long, is approximately 40°. If the lines of the rampart in the drawn section are projected then the N rampart was about 30 m wide; this may have

[48]Although Mallet identifies what he called a fosse cut into the second rampart ("glacis de terre noire" 1988:54f.), the sections do not reveal the existence of such a feature.

been the rampart's actual width since the interior of the rampart on the W side was never exposed.

The top of the N rampart featured a leveled base about 5.5 m wide upon which a *casemate wall* was constructed. The construction of this wall appears to have consisted of two brick walls, the inner of which was more than 1.5 m wide, while the outer wall was 0.5 m wide. The entire thickness of this wall appears to be more than 3.6 m. The wall around the N, W, and S sides of the site was at least 290 m long, and depending upon how much larger the E side was prior to its erosion, the fortifications on that side may have been at least another 200 m in length. Overall the rampart's top was situated between 6 and 8 m above the level of the surrounding plain.

At least one gate of the MB IIB–C settlement on the S side can be identified, though Petrie had suggested that another was located on the NW side of the site. The S Gate was aligned, undoubtedly, with the route to the SW on the coastal road to Egypt (MacDonald et al. 1932:29f. and fig. 77). Although only the W half of the gate was preserved, it appears to conform to the traditional *six-pier gate* for which two chambers were preserved in the W tower. Each of the 2.2 to 3.8 m wide piers was situated about 3 m apart with a passage between the piers about 3.6 m wide and 18 m long. The façade of the tower was about 24 m as measured along its N elevation. The gate was built of mudbricks directly on sand (ibid.). On the gate's W side remnants of the roughly 2.5 m wide town *fortification wall* abutting the gate were exposed. That a single wall of a varied thickness from the walls of the casemate on the N rampart (see above) was used in this part of the site supports, as at other sites, the identification of intra-site variability within fortification construction. The gate projected a little more than 4 m from the town wall. The interior threshold of the gate was paved with boulders and cobbles. The gate's mudbrick construction is dated to the MB IIB–C with renovations dated to the LB I.

An examination of the plan of the site with specific attention given to the orientation of the S Gate in relation to the remainder of the site suggests that much of the E half of the site has been destroyed by the meandering of the wadi. If the site was roughly circular in layout with a diameter of 200 m, then the inhabitable area within the Middle Bronze Age walls would have been limited to about 3.1 ha (cf. 6 ha, Broshi and Gophna 1986).

GATH (TELL ES-SAFI)*

MR 1359.1237
Refs.: Uziel and Maeir (2005).
Plans: Site (Uziel and Maeir 2005:fig. 2); MB settlement size (ibid., fig. 4).
Sects.: NA.
Photos: NA.
Tell es-Safi, identified with biblical *Gath* of the Philistines and Late Bronze Age *Gimtu* of the Amarna letters, is located 30 km ENE of Ashkelon on the S bank of the Wadi Elah in the Shephelah along the route leading E into the hill country. Although Gophna and Beck identify it as a fortified settlement during the MB IIA (1981:47), not until recent excavations by Aren Maier has any evidence of the settlement's Middle Bronze Age fortifications come to light. However, it is most likely that these fortifications date to the MB IIB–C.

Although the results of a survey of the site conducted by Aren Maier and Joe Uziel revealed a "surface scatter" of MB IIB–C pottery of about 10 hectares (Aren Maier, personal communication, 2006), they suggest that the site size was only about 8 ha in size at that time (Uziel and Maeir 2005:Table 3). During the July 2006 season the first remains of the site's MB IIB–C fortifications, including a well-built stone *tower* and whitish layers of the *glacis* that probably belonged to an earthen rampart, were encountered just to the NW of the summit of the tell in Area F (Aren Maier, personal communication, 2006). Unfortunately, the underlying shape of the area of the site occupied during the Middle Bronze Age is not suggestive of the site's morphology at that time (Uziel and Maeir 2005:figs. 2 and 4).

GERISA, TEL (TELL JERISHE)

MR 1320.1665
Refs.: Geva (1982); Herzog (1984a; 1984b; 1990; 1991; 1997a; 1997b).
Plans: Site (Herzog 1984a:29); SE (fig. 3).
Sects.: SE (Geva 1982:figs. 12, 13, 15–22).
Photos: SE (Geva 1982:pls. 7–10).
The mound of Gerisa is located about 6 km NE of Jaffa and 15 km W of Aphek. Shulamit Geva's publication of Eleazar L. Sukenik's excavations on the SE side of Gerisa serve as the primary basis for a detailed analysis of the Middle Bronze Age fortifications, which were first excavated between 1936 and 1951 (Geva 1982:10–18). However, this report must be interpreted in light of more recent excavations conducted by Ze'ev Herzog in Field A on the SSE side of Gerisa since 1981, which have sought to clarify the sequence of Gerisa's Middle Bronze Age fortifications (Herzog 1984a; 1984b; 1990; 1991). These later excavations have exposed much of the surface of the mound (2.6 ha) and revealed that the earliest

phase of MB IIA settlement was unfortified, like the EB III settlement, but was followed by two fortification systems the first dated to the MB IIA and the second to the MB IIB period (Herzog 1991:51), in contrast to the one system Sukenik had originally identified (see Geva 1982:14ff.).[49] Nevertheless, prior to final publication of Herzog's excavations, any understanding of the history of the fortifications can only be considered preliminary.

MB IIA, PHASE 3

Following a MB IIA phase of unfortified settlement (MBIIA, Phase 2 in Cohen 2002a:84), which was also detected in previous excavations (Geva 1982:14), Gerisa was first fortified during the MB IIA3 with a 1.7 m wide mudbrick *fortification wall* constructed along the edge of the mound at ca. 130 m ASL (Herzog 1991:51). Against the exterior of this wall a *glacis* was added, which was composed of hamra and small stones covered with a layer of crushed kurkar sandstone. The settlement associated with this wall seems to have experienced a conflagration since collapsed mudbrick and destruction debris had accumulated against its inner face during an early phase of its use. Levels associated with this glacis do not appear to have been encountered by former excavations at the site.

MB IIA, PHASE 4

During the next phase of MB IIA occupation the *fortification wall* (2.2 m wide in Field A) was rebuilt above the remains of the earlier wall and almost one meter within the perimeter of the earlier wall (Herzog 1990:61). This phase also featured the building of a second *glacis* which covered the slope with thirteen courses of mudbrick in Field A, which were then covered with a layer of crushed kurkar sandstone. This glacis corresponds with the glacis identified in Sukenik's excavations, which was also constructed of mudbricks though with far fewer courses; they appear to range from one course (Area IX) to as many as six courses (Area D). Mudbricks used in the glacis varied from 35 x 35 cm to 40 x 50–55 cm (Geva 1982:15). Geva described this glacis as interlocking with the contemporary town wall, which was reportedly 1.8 m wide (ibid., 15), though sections suggest that it was 4.6 m wide in Area D (ibid., fig. 12) and about 2.7 m wide in Area IX (ibid., fig. 17). The slope of this glacis was 20° in Field A according to Herzog (1990:61) but about 25° in Sukenik's section (Geva 1982:figs. 12f., 16, 20). Each of the two MB IIA town walls were associated with at least two floors (Herzog 1997a:113).

MB IIB–C

At some point the fortification wall was replaced by yet another *fortification wall* built of mudbricks 3 m wide during the MB IIB–C, which was itself accompanied by the building of a third *glacis* over the second one (Herzog 1997b:395). The site was destroyed in the transition between the MB IIC and LB I (Herzog 1993a:481).

LB I

The town wall may have continued in use during the LB I when small domestic structures were built against it, though later during the Late Bronze Age the site seems to have been unfortified (Herzog 1997b:395).

GEZER (TELL EL-JEZER)

MR 1425.1407

Refs.: Macalister (1912); Rowe (1935); Dever et al. (1970); Bullard (1970); Kempinski (1972); Dever (1973a; 1973b; 1974b); Seger (1975); Bunimovitz (1983); Hardin and Seger (2006).

Plans: Site (Dever et al. 1970:plan 1); Inner and Central walls (Macalister 1912, vol. 3, fig. 51); S Gate (ibid., vol. 1, fig. 125); W Migdol tower (Rowe 1935:General Plan); Field I: Tower 5017 (Dever et al. 1970:plan 8).

Sects.: Central Wall (Macalister 1912:fig. 119); Field I (Dever et al. 1970:plan 2).

Photos: Inner Wall (Macalister 1912: vol. 1, figs. 118, 120–4, 131); W Tower (Rowe 1935:pls. 1, 5f.); Field I: Tower 5017 (Dever et al. 1970:4b, 6a–b).

Gezer is situated on the route from the coastal road E into the hill country about 42 km to the NE of Ashkelon. Gezer was approximately 7 ha in size within the line of its MB IIC wall (Herzog 1997a:156). The fortifications of Gezer in the Middle Bronze Age appear to date entirely to the MB IIC, Strata 11a–10a in Field I (Figure 76). Excavations by the Hebrew Union College (HUC) expedition in Field I in the SW part of the site, in addition to previous excavations by Robert A. S. Macalister and Alan Rowe, revealed the remains of these fortifications, which included a fortification wall, rampart and glacis, towers, a bastion, and the S Gate. The MB IIC date assigned to these remains was established by the HUC expedition and, therefore, the inclusion of the results of the earlier expeditions follows their redating of these features. The correlations among the various features of Gezer's Middle Bronze Age defenses and the areas excavated by the different expeditions are summarized in Table 30.

[49] The three phases identified by the most recent expedition could be seen quite clearly in Geva's section of Area IX B from Sukenik's 1936 excavation season (Geva 1982:fig. 18).

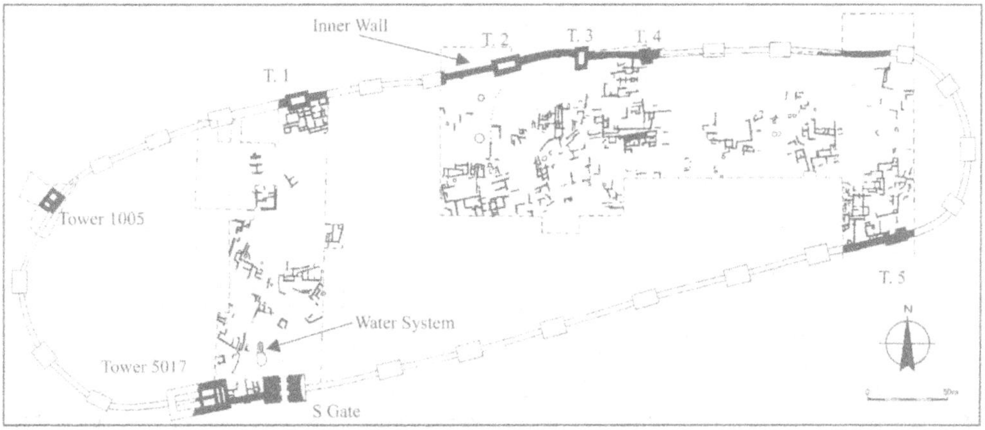

Figure 76. Fortifications of Gezer (after Herzog 1997a:fig. 4.20).

Table 30. The Middle Bronze Age stratigraphy of Gezer (after Dever 1986b:fig. 2).

Period	Stratum (HUC)	Macalister	Rowe	Field I	Field IV
MB IIA	XXII				
MB IIB	XXI			Str. 10	
MB IIB–C	XX			Str. 9	
MB IIC	XIX	"Inner Wall" (town wall), 5 towers	Tower 1005	Str. 8A–B Tower 5017	Str. 6
MB IIC/LB IA	XVIII	"Central Wall" (revetment)		Str. 7A–B Glacis 8012	Str. 5A–B Wall, S Gate, Glacis
LB IA				Glacis 8009/9007	

MB IIC: STRATUM XIX

Substantial evidence of the MB IIC fortification wall, to which Tower 5017 was apparently connected, and the wall's towers has been brought to light. The town wall, referred to as the Inner Wall, was first exposed by Macalister (1912:236ff.) along with two other town walls, which included the Central Wall and Outer Wall. Although the Outer Wall was recognized as post-Middle Bronze Age in date, Macalister suggested that the Central Wall, which he identified as "a rude earth bank lined inside and outside with a stone facing," preceded the construction of the Inner Wall (ibid., 236). However, as the HUC excavations revealed in Field I part of the so-called Central Wall actually consisted of a *revetment wall* (see Stratum XVIII below) below the *earthen rampart* (Macalister's "rude earth bank," Glacis 8012 of HUC excavations, discussed below). The revetment wall followed the construction of the Inner Wall with which Tower 5017 was associated.

The Inner Wall, which Macalister exposed around much of the site (see "First Semitic Period" in Macalister 1912: vol. 3, pl. 2) is to be identified, therefore, as the *fortification wall* of the MB IIB–C settlement. The wall was recorded by Macalister to be roughly 4 m wide with towers spaced about 27 m apart with dimensions of roughly 12.5 m long by 7.3 m wide (ibid., 238). The town wall's foundation was constructed of many courses of "hammer-dressed stones" ranging from very moderate size to some of the largest stones used in any construction at the site (Macalister 1912:238f.). The complete length of this wall would have been approximately 1,160 m around the entire site. Close examination by the HUC expedition of an original photo taken by Macalister's expedition of a tower on the Inner Wall revealed that much of its mudbrick superstructure remained intact (1912:vol. 2, fig. 121).

Five complete *towers* (see Table 31) and a part of another tower associated with the town wall were exposed by Macalister's expedition (1912: vol. 3, pl. 3). A seventh tower appears to have been exposed by Rowe on the N end of the W side of the site, though he claimed that the wall built along the structure's E side, which was 2.35 m wide, was not the town wall but that it was located "much further to the west" (1935:28). It is likely that Rowe was referring to Macalister's Outer Wall which was dated to the Iron Age and was located further W. Reconstructions of the plan of the Gezer's Middle Bronze Age fortifications since Dever et al. (1970:plan 1) have favored an identification of this structure as a tower on the town wall (see also Herzog 1997a:fig. 4.20). The tower (no. 1005) was 13 m long by 8 m wide with walls ranging in thickness from 1.8 to 2.3 m (Rowe 1935:28). Its foundations in contrast to those of the town wall in other areas or Tower 5017 appear to have been rather shallow, perhaps only one or two courses of large stones (ibid., pls. 1, 5–6), and no evidence of its mudbrick superstructure has survived.

Portions of Tower 5017 were exposed during the first season of the HUC expedition 1964–65 (Dever et al. 1970:18ff.). Although the building was identified as a tower, the structure is actually a *bastion* (cp. Ebla's bastions). The entire bastion was not excavated but projections of its dimensions suggest that it was 15.6 m wide and more than 22 m long (ibid., 18).[50] Macalister had originally associated this structure with the "Inner Wall," and Dever had once assumed that the structure was a "localized thickening of the wall" just to the W of the S gate (ibid., 19). The foundation of this bastion was built primarily of large stones (avg. 1 m long by 0.75–0.9 m wide) with smaller stones between them and stood about 4.25 m high; the mudbrick superstructure has presumably been robbed out (see also ibid., n. 3 and town wall above). The stones of the exterior (S) elevation of the building were dressed, while those used on the interior (N) elevation were not because, as the identification of a foundation trench in the section illustrated (cp. ibid., plan 2, p. 41), this face would never have been exposed. The earliest surface associated with this fortress was attributed to Field I's Stratum 8b (ibid., 19), and the latest pottery in the foundation trench associated with this building was dated to the MB IIC (ibid., 42). In light of this evidence Kempinski's attempt (1972) to redate this structure to the MB IIA cannot be accepted (see also Dever 1973b).

[50]Herzog suggests that the bastion was up to 30 m long (1997a:156), which is in keeping with the ratio of 2:1 (*l.*:*w.*) of other Middle Bronze Age towers (see p. 65). His illustration suggests that it was up to 40 m in length (see Figure 76).

Table 31. Dimensions of Gezer towers (from Dever 1986b:plan 1).

Tower	*l.* (m)	*w.* (m)
No. 1	12	8
No. 2	16	8
No. 3	8	12
No. 4	6	8
No. 5	13	6
Tower 1005	13	8

MB IIC–LB IA: Stratum XVIII

In addition to the evidence of the town wall and towers both Macalister and the HUC expedition brought to light evidence of an *earthen rampart* (Glacis 8012) built against the MB IIB wall during the MB IIC–LB IA (Dever et al. 1970:42ff.). As Dever has convincingly argued, the rampart revealed in Field I was supported in a later stage by a revetment wall (9011) located at its base (ibid., 42), which Macalister also identified but called the Central Wall, not realizing that it was constructed after and against the Inner Wall (see Macalister 1912:vol. 1, p. 236, esp. fig. 120 where the rampart can be seen sloping down from the Inner Wall to the Central Wall). However, Dever's suggestion that Macalister removed the glacis between the Inner Wall and Outer Wall being "unaware of what it was" ignores that he did consider the "rude earth bank" to be a part of the construction of the Central Wall.

Dever also assumes that the glacis had been constructed around the entire site as it appeared in Field I. However, Macalister had already observed that it was "by no means uniform throughout its length" (ibid.). This lack of uniformity in construction is attested at other Middle Bronze Age sites (e.g., Dan, Tel Haror, and Jericho). Nevertheless, as the section of the excavation of the rampart (Glacis 8012) in Field I by the HUC expedition revealed (Dever et al. 1970:plan 2), the rampart was 6.5 m high and 10 m wide (see Figure 7).[51] The section also reveals the very highly developed technique ("sandwich" method) which was employed in which alternating "dikes" of crushed chalk and concave layers of debris were heaped to achieve a 45° slope for the rampart.

[51]Although Dever was cautious not to offer figures concerning the size of the rampart, based upon evidence from other sites it would seem that the rampart probably continued at the same slope up to the foundation of the wall (Tower 5017), which may originally have stood a couple of meters higher. Furthermore, the rampart's dimensions and composition seem consistent between Dever's Field I and Macalister's exposure (Macalister 1912:vol. 1, fig. 120).

These layers of the rampart were sealed by a *glacis* of "impermeable chalk paste" (ibid., 43). According to Reuben Bullard it was necessary "to go to the base of the local Shephelah" to obtain the unweathered chalk used in the construction of the glacis (1970:119). At a later date the *revetment wall* (9011) was constructed out of sixteen courses of stone standing 4.5 m high and 3 m wide (Dever et al. 1970:43).

According to Dever the evidence for the rampart (Glacis 8012) and later revetment wall (9011) suggest that the rampart and its glacis should be dated to the MB IIC, while the revetment wall dated to the LB IA when another glacis (8009/9007) was added behind this wall and above the original glacis (Dever et al. 1970:44). Dever's interpretation of the relative sequence of these features and the assignment of Wall 9011 to a period following the construction of the rampart 8012 is corroborated by the section drawing (ibid., plan 2). Dever also suggests that a mudbrick structure was built atop Wall 9011 (ibid., 44) in this period, which supported the final glacis, though Macalister makes no reference to such a wall in his discussion of the Central Wall. Nevertheless, 9011 appears to have continued in use through the Iron Age and it became the foundation for the construction of the Outer Wall (Dever 1974a:35ff.).

Although Macalister attempted to identify two gates associated with the Middle Bronze Age wall, one on the NE (see Macalister 1912:vol. 3, pl. 51) and another on the W end of the S side (ibid., I, fig. 125), his identification of the NE gate has not been accepted. The S Gate, on the other hand, which was preserved to a height of 4.3 m, was a *six-pier gate* built between two mudbrick towers with dimensions of 8.7 (E-W) x 12.9 (N-S) m and 8.5 x 12.9 m (ibid., 241). The gate's façade measured 24 m across and it was 12.9 m deep. Bricks used in the gate's construction averaged 38 x 29 x 10 cm. Macalister also detected lime plaster coating on many of the bricks. Because the gate was not set "at right angles to the wall" the E tower projected slightly further from the wall. The passage between the towers, which descended slightly from within the town, was composed of "roughly laid" fieldstones and was 12.9 m long and 2.75 m wide. Three pairs of orthostats were added to the two sides of the passageway to create the six piers of the six-pier gate. The orthostats on one side of the gate were spaced approximately 1.97 m apart (ibid., 242). These orthostats were incredibly large with an average thickness between 57 and 60 cm; the length of those on the W side from S to N measured 2.7 m, 2.5 m, and 2.2 m, while those on the E from S to N 2.58 m, 2.65, and 2 m.

The date of the destruction of the site's fortifications, either at the end of the MB IIC or during the LB IA, has been a matter of debate. Nevertheless, there is evidence that the site may have been destroyed in a fire as suggested by the cracking of the orthostats in the town gate (Macalister 1912:242).

WATER SYSTEM

Recently Ronny Reich and Eli Shukron have made a convincing case for the identification of the Gezer water system as a MB IIC feature (Reich and Shukron 2003), which as I have noted in chapter 3 must be treated as part of the Middle Bronze Age fortification strategy (ibid., 72).

GIV'AT SHARETT*

MR 1483.1274
Refs.: Bahat (1993).
Plans: Site (ibid., 254).
Sects.: NA.
Photos: Site (ibid., 254).
The MB IIB–C village of Giv'at Sharett is located about 1.6 km to the SE of Beth-Shemesh. Like the contemporary village at Manahat, Giv'at Sharett was only enclosed by the *rear walls* of the houses that made up the settlement, as Dan Bahat suggests (1993:254). Although it is true that such defenses were probably inadequate against an organized army, the site's proximity to Beth-Shemesh probably meant that its inhabitants would have sought refuge there during a true crisis.

HAMMAH, TELL EL-

MR 1973.1977
Refs.: Cahill et al. (1988); Cahill and Tarler (1991; 1993).
Plans: Site (Cahill 2006:fig. 2).
Sects.: NA.
Photos: Site aerial (Cahill and Tarler 1993).
Tell Hammah is located 16 km S of Beth-Shean. A mudbrick *fortification wall* measuring 2.8 m wide featuring an offset 0.4 m wide and 1.75 m in length was revealed during the 1988 season in a trench (20 x 3 m) in Square H6 on the SE side of this small mound [2.2 ha] (Cahill et al. 1988:194). An *earthen rampart* of alternating layers of earth mixed with stones and stone chips was identified against the exterior of the wall in Squares H6, G6, and F6. Although none of the pottery found in the rampart was later than the Middle Bronze Age, a more precise date is not possible. The rampart was founded on bedrock around –120 m ASL at which the earliest layers of chipped stone were deposited. If the fortification line followed the –115

m contour line, then the intramural Middle Bronze Age settlement was about 1.3 ha (Cahill and Tarler 1993:562).[52]

HAROR, TEL (TELL ABU HUREIRA)

MR 11257.08795

Refs.: Oren and Morrison (1986); Oren (1993a; 1996; 1997b).

Plans: Site (Oren and Morrison 1986:fig. 2).

Sects.: E3 (Oren 1996:fig. 8); E4 (ibid., fig. 4); E5 (ibid., fig. 5); E7 (ibid., fig. 6).

Photos: Site aerial (Oren 1997b:fig. 8.2).

Tel Haror is located along the Nahal Gerar 23 km SE of Tell el-'Ajjul and 31.6 km to the SSE of Ashkelon. The site was first settled at the start of the MB IIB and grew to approximately 16.2 ha in size (Figure 77). During the sixth season of excavation it was confirmed that the site had obtained its full size during the MB IIB when its ramparts were constructed (Oren et al. 1991:73). Haror consisted of an upper town and a lower town that was surrounded by a fosse and earthen ramparts, which extended from the upper town to the S and W. On the S side Haror was protected by the Nahal Gerar. The rampart and fosse were explored in several areas: Trench E 3 on the NE side of the acropolis, Trenches E 4 and E 5 at the S and N ends, respectively, of the W rampart, and Trench E 7 at the W end of the N rampart.

The lower town was surrounded by 1,115 m long *earthen ramparts* which averaged 20 m wide and 7 to 8 m high and formed a trapezoidal shaped enclosure at the base of the upper town (Oren 1996:22; 1997a:474). While these dimensions are an average based on the areas where the rampart has been excavated, the actual dimensions of the ramparts were slightly different in each location (see Table 32 below). The ramparts were built without a core wall and they were consolidated by using alternating layers of different materials including kurkar gravel and sand, and earth and loess. These materials were all found within the rampart in the reverse sequence in which they occur in the region's geological sequence, suggesting that they were excavated primarily from the fosse (ibid., 17). Where kurkar was included in the rampart's construction, whether crushed or in large pieces, it appears to have been taken from the excavation of the fosse immediately outside the rampart (ibid., 22). Some evidence suggests that the ramparts were also covered by a *glacis* of flat kurkar stones and mud plaster. On the N side of the site it appears that a *revetment wall* of kurkar stones was also built to keep the rampart from eroding (Oren 1993a:582).

From the average dimensions of the ramparts it is possible to suggest that the combined volume of earth and stone incorporated into the construction of the ramparts around the lower town, which were about 1,115 m long, was approximately 80,175 m^3 (see Table 32 below), which can be compared with Oren and Yekutieli's estimate of 150,000 m^3 for a rampart 1,300 m long (Oren 1996:23). Even if we include the 185 m of rampart probably remaining around the N and E sides of the upper town, averaging 6 m high and 20 m wide (see E7; ibid., 19) for another 11,100 m^3, in order to obtain a total length of 1,300 m, we still fall considerably short of the estimate provided by the excavators. Even doubling the length of the upper town's fortifications, which were probably present based on the section of Trench E 7, results in a figure (102,375 m^3) that is still well below the excavators' estimate.

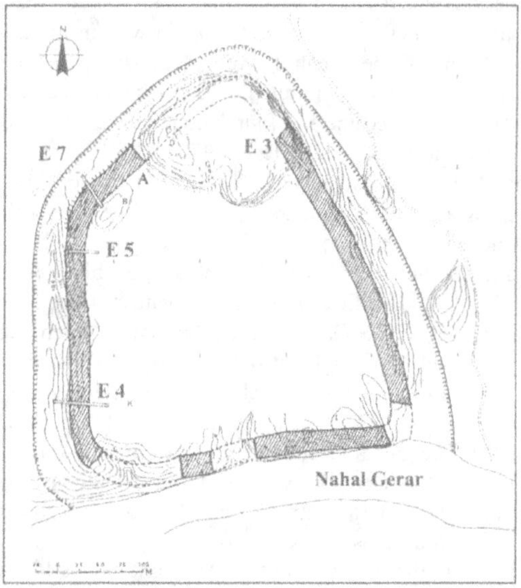

Figure 77. Ramparts of Tel Haror (after Oren 1996:fig. 1).

Erosion may have claimed the original mudbrick town wall since no evidence of it was found in the areas excavated, although Oren has recently suggested that it may never have been crowned by a wall (1996). However, mining of the ramparts for building materials in the late Iron Age may also have contributed to the absence of evidence for the town's fortification wall. Evidence from the SE part of the site suggests that during the course of the MB IIB–C the rampart was abandoned and a new mudbrick fortification wall was built at the base of the rampart while

52 Early reports mistakenly put the fortifications at–155 m ASL (Cahill et al. 1988:194; see also Tarler et al. 1991:135).

numerous pits were cut into the rampart's surface (Oren 1993a:582).

Outside of the rampart a *fosse* was excavated around three sides of the settlement (N, W, and E), while the Nahal Gerar flanked the site to the S. The fosse was mostly explored in trench E7 on the N side of the site where it appears to have been about 4 m deep and 12 to 15 m wide (Oren 1996). As noted above, the fosse has been considered the primary source for the material used in the construction of the rampart.

Table 32. Dimensions (in meters) of Haror's earthen ramparts (after Oren 1996).

Area	*h.* (m)	*w.* (m)	*l.* (m)	*vol.* (m^3)
East†	22	7	335	23,450
E 4 (W)	22	7	300	23,100
E 5 (W)	30	8	—	—
E 7 (N)	16	6	115	5,520
South†	22	7	365	28,105
Total			1,115	80,175

†Not excavated. Figures based on average dimensions of fortifications of each excavated section (E 3, 4, 5, and 7).

HAZOR (TELL EL-QEDAH)

MR 2032.2691

Refs.: Yadin et al. (1958; 1960; 1961; 1989); Yadin (1972b; 1975); Dunayevsky (1990); Ussishkin (1992); Mazar (1997a); Covello-Paran (1999).[53]

Plans: Site (Yadin et al. 1958:pl. 2; Ben-Tor and Bonfil 1997:plan V.30); Area A (Yadin et al. 1989:plan 16); Area C (1958:plan 179); Area G (1989:plan 30); Area H (ibid., plan 41); Area K (ibid., plans 42f., 45); Area P (A. Mazar 1997a:plans V.1–5, 7).

Sects.: Area A (Yadin et al. 1989:plan 16); Area C (Yadin et al. 1958:plan 179); Area G, Sections A-A, B-B, E-E, and F-F (ibid., plan 36); Area H (ibid., plan 41); Area K (ibid., plan 44); Area P, Sections A-A1 (A. Mazar 1997a:Section V.1) and B-B1 (ibid., V.8); Area E Lower town, Section A-A (Yadin 1972b:fig. 11).

Photos: Site (Yadin et al. 1958:pl. 1); Area A (Yadin et al. 1961:pls. 29); Area C (Yadin et al. 1958:pls. 26); Area G (Yadin et al. 1961:pls. 78–80, 94); Area H (ibid., pls. 103); Area K (ibid., pls. 130–45); Area P (A. Mazar 1997a:photos V.1–11, 13–27).

Hazor (*Hazura* of Mari texts) is located in the Huleh Valley about 27 km SSW of Tel Dan. The large settlement of Hazor, which included both an upper and lower town covering a gross area of about 74 ha (Yadin et al. 1958:2), reached this size no earlier than late MB IIA to early MB IIB (Ben-Tor and Bonfil 1997:198). The upper tell's surface is about 6.1 ha (10.6 ha at the base, Yadin et al. 1958:1).

Elements of the fortifications of Hazor dating to the Middle Bronze Age have been explored on the tell and around the lower settlement in Areas C, G, H, K, and P (Figure 78). A thorough series of final reports for the excavation of its Middle Bronze Age defenses serves as the basis for analyzing the site's fortifications.[54] Since the construction of the fortifications in each of the excavated areas was not identical, the results of the excavations are treated here by area rather than stratigraphically or chronologically. Although the archaeological evidence pertaining to these areas was provided during excavations, alternative interpretations of these features have been advanced (see principally comments in Yadin et al. 1989:xvii ff.).

Although the results of the excavations of the fortifications of the lower town are best known, excavations of the tell itself also revealed portions of the upper town's Middle Bronze Age defenses. Correlations between the stratigraphy of the upper and lower towns are provided in Table 33. Since the earliest Middle Bronze Age occupation at Hazor may have been limited to the upper town, this description begins with the remains of the fortifications of the upper town.[55]

UPPER TOWN

AREA A

The eastern *fortification wall* (L.375) of the upper town was excavated in squares O12/13 (Yadin et al. 1989:51f.).[56] This wall's mudbrick superstructure was preserved to a width of 8.8 m and a height of 3.5 m. The mudbricks in this wall were of two types, which are visible in the photographs (Yadin et al. 1961:pl. 29): a dark ("black") and a light-colored variety, which were said to have consisted of a greater quantity of lime. Darker bricks were used on the exteriors of the walls, while the lighter bricks were used for the core of the wall; they measured 40 to 50 cm square. Plaster was also preserved on the exterior faces of the wall (Yadin et al. 1989:52). The stone foundation of the wall was between 1 and 2 m high and the wall has been dated to the MB IIB.

[53]I would like to thank K. Covello-Paran for allowing me to preview the manuscript of the forthcoming report of the Area Q and N excavations, which are summarized in this entry.

[54]Additional details can be found in overviews of the results of the excavations published by Yadin (1972b:esp. 51–65, 106–09, and 12–28; 1975:129–41).

[55]The suggestion that the upper mound (tell) of Hazor was settled during the MB IIA is based upon the finds of MB IIA–B pottery on the tell (Yadin 1972b:121f.) and an MB IIA tomb in Area L excavated in 1971 (Ben-Tor 1997a:2f.).

[56]This feature was not correlated with either the stratigraphy of the upper or lower towns.

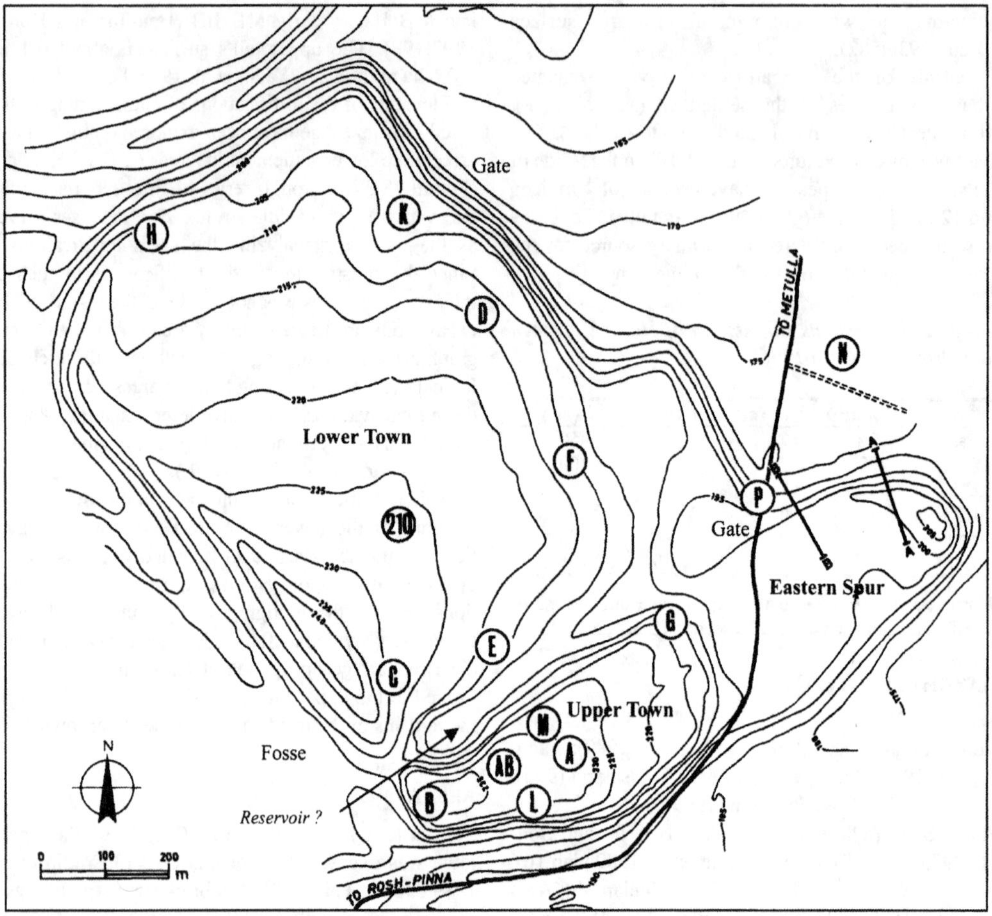

Figure 78. Plan of Hazor (after Yadin 1972b:fig. 3).

Table 33. Stratigraphic correlations between the upper and lower towns at Hazor.

Period	Upper Town	Lower Town	Area P
MB IIB	Stratum XVII	Stratum 4	Phase D
MB IIC	Stratum XVI	Stratum 3	Phase C
LB I	Stratum XV	Stratum 2	Phase C?–B?
LB IIA	Stratum XIV	Stratum 1b	Phase B
LB IIB	Stratum XIII	Stratum 1a	Phase A

Area G

Portions of the fortifications of the upper town were also exposed in Area G (Yadin et al. 1989:165ff.). Excavations here suggest that the tell's level during its earliest Middle Bronze Age settlement (late MB IIA to MB IIB) averaged 212 m ASL (ibid., 166). During the Middle Bronze Age (Str. XVII?) a stone *glacis* (10570) was built around the tell and in Area G it can be observed rounding the tell's NE corner, thus sealing its N and E slopes (ibid., 168). Its height was approximately 4 m and its base was built on a level anywhere from 204 to 208 m ASL (ibid., 169). The fortification wall built atop this glacis, which was not recovered, would have stood at about the same elevation throughout, ca. 209.8 m ASL based on the section (ibid., plan 36). The slope of this glacis was 68°. The glacis of the upper town was complimented by a small *fosse* that ran around the base of the glacis.[57]

[57]Although Yadin used both *fosse* and *moat* to refer to this feature (see Yadin et al. 1989:169, editor's note), in keeping

Ceramics from later occupation in the fosse confirm that both the fosse and the glacis were of Middle Bronze Age date, though it was not possible to determine whether or not the glacis belonged to Stratum XVII or XVI of the upper town.[58]

LOWER TOWN
AREA C

The *earthen rampart* on the W side of the lower town was first explored in Area C during the first season of excavation at Hazor (Yadin et al. 1958:71ff.). Unfortunately, its excavation cannot be considered complete since the trench did not completely bisect the rampart and thus a section of its fill layers was never revealed. Nevertheless, based on the limited excavation here Yadin considered the rampart to have been constructed from the debris taken from the construction of the fosse just to the W of the rampart (ibid., 2). The rampart's dimensions, which are roughly 90 m wide and 15 m high with an exterior slope of 18°, vary depending on the amount of material taken from the fosse directly outside the rampart, which consisted of earth and stones (ibid., 75). With a length of about 700 m the rampart's volume would have been about 472,500 m^3.

Evidence that the rampart was reinforced after its construction by walls built on its inner slopes was also revealed. The inner slope of the rampart, which features a slope of 16°, was also covered with a *glacis* of cobbles, which may have served to prevent erosion of the rampart into the settlement; it was in most places only one cobble thick, but in another place it measured as much as 3 m thick (Sq. F5). If a wall was constructed at the top of the rampart, then its base—taking erosion into consideration—would have rested at about 243.58 m ASL (ibid., 2). The rampart continued in use through the final phase of settlement in the lower town during the LB II (ibid., 92). Yadin suggested that one of the four gates was located at the S end of the W rampart, between the rampart in Area C and the upper town (Yadin et al. 1958:2), though this area has not yet been excavated.

Although no excavation trench was placed across the *fosse* on the W side of the site outside of the rampart (see Figure 78 above), Yadin determined it to be approximately 600 m long N-S, 40 m wide at the base, and 80 m wide at the top with an average depth of 15 m (ibid., 2). These dimensions yield an approximate volume of 540,000 m^3.

AREA H

Additional evidence for the town's rampart was revealed in Area H to the E of the temple excavated on the N side of the lower town (Yadin et al. 1989:212ff.). The *earthen rampart* (2193), which was determined to be a continuation of the W rampart, was built atop an underlying natural slope of 12° with a slope of 25°. It stood about 8.5 m high with a base about 55 m wide. The summit of the rampart was around 210 m ASL (ibid., 214). The inner slope of the rampart was about 32°. If this rampart possessed roughly the same dimensions along the 600 m stretch on the N side of the site, this rampart would have been about 140,250 m^3.

AREA K

In Area K on the NE side of the lower town a *six-pier gate*(?) was excavated by Moshe Dothan after its location had been established by John Garstang based on a depression in the site's contours (Yadin 1972b:58ff.; Yadin et al. 1989:276ff.). Only the S half of the gate was excavated to avoid completely exposing the feature (Yadin 1972b:59). Although Yadin concluded that the gate conformed to the six-pier plan like the gate in Area P (ibid., fig. 12, p. 59), no piers were revealed during excavations. Unlike the gate in Area P it also exhibited considerable modification during the main phases of its use. Its earliest phase coincided with the earliest Middle Bronze Age occupation of the lower town (Str. 4, MB IIB); it had been built on virgin soil. Unfortunately, only a small portion of this gate was exposed below the later gates and, therefore, its plan appears less complete. Nevertheless, the gate consisted of two 9 m square towers made of solid mudbrick built on a basalt fieldstone foundation up to 1.4 m high (Yadin et al. 1989:277).[59] The position of the gate was much further toward the settlement than were successive gates, which may be suggestive of the population pressure upon the settlement over the course of the Middle Bronze Age, which required movement outward of the gate and wall in order to provide additional living space within the settlement.

The gate's towers were abutted by a *casemate fortification wall* to the N and S, which consisted of 1.7 m wide outer walls (W.5530 and 5529) enclosing a 1.7 m wide space. Although these walls were only preserved about 1 m long, if the space between the walls were filled the wall's entire width would have been approximately 5 m (Yadin et al. 1989:279). The

---

with the conventions outlined in this work this feature is identified as a fosse.

[58]David Ussishkin's arguments for dating these structures to the Iron Age are unconvincing (1992:276ff.).

[59]Mudbrick dimensions could not be obtained due to the mudbrick superstructure's condition (Yadin et al. 1989:279).

destruction of the gate of Stratum 4 led to this gate's abandonment and the building of the Stratum 3 gate (ibid., 280).

The Stratum 3 (MB IIC) gate of Area K involved a considerable departure from the plan of the earlier gate, which was, in fact, of the *six-pier gate* type, in addition to a shift in the lower town's *fortification wall* almost 13 m out from the settlement (Yadin et al. 1989:280ff.). The projected size of the entire gate complex is 20.3 m wide by 16.5 m deep with towers 6.5 m wide by 16.5 m long (ibid., 281). The exterior wall of the tower (W.5501) on the E side was the widest of the tower's walls and measured 2.8 m wide while the remaining three walls (W.5505, 5511, 5506) averaged 2 m wide. These walls were preserved to a height of 3 m, the lowest 1.5 m of which appears to have served as the foundation for the gate. Along these walls the remains of mudbricks confirm that the superstructure was indeed built of mudbrick. Like the gate in Area P (see below), the towers of the Stratum 3 gate in Area K also featured two rooms (L.5005 and 5001) connected by a single passage 0.8 m wide. The E of the two rooms (L.5001) also provided evidence for a doorway 1 m wide leading into the gate's exterior chamber.

The passage of the gate between the towers, which was paved though its individual chambers were not, is thought to have measured about 3.1 m wide (see Yadin et al. 1989:plan 42).[60] The passage actually slopes as it leads outside the settlement about 1.5 m over a distance of 16.5 m (a grade of about 5°). Three piers with 3.4 m spaces between them projected 2.3 m from each of the gate's two towers. The piers were not of similar size, the widest being the most exterior (3 m), the second widest the center pier (2.3), and the narrowest the interior pier (2 m).

The town *fortification wall* associated with the Stratum 3 gate was moved 9 m to the E of the Stratum 4 wall (Yadin et al. 1989:282). It approached the center of the S tower and like the earlier wall featured a casemate construction of a combined width of 5.2 m, which included a 1.5 m wide outer wall (W.5534), a 1.4 m wide inner wall (W.5528), and a space of 2.3 m between them. Like the gate, these walls were built of mudbrick with stone foundations. Excavation of the casemate wall approaching the N side of the N tower of the gate indicates that the overall width of 5.2 m could be achieved with walls of varying width; in this case the outer wall (W.5502) was 1.9 m wide and the inner wall (W.5523) 1.2 m wide with a space of 2.1 m between them. It should be noted that the exterior wall had foundations more than 2 m deeper than those of the interior wall of the casemate and this was probably intended to compensate for the building of the casemate on the slope of the site in this area. Excavations to the N of the gate also suggest that the casemate was not continuous but was instead replaced by a solid wall after a distance (ibid., 283).

In addition to the wall of Stratum 3 it appears that a *revetment wall* (W.5503) of large basalt boulders was constructed 10 m E of the fortification wall (Yadin et al. 1989:284). A fifty-meter length of this wall preserved to a height of nearly 5 m was exposed. Although it could not be stratigraphically connected to the gate and town wall of this period, the presence of pottery below the revetment wall suggests that it was not constructed as part of the initial defenses of the settlement (ibid., Yadin 1972b:60f.). Furthermore, its distance from the Stratum 3 fortifications made it ideal for regulating entry to the town in this period, while it would have been too far to have served a similar purpose in Stratum 4. Yadin postulated that this wall also remained in use during later phases of the settlement's fortifications.[61]

In Stratum 2 (LB I) no major changes were made to the late Middle Bronze Age fortifications in Area K. Although the area was never explored, Yadin suggested that another gate may have served the lower town on the E side S of Area K (Yadin et al. 1958:3).

AREA P

Excavations in Area P by Amihai Mazar revealed that the modern road, which had been laid across this part of the site at a 45° angle (on a NE-SW line) to the embankment of the fortifications, used the depression in this area which corresponded with an ancient *six-pier gate* that led into the lower town on its E side (A. Mazar 1997a:353). Only the N half of this gate survived and for this reason it is not possible to derive the complete dimensions of this structure in its earliest phase

[60]Note that the figure of 7.3 m given in the text of *Hazor III–IV* as the width of "the gate's passageway" (Yadin et al. 1989:281) is incorrect but refers to the combined breadth of the gate's chambers from N to S. The correct figure of 3.1 m for the "width of the passageway" is provided at the end of the same paragraph.

[61]Ussishkin suggests that this wall was never exposed in antiquity, but had been "covered by earthen fills" and only served to "stabilize the constructional fills on which the gatehouse and the piazza were based" (1992:276). Unfortunately, Ussishkin's theory cannot be tested as it rests on the assumption that the deterioration that the wall had suffered within the thirty-four years since its excavation is the best indication that it could not have "stood exposed and in use for a few centuries without being affected." The counter-argument against this suggestion is that during this time the wall was never maintained, as it most certainly was in antiquity.

(D). The gate appears to have attained its layout during Stratum 4 (MB II), local phase D, when the structure was built with what is thought to have been a six-pier plan. Its piers (e.g., W.1449), which extended back into the structure to form its W and E elevations, as well as its inner dividing wall between its towers measured 3.3 m wide, while its passage walls measured 3.1 m wide [W.1447] (ibid., 354). The four chambers within the towers (L.1435 to S)—two on each side—measured 5 x 3.2 m; these appear to have been connected by a narrow passage (L.1410C) measuring 70 to 80 cm wide. Although the phase D gate was considerably damaged, no destruction layer could be associated with the gate's condition (ibid., 355). The floor of the gate was only exposed in a small area against its inner wall (L.1434). A part of the stone-paved ramp (L.1424) that led to the gate may have been exposed against the E elevation of the N half of the gate.

In Phase C, Stratum 3 (MB IIC or LB I) the gate appears to have been built along the lines of the earlier plan with the new floor (L.1429) about 1 m above the earlier floor (A. Mazar 1997a:355). The dimensions of the W tower of this gate (L.1414) are reconstructed as 20.6 x 9.5 m with roughly 1 m of its stone foundation preserved, which was probably its complete height. Based on the schematic plan of the Phase C gate it was about 25 m across and 20 m deep (ibid., plan V.3). No evidence of the mudbrick superstructure remained. The Phase C walls appear to have remained in use during Phase B (Str. 1b, LB I–IIA). The interior passage wall during these two phases was narrower, only 2.65 m wide, which contributed to the creation of deeper chambers between the gate's piers. A building built against the N tower of the gate (L.1418) may have been built as early as Phase C, though it was definitely in use during Phase B (ibid., 363). The exact purpose of this building is unknown. Ceramics found on the floor of the W tower (L.1414) date the destruction of the gate to the MB IIC (ibid., 357). When the Phase B gate was destroyed it was replaced by the final gate in Area P of Phase A and dated to the LB IIB (ibid., 360).

Also in Area P remains of stone-built *terraces* were excavated to the S of the gate (A. Mazar 1997a:368ff.). This set of six terraces bridged an elevation difference of 3.45 m over a distance of 10 m as they approached the S elevation of the S tower of the gate (see Section A-A[1]). Each terrace was between 40 and 60 cm high. The earliest of the terraces identified is dated to Phase C, although they may have first been constructed in Phase D; they continued in use into Phase B when the width of the terrace (and perhaps the associated wall) was widened to 15 m (ibid., 367). Probes excavated below these terraces demonstrated that they were originally dug into the earthen rampart. As was the case in the construction of the gate, the use of basalt stones in the construction of the terraces serves as an indication of the later date of the structure.

EASTERN SPUR: SECTIONS A-A & B-B

Before the excavation of the gate in Area P two sections were made through the N side of the fortifications of the Eastern Spur of the lower settlement in advance of construction by Kibbutz Ayeleth Ha-Shahar (Yadin 1972b:25, 54–57; Dunayevsky and Kempinski 1990). Although sections A-A and B-B were completed in 1965 (for locations, see Figure 78 above), final publications of these soundings remain forthcoming.

Section A-A is the E-most of the two sections (ibid., fig. 11) and serves to demonstrate that this portion of the rampart was constructed with a *core wall* built of mudbricks preserved over 7 m high, which was wider at the bottom (10.5 m) than at the top where it measured 8 m thick (ibid., 55). Yadin identified the construction of the core as a "structural casemate" noting that it featured a space "3 m wide and 5 m deep, filled with basalt and other stone pebbles and beaten earth." The construction of the core was completed with mudbricks measuring 40 x 30 x 15 cm and Yadin noted that often the vertical mortar joints between bricks could be as much as 10 cm thick. The build up of an *earthen rampart* occurred in sets of debris (I–III) that were piled against the front of the wall and plastered when complete, which appears to suggest that it functioned as an exterior surface prior to the addition of the layers of the interior side of the core wall. The final "block" of debris on the exterior was sealed by a *glacis* of 15 cm of plaster (Dunayevsky and Kempinski 1990:25). All of the ceramics recovered from the fortifications here were of MB IIA and IIB in date (ibid., 26).

As if to only demonstrate that construction of Hazor's defenses was not contiguous across the site, Section B-B revealed that no mudbrick core wall was used only 140 m to the W of Section A-A (Dunayevsky and Kempinski 1990). The core in Section B-B consisted only of earth that may have been a topographical necessity for the construction of the rampart as it approached the gate in Area P.

Despite irregularities in the nature of its construction it is possible to estimate the amount of fill used in the rampart of the Eastern Spur based upon Section A-A, which appears not to have lost a great deal of its height against the core wall. Its combined width minus the core wall was roughly 46 m, while the rampart's height was about 8 m. Presuming that these dimensions represent the average size of the Eastern Spur's rampart, which was 880 m long, the rampart's volume would have been approximately 161,920 m^3.

Table 34. Summary of Middle Bronze Age fortifications of Hazor.

Area	Str. XVII/4	Str. XVI/3
UPPER TOWN		
Area A	Wall?	→
Area G	Stone glacis and Fosse (date uncertain)	→
LOWER TOWN		
Area C	Rampart and Fosse	→
Area H	Rampart	→
Area K	Six-pier(?) gate 1st Town wall	Six-pier gate 2nd Town wall
	Revetment wall	→
Area P	Six-pier gate Terraces > town wall)	Six-pier gate Terraces > town wall
EASTERN SPUR		
Sects. AA & BB	Rampart and core wall	→

Table 35. Dimensions and volume of Hazor's earthen ramparts.

Section	*h.* (m)	*w.* (m)	*l.* (m)	*vol.* (m^3)
West (Area C)	15	90	700	472,500
North (Area H)	8.5	55	600	140,250
East†	8.5	55	800	187,000
Eastern Spur	8	46	880	161,920
TOTAL	—	—	2,980	961,670

†Height and width are based on rampart dimensions in Area H.

Exploration in Area Q during salvage excavations carried out in 1995 revealed that the Eastern Spur's *earthen rampart* was composed of earth, crushed limestone, and fieldstones (Covello-Paran 1999). Further excavation in Area Q3, 10 m E of Area P on the Eastern Spur, identified five construction layers within the rampart (see n. 53 above). They, comprised a large amount of occupational debris sandwiched between three plaster layers. These findings again confirm the lack of uniformity in the construction of Hazor's defenses.

AREA N2

In addition to the excavations of the Eastern Spur Covello-Paran suggests that an outer earthen rampart may have existed north of and parallel to the northern rampart of the Eastern Spur in Area N2 (see n. 53 above). Although this is possible and would be analogous to the ramparts surrounding Munbaqa, the results require further excavation to be confirmed.

SUMMARY

In all, the defenses of Hazor during the Middle Bronze Age were formidable (Table 34). Yadin noted that around the lower town the fosse, rampart, and glacis would have served to keep enemy troops at a distance of at least 130 m in any location (Yadin et al. 1958:3). Essentially these same features served as the defenses of Hazor during the LB I. The traditional figures for the size of Hazor range from 74 to 80 ha (ibid., 2; Broshi and Gophna 1986), however, these are only estimates of its gross size since they include the area occupied by the fortifications. A much more reasonable figure is to be found in Herzog's estimate of 63.1 ha (6.1 ha upper town, a 50 ha lower town, and 7 ha Eastern Spur, Herzog 1997a:120). As Table 35 illustrates the total volume of Hazor's ramparts was approximately 961,670 m^3 of earth and debris.

HEBRON (JEBEL RUMEIDAH)

MR 1597.1036

Refs.: Hammond (1968); Ofer (1993); Chadwick (1992; 2005).

Plans: Site and excavation areas (Hammond 1968:fig. 6; Ofer 1993).

Sects.: NA.

Photos: Area I3 (Hammond 1968:pl. XXXb; Chadwick 1992:167–69).

Hebron is located 30 km SSW of Jerusalem in the hill country. The area of Tel Hebron's Early and Middle Bronze Age settlements seems to have centered on a "low spur of Jebel Rumeida" according to Ofer (1993:607). The Early and Middle Bronze Age defenses were excavated by two expeditions: one from

1964 to 1966 by Philip Hammond (1968) and from 1983 to 1986 by Avi Ofer (1993). Evidence for the Early and Middle Bronze Age walls and Middle Bronze Age fortifications was revealed in Area I3 on the S side (1964 and 1966 seasons), Area G (also Area I7) on the SW—the suggested location of one gate—and Area F on the N side of the site.

MB IIC

More than 2 m of the remains of a 9.5 m thick cyclopean stone *fortification wall* that enclosed the settlement were exposed above the surface prior to Hammond's excavation in Area I3 on the S side of the site. Although Hammond had incorrectly identified another wall (of Roman date) in Area G as part of the Middle Bronze Age defenses, there is still ample evidence in Area G for the Middle Bronze Age wall, a stone *glacis* (and presumably an *earthen rampart*) built against the wall, a *fosse* cut into bedrock, and possibly one gate (Ofer 1993:608). The tell was no more than 4.9 ha (ibid., 607).

HEFER, TEL (TELL EL-IFSHAR)

MR 1415.1976
Refs.: Paley et al. (1985); Paley and Porath (1993; 1997).
Plans: NA.
Sects.: NA.
Photos: NA.
Tel Hefer is located on the southern coastal plain about 4 km from the Mediterranean Sea coast and about 8.8 km SW of Tel Zeror. While Hefer may have been unfortified through an early part of the MB IIA, as revealed in Area A the site was fortified at least by Phase 4 of the MB IIA. The challenge at Hefer is, however, correlating each of the phases in the Area A sequence with those in Area C, which are used to identify the site's Middle Bronze Age phases. While it appears that the fortifications have not been preserved in most areas (such as Area C) where they would have been expected (Paley et al. 1985), correlations between the sequences of Areas A and C suggest that Hefer was fortified during Phase 1 of the MB IIA (i.e., Phase B–C of Hefer).

The evidence for the Middle Bronze Age defenses of Tel Hefer consist of a 4 m thick mudbrick *fortification wall* in Area A belonging to local Phase A/14 (general Phase G) on the W side of the site (Paley et al. 1985:45f.; Paley and Porath 1993:611). This wall is interpreted as part of a large public building the outer face of which "had been destroyed when the slopes of the mound were eroded and mined in later periods" (Paley and Porath 1993:611). Since the wall was more than 2 m thick and was found in the step trench on the N side of the site, it probably belonged to a fortified structure. Furthermore, the excavators note that, "natural erosion as well as soil quarrying, principally during the Byzantine period, have reduced the overall size of the site and, therefore, the Middle Bronze remains are close to the surface" (Paley and Porath 1997:369).

Hefer is estimated to have been a little more than 4 ha in size during this period (Paley and Porath 1997:369). Three destructions, which may have occurred across the site, have been detected in Area C (Ph. B [=IIA1], C [=IIA1] and E [=IIA3]). These appear to correspond with phase 15 in Area A, which predates the construction of Hefer's fortifications (Paley and Porath 1993:611). The fortifications may have been built, therefore, in response to the cause(s) of these destructions.

IBLEAM (KHIRBET BAL'AME)*

MR 1777.2058
Refs.: Zertal (1993).
Plans: NA.
Sects.: NA.
Photos: NA.
Ibleam is located in the northern Samarian highlands about 18.5 km to the SE of Megiddo. The MB IIB fortifications of Ibleam are noted to include a fortification wall and "huge earthen rampart" (Zertal 1993:1312). According to Broshi and Gophna the site's gross size was 9 ha (1986).

IRBID, TELL

MR 229.218
Refs.: Lenzen (1989; 1992).
Plans: NA.
Sects.: NA.
Photos: NA.
Irbid is located 24 km NE of Pella in Jordan. It was fortified around the start of the MB IIA and later renovation of the fortification system during the MB IIB–C appears to have included a *fortification wall* built of basalt boulders (Lenzen and McQuitty 1989:300). Apparently many of these stones featured dimensions of more than 4 m square (Lenzen 1992:456). The site is circular, a layout that is attributed to its fortification wall and it may have been as much as 20 ha in size according to Falconer (2001), though dimensions of 350 x 400 m cited elsewhere suggest that the site's size was actually only about 14 ha during the Middle Bronze Age (Lenzen 1992:456).

Iskander, Khirbet [EB IV]

MR 2233.1072

Refs.: Parr (1960); Richard and Boraas (1984; 1988); Richard (1990; 1993; 1997).

Plans: Site (Richard 1990:fig. 2); Area B (Richard and Boraas 1988:fig. 3; Richard 1990:fig. 3); Area C (ibid., figs. 23f.).

Sects.: Area B (ibid., fig. 5).

Photos: Trench I (Parr 1960:pl. 13); Area B (Richard and Boraas 1988:fig. 2; Richard 1990:fig. 4; 1993:650); Area C (1990:figs. 25f.).

The 3 ha site of Khirbet Iskander is located on the N bank of the Wadi el-Wala in Transjordan about 50 km SSW of Amman (Richard 1993). Iskander, which had been settled during EB I, was resettled during the EB IV following an occupation gap during the EB II–III. The site is one of the only sites that is known to have been fortified during the EB IV in the southern Levant. Multiple phases of occupation are attested during the EB IV, the earliest of which (Area B Phase G) featured an unfortified village. This was followed by four phases of fortified EB IV settlement [Area B Phases E–B] (Richard 1990:36ff.). Sections of the fortifications were excavated in Areas A, B, and C, located on the SW, NW, and SE sides of the site respectively.

The fortifications on the N side of the site were exposed in Area B where four phases of fortified occupation were delineated. These were constructed in Phase E when the first *fortification wall*, the so-called "inner wall" (3016=4067), was built (Richard 1990:36). This wall was built with a stone foundation and mudbrick superstructure, the exterior face of which had collapsed to a 30° slope (1993:650), perhaps as a result of an earthquake (1990:36). In Phase D the "outer wall," which was the new *fortification wall*, was constructed just outside of the "inner wall" that now featured buttresses to keep if from collapsing further. The space between these walls was filled with rubble. The outer wall (4020, 3017, and 8024) was preserved to a width of 2.5 m and was supplemented with a *tower* preserved 3 m high (ibid.). When this system fell out of use in the middle of the EB IV the site appears not to have been refortified.

In Area C there is evidence of the SE entryway of the settlement dated to Phases C–A (Richard 1990:44; 1997). Though this *gate* is not of the postern type it does not appear to conform to a previously known gate type and its plan can probably be considered *ad hoc* since it was clearly a modification of earlier buildings in this part of the site (Richard 1990:fig. 23). However, the 2.25 m wide passage, aligned N-S and flanked by guardrooms, was plastered and lined with benches and may have been built with a roughly symmetrical plan in mind (1997:188). On its interior end steps led into an open courtyard before leading into the town. The S end of the gate is flanked by what appear to be the remains of the S *fortification wall* stretching to Area A on the W.

In Area A limited clearance of the S *fortification wall* of the late EB IV settlement (Ph. B) running between Areas A and C revealed that it was 0.8 m wide, the stone foundation having been preserved to a height of 1.25 m (Richard 1993:650).

Jaffa (Tel Yafo)

MR 1267.1625

Refs.: General (Kaplan 1967; 1972a; Kaplan and Ritter-Kaplan 1993); Area B (Kaplan 1960; 1961; 1964).

Plans: Site (Kaplan and Ritter-Kaplan 1993:655).

Sects.: NA.

Photos: NA.

Jaffa, located on the Mediterranean coast about 47 km N of Ashkelon, was originally situated at the mouth of the Ayalon River. Excavations at Jaffa by Jacob Kaplan from 1955 to 1974 revealed an MB IIB–C (Levels VIII and VII) earthen rampart in Areas A (SE) and B (NE), as well as the remains of a fortification wall. There is no evidence to support Kaplan's earlier claim that the enclosure could be dated to the MB IIA (1972a:75).

In Area B on the NE side of the tell a section of the *earthen rampart* aligned N-S was exposed under the floors of the Hammam in 1960 (Kaplan 1961:191f.). The rampart was constructed of "beaten earth" and kurkar. To the W of the Hammam a trench designated Area D provided a second exposure of the MB IIB–C rampart in Area B (Kaplan 1964:286), although its full dimensions remain uncertain (Figure 79). This feature was entirely overlaid by an earthen rampart and stone glacis dated to the eighth century BC (Kaplan 1961:191–92; Kaplan and Ritter-Kaplan 1993:658). In Area A Kaplan identified a mudbrick *fortification wall* (the southern of two walls there) that he assigned to Level VII (Kaplan 1960:122), which rested directly upon the *earthen rampart* made of "beaten earth and kurkar" which he assigned to Level VIII (Kaplan and Ritter-Kaplan 1993:657).

Based on the alignment of the rampart in both areas and undoubtedly on Kaplan's impressions regarding the ramparts of Yavneh-Yam, he suggested that the original enclosure was square (Kaplan and Ritter-Kaplan 1993:658). According to Kaplan, therefore, fifty percent of this enclosure had eroded on the seaward side of the site (Kaplan 1967). However, this conclusion cannot be substantiated and seems to be

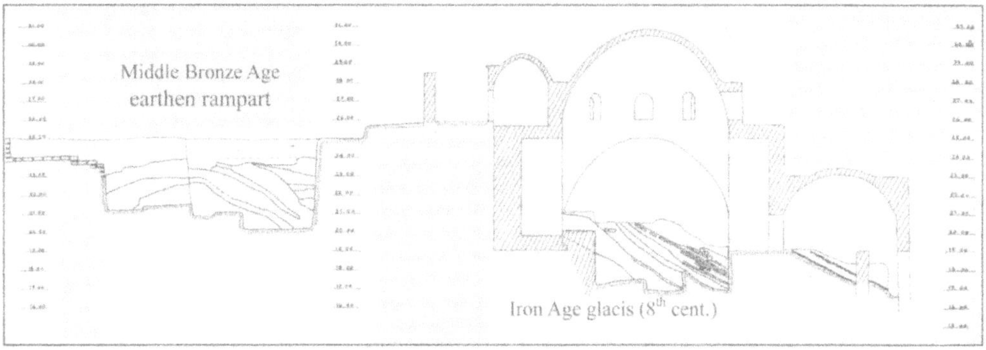

Figure 79. North section of Middle Bronze Age, Iron Age, and later fortifications in Areas D and B (left to right) at Jaffa from Kaplan's archive. Courtesy of the Jaffa Cultural Heritage Project.

entirely based on Kaplan's excavations at Yavneh-Yam. Preliminary examination of the ceramic evidence from Areas B and D suggest that since almost no diagnostic sherds were encountered in the lowest levels of the earthen rampart it must be dated primarily by its stratigraphic position below that of the Iron Age rampart. This interpretation is supported by archaeological parallels at 'Achzib, Beirut, and Byblos where a sequence of defensive construction from the Middle Bronze Age through the Iron Age has been revealed.

In light of disturbances caused by medieval and modern settlement at Jaffa it remains difficult to estimate the size of Middle Bronze Age Jaffa. Broshi and Gophna suggest a figure of around 3 ha for the tell itself (1986).

JATT*

MR 1540.2005

Refs.: Kotter (1986:138–40, 417–21).

Plans: NA.

Sects.: NA.

Photos: NA.

Jatt is located in northwestern Samaria about 100 m above the surrounding landscape. The site, which is dated to the late MB IIB–C, was approximately 8 ha in size (Kotter 1986:139). Kotter emphasizes the natural defensibility of the site on every side and that because of its location in the hill country it was supplied with sufficient limestone for the construction of its defenses (ibid.). Gophna and Beck identify it as fortified (see map on p. 47 in 1981), although the nature of these fortifications is unclear.

JEMMEH, TELL

MR 097.088

Refs.: Petrie (1928); Van Beek (1992; 1993).

Plans: Site (Petrie 1928:pl. 4).

Sects.: ST1 (Van Beek 1992:Ill. 1–3, 5f.).

Photos: Site aerial (Van Beek 1993:667); ST1 (Van Beek 1992:Ill. 4).

Tell Jemmeh, located about 32.7 km SSW of Ashkelon, is often identified with Late Bronze Age Yurza mentioned in New Kingdom sources. The site is located up the Nahal Besor from Tell el-'Ajjul with the wadi, which has severely cut into the site, on its N side. The site was established during the MB IIB and remained occupied through the Iron Age (Van Beek 1993:668).

Petrie had originally revealed part of a what he called a "rammed clay wall" but also referred to as a "glacis" (1928:22, pl. 25), which Van Beek identified as a "rammed earth revetment" (1992). A trench on the site's S side (ST1, exact location uncertain) provided evidence of this feature, which must be identified as an *earthen rampart* that provided the only evidence of Jemmeh's Middle Bronze Age defenses. This unimpressive rampart, which was reidentified with a bulldozer, was preserved about 1.6 m high and was more than 6 m wide since its interior face was not exposed (ibid., 4ff.). Since it was part of the site's earliest phase of continuous occupation it was constructed on virgin soil with alternating layers of sand (20–28 cm thick) and clay (1–1.5 cm thick). The rampart fell out of use and was subsequently buried by more than 2.15 m of debris (Van Beek 1993:668).

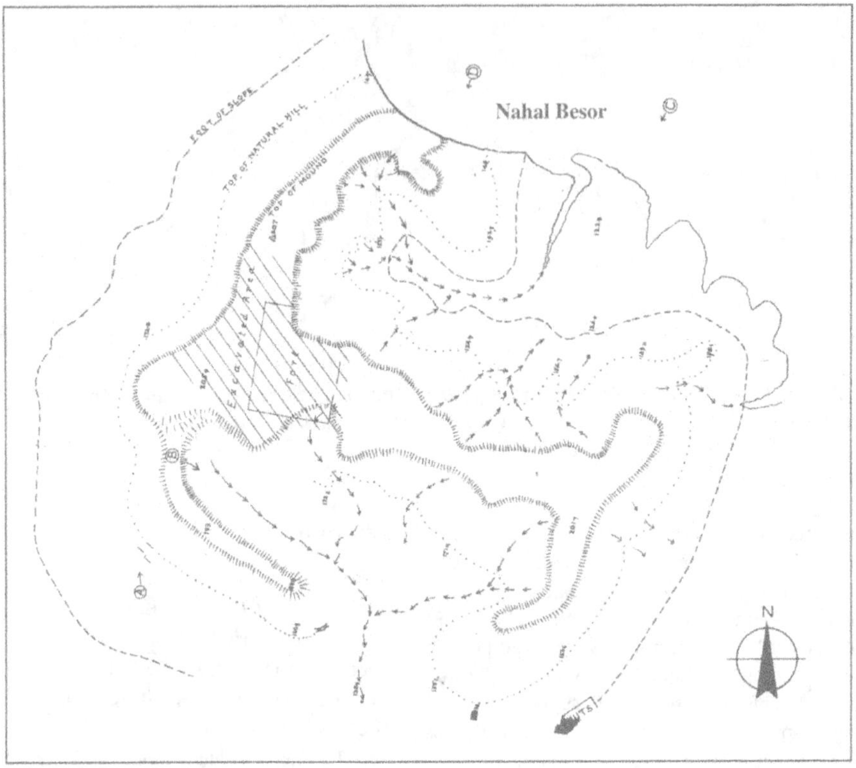

Figure 80. Plan of Jemmeh (after Petrie 1928:pl. 4).

The site's original plan was probably rectilinear as suggested by three pieces of evidence: the roughly square shape of the mound in Petrie's original plan (1928:pl. 4), the straight lines of the rampart and fosse evident in black and white aerial photographs of the site (Van Beek 1993:667), and the orientation of the Eighteenth Dynasty buildings excavated by Petrie on the W side of the mound—the W wall of this complex is parallel to the line of the W rampart and fosse (see Petrie 1928:pl. 6). Originally the site's N defenses were built along the Nahal Besor, but the wadi has now greatly eroded much of this side of the mound (Figure 80). If this reconstruction is accurate then the site was roughly square, 170 m on a side, with an inhabitable area of approximately 2.9 ha.

JENIN*

MR 1783.2074
Refs.: Zertal (1993).
Plans: NA.
Sects.: NA.
Photos: NA.

Jenin is located in the northern Samarian hills 10 km SE of Ta'anach. Its MB IIB fortifications are noted to include a fortification wall and a "huge earthen rampart " (Zertal 1993:1312). According to Broshi and Gophna the site's gross size was ca. 0.7 ha (1986).

JERICHO (TELL ES-SULTAN)

MR 196.141
Refs.: Sellin and Watzinger (1913); John Garstang (1930); John Garstang and J. B. E. Garstang (1940); Kenyon (1981); Ussishkin (1989); Marchetti and Nigro (1998; 2000; forthcoming).
Plans: Site (Kenyon 1981:figs. 1, 4; Marchetti and Nigro 2000:fig. 1); Site H (Kenyon 1981:pls. 328a, 329); Site A (ibid., pl. 343); Trench II (ibid., pl. 255b); Trench III (ibid., pls. 271f.); Area A (Marchetti 1998:figs. 4.15, 37; 2000:figs. 5:5, 7–9, 59); Area C (Sarie' 1998:fig. 3.4); Area D (Nigro 2000:figs. 3:1–2; Marchetti 2003:figs. 7f. and 10); Area E (Marchetti and Yasin 2000:fig. 4:6); Schematic of fortifications (F. Nigro 1998:figs. 4:39, 44f.).

Sects.: Site H (Kenyon 1981:pls. 339f.); Trench I (ibid., pl. 236); Trench II (ibid., pls. 259f.); Trench III (ibid., pls. 273–275); Area C (Sarie' 1998:fig. 3:1); Area A (Marchetti 1998:figs. 4:2, 4:12; 2000:5:1f.); Area E (Marchetti and Yasin 2000:fig. 4:1).
Photos: Trench I (Kenyon 1981:pls. 89b–93a); Trench II (ibid., pls. 108a–110a); Trench III (ibid., pls. 125b–128); Area A (Marchetti 1998:4:3–6, 9–11, 13f., 19, 35f., 38, 40–42; 2000:figs. 5:10–19, 24, 39, 60–64, 67); Area C (Sarie' 1998:figs. 3:2–3 and 3:5–10); Area E (Marchetti and Yasin 2000:figs. 4:2, 8–10); Area D (Nigro 2000:figs. 3:3–7).
Jericho is located about 22 km NE of Jerusalem and about 10 km N of the Dead Sea.[62] Four major expeditions have explored the Middle Bronze Age defenses of Jericho in various locations since 1907. These features include its ramparts, a massive stone revetment, and a mudbrick defensive wall. The first expedition by Ernst Sellin and Carl Watzinger between 1907 and 1909 exposed primarily the MB IIC stone revetment at the foot of the last rampart (see esp. plan 3 and figs. 34f., 1913).[63] John Garstang continued work at the site from 1930 to 1936 exposing more of the revetment (see pl. 6, 1930; 1931; see also plan of third town in pl. 5, 1940).

Kenyon's excavations at the site from 1952 to 1958 were successful in obtaining sections (on the N, W, and S sides of the tell) of what she had identified as three separate ramparts dated to the MB IIB–C (Kenyon 1981). She also created the first composite plan showing the areas that had been excavated by these projects (ibid., fig. 1). Her original impression of Jericho's Middle Bronze Age defenses seems to have come predominantly from the wall she discovered in Site H and the eroded remains of the ramparts which she had uncovered in Trenches I–III (see Table 36 below). She suggested that erosion along with the construction of the modern road on the E side of the site had taken a toll on both the rampart and mudbrick wall atop the rampart (Kenyon 1993:679). Firsthand examinations of the site and section drawings of trenches, as well as of the topography of the settlement, all illustrate the extent of this erosion. Following Kenyon's work, in a review of the data for the Middle Bronze Age defenses Ussishkin suggests identifying a fourth rampart (see esp. 1989:fig. 9). His suggestion, however, has not gained any following and, fortunately, our understanding of the Middle Bronze Age defenses is not restricted to the conclusions resulting from the three earlier expeditions.

The descriptions of Jericho's fortifications provided by the first three expeditions have been taken into account by the work of the Italian expedition from the University of Rome "La Sapienza" and the Palestinian Department of Antiquities led by Nicolò Marchetti and Lorenzo Nigro. Their work began in 1997 with the intent of renewing exploration of the Middle Bronze Age fortifications. They were successful not only in clarifying the phasing of the fortifications, but also in identifying a Middle Bronze Age mudbrick wall in Area C that crowned the rampart—something that had eluded previous expeditions. The stratigraphic phasing established by this expedition is used here to refer to the MB I–II (IIA–C) periods (Per. IVa–c) and for all previously excavated features of the Middle Bronze Age defenses, which have been reconsidered by the new expedition.[64] The synthesis produced by the new expedition has succeeded in explaining many of the discrepancies that existed between the results of the previous three expeditions (see also Table 37 below).[65]

MB IIA–B (Periods IVa–b) Fortification Walls

According to the Italian expedition's synthesis of finds, the fortifications of Area G or Kenyon's Site H in Period IVa1 consisted of a mudbrick fortification wall and tower. Based on her excavation of Site H Kenyon believed that the earliest evidence of the Middle Bronze Age defenses of Jericho were mudbrick *fortification walls* found in square HVI on the E side of the mound (Kenyon 1981:346ff.). According to her the earliest of these was wall HAH (phase xvi), which was unfortunately poorly preserved and was never completely excavated. It was succeeded by a 1.4 m thick mudbrick wall (HAJ, phase xviii) which was later widened by 0.5 m with the addition of wall HAK (phase xix). The passage of time between the constructions of these walls can be seen in the accumulation of debris against the W side of the walls in the N section (ibid., pl. 339). In phase xx wall HAL (0.4 m thick) was added to wall HAK. In phase xxi the composite

[62] I would like to thank Nicolò Marchetti for information regarding this entry.

[63] As a result of their participation in the early phases of archaeological exploration of Palestine they mistakenly attributed all Middle Bronze Age remains to the "Israelitische Period."

[64] The Italian expedition has established a complete stratigraphic sequence for all periods attested at Jericho (see Marchetti and Nigro 2000:9, Table 1; and Marchetti 2003:296, Table 1).

[65] Despite the clarity achieved thus far, *Quaderni di Gerico 3*, which will undoubtedly alter somewhat the conclusions presented here, remains forthcoming (Marchetti and Nigro forthcoming).

mudbrick wall built of these three units (HAJ-HAL), which attained a width of 2.3 m, was destroyed. Kenyon suggested that the mudbrick debris to the W (phase xxi) was the remains of these walls. After this destruction, according to Kenyon, there was no other evidence of the town wall in the excavated area since she presumed that it was moved further to the E. Nevertheless, the Italian expedition has interpreted this wall as a single wall 2.3 m thick (Marchetti 2003:304–06).

Also excavated in Site H were the remains of a mudbrick *tower* abutting the wall HAJ-HAL. Portions of three of its mudbrick walls (HBJ, HBK, HBL) were preserved to heights from 2 to 2.95 m (Kenyon 1981:351f., pl. 40). Kenyon had identified this structure as a defensive tower belonging in theory to a gateway that provided access to the spring. The tower's walls were very thick and were destroyed by fire on at least two occasions. There were, however, no finds suggesting that the tower had a military function.

The latest expedition's soundings in Area D near Kenyon's squares H II, III, and IV have yet to provide results confirming Kenyon's conclusions about the earliest Middle Bronze Age remains. However, they have identified a mudbrick *fortification wall* (W.7) over 3.5 m wide (Nigro 2000). This, Nigro suggests, corresponds with Kenyon's walls HCJ and HCP, about which little was known (Kenyon 1981:356f.). Approximately 20 m of the length of W.7 was exposed to a height of twelve courses (about 1.75 m high). The bricks of W.7 measured 42 x 36 x 15 cm. Based on pottery recovered the wall has been dated by Marchetti to Period IVa2 and IVb (2003:306–12).

First Rampart

The Italian expedition's work in Area C (N of Trench I) and Area A (at the S end of Trench III) has had significant implications for both the identification and dating of Jericho's Middle Bronze Age ramparts. Table 37 below summarizes the re-phasing of Kenyon's constructional phases including new features identified by the most recent expedition (cp. Kenyon's results, see Table 36 below). Marchetti emphasizes that the most useful basis for determining the actual number of ramparts at Jericho and the periods during which they were constructed is on the basis of the latest pottery recovered from within each of the layers assigned to the ramparts, thus establishing a *terminus post quem* for each rampart's construction. This evidence taken in combination with the stratigraphic relationships evident in the section drawings makes it possible to reconstruct each rampart's building phases. Based on this approach Marchetti suggests that instead of the three ramparts that had been identified by Kenyon, only two ramparts actually existed: one dated to the MB IIA–B, Periods IVa–b, and another to the MB IIC, Period IVc (Marchetti 1998).

Evidence of Jericho's first *earthen rampart* comes in the first place from Kenyon's excavation in Trenches I, II, and III. In Trench I, on the W side of the site she identified portions of the earliest Middle Bronze Age rampart as part of her Stage XLIII occupation (Kenyon 1981:16, 108ff.).[66] However, Kenyon had misidentified the EB IV remains and some indistinct Middle Bronze Age occupation debris below her first rampart as a berm, which was 5.4 m above the level of the plain. Thus, according to her interpretation the first rampart was only 5 m high and 14.5 m wide and was built of yellow and gray earth and sealed with huwwar plaster with a slope of 38°.[67] This plaster was securely fixed to the surface by tongues that were set into the rampart. To support her interpretation of the first rampart in Trench I Kenyon was led to suggest that a robbed out fieldstone retaining wall (KA) had once abutted the face of the "berm" below the first rampart. However, Marchetti's reexamination suggests that wall KC, in fact, functioned as the *retaining wall* for the first rampart in Trench I.[68] Furthermore, Marchetti also demonstrates that the EB IV and Middle Bronze Age fill identified by Kenyon below the first rampart actually formed the base of the original rampart (cp. Trench III, see Figure 84). This is established, according to Marchetti, by reexamining the section of Trench I in light of the results from Area C (see below). Therefore, according to Marchetti's reinterpretation of Kenyon's results the two ramparts can be distinguished by their revetment walls built at their bases and that the glacis of the original rampart was constructed of huwwar and plaster, which was locked into the rampart's surface (see Figure 82).

Contemporary with the life of this wall to the S of it by the Spring Hill were the remains of a complete *tower* discovered by Garstang (Marchetti 2003). The dimensions of the tower were approximately 10 x 16.25 m. It featured walls between 2 and 2.5 m thick.

---

[66] Kenyon's use of the term *stages* refers to "structural phases . . . indicating a main alteration in plan." This terminology is almost synonymous with the use by most other archaeologists of the terms *stratum*, and sometimes for the *phases* of those strata, though she suggested that *stages* here could not be "applied from site to site." In place of locus numbers she preferred *phase* designations (see Kenyon 1981:2f.).

[67] Kenyon cites its average width as 5 m (1981:108), but this can be disputed as too narrow even for an average width based on the N section of Trench I.

[68] Kenyon assigned the fill behind KC to the second rampart and its repair (Kenyon 1981:pl. 236).

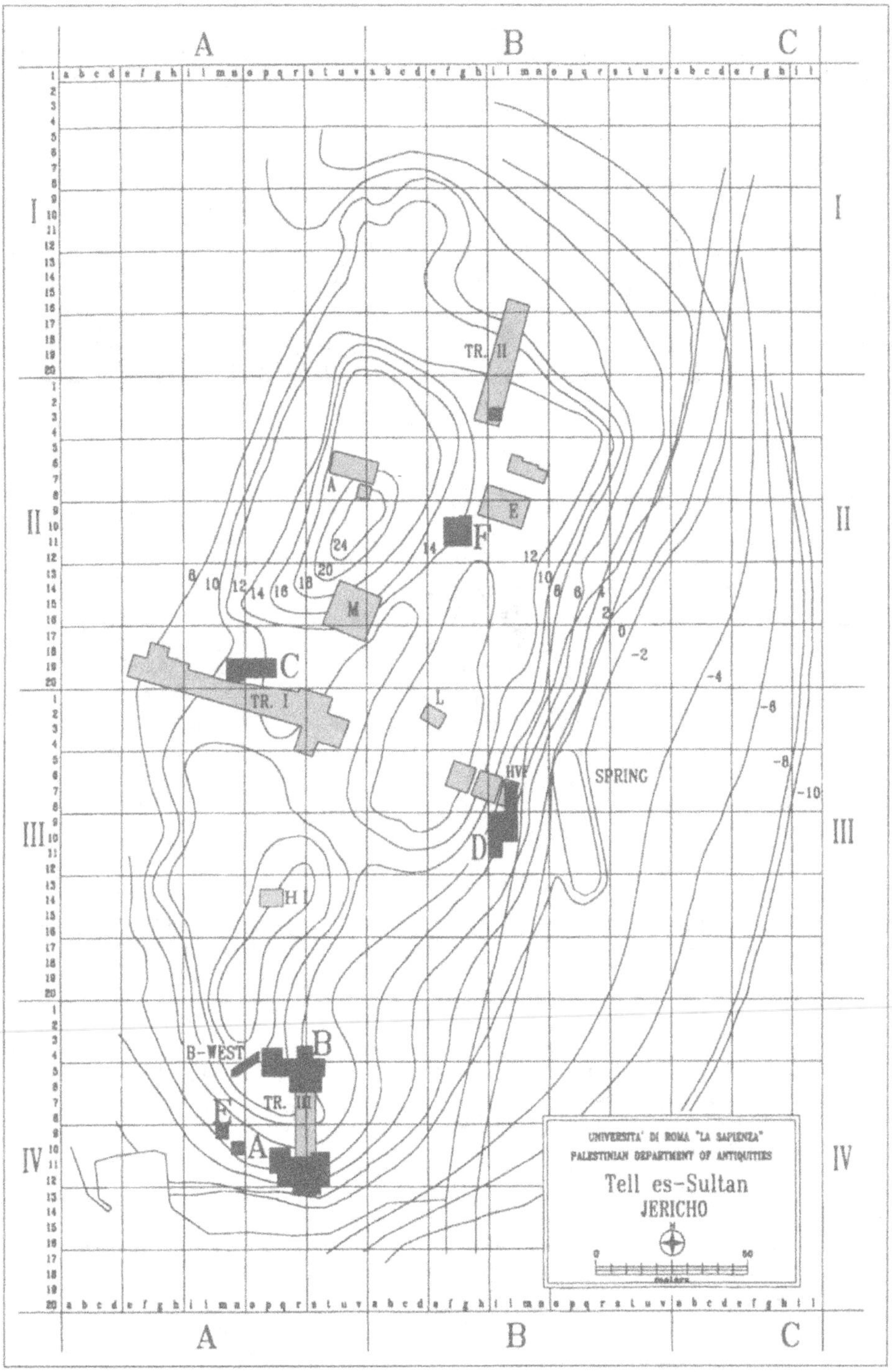

Figure 81. Plan of Jericho (after Marchetti and Nigro 2000:fig. 1).

Table 36. Jericho feature correlations with rampart phases according to Kenyon.

Area	Rampart 1	Rampart 2	Rampart 3	Post-Rampart 3 Wall	Post-Rampart 3 debris
Trench I Stage XLIII	*lvii*	*lviii–lx* KB, KC	*lxi-lxiiia* KD†, KE	Not detected	*lxiv-lxvi*
Trench II Site O Stage XXII	*lxix* OEK, OEL, OEM	*lxx*	*lxxi–lxxib* KD, OEN, OEO†, OEQ	*lxxii* OEP, OEQ, OER, OEX	*lxxii-lxxiii*
Trench III Site N Stage XXI	*lxxxi– lxxxii* NFK, NFL, NFM, NFN, NFO, NFP, NFQ, NFV	Not detected	Not detected NGK†	 NFR, NFS	
Site A			*Phase uncertain* Foundation(?) for town wall on rampart		

†Denotes the great revetment that was exposed around three-quarters of the site.

Table 37. Revised phasing of the fortifications of Jericho by Italian expedition.

Area	MB IIA–B: Rampart #1 (Periods IVa–b)	MB IIC: Rampart #2 (Period IVc)	Refs.
Trench I	*lvi–lvii*, lower *lx* KC	*lviii–lix*, upper *lx*, *lxi–lxv* KB, KD†, KE	Kenyon (1981:pl. 236); Marchetti (1998:fig. 4)
Trench II Site O	*lxix* OEK, OEL	*lxx–lxxi* OEM, OEO†, OEQ, OER	Kenyon (1981:pl. 259); Sellin and Watzinger (1913:fig. 34f.); Marchetti (1998:fig. 4)
Trench III Site N	*lxxx–lxxxi* *lower lxxxi–lxxxii,* *lxxxv–lxxxvi*; NFP (=W.3), NFK, NFL, NFM, NFN, NFO, NFP, NFQ, NFV	upper *lxxxi* NGK†	Kenyon (1981:pl. 273); Marchetti (1998:fig. 4)
Site H/ Area D	IVa1: Wall=HAJ+HAK+HAL; Tower=HBJ+HBK+HBL IVa2–Vb: Wall=HCJ+HCP (=W.7) Garstang "Tower"		Kenyon (1981:pls. 328a, 29); Nigro (2000:figs. 3); Marchetti (2003:fig. 8)
Area A-B	W.3 (=NFP), W.5 (=W.274), W.70, W.19, W.15, W.164, W.190	W.4† (P.154), W.174, W.71, W.72, F.60, F.63 P.14, P.17	Marchetti (1998:figs. 4; 2000:fig. 5)
Area C	W.93, W.98, W.84, W.112, L.111	W.113	Marchetti (1998:fig. 4)
Area E	Tower E1 (W.268, W.270), W.274 (=W.5)	F.269	Marchetti (2000:fig. 4)

†Denotes the "great revetment" exposed around three-quarters of the site.

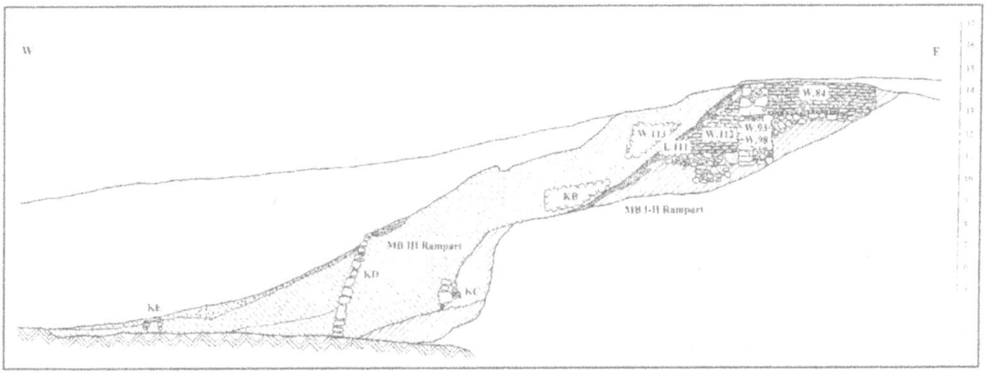

Figure 82. Schematic north section of Kenyon's Trench I and Area C (Marchetti 1998:fig. 4:44). Courtesy of the Italian-Palestinian Expedition at Tell es-Sultan/Jericho.

To the N of Trench I on the NW side of the mound, this rampart was also identified by Kenyon in the S section of Sellin and Watzinger's excavation area near Kenyon's Site A, which was cleaned and drawn by her (1981:374f., pl. 43). Here foundation stones rested at the summit of the rampart suggesting that a fortification wall crowned the rampart, though the superstructure remained elusive. The Italian expedition produced evidence of this same foundation and what was the mudbrick wall above it (see Figure 82) associated with the first rampart in the vicinity of Trench I (Sarie' 1998:103–15). Their E-W trench in Area C was laid N of and adjacent to the N section of Kenyon's Trench I. Here they discovered a mudbrick *fortification wall* (W.84) composed of nine courses of preserved mudbricks (30 x 30 x 10 cm) roughly 1.2 m high and, most impressively, 5.75 m wide (sixteen rows, see figs. 3:1, 3:4 in Sarie' 1998). This wall was identified as the mudbrick wall that crowned the first rampart (see Figure 82). Kenyon did not encounter it in her excavation of the area nor was it revealed in the section (1981:pl. 236). The wall was built upon a fieldstone foundation.[69]

The fortification wall was constructed behind both a small fieldstone *retaining wall* (W.98), nine courses high and four rows wide, and another fieldstone wall (W.93) that served as a footing to the mudbrick wall, three rows wide and three courses high. This footing would have reinforced the foot of the mudbrick fortification wall (W.84), which was built both behind it and upon it. W.93 and W.98 were in turn buttressed in front by the top portion of the *earthen rampart*, here composed of mudbrick (W.112) sixteen courses high (with a triangular section) resting on a narrow fieldstone wall (W.114; cp. Megiddo XIII–XII Area C Wall A). The entire core in front of the wall was then covered by a compact, crushed huwwar *glacis* (L.111), which is elsewhere evident in the E section of Trench II (see Figure 83 below). The actual elevation in Trench I of the wall and its associated features as found in Area C must have been higher than the top of Kenyon's Trench I, as is indicated by the lower elevations of the EB III walls K and M in Trench I, which appear to rise just slightly to the N as they enter Area C. This is, no doubt, in part due to the greater erosion that occurred in the area of Kenyon's Trench I, but it is more likely that it indicates the presence of a gate—at least an Early Bronze Age gate—to the S of Kenyon's Trench I. This would have meant that the preserved remains of the wall ended just to the N of Kenyon's trench.

The first rampart, which was also excavated in Trench II (see Figure 83 below), was covered with a plaster glacis surface with a slope of 35° (Kenyon's phase *lxix*). The plaster surface here featured the same construction technique of interlaced fingers of plaster in the fill of the rampart. Here the first rampart's dimensions, approximately 7 m high, were also very comparable to its dimensions in Trench I, thus seeming to further confirm the Italian expedition's reinterpretation of Kenyon's results.

[69] It should be observed that the trench laid for Area C by the Italian expedition is in strict conformance to the N-S grid, which affects even if only slightly any attempts to use the section drawings for measurements without mathematical correction. Area C, therefore, diverges 19° from a perpendicular cut through the fortification line. The following widths are taken from the section drawing (excluding W.84 measured in plan 3:4) and corrections are included in parentheses: W.93 1.54 m (1.46 m); W.98 1.98 m (1.87 m); W.112 3.3 m (3.12 m); W.113 2.75 m (2.6 m); and W.114 1.87 m (1.77 m).

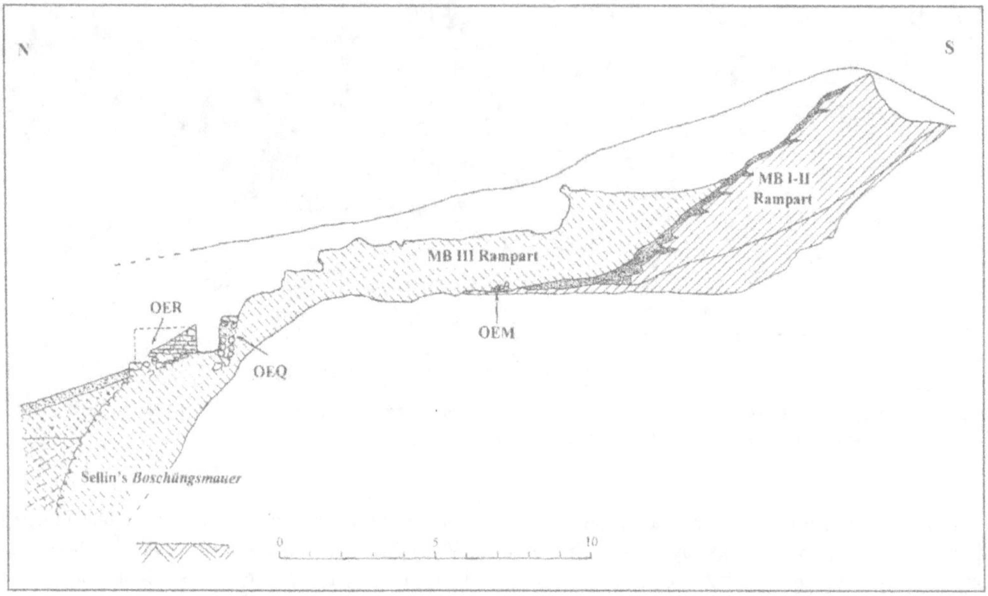

Figure 83. Schematic east section of Kenyon's Trench II (Marchetti 1998:fig. 4:45). Courtesy of the Italian-Palestinian Expedition at Tell es-Sultan/Jericho.

For Trench III Site N Kenyon suggested that only the first rampart (XXI phase *lxxxi*) survived (Kenyon 1981:215ff.).[70] However, here the situation is again complicated by even more erosion than in Trenches I and II (Note how different the W and E sections of Trench III are in figs. 273f.). Several rib or *retaining walls* incorporated into the construction of the first rampart in the area of Trench III are partially preserved (NFL, NFM, NFN, and NFO). It seems that repairs were also done during the life of this rampart in the course of which a fieldstone *revetment wall* NFP was constructed to control erosion at the base of the rampart. Some building seems to have taken place outside of wall NFP after the construction of the first rampart, which is evident from both Kenyon's and Marchetti's excavations.

While the earliest rampart featuring a plaster and huwwar *glacis* was identified in Trenches I, II, and III and Areas A and C, it was best preserved in Kenyon's Trench II. The glacis surface associated with its construction is one of the clearest indications that the earthen embankment that it sealed belonged to a single architectural feature, namely the first rampart as Kenyon had originally identified it (if only its upper portion). In addition to its uniform surface, in Trench I and III a low *revetment wall* was constructed at the foot of the slope, walls KC (see Figure 82) and NFP (see Figure 84; =W.3+W.5), respectively. In Area C, as mentioned earlier, excavation revealed the fortification wall, which was not, however, preserved in Kenyon's Trench I just to the S. It is possible to estimate the length of the wall and rampart as approximately 660 m. If the rampart's dimensions, about 24 m wide and 7 m high, as indicated in Trench I/Area C are accurate, then its volume would have been approximately 55,440 m^3.

Despite the clarity of this new analysis, a precise dating of the first rampart is difficult. The earliest deposition of earth for the building of the first rampart in Trenches I, II, and III contained primarily EB IV sherds (see Marchetti 1998: n. 11), indicating that there was probably no MB IIA settlement at the site until after the construction of the rampart had begun. The upper layers of the rampart featured mostly Middle Bronze Age sherds. Based on the sherds contained in the layers of the first rampart Marchetti has concluded that the rampart was constructed during

[70]Use of the section of Trench III to determine the width of the features illustrated must take into account that the trench conforms to the N-S axis of the grid. It is difficult, though, to determine how much the grid's alignment deviated from a line perpendicular to the fortifications since it is different at each contour elevation along the length of the trench. However, the following figures provide approximate corrections: 15° deviation for the Early Bronze Age city wall at the N end of the trench, and a 6° deviation for the first rampart and the stone revetment wall at the S end of the trench.

the late MB IIA and continued in use during the MB IIB–C (ibid., 118). Associated with the MB IIA–B rampart the Italian expedition has also revealed the continuation on the SW side of the tell of the revetment wall W.3=W.5 toward the NW from Area A and Trench III. Also built on the line of the wall was what appears to have been the foundation of a rectangular *tower* (W.280 and W.268) about 7.5 m long [width unknown] (Marchetti and Yasin 2000).

Although little context exists at present for a thorough discussion of the evidence for the *fortification wall* and *tower* belonging to the lower town excavated by the Italian expedition in Area A in 1998, it is still worth discussing here. The tower's full dimensions are not available, but it was slightly over 5 m wide (Marchetti 2000:fig. 5). Its attribution to Period IVb means that it was thought to have been built following the construction of the first rampart, but before the second began (Per. IVc). The exterior mudbrick wall (W.19) measures approximately 1.75 m wide and formed the outer wall of the tower. The tower's three other walls (W.15, W.164, W.190) were of similar thickness. If the context and orientation of the tower are understood correctly then the tower projected inward from the wall rather than outward (cp. Megiddo XIII).

MB IIC: PERIOD IVC: SECOND RAMPART

Although Kenyon originally identified a total of three ramparts, it is now clear that, in fact, only two ramparts were constructed during the Middle Bronze Age at Jericho (Marchetti 1998).[71] The second rampart, according to Marchetti, was built during the MB IIC and is most easily identified by the tall stone *revetment wall* that encircled the tell at the foot of the rampart. This so-called "Great revetment wall" was traced around nearly three quarters of the site by Sellin and Watzinger (1913:Taf. I), who referred to it as the "israelitische Boschüngsmauer." It should be recalled, however, that they had misattributed Middle Bronze Age remains to the Israelite period. The revetment wall was then again identified by Kenyon as wall KD in Trench I, OEO in Trench II, and NGK in Trench III. The Italian expedition only encountered the revetment wall anew in Area A-B and identified it as W.4 and dated it to Period IVc (see * items in Table 37 above). Founded on bedrock, this wall stood 4.65 m high and featured a 75° batter in Trench I. This revetment wall like its predecessors was no more than a single row wide and was probably intended to prevent erosion of the slope, but its stones were "slightly dressed boulders, with the crevices very carefully packed with small stones" (Kenyon 1981:110).

From the top of the revetment wall the slope of the second *earthen rampart* was a mere 22° and Kenyon detected several rows of mudbrick (1.25 m wide) just above the retaining wall laying on the surface's slope (ibid). These should be identified as the remains of a mudbrick *glacis* intended to protect the surface of the rampart. The composition of this rampart was more varied and included mudbrick debris, rubble, and stones in addition to earth. Although no remains of a separate fortification wall associated with the second rampart have been identified, this is not surprising since its fortification wall would most likely have overlaid or reused the wall of the first rampart.

At the foot of wall KD chipped stone and gravel was deposited up to 7 m from the base of the rampart against a 0.5 m high fieldstone wall. Based on the section drawing of Trench I (Kenyon 1981:pl. 236) we may presume that this was intended to cause water that would have rushed against the base of wall KD to be channeled away or filtered down in a controlled manner. Some time after the defenses went out of use fallen red mudbricks were deposited beyond wall KD and were then overlaid by mudbrick detritus. These are presumably the remains of the wall that crowned this rampart (Kenyon 1981:110). Schematic section drawings by Marchetti seem to identify the accumulated debris beyond the revetment wall as part of the MB IIC rampart, Period IVc (1998:figs. 4.39, 44f.). Whether this is intended to illustrate the deposition of eroded material from this rampart or that this material was part of the original rampart in situ is unclear. The dating of the material from this deposition will also, unfortunately, not resolve this issue if it in fact yields a contemporary date since this would be expected whether it was eroded material from the rampart or part of its original construction. In the final analysis the last rampart exhibited the greatest technical ability and a clear improvement in the approach to the defense of the settlement since the construction of the first rampart.

[71] The second rampart identified by Kenyon (in Trench I) was built against the first, resting on both the remainder of the berm and the level of the surrounding plain. Like the first, it too was initially constructed of earth against the small fieldstone retaining wall KB (phase lviii), but in a second phase (lix–lx), which has been interpreted as a repair of the collapsing rampart, KB was added and a new fieldstone wall (KC), described as a curb due to its small size, was also added beyond the berm. Plaster was again used on the surface, but no technique was employed for securing it on the surface (Kenyon 1981:109f.). According to Kenyon, a third and final rampart was then built against the second rampart in what she identified as phase lxi in Trench I; it was then faced with fieldstone retaining wall KD.

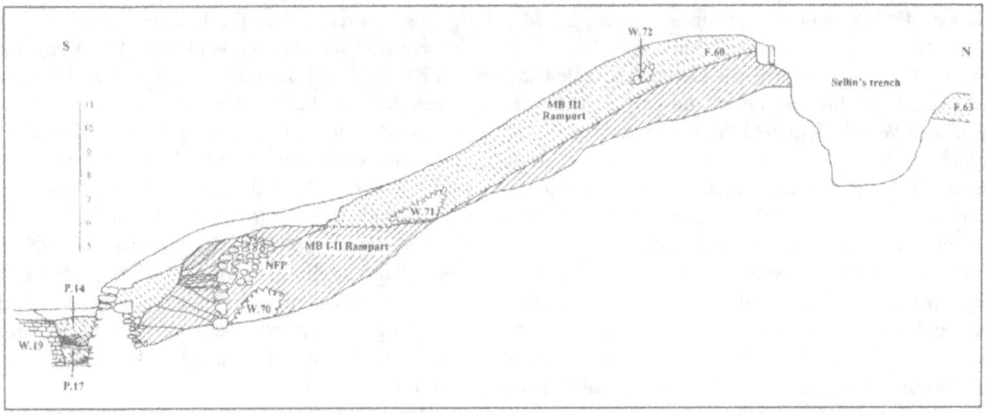

Figure 84. Schematic section of rampart in Kenyon's Trench III and Area A (Marchetti 1998:fig. 4:39). Courtesy of the Italian-Palestinian Expedition at Tell es-Sultan/Jericho.

GATES

Given the vast amount of excavation at Jericho which has led to exploration of over fifty percent of the settlement's surface it is curious that no gate structures dated to the Middle Bronze Age have been identified. An examination of the historical topography of Jericho's W slope seems to indicate that a depression has existed since excavations began just to the S of Kenyon's Trench I and Garstang's Area u. It is an area just to the S of all previous excavations, and it may have been the location of the W Gate leading into the hill country and the Jordan Valley road. The gate's location is suggested by a hollow bounded by the u-turns of the N and S 14 m contour lines (see Figure 81). This location, if correctly identified with the gate, is perfectly centered on the W side of the tell, approximately 130 m from both the N and S ends of the length of the first rampart here. The presence of this feature may also explain a shallow dive in the EB III walls K and M and the disappearance (i.e., erosion) of the Middle Bronze Age mudbrick fortification wall crowning the rampart as it neared the gate proceeding S from Area C into Trench I. This may also indicate that the Middle Bronze Age gate was, in fact, located above the EB II–III gate.

LATE BRONZE AGE?

To date there has been no evidence for the fortifications of Jericho during the Late Bronze Age. Yadin's suggestion that the last Middle Bronze Age wall continued in use through the Late Bronze Age until the thirteenth century to be identified with the wall of Jericho destroyed by the Israelites (1982:22) finds no support in the present study. No walls of any Early or Middle Bronze Age site lasted for three hundred years without needing to be rebuilt, despite the implications of Yadin's suggestion.

JERUSALEM

MR 1727.1310

Refs.: Kenyon (1974); Shiloh (1984); Reich (1987); Reich and Shukron (1998; 1999b; 1999a; 1999c; 2000; 2004); Steiner (2001).

Plans: Site (Kenyon 1974:fig. 18; Reich and Shukron 2000:329); Sects. of Middle Bronze Age wall (Reich 1987:figs. 1, 3); Macalister Field V (1926:pl. 5), cp. Parker section of Gallery XIX (Vincent 1911:pl. 6) and Kenyon Trench A (1974:fig. 16; Steiner 2001:fig. 3.4); Shiloh Area E1 (1984:fig. 14); Water System in Areas C and H (Reich and Shukron 2004:fig. 1).

Sects.: Kenyon Trench A (1974:figs. 15, 17; Steiner 2001:fig. 3.3); Schematic of Water System in Areas C and H (Reich and Shukron 2000:334).

Photos: Kenyon Trench A (1967:pls. 19f.; Steiner 2001:fig. 3.2); Shiloh's Area E wall (Cahill and Tarler 2000:33); Reich and Shukron, Area H (1999c:30; Reich and Shukron 2004:figs. 2–5).

Jerusalem (MB *Rušalimum*; LB *Urusalim*) is located in the hill country 30 km NNE of Hebron. Excavations along the E slope of the City of David have revealed various sections of the Middle Bronze Age (so-called "Canaanite" or "Jebusite") wall.

MB IIA–B FORTIFICATION WALL

The *fortification wall* (Kenyon Wall NB=Wall 3 Steiner) consisted of "cyclopean" construction. The wall, of which 13 m was exposed by Kathleen Kenyon, was 2 m wide and was preserved 1 m high (Steiner 2001:10f.). Kenyon and Macalister were the first excavators to propose an early date for the sections of this Middle Bronze Age town wall, but Kenyon only re-excavated the section that Parker had first exposed (Reich 1987:163). The section of the wall Kenyon re-exposed was in Trench A about two-thirds of the way down the E slope of the "City of David" (Kenyon 1967:24f.; 1974:81ff.). She suggested that the wall was located in this area in order to bring the means of accessing the Middle Bronze Age water source (see below) within the site's defenses (Kenyon 1967:25). The pottery in the wall was dated by her to the eighteenth century B.C. (ibid., 24). This date was upheld in Steiner's analysis of the ceramics (2001:11–12) and used to reiterate a transitional MB IIA–B date for the construction of wall 3. This section of the wall, as noted above, was previously excavated by Parker and can be seen in Vincent's publication of the wall (see 1911:pl. 6).

Kenyon appears to have exposed another section of the Middle Bronze Age wall running E-W to the NW of wall NB. The remains here were referred to as part of Structure J and they consisted of what appears to have been a section of the foundations of a *fortification wall* (Steiner 2001:12ff.). Its width, however, remains unknown. Although Steiner suggests that a natural depression to the N of this wall may have functioned as a fosse, improving the wall's defensibility in this location (ibid., 21), there is no evidence to support this interpretation and fosses are not attested at other settlements in the highlands in this period.

Excavations by Yigal Shiloh also revealed segments of the Middle Bronze Age *fortification wall* in Area E1 (1984:12). Here a 20 m long section of a 3 m wide "cyclopean" stone wall was exposed and assigned to Stratum 18B, the earliest Middle Bronze Age occupation dated to the transitional MB IIA–B. A buttress (W.285) was subsequently added to its interior during Stratum 18A and the entire structure served into Stratum 17 the final Middle Bronze Age occupation of the site. Shiloh placed Strata 18 through 17 within the eighteenth century (ibid.).

Water System and Towers

Ongoing excavations by Ronny Reich around the Gihon Spring down slope from Kenyon's Wall NB have revealed further remains of the Middle Bronze Age fortification system (Reich and Shukron 2004). These remains consist of the foundations of two massive towers, one of which enclosed the Gihon Spring while the other flanked the N side of a Middle Bronze Age pool cut into the bedrock to the SW of the spring. The spring and the pool are connected by what is identified as Channel II, which then proceeds S from the pool along the foot of the E slope of the City of David. During the Middle Bronze Age this channel was cut through bedrock from the surface of the slope and then covered with large boulders to obscure its location (ibid., 213).[72]

A tower located to the E has been dubbed the Spring Tower because it protects a space of about 6 m by 10 to 11 m where the Gihon Spring emerges from a vertical surface of the bedrock (Reich and Shukron 1999b:78*). The building's overall dimensions are 16.6 m E-W by 13.75 m N-S (2004:212). Its "cyclopean" stone walls were as much as between 4 and 7 m wide (ibid., 213) and were preserved to a height of more than 4 m (Reich and Shukron 1999b:78*).

To the W of this tower and just N of the pool was another small tower called the *Pool Tower* with walls of about 3 to 3.5 m amidst the stones of which MB IIB sherds were found (ibid., 77*; 1999c:30). The floor of this tower featured "flat stone slabs" (ibid.). The excavators have speculated that a similar tower protected the pool on its S side (Reich and Shukron 1999c:30).

Although the remains of Jerusalem's Middle Bronze Age occupation are substantial, they have been primarily limited to the E slope of the City of David and, therefore, a reconstruction of the extent of the town is not possible.[73] Nevertheless, we may suppose that its fortifications encompassed most of what is identified as the City of David, the peninsular hill S of the Ophel, which was flanked by the Tyropoeon Valley on the W and the Kidron Valley on the E. The remains thus far exposed indicate that the character of the construction of Jerusalem's Middle Bronze Age wall was similar to that of other hill country sites such as Shiloh and Hebron, relying heavily upon quarried stones for the construction of its wall.

Kabri, Tell

MR 1633.2682

Refs.: Kempinski (1987a; 1987b; 1991a; 1991b; 1993a; 2002a; 2002b); Kempinski and Niemeier (1991).

Plans: Site (Kempinski 2002b:fig. 1.3); Area C (fig. 4.19); Area T (fig. 4.24).

Sects.: Area C1, E section (Kempinski 1987a:figs. 5f.; 1991:figs. 10f.); Rampart schematic (2002a:4.20).

---

[72]I would like to thank Ronny Reich for a tour and detailed explanation of this system in December 2006.

[73]Despite the limited data, Steiner suggests that the settlement was between 4 and 5 ha in size during the Middle Bronze Age (2001:21), following Broshi and Gophna's estimate of 4 ha (1986).

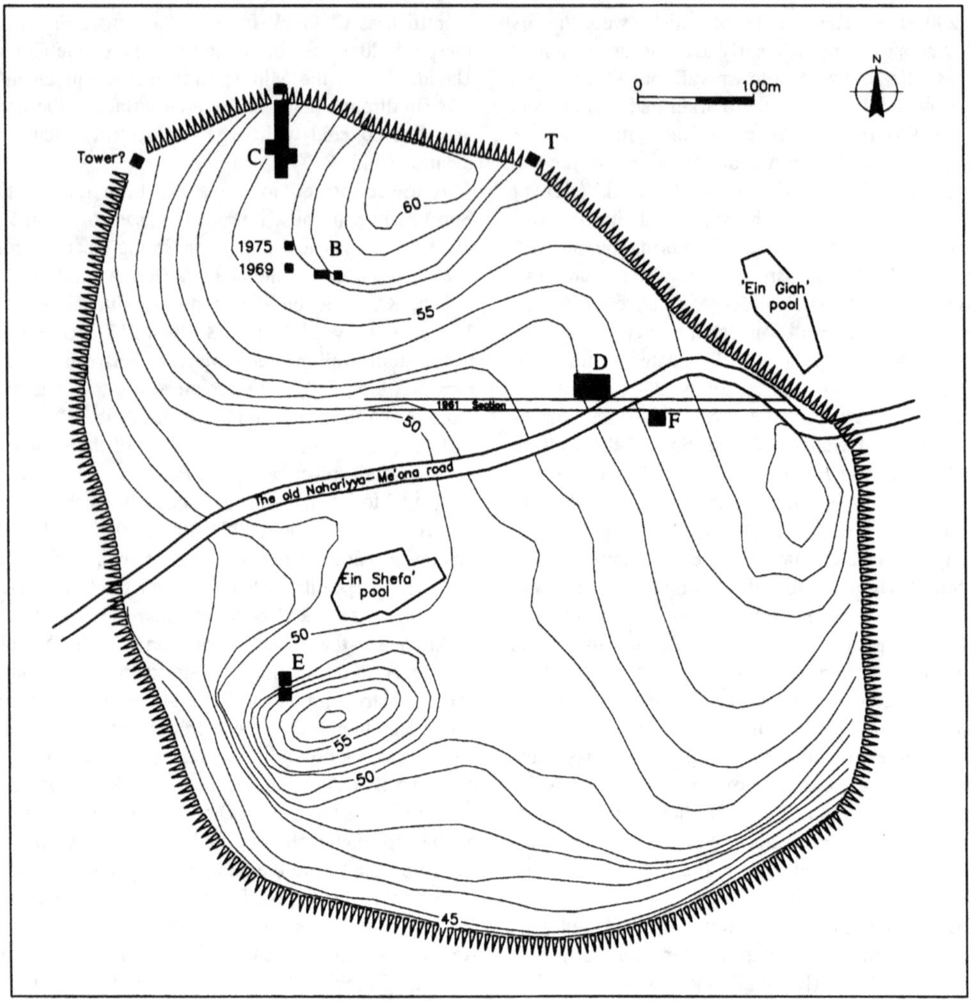

Figure 85. Plan of Kabri (after Kempinski 2002b:fig. 1.3).

Photos: Area C1 (Kempinski 2002a:4.21f.); Area T (fig. 4.25).

Kabri is located a little more than 4 km E of Nahariya and the cost along the Nahal Gaton. Its earliest Middle Bronze Age occupation, although limited primarily to tombs, is dated to Phase 2–3 of the MB IIA (Cohen 2002a:84f.) following a period of abandonment during the EB IV (Kempinski 2002b:5). Though remains of the stone foundation of the late MB IIA to MB IIB fortification wall have been traced along the surface of the site's earthen rampart during the course of normal excavations (Kempinski 2002a:35), the MB II fortifications of Kabri were first exposed in 1961 during the cutting of a 500 m long and 3.5 m deep trench across the site from E to W by a local contractor (1993a:839). This trench exposed a 35 m wide earthen rampart at the site's edge (Figure 85). Eventually excavations by Aharon Kempinski and Wolf-Dietrich Niemeier also exposed the fortifications in Areas C and T2, on the N and NE sides of the site respectively. Kempinski concluded that the remains of the Middle Bronze Age fortifications represented a single phase of construction dated to the late MB IIA or MB IIB (Kempinski 2002a:39).

In Area C parts of an earthen rampart, a revetment wall, and town wall were identified (Kempinski 2002a:37). The *earthen rampart* consisted of a mounded earth core, fills above this, and what appears to have been a *revetment wall* (W.404) located at the rampart's outer foot. The fills used in the rampart were deposited using the "sandwich method" and they included occupational debris from the EB III town, as well as pebbles and crushed kurkar. If the reconstruction of the slope below W.404 is accurate in the section drawing (ibid., fig. 4.20), then the slope of the rampart may have been as much as 45°, at least at the bottom. A fosse is suggested to have been built at the foot of the rampart (ibid.), but it has never been identified in excavations. Just inside the crest of the rampart in Area C a wall, referred to as the "inner wall" (W.400), was constructed (ibid., 37). It is probably to be identified as the *fortification wall* used with this rampart. This mudbrick wall was more than 4 m wide and was built upon a stone foundation and preserved about 1 m high.

Excavations in Area T2—so-called because the remains here were originally identified as part of a *tower* (Tower 2)—revealed what Kempinski identified as the 5.5 m wide stone foundations of a casemate wall [W.1601/1600] (2002a:37). In Area T2 the inner of the two walls (W.1601) was equated with the revetment wall (W.404) in Area C (Kempinski 1991a:fig. 11). While Kempinski's interpretation of the fortifications of Kabri is for the most part in keeping with the general understanding of the construction of Middle Bronze Age fortifications, his suggestion concerning the construction of a casemate wall on the front of the rampart cannot be accepted. In the first place, there are no parallels for such a construction on the front of a rampart. In the second place, an examination of the photograph of Wall 1600 (Kempinski 2002b:fig. 4.25) clearly shows that the structure projects several meters from the rampart (or main defenses whatever they might have been when this structure was built) and that the large stone masonry terminates on the S side. It is not possible to demonstrate that this feature functioned as a revetment wall (W.404) in Area C and we do not need to see a continuous and similarly constructed fortification system between Areas C and T, which are removed by more than 200 m. Therefore, Kempinski's original identification of the structure as a *tower*, tentatively dated to the MB IIB, which projected from the main line of defense, should be revived.

The Middle Bronze Age fortifications of Kabri have been provided with a *terminus post quem* date of late MB IIA to MB IIB (Kempinski 2002a:39). This is based upon MB IIB ceramics found in the rampart fill and the fills of the revetment walls (W.1600). These fortifications probably enclosed an area of roughly 32 ha within the nearly circular layout of the ramparts, which were approximately 2,130 m long (Kempinski and Niemeier 1994:fig. 2). If the ramparts were only 20 m wide then 4.26 ha of the settlement would have been occupied by the rampart, reducing the overall settlement to an area of less than 28 ha in size. This reflects considerable growth from the MB IIA when the settlement is estimated to have been between 15 and 20 ha (Kempinski 2002b:451).

The location of one of the Middle Bronze Age gates was identified on the NW side of the tell but was not excavated (Kempinski 1993a:841). The location of another gate on the E side of the site was also suggested (Kempinski 2002a:35). According to Kempinski, the site's destruction is to be dated to the end of the MB IIC.

KEBARA, EL-*

MR 1793.1967
Refs.: Zertal (2004).
Plans: NA.
Sects.: NA.
Photos: NA.
El-Kebara is located in the northern Samarian hills 17 km NNE of Shechem. The remains of the Middle Bronze Age *fortification wall* are probably those of a 50 m long wall built of large stones, which is located on the S side of the site and was identified during Zertal's survey (2004:237). This stretch of wall was built 5 m wide and preserved up to 4 m. There is a 6 m wide opening, perhaps a *gate*, in the wall, which is flanked by the remains of towers. According to Broshi and Gophna the site's gross size during the MB IIB–C was only 0.2 ha (1986).

KEDESH*

MR 199.279
Refs.: Aharoni (1993).
Plans: NA.
Sects.: NA.
There is limited evidence that Tel Kedesh, located 10.5 km to the NNW of Hazor, may have been fortified. However, the remains of a 5 m wide mudbrick wall found in section were ascribed to the Early Bronze Age by Yohanan Aharoni (1993:856). The site was about 10 ha in size during the Middle Bronze Age.

KEISAN, TELL

MR 164.253
Refs.: Humbert (1993); Seton-Williams (1980).
Plans: Site (Seton-Williams 1980:pl. 2); S trench (ibid., pl. 4).
Sects.: S trench (Seton-Williams 1980:pl. 4).
Photos: Site aerial (Humbert 1993:863); S trench (Seton-Williams 1980:pl. 126).
Tell Keisan is located some 9 km SE of Akko. Although the site was clearly fortified during the EB II (Str.

XVI) with a 5 m wide mudbrick wall (Humbert 1993:863), there is no relationship between these fortifications and those of the Middle Bronze Age. The remains of the Middle Bronze Age fortifications (Str. XV–XIV) were found in a trench on the S side of the site dug by John Garstang as part of the Nielson Expedition (Seton-Williams 1980:382f.). These remains included a core wall, two earthen ramparts, glacis, and revetment walls for the retention of the glacis.

MB IIB: STRATUM XVI

The Middle Bronze Age *core wall*, which probably also served as the foundation of the town wall, was preserved 1 m wide and to a height of 5 m, though its base was never reached (see Seton-Williams 1980:pl. 4 and p. 382). This part of the fortifications appears to have been constructed entirely of stones, which averaged about 45 to 50 cm on a side. Though the inner face of the wall was sloped, the outer face was vertical.[74] An *earthen rampart* (so-called "earth debris") 8 m high and roughly 19 m wide, and a *glacis* (so-called "lower revetment;" ibid, pl. 4) rested against the exterior face of this wall, thus suggesting that this wall was probably the core of the rampart and also served as the foundation for the mudbrick superstructure of the fortification wall. The ceramics in the fill of the earthen rampart were MB IIA and earlier, therefore suggesting that the fortifications were probably initiated no later than the start of the MB IIB. The glacis, which was identified around the entire tell (ibid.), stretched 19 m from the top of the rampart to the base of the tell and ranged from 0.5 to 2.5 m thick. It was composed of "white plaster and rubble."

MB IIB–C: STRATUM XV

A later renovation of the *earthen rampart* led to the addition of about 80 to 100 cm of earth above the original glacis and the construction of a second *glacis* about 70 cm thick on top of this (so-called "upper revetment," Seton-Williams 1980:pl. 4). This glacis was also dated to the MB IIB–C based on ceramics. Garstang also identified *retaining walls* (e.g., Walls X and Y) within the fills of both ramparts for "holding up the *debris* during construction" (ibid.). Both glacis appear to have featured *revetment walls* (Walls B1 and XA) between 0.5 and 1 m high at their base for preventing the erosion of the rampart.

The rampart would have raised the base of the wall about 10 m above the level of the plain. Given the known width of the two ramparts, 19 m and 21 m respectively, it is possible to estimate their respective slopes as 28° and 25°. If the top of the core wall can serve as a basis for suggesting the thickness of the town wall, then it would have been around 1 m thick. The circumference of the Middle Bronze Age fortifications is difficult to determine since they do not seem to parallel the site's present contour lines in Garstang's trench. Based upon the angle of the fortifications in this trench with respect to the contours of the tell it seems possible that the fortifications at this point represent a portion close to a corner in the town wall (see Seton-Williams 1980:pl. 2). If the site's plan was roughly oval as it appears today and may be supported by MB IIB–C *fortification wall* remains on the NW side of the site (ibid.), then the fortifications may have been about 810 m long, enclosing a space of about 4.5 ha.

Jean-Baptiste Humbert also suggests that a Middle Bronze Age gate may have been located on the N side of the site in Area D (1993:863), below the remains of Iron Age residences. The location of the access road to the W in the center of the N side of the site, however, might just as likely have been the location of one of the site's gates.

KHEIBAR, KHIRBET

MR 1764.1954
Refs.: Kotter (1986:140ff., 425ff.), Zertal (2004).
Plans: NA.
Sects.: NA.
Photos: Zertal (2004:fig. 180).
Sitting 45 m above its surroundings, Khirbet Kheibar is located in the northern Samarian hills about 6 km to the NW of Tell el-Far'ah North (Zertal 2004:241). The upper part of the mound was surrounded by a *fortification wall* preserved to a height of 3 m. Within this wall the site was 2 ha in size. Zertal suggests that Kheibar may be identified with *Milcah* mentioned in Josh 17:3ff. (ibid., 242).

KINNERET (TELL EL-'OREME)*

MR 200.252
Refs.: Fritz (2000).
Plans: NA.
Sects.: NA.
Photos: NA.
Kinneret is located on the NE shore of the Sea of Galilee approximately 20 km SSW of Hazor. Although final publication of the evidence for its Bronze Age fortifications remain forthcoming, Volkmar Fritz claimed that during its foundation in the sixteenth century B.C. the site featured a *fortification wall* and

[74]The slope cannot be determined with certainty as the section is schematic and the trench was not aligned perpendicularly to the slope of fortifications.

earthen rampart, which he called a "glacis" (2000:508). It continued to function until it was destroyed at the end of the LB I. Apparently, these fortifications served as a model for the construction of its fortifications during the Iron I period.

Kitan, Tel (Tell Musa)

MR 2043.2220
Refs.: Eisenberg (1993).
Plans: NA.
Sects.: NA.
Photos: NA.

Tel Kitan is situated on the W side of the Jordan Valley near the mouth of the Nahal Tavor approximately 12 km NNE of Beth-Shean. The site was occupied from the MB IIA through the LB I (Str. VI–III), but appears to have only been fortified during the MB IIC, Stratum IV (Eisenberg 1993:881). At that time the site featured a mudbrick *fortification wall*, which has been exposed on the S and W sides of the tell; on the S side the wall was 2 m thick and was built of square mudbricks resembling those used in the Middle Bronze Age temple, while on the W side domestic units were built against the wall. The site, which was approximately 0.8 ha in size (Broshi and Gophna 1986), appears to have consisted of little more than a handful of dwellings centered on the temple. The Stratum IV settlement was destroyed at the end of the Middle Bronze Age (Eisenberg 1993).

Kurdane, Tel (Aphek?)*

MR 160.249
Refs.: Broshi and Gophna (1986).
Plans: NA.
Sects.: NA.
Photos: NA.

Tell Kurdane (ancient *Aphek*?) is located 9 km to the SE of Akko. Recent exploration of the site's fortifications by Ezra Marcus has revealed a field stone *revetment* at the foot of an *earthen rampart*. However, due to a complete absence of ceramics in the excavated fill of the earthen rampart, it has not been possible to establish the Middle Bronze Age phase during which the rampart was constructed (personal communication, 2006). Marcus estimates the size of the Middle Bronze Age settlement to be approximately 12.4 ha based on the extent of the earthen rampart (cf. 1.75 ha est. by Broshi and Gophna 1986).

Lachish (Tell ed-Duweir)

MR 1357.1083
Refs.: Tufnell et al. (1940); Tufnell (1953; 1958); Ussishkin (1993; 2004c).
Plans: Site (Tufnell 1953:pl. 106).
Sects.: NW (Tufnell 1958:pl. 90); NE (ibid., pl. 96); Fosse (Tufnell et al. 1940:pl. 69).
Photos: NW side (Tufnell 1958:pl. 5); NE side (ibid., pl. 6); Fosse (Tufnell et al. 1940:pls. 69, 72; Tufnell 1953:pl. 125; 1958:pl. 6); so-called "sapper's tunnel" (Tufnell 1958:pl. 6).

Lachish (EA *Lakiša*) is located in the western foothills of the hill country approximately 30 km ESE of Ashkelon. Lachish is flanked on its E side by the Nahal Lachish (Wadi Ghafr). Although the site featured MB IIA occupation (Ussishkin 2004a:144ff.), the construction of the rampart during the MB IIB was most likely responsible for the shape of the mound today (Ussishkin 2004b:55), which is roughly rectilinear, almost square (335 m x 235 m), with its four corners oriented to the cardinal points of the compass (Figure 86). Excavation of Lachish revealed several elements of its defenses that date to the MB IIB–C. Among these were a plaster glacis, earthen rampart, fosse, and remains of a fortification wall (Tufnell 1958:45–48).

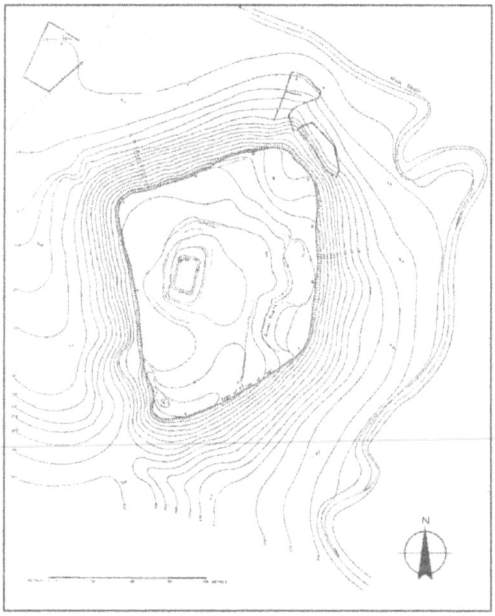

Figure 86. Plan of Lachish (after Tufnell 1953:pl. 106).

On the NW side of the mound excavation of an E-W trench revealed the site's Middle Bronze Age rampart, glacis, and fosse (see section AB in Tufnell 1958:pl. 90). The *earthen rampart* was composed in some places of "grey debris and brick wash" deposited in alternating layers, sometimes described as the "sandwich" technique, while in other places it consisted of earth (i.e., *terre pisée*) with limestone chips and dark clay (ibid., 45f.). Sherds recovered from the rampart's layers suggest a date within the MB IIB–C period. The slope of the earthen rampart probably resulted from the slope of the bedrock below the rampart. The rampart was about 30 m wide from its top to the bottom at the fosse and is preserved to height of about 7 m (ibid., 46), though it might originally have stood as much as 16.6 m above the valley floor. Its full length would have been about 1,020 m.

A lime-plastered *glacis* covering the rampart was exposed at the base of the rampart with a slope of 29°. A *fosse* also accompanied the rampart (see Tufnell et al. 1940:pl. 72). Originally cut from the bedrock (ibid., pl. 69), it was traced for almost 140 m on the NW side of the mound; it is 2 m deep and 9 m wide. Excavation of a circular pit in square C13 suggests that the fosse also served to collect water from the slopes of the mound in what have been a cistern in square B13 (ibid., 46). Ceramics retrieved from the fosse suggest that it was in use by at least the mid-MB IIB. No wall associated with this earthen rampart, its glacis, or fosse was ever identified.

The settlement of the MB IIC to LB I, which is attributed to pre-Level VII, appears to have featured a mudbrick *fortification wall* approximately 3.2 m wide built into the glacis on the slope of the rampart at approximately 244 m ASL (Tufnell 1958:48, pl. 96); five courses of this wall were preserved. This suggests that the settlement experienced some growth and the line of the town wall moved outward slightly (contra Ussishkin 2004d:55). Although Tufnell acknowledged that the wall was built after what she identified as a glacis (see Tufnell 1958:pl. 96), she noted that it probably did not constitute "a substantially later phase" (ibid., 48). The length of this wall would have been about 900 m enclosing an area of about 5 ha. Ussishkin suggests that the Middle Bronze Age gate may be located below the Iron Age gate on the W side of the mound (2004b:55). Tufnell noted that Level VIII was brought to an end by an intense conflagration (1958:48), which Ussishkin's excavations confirmed (2004a:154ff.).

Tufnell reported that she detected a "sappers' tunnel" dug into the site's W rampart (1958:pls. 6). However, because this tunnel parallels the summit of the mound and at no point along the 22 m of its excavated length did it undermine the wall, makes its difficult to accept its identification as a "sappers' tunnel." Furthermore, the location of the beginning of the tunnel also makes it difficult to accept that the sappers would have begun their work this close to the wall.

MALHATA, TEL (TELL EL-MILH)

MR 1525.0696
Refs.: Kochavi (1967; 1993); Beit-Arieh (2003).
Plans: NA.
Sects.: NA.
Photos: NA.

Tel Malhata is located on the S bank of the Nahal Malhata about 5 km ENE of Tel Masos. The MB IIB settlement at Tel Malhata, designated Period B, was represented by remains of the defenses of what has been interpreted as a fortress of approximately 1 ha in size (Kochavi 1993:935). The defenses consisted of an *earthen rampart* "built of alternating layers of loess and river pebbles" supported within the rampart by "three stone *retaining walls*." A stone *core wall* about 1 m thick was built to support the inside face of the rampart with structures built along its interior slope and against this wall. The rampart's *glacis* featured a plastered slope of 30°. The site, which was founded at the start of the MB IIB, was destroyed during the MB IIC. A 2.5 m thick wall, perhaps the town *fortification wall*, was revealed in section B (Kochavi 1967:272). Structures exposed inside this wall appear to have been destroyed during a fire that Kochavi dated to the MB IIC. The defenses of this period were accompanied by a *fosse*.

MANAHAT (NAHAL REPHAIM)*

MR 1679.1289
Refs.: Edelstein and Greenhut (1989); Edelstein and Milevski (1990); Edelstein (1993).
Plans: Site (Edelstein 1993:1281).
Sects.: NA.
Photos: Isometric reconstruction (Edelstein and Greenhut 1989:fig. 103).

Manahat is located in the Rephaim Valley about 5 km SW of Jerusalem. This small rural site was between 3 and 3.5 ha in size during the MB IIB–C (Edelstein and Greenhut 1989:121). It has been characterized as an undefended site, but the evidence from Area 100 on the S side of the site suggests that the perimeter enclosure was formed by the *rear walls* of houses around the settlement (cp. Giv'at Sharett). This does not, therefore, actually constitute a fortification wall.

MARJAMA, KHIRBET ('AIN SAMIYA)*

MR 1816.1554
Refs.: Zohar (1980); Finkelstein (1988; 1993a:1313).
Plans: NA.
Sects.: NA.
Photos: NA.

Khirbet Marjama is located 8 km SSE of Shiloh in the hills of southern Samaria (for location, see Finkelstein 1988:fig. 54). It was approximately 3 ha in size (ibid., 145). It is identified as a fortified settlement of the MB IIC (1993a:1313).

MASOS, TEL (KHIRBET EL-MASHASH)

MR 146.069

Refs.: Fritz and Kempinski (1983); Singer (1983); Kempinski (1993b).

Plans: Site (Fritz and Kempinski 1983:plan 27); Area E Trenches 1–3 (ibid., plan 28).

Sects.: Area E: Trench 1 (Singer 1983:fig. 22), Trench 2 (ibid., fig. 23)

Photos: Site aerial (Fritz and Kempinski 1983:pls. 1, 70–71); Area E Trench 1 (ibid., pls. 72a, 73–75).

Tel Masos is located 39 km SE of Haror and 6 km W of Malhata. The remains of an enclosure dated to the MB IIC were excavated in 1972 and 1974 about 600 m to the SW of Tell Masos on the S bank of the Nahal Beersheba in Area E (Singer 1983:187). The Middle Bronze Age settlement was about 1.7 ha in gross size (ibid., 186), though only 1.2 ha within the ramparts was probably inhabited, a square area approximately 110 m on a side. It is possible that the MB IIC settlement should be more specifically dated to the period during the Thirteenth Dynasty in Egypt (i.e., MB IIB), as Kempinski suggested, when it may have functioned as a "way station, controlled by the block of city-states crystallizing at that time in the southern Shephelah and along the southern coast of Canaan" (1993b:986). However, this date remains uncertain. While the settlement's Middle Bronze Age occupation was divided into two phases by Kempinski (1993b),[75] the original publication by Singer (1983) advances only a single settlement with several sub-phases of occupation during the late MB IIB–C but with no variation in settlement plan and no assertion that the site was a fortress prior to the construction of the rampart. The following description adheres to Singer's interpretation since the data he published appear to support a single phase of occupation. Furthermore, this interpretation has been accepted by A. Mazar who suggests that the site should be considered MB IIC in date (see A. Mazar 1990:196, Table 4).

Following what may have been earlier occupation during the first half of the MB IIB, an *earthen rampart* enclosure was constructed in the MB IIC of earth and debris, with a glacis, an inner *retaining wall*, and a *revetment wall* on the outside of the rampart (Singer 1983:187ff.). Today only the E and S sides of this four-sided enclosure are preserved, the N and W sides being "completely eroded away by winter floods," according to Singer (ibid., 186). The basal layers of the rampart as seen in Trench 2, where only half a meter of the rampart was preserved, appear to include in some places earlier occupational debris such as ash and burnt bricks (ibid., 187). Elsewhere the rampart was primarily constructed of loess intentionally deposited upon existing loess mounds up to 6 m high; the difference between the intentional and original deposits of similar material is apparent in pl. 74a of the final report (Fritz and Kempinski 1983). At least where existing loess mounds were present, the earliest layers, which were deposited as part of the rampart's construction, appear to have been initially deposited against structures built on the interior slope where the tip lines were steepest (35°). Further down the slope the tip lines of the fill were not as steep [ca. 30°] (Singer 1983:187).

Based upon the section of Trench 1 it appears that the entire rampart was as much as 50 m wide, though Singer's estimate of 30 m accounted specifically for the intentional buildup between the outer revetment and the S wall of building 806. The net effect of the buildup of the ramparts was to raise the site's defenses more than 7 m above the surrounding plain, even though the intentional buildup in this location constituted only the top 1.5 m of the complete buildup. The slope of the exterior of the rampart was approximately 17° when excavated, but may originally have been 25°. If the rampart was 440 m long, then its volume was approximately 77,000 m^3.

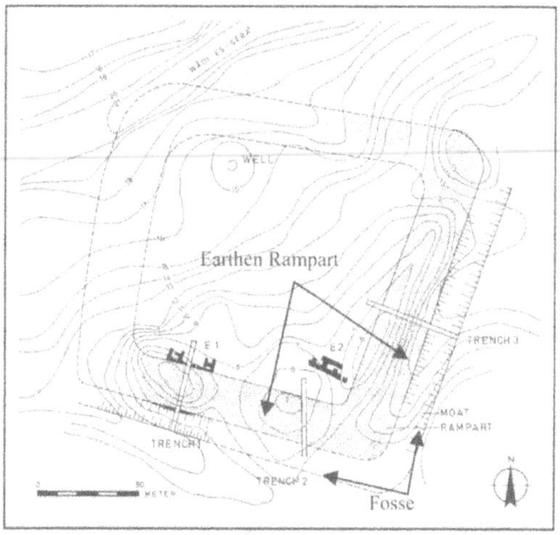

Figure 87. Plan of Masos (after Fritz and Kempinski 1983:plan 27).

[75]In Kempinski's analysis the earliest settlement was identified as a fortress that was then followed by the construction of a five-sided rampart enclosure built from the remains of the earlier fortress but which also featured a revetment wall and fosse.

It is difficult to be certain that the rampart featured a distinct *glacis* around the entire site, though Singer asserts that portions of a pebble glacis were detected (1983:188; see also fig. 22). Nevertheless, it is clear that the foot of the rampart was provided with a stone *revetment wall* to prevent erosion of the rampart into the fosse; this wall was preserved to a height of about 1.5 m. A *fosse* was also dug around the site into the soil, at least on the three sides where the wadi was not present, in order to improve its defensibility. Its extent can be seen in the aerial photograph of the site. It appears to have been about 5 m wide at its top and 2 m wide at its base and about 1 m deep (ibid., fig. 22), even though it is illustrated as nearly 20 m wide on the plan of the Middle Bronze Age settlement (Fritz and Kempinski 1983:plan 27). The excavators suggested that the earth taken from the fosse was used in the raising of the ramparts. The length of the fosse was about 150 m on a side, perhaps as much as 450 m long if it was not present on the N side where the wadi is located.

The excavations of Masos allow the following reconstruction of the constructional phases of the site's fortifications. Once the location of the rectilinear fortifications had been decided upon the initial steps would have included the building up of low spots between loess mounds probably with fill from the fosse. This continued until the rampart achieved a roughly 6 to 7 m height above the surrounding plain. In the second phase of construction structures (buildings 806, 819, etc.), perhaps to be identified as bastions and towers, were built along the crest of the rampart. In order to provide them with solid foundations more fill was added to the rampart's surface against the exterior walls of these buildings using occupational debris in certain areas (Trench 2) and excavated loess in others (Trench 1). This raised the rampart at least another 1.5 m. Finally, a pebble glacis may have been added to protect the surface of the rampart. There does not appear to be any evidence for positing that a separate wall crowned the ramparts. Instead, it appears that only the exterior walls of structures along the crest of the rampart provided walls.

ME'AMER, TEL (TELL EL-'AMR)*

MR 159.237

Refs.: Broshi and Gophna (1986); Prausnitz (1993a).

Plans: NA.

Sects.: NA.

Photos: NA.

Tel Me'amer, which is located 20 km S of Akko, is identified as a Middle Bronze Age site of approximately 0.7 ha featuring an *earthen rampart* (Broshi and Gophna 1986).

MEGIDDO (TELL EL-MUTESELLIM)

MR 1675.2213

Refs.: Schumacher (1908); Guy (1931); Loud (1948); Yadin (1972c); Kempinski (1989); Finkelstein et al. (2000).

Plans: Site (Loud 1948:figs. 376f.); Area AA (ibid., figs. 378–380); Area BB (ibid., figs. 396–400); Area CC (ibid., fig. 407); see also reconstructions of Strata XIII–X (Kempinski 1989:plans 2–5, fig. 38); Area F (Ilan et al. 2000:fig. 4.9); Lower town wall (Guy 1931:fig. 14).

Sects.: See Pflock 22 (Schumacher 1908); Areas AA, BB, and CC (Loud 1948:fig. 416); Area F (Ilan et al. 2000:figs. 4.3f.).

Photos: Area AA (Loud 1948:figs. 6–22, 25–27); Area BB (ibid., figs. 197–201); Area CC (ibid., figs. 264–267); Area F (Ilan et al. 2000:figs. 4.5–4.8).

Megiddo is located in the Jezreel Valley about 11 km SE of Tel Yoqne'am. Although extensive excavations were carried out at Megiddo by the Oriental Institute's expedition resulting in broad exposures of the town's architecture and the development of a ceramic sequence, several problems with those results have made it necessary to cautiously approach questions of Megiddo's stratigraphy.[76] Pertinent reexaminations of the Bronze Age stratigraphy have included work by Kenyon (1958; 1969), Müller (1970); Yadin (1972c), Gonen (1987), Kempinski (1989), Esse (1991:67–90), and most recently by the expedition led by Finkelstein and Ussishkin (Finkelstein et al. 2000). Yet despite these reassessments and changes to the phasing of archaeological loci, the stratigraphic understanding of the defenses remains largely as they were originally published. The Middle Bronze Age fortifications were excavated by the Oriental Institute expedition in three successive seasons, 1936 (Area CC on the S), 1937 (Area BB on the E), and 1938 (Area AA on the N). The dates assigned to the Middle Bronze Age stratigraphy used here are, however, those put forward by the most recent expedition to Megiddo (Ilan et al. 2000).

The first phase of Megiddo's Middle Bronze Age settlement belonging to the MB IIA was Stratum

[76]These problems among others outlined by Kenyon (1958:51*f.; 1969) include the archaeological methodology employed, hasty and limited publication, interruptions in work during the course of the excavation, and a ceramic assemblage of mixed reliability. Nevertheless, one cannot diminish the influence of this early project upon the field, due to its methodological improvements on the work of even earlier archaeologists such as Sir Flinders Petrie and the important role that the site's stratigraphy has played in the years since its excavation.

XIV. No remains of the fortifications or perimeter of the settlement dated to this period have been revealed.

MB IIA: STRATUM XIIIA

The earliest evidence for the Middle Bronze Age fortifications of Megiddo is associated with Stratum XIII. Close examination of the remains of this stratum in Area BB has enabled further subdivision of its remains into phases A and B (Loud 1948:84ff.). It does not appear, however, that Megiddo was fortified until Stratum XIIIA, the first time since the EB III. In this phase a 1.8 m thick mudbrick *fortification wall* of offset-inset construction was built in Area AA (Figure 88) where a two-chambered, bent-axis gate complex (see discussion below) was also excavated (ibid., 6ff.). The insets in the town wall in Area AA were 2 m apart. Because Stratum XIII was the earliest stratum excavated in Area AA, Loud postulated that the fieldstone foundations for the gate, which were wider than their mudbrick walls may have belonged to an earlier stratum, such as Stratum XV, which was not excavated in this area (Loud 1948:6f.). Penetrating the foundation of the town wall to the E of the gate was a drain over 15 m in length that channeled water from the town out onto the steps of the town's approach (4103).

On the n side of this stepped approach the battered stone retaining wall for the n wall of the citadel gate was preserved 3.7 m high. The gate is essentially a four-pier gate featuring a 90° turn between the main chamber and the stairs leading to the lower town. Passages between the gate's piers were 1.5 m at their narrowest. Outside the gate nine steps 10 to 35 cm high led down the upper tell's slope to the lower town or "terrace." The width of the gate and the presence of these steps seem to argue against the use of this gate for chariot or wagon access, without negating the possibility that pack animals could have come up these steps. For this reason and in keeping with the results of excavation in the lower town, it is possible to postulate that this gate served only as a citadel gate leading down to the lower town (discussed below). It was suggested by Loud that the gate was not covered and no evidence for benches was discovered. Eleven

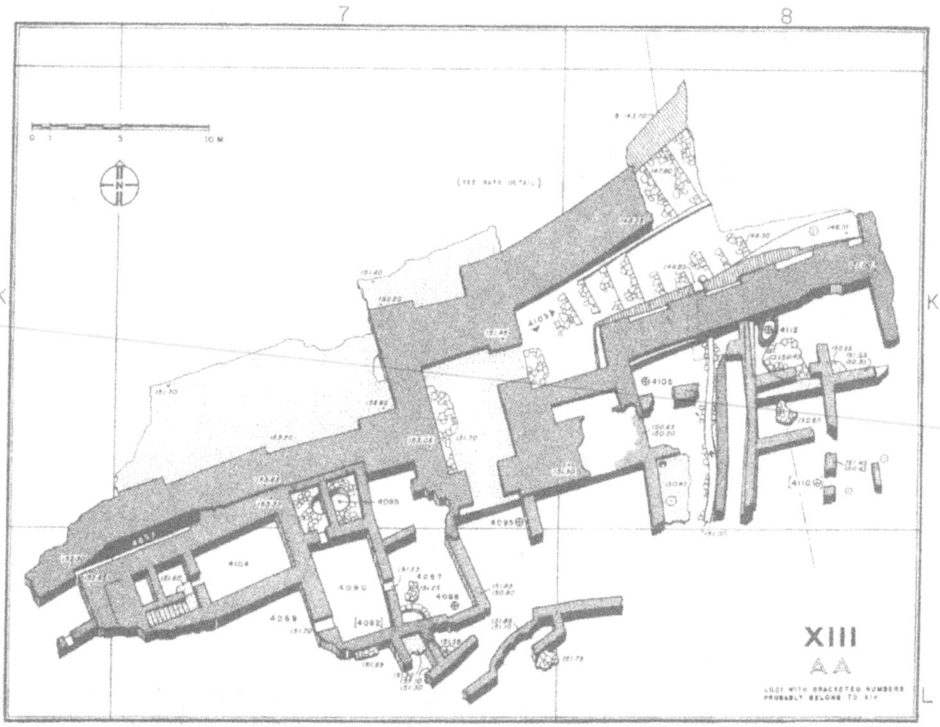

Figure 88. Gate and wall of Stratum XIIIA in Area AA (Loud 1948:fig. 378, top). Courtesy of the Oriental Institute of the University of Chicago.

courses of mudbrick were preserved in the citadel gate.[77] A *supplemental earthen rampart*, which was identified as a glacis, made of chipped limestone and less than 2 m wide, was also found abutting the walls of the earliest gate and town wall (Loud 1948:6–8, figs. 6–20). Kenyon determined its slope to be about 16° (1969:55).

Although the identification of structure 4104 W of the gate and just inside the town wall (though not abutting it) is uncertain, its rectangular layout (13 x 6.5 m) and partially preserved staircase suggest that it was a *tower* that provided access to the top of the town wall (Loud 1948:7f.). It may have been constructed on the interior of the wall owing to the lack of flat ground upon which to build the tower outside the town wall (cp. Tel Zeror and Tel Poleg). Further confirmation of this identification is provided by a mudbrick tower excavated in Area BB in Stratum XII (see discussion below). No finds are reported as associated with structure 4104, but in the same room as its well-preserved staircase a mudbrick storage bin and bench evoke how this structure served the town's inhabitants.

The earliest fortification wall found in Area AA is probably equated with the mudbrick town wall 3182 in Area BB, though it is 30 cm thinner in Area BB. In Area BB (Figure 89) the *fortification wall* was only 1.5 m thick, and the insets, which were one brick deep (35 cm), were regularly spaced, slightly less than 3 m apart (Loud 1948:87). Wall 3182 was built of 35 x 35 x 10 cm mudbricks and the entire structure was founded upon a stone foundation (2149) 3 m wide. Although this wall is described as made of brown mudbricks, close examinations of fig. 398 (Loud 1948; note the line dividing N from the S phase in the construction of 3182) and Gordon Loud's field diary reveal that at the S end of Area BB wall 3182 was constructed slightly differently. Here it was found to be constructed of gray mudbricks over 2.1 m thick with offsets spaced only 2 m apart. This difference led Loud to suggest in his field diary, but not in the publication, that there may have been a gate here.[78] Although a gate seems improbable here given the lack of evidence for other gates or an approaching ramp in this area, this change in construction style may simply characterize the lack of continuity in wall construction around the site, which is also attested at other sites. The fieldstone foundation, which was twice as wide as the wall also followed the offset-inset layout of the mudbrick wall. The remaining excess foundation protruded on the interior of the town wall (Loud 1948:84–87, figs. 197–201). This was perhaps because in Area BB the citadel's wall formed the outer defense of the settlement where in Area AA the wall primarily functioned to separate the citadel from the lower town.

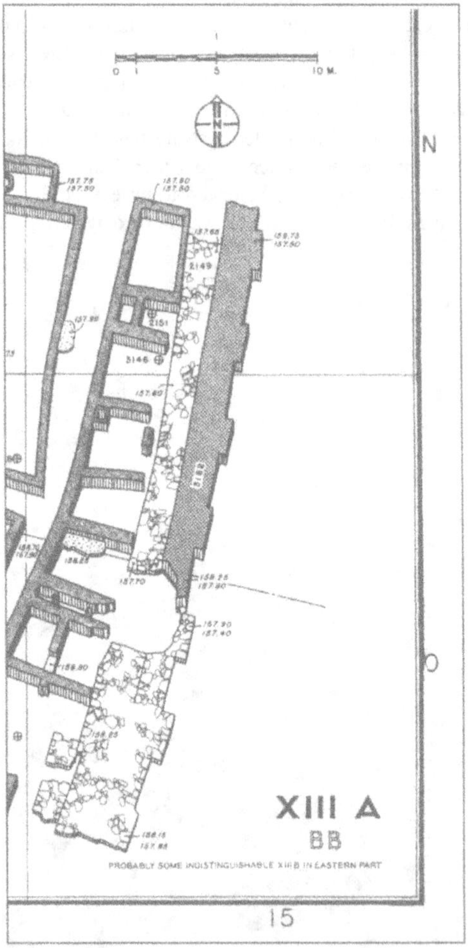

Figure 89. Wall of Stratum XIII in Area BB (Loud 1948:fig. 397). Courtesy of the Oriental Institute of the University of Chicago.

The Stratum XIII defenses were also revealed in Area CC, though the results are not as clear as those

[77] This observation is based on unpublished photo A4636 of the Oriental Institute's expedition (Archives of the Oriental Institute).

[78] This is based on Loud's field diary entry for March 18, 1937 for the East Area (BB): "The mudbrick wall [3182] is puzzling near the south end, for here at what may be part of a gate the two kinds of brick are reversed—the brown brick [identified with 3182] continues in the line of the gray [previously identified with 3181], while the gray is behind [3182]—on the town side of the brown" (Oriental Institute Archives). This observation is a good lesson in avoiding equation of walls based simply on brick size or color.

in Areas AA and BB (Loud 1948:105ff.). In this part of the site it was more difficult to assign remains to Stratum XIII or XII due to a limited exposure in a 5 m wide trench. Nevertheless, two complexes of walls, referred to as A and B, were exposed and have been associated with the Strata XIII and XII defenses. The mudbrick *fortification wall* within the wall A complex, including the outermost wall on the edge of the slope and its "glacis," was preserved 4 m high and 2 m wide. It featured two phases in the course of its construction (ibid., fig. 416). It appears that wall A formed the interior buttress of an early *earthen rampart* built at a distance from the town wall itself (i.e., complex B). This is suggested by the 3 m high, N elevation of wall A, which slants at 70° against a mudbrick core construction 3.5 m wide and 2.5 m high with a triangular section. The face of this structure is covered on the S by what was identified as a "glacis" (4 m wide x 1 m thick of unknown composition) seen in the section drawing with an approximate slope of 40°. However, the reconstruction Loud proposed for the defenses of Strata XIII–XII in Area CC (ibid., fig. 267) is likely incorrect since the interior face of the lower portion of the wall A complex probably did not remain exposed nor was its mudbrick superstructure suspended in mid-air over its lower courses. It would have been necessary to build up the interior against the rampart if a wall was to be built on top of it.

It was perhaps for this reason that Kempinski suggested that the wall A complex related to neither 3181 or 3182, as found in Areas AA and BB (1989:109). Instead he suggested that the wall A complex was a part of a greater enclosure wall and rampart that encircled both the upper tell and lower town, and for this reason it was not detected in Areas AA and BB. It is also possible that the wall A complex should be dated to Stratum XIIIA and continued in use during Stratum XII when the wall B complex was constructed. Nevertheless, the temporal relationship between the wall A and wall B complexes remains uncertain.

The wall B complex, which comprised part of the defenses in Area CC, consisted of two walls, perhaps the remains of a tower (dimensions unknown) like the one in Area AA but larger. Unfortunately, although a doorway was preserved to a height of 2.15 m between the N and S mudbrick walls within the tower, no finds from inside the structure help to clarify its function.

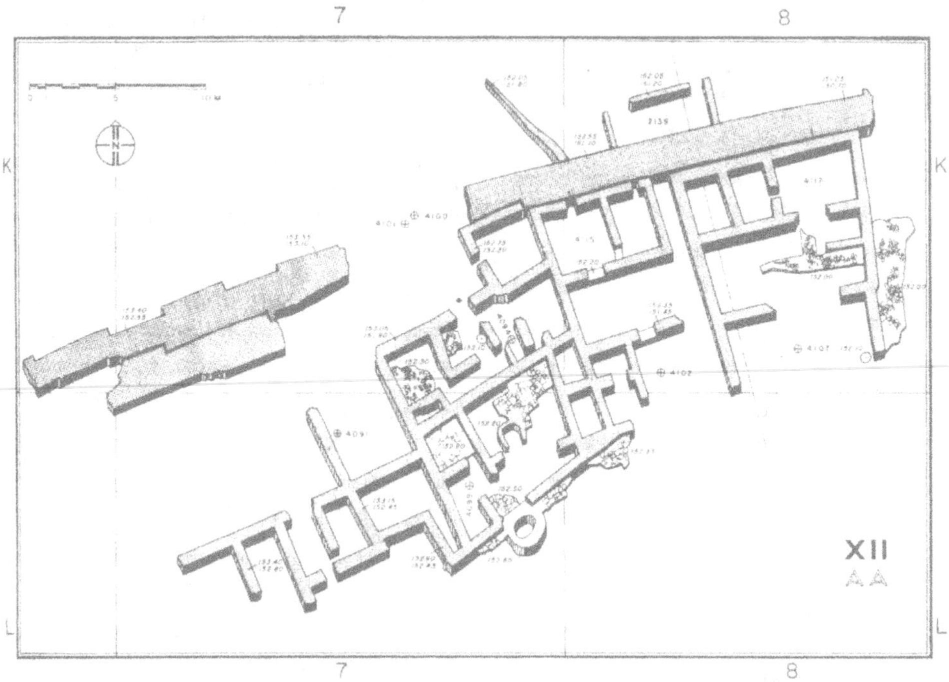

Figure 90. Gate and wall of Stratum XII in Area AA (Loud 1948:fig. 378, bottom). Courtesy of the Oriental Institute of the University of Chicago.

Both of these walls were 1.75 m thick and made of mudbricks 35–40 x 35–40 x 10 cm, dimensions similar to bricks used in wall (3182) of Stratum XIIIA in Area BB. Although the S wall was preserved to a height of 1.75 m, the N wall was only preserved to a height of 1 m. Perhaps the S wall served as the extension of 3182/3181 from Area BB (Loud 1948:105–13). However, if Kempinski's reconstruction is correct, it would seem that no reinforcement of the wall in this area was undertaken during Stratum XII as in Area BB after its initial construction, which was perhaps due to the continued serviceability of the rampart (wall A complex). Given that the wall B structure was constructed on a deliberately packed surface at a distance of 10.5 m from the wall A complex with no structures of consequence between them, Kempinski's interpretation is plausible.

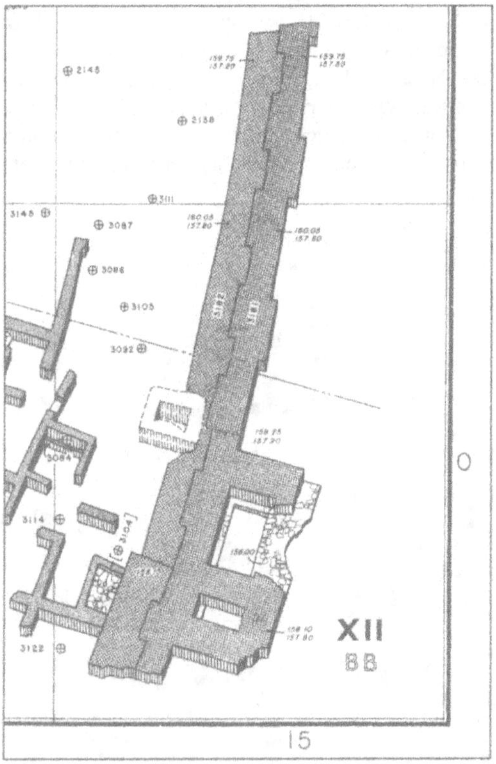

Figure 91. Area BB Strata XIIIA and XII walls (Loud 1948:fig. 398). Courtesy of the Oriental Institute of the University of Chicago.

MB IIA–B: Stratum XII

During Stratum XII, as mentioned above, Megiddo's *fortification wall* was widened around the entire site as evidenced in Areas AA and BB (Loud 1948:8ff., 87ff.). In Area AA only the wall to the W of the location of the gate, which was not preserved, was thickened with an addition of nearly 2 m (Figure 91), which was also plastered.[79] Although here the wall retained the offset-inset design of the earlier wall, the portion of the wall to the E of the gate's location lacked buttressing and was not built with an offset-inset design. No remains of the Stratum XII citadel gate were preserved since it was presumed to have shifted eastward in the tradition of later gates (Loud 1948:8–15, figs. 21f.). Kempinski has posited the continued use of the Stratum XIII gate in the first phase of Stratum XII (his stratum XIIB),[80] an interpretation with which Herzog agrees (1997a:104). According to Kempinski, this gate was replaced in Stratum XIIA by a direct access gate of which nothing has remained (1989:46f., 110, plan 3).

In Area BB *fortification wall* 3181 was added to the exterior of 3182 doubling its width and improving its construction (see Figure 91 above). Evidence of this improvement in construction can be seen in the plastering of the face of the wall,[81] which featured a more careful construction, the use of better bricks, and the addition of a *tower* (9.5 x 6 m) constructed of mudbrick on the exterior of the wall (Loud 1948:87–92, figs. 198–201).

Excavation in Area F of the Tel Aviv University expedition has revealed part of a rampart around the lower town to the NE of the upper town's gate (Ilan et al. 2000:78ff.). Unfortunately, limited soundings in this trench, which did not reach bedrock or virgin soil, make it difficult to interpret the results. Nevertheless, in Area F a wall (94/15), which is probably to be identified as a *revetment wall*, was excavated. Against the lower part of this wall layers of an *earthen rampart* were deposited, leaving the upper 2 m or more of the wall exposed (ibid., 80). It was apparently built around the lower town and probably corresponds, as suggested

---

[79]This plastering on the addition to the mudbrick wall during Stratum XII, which was preserved in an identical fashion as that found in Area BB (i.e., between the bricks' joints), is only evident on the N face of the wall in unpublished photo A4528 (Oriental Institute Archives).

[80]The division of Stratum XII in Area AA into A and B, which is not original to the Megiddo stratigraphy, was arrived at by Kempinski (1989). This division was not, however, suggested to apply to any of the other excavation areas.

[81]Loud has referred to this as plaster "pointing" of the joints between the mudbricks. Although this is clearly what remains of the plaster after excavation (see Loud 1948:fig. 199), it is more likely that the entire exterior of the wall had been plastered and that only small sections of the plaster have been preserved where recesses in the surface, such as between bricks, held the plaster longest.

by the Tel Aviv expedition, with Wall K identified by Guy, though misdated to the tenth century B.C. (see Guy 1931:fig. 14, for the course of this wall). Following some limited soundings of the lower town in 1971 and 1972 Yadin had suggested that the "rampart" only consisted of "building remnants and the accumulation after the MB II" and for this reason could not be considered part of the site's defenses (1972c:162). However, the results from the Tel Aviv expedition appear to confirm the existence of defenses around the lower town, the earliest of which can now be placed at least at the start of the MB IIB.

MB IIB: STRATUM XI

In Stratum XI the town's defenses were modified yet again. This time in Area AA a much thinner, though buttressed *fortification wall* was built upon an earthen rampart about 7 m out from the location of the earlier town wall. The 1.25 m thick wall, presumably composed of mudbrick, was built on a rubble foundation with interior buttresses (1 x 1 m) about 1.5 m apart. A *glacis* was found running up the rampart and against the lower, exterior part of the wall though its composition was not described. While a supplemental *earthen rampart* can be posited from the section (Loud 1948:fig. 416), though it is not explicitly described by Loud, it is impossible to be certain of the embankment's original height since the section drawing is only schematic. Although estimates for the rampart's height range from 3 m (Kempinski 1989:110) to 5 m (Kenyon 1969:56, fig. 27), an examination of the section seems to support Kempinski's figure. The outer slope of the glacis and rampart was 45° based on an examination of the same section. Although Kenyon's reconstruction of the interior slope of this embankment is speculative, it is probably a reliable reconstruction of the remains found to the S of the wall but not displayed in section.

Only the W remnant of the gate was found on the E side of Area AA, but the remains are insufficient to reconstruct its layout (Loud 1948:15, fig. 26). No remains of this period's defenses were found in Area BB or CC by the Oriental Institute Expedition. However, there is a conspicuous alignment of the E elevation of structures in Square O15 in Area BB with the inner elevation of Wall 3181 of Stratum XII, suggesting that perhaps the earlier wall continued in use during Stratum XI.

MB IIC: STRATUM X

Remains of the defenses of Megiddo in Stratum X are very poorly preserved. Nothing is left of the fortification walls in Areas AA, BB, or CC. Although Loud speculated that the Stratum IX *six-pier gate* may have been in use at the end of Stratum X (see Loud 1948:15f., fig. 380), it is not possible to demonstrate this stratigraphically. There is also no evidence of a uniformly constructed wall associated with this gate in any of the areas excavated. Nevertheless, based upon the uncertain stratigraphic reconstructions of Area AA it is likely that the six-pier gate associated with Strata IX to VIIA was constructed during the later half of the MB II. If the Stratum IX gate was actually constructed during Stratum X, then it was either accompanied by the fortification wall identified with Stratum IX or the gate was built into the fortifications of Stratum XI. There are no comparanda to suggest that a late Middle Bronze Age gate such as this one stood without a fortification wall.

LB I: STRATUM IX

The only fortifications usually identified with the Late Bronze Age settlement of Megiddo are those of the six-pier gate discussed in the previous section. However, it appears that the *fortification wall* of the first phase of the LB was 2.5 m thick and extended westward from the palace of Stratum IX (see Loud 1948:fig. 381). It featured a saw-tooth construction along its outer elevation with recesses between 0.5 and 1 m deep.

No fortifications associated with subsequent Late Bronze Age strata have been identified.

MEVORAKH, TEL (TELL MUBARAK)

MR 1433.2155
Refs.: Stern (1984).
Plans: Fortress (figs. 21, 26–28).
Sects.: Fortress (fig. 32f.).
Photos: Fortress (pls. 18–19).

Tel Mevorakh is 4.4 km downstream from Burga, which is the nearest major Middle Bronze Age settlement along the Nahal Tanninim. The Middle Bronze Age settlement was identified as a fortress by Ephraim Stern and was said to have been destroyed at the end of the Middle Bronze Age (Stern 1984:36). Its location in the center of the coastal plain between the plain of Sharon to the S and the Carmel coastal plain must certainly be considered strategic.

MB IIA, PHASE 2: STRATUM XV

The earliest occupation on Mevorakh was a well-planned "road-fortress" on virgin soil during the middle of MB IIA, Stratum XV (Stern 1984:69). Unfortunately, remains from this phase are limited to the central excavation area where only three mudbrick walls on stone foundations (W.182, 188, and 191) of fortress B-349 were revealed. While Stern identifies this structure as part of the fortress, its location at the NW corner of the site suggests that it was more specifically the NW *tower* of the fortress that must have encompassed the entirety of the natural hill. The walls of this structure average 1 m wide. Preserved to almost 3 m in height in one location, the walls are the most rectilinear in plan of those found during the course of excavations by Stern.

MB IIA, PHASE 3–4: STRATUM XIV

In the second phase of MB IIA occupation (Str. XIV), it is clear that the site experienced a population growth that forced settlement beyond the walls of this fortress (Stern 1984:52). This was indicated by the building of a mudbrick *fortification wall* (186) of a similar character to the W and against the tower's exterior wall (182), as well as the construction of a house dated to the MB IIA. It should also be noted that no rampart or glacis was built during either of these phases.

MB IIB: STRATUM XIII

During the first phase of MB IIB settlement at Mevorakh (Str. XIII) an *earthen rampart* of sand, hamra, and kurkar (so-called *terre pisée*) was deposited over the earlier, perimeter walls of the Stratum XIV–XV settlement. This formed a small crater, not unlike the craters of larger settlements (e.g., Ashkelon, Tel Dan, Hazor, Timnah) within which the settlement was located. Its inner slope was covered by a seeming *glacis* of "thin layers of hard clay, mixed with a large number of potsherds" (Stern 1984:49). This undoubtedly served to protect the rampart against erosion.

Stern has characterized the site during this period as a border fortress, serving Tel Dor (ibid., 69). The gate entrance to the site must have been located on the E side where a slope feasible for an entrance was found and since the N, W, and S sides of the site form the steep sides of the rectangular enclosure. During this period the site was about 0.1 ha (ibid., 68). It is not possible to estimate the quantity of earth and materials used in the ramparts enclosing Mevorakh in this period since insufficient sections are available from which to establish their dimensions.

MB IIC: STRATUM XII

The final stage of Middle Bronze Age occupation at Mevorakh during the MB IIC (Str. XII) continued with the occupation of the MB IIB fortress. The site is thought to have been no more than about 0.1 ha in size (Stern 1984:1). The strategically important role of this site along the coastal plain throughout the Middle Bronze Age is attested by its nearly continuous occupation despite the abrupt interruption of occupation at other Middle Bronze Age sites such as nearby Burga, which was abandoned at the start of the MB IIB. At the start of the Late Bronze Age Mevorakh was leveled for the construction of the Late Bronze Age temple, which required more than a meter of buildup above the Middle Bronze Age fort.

MICHAL, TEL (DHAHRAT MAKMISH)*

MR 131.174
Refs.: Herzog et al. (1989).
Plans: Site (Herzog et al. 1989:1.2; Herzog 1989:4.2); N end of high tell (ibid., fig. 4.7).
Sects.: W (Herzog 1989:fig. 4.3); E (fig. 4.4); S (fig. 4.5); Composite A-A (fig. 4.1).[82]
Photos: Site aerials (Herzog et al. 1989:pls. 2, 5), W trench (pl. 7); S trench (pls. 8f.).
Tel Michal is located on the Mediterranean coast about 12 km NNE of Jaffa. The Middle Bronze Age settlement at Tel Michal was limited to the high tell (0.2–0.3 ha) during the MB IIC (Herzog 1997a:129). Although a composite drawing of the E-W trench across the excavation area is published, the step trench revealed only a very limited picture of an embankment which is conjectured to have been built during the MB IIC. According to the excavator much of the settlement's W side has been eroded, a process which was occurring even in antiquity as evident from the eastward shift in the LB I (Str. XVI) and II (Str. XV) settlements (Figure 92).[83]

MB IIC: STRATUM XVII

While a substantial part of the picture of the MB IIC settlement has been reconstructed by the excavator, MB IIC remains (Str. XVII) consisted mostly of fills and were encountered in the W step trench (Sq. M16–18) and in squares M8, O13–15, S15, and R15 (Herzog 1989:29ff.). These fills, which contain only MB IIC sherds, formed a supposed platform (782) over 4 m high expanding the area of the original kurkar hill for the construction of the settlement. The platform on the NE and E sides was purportedly delimited from the rampart and glacis by mudbrick walls S156 (which was 0.64 m thick, preserved to a height of 3.27 m, and was truncated on the W where the mound has since collapsed) and N156, respectively. But the identification of Locus 782 as a platform is

[82]It is important to note that in addition to the limited information that these sections actually provide concerning the Middle Bronze Age deposits at T. Michal they were drawn from photographs and, therefore, have undoubtedly introduced some inaccuracy (see Herzog et al. 1989:7).
[83]Although it is doubtful that T. Michal possessed an earthen rampart or glacis, this site is included in this catalog because of its former association with Middle and Late Bronze Age fortification strategies.

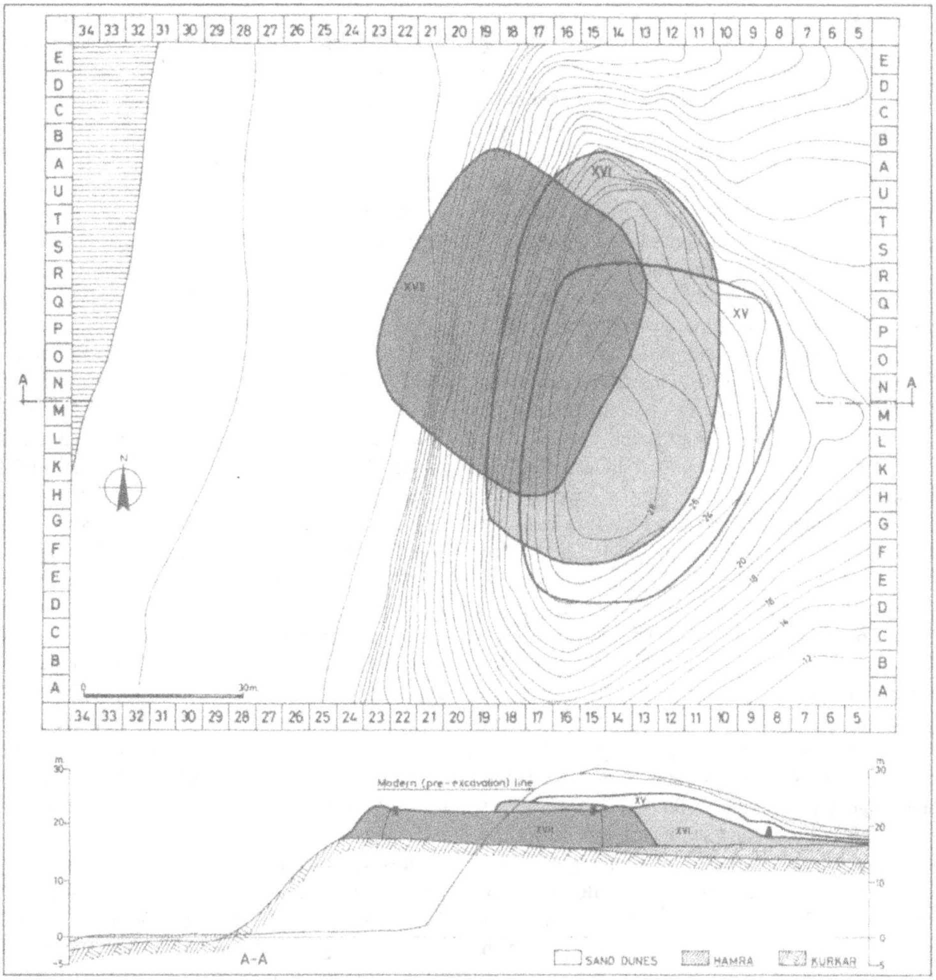

Figure 92. The evolution of Tel Michal's embankments (after Herzog 1989:fig. 4.2).

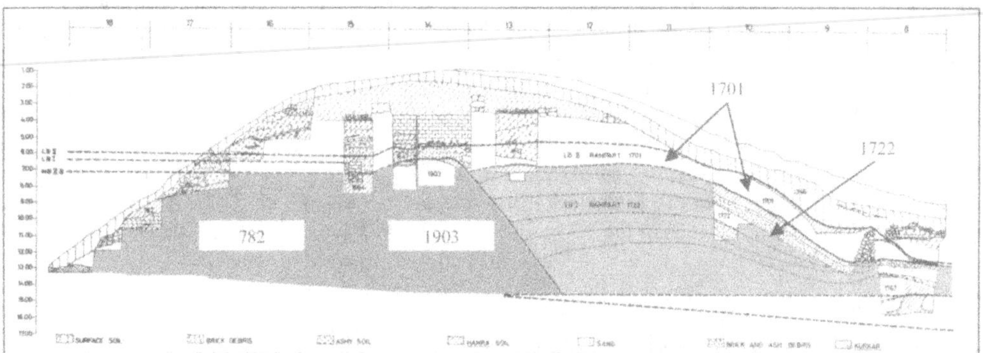

Figure 93. Schematic of section through Tel Michal's embankments (after Herzog 1989:fig. 4.1).

very speculative; since fig. 4.3 shows that it was no more than a debris layer about 1 m thick (see Herzog 1989). The nature of this layer, if it reached the center of the mound, can only be guessed at. Therefore, there can be no certainty with regard to its identification as a platform.

An equally dubious "sand rampart," consisting of Locus 719 on the N and Loci 1657, 1663, and 1903 on the E, was supposedly constructed around the entire site and against this platform. However, this cannot be verified as a rampart since it is nowhere demonstrated to have abutted the platform (Locus 782) and was also never excavated to a sufficient depth to verify its thickness and hence its identity as a rampart. This sand may be nothing more than wind-blown accumulation against the kurkar ridge atop which the settlement was founded. Equally doubtful is the identification of hamra covering the sand of this supposed rampart, which was exposed on the NE side of the tell in square T14 (Locus 716) and on the E in square N13–14 (Loci 1665 and 1901), as a glacis intended to keep the sand rampart from eroding. Therefore, based upon the evidence available at present there is insufficient reason for the identification of a platform, rampart, or glacis at this site. The lack of evidence for a fortification wall at Tel Michal would only seem to support this revised interpretation, as such features are frequently attested together. Although Herzog has estimated that 31,500 m^3 of earth and sand were moved for the construction of the platform and ramparts at Tel Michal (1989:32), no soundings penetrated the core of the tell to a sufficient depth and therefore there is no reliable basis for calculating the quantity of earth used in these features.

Based on a projection of the building walls identified around the site's edge, the site is estimated to have been 2.5 ha in gross size (cp. with figure below) during the MB IIC, of which only 20% remains today. However, based upon the defined limits of the site identified for the MB II its inhabitable area appears to have measured only approximately 30 x 30 m or roughly 0.1 ha in size. Herzog has attributed the late Middle Bronze Age destruction to tectonic activity, downplaying the presence of considerable amounts of brick and ash from the settlement that were present during the construction of the LB I rampart (Locus 1722) (1989:38).

LB I–II: Strata XVI–XV

Two successive settlements (Str. XVI and XV), neither of which was enclosed by a fortification wall, were built during the LB I and II (Herzog 1989:38ff.). Both of these featured what were referred to as ramparts built of sand, and are compared to the "rampart" of the MB IIC. Although these features were probably intentionally added to the E, N, and S sides of the site changing the shape of the tell over time, their identification as earthen ramparts is unnecessary as they bear no functional similarity to Middle Bronze Age fortifications. Even though the site's inhabitable area does not appear to have increased significantly (see Herzog 1989:figs. 4.2 and 4.8), these features were probably no more than attempts to expand the settlement to accommodate modest growth during this period.

The Stratum XVI embankment, referred to as a rampart (1722 on the E, 1562 on the S, and 1381 on the N), was composed of sand (Loci 761, 1722, 1562), ash (1764, 1750, 1727, 1746), brick and occupation debris (1750, 1730, 1556, 1559), and some hamra soil which was used to stabilize the sand (1562, 1558). This embankment was covered with a hamra surface, which Herzog identifies as a glacis (1718 on the E, 1555 on the S, and 1383 on the N). Although the preservation of the embankment is not identical on the E and S sides, both locations reflect the same general sequence of deposition: sand, brick and occupation debris, and a hamra surfacing. While the correspondence between embankment layers at different locations within a site is unusual, here it can be explained by the small size of Tel Michal, which probably meant that each of its embankments were constructed in a very short period of time. The slope of this embankment was preserved to about 30°. Herzog suggests that the remains of building 873 on the N side of the site during Stratum XVI may have been a "fort guarding the approaches from the north" (ibid., 41), but again no clear evidence to support this identification is available.

The Stratum XV embankment, which was also identified as an earthen rampart (1701/1736 on the E, 971 on the S, and 1512 in the center), was also constructed mostly of sand, which was held in place by a retaining wall [M.91] (Herzog 1989:39f.). Not as much mudbrick debris (e.g., 1554) and ash (e.g., 1736, 1512) appear to have been used in its construction as in the previous embankments, yet like them it too was covered with a hamra surface, again referred to as a glacis (1702/1737 on the E and 956/963/961/1262 on the S). The slope of this rampart was preserved to about 25°.

Nagila, Tel

MR 127.101

Refs.: Amiran and Eitan (1965; 1993).

Plans: Site (1965:fig. 11; 1993:1080).

Sects.: Area C (1965:fig. 11).
Photos: NA.
Nagila is located along the Nahal Shiqma approximately 27 km SE of Ashkelon in the coastal plain. During the MB IIB Tel Nagila (Str. XI) was resettled after an occupational gap during the EB IV and MB IIA. The fortification system was excavated in Areas C and F, though it is clear that settlement had spread outside the fortifications during part of the MB IIB–C. Area C on the NW side of the site provided the most information about the site's fortifications, which consisted of a rampart, glacis, fortification wall, and a fosse at the foot of the rampart. In addition to these defenses the site was flanked on the E by the Nahal Shiqma.

Its *earthen rampart* consisted of a core of mounded earth on the interior slope of which a mudbrick *fortification wall* 2.3 to 2.5 m thick was built. About 3 m of the wall's elevation was preserved. The lower portion of the exterior of the wall and the rampart's surface were covered by alternating layers of earth, caped by a crushed chalk *glacis*. The rampart and glacis sloped at an angle of about 35° (see Amiran and Eitan 1965:fig. 11). The excavation of Area C on the inner slope of the rampart revealed not only the town wall but another parallel wall about 3 m behind it (cp. Achzib). This wall of similar thickness probably belonged to a rectangular *tower* that was part of the defenses at this point along the wall (see reconstruction in Herzog 1997a:fig. 4.21). That this tower was involved in a site-wide fiery destruction is evident from the thick layer of ash excavated above its floor. Amiran and Eitan have noted that the level of the floor inside the perimeter wall was about 1.5 m lower than the level at which the rampart met the exterior of the wall. This perhaps illustrates that the wall must have been built to a considerable height if it was to serve as an effective defensive wall above the rampart. The estimated length of the rampart is approximately 685 m.

From an examination of the site's plan and section in Area C it is possible to also add some details concerning the size of the *fosse* that encircled the site on at least three sides. The fosse was apparently carved into the bedrock about 3 m deep and about 8.5 m wide at its base (almost 14 m at its top) along the 850 m perimeter of the base of the rampart (Amiran and Eitan 1965; 1993). The E side was probably protected by the Nahal Shiqma.

The site's perimeter wall was also excavated in Area F where fallen bricks and additional evidence of the conflagration that brought its Middle Bronze Age occupation to a close were exposed. Despite this terminal date for the fortifications, the excavators were unable to conclude during which phase of the site's MB IIB–C occupation (Str. XI–VII) the fortifications were constructed.

During the LB I when the site was unfortified a flagstone pavement was laid over the fortification wall in Area F.

NAHARIYA, TEL

MR 159.268
Refs.: Yogev (1993).
Plans: Site (p. 1089).
Sects.: NA.
Photos: Site aerial (p. 1089).
Nahariya is located on the N coast of Israel about 10 km N of Akko and is situated on the S bank of the Nahal Gaton. Although the full extent of the original mound is uncertain it is possible that the site was as large as 3.75 ha during the Middle Bronze Age within a space of roughly 250 by 150 m. The remains of the Middle Bronze Age belong to Strata III and II on the main mound. Stratum III, which has been divided into three sub-phases (IIIC–IIIA), dates to the MB IIA but did not feature any fortifications (ibid., 1089).

During the early MB II (IIB, Stratum II), the site was, however, fortified with a 3.8 m wide *fortification wall*, 12 m of which were exposed in a sounding excavated in 1980 on the N side of the site. The foundation of this wall was built of three to four courses of kurkar stones. If this wall enclosed the entire Middle Bronze Age mound it would have been about 800 m long. The *N Gate* of the settlement was also located in this sounding on the N side of the site, though only the 3 m wide passageway and part of the tower on the W side of the gate were exposed. Both of these structures have been provisionally dated to the MB IIB.

Nothing is known of the defenses of the site in the periods immediately following.

NAJJAR, KHIRBET EN-*

MR 1782.2056
Refs.: Zertal (2004).
Plans: NA.
Sects.: NA.
Photos: NA.
Khirbet en-Najjar sits 65 m above its surroundings in the northern Samarian hills about 1 km SE of Ibleam (Zertal 2004:160). The only access to the 3.2 ha tell is from the SE where there are remains of a 4 m wide stone *fortification wall* ascribed to the MB IIB that is still preserved to a height of 2 to 3 m (ibid., 160f.). The remains of what was probably a *tower* located S of the site's entrance are also visible.

NAMI, TEL*

MR 1433.2296
Refs.: Artzy and Marcus (1991); Artzy (1993).
Plans: Site (Artzy and Marcus 1991:fig. 4).
Sects.: NA.
Photos: NA.
The site of Tel Nami is located about 8 km north of Dor (Artzy and Marcus 1991:14). Middle Bronze Age occupation at Nami was limited to the MB I (IIA), during which both the peninsula and an area now designated Nami East were occupied. The fortifications, probably part of the earthen rampart, have been observed only in an excavated section (Michal Artzy, personal communication, 2006). At the start of the MB IIB the settlement was abandoned following "an intense conflagration" detected in Area D (Artzy and Marcus 1991:13*), and the site was not resettled until the LB II.

NEBI RUBIN*

MR 1245.1485
Refs.: Gophna and Beck (1981); Gophna and Broshi (1986).
Plans: NA.
Sects.: NA.
Photos: NA.
Nebi Rubin is located about 3 km ENE of Yavneh-Yam. The site is listed on Gophna and Beck's settlement map as fortified (1981:47), as well as in Broshi and Gophna's catalog where it is designated as having an earthen rampart and noted as approximately 10 ha in size (1986:84). It was probably fortified during the MB IIB–C after being initially occupied during the MB IIA2/3 phase (Cohen 2002:88).

NIMRIN, TELL

MR 2094.1451
Refs.: Flanagan et al. (1994).
Plans: Site (fig. 1).
Sects.: N40/W20 (figs. 13 and 15).
Photos: N40/W20 (figs. 12 and 14).
Tell Nimrin is located on the E side of the Jordan Valley approximately 14 km ENE of Jericho on the S bank of the Wadi Shuaib at its mouth. The site may have been unfortified during the MB IIA, however, clear evidence of the site's MB IIB–C fortifications has come to light in square N40/W20 on the NW flank of the site. Despite uncertainty regarding the identification of these remains as part of the site's fortifications, and not as some other type of monumental construction (Flanagan et al. 1994:217ff.), it is reasonably certain that these remains constitute the site's defenses (see also Falconer 2001:278).

The remains consist of two phases of an E-W stretch of the town's *fortification wall* (Flanagan et al. 1994:218) that have been attributed to the MB IIB and IIC. Of the first phase of the fortifications [MB IIB] the remains of the stone built foundation (L.222) were preserved to about 4 m high (over ten courses). The second phase (MB IIC) was at least 3 m wide and consisted of about six courses of "small boulders" (L.201) and a mudbrick superstructure (L.143) of which ten courses were preserved; the entire structure including both phases stood upon excavation more than 4 m high (ibid.).

There is no clear evidence of a violent destruction to coincide with the abrupt abandonment of the settlement. The 1 ha settlement was abandoned during the late sixteenth or early fifteenth century B.C. (Flanagan et al. 1994:217ff.).

PELLA (TABAQAT FAHL)

MR 2075.2065
Refs.: Smith and Potts (1992); Knapp (1993); Bourke (1997); Bourke et al. (1998); Falconer (2001); McLaren (2003).
Plans: Site (Bourke 1997:fig. 1; see also McLaren 2003:fig. 7); Area III (Smith and Potts 1992:fig. 7; see also McLaren 2003:figs. 8a, 15); Area XXVIIIC (Bourke et al. 1998:fig. 18; see also McLaren 2003:figs. 17, 22a, 24a, 25a).
Sects.: Area IIIC (McLaren 2003:fig. 8b); Area IIIF (ibid., figs. 9–12); Area XXVIIIC (ibid., figs. 18f., 22b, 23, 24b, 25b).
Photos: Area IIID (R. H. Smith and Potts 1992:pl. 29a), Area IIIF (Bourke et al. 1998:figs. 15f.); Area XXVIIIC (ibid., fig. 19).
Pella is located on the E side of the Jordan Valley about 12.5 km to the SE of Beth-Shean. Excavations since 1983 have revealed extensive remains of the site's Middle Bronze Age defenses. These defenses, though already known to date to the Middle Bronze Age (Str. XI), now appear to date more specifically to the MB IIA–B based upon ceramics recovered from the wall in Area III and the tower in Area XXVIII (Bourke et al. 1998:189ff.). These fortifications continued in use through the end of the Middle Bronze Age (Bourke 1997:104ff.).

MB IIA–B
AREA IIIF, PHASES B–C[84]

The earliest Middle Bronze Age fortifications of Pella consisted of the remains of a large mudbrick *fortifica-*

[84] Local phases were employed for each excavation area at Pella. For correlations between phases in each area, see Table 2 in McLaren (2003:14).

tion wall excavated in Area IIIC, D, and F [Wall 41] (R. H. Smith and Potts 1992:40ff.). This wall fortified the settlement's SE side where the slope of the terrain drains toward the Wadi Jirim to the S. Perhaps for this reason the total elevation of the N-S stretch of the mudbrick wall exposed in Area III was more than 11 m across the 10 m of its exposed length, although the wall was only preserved to a height of 6 m (thirty-five courses of brick) in any one place (McLaren 2003:15). Wall 41, the earliest of these two walls, was 3.5 m wide and was constructed of mudbricks on a fieldstone foundation. Three distinct bands of brick were formed by the use of different types of mudbrick during this wall's construction.

In one place on its E side the wall appears to have featured a buttress (Buttress 100) about 90 cm thick, which was added to the wall's exterior. Based upon the ceramics recovered from the foundation trench of Wall 41, the construction of the wall is to be placed in the transitional MB IIA–B period (ibid., 13). Wall 41 was also accompanied by a parallel interior fortification wall (Wall 52) and a "roadway" that existed between these walls.

The main feature used to distinguish Phase C from Phase B is the presence of what was identified as a *glacis* built above the slope outside the wall and against the exterior face of the mudbricks in Wall 41 (McLaren 2003:15). Since it consists of "multiple layers of plaster and stone" it can be unequivocally identified as a glacis. Its slope, however, is barely noticeable (ibid., fig. 9). By the end of Phase C the roadway between the walls had fallen out of use as domestic architecture was built over it (ibid., 14).

AREA XXVIIIC, PHASE D

A second section of the Middle Bronze Age fortification wall on the S side of the site along with a tower were identified in Area XXVIIIC (Bourke et al. 1998:189ff.). These features have been identified as contemporaneous with Phases B–C in Area IIIF. The W section of the S town *fortification wall* (Wall 9) was approximately 11 m long, 2.5 to 3.0 m thick, and was coated on its exterior with lime plaster 2 to 3 cm thick (Bourke et al. 1998:193; McLaren 2003:18). It was built on a stone foundation and its superstructure featured 2 m (eighteen courses) of preserved mudbrick.

This wall was reinforced on its interior and exterior sides by Buttresses 16 and 12, respectively (McLaren 2003:18). Buttress 16 was 2.5 m wide and 1 m thick and featured a composition identical to Wall 9. Buttress 12 built in the corner between Tower 1 and Wall 9 was triangular in shape; it was 1 m thick and was built of stone and coated with lime plaster like Wall 9.

The *fortification wall* stretching 16 m to the E (Wall 10) from Tower 1 was 2.5 to 3.0 m thick and was also built on a stone foundation (McLaren 2003:18). It is said to have featured a saw-tooth design, although only one such "tooth" was revealed. Wall 10 was supported on its interior by a perpendicular wall (Wall 15) with stone foundation and mudbrick superstructure. It is, however, unlikely to have functioned as part of a casemate construction (contra McLaren 2003:18). Instead it may have functioned to level the area inside the wall for other uses. Ceramics from the tower's foundations suggest an MB IIA–B date (Bourke et al. 1998:194).

Both sections of the S wall flanked the *tower* (Tower 1), which was built mostly on the inside of the town wall (cp. Megiddo). It measured 8 m long (E-W) and more than 12 m (N-S) wide (ibid., 191). It was built with a foundation of five courses of stone cut into the bedrock and a mudbrick superstructure of which more than forty courses of mudbrick were preserved; facing bricks measured 40–50 x 40 x 10 cm, while interior bricks were rectangular, 70 x 40 x 10 cm (McLaren 2003:17). The tower featured at least one chamber (Locus 30) the interior faces and floor of which were lime plastered; the chamber also possessed one bench.

MB IIB(?): AREA IIIF, PHASE D

During Phase D in Area IIIF structural damage caused Wall 41 to fall out of use (McLaren 2003:15). This phase corresponds with an occupational gap between Phases VIII and VII (Area IIIC-D).

MB IIB: AREA IIIF, PHASE E

In Phase E the fortifications were reconstructed with the addition of buttresses and reinforcing walls (McLaren 2003:15). Wall 7, preserved to a height of 4.5 m (forty-six courses of brick) and 3.58 m wide, was added during this phase, which was identified in Square IIIF to the NE (R. H. Smith and Potts 1992:42f.). This wall abutted the exterior face of Wall 41 running E; it did not feature a stone foundation. Although the function of this wall has been debated, it is likely that it at least served in some way to buttress Wall 41 on its exterior, as McLaren suggests (2003:16). Buttress 37 was also added to the exterior face of Wall 41 in Area IIID during this phase, while Buttress 83 was added to its interior face in Area IIIC (ibid.).

The fortifications in Area XXVIII may have continued in use during this phase.

AREA IIIF, PHASE F

This phase represents the first phase following the abandonment of the Middle Bronze Age fortification system.

If occupation was limited to the mound during the Middle Bronze Age then the walls in Areas III and XXVIII would delimit an area approximately 265 m long and 135 m wide (about 3.6 ha). Thus the length of the wall would have been about 800 m.

POLEG, TEL

MR 1352.1849
Refs.: Gophna (1973; 1993b); Kochavi et al. (1979).
Plans: Site (Gophna 1973:fig. 1); Area A (ibid., fig. 2); Area C (ibid., fig. 3); Area E (ibid., fig. 6).
Sects.: Area C (Gophna 1973:fig. 4).
Photos: Area C (Gophna 1993b:1194).
Tel Poleg is located along the Nahal Poleg about 22 km SW of Tel Zeror. Information concerning the fortifications from the MB IIA fortress at Poleg was obtained from salvage excavations conducted in 1959 and 1964 (Gophna 1973; Kochavi et al. 1979:133–39). The small size of the site, 1.2 ha, suggests that it most likely functioned as a fortress on the coastal road (Figure 94). Excavations in Areas A, C, D, and E, on the E, S, and W parts of the mound revealed portions of the fortification wall. Although it was impossible to establish stratigraphic links between these excavated parts of the wall since the site had served as a quarry, the ceramic remains have made it possible to date each of the areas to the MB IIA.

In Area A the mudbrick *fortification wall* (Locus 21) was 2.7 m wide. It was perhaps not wider here because the steep slope on this side of the site also aided its defense. Bricks used in this structure measured 60 x 50 x 12 cm (Gophna 1973:111).

In Area C a mudbrick *tower* (Locus 24) measuring 15 x 8 m along with a portion of the wall measuring 5 m thick, which ran N toward Area D, were excavated (ibid., 115). From the published photographs it is possible to discern that about twenty rows and twenty courses of mudbrick were preserved to a height of more than 2 m (Gophna 1993b; also Gophna 1973:115). The average dimensions of the bricks used in this tower were 55 x 35 x 12 cm in size (ibid., 115). The slope on the W side of the site was considered more moderate, perhaps explaining the disparity in thickness between the walls found in Areas C and D, and the much thinner wall found in Area A, where the slope was considerably steeper.

Some indication of the slope, roughly 20°, leading up to the wall is provided in the section drawing of the W section of Area C (Gophna 1973:fig. 4); this was identified as the *earthen rampart* built of layers of crushed kurkar. The excavation of Area E revealed the S half of the *gate complex* of the "indirect approach type" (Kochavi et al. 1979:132). Aside from this gate it is impossible given the state of preservation of the site and its immediate environs to identify other gates. Its proximity to the Nahal Poleg would have provided it with a source of water and would have improved the defensibility of the S side of the mound where it is thought to have flowed in antiquity.

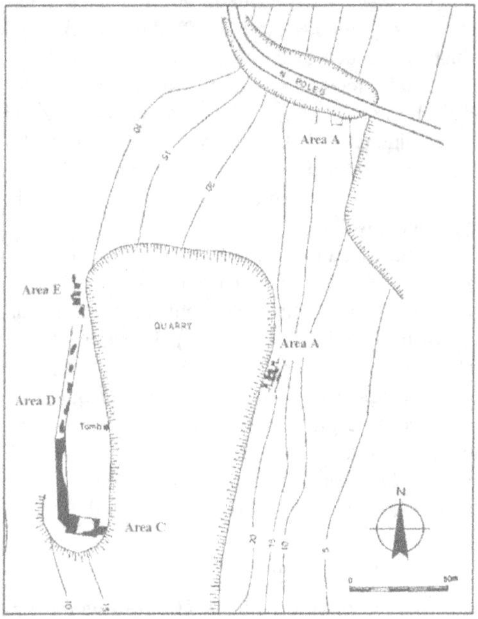

Figure 94. Plan of Tel Poleg (after Kochavi et al. 1979:fig. 5).

PORAN, TEL (TELL EL-FARANI)

MR 1137.1242
Refs.: Gophna (1992a).
Plans: Site (fig. 2).
Sects.: Salvage cut (fig. 4).
Photos: NA.
Tel Poran, a site of some 10 to 12 ha, was located along the coastal plain about 7 km S of Ashdod and 8 km N of Ashkelon. It was among a number of sites discovered in the process of salvage archaeology in Israel, which were excavated in the 1960s and 1970s

(Gophna 1977; 1992a). Like many sites founded during the Middle Bronze Age, it was situated along a wadi, the Nahal Ivtah, although this fact seems to have played no role in the site's selection during the Middle Bronze Age since it had already been occupied during the EB II–III.

Reoccupation of the site during the Middle Bronze Age was probably concerned with reusing the existing, though dilapidated, remains of a 5 m thick Early Bronze Age mudbrick fortification wall. It appears to have served as the *core wall* for the building of the Middle Bronze Age *earthen rampart*. Nevertheless, it is not possible to provide a specific date for the rampart's construction. During the process of building the rampart over the EB II–III wall remains, layers of crushed kurkar, bricky material, and ash were heaped outside the wall behind a small *retaining wall* 9 m from the base of the wall. Gophna is probably correct in identifying the mudbrick layer on the rampart's surface as a *glacis*. Subsequent natural accumulation accounts for another 10 m added to the slope outside of the retaining wall as well as on the Middle Bronze Age rampart slope.

Greatest attention has been paid to the absence of evidence for a wall on top of the rampart. Gophna suggests that this conclusion is conclusively demonstrated by the remains of a pit dug into the top of the wall's surface, which contained Early Bronze Age pottery (Gophna 1992a:fig. 5). Nevertheless, the pit could have been dug before a Middle Bronze Age wall was built over it, which has subsequently eroded away in this location. It is unfortunate that the 33 m section provided does not include the stratigraphic relationship between the wall and its rampart and the remains within the Early Bronze Age wall.

Final conclusions regarding the site's fortifications must await more complete publication of the excavator's findings.

Qana, Tel*

MR 1397.1707
Refs.: Broshi and Gophna (1986); Gophna and Beck (1988).
Plans: NA.
Sects.: NA.
Photos: NA.

Tel Qana is located about 5 km NW of Aphek in the southern coastal plain. The MB IIA settlement was approximately 0.7 ha (Gophna and Portugali 1988:26). It was identified as a site with an earthen rampart by Broshi and Gophna (1986).

Qarqaf, Khirbet*

MR 1643.1859
Refs.: Zertal (1993:1312; 2004:381–83).
Plans: NA.
Sects.: NA.
Photos: Site (2004:fig. 317).

The 0.8 ha site of Khirbet Qarqaf is located in the northern Samarian hills about 15 km NW of Shechem. It sits 70 m above the road between Shechem and Dothan (Zertal 2004:381). The MB IIB fortifications of Khirbet Qarqaf are said to include a substantial stone *fortification wall* (1993:1312).

Qashish, Tel (Tell Qasis)

MR 160.232
Refs.: Ben-Tor et al. (2003).
Plans: Site (Bonfil 2003:plan 1); Area A (ibid., plans 24–25).
Sects.: Area A, sect. C–C' (Ben-Tor and Bonfil 2003:191, see plan 32 on p. 231 for location).
Photos: Site aerial (Ben-Tor et al. 2003:photos 1 and 2); Area A (ibid., photos 75–76).

Tel Qashish is situated on the N bank of the Nahal Kishon 2 km NW of Tel Yoqne'am (Ben-Tor et al. 2003:1ff.). The earliest phase of MB IIA settlement detected, Stratum X, provided no architectural evidence and thus whether or not it was fortified remains unknown (Bonfil 2003:185). The Stratum IX settlement seems, therefore, to be the first fortified, which occurred during the transition between the MB IIA and IIB (MB IIA, Phase 4 in Cohen 2002a:90).

MB IIA, Phase 4: Strata IXC–B

A 15 m stretch of the stone foundation of the town *fortification wall* (W.20 & 242) survived to a width of approximately 1.7 m and a height of 1.4 m in Area A on the N side of the site (Ben-Tor and Bonfil 2003:190).The fortification walls were founded upon the remains of Early Bronze Age walls, which may have been visible when the site was resettled following a hiatus during the EB IV.

The wall featured a rectangular *tower*, measuring 5.5 x 4.5 m (W.42, 43, & 47), that projected inward from the fortification wall into which it was constructed. West of the tower the wall was flush with its outer face, however, the eastern portion (W.242) appears to have abutted the center of the wall, which resulted in an offset-inset construction. The excavators have suggested the possibility, based on traces of plaster on the inner face of the wall that the tower was a later addition; however, this remains uncertain.

A feature probably to be identified as an *earthen rampart*, of the supplemental type though identified as a "glacis" by Ben-Tor, may have been built against the outside of the wall. It consisted of "small stones and crushed stones mixed with earth" (Ben-Tor and Bonfil 2003:190). Nevertheless, its identification as such is far from certain given its appearance in section and photos, which suggests its identification as collapsed material down the slope.

Although the Middle Bronze Age settlement has been described as 1 ha in size, if the enclosure was only about 130 m E to W by 50 m N to S then its inhabitable area was no more than about 0.7 ha. The site appears to have been unfortified during the Late Bronze Age (Ben-Tor 1993b:1203).

QIRYAT SHEMONA SOUTH

MR 204.292
Refs.: NA.
Plans: NA.
Sects.: NA.
Photos: NA.

Salvage excavations in advance of the expansion of a traffic circle in the town of Qiryat Shemona revealed a Middle Bronze Age fortress west of Tel Dan in the northern Huleh Valley (Yassur-Landau, personal communication, 2006). The site consisted of a roughly 0.25 ha multiperiod Middle Bronze Age settlement founded during the MB IIA.

MB IIA, PHASE 4: STRATUM VII

During the mid-MB IIA the site was fortified by a mudbrick *fortification wall* (W.522) on stone foundations. Twenty-five meters of the length of this wall, which was approximately 2.5 m thick, were exposed on the north side of the fortress. At the eastern end of this section an almost square *tower* (268) measuring 6 x 5 m was also exposed. Inside the tower a hearth was identified. Occupation consisting of domestic structures was revealed inside the wall. The ceramics of his stratum are noted to be contemporary with the Dan XII ceramic assemblage. This settlement was deliberately destroyed, though not by fire, as suggested by various destruction deposits filled with smashed ceramics.

MB IIA–B: STRATA VI–V

During Stratum VI, which was contemporary with Dan XI, the site continued as a small unfortified village. The village of Stratum V was contemporary with Hazor Stratum 4 of the lower town.

QUMEI, KHIRBET*

MR 1707.1834
Refs.: Zertal (2004).
Plans: NA.
Sects.: NA.
Photos: NA.

Khirbet Qumei sits in the northern Samarian hills less than 5 km NW from Shechem and 35 m above its surroundings (Zertal 2004:403f.). The MB IIB fortifications of Khirbet Qumei are noted to include a stone-built *fortification wall* (Zertal 1993:1312). The site is reported to have been 2.2 ha in Zertal's survey (2004:403), though what portion of this was occupied during the MB IIB is unclear.

REHOV, TEL (TELL ES-SAREM)*

MR 197.207
Refs.: Broshi and Gophna (1986); A. Mazar (1999).
Plans: Site (A. Mazar 1999).
Sects.: NA.
Photos: NA.

Rehov is located on the W side of the Jordan Valley 5 km S of Beth-Shean (A. Mazar 1999:2). According to Broshi and Gophna (1986) Rehov features an *earthen rampart* datable to the Middle Bronze Age. However, excavations led by Amihai Mazar from 1997 through 2001 have not yet exposed remains of the Middle Bronze Age fortifications of the site. Nevertheless, a cemetery to the S of the site (A. Mazar 1999:7) as well as sherds belonging to phase D-11 that may date to the MB IIC (ibid., 11) suggest that Rehov was occupied during the Middle Bronze Age. A precise estimate of the size of the Middle Bronze Age settlement is not yet possible.

RUKAIS

MR 2979.1946
Refs.: Betts et al. (1996); Eames (1997); Betts (1998); McLaren (2003).
Plans: Site (McLaren 2003:fig. 29a); Area 8 (ibid., fig. 36); Area 14 (ibid., fig. 41); Area 16 (ibid., figs. 44b–c); Gate (ibid., figs. 45a–b).
Sects.: Area 5 (Betts et al. 1996:fig. 5; McLaren 2001; 2003:figs. 30–32); Area 8 (McLaren 2003:figs. 36f.); Area 14 (ibid., fig. 41).
Photos: Site aerial (McLaren 2003:fig. 29b); Area 16 (ibid., fig. 44a).

The MB IIC site of Rukais is located 25 km to the SE of Bosra along the Wadi al-'Ajib in the Hauran on the border between Jordan and Syria (Betts et al. 1996). The site is approximately 0.8 ha in size, though it may prove to be larger once the lower town to the SW is included (Eames and Schroder 1997:150). Although the site had been occupied during the EB IB, it was only resettled during the MB IIB and only fortified during the MB IIC.

MB IIB: PHASE II.1

This phase of settlement at Rukais consisted of an unwalled MB IIB village (McLaren 2003:21 ff.). No architectural remains were found associated with this phase in any area where the fortifications of the following phase were exposed.

MB IIB: PHASE II.2.I

The settlement in this period was about 0.8 ha in size (100 m in diameter) and was enclosed by a "massive" *fortification wall* built of basalt fieldstones (wall 200 in Area 5, 000 in Area 8, and 14000 in Area 14, McLaren 2003). It was preserved 4 m high and between 2.5 and 3 m in width in Area 5, but to 4.5 m high and between 3 and 3.5 m wide in Area 8, and 3 m thick in Area 14 (ibid., 21ff.). In Area 14 the wall was buttressed on its interior by Wall 14033, a 3 m thick red mudbrick wall (bricks 45–50 x 45–50 x 12–15 cm).

Further evidence for the site's fortifications during this phase consist of the only evidence thus far revealed in Jordan for a *four-pier gate* during the Middle Bronze Age. It was excavated in Area 16 on the N side of the settlement (ibid., 24f.). Its exterior façade was flush with the outer face of the fortification wall. Due to the site's small circumference the exterior (N) elevation of the gate, measuring 15 m wide, was wider than its interior (S) elevation, which measured 13.5 m wide (cp. Shechem E Gate). The towers flanking its axis were thus each 6.5 m wide on the N while only 5 m wide on the S, and the E tower was approximately 11 m long, while the W was about 10.5 m long. Both towers featured rooms that have been identified as guardrooms. A stone-covered drain ran beneath the passage of the gate, which at its narrowest was approximately 2 m wide. However, where the gate's lintel, which featured a door socket, protruded into the passage it was no more than 1.5 m wide. The structure was preserved to a height of 2.5 m (McLaren 2001:98). Although the gate can only be dated by the stratigraphy deposited following its use, direct connection with the main fortification wall is indisputable as is its placement in this phase.

PHASE II.2.II

This phase, which followed the phase characterized by the fortifications described above, represents a return to an unfortified settlement as was exposed in each of the areas excavated (ibid., 21ff.).

SAHAB

MR 245.142
Refs.: Ibrahim (1989; 1997).
Plans: Site (Ibrahim 1989:519).
Sects.: NA.
Photos: NA.

Sahab is located 12 km SE of Amman. The site features Middle Bronze Age fortifications including a rampart excavated in Area H on the SE corner of the site and the remains of a Late Bronze Age wall (possibly also of late Middle Bronze Age date) in Areas G II and III on the SW corner of the site. The Middle Bronze Age fortifications in Area H consisted of an *earthen rampart* and *fortification wall* (Ibrahim 1989:518). The identification of the glacis mentioned in connection with the rampart cannot, however, be confirmed. The wall in Area G II may be associated with a Middle Bronze Age wall excavated in Area H III, but this has not yet been established (Ibrahim 1997).

The site was definitely walled during the Late Bronze Age as evidenced in Areas H III–IV on the SE, H II on the E, and H II, B019 on the N (Ibrahim 1997; 1989:519). Based on the reconstructed extent of the Late Bronze Age wall (which appears to trace the course of the Middle Bronze Age wall, Ibrahim 1989:519), the site appears to have been roughly square and about 125 m on a side for a net area within its wall of about 1.6 ha. The complete length of the fortifications was more than 500 m.

Late Bronze Age fortifications were exposed at Sahab (for references, see Strange 2001:310).

SERA', TEL (TELL ESH-SHARI'A)

MR 119.088
Refs.: Oren (1993b; 1997c).
Plans: Site (Oren 1993b:1329).
Sects.: NA.
Photos: NA.

Tel Sera' is located about 32 km to the SSE of Ashkelon. Recent excavations by Eliezer Oren at the site identified late Middle Bronze Age fortifications at the site (Oren 1997c:1). The site was probably settled during the MB IIC (Str. XII) on a natural mound of about 4 to 5 m in height along the Nahal Gerar, which provided additional defense of the site's SE side (1993b:1329). Evidence for the site's defenses was encountered in Area B (ibid., 1330). However, no regularity of the site's plan can be inferred from its contours. The site and probably its fortifications do not appear to have been altered during the LB I.

Shechem (Tell el-Balatah)

MR 1755.1805

Refs.: Sellin (1941); G. E. Wright (1964); Dever (1974b); Seger (1974; 1975); G. R. H. Wright (1985b); Lederman (1985); Ussishkin (1989); Campbell (2002); G. R. H. Wright (2002).

Plans: Site (G. R. H. Wright 2002:ills. 2, 7), Contours (ibid., ill. 3); Schematics (Campbell 2002:figs. 78, 100, 22–24, 29); Field I (G. R. H. Wright 2002:ills. 12–15); Field III (ibid., ills. 38f.); Field IV (Campbell 2002:fig. 144; G. R. H. Wright 2002:ill. 52); NW Gate (Campbell 2002:fig. 129; G. R. H. Wright 2002:ills. 48f.).

Sects.: E Area: Field I (Campbell 2002:figs. 127f., 31f.; G. R. H. Wright 2002:ills. 20–22, 24–27); NW area (Campbell 2002:fig. 99; G. R. H. Wright 2002:ill. 10); Field III (G. R. H. Wright 2002:ills. 41–44); Field IV (ibid., ills. 54–59); Field V (Campbell 2002:fig. 17; G. R. H. Wright 2002:ill. 68); Section AA (G. R. H. Wright 2002:ill. 8); Field VI (figs. 11, 14); Field XIII (Campbell 2002:fig. 35).

Photos: Field I (Campbell 2002:fig. 130); Field III (ibid., figs. 112, 114); Field VI (ibid., fig. 18).

Shechem (Tell Balatah) is located in the hill country about 49 km N of Jerusalem. The excavations of Shechem were conducted by several expeditions over a number of years. The cumulative results of these excavations have been recently presented by Edward F. Campbell and G. R. H. Wright (2002; 2002) and they include treatment of the excavations of the Middle Bronze Age defenses on the NW (Fields IV–VI, XIII), N (Field VIII), NE (Field III), and the E (Field I) sides of the site (Figure 95).

MB IIA/B: Stratum XX

There are no data to suggest that Shechem was fortified during the early MB IIA (Campbell 2002:26). It was first fortified, therefore, at the start of the MB IIB, Stratum XX, with Wall D, the first *fortification wall*, of about 2.65 to 2.85 m wide (ibid., 27). Earlier MB IIA settlement was also present outside this wall (ibid., 25). The faces of the wall were built of dressed or specially selected stones with rubble piled between these (ibid., 28). Along the 52 m of its exposed length it was preserved to a maximum height of 1.7 m and some of its mudbrick superstructure was found during Austro-German excavations, though the Joint Expedition did not find any similar remains in an extension within Area 9. The latest ceramics recovered from a dismantling of this wall were determined to be of transitional MB IIA–B date. Unfortunately, this wall has only been identified on the NW side of the site, although the C Rampart which was built against this wall was itself identified in the other areas excavated. The location of Wall D, 40 m inside the line of the later Middle Bronze Age fortifications in the NW area, suggests that the site was considerably smaller than it was at the beginning of the MB IIB.

MB IIB: Strata XIX–XVII

The Joint Expedition's excavations in Field V also revealed the defenses of Shechem, which followed those of Wall D during the MB IIB (Campbell 2002:44f.). Some time after the construction of Wall D the fortifications of Shechem were modified with the addition of the first *earthen rampart* of the supplemental type (Str. XIX), 37 m outside the line of Wall D.[85] The actual height of this rampart and its slope cannot be established—though J. Seger (1975) suggests that it was 15 m high—and for this reason this rampart was also identified as the "truncated rampart." Its composition nearest revetment Wall C at least partly included interlaced tongues of huwwar limestone and dark soil. This rampart's exterior was held in place by a *revetment wall* (Wall C) built of fieldstones with a slight batter that was preserved up to 5 m high (Campbell 2002:44). Although it was not completely excavated, its function can be compared to similar revetment walls at Jericho and Ebla. Because Wall C is associated with this rampart, the rampart is also referred to as the "C rampart."

A second earthen rampart, the so-called "augmented rampart" (Str. XVIII or XVII), was constructed following the first rampart also during the MB IIB. This feature corresponds with Dever's "Glacis B," L.72.125 (Campbell 2002:109). The rampart appears to have been detected in excavations in Area A-B in Fields IV and III (ibid., 45), though its slope was only about 15° in Dever's Field IV Section EE (G. R. H. Wright 2002:ill. 58). The basis for its identification in Field IV is that, even though revetment Wall C was not detected, a rampart appears to have been constructed with a surface that terminated at the base of the revetment wall of the latest Middle Bronze Age fortification system (i.e., Wall A, Campbell 2002:45). A sloped fill would have been unnecessary behind Wall A if it was intended only to provide the base for a level extension of the town to the line formed by Wall A. Sherds from the fill of this rampart date no later than the MB IIB. The date is furthermore confirmed by sherds found in the layers

[85] It is important to note, however, that no complete section of the fortifications was ever undertaken on the NW side of the site and, therefore, the actual nature of their construction must remain uncertain even if the major elements were partly exposed.

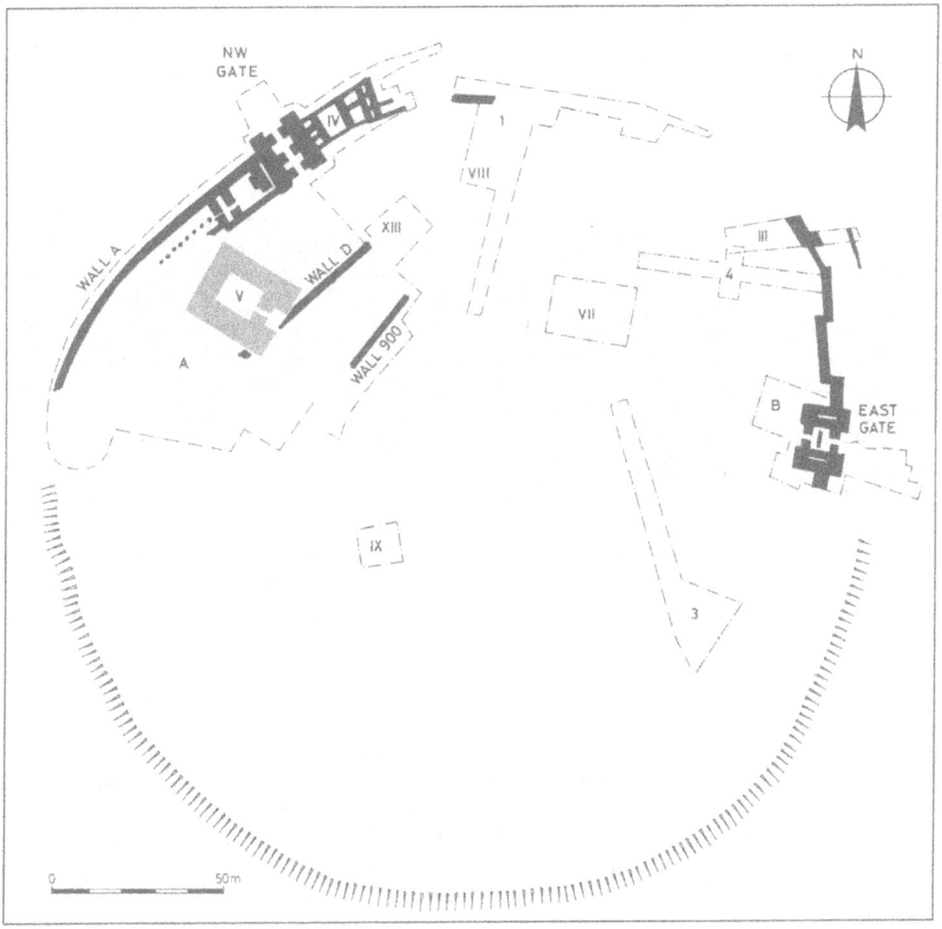

Figure 95. Plan of Shechem and its Middle Bronze Age fortifications (after Herzog 1997a:fig. 4.14).

above this rampart dated to the MB IIC. Both of these ramparts appear to have been constructed of materials taken from outside of the site. This is indicated by the small amount of pottery recovered from the ramparts' fills, the reddish earth used throughout, and the lack of ash and charcoal detected (Campbell 2002:49).

MB IIC: STRATUM XVI

A violent destruction at the end of Stratum XVII appears to have brought about "massive earth movement and filling at the fortifications and in the public precinct" that also "removed much of the evidence of destruction," though it was very evident in the domestic quarter (Campbell 2002:103).[86] The first phase of renovation of the site's fortifications in this period (Str. XVI) involved the construction of a "Cyclopean" *revetment wall* (Wall A) around the N half of the site (about 380 m long) at the base of the previous "augmented" rampart ("Glacis B"), standing as much as 10 m high in places (ibid., 105ff.). Its construction involved the placing of boulders, probably unquarried, with dimensions greater than 1 m to a side (some as much as 2.2 m long) on a thin layer of earth resting above bedrock. As each course of boulders was set in place fill was added behind them (i.e., the earthen rampart, see below). Although its dimensions vary throughout its construction, it was determined by the Austro-German expedition to have been as much as 2 m wide at the top and 4 m wide at its base and built

[86]It was the massive change brought on after this destruction that served as the basis for designating the following Middle Bronze Age strata as MB IIC (Campbell 2002:105). Thus the last two of six strata associated with the post-MB IIA occupation at Shechem were designated MB IIC.

eight to ten courses high. The exterior face of the revetment wall was sufficiently dressed "so as to present a relatively smooth and flush face," while its interior face was "uneven, and slants inward" (ibid., 106f.). Though large boulders formed the bulk of the wall smaller stones and cobbles were used to fill smaller spaces between them. The slope of the revetment appears to have been almost vertical along its lowest courses but about 15° toward the top (ibid., 107f.).

The Austro-German expedition also revealed two drains on the NW side of the site that fed water from this quarter out two openings more than 5 m above the base of revetment wall A, which were about 0.45 m wide and 1 m high. In certain places, such as to the NE of the W gate, the wall features a saw-tooth design similar to that of Shiloh's MB IIC fortifications (Campbell 2002:109). Evidence for the superstructure of Wall A was recovered from Field I during salvage work in the late 1980s (ibid., 115). This evidence included mudbricks measuring 55 cm long and 12 cm thick. Despite its incredible masonry, the length of Wall A around the entire site was only about 740 m.

The *earthen rampart* that was associated with this revetment wall was designated "Glacis A" (L.7276) by Dever (Campbell 2002:109). Although it was not traced across its entire width, it is thought to have abutted revetment Wall A at its upper, interior edge. It was composed of "chalk and rubbish" with a slope of approximately 21° in Field IV Section EE N of the NW Gate (G. R. H. Wright 2002:ill. 58). This rampart was not identified to the S of the NW Gate as seen in Field IV Section FF (ill. 59), but instead has been suggested to consist simply of a filling on top of the "augmented rampart" or "Glacis B" behind Wall A.[87] At the same time as construction of the "Glacis A" rampart a *fortification wall* (Wall B/2) was constructed, which was probably set into the rampart if its lower courses did not themselves rest upon a stone core. Wall B/2 (L.72.106B) was built approximately 3 m wide of stone. Although it has been suggested that wall 7287 seen in Section EE was part of a tower associated with this fortification wall, the basis for this remains unclear as no floors can be seen in section behind the wall; it is instead more likely that the small remains of this wall constituted a *retaining wall* within the rampart, such as at Jericho. Wall B/2 appears to have experienced the site-wide destruction at the end of Stratum XVI after which time it was replaced with Wall B/1 (L.72.106A), a 1.75 m wide stone wall (Campbell 2002:109).

The gate, referred to as the N or NW Gate, was another prominent feature of the MB IIC defenses of Shechem (Campbell 2002:110ff.). It was revealed by Sellin's expedition to have been constructed with fieldstone foundations and a superstructure of red and pinkish-white mudbricks, which measured 38 x 38 x 10–12 cm. Although these remains were once preserved to a combined height of 2 m, the mudbrick superstructure has not survived during the past century of exposure. The gate conformed to the well-known *six-pier gate* plan and had overall dimensions of 18.3 m wide (with an accompanying passage of equal length) and was 16.8 m wide along its exterior (NW) elevation while only 16.2 m wide on the interior (SE). Campbell notes that the difference in the width of these two elevations resulted from the gate conforming to the curve of the wall (ibid., 110). If this was the basis for its construction in this manner, it was wholly a subjective decision on the part of the builders given that it would have made no difference to the integrity and/or function of the gate if it had been inserted with a square plan into the curving wall, as it clearly was not significant at other sites (e.g., Ashkelon, Megiddo). Ironically, the spaces between the piers conform to exactly the opposite principle with the widest being the most interior of the three; their dimensions from the interior to exterior are 3.0, 2.9, and 2.8 m respectively (see G. R. H. Wright 2002:ill. 49); one would have expected the widest space to have been between the exterior piers. Each of these piers projected about 2 m and they were lined with massive limestone orthostats. Although no door sockets were recovered, it is not necessary to suggest as Campbell does that sliding doors were used, which he also suggests was the case even for doors with evidence of door posts such as at Ebla (Campbell 2002:111).

Although the gate in its final state (i.e., at the end of the MB IIC) appeared to have been symmetrical, the results of the Joint Expedition's work have demonstrated that the gate was probably not originally built like this (Campbell 2002:110). This realization was arrived at following the discovery of a flagstone surface that was associated with the gate's earliest construction, but was built over by a lateral wall on the N side of the passage which made the recesses

[87] Section FF does not provide a complete record of the fortifications in this area. However, the excavators have not always treated this section in this light throughout their report (i.e., as secondary to the more complete representations of the site's sequence of fortifications). For this reason there has been confusion concerning the exact sequence of fortifications, particularly in Stratum XV, in their report. In my opinion the assertion of the existence of only a single rampart based on Section FF is misleading, since it is possible that two ramparts with slopes comparable to those seen in Section EE were actually present below the levels reached (G. R. H. Wright 2002:ill. 58).

between the piers on the N side as deep as those on the S side of the gate. This, Campbell suggests, was part of the renovation of the gate when the refurbished building 7200 was bonded to the gate. However, given the numerous examples of six-pier gates where no such sequence of events is attested, and that the duration of elapsed time before the construction of the lateral wall cannot be established, it is possible that these are merely constructional subphases within the original gate's main building phase in Stratum XVI.

Although the evidence for a ramp leading up to the gate, which rested 5 m above the bottom of revetment Wall A, from outside the settlement remains unclear (Campbell 2002:112f.), the suggestion that some ramp did lead up to the gate's entrance is warranted.

Contrary to David Ussishkin's proposal (1989), the exterior of Wall A was completely exposed in antiquity and the disparity in elevation between the gate and the bottom of the "fosse" around the site was considerable (about 5 m). As Campbell himself notes, there are several reasons why Ussishkin's suggestion that Wall A was merely intended to support the fill around the site cannot be maintained. First, the original excavator noted a ramp "along the wall," which it may be deduced was intended to account for disparity between the height of the town and the elevation outside. Second, there is no ceramic evidence to support a fill completely burying Wall A, which would date to the late Middle Bronze Age and no later. Third, as Campbell observes, Wall A is capable of supporting the fill behind it and standing as it does despite its steep face and, therefore, a fill was not structurally necessary. Fourth, the dressing of the face of the stones used in Wall A is further proof that they were exposed in antiquity. Finally, if a fill existed up to the elevation of the drains which pour from the wall in the NW area, as Ussishkin suggests, then these very large drains (see above) would have compromised the defenses in this area.

Terminal MB IIC: Stratum XV

Following what appears to have been a site-wide destruction, which brought Stratum XVI to a close, the site was refortified with an offset-inset *fortification wall* (Wall B), which was identified in Fields I, III, and IV (Campbell 2002:118). It was previously identified around the site by Welter and Sellin who traced the bricks on top of the wall (ibid., 123). The bricks in this wall appear to measure from 35–43 x 35–43 x 15–16 cm in size (ibid., 150). This fieldstone wall, which ranged in width from 3.25 to 3.75 m, was built in most places at the crest of the remains of the earlier "augmented" rampart (ibid., 123). In Fields I and III Wall B was accompanied by a thinner wall running parallel to its interior face at distances ranging from 1.75 to 7 m from it (ibid., 133). Campbell suggests that these walls were intended to support battlements on top of Wall B.

In Field IV Wall B, designated Wall B1, was built atop the ruins of Wall B2 with its foundation trench cutting into "Glacis A" (Campbell 2002:118ff.); it has been suggested that much of the interior face of Wall B1 to the N of the gate had "fallen away," thus accounting for its being half as wide as Wall B elsewhere on the tell (ibid., 119). Wall B was never reached S of the gate. Contemporary with Wall B/1, according to the Joint Expedition, was the construction of a 4 m fieldstone extension of the elevation of revetment Wall A (L.7210), which was identified to the N of the gate. This was then followed almost immediately by a massive leveling of the space between Wall A and Wall B, upon which complexes 7200 to the N of the gate and 7300 to the S of the gate were said to have been constructed.[88]

The first phase of Wall B was brought to a close by a destruction, evident in Field IV, which resulted in a meter of destruction debris on the inside of Wall B1 (L.72.106). The first phase of the fortifications of Stratum XV, characterized by Wall B, was followed by the construction during the second phase of a new wall (L.7275B) detected both to the N and S of the NW Gate (ibid., 121). This stone wall was only 0.85 m wide and, at least in Section EE, appears to lean inward. Some time after its construction the building complexes 7200 and 7300 were abandoned and following this abandonment a new wall (L.7275A) was added parallel to Wall E (L.7275B) about 2.5 m away from it. These two walls were also detected to the N and S of the NW Gate, thus constituting a casemate system of defense (Wall E) in the third and final phase of Stratum XV's defenses (ibid., 123). The three destructions that followed each of these three sub-phases in Field IV were identified securely within the latest phase of the MB IIC, Stratum XV,

[88] Although Campbell suggests that the leveling of the area inside Wall A in Field IV S of the gate occurred in Stratum XVI and not XV, as he claimed it did to the N of the gate (ibid., 123), the basis for this is unclear. Both strata are designated as MB IIC and ceramics are insufficiently distinguished to provide further subphasing. Furthermore, the reconstruction of the line of the original rampart ("Glacis 7333A") behind Wall A is based only on the extrapolation of sloping lines encountered in two probes at a distance of more than 6 m from the inside of Wall A. Yet, a rampart with a slope comparable to that revealed in Dever's excavations would have resulted in Wall A projecting no more than 3 m above the rampart.

without any ceramic evidence for placing them within the Late Bronze Age. The results concerning Wall B in Field III are confusing largely because the Joint Expedition attempted to continue the work of earlier exploration in this area. Since the identification of Wall A was uncertain in this field,[89] it is impossible to assert with confidence the relationship of Wall B to the earlier rampart's construction.

The E Gate built on the E side of the site was a *four-pier gate* and has been dated to a late phase of the Middle Bronze Age (Campbell 2002:131ff.).[90] The E elevation (outside) of the gate was 18.2 m across and the W elevation (inside) was 17.8 m wide. The N elevation of the N tower measured 13.3 m while the S elevation of the S tower was 13.8 m (ibid., 133). Based on ill. 13 (G. R. H. Wright 2002) the gate's inner piers project 1.8 m into the passage and the outer piers 1.6 m with passages between both sets of piers narrowing at the most to approximately 3.35 m in width. The structure was preserved to a height of 5.64 m. As expected both of the gate's towers featured chambers designated guardrooms by Campbell. Two phases of this gate are attributed to the Middle Bronze Age. The earliest phase of this gate, which was built into the earliest rampart ("C rampart"), was followed, presumably, by the raising of the original limestone orthostats approximately 2 m to the level of the gate after renovations were completed. All pottery found in association with the E Gate was MB IIC in date.

Like the rest of the site the gate was destroyed in the MB IIC. Campbell even suggests that a chunk from the surface of "the orthostat at the front of the south inner pier" which was "tilted 29 degrees off vertical" may suggest that it was "inflicted by a battering ram" (Campbell 2002:137). Two articulated skeletons and the bones of four others, which were related to the final Middle Bronze Age destruction of Shechem, were discovered amidst the ash and fallen bricks on the steps inside the gate (ibid., 137, 139).

LB Fortifications

According to Campbell the Late Bronze Age fortification system (Str. XIV), which is evident in Fields I and IV, appears to have relied upon reuse of the MB IIC fortifications, but evidently decreased the overall size of the settlement (2002:169ff.). Wall A was reused and a new wall was constructed above the remains of the Wall B system. The E Gate was reused with new guardrooms being added.

Shiloh (Khirbet Seilun)

MR 1775.1626

Refs.: Buhl (1969); Finkelstein et al. (1993).

Plans: Site (Finkelstein et al. 1993:fig. 1.5); Area C (Lederman 1993:figs. 2.1, 2.3); Area D (Lederman and Finkelstein 1993:figs. 3.1, 3.3); Area H-F (Finkelstein and Lederman 1993a:fig. 4.10); Area K (ibid., fig. 5.7); Schematic (Finkelstein 1988:fig. 66).

Sects.: Area C (Lederman 1993:fig. 2.5); Area D (Lederman and Finkelstein 1993:fig. 3.9); Areas H-F (Finkelstein and Lederman 1993a:fig. 4.11); Area K (1993b:fig. 5.8).

Photos: Area C (Lederman 1993:figs. 2.8, 13–14); Area D (Ben-Tor 1992a:pl. 28; Lederman and Finkelstein 1993:figs. 3.2, 4–8); Areas H-F (Finkelstein and Lederman 1993a:figs. 4.2–4, 7–8, 12–13, 18).

Shiloh is located in the hill country about 17 km SSE of Shechem. The most recent results are those of the last expedition to Shiloh by Finkelstein, Bunimovitz, and Lederman (Finkelstein et al. 1993). While Shiloh was occupied during the MB IIB, a conclusion based on ceramic evidence, the site was only fortified during the MB IIC. Remains of the MB IIC fortifications of Shiloh were exposed in trenches in Areas C, D, H-F, J, K, and M.

Excavation in Area C on the W side of the site revealed portions of an offset-inset *fortification wall* (E381) built of fieldstones on bedrock over 2 m wide. Preserved to a height of 2.5 m, the wall was dated to the MB IIC [Str. VII] (Lederman 1993:17ff.). A clay *glacis* was laid between the outer face of the fortification wall and a *retaining wall* of a single course of fieldstones set directly on the bedrock.

The construction of the wall and glacis were further examined in Area D on the site's NE side where a section trench over 20 m long was excavated (Lederman and Finkelstein 1993:35ff.). Here the *core wall* of the rampart constructed of large fieldstones directly on bedrock (L.301) was exposed. This wall, preserved 6.7 m high and 3.8 m wide, featured a saw-tooth design. No remains of a mudbrick superstructure were found on its top surface, though the employment of such a building technique in a wall which was not intended to be seen, suggests that this wall served as the foundation for a mudbrick superstructure with presumably the same saw-tooth construction. It may, therefore, be concluded that this mudbrick wall may have been as much as 3.8 m wide in Area D. The interior of the wall seems to have served as a fill (417) for subsequent construction in-

[89]It has been suggested that Wall A should be identified with wall 655 (G. R. H. Wright 2002:ill. 44), but this cannot be confirmed.

[90]The E Gate was mostly excavated by Sellin and Welter, though the Joint Expedition (1956) was able to excavate part of the S tower of the gate (Campbell 2002:131).

side the town wall. This fill was bounded on the SE by another large wall (M.332) that connected to the interior of L.301 and the continuation of the perimeter wall, here designated N.321. It was only about 3 m thick and did not continue the saw-tooth technique. This portion of the wall connected to the outer face of L.301 creating an offset like that found in Area C.

The rampart and glacis were the largest outside wall L.301 in Area D (Lederman and Finkelstein 1993:35ff.). Although Finkelstein has not distinguished the rampart from the glacis, it is possible to suggest that the lower three layers of the five that he designated part of Glacis 723 should be identified as a *supplemental earthen rampart* upon which the glacis was then laid. The first step in the construction of the rampart was the erection of a *retaining wall* (M.291) along the existing 15° slope of the bedrock outside the wall. Clay was then deposited on both sides and above this retaining wall. Above this a layer of gray ash was then added with another layer of reddish-brown earthen fill nearly one meter thick deposited on this.

A *glacis*, which was also attested in other areas and consisted of a chalky material and small stones, was added to the rampart. Alternating layers of this material and a *terra rosa* soil created fingers of the chalky material which were imbedded within these soil layers. The glacis featured a slope of about 28° and a width of about 20 m. Thus, along the 175 m of the E side, if the rampart extended this entire length (with a cross section of 20 x 7 m), its volume would have been approximately 12,250 m^3 (cp. 14,000 m^3 in, Finkelstein et al. 1993:379). The volume of the core wall in this same area would have been 4,455 m^3.

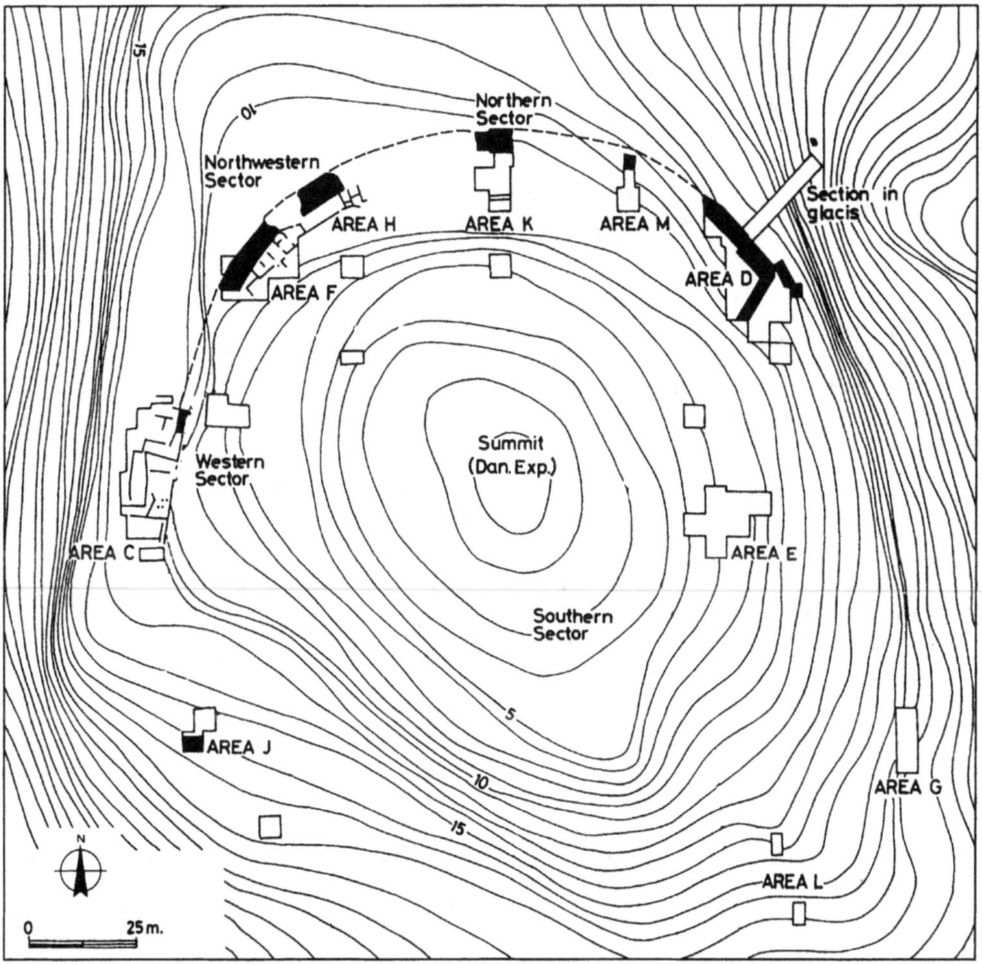

Figure 96. Plan of Shiloh showing Middle Bronze Age fortifications (after Finkelstein et al. 1993:fig. 1.5).

On the NW side of the site the MB IIC casemate-type *fortification wall* was excavated in Areas H and F, L281 and J291 respectively, where it was found to be 4.8 to 5 m thick (Finkelstein and Lederman 1993a:49–62). Here again the wall (i.e., its foundation) was constructed of fieldstones on bedrock, though its outside line is hard to follow due to later robbing of the wall. Inside the wall in Area H a number of store rooms were excavated. The only evidence for a glacis was detected in a small part of Area F. In Area F ash was found that indicates that the site was destroyed at the end of the Middle Bronze Age.

On the SW side of the site in Upper Area J the MB IIC *fortification wall* was again detected, built in the same fashion and about the same width as the wall in Areas H and F (Finkelstein and Lederman 1993b). While no Middle Bronze Age gate was identified, the excavators have suggested that it may have been located in lower Area J on the SW side of the site where another portion of the MB IIC *glacis* was revealed. The composition of this glacis was similar to that of the glacis in Area D (ibid., 69).

In Area K the MB IIC *fortification wall* (U251) was again located and seems to have been nearly 5.25 m wide, though much of its E end was robbed. Here the wall featured an offset-inset construction. No remains of the glacis were detected in this part of the mound, perhaps because bedrock was just below the surface. As in Areas F and M store rooms were built inside the town wall in Area K.

Between Areas D and K, the same *fortification wall* was also partially exposed in Area M. The defenses of Shiloh for the remaining 390 m of its perimeter may, therefore, have consisted of such defenses with an average width between 2 and 3.8 m.

Perhaps the most important observation concerning the Middle Bronze Age defenses of Shiloh is that they were not of uniform construction around the entire site. Instead, it is clear that certain areas were considered more vulnerable than others to attack and, therefore, the construction of its defenses varied substantially among these areas.

SHIMRON, TEL (SEMUNIEH)

MR 1699.2340

Refs.: Raban (1982); Portugali (1982); Kotter (1986:126f., 381ff.).

Plans: Site (Raban 1982:fig. g; Portugali 1982:figs. 3–6, 9); S Gate (Raban 1982:fig. k); N Gate (ibid., fig. j).

Sects.: NA.

Photos: Ramparts (Raban 1982:figs. a–b); S Gate (fig. h); N Gate (fig. f); SE Rampart (fig. i).

Tel Shimron is located 10 km to the NE of Tel Yoqne'am in the Jezreel Valley. The site was between 15 and 20 ha during the MB IIA–B (Raban 1982; Portugali 1982:183; see also Kotter 1986:126f.). It features a high upper town on its E side and a large lower town encircled by ramparts to the W. Two gates were identified on the N and S sides of the lower town (Figure 97 below).

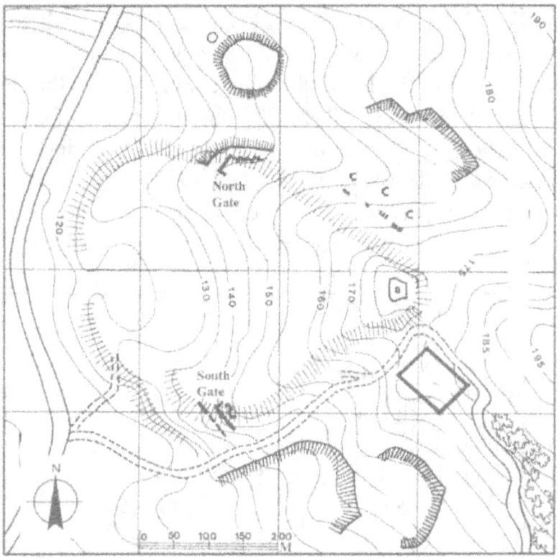

Figure 97. Plan of Shimron (after Raban 1982:fig. g).

The *earthen ramparts* run a clockwise course from the S edge of the upper town to the W around to the N side of the upper town, a circuit of about 1,200 m (Raban 1982:fig. g). According to Raban these ramparts may have been built against stone walls, which were identified in places along the ramparts, or these may have simply been the foundations of the fortification wall that crowned the ramparts (ibid., 71). It is also possible that these remains functioned as both the core of the rampart and the fortification wall's foundation. In any case, it is clear that in many places the rampart was laid directly upon the limestone bedrock and was on average raised to a height of 20 m. On the S side of the site the core of the rampart was made of red bricks covered with crushed basalt and black earth (ibid., 74). In most places it appears that the rampart was protected by a *glacis* of crushed limestone. The acropolis also featured its own *glacis* composed of chipped limestone (ibid., 71), which was no more than 150 m long.

The locations of two gates were established from a survey of the site. On the S side of the site the E half of a *six-pier gate* was exposed on the surface at the base of the rampart (Raban 1982:73, fig. k). It was approximately 20 m long. Attached to it on the E were the remains of its E tower. A 25 m stretch of the town wall with interior buttresses, which were 2 m wide and 1 m apart, was attached to the gate to its E, but unfortunately its width was not reported (ibid., 73). The N Gate was not really preserved; it would have necessitated a sharp turn to the E upon exit due to the surrounding topography.

According to Kempinski the fortifications of this site date to the MB IIB–C (1992d:183).

SHUSH, TEL (TELL ABU SHUSHA)*

MR 1634.2246
Refs.: Raban (1999).
Plans: NA.
Sects.: NA.
Photos: NA.
Tel Shush is located less than 3 km to the NW of Megiddo (Raban 1999:61*). Ceramics collected at the site included MB IIA sherds. According to the Israel Antiquities Authority survey the site's gross size was 4 ha and it was enclosed during the Middle and Late Bronze Ages by an earthen rampart.

SHUWEIKET ER-RAS, KHIRBET*

MR 1536.1942
Refs.: Zertal (1993).
Plans: NA.
Sects.: NA.
Photos: NA.
Shuweiket er-Ras is located in the western foothills approximately 27 km NW of Shechem. It was identified as a fortified MB IIB settlement with a fortification wall and "huge earthen rampart" (Zertal 1993:1312); it is identified as approximately 3 ha in size (Broshi and Gophna 1986).

TA'ANACH (TELL TA'ANNEK)

MR 171.214
Refs.: Lapp (1964; 1967; 1969).
Plans: Site (1969:fig. 1); SW 7–7 (ibid., fig. 11).
Sects.: SW 1–28 and 29 (1964:fig. 2); SW 7–1 to 9–1 (1967:fig. 6); SW 7–7 (1969:fig. 10); NW 9–1 (1964:fig. 4).
Photos: NW 9–1 (1964:fig. 5); SW 7–7 (1969:figs. 12–14).
Ta'anach is located about 8 km to the SE of Megiddo. The site appears to have been well-fortified during the EB II–III. The fortifications of this phase have been identified on the S and W sides of the site and belong to four major phases (Lapp 1969:5ff.). Although there may have been some occupation of the site during the MB IIB, the site does not appear to have been fortified until the MB IIC. These defenses were first exposed during excavations led by Paul Lapp when they were excavated in NW 9–1 on the W side of the site (1964:15), in both SW 1–28 and 29 on the S side of the site (ibid.), and in SW 7–7 and SW 8–1 also on the W side of the site (1969:4).

MB IIC

The first phase of the defenses of Ta'anach have been dated roughly to the early sixteenth century B.C. by Lapp (1969:16ff.). They were built above the remains of the EB II–III fortifications and consisted of two main phases with which a series of glacis were associated. The defenses in this period included a poorly constructed fieldstone *fortification wall* with perhaps some exterior buttresses (118) built directly atop the remains of the Early Bronze Age wall as revealed in SW 7–7 and 8–7. Whether or not the wall featured a mudbrick superstructure remains uncertain. Associated with this wall was the earliest *earthen rampart*, which was built of "strips of huwwar and clayey brown earth . . . representing the 'sandwich method'" identified with the poorly preserved remains of Glacis 29 in NW 9–1 (1964:15, fig. 4) but better preserved as Glacis 69 in SW 9–1 (1967:13).

The second main phase of the MB IIC fortifications, which dated roughly to the end of the sixteenth century B.C., saw the construction of a *casemate wall* (1969:16, 19f.), with which two later *glacis* were associated. The casemate wall was a total of 4.6 m wide with exterior and interior walls of 1.75 m and 1.5 to 2.25 m respectively (ibid., 20). Associated with this casemate wall was the construction of a second *earthen rampart* consisting of "brown stony rubble" covered by horizontally deposited layers of earth (1964:15). This was identified as Glacis 15 in NW 9–1, Glacis 67 in SW 9–1 (1967:13), and perhaps layer 53 in SW 8–7 (1969:19). This glacis was retained by a small fieldstone *revetment wall* preserved in NW 9–1 (no. 33 in fig. 4, 1964).

A third and final *glacis* was built consisting of a one-meter thick "layer of yellow-white huwwar" limestone deposited over top of the second glacis and perhaps built against the outer wall of the West Building (ibid., 15). This was identified with Glacis 64 in SW 9–1 (1967:13). This glacis was later damaged by Late Bronze Age occupation (1964:15). On the S side of the site (SW 1–28 and 1–29) two glacis, which probably correspond with the first and second glacis, could also be identified (ibid.).

The circumference of the MB IIC fortifications may have been as much as 840 m around the site enclosing an area of approximately 3.8 ha. No gates have thus far been identified.

LATE BRONZE AGE

The *casemate wall* of the MB IIC continued in use into the LB I, though only after a site-wide destruction. No additional defenses were afterward constructed until the Iron Age (1969:33).

TIMNAH (TEL BATASH)

MR 141.132
Refs.: A. Mazar (1997b).
Plans: Site (fig. 2); Area A (P/S 1); Area B (P/S 9).
Sects.: Area A (P/S 8); Area B (P/S 24–27 and figs. 6–8).
Photos: Site aerial (photos 1, 3); Area A (photo 7); Area B (P/S 23–26).

Timnah (Tel Batash) is located 5.4 km east of Ekron at the mouth of the Sorek Valley. Amihai Mazar suggests that the site took its shape from the initial rectilinear plan of the Middle Bronze Age fortifications that were first built during MB IIB on the S bank of the Nahal Sorek (A. Mazar 1997b). The rampart and wall dated to MB IIB–C, Strata XII–XI, at Tel Batash were excavated in areas A and B on the N side of the mound (Figure 98).

AREA A

In Area A a step trench 40 m x 5 m was excavated on the N slope. This step trench revealed only three layers belonging to the construction of the *earthen rampart* (183, 186, 191), which can be seen in the section. On one of these layers traces of lime were detected indicating that it was probably plastered. Excavation of the *fosse* detected at the foot of the slope did not reveal its full dimensions since work stopped 2.1 m below the level of the plain (A. Mazar 1997b:21–23).

AREA B

A larger exposure of the rampart was made in Area B where an initial trench was lengthened to the N with mechanical equipment. Mazar suggests that "the rampart had suffered from severe erosion" leaving only the lower portion, which was preserved to a height of 4.5 m. The majority of the rampart was composed of alternating layers of pebbles and soil employed in what Mazar refers to as "constructional 'triangles'" deposited against the mudbrick town wall (B863). This technique, which was also employed at Gezer, was intended to reinforce the earthwork by sloping the layers back toward the wall instead of away from it. That the rampart was built after a short-lived settlement on the site during the MB IIB is evident from the Middle Bronze Age sherds found in the rampart's fill (A. Mazar 1997b:34–38). The entire circuit of the ramparts was about 600 m long. Assuming a maximal height of 6 m and a possible width of up to 25 m of exterior rampart and another 25 m as the width of the interior rampart then the volume of the earthen rampart would have been no more than 90,000 m^3 (cp. figure of 140,000 cubic m Mazar 1997:250).

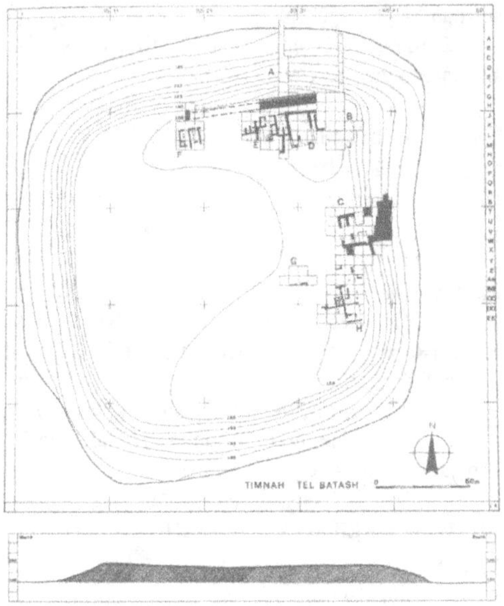

Figure 98. Plan and profile of Timnah. Reprinted, by permission, from A. Mazar (1997b:fig. 2).

In addition to the rampart, in Area B a portion of a well-fortified mudbrick building termed the "citadel" was uncovered. Two rooms of this structure (725 and 1205), built against the inside of the 2.4 m wide mudbrick *fortification wall* were excavated. The W wall of the structure (B922) was exposed with a width of 3.5 m (wider than the fortification wall) and preserved 4 m high revealing why Mazar has referred to this structure as a citadel. The two rooms exhibited two phases of occupation associated with two separate floors, assigned to Strata XII (1114 and 1211) and XI (725 and 1205) respectively. The plaster floor (B725) in the N room was traced up the inside of the fortification wall (here designated B774). Unfortunately, no significant material culture was found in this structure, which Mazar concludes is evidence that the inhabitants had time to remove their possessions. As the architecture would seem to suggest, a citadel or tower seems to have occupied the NE corner of the settlement overlooking the Nahal Sorek. The citadel fell out of use during the LB I when floors were laid over the stumps of its walls (A. Mazar 1997b:249–52).

No gates have been excavated at Batash nor can their locations be determined based on the topography of the site, despite Kempinski's suggestion that four gates, one on each side, provided access to the site as at Qatna (1992a).

'Umayri, Tell el-

MR 2342.1420
Refs.: Clark (2000; 2002)
Plans: Site (Herr 2000:fig. 2.1); Field B (Clark 2000:fig. 4.9).
Sects.: Field B schematic section (Clark 2000:figs. 4.3, 4.11; 2002:fig. 4.4).
Photos: Site aerial (Herr 2000:fig. 2.2); Field B (Clark 2000:figs. 4.1, 4.8, 4.10, 4.12; 2002:figs. 4.5–4.7).
Tell el-'Umayri is located in the highlands of Transjordan about 38 km E of Jericho. Recent excavations at Tell el-'Umayri have revealed that the settlement was fortified during the MB IIC, Phase 12 at the site (Clark 2000:64ff.; 2002:49ff.). Based on the size of the tell, the site may have been approximately 6.5 ha during this period (Falconer 2001:281).

The fortifications included an *earthen rampart* with a 25° to 30° slope perhaps originally 35°, built directly on bedrock on the W side of the site. Excavation of the rampart revealed that bedrock was leveled in preparation for its building. The rampart raised the defenses 5 m above the area to the site's W with the addition of 3 m of earth atop the slope (Clark 2002:50). The surface of the rampart may have been protected with a *glacis* composed of crushed nari (Clark 2000:66). About 8 m down the slope from the foot of the rampart a 6 m wide and 5 m deep *fosse* was cut into the bedrock. The farthest face of this fosse was not worked. It is not known how long this fosse stretched to the N or S (ibid., 65). No remains of these elements of the fortification system have been found on the E side of the site where the slope is thought to have provided greater defense of the settlement.

It appears that an earthquake ca. 1200 B.C. may have damaged the Middle Bronze Age rampart (Clark 2002:51). No Late Bronze Age fortifications have been identified. The next fortifications are those of the Iron I period (Phase 11B).

'Urma, Khirbet*

MR 18050.17265
Refs.: Finkelstein (1988; 1993a).
Plans: NA.
Sects.: NA.
Photos: Site (Finkelstein 1988:fig. 44).
Khirbet 'Urma (biblical *'Arumah)* is located in the hills of southern Samaria 8 km SE of Shechem (see Finkelstein 1988:fig. 54). It was identified as a fortified MB IIC settlement (1993a:1313), ca. 1.5 ha in size (1988:159).

Yavneh-Yam

MR 1212.1479
Refs.: Kaplan (1969; 1993).
Plans: Site (Kaplan 1993:1504).
Sects.: Area A (Kaplan 1993:1506; Herzog 1986:fig. 48).
Photos: Area A (Kaplan 1993:1506); Area H (ibid., 1505).

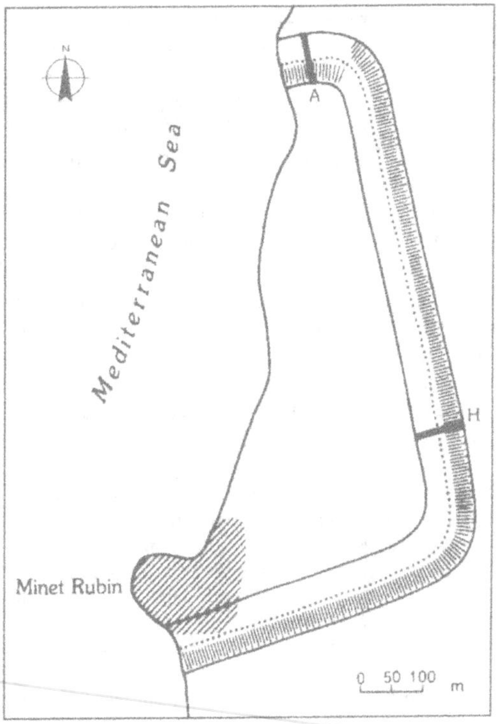

Figure 99. Plan of Yavneh-Yam (after Kaplan 1969:1504).

Yavneh-Yam is located 32 km N of Ashkelon on the Mediterranean coast (Figure 99). Although there is some question regarding how large the settlement was during the MB IIA, it appears to have featured occupation at least starting in Phase 2 or 3 (Cohen 2002a:94). The site expanded during the MB IIB–C with the construction of a massive earthen rampart, parts of which survived on the N, E, and S sides of the site. Excavations were conducted at Yavneh-Yam from 1967 to 1969 by Jacob Kaplan with the intention of studying "the method of construction of the enclosure and its ramparts" (1993).[91]

Although Kaplan reconstructed the site's layout as a square enclosure (800 x 800 m) comparing it with the square enclosure at Qatna, the original enclosure was probably akin to the enclosures of nearby sites such as Ekron and Tell el-'Ajjul. It must be assumed, therefore, that most of the length of the S rampart has survived and that we lack only a slight portion of the W ends of the N and S ramparts; it is very likely that as at Ashkelon no rampart ran N-S on the W side of the site. Presuming that far less erosion has occurred than has been suggested, the original enclosure featuring straight sides would have been a rectangle transected by the coast; it would have measured 800 m N-S on its E side, with 575 m E-W on its S side, and a small portion of rampart of approximately 125 m long on its N, for a total length of about 1,500 m enclosing a gross area of no more than 28 ha (rather than the 64 ha originally cited by Kaplan 1993). After subtracting the area occupied by the exterior slope of the ramparts the inhabitable area was approximately 16.9 ha.

Two areas of Yavneh-Yam's earthen rampart were excavated by Kaplan: Area A where a section was cut through the rampart's N side and Area H, a depression in the rampart toward the S end of the E side where three superimposed gates were revealed. Since no final publications of Kaplan's excavations are available, his brief summary of the rampart is retained:

> The rampart was found to be constructed as follows: first, the whitish sand covering virgin soil was leveled along the rampart's proposed alignment. A layer of hamra, 12 cm thick, was then laid as a bedding for the rampart core, constructed of light-brown, pounded earth. In the final stage, the core was encased by a sheath of hamra, built up from the base on both sides of the core, toward the top. The glacis was in two layers: the lower layer, of heavy clay soil approximately 60–70 cm thick, extends from the top of the rampart down to virgin soil. The second, covering layer, of crushed kurkar 50 cm thick, was probably intended to prevent the damp clay soil from desiccating and pulverizing. At a later stage, an additional glacis was laid over the crushed kurkar layer. This new glacis was constructed in two parts: the lower, about 3 m high was in the form of a retaining wall inclined about 45 degrees; from that point, and up to the top, the rampart was paved with stone and raked to approximately a 30-degree angle (Kaplan 1993:1505).

Although Kaplan suggested that two separate glacis were constructed, in the absence of published sections and a detailed report we have no basis for distinguishing any phases in the construction of this rampart. It is possible, therefore, to argue that this was a single *glacis* that consisted of three separate elements, each of which were added in succession: the first of clay, the second of crushed kurkar, and the last of stone. From Kaplan's schematic plans it appears that the rampart measured about 75 m wide at its base.

In Area H three successive gates belonging to the MB IIA (though probably dated to late MB IIA or MB IIB) through LB I were excavated (for plan, see Herzog 1986:fig. 48). Gate III, the earliest gate, was a typical *six-pier gate* built of mudbricks (see Table 9 for dimensions). Gate II was built above Gate III but was only a *four-pier gate*. However, Gate II also featured 2.4 m thick walls and a staircase that led to an upper story in the E tower. The exteriors of Gates III and II both of which are dated to the Middle Bronze Age were protected by rubble walls.

A final gate, Gate I, was built during the LB I above Gate II. Only half of this structure was exposed, which consisted of two rooms and a supporting wall. Several "anchor" walls, three to the N and two to the S, were found outside of the gate towers. Kaplan suggested that these walls, the earliest of which were contemporary with Gate III, were intended to anchor the gate in the ramparts and he concluded from this that the gate was a later addition to a preexisting rampart (1969:121). However, these walls can be understood as terrace walls attaching the fortification wall to the gate system. Another gate system was encountered in the middle of the E rampart N of Area H during a sounding made there, but no further details have been published concerning this structure.

Though the presence of MB IIA sherds on the "lowest floor in one of the [gate] chambers" indicates that its earliest use could have been in the MB IIA, the existence of only three phases of gate construction, which were in use from the MB IIA to the LB I are suggestive of a shorter (i.e., ca. 1700–1530 B.C.) rather than longer period of use (i.e., ca. 1900–1530 B.C.). Furthermore, based upon the rectilinear layout of the fortifications and the six-pier gate built with this rampart, the structure should probably be dated to the late MB IIA at the earliest, and more probably to the transitional MB IIA–B,[92] as at Tel Dan and possibly Hazor. However, only publication and analysis of the ceramic evidence from the rampart

[91]Despite discussion of the ramparts of Yavneh-Yam in three separate publications, this summary, taken from the *NEAEHL*, is identical to earlier summaries published by Kaplan (1975:4–6; 1978).

[92]A. Mazar has also suggests an MB IIB date for the gate and rampart (1990:227f., n. 7).

itself will allow us to be more certain of the date of the gate and the enclosure.

YOQNE'AM, TEL

MR 1604.2289
Refs.: Ben-Tor and Zarzecki (1988–1989); Ben-Tor (1993a); Ben-Tor et al. (1996).
Plans: Site (Ben-Tor et al. 1996:plan III.2).
Sects.: NA.
Photos: Site aerial (Ben-Tor et al. 1996:photo III.1); Area A (Ben-Tor and Zarzecki 1988–1989:fig. 164).
Tel Yoqne'am is located in the Jezreel Valley about 10 km NW of Megiddo. Ceramic evidence from Tel Yoqne'am suggests that the site was continuously occupied from at least EB I through the start of the Middle Bronze Age (Ben-Tor 1993a:2). A cave burial dated to the early MB IIA suggests that the site was also occupied during an early phase of this period designated Stratum XXIIIC (Ben-Tor 1997b:811). However, the earliest evidence for the fortifications of the site are the remains of two fortification systems, which based upon their ceramics have been ascribed to Strata XXIIIB–A (MB IIA, Phases 3–4 Cohen 2002a:95), which were founded upon bedrock in Area A on the NW side of the site (Ben-Tor et al. 1996:2).

MB IIA, PHASE 3–4: STRATA XXIIIB–A

The first Middle Bronze Age *fortification wall* (Str. XXIIIB) was built 2.4 m wide with mudbricks (Ben-Tor and Zarzecki 1988–1989:195). This first wall was determined to have been founded on bedrock, which led Ben-Tor to suggest that the construction of the rampart eradicated the architectural remains of earlier periods on the site (1993a:811). It is noteworthy that the first wall, though built on bedrock, also featured a stone foundation.

During a later renovation of the defenses another *fortification wall*, 3 m wide and featuring a *tower*, was built during Stratum XXIIIA. This wall overlaid the Stratum XXIIIB wall in Area A1 and like the earlier wall featured a stone foundation of a single course and a mudbrick superstructure. This new, wider wall was built to accommodate the settlement's size and therefore was built further out so that its outer face was founded upon the earlier glacis outside the original wall. It was preserved to a height of almost 3 m and featured a unique construction in that stone courses were used between brick courses. This wall was also identified in Area A4 where it was preserved to a height of more than 2.5 m (1988–1989:196). Based upon the published plans of the site and the location of the Late Bronze Age remains in Area A1 (see Ben-Tor 1993a:fig. on p. 808) it is possible to suggest that these Stratum XXIIIA fortification walls were located at about the 90 m contour line, or at least 10 m above the plain to the N. If this is a reliable reconstruction of the location of the MB II walls, then it is possible that this wall was 830 m long and enclosed a space of about 4.7 ha.

In addition to the wall, the slope of the site was reinforced with a *glacis* built upon bedrock and detected in Areas A1 and A4 (Ben-Tor 1993a:811). This glacis abutted the foot of the first MB IIB town wall at the top of the slope and was retained at its foot by a mudbrick *revetment wall* built on bedrock 4.5 m thick (Ben-Tor and Zarzecki 1988–1989:196).

MB IIB: STRATUM XXII

During the start of MB IIB the Stratum XXIIIA wall was replaced by a 1.5 m wide mudbrick *fortification wall* in Stratum XXII (Ben-Tor 1993a:810). A *tower* was apparently built against its interior face, though the exterior face of the wall was badly damaged due to erosion on the slope of the site.

MB IIC: STRATUM XXI

The following settlement at Yoqne'am during the MB IIC was, interestingly enough, unfortified according to Ben-Tor (1993a:810).

ZEROR (KHIRBET ET-TELL DHURUR)

MR 1476.2038
Refs.: Ohata (1967; 1970); Kochavi et al. (1979).
Plans: Site (Ohata 1970:pl. 1; Kochavi et al. 1979:fig. 13); Area A IV (Ohata 1970:pl. 9); Area D (ibid., pl. 11; Kochavi et al. 1979:figs. 15f.).
Sects.: Area A IV (Ohata 1970:pl. 9); Areas C and D (ibid., pl. 12).
Photos: Area C (ibid., pl. 45), Area D (pls. 49f.).
Tel Zeror is located on the coastal plain about 12 km SSE from Tel Mevorakh. The site was founded during the MB IIA (Ph. 2–3, Cohen 2002a:96). The Middle Bronze Age settlement was limited to this period, as no remains dated to the MB IIB–C have been identified. The site's MB IIA defenses consisted of the remains of two fortification walls, a tower, a rampart, and a fosse. Prior to excavation it was not possible to discern any of these elements from the site's contours. But fortification walls were discovered in Areas A, C, and D. Trench IV in Area A was laid across the defenses on the E side of the acropolis. In Area C Trench D1–5 ran N-S across the S fortifications. In Area D an E-W trench was excavated across the wall and fosse on the W side of the site.

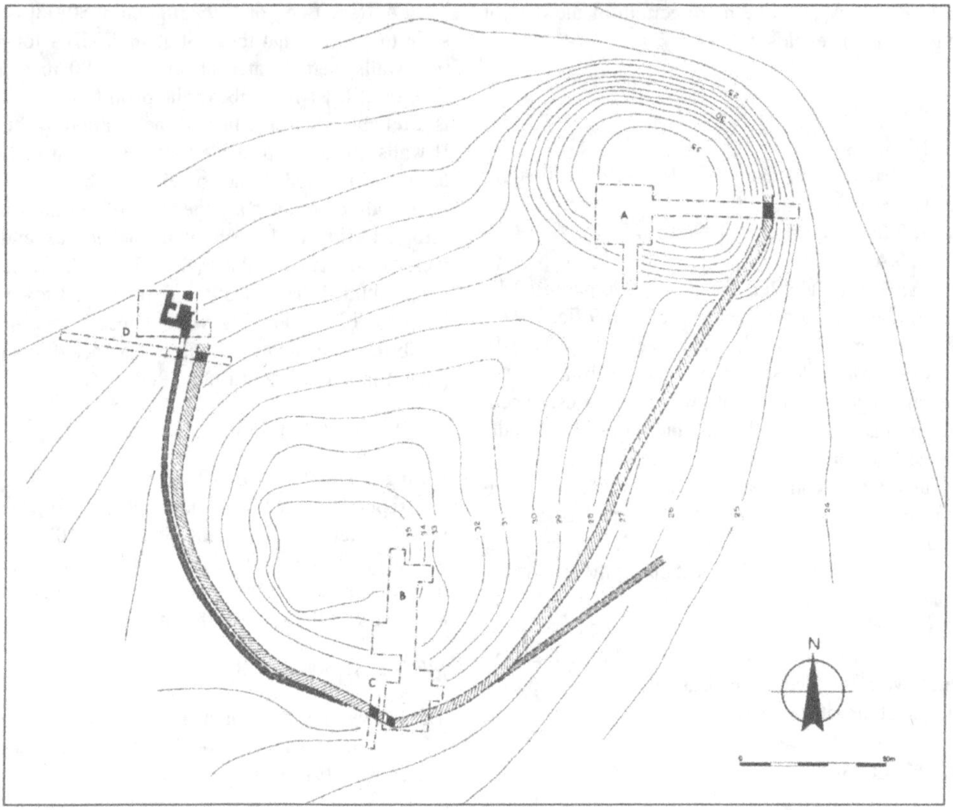

Figure 100. Plan of Tel Zeror (after Kochavi et al. 1979:fig. 13).

Two walls are attested for the MB IIA town defenses. Both of these phases were revealed in Areas C and D. The first *fortification wall* (W.1) in Area C was 2.5 m wide and seems to have been constructed on an *earthen rampart* at 26.2 m ASL (Ohata 1970:49ff.).[93] The rampart was 20 m wide, but since its excavation was not completed and the trench was not extended far enough N, a complete description of the rampart in this area is not possible (see section; ibid., pl. 12). It was also detected in Area A, though the first wall was not exposed there (ibid., 43ff.). In Areas A and C earthen piles, which were pits dug into the rampart and filled with distinct material, were, according to the excavator, intended to stabilize the rampart. However, given the miniscule slope the rampart in both areas and that the pits were dug into a flat surface, it is difficult to accept this suggestion. In Area D the evidence that the first fortification wall (W.1) was built on a small earthen rampart is much clearer than in Area C (ibid., 57f.). Here the rampart was composed of reddish sandy soil about 1.9 m thick and it was preserved about 1.4 m high. The wall itself was almost 2 m wide. Along with the rampart the remains of a rectangular *tower* can also be associated with the first town wall in Area D (ibid., 60f.). Unfortunately, it was not completely excavated so its dimensions remain unknown.

The earth for the rampart upon which the fortification wall (W.1) was built in Area D is assumed to have been taken from the *fosse* dug about 18 m in front of it (Ohata 1970:59f., pl. 12). The fosse was approximately 10 to 12.5 m wide with an outer scarp of 58° and an inner slope of 43°. It was excavated down to the water table at about 20 m ASL, which would make the fosse between 4 and 5 m deep. The fosse did not seem to have been cut into bedrock, but without excavation of the lowest levels of the fosse it is not possible to be certain whether or not this is true. It is also uncertain whether or not the fosse ran around the entire settlement.

[93] In the absence of consistent loci references in the final reports for the Zeror excavations, loci references for Area D have been adapted from Kochavi et al. (1979:fig. 15).

The remains of the first wall in Area C were used to level the area for the construction of a second *fortification wall* at a later stage of the MB IIA, raising the base of the new wall to 28 m ASL and infilling the fosse below with remaining rubble (Ohata 1970:49ff.). Ohata has speculated that a fosse associated with the second wall might have existed farther to the S of the excavated trench but this has not been verified. It was impossible to determine the complete width of the second wall in this area, though it was preserved as much as 5 m wide. The second wall was also identified in Area A (ibid., 43ff.). It was built of mudbrick, eight courses of which were preserved to a height of 1.5 m, on a foundation of a single course of stones. It was estimated to be about 3 m thick. This wall fell out of use at the start of the MB IIB, when the site was abandoned (ibid., 52f.). In Area D the second wall was built about 8 m inside the first wall, which is clear from the small structure on the W side of the wall into which the wall was cut when it was built (ibid., 66, pl. 12:). Having perhaps realized the insufficient protection provided by the first wall, the second wall was built much thicker, 4.3 m wide, and was preserved to a height of 3.6 m. None of the walls in Area D were built upon stone foundations like those in Area A and C.

ZUREKIYEH, 'AIN (POLEG EAST FORT)

MR 1382.1833
Refs.: Gophna and Ayalon (1982); Gophna (1993b:1195f.).
Plans: Site (Gophna 1993b:fig. 2); Area A (fig. 3); Area B (fig. 4).
Sects.: Area B (fig. 5).
Photos: NA.

Survey excavations of 'Ain Zurekiyeh, located only 3.5 km ESE from Tel Poleg along the Nahal Poleg, have revealed two phases of an early MB IIA fort occupying approximately 0.4 ha. The plan of the fort appears to have been approximately square, although the lines of the walls conform to the contours of the kurkar ridge. Its *fortification walls* were 3.2 m wide and were constructed of mudbrick on stone foundations. Excavations in Areas A and B during salvage operations exposed 18 m and 16 m sections of this wall on the N and E sides of the site, respectively (Gophna and Ayalon 1980:147; 1993b:1194f.). In Area A the stone foundations were built directly on the kurkar bedrock with its inner and outer elevations faced with large kurkar blocks and the space between them filled with kurkar rubble and smaller stones (Gophna and Ayalon 1982:69). Two phases of this wall, W.1012 and W.1011, were identified, but no occupational surfaces associated with either phase could be identified (ibid., 69ff.). In Area B a single phase of a wall (W.1014) of comparable thickness and construction to that of the earlier phase in Area A (W.1012) was identified (ibid., 72). But here the mudbrick superstructure survived to a height of five courses (60 cm) built of hamra mudbricks without mortar. Occupational levels on the inside of the wall remained intact in Area B.

The estimated circumference of the walls is about 250 m. Based upon ceramics from the site the excavators have suggested that the site was occupied in the earliest phase of the MB IIA contemporary with the "Pre-Palace" phase at Aphek but prior to the occupation of Tel Poleg (ibid., 77). No evidence of a rampart has been revealed, although the slope outside the wall in Area B to the E was approximately 32°.

3. Egypt

Avaris (Tell ed-Dab'a)

Lat. Lon.: 30°47'15" N, 31°49'20" E

Refs.: Bietak et al. (1994); Bietak (1996:esp. 63ff.; 2001:esp. 352f.); Bietak et al. (2001).

Plans: Site (Bietak et al. 2001:fig. 1); 'Ezbet Helmi (Bietak et al. 2001:figs. 12, 33); Area H/I (Bietak et al. 1994:figs. 7f.).

Sects.: NA.

Photos: Area H/I (Bietak et al. 1994:pls. 2–3A).

Tell ed-Dab'a, identified with *Avaris* of the Fifteenth Dynasty, is located along the Pelusiac branch of the Nile in the eastern Egyptian Delta. Although excavations provide a thorough understanding of the settlement from the Middle Kingdom through the Eighteenth Dynasty, only one sounding has been made of the site's fortifications. Furthermore, the remains of the fortification wall of Avaris are limited to the end of the Hyksos period when the settlement had expanded westward into this area and, therefore, nothing is known of the site's fortifications during the earlier part of the Middle Bronze Age.

Dynasty XV: Stratum D/2 (Area H/I Stratum V)

The town's *fortification wall* was excavated in 'Ezbet Helmi on the W side of Tell ed-Dab'a on the W bank of the present river course (Bietak et al. 1994:27–31). This mudbrick town wall dates to the late Hyksos period (Str. D/2, local Stratum V) and featured two building phases (Bietak et al. 1994:28). In the earliest phase the wall was 6.2 m (12 cubits) wide and its exterior face featured a sloped footing at the base of the wall of 78° to 80°. The wall was buttressed on the exterior at regular intervals of approximately 18.1 m (35 cubits) by square towers 5.3 m long (10 cubits), which projected approximately 3.6 m (7 cubits) from the wall. The towers were clearly built as part of the original phase of construction with solid mudbrick foundations. The wall was preserved up to 1.3 m high. During a second phase of Stratum V, this wall was expanded to a width of 8.5 m [16 cubits] (ibid., 29).

During this period an orchard was located just inside the wall as suggested by the tree pits excavated there, and it is believed that the Pelusiac branch of the Nile ran on the N side of the wall (Bietak et al. 1994:30). Although only a 50 m stretch of this wall has been exposed (ibid., 28), based on a magnetometry survey the wall can be traced for about 280 m to the NE (Bietak et al. 2001:48, fig. 33). There is no evidence to suggest that the wall rested upon an earthen rampart. In fact, the original wall appears to have been built upon a foundation of yellow sand (Bietak et al. 1994:28), a typical Egyptian construction technique.

In addition to the wall a *bastion* whose dimensions appear to have been 39.5 x 26 m (75 x 50 cubits) was built in Area H/II 100 m to the S (Bietak 1996:63f.). It may have been comparable in function to the later Eighteenth Dynasty bastion constructed in Area H/I. This structure is referred to as the Hyksos Citadel.

Dynasty XVIII

At the start of the Eighteenth Dynasty the fortification wall's function appears to have been slightly altered (Bietak et al. 1994:30f.). The construction of a large structure, referred to as the "Große Platform," which served as the foundation for a large building, cut into the inner face of the wall, though not destroying it (Bietak 1996:68). Bietak identifies this structure as a *bastion* (70.5 x 47 m) comparable with those along the ramparts at Avaris (ibid., 70 and fig. 58).

BIBLIOGRAPHY

Adams, Robert McCormick

1965 *Land behind Baghdad: A History of Settlement on the Diyala Plains*. Chicago: University of Chicago Press.

1981 *Heartland of Cities: Surveys of Ancient Settlement and Land Use on the Central Floodplain of the Euphrates*. Chicago: University of Chicago Press.

Aeneas Tacticus (Aineias the Tactician)

1990 *How to Survive under Siege*. Translated by D. Whitehead. Clarendon Ancient History Series. Oxford: Clarendon.

Aharoni, Yohanan

1993 Kedesh (in Upper Galilee): Tel Kedesh. In *NEAEHL*, vol. 3, ed. E. Stern, 855–56. Jerusalem: Israel Exploration Society.

Akkermans, Peter M. M. G., and Glenn M. Schwartz

2003 *The Archaeology of Syria: From Complex Hunter-Gatherers to Early Urban Societies (ca. 16,000–300 BC)*. Cambridge: Cambridge University Press.

Albright, William Foxwell

1922 Palestine in the earliest historical period. *JPOS* 2:110–38.

1924 The town of Selle (Zaru) in the 'Amarna tablets. *JEA* 10:6–8.

1926 The Historical Background of Genesis XIV. *JSOR* 10:231–69.

1935 Palestine in the earliest historical period. *JPOS* 15:193–234.

1938a The Chronology of a South Palestinian City, Tell el-'Ajjul. *American Journal of Semitic Languages and Literatures* 55:337–59.

1938b *The Excavation of Tell Beit Mirsim II: The Bronze Age*. AASOR 17. New Haven: American Schools of Oriental Research.

1949 *The Archaeology of Palestine*. 1st ed. Harmondsworth, UK: Penguin.

1960 *The Archaeology of Palestine*. 2nd ed. Harmondsworth, UK: Penguin.

1964 The Eighteenth-Century Princes of Byblos and the Chronology of Middle Bronze. *BASOR* 176:38–46.

1965 Further Light on the History of Middle-Bronze Byblos. *BASOR* 179:38–43.

1966 Remarks on the Chronology of Early Bronze IV-Middle Bronze IIA in Phoenicia and Syria-Palestine. *BASOR* 184:26–35.

1973 The Historical Framework of Palestinian Archaeology between 2100 and 1600 BC (EB IV, MB I, MB IIA–B). *BASOR* 209:12–18.

Alex, Michael, and Werner Wolfner

1984 Mittlere Jahrensniederschläge und Variabilität. In *TAVO*. vol. A IV 4. Wiesbaden: Reichert Verlag.

Alt, Albrecht

1954 Beziehungen zu Ägypten in den Briefen von Mari? *ZDPV* 70:130–34.

Amiran, Ruth

1970 *Ancient Pottery of the Holy Land: From Its Beginnings in the Neolithic Period to the End of the Iron Age*. English ed. [n.p.]: Rutgers University Press.

Amiran, Ruth, and Abraham Eitan

1965 A Canaanite-Hyksos City at Tell Nagila. *Archaeology* 18:113–23.

1993 Nagila, Tel. In *NEAEHL*, vol. 3, ed. E. Stern, 1079–81. Jerusalem: Israel Exploration Society.

Archi, Alfonso

1998 The Regional State of Nagar according to the Texts of Ebla. In *About Subartu: Culture, Society, Image*, ed. M. Lebeau, 1–15. Subartu 4/2. Turnhout: Brepols.

Arns, Raimund, et al.

1984 Ausgrabungen in Tall Bi'a 1982 und 1983. *MDOG* 116:15–63.

Artzy, Michal

1993 Nami, Tel. In *NEAEHL*, vol. 3, ed. E. Stern, 1095–98. Jerusalem: Israel Exploration Society.

Artzy, Michal, and Ezra Marcus

1991 The MBIIA Coastal Settlement at Tel Nami. *Michmanim* 5:5*–16*.

Ashbee, Paul, and Ian W. Cornwall

1961 An Experiment in Field Archaeology. *Antiquity* 35:129–34.

Astour, Michael C.

1992 An Outline of the History of Ebla (Part 1). In *Eblaitica: Essays on the Ebla Archives and Eblaite Language 3*, eds. C. H. Gordon and G. A. Rendsburg, 3–82. Winona Lake, IN: Eisenbrauns.

2002 A Reconstruction of the History of Ebla (Part 2). In *Eblaitica: Essays on the Ebla Archives and Eblaite Language 4*, eds. C. H. Gordon and G. A. Rendsburg, 57–195. Winona Lake, IN: Eisenbrauns.

Atkinson, Richard J. C.

1961 Neolithic Engineering. *Antiquity* 35:292–99.

Avi-Yonah, Michael, ed.

1975 *Encyclopedia of Archaeological Excavations in the Holy Land*. 4 vols. Jerusalem: Israel Exploration Society.

Badawy, Alexander M.

1964 Preliminary Report on the Excavations by the University of California at Askut. *Kush* 12:47–53.

1965 Askut: A Middle Kingdom Fortress in Nubia. *Archaeology* 18:124–31.

1966 Archaeological Problems Relating to the Egyptian Fortress at Askut. *JARCE* 5:23–27.

Badre, Leila

1997a BEY 003 Preliminary Report: Excavations of the American University of Beirut Museum. *BAAL* 2:6–94.

1997b Kamid el-Loz. In *OEANE*, vol. 3, ed. E. M. Meyers, 265–66. New York: Oxford University Press.

2002 The Bronze Age of Beirut: Major Results. *ARAM* 13–14:1–26.

Badre, Leila, and Eric Gubel

1999–2000 Tell Kazel, Syria: Excavations of the AUB Museum 1993–1998, Third Preliminary Report. *Berytus* 44.

Badre, Leila, et al.

1990 Tell Kazel, Syria. Excavations of the AUB Museum 1985–1987. Preliminary Reports. *Berytus* 38:10–124.

Badre, Leila, and Jean-Paul Thalmann

1996 Les découvertes archéologiques du centre-ville de Beyrouth. *CRAIBL* 1996:87–97.

Bahat, Dan

1993 Beth-Shemesh (Giv'at Sharett). In *NEAEHL*, vol. 1, ed. E. Stern, 253–54. Jerusalem: Israel Exploration Society.

Baker, Jill L.

2003 The Middle and Late Bronze Age Tomb Complex at Ashkelon, Israel: The Architecture and the Funeral Kit. Unpublished Ph.D. Dissertation. Department of Old World Archaeology and Art, Brown University.

Bar-Yosef, Ofer, et al.

1993 The Kefar Monash Hoard. In *NEAEHL*, vol. 2, ed. E. Stern, 614. Jerusalem: Israel Exploration Society.

Baramki, Dimitri C.

1966 *The Archaeological Museum of the American University of Beirut*. Beirut: American University of Beirut.

Barnett, Richard D.

1963 Xenophon and the Wall of Media. *JHS* 83:1–26.

Beit-Arieh, Itzhaq, ed.

2003 *Map of Tel Malhata (144)*. ASI. Jerusalem: Israel Antiquities Authority.

Ben-Tor, Amnon

1992b The Early Bronze Age. In *The Archaeology of Ancient Israel*, ed. A. Ben-Tor, 81–125. Translated by R. Greenberg. New Haven: Yale University Press.

1993a Jokneam. In *NEAEHL*, vol. 3, ed. E. Stern, 805–11. Jerusalem: Israel Exploration Society.

1993b Qashish, Tel. In *NEAEHL*, vol. 4, ed. E. Stern, 1200–03. Jerusalem: Israel Exploration Society.

1997a Hazor. In *OEANE*, vol. 3, ed. E. M. Meyers, 1–5. New York: Oxford University Press.

1997b Yoqne'am. In *OEANE*, vol. 5, ed. E. M. Meyers, 381–83. New York: Oxford University Press.

2004 Hazor and Chronology. *ÄuL* 14:45–67.

Ben-Tor, Amnon, ed.

1992a *The Archaeology of Ancient Israel*. New Haven: Yale University Press.

Ben-Tor, Amnon, et al., eds.

1996 *Yoqne'am I, The Late Periods: Final Report of the Archaeological Excavations (1977–1988)*. Qedem Reports 3. Jerusalem: Institute of Archaeology, Hebrew University of Jerusalem.

2003 *Tel Qashish: A Village in the Jezreel Valley: Final Report of the Archaeological Excavations (1978–1987)*. Qedem Reports 5. Jerusalem: Israel Exploration Society.

Ben-Tor, Amnon, and Ruhama Bonfil

2003 The Stratigraphy and Pottery Assemblages of the Middle and Late Bronze Ages in Area A. In *Tel Qashish: A Village in the Jezreel Valley: Final Report of the Archaeological Excavations (1978–1987)*, eds. A. Ben-Tor et al., 185–276. Qedem Reports 5. Jerusalem: Israel Exploration Society.

Ben-Tor, Amnon, and Ruhama Bonfil, eds.

1997 *Hazor V: An Account of the Fifth Season of Excavations, 1968*. Jerusalem: Israel Exploration Society.

Ben-Tor, Amnon, and A. Zarzecki

1988–1989 Tel Yoqne'am 1988. *ESI* 7–8:195–96.

Bergman, C. A. et al.

1988 Experimental archery: projectile velocities and comparison of bow performances. *Antiquity* 62: 658–70.

Bertemes, Frenz

1986 Die „Mittelbronzezeitliche" Stadtmauer am Osthand des Tell Kamid el-Loz. In *Kamid el-Loz 1977–81*, 77–100. Saarbrücker Beiträge zur Altertumskunde 36.

Betts, Alison V. G., ed.

1998 *The Harra and the Hamad: Excavations and Explorations in Eastern Jordan, Volume 1*. Sheffield Archaeological Monographs 9. Sheffield: Collis.

Betts, Alison V. G., et al.

1996 Studies of the Bronze Age Occupation in the Wadi al-'Ajib, Southern Hauran. *Levant* 28:27–39.

Bienkowski, Piotr

1989 The Division of Middle Bronze IIB–C in Palestine. *Levant* 21:169–79.

Bietak, Manfred

1984 Problems of Middle Bronze Age Chronology: New Evidence From Egypt. *AJA* 88:471–85.

1996 *Avaris: The capital of the Hyksos, recent excavations at Tell el-Dab'a*. London: British Museum.

2001 Dab'a, Tell ed-. In *The Oxford Encyclopedia of Ancient Egypt*, vol. 1, ed. D. B. Redford, 351–54. New York: Oxford University Press.

2002 Relative and Absolute Chronology of the Middle Bronze Age: Comments on the Present State of Research. In *The Middle Bronze Age in the Levant: Proceedings of an International Conference on MB IIA Ceramic Material, Vienna, 24th–26th of January 2001*, ed. M. Bietak, 29–42. CCEM 3. Vienna: Österreichischen Akademie der Wissenschaften.

2003 *The Synchronisation of Civilisations in the Eastern Mediterranean in the Second Millennium B.C. II.* CCEM 4. Vienna: Österreichischen Akademie der Wissenschaften.

Bietak, Manfred, et al.
1994 Neue Grabungsergebnisse aus Tell el-Dab'a und 'Ezbet Helmi im östlichen Nildelta (1989–1991). *ÄuL* 4:9–80.

Bietak, Manfred, Josef Dorner; and Peter Jánosi
2001 Ausgrabungen in dem Palastbezirk von Avaris. Vorbericht Tell el-Dab'a/'Ezbet Helmi 1993–2000. *ÄuL* 11:27–119.

Biran, Avraham
1980 Two Discoveries at Tel Dan. *IEJ* 30:89–98.
1981 The Discovery of the Middle Bronze Age Gate at Dan. *BA* 44:139–44.
1984 The Triple-Arched Gate of Laish at Tel Dan. *IEJ* 34:1–19.
1990 The Middle Bronze Age Ramparts of Tel Dan. *EI* 21:56–65.
1993 Dan. In *NEAEHL*, vol. 1, ed. E. Stern, 323–32. Jerusalem: Israel Exploration Society.
1994 *Biblical Dan.* Jerusalem: Israel Exploration Society.
2000 Tel Dan—1996. *HA—ESI* 20:1*–2*.

Biran, Avraham, et al.
1996 *Dan I: A Chronicle of the Excavations, the Pottery Neolithic, the Early Bronze Age and the Middle Bronze Age Tombs.* Annual of the Nelson Glueck School of Biblical Archaeology. Jerusalem: Hebrew Union College-Jewish Institute of Religion.

Bluard, Christine
1997 Recherches sur le périmètre externe (chantier H). In *Tell Beydar, Three Seasons of Excavations (1992–1994)*, eds. M. Lebeau and A. Suleiman, 179–91. Subartu 3. Turnhout: Brepols.

Boas-Vedder, Ir. Diny
2001 Appendix 2. Jerusalem: The Wall from the MB II Period. In *Excavations by Kathleen M. Kenyon in Jerusalem 1961–1967, Volume III: The Settlement in the Bronze and Iron Ages*, ed. M. L. Steiner, 137–53. Copenhagen International Series 9. London: Sheffield.

Boehmer, Rainer Michael
1972 *Die Kleinfunde von Boghazköy.* WVDOG 87. Berlin: Mann Verlag.

Bonfil, Ruhama
2003 Characteristics of Settlement Plans in the Middle Bronze Age II. In *Tel Qashish: A Village in the Jezreel Valley: Final Report of the Archaeological Excavations (1978–1987)*, eds. A. Ben-Tor et al., 324–26. Qedem Reports 5. Jerusalem: Israel Exploration Society.

Borger, Rykle
1981 *Assyrische-babylonische Zeichenliste.* 2nd ed. AOAT 33A. Vluyn: Verlag Butzon und Bercker Kevelaer.

Bounni, Adnan
1997 Kazel, Tell al-. In *OEANE*, vol. 3, ed. E. M. Meyers, 275–76. New York: Oxford University Press.

Bourke, Stephen J.
1989 The Transition from the Middle to Late Bronze Age in Syria: The Evidence from Tell Nebi Mend. *Levant* 25:155–95.
1997 Pre-classical Pella in Jordan: A Conspectus of Ten Year's Work (1985–1995). *PEQ* 129:94–115.

Bourke, Stephen J., et al.
1998 Preliminary Report on the University of Sydney's Sixteenth and Seventeenth Seasons of Excavations at Pella (Tabaqat Fahl) in 1994/95. *ADAJ* 42:179–211.

Bradbury, Jim
1985 *The Medieval Archer.* New York: St. Martin's.

Braemer, Frank
1984 Prospections archéologiques dans le Hawran (Syrie). *Syria* 61:219–50.
1993 Prospections archéologiques dans le Hawran (Syrie) III. *Syria* 70:117–65.
2002 Le rempart de Bosra au IIe millénaire avant notre ère. *Syria* 79:65–74.

Braidwood, Robert J.
1937 *Mounds in the Plain of Antioch: An Archeological Survey.* OIP 48. Chicago: University of Chicago Press.

Braudel, Fernand
1972 *The Mediterranean and the Mediterranean World in the Age of Philip II.* New York: Harper & Row.

Bretschneider, Joachim
1997 Untersuchungen an der äußeren Wallanlage (Feld H). In *Tell Beydar, Three Seasons of Excavations (1992–1994)*, eds. M. Lebeau and A. Suleiman, 193–207. Subartu 3. Turnhout: Brepols.

Bretschneider, Joachim et al.
2004 Tell Tweini, Ancient Gabala, in the Bronze Age. *ÄuL* 14:215–30.

Bretschneider, Joachim, and Karel van Lerberghe
1997 Kranzhügel-Städte und Keilschrift-Archive. *AW* 28:289–96.

Broshi, Magen, and Ram Gophna
1986 Middle Bronze Age II Palestine: Its Settlement and Population. *BASOR* 261:73–90.

Brown, F., et al.
1906 *The Brown-Driver-Briggs Hebrew and English Lexicon.* Peabody, MA: Hendrickson.

Bryce, Trevor R.
1998 *The Kingdom of the Hittites.* 1st ed. Oxford: Clarendon.

Buccellati, Giorgio
1979 The Fourth Season: Introduction and the Stratigraphic Record. In *Terqa Preliminary Report 10.* BiMes 10. Malibu, CA: Undena.

1998 Urkesh as Tell Mozan: Profiles of the Ancient City. In *Urkesh and the Hurrians: Studies in Honor of Lloyd Cotsen. Urkesh/Mozan Studies 3*, eds. G. Buccellati and M. Kelly-Buccellati, 11–34. BiMes 26. Malibu, CA: Undena.

Buccellati, Giorgio, and Marilyn Kelly-Buccellati

1978 Terqa Preliminary Report No. 6: The Third Season—Introduction and the Stratigraphic Record. SMS 2.6:1–36.

1983 Terqa: The First Eight Seasons. *AAS* 33:47–67.

1988 *Mozan 1: The Soundings of the First Two Seasons*. BiMes 20. Malibu, CA: Undena.

1997a Mozan, Tell. In *OEANE*, vol. 4, ed. E. M. Meyers, 60–63. New York: Oxford University Press.

1997b Terqa. In *OEANE*, vol. 5, ed. E. M. Meyers, 188–90. New York: Oxford University Press.

1997c Urkesh: The First Hurrian Capital. *BA* 60:77–94.

Buchanan, Briggs

1966 *Catalogue of Ancient Near Eastern Seals in the Ashmolean Museum, Vol. 1: Cylinder Seals*. Oxford: Clarendon.

1981 *Early Near Eastern Seals in the Yale Babylonian Collection*. New Haven: Yale University Press.

Buhl, Marie-Louise, and Svend Holm-Nielson

1969 *Shiloh, the Danish Excavations at Tall Sailun, Palestine in 1926, 1929, 1932, and 1963: The Pre-Hellenistic Remains*. Archaeological-Historical Series I. Copenhagen: National Museum of Denmark.

Bullard, Reuben G.

1970 Geological Studies in Field Archaeology. *BA* 33:98–132.

Bunimovitz, Shlomo

1983 Glacis 10014 and Gezer's Late Bronze Age Fortifications. *TA* 10:61–70.

1989 The Land of Israel in the Late Bronze Age: A Case Study of Socio-Cultural Change in a Complex Society (Hebrew). Unpublished Ph.D. Dissertation, Tel Aviv University.

1992 Middle Bronze Age Fortifications in Palestine as a Social Phenomenon. *TA* 19:221–33.

1993 The Changing Shape of Power in the Bronze Age. In *Biblical Archaeology Today, 1990 Supplement: Proceedings of the Second International Congress on Biblical Archaeology*, eds. A. Biran and J. Aviram, 142–49. Jerusalem: Israel Exploration Society.

1995 On the Edge of Empires—Late Bronze Age (1500–1200 BCE). In *The Archaeology of Society in the Holy Land*, ed. T. E. Levy, 320–31. New York: Facts on File.

Bunimovitz, Shlomo, and Zvi Lederman

1993 Beth-Shemesh. In *NEAEHL*, vol. 1, ed. E. Stern, 249–53. Jerusalem: Israel Exploration Society.

2003 Tel Beth Shemesh, 2001–2003. *IEJ* 53:233–37.

Bunnens, Guy L., and Arlette Roobaert

1988 Le mur de enceinte (Area K). In *Mozan 1: The Soundings of the First Two Seasons*, eds. G. Buccellati and M. Kelly-Buccellati, 61–64. BiMes 20. Malibu, CA: Undena.

Burke, Aaron A.

2004 The Architecture of Defense: Fortified Settlements of the Levant during the Middle Bronze Age. Unpublished Ph.D. Dissertation. Department of Near Eastern Languages and Civilizations, University of Chicago.

2007 Magdalūma, Migdālîm, Magdoloi, and Majādīl: The Historical Geography and Archaeology of the Magdalu (Migdāl). *BASOR* 346:29–57.

forthcoming More Light on Old Reliefs: New Kingdom Egyptian Siege Tactics and Asiatic Resistance. In *Festschrift in Honor of Lawrence E. Stager*, ed. J. D. Schloen. Winona Lake, IN: Eisenbrauns.

Cahill, Jane M.

2006 The Excavations at Tell el-Hammah: A Prelude to Amihai Mazar's Beth-Shean Valley Regional Project. In *"I Will Speak the Riddles of Ancient Times": Archaeological and Historical Studies in Honor of Amihai Mazar on the Occasion of His Sixtieth Birthday*, vol. 2, eds. A. M. Maeir and P. de Miroschedji, 429–59. Winona Lake, IN: Eisenbrauns.

Cahill, Jane M., et al.

1988 Tell el-Hammah, 1988. *IEJ* 38:191–94.

Cahill, Jane M., and David Tarler

1993 Hammah, Tell el-. In *NEAEHL*, vol. 2, ed. E. Stern, 561–62. Jerusalem: Israel Exploration Society.

2000 Excavations Directed by Yigal Shiloh at the City of David, 1978–1985. In *Ancient Jerusalem Revealed*, ed. H. Geva, 31–45. Jerusalem: Israel Exploration Society.

Callaway, Joseph

1993 Ai. In *NEAEHL*, vol. 1, ed. E. Stern, 39–45. Jerusalem: Israel Exploration Society.

Callot, Olivier

1986 La région nord du Palais Royal d'Ougarit. In *CRAIBL*, 735–55. Paris.

Calvet, Yves

2002 Early Middle Bronze Age Sites in the Area of Salamiyah (Syria). In *The Middle Bronze Age in the Levant: Proceedings of an International Conference on MB IIA Ceramic Material, Vienna, 24th–26th of January 2001*, ed. M. Bietak, 103–12. CCEM 3. Vienna: Österreichischen Akademie der Wissenschaften.

2006 Le rempart d'Ougarit. In *Les espaces syro-mésopotamiens. Dimensions de l'expérience humaine au Proche-Orient ancien*, eds. P. Butterlin et al., 191–200. Subartu 17. Turnhout: Brepols.

Campbell, Edward F.

2002 *Shechem III, The Stratigraphy and Architecture of Shechem/Tell Balatah, Volume I: Text*. American Schools of Oriental Research Archaeological Reports. Atlanta: Scholars.

Carmi, Israel, and David Ussishkin
2004 ^{14}C Dates. In *The Renewed Archaeological Excavations at Lachish (1973–1994)*, vol. 1, ed. D. Ussishkin, 2508–13. Monograph Series 22. Tel Aviv: Emery and Claire Yass Publications in Archaeology.

Casana, Jesse
2003 From Alalakh to Antioch: Settlement, Land Use and Environmental Change in the Amuq Valley of Southern Turkey. Unpublished Ph.D. Dissertation. Department of Near Eastern Languages and Civilizations, University of Chicago.

Casana, Jesse, and Tony J. Wilkinson
2005a Appendix A: Gazetteer of Sites. In *The Amuq Valley Regional Projects, Volume 1: Surveys in the Plain of Antioch and Orontes Delta, Turkey, 1995–2002*, ed. K. A. Yener, 203–80. OIP 131. Chicago: Oriental Institute of the University of Chicago.
2005b Settlement and Landscapes in the Amuq Region. In *The Amuq Valley Regional Projects, Volume 1: Surveys in the Plain of Antioch and Orontes Delta, Turkey, 1995–2002*, ed. K. A. Yener, 25–65. OIP 131. Chicago: Oriental Institute of the University of Chicago.

Cecchini, S. M., and Stefania Mazzoni
1998 *Tell Afis (Siria): The 1988–1992 Excavations at the Acropolis*. Seminari di Orientalistica. Pisa: Gardini Editori e Stampatori.

Cerulli, Tiziano
2000 Appendix G: Analysis of Mudbrick Samples from Tell es-Sultan (Jericho). In *Excavations at Jericho, 1998: Preliminary Reports on the Second Season of Archaeological Excavations and Surveys at Tell es-Sultan, Palestine*, eds. N. Marchetti and L. Nigro, 349–54. Quaderni di Gerico 2. Rome: Università di Roma «La Sapienza».

Chadwick, Jeffrey R.
1992 The Archaeology of Biblical Hebron in the Bronze and Iron Ages: An Examination of the Discoveries of the American Expedition to Hebron. Unpublished Ph.D. Dissertation. Department of Languages and Literatures, University of Utah.
2005 Discovering Hebron: The City of the Patriarchs Slowly Yields Its Secrets. *BAR* 31:24–33, 70.

Chang, Kwang-Chih
1986 *The Archaeology of Ancient China*. 4th revised ed. New Haven: Yale University Press.

Charlton, Thomas H., and Deborah L. Nichols
1997 The City-State Concept. In *The Archaeology of City-States: Cross-Cultural Approaches*, eds. D. L. Nichols and T. H. Charlton, 1–14. Washington, DC: Smithsonian Institution.

Clark, Douglas R.
2000 Field B: The Western Defense System. In *Madaba Plains Project 4: The 1992 Season at Tall al-'Umayri and Subsequent Studies*, eds. L. G. Herr et al., 59–94. Berrien Springs, MI: Andrews University Press.
2002 Field B: The Western Defense System. In *Madaba Plains Project 5: The 1994 Season at Tall al-'Umayri and Subsequent Studies*, eds. L. G. Herr et al., 48–116. Berrien Springs, MI: Andrews University Press.

Clarke, Sommers
1916 Ancient Egyptian Frontier Fortresses. *JEA* 3:155–79.

Clausewitz, Carl von
1976 *On War*. Princeton: Princeton University Press.

Cohen, Susan L.
2002a *Canaanites, Chronology, and Connections: The Relationship of Middle Bronze IIA Canaan to Middle Kingdom Egypt*. SAHL 3. Winona Lake, IN: Eisenbrauns.
2002b Middle Bronze Age IIA Ceramic Typology and Settlement in the Southern Levant. In *The Middle Bronze Age in the Levant: Proceedings of an International Conference on MB IIA Ceramic Material, Vienna, 24th–26th of January 2001*, ed. M. Bietak, 113–31. CCEM 3. Vienna: Österreichischen Akademie der Wissenschaften.

Coleson, Joseph Edward
1982 Yaqqim-Addu of Sagaratum: The Correspondence of a District Governor in the Kingdom of Mari. Unpublished Ph.D. Dissertation. Department of Near Eastern and Judaic Studies, Brandeis University.

Collon, Dominique
1982 *The Alalakh Cylinder Seals: A New Catalogue of the Actual Seals Excavated by Sir Leonard Woolley at Tell Atchana, and from neighbouring sites on the Syrian-Turkish Border*. BAR International Series 132.

Cooley, Robert E., and Gary D. Pratico
1997 Dothan. In *OEANE*, vol. 2, ed. E. M. Meyers, 171. New York: Oxford University Press.

Cooper, Jerrold S., ed.
1986 *Sumerian and Akkadian Royal Inscriptions I: Presargonic Inscriptions*. Translation Series 1. New Haven: American Oriental Society.

Cotterell, B., and J. Kamminga
1990 *Mechanics of Pre-Industrial Technology*. New York: Cambridge University Press.

Courtois, Jacques-Claude
1973 Prospection archéologique dans la moyenne vallée de l'Oronte. *Syria* 50:53–99.

Covello-Paran, Karen
1999 Tel Hazor, the Eastern Spur. *HA—ESI* 107:1*–2*.

Cremaschi, Mauro, et al.
2002 The Geoarchaeology of Tell Mishrife. In *Excavating Qatna I: Preliminary Report on the 1999 and 2000 Campaigns of the Joint Syrian-Italian-German Archaeological Research Project at Tell Mishrife*, eds. M. al-Maqdissi et al., 17–24. Documents d'archaéologie syrienne 4. Damascus: Salhani.

Culican, W., and Thomas L. McClellan
1983 El-Qitar 1982–1983: A Preliminary Report. *AbrN* 22:29–63.

Cunliffe, Barry W.
1983 *Danebury: Anatomy of an Iron Age Hill Fort*. London: Batsford.

Curvers, Hans H., and Glenn M. Schwartz
1997 Umm el-Marra: A Bronze Age Urban Center in the Jabbul Plain, Western Syria. *AJA* 101:201–27.

Czichon, Rainer M., and Peter Werner
1998 *Tall Munbaqa-Ekalte I: Die Bronzezeitlichen Kleinfunde*. WVDOG 97. Saarbrücken: Saarbrücker Druckerei und Verlag.

Dalley, Stephanie
1984 *Mari and Karana: Two Old Babylonian Cities*. New York: Longman.

Dalokay, Y.
1969 Lehmflachdachbauten in Anatolien. Unpublished Ph.D. Dissertation, Fakultät für Bauwesen der Technischen Universität Carolo-Wilhelmina zu Brauschweig.

Damerji, Muayad Said Basim
1987 *The Development of the Architecture of Doors and Gates in Ancient Mesopotamia*. Tokyo: Kokushikan University Press.

Damluji, Salma Samar, ed.
1992 *The Valley of Mud Brick Architecture: Shibam, Tarim & Wadi Hadramut*. Reading, UK: Garnet.

Dayan (Nevo), Y.
1963 *Archaeological Survey of the Hulah Valley, Dan* (Hebrew).

Debruyne, Michel
2003 Research on the Outer Rampart (Area K). In *Tell Beydar, 1995–1999 Seasons of Excavations. A Preliminary Report*, eds. M. Lebeau and A. Suleiman, 387–94. Subartu 10. Turnhout: Brepols.

Delougaz, Pinhas
1990 Khafajah Mound B. In *Old Babylonian Public Buildings in the Diyala Region*, eds. H. Hill et al., 207–12. OIP 98. Chicago: University of Chicago Press.

Deshayes, Jean
1960 *Les outils de bronze de l'Indus au Danube*. BAH 71. Paris: Institut Français d'Archéologie du Proche-Orient.

Dever, William G.
1973a The Gezer Fortifications and the "High Place": An Illustràtion of Stratigraphic Methods and Problems. *PEQ* 105:61–70.
1973b Tower 5017 at Gezer—A Rejoinder. *IEJ* 23:23–26.
1974a *Gezer II: Report of the 1967–70 Seasons in Fields I and II*. Annual of the Nelson Glueck School of Biblical Archaeology, 2. Jerusalem: Hebrew Union College.
1974b The MB IIC Stratification in the Northwest Gate Area at Shechem. *BASOR* 216:31–51.
1976 The Beginning of the Middle Bronze Age in Syria-Palestine. In *Magnalia Dei, The Mighty Acts of God: Essays on the Bible and Archaeology in Memory of G. Ernest Wright*, eds. F. M. Cross et al., Jr., 3–38. Garden City, NY: Doubleday.
1977 The Patriarchal Traditions: Palestine in the Second Millennium BCE: The Archaeological Picture. In *Israelite and Judaean History*, eds. J. H. Hayes and J. M. Miller, 70–120. Philadelphia: Trinity Press International.
1980 New Vistas on the EB IV ("MB I") Horizon in Syria-Palestine. *BASOR* 237:35–64.
1985 Relations between Syria-Palestine and Egypt in the "Hyksos" Period. In *Palestine in the Bronze and Iron Ages: Papers in Honour of Olga Tufnell*, ed. J. N. Tubb. Occasional Publications 11. London: Institute of Archaeology.
1986a Abel Beth-Ma'acah: Northern Gateway of Ancient Israel. In *The Archaeology of Jordan and Other Studies: Presented to Siegfried H. Horn, Professor Emeritus of Archaeology and History of Antiquity, Andrews University, Berrien Springs, MI*, eds. L. T. Geraty and L. G. Herr, 207–22. Berrien Springs, MI: Andrews University Press.
1986b *Gezer IV: The 1969–1971 Seasons in Field VI, the "Acropolis"*. Annual of the Nelson Glueck School of Biblical Archaeology, 4. Jerusalem: Hebrew Union College.
1987 The Middle Bronze Age: The Zenith of the Urban Canaanite Era. *BA* 50:148–77.
1990a 'Hyksos,' Egyptian destructions, and the end of the Palestinian Middle Bronze Age. *Levant* 22: 75–81.
1990b Review of *Tell el-Fâr'ah II, 1; 2. Le Bronze Moyen stratigraphie des vestiges du Bronze Moyen II*. *Pal* 15:154–58.
1992a The Chronology of Syria-Palestine in the Second Millennium BCE: A Review of Current Issues. *BASOR* 288:1–26.
1992b Palestine, Archaeology of (Bronze–Iron Ages). In *ABD*, vol. 5, ed. D. N. Freedman, 109–14. New York: Doubleday.
1993a Gezer. In *NEAEHL*, vol. 2, ed. E. Stern, 496–506. Jerusalem: Israel Exploration Society.
1993b The Rise of Complexity in the Land of Israel in the Early Second Millennium B.C.E. In *Biblical Archaeology Today, 1990 Supplement: Proceedings of the Second International Congress on Biblical Archaeology*, eds. A. Biran and J. Aviram, 98–109. Jerusalem: Israel Exploration Society.
1995 Social Structure in the Early Bronze IV Period in Palestine. In *The Archaeology of Society in the Holy Land*, ed. T. E. Levy, 282–96. New York: Facts on File.
1997a Akko. In *OEANE*, vol. 1, ed. E. M. Meyers, 54–55. New York: Oxford University Press.
1997b Gezer. In *OEANE*, vol. 2, ed. E. M. Meyers, 396–400. New York: Oxford University Press.

Dever, William G., et al.
1970 *Gezer I: Preliminary Report of the 1964–66 Seasons*. Nelson Glueck School of Biblical Archaeology 1. Jerusalem: Hebrew Union College.

Dohmann-Pfälzner, Heike, and Peter Pfälzner
2002 Ausgrabungen der Deutschen Orient-Gesellschaft in der zentralen Oberstadt von Tall Mozan/Urkeš: Bericht über die in Kooperation mit dem IIMAS durgeführte Kampagne 2001. *MDOG* 134:149–92.

Dornemann, Rudolph H.
1978 Tell Hadidi: Bronze Age City on the Euphrates. *Archaeology* 31:20–26.
1979a Tell Hadidi: A Millennium of Bronze Age City Occupation—Euphrates Valley, Syria. In *Archaeological Reports from the Tabqa Dam Project*, ed. D. N. Freedman, 113–51. AASOR 44. Cambridge, MA: American Schools of Oriental Research.
1979b Tell Hadidi: An Important Center of the Mitannian Period and Earlier. In *Le moyen Euphrate: actes du colloque de Strasbourg 10–12 mars 1977*, ed. J.-C. Margueron, 217–34. Leiden: Brill.
1983 *The Archaeology of Transjordan in the Bronze and Iron Ages*. Milwaukee: Milwaukee Public Museum.
1985 Salvage Excavations at Tell Hadidi in the Euphrates Valley. *BA* 48:49–59.

Dorsey, David A.
1991 *The Roads and Highways of Ancient Israel*. Baltimore: Johns Hopkins University.

Dossin, Georges
1938 Signaux lumineux au pays de Mari. *RA* 35:174–86.
1972 *Adaššum* et *kirḫum* dans les textes de Mari. *RA* 66:111–30.

Dothan, Moshe
1973 The Foundation of Tel Mor and Ashdod. *IEJ* 23: 1–17.
1976 Akko: Interim Excavation Report First Season, 1973/4. *BASOR* 224:1–48.
1982 *Ashdod IV: Excavation of Area M: Fortifications of the Lower City*. ʿAtiqot 15. Jerusalem: Israel Exploration Society.
1993a Acco (Tel Acco). In *NEAEHL*, vol. 1, ed. E. Stern, 16–24. Jerusalem: Israel Exploration Society.
1993b *Ashdod V Excavation of Area G: The Fourth through Sixth Seasons of Excavations, 1968–1970*. ʿAtiqot 23.
1993c Esur, Tel. In *NEAEHL*, vol. 1, ed. E. Stern, 426–28. Jerusalem: Israel Exploration Society.

Dothan, Moshe, and Avner Raban
1980 The Sea Gate of Ancient Akko. *BA* 43:35–39.
1984 Tel Akko. *IEJ* 34:189–91.

Dothan, Trude, and Seymour Gitin
1993 Miqne, Tel (Ekron). In *NEAEHL*, vol. 3, ed. E. Stern, 1051–59. Jerusalem: Israel Exploration Society.
1997 Miqne, Tel. In *OEANE*, vol. 4, ed. E. M. Meyers, 30–35. New York: Oxford University Press.

Drews, Robert
1988 *The Coming of the Greeks: Indo-European Conquests in the Aegean and the Near East*. Princeton: Princeton University Press.

Dunand, Maurice
1939 *Les fouilles de Byblos I: 1926–1932*. Paris: Paul Geuthner.
1949–1950 Rapport préliminaire sur les fouilles de Byblos en 1948. *BMB* 9:53–64.
1954 *Les fouilles de Byblos II: 1933–1938*. Paris: Paul Geuthner.
1955a Rapport préliminaire sur les fouilles de Byblos en 1950. *BMB* 12:7–12.
1955b Rapport préliminaire sur les fouilles de Byblos en 1951. *BMB* 12:13–20.
1955c Rapport préliminaire sur les fouilles de Byblos en 1952. *BMB* 12:21–23.
1956a Rapport préliminaire sur les fouilles de Byblos en 1954. *BMB* 13:73–78.
1956b Rapport préliminaire sur les fouilles de Byblos en 1955. *BMB* 13:79–86.
1961 Rapport préliminaire sur les fouilles de Byblos en 1959. *BMB* 16:81–85.
1964 Rapport préliminaire sur les fouilles de Byblos en 1961. *BMB* 17:25–28.
1967 Rapport préliminaire sur les fouilles de Byblos en 1965. *BMB* 20:21–26.
1968 *Byblos: Its History, Ruins and Legends*. 2nd ed. Beirut (orig. French 1963 ed.).

Dunayevsky, Immanuel, and Aharon Kempinski
1990 The Eastern Rampart of Hazor. *ʿAtiqot* 10:23–28.

Dunham, Dows, ed.
1967 *Second Cataract Forts II: Uronarti Shalfak Mirgissa*. Boston: Museum of Fine Arts.

Dunham, Dows, and Jozef M. A. Janssen, eds.
1960 *Second Cataract Forts I: Semna Kumma*. Boston: Museum of Fine Arts.

Eames, Samantha, and Maria Schroder
1997 A Middle Bronze Age Landscape: Occupation and Resource Management in Wadi el-ʿAjib, al-Mafraq. In SHAJ 6, 149–58. Amman: Department of Antiquities.

Edelstein, Gershon
1993 Rephaim, Nahal (The Manahat Excavations). In *NEAEHL*, vol. 4, ed. E. Stern, 1281–82. Jerusalem: Israel Exploration Society.

Edelstein, Gershon, and Zvi Greenhut
1989 Manahat—1987/1988. *ESI* 7/8:117–23.

Edelstein, Gershon, and Yanir Milevski
1990 Manahat—1989. *ESI* 9:148–50.

Edzard, Dietz O.
1981 Neue Erwägungen zum Brief des Enna-Dagan von Mari (TM.75.G.2367). *StEb* 4:89–97.

Eichler, Seyarre, et al.
1984 Ausgrabungen in Tall Munbaqa 1983. *MDOG* 116:65–93.

Einwag, Berthold
2002 The Early Middle Bronze Age in the Euphrates Valley: The Evidence from Tuttul/Tell Biʿa. In *The Middle Bronze Age in the Levant: Proceedings of an International Conference on MB IIA Ceramic Material, Vienna, 24th–26th of January 2001*, ed. M. Bietak, 143–61. CCEM 3. Vienna: Österreichischen Akademie der Wissenschaften.

Eisenberg, Emanuel
1993 Kitan, Tel. In *NEAEHL*, vol. 3, ed. E. Stern, 878–81. Jerusalem: Israel Exploration Society.

Eisenberg, Emanuel, and Gershon Edelstein
1993 Rephaim, Nahal. In *NEAEHL*, vol. 4, ed. E. Stern, 1277–80. Jerusalem: Israel Exploration Society.

Eitan, Abraham
1972 Tell Beit Mirsim G–F—The Middle Bronze IIA Settlement. *BASOR* 208:19–24.

Emery, Walter B., ed.
1965 *Egypt in Nubia*. London: Hutchinson.

Emery, Walter B., et al.
1979 *The Fortress of Buhen: the Archaeological Report*. London: Egypt Exploration Society.

Engberg, Robert M.
1939 *The Hyksos Reconsidered*. SAOC 18. Chicago: University of Chicago Press.

Eph'al, Israel
1996 *Siege and Its Ancient Near Eastern Manifestations*. Jerusalem: Magnes (Hebrew).

Erkanal, Hayat
1977 *Die Äxte und Beile des 2. Jahrtausends in Zentralanatolien*. Prähistorische Bronzefunde, pt. IX, vol. 8. Munich: Franz Steiner.

Esse, Douglas L.
1991 *Subsistence, Trade, and Social Change in Early Bronze Age Palestine*. SAOC 50. Chicago: University of Chicago Press.

Falconer, Steven E.
1987 Heartland of Villages: Reconsidering Early Urbanism in the Southern Levant. Unpublished Ph.D. Dissertation, University of Arizona.
1994a The Development and Decline of Bronze Age Civilisation in the Southern Levant: A Reassessment of Urbanism and Ruralism. In *Development and Decline in the Mediterranean Bronze Age*, eds. C. Mathers and S. Stoddart, 305–33. Sheffield Archaeological Monographs, 8. Sheffield: J. R. Collins.
1994b Village Economy and Society in the Jordan Valley: A Study of Bronze Age Rural Complexity. In *Archaeological Views from the Countryside: Village Communities in Early Complex Societies*, eds. G. M. Schwartz and S. E. Falconer, 121–42. Washington, DC: Smithsonian Institution.
2001 The Middle Bronze Age. In *The Archaeology of Jordan*, eds. B. MacDonald et al., 271–89. Sheffield: Sheffield Academic Press.

Falconer, Steven E., et al.
2003 Bronze Age Rural Economic Transitions in the Jordan Valley. In *The Near East in the Southwest: Essays in Honor of William G. Dever*, ed. B. A. Nakhai, 1–17. AASOR 58. Boston: American Schools of Oriental Research.

Falconer, Steven E., and Bonnie Magness-Gardiner
1993 Hayyat, Tell el-. In *NEAEHL*, vol. 2, ed. E. Stern, 591–93. Jerusalem: Israel Exploration Society.

Faris, Nabih Amin, and Robert Potter Elmer, eds.
1945 *Arab Archery: An Arabic Manuscript of About A.D. 1500 "A Book on the Excellence of the Bow & Arrow" And the Description Thereof*. Princeton: Princeton University Press.

Fathy, Hassan
1969 *Gourna: A Tale of Two Villages*. Cairo: Ministry of Culture.

Figulla, Hugo Heinrich
1967 *Old Babylonian Naditu Records*. Cuneiform Texts from Babylonian Tablets in the British Museum. London: British Museum.

Finkbeiner, Uwe
1997 Fortifications of Prehellenistic Beirut; The Archaeological Site BEY 020. *National Museum News*: 7–9.
2001 Emar 1999: Bericht über die 3. Kampagne der syrisch-deutschen Ausgrabungen. *BM* 32:41–120.

Finkbeiner, Uwe, and Thomas Leisten
1999–2000 Emar and Balis 1996 and 1998: A Preliminary Report of the Joint Syrian-German Excavations with the Collaboration of Princeton University. *Berytus* 44:5–57.

Finkbeiner, Uwe, and Hélène Sader
1997 BEY 020: Preliminary Report of the Excavations 1995. *BAAL* 2:114–66.

Finkbeiner, Uwe, and Ferhan Sakal
2003 Emar 2002: Bericht über die 5. Kampagne der syrisch-deutschen Ausgrabungen. *BM* 34:10–117.

Finkelstein, Israel
1988 *The Archaeology of the Israelite Settlement*. Jerusalem: Israel Exploration Society.
1992 Middle Bronze Age 'Fortifications': A Reflection of Social Organization and Political Formations. *TA* 19:201–20.
1993a Samaria (Region): The Southern Samarian Hills Survey. In *NEAEHL*, vol. 4, ed. E. Stern, 1313–14. Jerusalem: Israel Exploration Society.
1993b The Sociopolitical Organization of the Central Hill Country in the Second Millennium B.C.E. In *Biblical Archaeology Today, 1990 Supplement: Proceedings of the Second International Congress on Biblical Archaeology*, eds. A. Biran and J. Aviram, 292–306. Jerusalem: Israel Exploration Society.
1996a The Philistine Countryside. *IEJ* 46:225–42.
1996b The Territorial-Political System of Canaan in the Late Bronze Age. *UF* 28:221–55.
2002 Gezer Revisited and Revised. *TA* 29:262–82.

Finkelstein, Israel, and Zvi Lederman
1993a Areas H-F: Middle Bronze III Fortifications and Storerooms. In *Shiloh: The Archaeology of a Biblical Site*, eds. I. Finkelstein et al., 49–64. Monograph Series 10. Tel Aviv: Emery and Claire Yass Archaeology.
1993b Excavation Results in Areas E, G, J, K, L and M. In *Shiloh: The Archaeology of a Biblical Site*, eds. I. Finkelstein et al., 65–78: Monograph Series 10. Tel Aviv: Emery and Claire Yass Archaeology.

Finkelstein, Israel, et al.
1993 *Shiloh: The Archaeology of a Biblical Site.* Monograph Series 10. Tel Aviv: Emery and Claire Yass Archaeology.

Finkelstein, Israel, et al., eds.
2000 *Megiddo III: The 1992–1996 Seasons.* The Emery and Claire Yass Publications in Archaeology. Jerusalem: Graphit.

Fischer, Peter M.
1997 Tall Abu al-Kharaz: The Swedish Jordan Expedition 1995–1996, Sixth and Seventh Season Preliminary Excavation Report. *ADAJ* 41:129–44.
1998 Tall Abu al-Kharaz: The Swedish Jordan Expedition 1997, Eighth Season Preliminary Excavation Report. *ADAJ* 42:213–23.
2000 A Synthesis of Ten Campaigns at Tell Abu al-Kharaz, Jordan Valley: The Early, Middle, Late Bronze and Iron Ages. In *Proceedings of the First International Congress on the Archaeology of the Ancient Near East, Rome May 18th–23rd 1998,* vol. 1, eds. P. Matthiae et al., 447–70. Rome: Università degli Studi di Roma "La Sapienza".

Fischer, Peter M., and Moain Sadeq
2000 Tell el-'Ajjul 1999: A Joint Palestinian-Swedish Field Project, First Season Preliminary Report. *ÄuL* 10:211–26.

Fisher, Clarence S.
1905 *Excavations at Nippur, Part I: The Topography and City Walls of Nippur.* New York: G. E. Stechert & Co.

Flanagan, James W., et al.
1994 Tell Nimrin: Preliminary Report on the 1993 Season. *ADAJ* 38:205–44.

Fleming, Daniel E.
2004 *Democracy's Ancient Ancestors: Mari and Early Collective Governance.* New York: Cambridge University Press.

Fortin, Michel
1981 Military Architecture in Cyprus during the Second Millennium B.C. Unpublished Ph.D. Dissertation. Department of Bedford College, University of London.
1983 Recherches sur l'architecture militaire de l'âge du bronze à Chypre. *Échos du monde classique (Classical Views)* 27:206–19.
1989 La soi-disant forteresse d'Enkomi I (Chypre) à la fin du Bronze Moyen et au début du Bronze Récent. In *Transition: le monde Égéen du Bronze Moyen au Bronze Récent, Actes de la deuxième Rencontre égéenne internationale de l'Université de Liège,* ed. R. Laffineur, 239–49. Aegaeum 3. Liège: Université de l'État à Liège.
1995 A Cluster of Middle Cypriot III-Late Cypriot I Fortified Sites on the Ayios Sozomenos Plateau. In *Klados: Essays in Honour of J. N. Coldstream,* ed. C. Morris, 87–102. London: University of London Press.
2006a À la recherche de l'ancienne Tunip à Tell 'Acharneh, dans la vallée de l'Oronte (Syrie). In *Les espaces syro-mésopotamiens: dimensions de l'expérience humaine au Proche-Orient ancien,* eds. P. Butterlin et al., 213–21. Subartu 17. Turnhout: Brepols.
2006b Tell 'Acharneh: localisation, description et recherches antérieures. In *Tell 'Acharneh 1998–2004: rapports préliminaires sur les campagnes de fouilles et saison d'études,* ed. M. Fortin, 3–23. Subartu 18. Turnhout: Brepols.
2006c Tell 'Acharneh: rapport préliminaire sur les campagne de fouilles de 2002. In *Tell 'Acharneh 1998–2004: rapports préliminaires sur les campagnes de fouilles et saison d'études,* ed. M. Fortin, 120–39. Subartu 18. Turnhout: Brepols.

Foster, Benjamin R.
1982 The Siege of Armanum. *JANES* 14:27–36.

Franken, Hendricus J.
1961 The Excavations at Deir 'Allā in Jordan, Second Season. *VT* 11:361–72.

Franken, Hendricus J., and Moawiyah M. Ibrahim
1978 Two Seasons of Excavations at Tell Deir 'Allā, 1976–1978. *ADAJ* 22:57–80.

Frayne, Douglas
1990 *Old Babylonian Period (2003–1595 BC).* The Royal Inscriptions of Mesopotamia: Early Periods, 4. Toronto: University of Toronto Press.
1993 *Sargonic and Gutian Periods (2334–2113 BC).* The Royal Inscriptions of Mesopotamia: Early Periods, 2. Toronto: University of Toronto Press.
1997 *Ur III Period (2112–2004 BC).* The Royal Inscriptions of Mesopotamia: Early Periods 3/2. Toronto: University of Toronto Press.
2006 Tell 'Acharneh and the Orontes Valley: Early Textual Sources. In *Tell 'Acharneh 1998–2004: rapports préliminaires sur les campagnes de fouilles et saison d'études,* ed. M. Fortin, 25–42. Subartu 18. Turnhout: Brepols.

Free, Joseph P.
1958 The Fifth Season at Dothan. *BASOR* 152:10–18.

Fritz, Volkmar
1995 *The City in Ancient Israel.* Sheffield: Sheffield.
2000 Bronze Age Features in Cities of the Early Iron Age in the Southern Levant. In *Proceedings of the First International Congress on the Archaeology of the Ancient Near East, Rome May 18th–23rd 1998,* vol. 1, eds. P. Matthiae et al., 507–11. Rome: Università degli Studi di Roma "La Sapienza".

Fritz, Volkmar, and Aharon Kempinski
1983 *Ergebnisse der Ausgrabungen auf Hirbet el-Mšaš (Tel Masos) 1972–1975.* Abhandlung des Deutschen Palästinavereins. Wiesbaden: Harrassowitz.

Fugmann, Ejnar
1958 *Hama, fouilles et recherches 1931–1938, vol. II/1, L'architecture des périodes pré-hellénistiques.* Copenhagen: Foundation Carlsberg.

Funk, Robert W.
1968 The History of Beth-Zur with Reference to its Defenses. In *The 1957 Excavations at Beth-Zur,* eds. O. R. Sellers, et al., 4–17. AASOR 38. New Haven: American Schools of Oriental Research.

1993 Beth-Zur. In *NEAEHL*, vol. 1, ed. E. Stern, 259–61. Jerusalem: Israel Exploration Society.

Gaballa, G. A.
1976 *Narrative in Egyptian art*. Deutsches Archäologisches Institut, Abteilung Kairo. Mainz am Rhein: Zabern.

Gal, Zvi
2000 A Sounding at Tel Par. *'Atiqot* 39:83–103.

Gal, Zvi, and Moshe Kochavi
2000 Area B Stratigraphy, Architecture and Tombs. In *Aphek-Antipatris I: Excavation of Areas A and B, The 1972–1976 Seasons*, eds. M. Kochavi, et al., 59–92. Tel Aviv: Emery and Claire Yass Publications in Archaeology.

Garfinkel, Yosef
2001 Warrior Burial Customs in the Levant during the Early Second Millennium. In *Studies in the Archaeology of Israel and Neighboring Lands in Memory of Douglas L. Esse*, ed. S. R. Wolff, 143–61. SAOC 59. Winona Lake, IN: Eisenbrauns.

Garstang, John
1930 Jericho. Sir Charles Marston's expedition of 1930. *PEFQS*:123–32.
1931 The walls of Jericho: The Marston-Melchett Expedition of 1931. *PEFQS*:186–96.

Garstang, John, and J. B. E. Garstang
1940 *The Story of Jericho*. London: Hodder and Stoughton.

Gasche, Hermann, et al.
1998 *Dating the Fall of Babylon: A Reappraisal of Second-Millennium Chronology*. Mesopotamian History and Environment II:4. Ghent: University of Ghent/Oriental Institute of the University of Chicago.

Gasche, Hermann, and Roland Paepe
1980 The Surrounding Wall of Abu Habbah (Sippar) in Relation to that of Tell ed-Der and to the Regional Fluviatile System. In *Tell ed-Der III: Sounding at Abu Habbah (Sippar)*, ed. L. de Meyer, 37–52. Leuven: Peeters.

Gershuny, Lilly
2002 A Middle Bronze Age Cemetery in Ashqelon. In *The Middle Bronze Age in the Levant: Proceedings of an International Conference on MB IIA Ceramic Material, Vienna, 24th–26th of January 2001*, ed. M. Bietak, 185–88. CCEM 3. Vienna: Österreichischen Akademie der Wissenschaften.

Geva, Shulamit
1982 *Tell Jerishe: The Sukenik Excavations of the Middle Bronze Age Fortifications*. Qedem 15. Jerusalem: Institute of Archaeology, Hebrew University of Jerusalem.

Gianessi, Deborah
1998 Architecture and Stratigraphy. In *Tell Afis (Siria): The 1988–1992 Excavations at the Acropolis*, vol. 2, eds. S. M. Cecchini and S. Mazzoni, 101–21. Seminari di Orientalistica. Pisa: Gardini Editori e Stampatori.

Gibson, McGuire
1964 The Mace, the Axe, and the Dagger in Ancient Mesopotamia. Unpublished M.A. Thesis. Department of Oriental Languages and Civilizations, University of Chicago.

Gibson, McGuire, et al.
1983 The Southern Corner of Nippur: Excavations during the 14th and 15th Seasons. *Sumer* 39:170–90.
2002 First Season of Syrian-American Investigations at Hamoukar, Hasekeh Province. *Iraq* 64:45–68.

Gittlen, Barry Melvin
1992 The Late Bronze Age "City" at Tel Miqne/Ekron. *EI* 23:50*–53*.

Glock, Albert Ernest
1968 Warfare in Mari and Early Israel. Unpublished Ph.D. Dissertation, University of Michigan.
1993 Taanach. In *NEAEHL*, vol. 4, ed. E. Stern, 1428–33. Jerusalem: Israel Exploration Society.

Glock, Alice, ed.
1988 *Miniscule Monuments of Ancient Art: Catalogue of Near Eastern Stamp and Cylinder Seals Collected by Virginia E. Bailey*. Madison, NJ: New Jersey Museum of Archaeology.

Goetze, Albrecht
1962 Two Ur-Dynasty Tablets Dealing with Labor. *JCS* 16:13–16.

Goldberg, Paul
2004 Section F: The Micromorphology of Late Bronze Age Mudbrick. In *The Renewed Archaeological Excavations at Lachish (1973–1994)*, vol. 5, ed. D. Ussishkin, 2593–603. Monograph Series 22. Tel Aviv: Emery and Claire Yass Publications in Archaeology.

Gonen, Rivka
1975 *Weapons of the Ancient World*. Cassell's Introducing Archaeology Series. Jerusalem: Peli.
1984 Urban Canaan in the Late Bronze Period. *BASOR* 253:61–74.
1987 Megiddo in the Late Bronze Age: Another Reassessment. *Levant* 19:83–100.
1992 The Late Bronze Age. In *The Archaeology of Ancient Israel*, ed. A. Ben-Tor, 211–57. Translated by R. Greenberg. New Haven: Yale University Press.

Gophna, Ram
1973 The Middle Bronze Age II Fortifications at Tel Poleg. *EI* 11:111–19.
1977 Fortified Settlements from the Early Bronze and Middle Bronze II at Tel Poran. *EI* 13:87–90, 293* (Hebrew).
1979 A Middle Bronze Age II Village in the Jordan Valley. *TA* 6:28–33.
1992a Early Bronze Age Fortification Wall and Middle Bronze Age Rampart at Tel Poran. *TA* 19:267–73.
1992b The Intermediate Bronze Age. In *The Archaeology of Ancient Israel*, ed. A. Ben-Tor, 126–58. New Haven: Yale University Press.
1993a Far'ah, Tell el- (South). In *NEAEHL*, vol. 2, ed. E. Stern, 441–44. Jerusalem: Israel Exploration Society.

1993b Poleg, Nahal. In *NEAEHL*, vol. 4, ed. E. Stern, 1193–95. Jerusalem: Israel Exploration Society.

Gophna, Ram, and Etan Ayalon
1980 Survey of the Central Coastal Plain, 1978–1979: Settlement Pattern of the Middle Bronze Age IIA. *TA* 7:147–51.
1982 A Fortified Middle Bronze Age IIA Site at ʻAin Zurekiyeh in the Poleg Basin. TA 9:69–78.

Gophna, Ram, and Pirhiya Beck
1981 The Rural Aspect of the Settlement Pattern of the Coastal Plain in the Middle Bronze Age II. *TA* 8:45–80.

Gophna, Ram, and Juval Portugali
1988 Settlement and Demographic Processes in Israel's Coastal Plain from the Chalcolithic to the Middle Bronze Age. *BASOR* 269:11–28.

Gore, Rick
2001 Ancient Ashkelon. *NG* 199:66–93.

Goren, Yuval et al.
2004 *Inscribed in Clay: Provenance Study of the Amarna Tablets and Other Ancient Near Eastern Texts*. Tel Aviv: Emery and Claire Yass Publications in Archaeology.

Grant, Elihu
1931 *Ain Shems Excavations (Palestine) 1928–1931: Part I*. Haverford, PA: Haverford College.
1934 *Rumeileh being Ain Shems Excavations (Palestine): Part III*. Haverford, PA: Haverford College.

Grant, Elihu, and George Ernest Wright
1938 *Ain Shems Excavations (Palestine): Part IV (Pottery)*. Haverford, PA: Haverford College.
1939 *Ain Shems Excavations (Palestine): Part V (Text)*. Biblical and Kindred Studies. Haverford, PA: Haverford College.

Grayson, A. Kirk
1987 *Assyrian Rulers of the Third and Second Millennia BC (to 1115 BC)*. The Royal Inscriptions of Mesopotamia, Assyrian Periods. Toronto: University of Toronto Press.

Greene, Joseph A., and Khairieh ʻAmr
1992 Deep Sounding on the Lower Terrace of the Amman Citadel: Final Report. *ADAJ* 36:113–44.

Gregori, Barbara
1986 "Three-Entrance" City-Gates of the Middle Bronze Age in Syria and Palestine. *Levant* 18:83–102.

Guardata, F. Baffi
1990 Tell Touqan, une ville paléosyrienne. *AAS* 40:64–75.
1994 Tell Tuqan (Siria): campagna di scavi 1993. La porta urbica di Sud-Est. *OE* 1994:14–15.
2000 Les tombes du Bronze Moyen dans le secteur des fortifications à Ebla. In *Proceedings of the First International Congress on the Archaeology of the Ancient Near East, Rome May 18th–23rd 1998*, vol. 2, eds. P. Matthiae, et al., 55–78. Rome: Università degli Studi di Roma "La Sapienza".

Guy, Philip L. O.
1931 *New Light from Armageddon: Second Provisional Report (1927–29) on the Excavations at Megiddo in Palestine*. Oriental Institute Communications 9. Chicago: University of Chicago Press.

Hachmann, Rolf
1991 Kamid el-Loz 1963–1981: German Excavations in Lebanon, Part I. *Berytus* 37.

Haines, Richard C.
1971 *Excavations in the Plain of Antioch II*. OIP 95. Chicago: University of Chicago Press.

Hall, Gerald, et al.
1973 Architectural Study (Asvan 1968–1972 Interim Report). *AnSt* 23:245–69.

Hall, Harry Reginald
1923 The Hittites and the Egyptians. In *Anatolian Studies Presented to Sir William Mitchell Ramsay*, 165–85. London: Manchester University Press.

Hallo, William W., ed.
1997–2002
The Context of Scripture. 3 vols. Leiden: Brill.

Hallote, Rachel S.
1994 Mortuary Practices and Their Implications for Social Organization in the Middle Bronze Southern Levant. Unpublished Ph.D. Dissertation. Department of Near Eastern Languages and Civilizations, University of Chicago.
2002 Real and Ideal Identities in Middle Bronze Age Tombs. *NEA* 65:105–11.

Hammond, Philip C.
1968 Hebron. *RB* 75:253–58.

Hardin, James W., and Joe D. Seger
2006 Gezer Rectified: The Dating of the South Gate Complex. In *Confronting the Past: Archaeological and Historical Essays on Ancient Israel in Honor of William G. Dever*, eds. S. Gitin et al., 51–60. Winona Lake, IN: Eisenbrauns.

Harif, Amos
1978 Middle Kingdom Architectural Elements in Middle Bronze Age Megiddo. *ZDPV* 94:24–31.

Harvey, Stephen P.
2001 Tribute to a Conquering King. *Archaeology* 54: 52–55.

Hawkins, John David
1997 Carchemish. In *OEANE*, vol. 1, ed. E. M. Meyers, 423–24. New York: Oxford University Press.

Heltzer, Michael
1982 *The Internal Organization of the Kingdom of Ugarit*. Wiesbaden: Reichert Verlag.

Herbordt, Suzanne, et al.
1982 Ausgrabungen in Tall Biʻa 1981. *MDOG* 114:79–101.

Herr, Larry G., ed.
2000 *Madaba Plains Project 4: The 1992 Season at Tall al-ʻUmayri and Subsequent Studies*. Madaba Plains Project. Berrien Springs, MI: Andrews University Press.

Herzog, Ze'ev
1984a Tel Gerisa—1981/1982. *ESI* 1 (1982):28–31.

1984b Tel Gerisa, 1983. *IEJ* 34:55–56.

1986 *Das Stadttor in Israel und in der Nachbarländern*. Mainz: Philipp von Zabern.

1989 Middle and Late Bronze Age Settlements (Strata XVII–XV). In *Excavations at Tel Michal, Israel*, eds. Z. Herzog et al., 29–42. Publications of the Institute of Archaeology 8. Minneapolis: University of Minnesota Press.

1990 Tel Gerisa—1986. *ESI* 7–8 (1988/89):60–62.

1991 Tel Gerisa—1988. *ESI* 9 (1989/1990):51–52.

1992 Cities. In *ABD*, vol. 1, ed. D. N. Freedman, 1031–43. New York: Doubleday.

1993a Gerisa, Tel. In *NEAEHL*, vol. 2, ed. E. Stern, 480–84. Jerusalem: Israel Exploration Society.

1993b Michal, Tel. In *NEAEHL*, vol. 3, ed. E. Stern, 1036–41. Jerusalem: Israel Exploration Society.

1997a *Archaeology of the City: Urban Planning in Ancient Israel and Its Social Implications*. Monograph Series 13. Jerusalem: Emery and Claire Yass Archaeology.

1997b Gerisa, Tel. In *OEANE*, vol. 2, ed. E. M. Meyers, 394–96. New York: Oxford University Press.

Herzog, Ze'ev, et al.

1989 *Excavations at Tel Michal, Israel*. Publications of the Institute of Archaeology, 8. Minneapolis: University of Minnesota Press.

Heusch, Jan-Christoph

1979 Tall Habuba Kabira im 3. und 2. Jahrtausend—die Entwicklung der Baustuktur. In *Le moyen Euphrate: actes du colloque de Strasbourg 10–12 mars 1977*, ed. J.-C. Margueron, 159–99. Leiden: Brill.

Holland, Thomas A.

1976 Preliminary Report on Excavations at Tell es-Sweyhat, Syria, 1973–4. *Levant* 8:36–70.

1977 Preliminary Report on Excavations at Tell es-Sweyhat, Syria, 1975. *Levant* 9:36–65.

Hölscher, Uvo

1941 *The Excavation of Medinet Habu III: The Mortuary Temple of Ramses III, Part 1*. OIP 54. Chicago: University of Chicago Press.

1951 *The Excavation of Medinet Habu IV: The Mortuary Temple of Ramses III, Part 2*. OIP 55. Chicago: University of Chicago Press.

Horsnell, Malcolm J. A., ed.

1999 *The Year-Names of the First Dynasty of Babylon II, The Year-Names Reconstructed and Critically Annotated in the Light of Their Exemplars*. Hamilton, Ontario: McMaster University Press.

Hrouda, Barthel

1997 Isin. In *OEANE*, vol. 3, ed. E. M. Meyers, 186–87. New York: Oxford University Press.

Huehnergard, John

1987 *Ugaritic Vocabulary in Syllabic Transcription*. Harvard Semitic Studies 32. Atlanta: Scholars.

Huehnergard, John, and Wilfred H. van Soldt

1999 A Cuneiform Lexical Text from Ashkelon with a Canaanite Column. *IEJ* 49:184–92.

Humbert, Jean-Baptiste

1993 Keisan, Tell. In *NEAEHL*, vol. 3, ed. E. Stern, 862–67. Jerusalem: Israel Exploration Society.

Ibrahim, Moawiyah M.

1989 Sahab. In *Archaeology of Jordan II. 2, Field Reports: Surveys and Sites L–Z*, eds. D. Homès-Fredericq and J. B. Hennessy, 516–20. Akkadica Supplementum 7. Leuven: Peeters.

1997 Sahab. In *OEANE*, vol. 4, ed. E. M. Meyers, 450–52. New York: Oxford University Press.

Ilan, David

1995 The Dawn of Internationalism—The Middle Bronze Age. In *The Archaeology of Society in the Holy Land*, ed. T. E. Levy, 297–319. New York: Facts on File.

Ilan, David, et al.

2000 Area F. In *Megiddo III: The 1992–1996 Seasons*, eds. I. Finkelstein et al., 75–103. Jerusalem: Graphit.

Ingholt, Harald, ed.

1940 *Rapport préliminaire sur sept campagnes de fouilles à Hama en Syrie (1932–1938)*. Copenhagen: Ejnar Munksgaard.

Invernizzi, Antonio

1992 *Dal Tigri all'Eufrate II: Babilonesi e Assiri*. Studi e Materiali di Archeologia 6. Turin: Casa Editrice Le Lettere.

Ismail, Farouk, et al., eds.

1996 *Administrative Documents from Tell Beydar (Seasons 1993–1995)*. Subartu 2. Turnhout: Brepols.

Jaroš-Deckert, B.

1984 *Grabung in Asasif 1963–1970. Das Grab des Jnjjtf.f. Die Wandmalereien der XI. Dynastie*. Archäologische Veröffentlichungen 12. Mainz am Rhein: Philipp von Zabern.

Jarry, Marc

1939 Sur une blessure mortelle causée par une fléche de bronze à Ugarit. *Syria* 20:293–95.

Jasmin, Michaël

2006 The Political Organization of the City-States in Southwestern Palestine in the Late Bronze Age IIB (13th Century BC). In *"I Will Speak the Riddles of Ancient Times": Archaeological and Historical Studies in Honor of Amihai Mazar on the Occasion of His Sixtieth Birthday*, vol. 2, eds. A. M. Maeir and P. de Miroschedji, 161–91. Winona Lake, IN: Eisenbrauns.

Joffe, Alexander H.

1997 Palestine: Palestine in the Bronze Age. In *OEANE*, vol. 4, ed. E. M. Meyers, 212–17. New York: Oxford University Press.

Kagan, Kimberly

2006 *The Eye of Command*. Ann Arbor, MI: University of Michigan Press.

Kamlah, Jens, and Hélène Sader

2004 Deutsch-libanesische Ausgrabungen auf *Tell el-Burak*, südlich von Sidon. *ZDPV* 120:123–40.

Kaplan, Jacob
1956 Jaffa. *IEJ* 6:259–60.
1960 Jaffa. *IEJ* 10:121–22.
1961 Jaffa. *IEJ* 11:191–92.
1964a Jaffa's History Revealed by the Spade. *Archaeology* 17:270–76.
1964b Jaffa. *IEJ* 14:285–86.
1967 The Fortifications of Jaffa in the Hyksos Period. Paper presented at the Fourth World Conference of Jewish Studies, Jerusalem.
1969 Yavne-Yam. *IEJ* 19:120–21.
1971 Mesopotamian Elements in the Middle Bronze II Culture of Palestine. *JNES* 30:293–307.
1972a The Archaeology and History of Tel-Aviv—Jaffa. *BA* 35:66–95.
1972b *The Fortifications of Palestine in the MB II Period.* Publications of the Jaffa Museum of Antiquities 4. Tel Aviv-Jaffa: Jaffa Museum.
1975 Further Aspects of the Middle Bronze Age II Fortifications in Palestine. *ZDPV* 91:1–17.
1993 Yavneh-Yam. In *NEAEHL*, vol. 4, ed. E. Stern, 1504–06. Jerusalem: Israel Exploration Society.

Kaplan, Jacob, and Haya Ritter-Kaplan
1993 Jaffa. In *NEAEHL*, vol. 2, ed. E. Stern, 655–59. Jerusalem: Israel Exploration Society.

Karam, Naji
1997 BEY 013 rapport préliminaire. *BAAL* 2:107–09.

Keegan, John
1976 *The Face of Battle.* New York: Viking.
1993 *A History of Warfare.* New York: Vintage.

Kelly-Buccellati, Marilyn
1990 Three Seasons of Excavation at Tell Mozan. In *Tall al-Hamidiya 2*, eds. S. Eichler et al., 119–32. Orbis biblicus et orientalis, series archaeologica 6. Göttingen: Vandenhoeck & Ruprecht.

Kelso, James Leon
1968 *The Excavation of Bethel.* AASOR 39. Cambridge, MA: American Schools of Oriental Research.
1993 Bethel. In *NEAEHL*, vol. 1, ed. E. Stern, 192–94. Jerusalem: Israel Exploration Society.

Kemp, Barry J.
1972 Fortified Towns in Nubia. In *Man, Settlement and Urbanism*, eds. P. J. Ucko et al., 651–56. Cambridge, MA: Schenkman.
1983 Old Kingdom, Middle Kingdom and Second Intermediate Period c. 2686–1552 BC. In *Ancient Egypt: A Social History*, eds. B. G. Trigger et al., 71–182. Cambridge: Cambridge University Press.
1991 *Ancient Egypt: Anatomy of a Civilization.* New York: Routledge.
2000 Soil (including mud-brick architecture). In *Ancient Egyptian Materials and Technology*, eds. P. T. Nicholson and I. Shaw, 78–103. Cambridge: Cambridge University Press.

Kempinski, Aharon
1972 Review of *Gezer I, Preliminary Report of the 1964–66 Seasons* by W. G. Dever et al. *IEJ* 22: 183–86.
1974 Tell el-'Ajjûl—Beth-Aglayim or Sharuhen? *IEJ* 24:145–52.
1983 *Syrien und Palästina (Kanaan) in der letzten Phase der Mittlebronze IIB-Zeit (1650–1570 v. Chr.).* ÄAT 4. Wiesbaden: Harrassowitz.
1987a *Excavations at Kabri: Preliminary Report of the 1986 Season.* Tel Aviv: Institute of Archaeology, Dept. of Antiquities, Israel Exploration Society.
1987b Middle Bronze Age Fortifications. In *Excavations at Kabri: Preliminary Report of the 1986 Season*, ed. A. Kempinski. Tel Aviv: Institute of Archaeology, Dept. of Antiquities, Israel Exploration Society.
1989 *Megiddo: A City-State and Royal Centre in North Israel.* Materialien zur Allgemeinen und Vergleichenden Archäologie 40. Munich: C. H. Beck.
1991a The Earthen Rampart and Fortification. In *Excavations at Kabri: Preliminary Report of the 1990 Season 5*, eds. A. Kempinski and W.-D. Niemeier, 24–26. Tel Aviv: Tel Aviv University.
1991b Kabri—1990. *ESI* 10:76–77.
1992a Dan and Kabri—A Note on the Planning of Two Cities. *EI* 23:76–81 (Hebrew).
1992b Fortifications, Public Buildings, and Town Planning in the Early Bronze Age. In *The Architecture of Ancient Israel: From the Prehistoric to the Persian Periods*, eds. A. Kempinski and R. Reich, 68–80. Jerusalem: Israel Exploration Society.
1992c Middle and Late Bronze Age Fortifications. In *The Architecture of Ancient Israel: From the Prehistoric to the Persian Periods*, eds. A. Kempinski and R. Reich, 127–42. Jerusalem: Israel Exploration Society.
1992d The Middle Bronze Age. In *The Archaeology of Ancient Israel*, ed. A. Ben-Tor, 159–210. New Haven: Yale University Press.
1992e Urbanization and Town Plans in the Middle Bronze Age II. In *The Architecture of Ancient Israel: From the Prehistoric to the Persian Periods*, eds. A. Kempinski and R. Reich, 121–26. Jerusalem: Israel Exploration Society.
1993a Kabri. In *NEAEHL*, vol. 3, ed. E. Stern, 839–41. Jerusalem: Israel Exploration Society.
1993b Masos, Tel. In *NEAEHL*, vol. 3, ed. E. Stern, 986–89. Jerusalem: Israel Exploration Society.
2002a Area C. In *Tel Kabri: The 1986–1993 Excavation Seasons*, ed. A. Kempinski, 35–54. Monograph Series 20. Tel Aviv: Emery and Claire Yass Archaeology.
2002b *Tel Kabri: The 1986–1993 Excavation Seasons.* Monograph Series 20. Tel Aviv: Emery and Claire Yass Archaeology.

Kempinski, Aharon, and Wolf-Dietrich Niemeier, eds.
1991 *Excavations at Kabri: Preliminary Report of the 1990 Season 5.* Tel Aviv: Tel Aviv University.
1994 *Excavations at Kabri: Preliminary Report of the 1992–1993 Seasons, 7–8.* Tel Aviv: Tel Aviv University.

Kendall, Timothy
1975 Warfare and Military Matters in the Nuzi Tablets. Unpublished Ph.D. Dissertation. Department of Mediterranean Studies, Brandeis University.

Kenyon, Kathleen M.
1952 Excavations at Jericho, 1952. *PEQ* 1952:62–82.
1958 Some Notes on the Early and Middle Bronze Age Strata at Megiddo. *EI* 5:51*–60*.
1966 *Amorites and Canaanites*. The Schweich Lectures of the British Academy, 1963. Jerusalem: Israel Exploration Society.
1967 *Jerusalem: Excavating 3000 Years of History*. London: Thames & Hudson.
1969 The Middle and Late Bronze Age Strata at Megiddo. *Levant* 1:25–60.
1973 Palestine in the Middle Bronze Age. In *History of the Middle East and the Aegean Region c. 1800–1380 B.C.*, eds. I. E. S. Edwards et al., 77–116. The Cambridge Ancient History II:1. Cambridge: Cambridge University Press.
1974 *Digging Up Jerusalem*. London: Ernest Benn.
1979 *Archaeology in the Holy Land*. 4th ed. New York: Norton.
1981 *Excavations at Jericho III: The Architecture and Stratigraphy of the Tell*. London: British School of Archaeology in Jerusalem.
1993 Jericho. In *NEAEHL*, vol. 2, ed. E. Stern, 674–81. Jerusalem: Israel Exploration Society.

Kepinski-Lecomte, Christine, ed.
1992 *Haradum I: Une ville nouvelle sur le Moyen-Euphrate (XVIIIe–XVIIe siècles av. J.-C.)*. Paris: Éditions Recherches sur les Civilisations.

Kern, Paul Bentley
1999 *Ancient Siege Warfare*. Bloomington, IN: Indiana University Press.

Khadour, Muhammad, et al.
1997 Gindaris. *AJA* 101:118–19.

Kienast, Burkhart
1980 Der Feldzugsbericht des Ennadagan in literarhistorischer Sicht. *OrAnt* 19:247–61.

Kitchen, Kenneth A.
2000 Regnal and Genealogical Data of Ancient Egypt (Absolute Chronology I). The Historical Chronology of Ancient Egypt: A Current Assessment. In *Synchronisation of Civilisations in the Eastern Mediterranean in the Second Millennium B.C.: Proceedings of an International Symposium at Schloss Haindorf, 15th–17th November 1996 and at the Austrian Academy, Vienna, 11th–12th of May 1998*, ed. M. Bietak. Contributions to the Chronology of the Eastern Mediterranean 1. Vienna: Österreichischen Akademie der Wissenschaften.

Klengel, Horst
1965–1970 *Geschichte Syriens im 2. Jahrtausend v. u. Z.* Berlin: Akademie.
1987 Non-Slave Labour in the Old Babylonian Period: The Basic Outlines. In *Labor in the Ancient Near East*, ed. M. A. Powell, 159–66. New Haven: American Oriental Society.
1992 *Syria 3000 to 300 BC: A Handbook of Political History*. Berlin: Akademie.
1995 Tunip und andere Probleme der historischen Geographie Mittlesyriens. In *Immigration and Emigration within the Ancient Near East. Festschrift E. Lipinski*, eds. K. van Lerberghe and A. Schoors, 125–34. Leuven: Uitgeverij Peeters en Departement Oriëntalistiek.

Knapp, Arthur Bernard
1989 Complexity and Collapse in the North Jordan Valley: Archaeometry and Society in the Middle-Late Bronze Ages. *IEJ* 39:129–48.
1993 *Society and Polity at Bronze Age Pella: An Annales Perspective*. JSOT/ASOR Monograph Series 6. Sheffield: Sheffield Academic Press.

Knudstad, James
1966 Serra East and Dorginarti. *Kush* 14:165–86.

Kochavi, Moshe
1967 Tel Malhata. *IEJ* 17:272–73.
1974 Tel Aphek. *IEJ* 24:261–62.
1975 The First Two Seasons of Excavations at Aphek-Antipatris. *TA* 2:17–42.
1993 Malhata, Tel. In *NEAEHL*, vol. 3, ed. E. Stern, 934–36. Jerusalem: Israel Exploration Society.

Kochavi, Moshe, et al.
1979 Aphek-Antipatris, Tel Poleg, Tel Zeror and Tel Burga: Four Fortified Sites of the Middle Bronze Age IIA in the Sharon Plain. *ZDPV* 95:121–65.

Kochavi, Moshe, et al., eds.
2000 *Aphek-Antipatris I: Excavation of Areas A and B, The 1972–1976 Seasons*. Tel Aviv: Emery and Claire Yass Publications in Archaeology.

Koehler, Ludwig and Walter Baumgartner
1994 *The Hebrew and Aramaic Lexicon of the Old Testament*. Leiden: Brill.

Koliński, Rafał
2001 *Mesopotamian dimātu of the Second Millennium BC*. BAR International Series 1004. Oxford: Archaeopress.

Kooij, G. van der, and Moawiyah M. Ibrahim
1989 *Picking Up the Threads...A Continuing Review of the Excavations at Deir Alla, Jordan*. Leiden: University of Leiden.

Korfmann, Manfred
1972 *Schleuder und Bogen in Südwestasien; von den frühesten Belegen bis zum Beginn der historischen Stadtstaaten*. Antiquitas. Reihe 3, Abhandlungen zur Vor- und Frühgeschichte, zur klassischen und provinzial-römischen Archäologie und zur Geschichte des Altertums 13. Bonn: Habelt.
1973 The Sling as a Weapon. *Scientific American* 229:34–42.

Kotter, Wade Ralph
1986 Spatial Aspects of the Urban Development of Palestine during the Middle Bronze Age. Unpublished Ph.D. Dissertation. Department of Anthropology, University of Arizona.

Kouchoukos, Nicholas
1999 Landscape and Social Change in Late Prehistoric Mesopotamia. Unpublished Ph.D. Dissertation, Yale University.

Koucky, Frank L.
n.d. Early Maps and Records of Ashkelon. In *Ashkelon I*, eds. L. E. Stager and J. D. Schloen.

Kraus, Fritz Rudolph
1948 Ein Altakkadisches Festungsbild. *Iraq* 10:81–92.

Kühne, Hartmut, and Hans Steuerwald
1979 Das Nordost-Tor von Tell Mumbaqat. In *Le moyen Euphrate: actes du colloque de Strasbourg 10–12 mars 1977*, ed. J.-C. Margueron, 203–15. Leiden: Brill.

Kupper, Jean Robert
1951 Note lexicographiques. *RA* 45:120–30.

Kuschke, A., et al.
1976 *Archaeologische Survey in der nordlichen Biga, 1972*.

Lagarce, Jacques
1984 Remarques sur des ouvrages de soutènement et de défense à Ras Shamra et à Ras Ibn Hani. *Syria* 61:153–79.

Lambert, Wilfred G.
1990 Halam, Il-Halam and Aleppo. *MARI* 6:641–43.

Lapp, Paul W.
1964 The 1963 Excavations at Tell Ta'annek. *BASOR* 173:4–44.
1967 The 1966 Excavations at Tell Ta'annek. *BASOR* 185:2–39.
1969 The 1968 Excavations at Tell Ta'annek. *BASOR* 195:2–49.

Lavee, Hanoch, et al.
1993 Micromorphological Investigation of the Middle Bronze Age Glacis. In *Shiloh: The Archaeology of a Biblical Site*, eds. I. Finkelstein et al., 294–302. Monograph Series 13. Tel Aviv: Emery and Claire Yass Archaeology.

Lawrence, Arnold Walter
1965 Ancient Egyptian Fortifications. *JEA* 51:69–94.

Lazzeri, Simona, and Nicola Macchioni
2000 Appendix E: Identification of the Wooden Species of Some Charcoals from the 1998 Excavations at Jericho. In *Excavations at Jericho, 1998: Preliminary Reports on the Second Season of Archaeological Excavations and Surveys at Tell es-Sultan, Palestine*, eds. N. Marchetti and L. Nigro, 333–44. Quaderni di Gerico 2. Rome: Università di Roma «La Sapienza».

Lebeau, Marc
1997a Beydar. *AJA* 101:111–12.
1997b La situation géographique, la topographie et les périodes d'occupation de Tell Beydar. In *Tell Beydar, Three Seasons of Excavations (1992–1994)*, eds. M. Lebeau and A. Suleiman, 7–19. Subartu 3. Turnhout: Brepols.

Lebeau, Marc, and Antoine Suleiman, eds.
1997 *Tell Beydar, Three Seasons of Excavations (1992–1994)*. Subartu 3. Turnhout: Brepols.

LeBlanc, Steven A.
2003 Prehistory of Warfare. *Archaeology* 56:18–25.

Lederman, Zvi
1985 The Building Technique at Shiloh in the Middle Bronze Age II and the "Millo" of Shechem. Paper presented at the Eleventh Archaeological Conference in Israel: Abstracts of Lectures, Jerusalem.
1993 Area C: The Iron Age I Pillared Buildings and Other Remains. In *Shiloh: The Archaeology of a Biblical Site*, eds. I. Finkelstein et al., 15–34. Monograph Series 10. Tel Aviv: Emery and Claire Yass Archaeology.

Lederman, Zvi, and Israel Finkelstein
1993 Area D: Middle Bronze Age Stone and Earth Works, Late Bronze Age Dumped Debris and Iron Age I Silos. In *Shiloh: The Archaeology of a Biblical Site*, eds. I. Finkelstein et al., 35–48. Monograph Series 10. Tel Aviv: Emery and Claire Yass Archaeology.

Leemans, William F.
1982 The Pattern of Settlement in the Babylonian Countryside. In *Societies and Languages of the Ancient Near East: Studies in Honour of I. M. Diakonoff*, ed. M. A. Dandamayev, 246–49. Warminster: Aris & Phillips.

Lehmann, Gunnar
2002 *Bibliographie der archäologischen Fundstellen und Surveys in Syrien und Libanon*. Orient-Archäologie 9. Berlin: Leidorf.

Lehmann, Gunnar, et al.
1991 The 1990 Excavations at Abu Snesleh: Preliminary Report. *ADAJ* 35:41–62.

Lehmann, Gunnar, and Tammi J. Schneider
2000 Tell el-Far'ah (South), 1999 and 2000. *IEJ* 50:258–61.

Lenzen, Cherie J.
1992 Irbid, Tell. In *ABD*, vol. 3, ed. D. N. Freedman, 456–57. New York: Doubleday.

Lenzen, Cherie J., and A. M. McQuitty
1989 Irbid (Tell). In *Archaeology of Jordan II. 1, Field Reports: Surveys and Sites A–K*, eds. D. Homès-Fredericq and J. B. Hennessy, 298–300. Akkadica Supplementum 7. Leuven: Peeters.

Leonard, Albert
1989 The Late Bronze Age. *BA* 52:4–39.

Lerberghe, Karel van, and Gabriela Voet
2000 *Tell Beydar: Environmental and Technical Studies*. Subartu 6. Turnhout: Brepols.

Lichtheim, Miriam
1973–80 *Ancient Egyptian Literature*. Berkeley, CA: University of California.

Littauer, Mary Aiken, ed.
2002 *Selected Writings on Chariots and Other Early Vehicles, Riding and Harness*. CHANE 6. Leiden: Brill.

Littauer, Mary Aiken, and Joost H. Crouwel
1979 *Wheeled Vehicles and Ridden Animals in the Ancient Near East*. HdO, Kunst und Archäologie I/B2 L.1. Leiden: Brill.

1985 *Chariots and Related Equipment from the Tomb of Tut'ankhamun.* Tut'ankhamun's Tomb Series 3. Oxford: Griffith Institute.

Liverani, Mario
1982 Ville et campagne dans le royaume d'Ugarit. Essai d'analyse économique. In *Societies and Languages of the Ancient Near East: Studies in Honour of I. M. Diakonoff*, ed. M. A. Dandamayev, 250–58. Warminster: Aris & Phillips.

Lönnquist, Minna Angelina
2000 *Between Nomadism and Sedentism: Amorites from the Perspective of Contextual Archaeology.* Helsinki: Juutiprint.

Loon, Maurits N. van, ed.
1988 *Hammam et-Turkman I: Report on the University of Amsterdam's 1981–1984 Excavations in Syria.* PIHANS 63. 2 vols. Leiden: Nederlands Instituut voor Het Nabije Oosten.
2001 *Selenkahiye: Final Report on the University of Chicago and University of Amsterdam Excavations in the Tabqa Reservoir, Northern Syria, 1967–1975.* PIHANS 91. Leiden: Nederlands Instituut voor Het Nabije Oosten.

Loon, Maurits N. van, and Diederik J. W. Meijer
1987 Hammam et-Turkman on the Balikh: First Results of the University of Amsterdam's 1986 Excavation. *Akk* 52:1–9.

Loud, Gordon
1948 *Megiddo 2: Seasons of 1935–39.* OIP 62. Chicago: University of Chicago Press.

Ludwig, Wido, et al.
1970 Die Grabung auf dem Tell Habuba Kabira. *MDOG* 102:30–58.

Lyonnet, Bertille
1998 Le peuplement de la Djéziré occidentale au début du 3e millénaire, villes circulaires et pastoralisme: questions et hypothèses. In *About Subartu: Landscape, Archaeology, and Settlement*, ed. M. Lebeau, 179–93. Subartu 4/1. Turnhout: Brepols.

Ma'ayeh, Farah S.
1960 Recent Archaeological Discoveries in Jordan: Tell Safut (Sweileh). *ADAJ* 4–5:115.

Macalister, Robert Andrews Stewart
1912 *The Excavation of Gezer 1902–1905 and 1907–1909.* London: Palestine Exploration Fund.
1926 *Excavations on the Hill of Ophel, Jerusalem 1923–1925.* Annual of the Palestine Exploration Fund 4. Manchester, England: Palestine Exploration Fund.

MacDonald, Eann, et al.
1932 *Beth-Pelet II: Prehistoric Fara, Beth-Pelet Cemetery.* British School of Archaeology in Egypt 52. London: British School of Archaeology in Egypt.

Machiavelli, Niccolò
1965 *The Art of War.* New York: Da Capo.

Machule, Dittmar, and Dieter Rentschaler
1971 Arbeiten in Mumbaqat und Beobachtungen in der Umgebung. *MDOG* 103:48–58.

Machule, Dittmar, et al.
1992 Ausgrabungen in Tall Munbaqa/Ekalte 1990. *MDOG* 124:11–40.
1993 Ausgrabungen in Tall Munbaqa/Ekalte 1991. *MDOG* 125:69–101.
1996 Tall Munbaqa/Ekalte 1994. *MDOG* 128:11–32.
1986 Ausgrabungen in Tall Munbaqa 1984. *MDOG* 118:67–146.

Mackenzie, Duncan
1911 *Excavations at Ain Shems (Beth-Shemesh).* Annual of the Palestine Exploration Fund 1911, 41–94. Manchester, England: Palestine Exploration Fund.
1913 *Excavations at Ain Shems (Beth-Shemesh).* In Annual of the Palestine Exploration Fund 1912–1913. Manchester, England: Palestine Exploration Fund.

Maeir, Aren M.
2000 Hamadiya North: A Rural MB II Settlement in the Beth She'an Valley. *'Atiqot* 39:31–42.
2003 Does Size Count? Urban and Cultic Perspectives on the Rural Landscape during the Middle Bronze II. In *The Rural Landscape of Ancient Israel*, eds. A. M. Maeir et al., 61–69. BAR International Series 1121. Oxford: Archaeopress.

Magness-Gardiner, Bonnie
1994 Urban-Rural Relations in Bronze Age Syria: Evidence from Alalakh Level VII Palace Archives. In *Archaeological Views from the Countryside: Village Communities in Early Complex Societies*, eds. G. M. Schwartz and S. E. Falconer, 37–47. Washington, DC: Smithsonian Institution.

Mallet, Joël
1973 *Tell el-Fâr'ah (Région de Naplouse): L'installation du Moyen Bronze antérieure au rempart.* Cahiers de la Revue Biblique 14. Paris: J. Gabalda et Cie, Éditeurs.
1988 *Tell el-Fâr'ah II: Le Bronze Moyen, stratigraphie des vestiges du Bronze Moyen II.* Recherche sur les Civilisations 66. Paris: Éditions Recherches sur les Civilisations.

Mallowan, Max E. L.
1966 *Nimrud and Its Remains.* New York: British School of Archaeology in Iraq.

al-Maqdissi, Michel
2001 Kurzbericht über die syrischen Ausgrabungen in Mišrife-Qatna. *MDOG* 133:141–55.

al-Maqdissi, Michel, et al., eds.
2002 *Excavating Qatna I: Preliminary Report on the 1999 and 2000 Campaigns of the Joint Syrian-Italian-German Archaeological Research Project at Tell Mishrife.* Documents d'archaéologie syrienne 4. Damascus: Salhani.

Marchetti, Nicolò
1998 L'Area A. Le fortificazioni e l'insediamento del bronzo medio II–III. In *Scavi a Gerico, 1997. Relazione preliminaire sulla prima campagna di scavi e prospezioni archeologiche a Tell es-Sultan, Palestina*, eds. N. Marchetti and L. Nigro,

118–97. Quaderni di Gerico 1. Rome: Università di Roma «La Sapienza».

2000 Area A. The Middle Bronze II Public Building and Residential Quarter in the Lower Town. In *Excavations at Jericho, 1998: Preliminary Reports on the Second Season of Archaeological Excavations and Surveys at Tell es-Sultan, Palestine*, eds. N. Marchetti and L. Nigro, 193–281. Quaderni di Gerico 2. Rome: Università di Roma «La Sapienza».

2003 A Century of Excavations on the Spring Hill at Tell es-Sultan, Ancient Jericho: A Reconstruction of Its Stratigraphy. In *The Synchronisation of Civilisations in the Eastern Mediterranean in the Second Millennium B.C. II*, ed. M. Bietak, 295–321. CCEM 4. Vienna: Österreichischen Akademie der Wissenschaften.

Marchetti, Nicolò, and Lorenzo Nigro, eds.

1998 *Scavi a Gerico, 1997. Relazione preliminaire sulla prima campagna di scavi e prospezioni archeologiche a Tell es-Sultan, Palestina.* Quaderni di Gerico 1. Rome: Università di Roma «La Sapienza».

2000 *Excavations at Jericho, 1998: Preliminary Reports on the Second Season of Archaeological Excavations and Surveys at Tell es-Sultan, Palestine.* Quaderni di Gerico 2. Rome: Università di Roma «La Sapienza».

forthcoming *Excavations at Jericho, 1999 and 2000: Preliminary Reports on the Third and Fourth Seasons of Archaeological Excavations and Surveys at Tell es-Sultan, Palestine.* Quaderni di Gerico 3. Rome: Università di Roma «La Sapienza».

Marchetti, Nicolò, and Jehad Yasin

2000 Area E. The South-Western Middle Bronze I–II Fortifications. In *Excavations at Jericho, 1998: Preliminary Reports on the Second Season of Archaeological Excavations and Surveys at Tell es-Sultan, Palestine*, eds. N. Marchetti and L. Nigro, 181–92. Quaderni di Gerico 2. Rome: Università di Roma «La Sapienza».

Marcus, Ezra S., ed.

2003 *Dating the Early Middle Bronze Age in the Southern Levant: A Preliminary Comparison of Radiocarbon and Archaeo-Historical Synchronisations.* The Synchronisation of Civilisations in the Eastern Mediterranean in the Second Millennium B.C. II. Vienna: Österreichischen Akademie der Wissenschaften.

Marfoe, Leon

1995 *The Prehistoric and Early Historic Context of the Site: Catalogue and Commentary.* Kamid el-Loz 13, Saarbrücker Beiträge zur Altertumskunde 41. Bonn: Habelt.

1998 *Settlement History of the Biqa' up to the Iron Age.* Kamid el-Loz. Bonn: Habelt.

Margueron, Jean-Claude

1982 Mari: rapport préliminaire sur la campagne de 1979. *MARI* 1:9–30.

1983 Mari: rapport préliminaire sur la campagne de 1980. *MARI* 2:9–35.

1984 Mari: rapport préliminaire sur la campagne de 1982. *MARI* 3:7–39.

1987 État présent des recherches sur l'urbanisme de Mari–I. *MARI* 5:483–98.

1995 Emar, Capital of Ashtata in the Fourteenth Century BCE. *BA* 58:126–38.

1997a Larsa. In *OEANE*, vol. 3, ed. E. M. Meyers, 331–33. Translated by N. W. Leinwand. New York: Oxford University Press.

1997b Mari. In *OEANE*, vol. 3, ed. E. M. Meyers, 413–17. Translated by N. W. Leinwand. New York: Oxford University Press.

2000 Mari: derniers développements des recherches conduites sur le tell Hariri. In *Proceedings of the First International Congress on the Archaeology of the Ancient Near East, Rome May 18th–23rd 1998*, vol. 2, eds. P. Matthiae et al., 909–28. Rome: Università degli Studi di Roma "La Sapienza".

2004 *Mari, métropole de l'Euphrate au IIIe et au début IIe millénaire av. J.-C.* Paris: Picard.

Marín, Juan Antonio Belmonte

2001 *Répertoire Géographique des Textes Cunéiformes, Band 12/2: Die Orts- und Gewässernamen der Texte aus Syrien im 2. Jt. v. Chr.* Beihefte zum Tübinger Atlas des Vorderen Orients. Wiesbaden: Reichert Verlag.

Marquis, Philippe

1995 Les fouilles du centre-ville, état d'avancement des travaux: mars à septembre 1995. *National Museum News*:10–11.

Marrassini, Paolo

1971 Sui «campi fortificati» nell'età di Mari. *OrAnt* 10:107–22.

Marsden, Eric William

1969 *Greek and Roman Artillery: Historical Development.* Oxford: Oxford University Press.

1971 *Greek and Roman Artillery: Technical Treatises.* Oxford: Oxford University Press.

Marzal-Gonzalez, Angel

1969 The Organization of the Mari State. Unpublished Ph.D. Dissertation. Department of Near Eastern Languages and Civilizations, University of Chicago.

Master, Daniel M. et al., eds.

2005 *Dothan I: Remains from the Tell (1953–1964).* The Excavations of Joseph P. Free at Dothan (1953–1964). Winona Lake, IN: Eisenbrauns.

Matthiae, Paolo

1967 Mission archéologique de l'Université de Rome à Tell Mardikh. *AAS* 17:25–43.

1968 Mission archéologique de l'Université de Rome à Tell Mardikh (1966). *AAS* 18:5–20.

1970 Mission archéologique de l'Université de Rome à Tell Mardikh: rapport sommaire sur la quatrième et la cinquième campagnes, 1967 et 1968. *AAS* 20:55–72.

1977a *Ebla. Un impero ritrovato. Dai primi scavi alle ultime scoperte*. 1st ed. Turin: Einaudi.
1977b Le palais royal protosyrien d'Ebla: nouvelles recherches archéologiques à Tell Mardikh en 1976. *CRAIBL* 1977:148–74.
1979 Sondages à Tell Touqan (Syrie), 1978. *Akk* 14:6–10.
1980a Campagne de fouilles à Tell Mardikh-Ebla en 1979: les tombes princières et le Palais de la ville basse à l'époque amorrhéene. *CRAIBL* 1980:94–118.
1980b *Ebla: An Empire Rediscovered*. London: Hodder and Stoughton.
1982a Fouilles de 1981 à Tell Mardikh-Ebla et à Tell Touqan: nouvelles lumières sur l'architecture paléosyrienne du Bronze Moyen I–II. *CRAIBL* 1982:299–331.
1982b The Western Palace of the Lower City of Ebla: A New Administrative Building of Middle Bronze I–II. *Archiv für Orientforschung. Beiheft* 19:121–29.
1983 Tell Tuqan bei Ebla. Eine neue, grosse altsyrische Stadt. *AW* 14:40–50.
1989–1990
Tell Tuqan. *AfO* 36–37:335–39.
1995 *Ebla. Un impero ritrovato. Dai primi scavi alle ultime scoperte*. 3rd ed. Turin: Einaudi.
1996 Tell Mardikh-Ebla (Siria), campagna di scavi 1995. *OE* 1996:84–87.
1997a Ebla. In *OEANE*, vol. 2, ed. E. M. Meyers, 180–83. New York: Oxford University Press.
1997b Ebla and Syria in the Middle Bronze Age. In The Hyksos: New Historical and Archaeological Perspectives, ed. E. D. Oren, 379–414. University Museum Monograph 96. Philadelphia: The University Museum, University of Pennsylvania.
1997c Tell Mardikh 1977–1996: vingt ans de fouilles et de découvertes. La renaissance d'Ebla amorrhéenne. *Akk* 101:1–29.
1998 Les fortifications de l'Ébla paléo-syrienne: fouilles à Tell Mardikh, 1995–1997. *CRAIBL* 1998: 557–88.
2000 Nouvelles fouilles à Ébla (1998–1999): forts et palais de l'enceinte urbaine paléo-syrienne: fouilles Tell Mardikh. *CRAIBL* 2000:567–610.
2002 A Preliminary Note on the MB I–II Fortification System at Ebla. *DM* 13:29–51.
2006 Archaeology of a Destruction: The End of MB II Ebla in the Light of Myth and History. In *Timelines: Studies in Honour of Manfred Bietak*, vol. 3, eds. E. Czerny et al., 39–51. Orientalia Lovaniensia Analecta 149. Leuven: Peeters.
in press
Assedi divini e ostaggi mortali da Ebla a Troia. Sulla tradizione epica prima dell'Illiade. In *Rendiconti dell'Accademica Nazionale dei Lincei, Classe di Scienze Morali Storiche e Filologiche* 9.

Matthiae, Paolo, et al., eds.
1995 *Ebla. alle origini della civiltà urbana*. Milan: Electa.

Mazar, Amihai
1990 *Archaeology of the Land of the Bible 10,000–586 BCE*. 1st ed. Anchor Bible Reference Library. New York: Doubleday.
1997a Area P. In *Hazor V: An Account of the Fifth Season of Excavations, 1968*, eds. A. Ben-Tor and R. Bonfil, 353–86. Jerusalem: Israel Exploration Society, The Hebrew University.
1997b *Timnah (Tel Batash) I: Stratigraphy and Architecture*. Qedem 37. Jerusalem: Institute of Archaeology, Hebrew University of Jerusalem.
1999 The 1997–1998 Excavations at Tel Rehov: Preliminary Report. *IEJ* 49:1–42.

Mazar, Benjamin
1968 The Middle Bronze Age in Palestine. *IEJ* 18:65–97.
1970 Canaan in the Patriarchal Age. In *The World History of the Jewish People, Vol. 2*, ed. B. Mazar, 169–87. Ancient Times (Patriarchs and Judges). Tel Aviv: Massada.

Mazzoni, Stefania
1992 *Tell Afis e l'età del ferro*. Seminari di Orientalistica 2. Pisa: Gardini Editori e Stampatori.
1994 Afis. *AJA* 98:146–49.
2002 Late Bronze Age Pottery Production in Northwestern Central Syria. In *Céramique de L'âge du Bronze en Syrie, I: La Syrie du sud et la vallée de l'Oronte*, eds. M. al-Maqdissi et al., 129–51. BAH 161. Beirut: Institut Français d'Archéologie du Proche-Orient.

McClellan, Thomas L.
1984–1985
El-Qitar: Second Season of Excavations, 1983–1984. *AbrN* 23:39–72.
1986 El-Qitar: Third Season of Excavation, 1984–1985. *AbrN* 24:83–106.

McLaren, Peter Bruce
2001 A Middle Bronze Age Gate at Tell Rukeis, Jordan. In *Australians Uncovering Ancient Jordan: Fifty Years of Middle Eastern Archaeology*, ed. A. G. Walmsley, 95–106. Sydney: University of Sydney.
2003 *The Military Architecture of Jordan during the Middle Bronze Age: New evidence from Pella and Rukeis*. British Archaeology Reports 1202. Oxford: Archaeopress.

McLeod, Wallace
1965 The Range of the Ancient Bow. *Phoenix* 19:1–14.
1970 *Composite Bows From the Tomb of Tut'ankhamun*. Tut'ankhamun's Tomb Series 3. Oxford: Griffith Institute.
1972 The Range of the Ancient Bow: Addenda. *Phoenix* 26:78–82.

Meijer, Diederik J. W.
1988 Architecture and Stratigraphy. In *Hammam et-Turkman I: Report on the University of Amsterdam's 1981–1984 Excavations in Syria*, ed. M. N. van Loon, 69–127. PIHANS 63.
1989 Hammam et-Turkman on the Balikh: First Results of the Fifth Campaign, 1988. *Akk* 64/65:1–12.
1997 Hammam al-Turkman. *AJA* 101:121–22.
2001 Architecture and Stratigraphy. In *Selenkahiye: Final Report on the University of Chicago and University of Amsterdam Excavations in the Tabqa Reservoir, Northern Syria, 1967–1975*, ed.

M. N. van Loon, 25–105. PIHANS 91. Istanbul: Nederlands Instituut voor Het Nabije Oosten.

Mendelssohn, Kurt
1977 *The Riddle of the Pyramids*. New York: Thames and Hudson.

Mendenhall, George E.
1992a The Amorite Migrations. In *Mari in Retrospect: Fifty Years of Mari and Mari Studies*, ed. G. D. Young, 233–41. Winona Lake, IN: Eisenbrauns.
1992b Amorites. In *ABD*, vol. 1, ed. D. N. Freedman, 199–202. New York: Doubleday.

Meshel, Ze'ev
1992 The Architecture of the Israelite Fortresses in the Negev. In *The Architecture of Ancient Israel: From the Prehistoric to the Persian Periods*, eds. A. Kempinski and R. Reich, 294–301. Jerusalem: Israel Exploration Society.

du Mesnil du Buisson, Robert
1926 Les ruines d'el Mishrifé au nord-est de Homs (Émèse). *Syria* 7:289–325.
1927 L'ancienne Qatna ou les ruines d'el Mishrifé au nord-est de Homs (Émèse), deuxième campagne de fouilles, 1927. *Syria* 8:277–301.
1932 Une campagne de fouilles à Khan Sheikhoun. *Syria* 13:171–88.
1935a Le site archéologique de Mishrifé-Qatna. Collection de textes et documents d'orient. Paris: Boccard.
1935b Souran et Tell Masin. *Berytus* 2:121–34.

Meyer, Jan-Waalke
1996 Offene und geschlossene Siedlungen: Ein Beitrag zur Siedlungsgeschichte und historischen Topographie in Nordsyrien während des 3. und 2. Jts. v. Chr. *AoF* 23:132–70.

Meyer, Léon de, et al. eds.
1971 *Tell ed-Dēr I: rapport préliminaire sur la première campagne (février 1970)*. Leuven: Peeters.

Miglus, Peter, and Eva Strommenger
2002 *Tall Bi'a-Tuttul VIII: Stadtbefestigungen, Häuser und Tempel*. WVDOG 103. Saarbrücken: Saarbrücker Druckerei und Verlag.

Miller, Robert L.
1983 Chisel-ended arrowheads from Tell Hadidi, Syria. *BIA* 20:187–90.

Miller, Robert L., et al.
1986 Experimental Approaches to Ancient Near Eastern Archery. *WoAr* 18:178–95.

Minoff, Eli
1992 A Mace Head Mould from Tel Dan. *EI* 23:87*–89*.

Miron, Eli
1992 *Axes and Adzes from Canaan*. Prähistorische Bronzefunde, pt. IX, vol. 19. Stuttgart: Steiner.

Miroschedji, Pierre de
1993 Far'ah, Tell el- (North). In *NEAEHL*, vol. 2, ed. E. Stern, 433–38. Jerusalem: Israel Exploration Society.

Moorey, P. R. S.
1986 The Emergence of the Light, Horse-drawn Chariot in the Near East, c. 2000–1500 BC. *WoAr* 18:196–215.

Moortgat-Correns, Ursula
1972 *Die Bildwerke vom Djebelet el Beda in Ihrer räumlichen und zeitlichen Umwelt*. Berlin: Walter de Gruyter.

Moran, William L.
1992 *The Amarna Letters*. Baltimore: Johns Hopkins University.

Morris, Ian
1997 An Archaeology of Equalities? The Greek City-States. In *The Archaeology of City-States: Cross-Cultural Approaches*, eds. D. L. Nichols and T. H. Charlton, 91–105. Washington, DC: Smithsonian Institution.

Mousli, Majed
1989/1990
Tall Safinat-Nouh. *AfO* 36–37:300–08.

Müller-Karpe, Andreas
2003 Remarks on Central Anatolia Chronology of the Middle Hittite Period. In *The Synchronisation of Civilisations in the Eastern Mediterranean in the Second Millennium B.C. II*, ed. M. Bietak, 383–94. CCEM 4. Vienna: Österreichischen Akademie der Wissenschaften.

Müller, Uwe
1970 Kritische Bemerkungen zu den Straten XIII bis IX in Megiddo. *ZDPV* 86:50–86.

Na'aman, Nadav
1975 The Political Disposition and Historical Development of Eretz-Israel according to the Amarna Letters (Hebrew). Unpublished Ph.D. Dissertation, Tel Aviv University.
1988 Biryawaza of Damascus and the Date of Kamid el-Loz 'Apiru Letters. *UF* 20:179–94.
1997 The Network of Canaanite Late Bronze Kingdoms and the City of Ashdod. *UF* 29:599–625.
1999 Four Notes on the Size of Late Bronze Age Canaan. *BASOR* 313:31–37.

Naumann, Rudolf
1971 *Architektur Kleinasiens von ihren Anfängen bis zum Ende der hethitischen Zeit*. 2nd ed. Tübingen: Verlag Ernst Wasmuth.

Neev, David, et al.
1987 *Mediterranean Coasts of Israel and Sinai: Holocene Tectonism from Geology, Geophysics, and Archaeology*. New York: Taylor & Francis.

Neu, Erich
1996a *Das hurritische Epos der Freilassung, I. Untersuchungen zu einem zweisprachigem Textensemble aus Hattusa*. Studien zu den Bogazköy-Texten 32. Wiesbaden: Harrassowitz Verlag.
1996b La bilingue hourro-hittite: Contenu et sens. In *Mari, Ébla et les Hourrites: dix ans de travaux*, ed. J.-M. Durand, 363–76. Amurru 1. Paris: Éditions Recherches sur les Civilisations.

Newberry, Percy E.
1893 *Beni Hasan*. Archaeological Survey of Egypt. London: Kegan Paul International.
1894 *Beni Hasan II*. Archaeological Survey of Egypt. London: Kegan Paul International.

Newberry, Percy E., and F. L. Griffith
1895 *El Bersheh I*. Archaeological Survey of Egypt. London: Kegan Paul and Co.

Nigro, Francesco
1998 Appendice A. Il parco archeologico per la conservazione e la valorizzazione di Tell es-Sultan, antica Gerico. In *Scavi a Gerico, 1997. Relazione preliminaire sulla prima campagna di scavi e prospezioni archeologiche a Tell es-Sultan, Palestina*, eds. N. Marchetti and L. Nigro, 205–29. Quaderni di Gerico 1. Rome: Università di Roma «La Sapienza».

Nigro, Lorenzo
1997–1999
Yamkhad/Aleppo: Investigating the Second Millennium B.C. Capital of Northern Syria through Islamic, Byzantine and Classical Towns. In *Trails to the East: Essays in Memory of Paolo Cuneo*, vol. 1/2, eds. M. Calia et al. Environmental Design.
2000 Area D. The Tell Fortifications in Front of the Spring. In *Excavations at Jericho, 1998: Preliminary Reports on the Second Season of Archaeological Excavations and Surveys at Tell es-Sultan, Palestine*, eds. N. Marchetti and L. Nigro, 165–79. Quaderni di Gerico 2. Rome: Università di Roma «La Sapienza».
2002a The MB Pottery Horizon of Tell Mardikh/Ancient Ebla in a Chronological Perspective. In *The Middle Bronze Age in the Levant: Proceedings of an International Conference on MB IIA Ceramic Material, Vienna, 24th–26th of January 2001*, ed. M. Bietak, 296–328. CCEM 3. Vienna: Österreichischen Akademie der Wissenschaften.
2002b The Middle Bronze Age Pottery Horizon of Northern Inner Syria on the Basis of the Stratified Assemblages of Tell Mardikh and Hama. In *Céramique de L'âge du Bronze en Syrie, I: La Syrie du sud et la vallée de l'Oronte*, eds. M. al-Maqdissi et al., 97–128. BAH 161. Beirut: Institut Français d'Archéologie du Proche-Orient.

Northedge, Alastair
1983 The Fortifications of Qal'at 'Amman (Amman Citadel). *ADAJ* 27:436–60.

Novák, Mirko
1995 Die Stadtmauergrabung. In *Ausgrabungen in Tell Chuera in Nordost-Syrien I: Vorbericht über die Grabungskampagnen 1986 bis 1992*, vol. 2, eds. W. Orthmann et al., 173–82. Vorderasiatische Forschungen der Max Freiherr von Oppenheim-Stiftung. Saarbrücken: Saarbrücker Druckerei und Verlag.

Novák, Mirko, and Peter Pfälzner
2001 Ausgrabungen in Mišrife-Qatna 2000: Vorbericht der deutschen Komponente des internationalen Kooperationsprojektes. *MDOG* 133:157–98.
2002 Excavations in the Western Part of the Bronze Age Palace (Operation G). In *Excavating Qatna I: Preliminary Report on the 1999 and 2000 Campaigns of the Joint Syrian-Italian-German Archaeological Research Project at Tell Mishrife*, eds. M. al-Maqdissi et al., 17–24. Documents d'archaéologie syrienne 4. Damascus: Salhani.

Oates, David
1973 Early Vaulting in Mesopotamia. In *Archaeological Theory and Practice*, ed. D. E. Strong, 183–91. London: Seminar.
1985 Walled Cities in Northern Mesopotamia in the Mari period. *MARI* 4:585–94.
1990 Innovations in Mud-Brick: Decorative and Structural Techniques in Ancient Mesopotamia. *WoAr* 21:389–406.

Oates, David, et al.
1997 *Excavations at Tell Brak, Vol. I: The Mitanni and Old Babylonian Periods*. Oxford: Oxbow.
2001 *Excavations at Tell Brak, Vol. II: Nagar in the 3rd Millennium BC*. Oxford: Oxbow.

Oates, David, and Joan Oates
2001 Archaeological Reconstruction and Historical Commentary. In *Excavations at Tell Brak, Vol. II: Nagar in the 3rd Millennium BC*, eds. D. Oates et al., 379–96. Oxford: Oxbow.

Ofer, Avi
1993 Hebron. In *NEAEHL*, vol. 2, ed. E. Stern, 606–09. Jerusalem: Israel Exploration Society.

Ohata, Kiyoshi
1967 *Tel Zeror II: Report of the Excavation, Second Season, 1965*. Tokyo: Society for Near Eastern Studies.
1970 *Tel Zeror III: Report of the Excavation, Third Season, 1966*. Tokyo: Society for Near Eastern Studies.

Olmo Lete, Gregorio del, and Joaqín Sanmartín
2003 *A Dictionary of the Ugaritic Language in the Alphabetic Tradition*. HdO, Section I: The Near and Middle East. Leiden: Brill.

Opificius, Ruth
1964 Befestigungen des Zweistromlandes im Beginn des zweiten Jahrtausends. *BM* 3:78–90.

Oppenheim, A. Leo
1954 The Archives of the Palace of Mari II: A Review Article. *JNES* 13:141–48.
1977 *Ancient Mesopotamia: Portrait of a Dead Civilization*. Revised ed. Chicago: University of Chicago Press.

Oredsson, Dag
2000 *Moats in Ancient Palestine*. Coniectanea Biblica Old Testament Series. Stockholm: Almqvist & Wiksell International.

Oren, Eliezer D.
1975 The Pottery from the Achzib Defence System, Area D: 1963 and 1964 Seasons. *IEJ* 25:211–25.
1993a Haror, Tel. In *NEAEHL*, vol. 2, ed. E. Stern, 580–84. Jerusalem: Israel Exploration Society.
1993b Sera', Tel. In *NEAEHL*, vol. 4, ed. E. Stern, 1329–35. Jerusalem: Israel Exploration Society.
1996 The Middle Bronze Age Defence System at Tel Haror. *EI* 25:15–26.
1997a Haror, Tel. In *OEANE*, vol. 2, ed. E. M. Meyers, 474–76. New York: Oxford University Press.

1997b The "Kingdom of Sharuhen" and the Hyksos Kingdom. In *The Hyksos: New Historical and Archaeological Perspectives*, ed. E. D. Oren, 253–83. University Museum Monograph 96. Philadelphia: The University Museum, University of Pennsylvania.
1997c Sera', Tel. In *OEANE*, vol. 5, ed. E. M. Meyers, 1–2. New York: Oxford University Press.

Oren, Eliezer D., and M. A. Morrison
1986 Land of Gerar Expedition: Preliminary Report for the Seasons of 1982 and 1983. In *Preliminary Reports of ASOR-Sponsored Excavations 1980–1984*, ed. W. E. Rast, 57–87. Winona Lake, IN: Eisenbrauns.

Oren, Eliezer D., et al.
1991 Tel Haror–1988. *ESI* 9:69–73.

Orthmann, Winfried
1976 Mumbaqat 1974: Vorläufiger Bericht über die von der Deutschen-Orient-Gesellschaft mit Mitteln der Stiftung Volkswagenwerk unternommenen Ausgrabungen. *MDOG* 106:25–44.
1997 Chuera, Tell. In *OEANE*, vol. 1, ed. E. M. Meyers, 491–92. New York: Oxford University Press.

Orthmann, Winfried, and Hartmut Kühne
1974 Mumbaqat 1973: Vorläufiger Bericht über die von der Deutschen-Orient-Gesellschaft mit Mitteln der Stiftung Volkswagenwerk unternommenen Ausgrabungen. *MDOG* 106:53–97.

Orthmann, Winfried, et al.
1995 *Ausgrabungen in Tell Chuera in Nordost-Syrien I: Vorbericht über die Grabungskampagnen 1986 bis 1992*. Vorderasiatische Forschungen der Max Freiherr von Oppenheim-Stiftung 2. Saarbrücken: Saarbrücker Druckerei und Verlag.

Osten, Hans Henning von der
1937 *The Alishar Hüyük: Seasons of 1930–1932 Part II*. OIP 29. Chicago: University of Chicago Press.
1956 *Svenska Syrienexpeditionen, 1952/3 I, Tell es-Salihiye*. Skrifter Utgivna av Svenska Institutet I Athen. Lund: CVK Gleerup.

Özgüç, Tahsin
1997 Kaneš. In *OEANE*, vol. 3, ed. E. M. Meyers, 266–68. New York: Oxford University Press.

Paepe, Roland, et al.
1978 Sounding B: The Surrounding Wall of Tell ed-Der in Relation to the Regional Fluviatile System. In *Tell ed-Der II: Progress Reports (First Series)*, ed. L. de Meyer, 1–35. Leuven: Peeters.

Paley, Samuel M., et al.
1985 Tel Hefer—1984. *ESI* 3:44–45.

Paley, Samuel M., and Yosef Porath
1993 Hefer, Tel. In *NEAEHL*, vol. 2, ed. E. Stern, 609–14. Jerusalem: Israel Exploration Society.
1997 Early Middle Bronze Age IIA Remains at Tel el-Ifshar, Israel: A Preliminary Report. In *The Hyksos: New Historical and Archaeological Perspectives*, ed. E. D. Oren, 369–78. University Museum Monograph 96. Philadelphia: The University Museum, University of Pennsylvania.

Palumbo, Gaetano
1990 *The Early Bronze Age IV in the Southern Levant: Settlement Patterns, Economy, and Material Culture of a "Dark Age."* Contributi e Materiali di Archeologia Orientale 3. Rome: Università di Roma «La Sapienza».

Pardee, Dennis
1997a The Kirta Epic. In *The Context of Scripture I*, eds. W. W. Hallo and K. L. Younger, 333–43. Leiden: Brill.
1997b Ugaritic Prayer for a City Under Siege. In *The Context of Scripture I*, eds. W. W. Hallo and K. L. Younger, 283–85. Leiden: Brill.

Parr, Peter J.
1960 Excavations at Khirbet Iskander. *ADAJ* 4–5:128–36.
1968 The Origin of the Rampart Fortifications of Middle Bronze Age Palestine and Syria. *ZDPV* 84:18–45.
1972 Settlement Patterns and Urban Planning in the Ancient Levant: The Nature of the Evidence. In *Man, Settlement and Urbanism*, eds. P. J. Ucko et al., 805–10. Cambridge, MA: Schenkman.
1983 The Tell Nebi Mend Project. *AAS* 33:99–117.
1991 The Tell Nebi Mend Project: A Progress Report on the Institute of Archaeology's Excavations at Ancient Kadesh-on-the-Orontes in Syria. *Journal of the Ancient Chronology Forum* 4 (1990–1991):78–85.
1993 Tell Nebi Mend 1992. *PEQ* 125:88.
1997 Nebi Mend, Tell. In *OEANE*, vol. 4, ed. E. M. Meyers, 114–15. New York: Oxford Univ. Press.

Parrot, André
1958 *Mission archéologique de Mari II: Le palais*. BAH 68. Paris: Institut Français d'Archéologie de Beyrouth.
1964 Les fouilles de Mari: treizième campagne (printemps 1963). *Syria* 41:3–24.

Peilstöcker, Martin
2003 The Akko Plain (Israel) during the Middle Bronze Age—Urbanism in a Mediterranean Coastal Plain. Paper presented at the Annual Meeting of the American Schools of Oriental Research, Atlanta.

Pennells, Ernest
1983 Middle Bronze Age Earthworks: A Contemporary Engineering Evaluation. *BA* 46:57–61.

Peregrine, Peter N., et al.
1997 Geomagnetic Mapping of the Outer Town. In *Subsistence and Settlement in a Marginal Environment: Tell es-Sweyhat, 1989–1995 Preliminary Report*, ed. R. L. Zettler. MASCA Research Papers in Science and Archaeology 14. Philadelphia: The University Museum, University of Pennsylvania.

Petrie, William M. Flinders
1908 *Hyksos and Israelite Cities*. British School of Archaeology in Egypt 12. London: British School of Archaeology in Egypt.
1928 *Gerar*. British School of Archaeology in Egypt 43. London: British School of Archaeology in Egypt.
1931 *Ancient Gaza I*. British School of Archaeology in Egypt 53. London: British School of Archaeology in Egypt.

1932 *Ancient Gaza II.* British School of Archaeology in Egypt 54. London: British School of Archaeology in Egypt.

1934 *Ancient Gaza IV.* British School of Archaeology in Egypt 56. London: British School of Archaeology in Egypt.

Petrie, William M. Flinders, and Olga Tufnell

1930 *Beth-Pelet I (Tell Fara).* British School of Archaeology in Egypt 48. London: British School of Archaeology in Egypt.

Pettinato, Giovanni

1980 Bolletino militare della campagna di Ebla contro la città di Mari. *OrAnt* 19:231–45.

1981 *The Archives of Ebla: An Empire Inscribed in Clay.* Garden City, NY: Doubleday.

1991 *Ebla: A New Look at History.* Maryland: Johns Hopkins University.

Peyronel, Luca

2000 The MB II Fortress V at Tell Mardikh-Ebla (Syria): Preliminary Analysis of Architectural Context and Archaeological Materials. In *Proceedings of the First International Congress on the Archaeology of the Ancient Near East, Rome May 18th–23rd 1998*, vol. 2, eds. P. Matthiae et al., 1353–77. Rome: Università degli Studi di Roma "La Sapienza".

Pézard, Maurice

1931 *Qadesh, mission archéologique à Tell Nebi Mend, 1921–1922.* Paris: Librairie Orientaliste Paul Geuthner.

Pfälzner, Peter

1987a A Short Account of the Excavation of Tell Bderi 1985. *AAS* 36–37:276–91.

1987b A Short Account of the Excavation of Tell Bderi 1986. *AAS* 36–37:292–303.

1989–1990 Tall Bderi 1985–1987. *AfO* 36–37:212–21.

1990 Tell Bderi: The development of a Bronze Age Town. In *The Near East in Antiquity: German Contributions to the Archaeology of Jordan, Palestine, Syria, Lebanon and Egypt*, ed. S. Kerner, 63–79. Amman: Goethe-Institute.

Philip, Graham

1989 *Metal Weapons of the Early and Middle Bronze Ages in Syria-Palestine.* BAR International Series 526. Oxford: Archaeopress.

Pinnock, Frances

2001 The Urban Landscape of Old Syrian Ebla. *JCS* 53:13–33.

Pitard, Wayne T.

1987 *Ancient Damascus: A Historical Study of the Syrian City-State from Earliest Times until Its Fall to the Assyrians in 732 B.C.E.* Winona Lake, IN: Eisenbrauns.

1998 Before Israel: Syria Palestine in the Bronze Age. In *The Oxford History of the Biblical World*, ed. M. D. Coogan, 33–77. New York: Oxford University Press.

Poidebard, R. P. Antoine

1934 *Le trace de Rome dans le dèsert de Syrie: le limes de Trajan à la conquête arabe; recherches aériennes (1925–1932).* BAH 18. Paris: Paul Geuthner.

Portugali, Yuval

1982 A Field Methodology for Regional Archaeology. *TA* 9:170–88.

Posener, G., et al.

1971 Syria and Palestine c. 2160–1780 B.C. In *Early History of the Middle East*, 532–94. The Cambridge Ancient History I:2. Cambridge: Cambridge University Press.

Postgate, J. Nicholas

1992 *Early Mesopotamia: Society and Economy at the Dawn of History.* London: Routledge.

Powell, Marvin A., ed.

1987 *Labor in the Ancient Near East.* New Haven: American Oriental Society.

1990 Maße und Gewichte. In *Reallexikon der Assyriologie und Vorderasiatischen Archäologie*, vol. 7, ed. D. O. Edzard, 457–517. Berlin: Walter de Gruyter.

Prausnitz, Moshe W.

1973 Akziv and Avdon: On the Planning of a Port and a Fortress City in the Plain of Acre. *EI* 11:219–23.

1975 The Planning of the Middle Bronze Age Town at Achzib and Its Defences. *IEJ* 25:202–10.

1993a Acco (Plain of Acco). In *NEAEHL*, vol. 1, ed. E. Stern, 31. Jerusalem: Israel Exploration Society.

1993b Bira, Tel. In *NEAEHL*, vol. 1, ed. E. Stern, 262–63. Jerusalem: Israel Exploration Society.

Prausnitz, Moshe W., and Eilat Mazar

1993 Achzib. In *NEAEHL*, vol. 1, ed. E. Stern, 32–36. Jerusalem: Israel Exploration Society.

Pritchard, James Bennett, ed.

1969a *The Ancient Near East in Pictures Relating to the Old Testament.* 2nd ed. Princeton: Princeton University Press.

1969b *Ancient Near Eastern Texts Relating to the Old Testament.* 3rd ed. Princeton: Princeton University Press.

Pumpelly, Raphael, ed.

1908 *Explorations in Turkestan, Expedition of 1904: Prehistoric Civilizations of Anau, Origins, Growth, and Influence of Environment.* Washington, DC: Carnegie Institution.

Quenet, Philippe

1997 La séquence stratigraphique du IIIe millénaire (chantier G). In *Tell Beydar, Three Seasons of Excavations (1992–1994)*, eds. M. Lebeau and A. Suleiman, 169–77. Subartu 3. Turnhout: Brepols.

Raban, Avner, ed.

1982 *Map of Nahalal (28) 16–23.* ASI. Jerusalem: Israel Antiquities Authority.

1991 The Port City of Akko in the MBII. *Michmanim* 5:17*–34*.

1999 *Map of Mishmar Ha-'Emeq (32).* ASI. Jerusalem: Israel Antiquities Authority.

Randall-Maciver, David, and Charles Leonard Woolley
1911 *Buhen*. Eckley B. Coxe Junior Expedition to Nubia. Philadelphia: The University Museum, University of Pennsylvania.

Reade, Julian E.
1964 El-Mutabbaq and Umm Rus. *Sumer* 20:83–89.
1968 Tell-Taya (1967): Summary Report. *Iraq* 30:234–64.

Redford, Donald B.
1992 *Egypt, Canaan, and Israel in Ancient Times*. Princeton: Princeton University Press.

Reich, Ronny
1987 Four Notes on Jerusalem. *IEJ* 37:158–67.

Reich, Ronny, and Hannah J. Katzenstein
1992 Glossary of Architectural Terms. In *The Architecture of Ancient Israel: From the Pre-historic to the Persian Periods*, eds. A. Kempinski and R. Reich, 311–22. Jerusalem: Israel Exploration Society.

Reich, Ronny, and Eli Shukron
1998 Jerusalem, the Gihon Spring. *HA* 108:136–37.
1999a Jerusalem, the Gihon Spring. *HA—ESI* 110:77–8*.
1999b Jerusalem, the Gihon Spring. *HA—ESI* 109:63–4*.
1999c Light at the End of the Tunnel: Warren's Shaft Theory of David's Conquest Shattered. *BAR* 25:22–33, 72.
2000 The Excavations at the Gihon Spring and Warren's Shaft System in the City of David. In *Ancient Jerusalem Revealed*, ed. H. Geva, 327–39. Jerusalem: Israel Exploration Society.
2003 Notes on the Gezer Water System. *PEQ* 135:22–29.
2004 The History of the Gihon Spring in Jerusalem. *Levant* 36:211–23.

Reisner, George A., ed.
1931 *Mycerinus: The Temple of the Third Pyramid at Egypt*. Cambridge, MA: Harvard University.
1960 Egyptian Forts from Halfa to Semna. *Kush* 8:11–24.

Renfrew, Colin
1975 Trade as Action at a Distance: Questions of Integration and Communication. In *Ancient Civilization and Trade*, eds. J. A. Sabloff and C. C. Lamberg-Karlovsky, 3–59. Albuquerque, NM: University of New Mexico.
1984 *Approaches to Social Archaeology*. Edinburgh: Edinburgh University Press.

Richard, Suzanne
1990 The 1987 Expedition to Khirbet Iskander and Its Vicinity: Fourth Preliminary Report. In *Preliminary Reports of ASOR-Sponsored Excavations 1983–87*, ed. W. E. Rast, 33–58. BASOR Supplement 26. Baltimore: Johns Hopkins University.
1993 Iskander, Khirbet. In *NEAEHL*, vol. 2, ed. E. Stern, 649–52. Jerusalem: Israel Exploration Society.
1997 Iskander, Khirbet. In *OEANE*, vol. 3, ed. E. M. Meyers, 188–89. New York: Oxford University Press.

Richard, Suzanne, and Roger S. Boraas
1984 Preliminary Report of the 1981–82 Seasons of the Expedition to Khirbet Iskander and its Vicinity. *BASOR* 254:63–87.
1988 Khirbet Iskander, Jordan: 1984 Season. In *Preliminary Reports of ASOR-Sponsored Excavations 1982–85*, ed. W. E. Rast, 107–30. BASOR Supplement 25. Baltimore: Johns Hopkins University.

Ristvet, Lauren, and Harvey Weiss
2004 Tell Leilan Excavations 2002. *OE* 2004.

Ritner, Robert K.
1997 Execration Texts. In *The Context of Scripture I: Canonical Compositions from the Biblical World*, ed. W. W. Hallo, 50–52. Leiden: Brill.

Ronzevalle, P. L.
1914 Le campe retranché d'El Mišrifé. *Mélanges de la Faculté Orientale (Beyrouth)* 7:109–26.

Rosen, Arlene Miller
1986 *Cities of Clay: The Geoarchaeology of Tells*. Prehistoric Archaeology and Ecology Series. Chicago: University of Chicago Press.

Rowe, Alan
1935 The 1934 Excavations at Gezer. *PEFQS* 1935: 19–33.

Rowton, Michael B.
1974 Enclosed Nomadism. *JESHO* 17:1–30.

Saade, Gabriel
1979–1980 À la recherche des villes et villages du royaume ougaritien. *AAS* 29–30:215–30.

Saghieh, Muntaha
1983 *Byblos in the Third Millennium B.C.* Warminster: Aris & Phillips.

Sallaberger, Walther, and Jason Ur
2004 Tell Beydar/Nabada in its Regional Setting. In *Third Millennium Cuneiform Texts from Tell Beydar (Seasons 1996–2002)*, eds. L. Milano et al., 51–71. Subartu 12. Turnhout: Brepols.

Salonen, Erkki
1965 *Die Waffen der alten Mesopotamier, eine lexikalische und kulturgeschichtliche Untersuchung*. Studia Orientalia 33. Helsinki: Societas Orientalis Fennica.

Sarie', Issa
1998 Area C. The Middle Bronze I–II Earthen Rampart. In *Scavi a Gerico, 1997. Relazione preliminaire sulla prima campagna di scavi e prospezioni archeologiche a Tell es-Sultan, Palestina*, eds. N. Marchetti and L. Nigro, 103–15. Quaderni di Gerico 1. Rome: Università di Roma «La Sapienza».

Sarzec, E. de
1894–1912 *Découvertes en Chaldée*. Paris: E. Leroux.

Sasson, Jack M.
1969 *The Military Establishment at Mari*. Studia Pohl 3. Rome: Pontifical Biblical Institute.

Sauer, James A.
1986 Transjordan in the Bronze and Iron Ages: A Critique of Glueck's Synthesis. *BASOR* 263:1–26.

Sauvaget, Jean
1929 L'enceinte primitive de la ville d'Alep. *Mélanges de l'Institut Français de Damas* 1:133–59.

Schaeffer, Claude F. A.
1939 Les fouilles de Ras Shamra-Ugarit 10e et 11e campagnes (automne et hiver 1938–1939). *Syria* 20:277–92.
1948 *Stratigraphie comparée et chronologie de l'Asie Occidentale (IIIe et IIe millénaires), Vol. I: Syrie, Palestine, Asie Mineure, Chypre, Perse et Caucase.* Oxford.
1949 *Ugaritica II: nouvelles études relatives aux découvertes de Ras Shamra.* MRS 5. Paris: Institut français d'archéologie de Beyrouth.
1951 Reprise des recherches archéologiques à Ras Shamra-Ugarit sondages de 1948 et 1949 et campagne de 1950. *Syria* 28:1–21.

Schloen, J. David
1997 Ashkelon. In *OEANE*, vol. 1, ed. E. M. Meyers, 220–23. New York: Oxford University Press.
2001 *The House of the Father as Fact and Symbol: Patrimonialism in Ugarit and the Ancient Near East.* SAHL 2. Winona Lake, IN: Eisenbrauns.

Schmid, Hansjörg, et al.
1971 Die Grabung auf dem Tell Habuba Kabira. *MDOG* 103:8–37.
1973 Habuba Kabira-Tall. *MDOG* 105:38–67.

von Schuler, Einar, et al.
1969 Die Frühjahrskampagne 1969 in Habuba Kabira. *MDOG* 101:37–64.

Schulman, Alan R.
1964 Siege Warfare in Pharaonic Egypt. *Natural History* 73:12–21.
1982 The Battle Scenes of the Middle Kingdom. *JSSEA* 12:165–83.

Schumacher, Gottlieb
1908 *Tell el-Mutesellim I: Bericht über die 1903–1905.* Leipzig.

Schwartz, Glenn M.
1988 *A Ceramic Chronology from Tell Leilan, Operation 1.* Yale Tell Leilan Research 1. New Haven: Yale University Press.

Schwartz, Glenn M., et al.
2000 Excavation and Survey in the Jabbul Plain, Western Syria: The Umm el-Marra Project 1996–1997. *AJA* 104:419–62.
2003 A Third-Millennium B.C. Elite Tomb and Other New Evidence from Tell Umm el-Marra, Syria. *AJA* 107:325–61.
2006 A Third-Millennium B.C. Elite Mortuary Complex at Umm el-Marra, Syria: 2002 and 2004 Excavations. *AJA* 110:603–41.

Schwartz, Glenn M., and Harvey Weiss
1992 Syria, ca. 10,000–2,000. In *Chronologies in Old World Archaeology*, ed. R. W. Ehrich, Vol I:221–43; Vol II:185–202. Chicago: University of Chicago Press.

Scurlock, Jo Anne
1989 Assyrian Battering Rams Revisited. *SAAB* 3:129–31.

Seeden, Helga
1986 Bronze Age Village Occupation in Busra: AUB Excavations on the Northwest Tell, 1983–84. *Berytus* 34:11–81.

Seger, Joe D.
1974 The Middle Bronze IIC Date of the East Gate at Shechem. *Levant* 6:117–30.
1975 The MB II Fortifications at Shechem and Gezer: A Hyksos Retrospective. *EI* 12:34–45.

Sellers, Orid Rogers
1933 *The Citadel of Beth-Zur.* Philadelphia: Westminster.

Sellers, Orid Rogers, et al.
1968 *The 1957 Excavation at Beth-Zur.* AASOR 38. Cambridge, MA: American Schools of Oriental Research.

Sellin, Ernst, and Hans Steckeweh
1941 Sichem. *ZDPV* 64:1–20.

Sellin, Ernst, and Carl Watzinger
1913 *Jericho: die Ergebnisse der Ausgrabungen.* Leipzig.

Serangeli, Flavia
1978 Le liste di censo di Alalah IV. *Vicino Oriente* 1:99–131.

Seton-Williams, M. V.
1980 Appendice: Tell Keisan, 1935–1936. In *Tell Keisan (1971-1976): une cité phénicienne en Galilée*, eds. J. Briend and J.-B. Humbert, 381–92. Orbis biblicus et orientalis, series archaeologica 1. Freiburg: Éditions universitaires.

Shea, William H.
1979 The Conquests of Sharuhen and Megiddo Reconsidered. *IEJ* 29:1–5.

Shiloh, Yigal
1984 *Excavations at the City of David I, 1978–1982: Interim Report of the First Five Seasons.* Qedem 19. Jerusalem: Hebrew University.

Sigrist, Marcel, ed.
1988 *Isin Year Names.* Institute of Archaeology Publications, Assyriological Series 2. Berrien Springs, MI: Andrews University Press.
1990 *Larsa Year Names.* Institute of Archaeology Publications, Assyriological Series 3. Berrien Springs, MI: Andrews University Press.

Sigrist, Marcel, and Tohru Gomi
1991 *The Comprehensive Catalogue of Published Ur III Tablets.* Bethesda, MD: CDL.

Singer, Itamar
1983 The Middle Bronze Age Fortified Enclosure. In *Ergebnisse der Ausgrabungen auf Hirbet el-Mšaš (Tel Masos) 1972–75*, eds. V. Fritz and A. Kempinski, 186–97. Abhandlung des Deutschen Palästinavereins. Wiesbaden: Harrassowitz.

Smith, G. Elliot
1912 *The Royal Mummies.* Catalogue Général des Antiquités Égyptiennes du Musée du Caire 72. Paris.

Smith, H. S.
1966 Kor: Report on the Excavations of the Egypt Exploration Society at Kor, 1965. *Kush* 14:187–243.

Smith, Monica L.
2005 Networks, Territories, and the Cartography of Ancient States. *Annals of the Association of American Geographers* 95:832–49.

Smith, Robert H., and Timothy F. Potts
1992 The Middle and Late Bronze Ages. In *Pella in Jordan 2: The Second Interim Report of the Joint University of Sydney and College of Wooster Excavations at Pella 1982–1985*, eds. A. W. McNicoll et al., 35–81. Mediterranean Archaeology Supplement 2. Sydney: MEDITARCH.

Smith, Stuart Tyson
1995 *Askut in Nubia: The Economics and Ideology of Egyptian Imperialism in the Second Millennium B.C.* London: Kegan Paul International.

Soden, Wolfram, von, ed.
1965–1981 *Akkadisches Handwörterbuch*. 3 vols. Wiesbaden: Harrassowitz.

Soldt, Wilfred H. van
1996 Studies in the Topography of Ugarit (1): The Spelling of Ugaritic Toponyms. *UF* 28:653–92.
1997 Studies in the Topography of Ugarit (2): The Borders of Ugarit. *UF* 29:683–703.
1998 Studies in the Topography of Ugarit (3): Groups of Towns and Their Locations. *UF* 30:703–44.
1999 Studies in the Topography of Ugarit (4): Town Sizes and Districts. *UF* 31:749–76.
2005 *The Topography of the City-State of Ugarit*. AOAT 324. Münster: Ugarit-Verlag.

Sollberger, Edmond
1980 The So-Called Treaty between Ebla and 'Ashur'. *StEb* 3:129–55.

Sollberger, Edmond, and Jean Robert Kupper
1971 *Inscriptions royales sumeriennes et akkadienes*. Paris: Les editions du Cerf.

Spencer, A. Jeffrey
1979 *Brick Architecture in Ancient Egypt*. Warminster: Aris & Phillips.
1994 Mud Brick: Its Decay and Detection in Upper and Lower Egypt. In *The Unbroken Reed: Studies in the Culture and Heritage of Ancient Egypt in Honour of A. F. Shore*, vol. 11, eds. C. Eyre et al., 315–20. Occasional Publications. London: Egypt Exploration Society.

Stager, Lawrence E.
1985 The Archaeology of the Family in Ancient Israel. *BASOR* 260:1–35.
1991a *Ashkelon Discovered: From Canaanites and Philistines to Romans and Moslems*. Washington, DC: Biblical Archaeology Society.
1991b The Massive Middle Bronze Fortifications—How did they Work? *BAR* 17:30.
1992 The Periodization of Palestine from Neolithic through Early Bronze Times. In *Chronologies in Old World Archaeology*, ed. R. W. Ehrich, vol. I:22–41; vol. II:46–60. Chicago: University of Chicago Press.
1993 Ashkelon. In *NEAEHL*, vol. 1, ed. E. Stern, 103–12. Jerusalem: Israel Exploration Society.
1995 The Impact of the Sea Peoples in Canaan (1185–1050 BCE). In *The Archaeology of Society in the Holy Land*, ed. T. E. Levy, 332–48. New York: Facts on File.
1999 The Fortress-Temple at Shechem and the "House of El, Lord of the Covenant". In *Realiea Dei: Essays in Archaeology and Biblical Interpretation in Honor of Edward F. Campbell, Jr. at His Retirement*, eds. P. H. Williams, Jr. and T. Hiebert, 228–49. Atlanta: Scholars.
2001 Port Power in the Early and Middle Bronze Age: the Organization of Maritime Trade and Hinterland Production. In *Studies in the Archaeology of Israel and Neighboring Lands in Memory of Douglas L. Esse*, ed. S. R. Wolff, 625–38. SAOC 59. Winona Lake, IN: Eisenbrauns.
2002 The MB IIA Ceramic Sequence at Tel Ashkelon and Its Implications for the "Port Power" Model of Trade. In *The Middle Bronze Age in the Levant: Proceedings of an International Conference on MB IIA Ceramic Material, Vienna, 24th–26th of January 2001*, ed. M. Bietak, 353–62. CCEM 3. Vienna: Österreichischen Akademie der Wissenschaften.
2006 The House of the Silver Calf of Ashkelon. In *Timelines: Studies in Honour of Manfred Bietak*, eds. E. Czerny et al., 403–10. Orientalia Lovaniensia Analecta 149. Leuven: Peeters.

Starkey, James Leslie
1934 Excavations at Tell el-Duweir, 1933–1935. *PEFQS* 1934:164–75.

Stein, Diana L.
1997 Alalakh. In *OEANE*, vol. 1, ed. E. M. Meyers, 55–59. New York: Oxford University Press.

Stein, Gil J.
1991 Operation 4: The City Wall. *AJA* 94:547–55.

Steiner, Margreet L., ed.
2001 *Excavations by Kathleen M. Kenyon in Jerusalem 1961–1967, Volume III: The Settlement in the Bronze and Iron Ages*. Copenhagen International Series 9. London: Sheffield.

Steinkeller, Piotr
1987 Battering Rams and Siege Engines at Ebla. *NABU* 1987:14.

Stern, Ephraim
1984 *Excavations at Tel Mevorakh (1973–1976), Pt. 2: The Bronze Age*. Qedem 18. Jerusalem: Institute of Archaeology, Hebrew University of Jerusalem.
1993 *The New Encyclopedia of Archaeological Excavations in the Holy Land*. English ed. 4 vols. New York: Simon & Schuster.

Stewart, James R.
1974 *Tell El 'Ajjul: The Middle Bronze Age Remains*. SMA 38. Göteborg: Åström.

Stiebing, William H., Jr.
1970 Burial Practices in Palestine during the Bronze Age. Unpublished Ph.D. Dissertation, University of Pennsylvania.

Stone, Elizabeth C.

1997 City-States and Their Centers: The Mesopotamian Example. In *The Archaeology of City-States: Cross-Cultural Approaches*, eds. D. L. Nichols and T. H. Charlton, 15–26. Washington, DC: Smithsonian Institution.

Stout, Margaret E.

1977 Clay Sling-Bullets from Tell Sweyhat. *Levant* 9:63–65.

Strange, John

2001 The Late Bronze Age. In *The Archaeology of Jordan*, eds. B. MacDonald et al., 271–89. Sheffield: Sheffield Academic Press.

Strommenger, Eva

1976 Fünfter vorläufiger Bericht über die von der Deutschen Orient-Gesellschaft mit Mitteln der Stiftung Volkswagenwerk in Habuba Kabira unternommenen archäologischen Untersuchungen (Kampagnen 1974, 1974, 1975). *MDOG* 108:5–22.

1979 Ausgrabungen der Deutschen Orient-Gesellschaft in Habuba Kabira. In *Archaeological Reports from the Tabqa Dam Project*, ed. D. N. Freedman, 63–78. AASOR 44. Cambridge, MA: American Schools of Oriental Research.

1980 *Habuba Kabira: Eine Stadt vor 5000 Jahren.* Mainz am Rhein: Verlag Philipp von Zabern.

1981–1982
Tall Bi'a. *AfO* 28:210–12.

1991 Ausgrabungen in Tall Bi'a 1990. *MDOG* 123:7–34.

Strommenger, Eva, et al.

1986 Ausgrabungen in Tall Bi'a 1984. *MDOG* 118:7–44.

Suleiman, Antoine

1983 Excavations at Ansari-Aleppo for the Seasons 1973–1980. *AAS* 33:179–214.

1985 Fouilles d'Alep (Al-Ansari). *Syria* 62:135.

2003 The Northern Stratigraphical Sounding (Field G). In *Tell Beydar, 1995–1999 Seasons of Excavations. A Preliminary Report*, eds. M. Lebeau and A. Suleiman, 301–13. Subartu 10. Turnhout: Brepols.

Sun Tzu

1963 *The Art of War*. Trans. S. B. Griffith. London: Oxford University Press.

Sürenhagen, Dietrich

1999 Tell Gindaris and the Western Afrin Valley. *AAS* 43:159–67.

Tallon, Maurice, S. J.

1956 Une nouvelle enceinte antique en Émésêne. *AAS* 6:51–62.

Tarler, David, et al.

1991 Tell el-Hammah 1988. *ESI* 9:134–35.

Tefnin, Roland

1978–1979
Tall Abu Danna. *AfO* 26:145–47.

1979a Deux campagnes de fouilles au Tell Abou Danné (1975–1976). In *Le moyen Euphrate: actes du colloque de Strasbourg 10–12 mars 1977*, ed. J.-C. Margueron, 179–99. Leiden: Brill.

1979b Exploration archéologique au nord du lac de Djabboul (Syrie): une campagne de sondages sur le site d'Oumm el-Marra, 1978. *Annuaire de l'Institut de Philologie et d'Histoire orientales et slaves* 23:71–94.

1979c Tell Abou Danné, une ville fortifiée de l'âge du bronze du Syrie du nord. *Archéologia* 129:42–50.

1981–1982
Tall Abu Danna. *AfO* 28:201–02.

1983 Aperçu sur neuf campagnes de fouilles belges aux Tells Abou Danne et Oumm el-Marra (1975–1983). *AAS* 33:141–52.

Thalmann, Jean-Paul

1978 Tell 'Arqa (1978–1979) Rapport Provisoire. *BMB* 30:61–75.

1991 L'âge du Bronze à Tell 'Arqa (Liban): Bilan et perspectives (1981–1991). *Berytus* 39:21–38.

2002 Pottery of the Early Middle Bronze Age at Tell Arqa and in the Northern Levant. In *The Middle Bronze Age in the Levant: Proceedings of an International Conference on MB IIA Ceramic Material, Vienna, 24th–26th of January 2001*, ed. M. Bietak, 363–77. CCEM 3. Vienna: Österreichischen Akademie der Wissenschaften.

Trigger, Bruce G.

1982 The reasons for the construction of the second cataract forts. *JSSEA* 12:1–6.

Tubb, Jonathan N.

1985 Some Observations on Spearheads in Palestine in the Middle and Late Bronze Ages. In *Palestine in the Bronze and Iron Ages: Papers in Honour of Olga Tufnell*, ed. J. N. Tubb, 189–96. Occasional Publication 11. London: Institute of Archaeology.

1993 Sa'idiyeh, Tell es-. In *NEAEHL*, ed. E. Stern, 1295–300. Jerusalem: Israel Exploration Society.

Tufnell, Olga

1953 *Lachish III (Tell ed-Duweir): The Iron Age*. The Wellcome-Marston Archaeological Research Expedition to the Near East, Vol. III. London: Oxford University Press.

1958 *Lachish IV (Tell ed-Duweir): The Bronze Age.* The Wellcome-Marston Archaeological Research Expedition to the Near East, Vol. IV. London: Oxford University Press.

Tufnell, Olga, et al.

1940 *Lachish II: The Fosse Temple*. The Wellcome-Marston Archaeological Research Expedition to the Near East, Vol. II. London.

Tufnell, Olga, and Aharon Kempinski

1993 'Ajjul, Tell el-. In *NEAEHL*, vol. 1, ed. E. Stern, 49–53. Jerusalem: Israel Exploration Society.

Ussishkin, David

1982 *The Conquest of Lachish by Sennacherib*. Tel Aviv: Tel Aviv University, The Institute of Archaeology.

1989 Notes on the Fortifications of the Middle Bronze II Period at Jericho and Shechem. *BASOR* 276:29–53.

1991 Fortifications at Hazor. *EI* 21:1–5 (Hebrew).

1992 Notes on the Middle Bronze Age Fortifications of Hazor. *TA* 19:274–81.

1993 Lachish. In *NEAEHL*, vol. 3, ed. E. Stern, 897–911. Jerusalem: Israel Exploration Society.

2004a Area P: The Middle Bronze Age Palace. In *The Renewed Archaeological Excavations at Lachish (1973–1994)*, vol. 1, ed. D. Ussishkin, 140–87. Monograph Series 22. Tel Aviv: Emery and Claire Yass Publications in Archaeology.

2004b The Mound and Excavation Strategy. In *The Renewed Archaeological Excavations at Lachish (1973–1994)*, vol. 1, ed. D. Ussishkin, 23–49. Monograph Series 22. Tel Aviv: Emery and Claire Yass Publications in Archaeology.

2004c *The Renewed Archaeological Excavations at Lachish (1973–1994)*. Monograph Series 22. 5 vols. Tel Aviv: Emery and Claire Yass Publications in Archaeology.

2004d A Synopsis of the Stratigraphical, Chronological and Historical Issues. In *The Renewed Archaeological Excavations at Lachish (1973–1994)*, vol. 1, ed. D. Ussishkin, 50–122. Monograph Series 22. Tel Aviv: Emery and Claire Yass Publications in Archaeology.

Uziel, Joe, and Aren M. Maeir

2005 Scratching the Surface at Gath: Implications of the Tell es-Safi/Gath Surface Survey. *TA* 32:50–75.

Van Beek, Gus W.

1992 The Rammed Earth Revetment at Tell Jemmeh. *EI* 23:4*–9*.

1993 Jemmeh, Tell. In *NEAEHL*, vol. 2, ed. E. Stern, 667–74. Jerusalem: Israel Exploration Society.

Van de Mieroop, Marc

1997 *The Ancient Mesopotamian City*. Oxford: Clarendon.

van Seters, John

1966 *The Hyksos: A New Investigation*. New Haven: Yale University Press.

Vaux, Roland de

1960 Chronique: Exploration de la région de Salt. *RB* 47:398–425.

1962 Les fouilles de Tell el-Far'ah. *RB* 69:212–63.

Vegetius Renatus, Flavius

1993 *Epitome of Military Science*. Translated by N. P. Milner. Translated Texts for Historians 16. Liverpool: Liverpool University Press.

Vercoutter, Jean

1955 Kor est-il Iken? Rapport préliminaire sur les fouilles françaises de Kor (Bouhen sud), Sudan, en 1954. *Kush* 3.

1966 Semna South Fort and the Records of the Nile Levels at Kumma. *Kush* 14:125–64.

1970 *Mirgissa I*. Mission Archéologique Française au Soudan 1. Paris: CNRS.

Vila, Andre

1970 L'armement de la forteresse de Mirgissa-Iken. *Revue d'Égypt* 22:171–99.

Villard, Pierre

2001 Les administrateurs de l'époque de Yasmah-Addu. In *Amurru 2: Mari, Ébla et les Hourrites, dix ans de travaux*, eds. J.-M. Durand and D. Charpin, 9–140. Paris: Éditions Recherches sur les Civilisations.

Vincent, Louis-Hugues

1911 *Underground Jerusalem: Discoveries on the Hill of Ophel (1909–1911)*. London: H. Cox, "Field" Office.

Voss, Ross Joseph

2002 A Sequence of Four Middle Bronze Age Gates in Ashkelon. In *The Middle Bronze Age in the Levant: Proceedings of an International Conference on MB IIA Ceramic Material, Vienna, 24th–26th of January 2001*, ed. M. Bietak, 379–84. CCEM 3. Vienna: Österreichischen Akademie der Wissenschaften.

Walters, Stanley D.

1970 *Water for Larsa: An Old Babylonian Archive dealing with Irrigation*. Yale Near Eastern Researches 4. New Haven: Yale University Press.

Warmenbol, Eugene

1980 Oumm el-Marra à l'époque des royaumes d'Ebla et de Iamkhad. *Akk* 19:58–59.

Weinstein, James M.

1975 Egyptian Relations with Palestine in the Middle Kingdom. *BASOR* 217:1–16.

1981 The Egyptian Empire in Palestine: A Reassessment. *BASOR* 241:1–28.

1984 Radiocarbon Dating in the Levant. *Radiocarbon* 26:297–366.

Weippert, Helga

1988 *Palästina in vorhellenistischer Zeit*. Handbuch der Archäologie, Vorderasien II.1. Munich: C. H. Beck'sche Verlagsbuchhandlung.

Weiss, Harvey

1985 Tell Leilan on the Habur Plains of Syria. *BA* 48:6–34.

1990 Third Millennium Urbanization: A Perspective from Tell Leilan. In *Tall al-Hamidiya 2*, eds. S. Eichler et al., 159–66. Orbis biblicus et orientalis, series archaeologica 6. Göttingen: Vandenhoeck & Ruprecht.

1991 Archaeology in Syria. *AJA* 95:683–740.

1997 Leilan, Tell. In *OEANE*, vol. 3, ed. E. M. Meyers, 341–47. New York: Oxford University Press.

2000 Beyond the Younger Dryas: Collapse as Adaptation to Abrupt Climate Change in Ancient West Asia and the Eastern Mediterranean. In *Environmental Disaster and the Archaeology of Human Response*, eds. G. Bawden and R. M. Reycraft, 75–98. Anthropological Papers 7. Albuquerque, NM: Maxwell Museum of Anthropology.

Weiss, Harvey, et al.

1991 1985 Excavations at Tell Leilan, Syria. *AJA* 94:529–81.

Werner, Peter, ed.

1998 *Tall Munbaqa: Bronzezeit in Syrien*. Veröffentlichungen des Hamburger Museums für Archäologie und die Geschichte Harburgs—Helms-Museum, Nr. 80. Hamburg: Wachholtz Verlag Neumünster.

Wheeler, R. E. Mortimer
1947 Harappan Chronology and the Rigveda. *Ancient India* 3:78–82.

Wilcke, Claus
1970 Zur Geschichte der Amurriter in der Ur-III Zeit. *Die Welt des Orients* 5:1–31.

Wilkinson, Tony J.
1993 Linear Hollows in the Jazira, Upper Mesopotamia. *Antiquity* 67:548–62.
2000 Regional Approaches to Mesopotamian Archaeology: The Contribution of Archaeological Surveys. *JAR* 8:219–67.
2003 *Archaeological Landscapes of the Near East.* Tucson: University of Arizona.

Wilkinson, Tony J., and D. J. Tucker
1995 *Settlement Development in the North Jazira, Iraq: A Study of the Archaeological Landscape.* Iraq. Baghdad: British School of Archaeology in Iraq.

Wimmer, Donald H.
1997 Safut, Tell. In *OEANE*, vol. 4, ed. E. M. Meyers, 448–50. New York: Oxford University Press.

Winlock, Herbert Eustis
1945 *The Slain Soldiers of Neb-hepet-Re' Mentu-hotpe.* Metropolitan Museum of Art Egyptian Expedition 16. New York: Metropolitan Museum of Art.

Winter, Irene J.
1973 North Syria in the Early First Millennium B.C. with Special Reference to Ivory Carving. Unpublished Ph.D. Dissertation, Columbia University.

Wiseman, Donald J.
1953 *The Alalakh Tablets.* Occasional Papers of the British Institute of Archaeology at Ankara. London.

Wolff, Samuel R.
1991 Archaeology in Israel. *AJA* 95:489–538.

Woolley, Charles Leonard
1921 *Carchemish: Report on the Excavations at Djerabis, Part II, The Town Defenses.* London: Trustees of the British Museum.
1953 *A Forgotten Kingdom, being a record of the results obtained from the excavation of two mounds, Atchana and Al Mina, in the Turkish Hatay.* London: Max Parish.
1955 *Alalakh: An Account of the Excavations at Tell Atchana in the Hatay, 1937–1949.* Reports of the Research Committee of the Society of Antiquaries of London, XVIII. Oxford: Oxford University Press.

Woolley, Charles Leonard, and Richard D. Barnett
1952 *Carchemish: Report on the Excavations at Djerabis, Part III, The Excavations in the Inner Town; The Hittite Inscriptions.* London: Trustees of the British Museum.

Wreszinksi, Walter, ed.
1935 *Atlas zur Altaegyptischen Kulturgeschichte II.* Leipzig: Hinrich.

Wright, George Ernest
1964 *Shechem: The Biography of a Biblical City.* London: Gerald Duckworth.
1971 The Archaeology of Palestine from the Neolithic through the Middle Bronze Age. *JAOS* 91:276–93.

Wright, G. R. H.
1968 Tell el-Yehudiyah and the Glacis. *ZDPV* 84:1–17.
1969 Iran and the Glacis. *ZDPV* 85:24–34.
1984 The Monumental City Gate in Palestine and Its Foundations. *ZA* 74:267–89.
1985a *Ancient Building in South Syria and Palestine.* HdO, 7:2, b. Leiden: Brill.
1985b The City Gates at Shechem, Simple Reconstruction Drawings. *ZDPV* 101:1–8.
1997 Building Materials and Techniques: Materials and Techniques of the Bronze and Iron Ages. In *OEANE*, vol. 1, ed. E. M. Meyers, 363–67. New York: Oxford University Press.
2002 *Shechem III, The Stratigraphy and Architecture of Shechem/Tell Balatah, Volume 2: The Illustrations.* American Schools of Oriental Research Archaeological Reports. Atlanta: Scholars.

Wulff, Hans E.
1966 *Traditional Crafts of Persia: Their Development, Technology and Influence on Eastern and Western Civilizations.* Cambridge, MA: Massachusetts Institute of Technology.

Yadin, Yigael
1955a The Earliest Record of Egypt's Military Penetration into Asia? *IEJ* 5:1–16.
1955b Hyksos Fortifications and the Battering Ram. *BASOR* 137:23–32.
1963 *The Art of Warfare in Biblical Lands in the Light of Archaeological Study.* Jerusalem: McGraw-Hill.
1970 Warfare in the Second Millennium B.C.E. In *The World History of the Jewish People, Vol. 2*, ed. B. Mazar, 127–59. Ancient Times (Patriarchs and Judges). Tel Aviv: Massada.
1972a The Earliest Representation of a Siege from Mari and a 'Scythian Bow'. *IEJ* 22:91–94.
1972b *Hazor: The Head of All Those Kingdoms.* The Schweich Lectures of the British Academy. London: Oxford University Press.
1972c Megiddo. *IEJ* 22:161–64.
1973 The Tell Beit Mirsim G-F Alleged Fortifications. *BASOR* 212:23–25.
1975 *Hazor: The Rediscovery of a Great Citadel of the Bible.* London: Weinfeld and Nicolson.
1977 The Nature of the Settlements during the Middle Bronze IIA and the Problem of the Apheq Fortifications. *EI* 13:91–105 (Hebrew), 293*–94* (English summary).
1978 The Nature of the Settlements during the Middle Bronze IIA Period in Israel and the Problem of the Aphek Fortifications. *ZDPV* 94:1–23.
1982 Is the Biblical Account of the Israelite Conquest of Canaan Historically Reliable? *BAR* 8:16–23.

Yadin, Yigael, et al.
1958 *Hazor I: An Account of the First Season of Excavations, 1955.* Jerusalem: Magnes.
1960 *Hazor II: An Account of the Second Season of Excavations, 1956.* Jerusalem: Magnes.

1961 *Hazor III–IV: An Account of the Third and Fourth Seasons of Excavations, 1957–1958, Plates.* Jerusalem: Magnes.

1989 *Hazor III–IV: An Account of the Third and Fourth Seasons of Excavations, 1957–1958, Text.* Jerusalem: Magnes.

Yeivin, Shemuel

1951 The Date of the Tunnel in the Lachish Glacis. *EI* 1:29–31.

Yeivin, Zeev

1993 Makhruq, Khirbet el- (Area B). In *NEAEHL*, vol. 3, ed. E. Stern, 930–31. Jerusalem: Israel Exploration Society.

Yener, Kutlu Aslihan

1998 A View from the Amuq in South-Central Turkey: Societies in Transformation in the Second Millennium B.C. In *The Aegean and the Orient in the Second Millennium: Proceedings of the 50th Anniversary Symposium, Cincinnati, 18–20 April 1997*, eds. E. H. Cline and D. Harris-Cline, 366–74. Aegaeum 18. Liège: Université de Liège.

2005 *The Amuq Valley Regional Projects, Volume 1: Surveys in the Plain of Antioch and Orontes Delta, Turkey, 1995–2002.* OIP 131. Chicago: Oriental Institute of the University of Chicago.

Yener, Kutlu Aslihan, et al.

1996 The Oriental Institute Amuq Valley Projects, 1995. *Anatolica* 22:49–83.

2000 The Amuq Valley Regional Project, 1995–1998. *AJA* 104:163–220.

Yogev, Ora

1993 Nahariya (Excavations on the Mound). In *NEAEHL*, vol. 3, ed. E. Stern, 1088–90. Jerusalem: Israel Exploration Society.

Yon, Marguerite

1992 Ugarit: The Urban Habitat, The Present State of the Archaeological Picture. *BASOR* 286:19–34.

1997a *La cité d'Ougarit sur le tell de Ras Shamra.* Paris: Éditions Recherches sur les Civilisations.

1997b Ugarit. In *OEANE*, vol. 5, ed. E. M. Meyers, 255–62. New York: Oxford University Press.

Yuhong, Wu

1994 *A Political History of Eshnunna, Mari and Assyria during the Early Old Babylonian Period (From the End of Ur III to the Death of Šamši-Adad).* Periodic Publications on Ancient Civilizations 2. Changchun, China: Institute for History of Ancient Civilizations.

Yurco, Frank J.

1986 Merneptah's Canaanite Campaign. *JARCE* 23:189–216.

Žabkar, L.

1982 Semna South: A Preliminary Report on the 1966–1968 Excavations of the University of Chicago Oriental Institute Expedition to Sudanese Nubia. *JARCE*. 19:7–50.

Zaccagnini, Carlo

1979 *The Rural Landscape of the Land of Arraphe.* Rome: Università di Roma, Istituto di studi del Vicino oriente.

Zadok, Ran

2004 Appendix: The Name 'Lachish'. In *The Renewed Archaeological Excavations at Lachish (1973–1994)*, vol. 1, ed. D. Ussishkin, 120–22. Monograph Series 22. Tel Aviv: Emery and Claire Yass Publications in Archaeology.

Zayadine, Fawzi, et al.

1989 The 1988 Excavations on the Citadel of Amman: Lower Terrace, Area A. *ADAJ* 33:357–63.

Zertal, Adam

1993 Samaria (Region): The Mount Manasseh (Northern Samarian Hills Survey). In *NEAEHL*, vol. 4, ed. E. Stern, 1311–12. Jerusalem: Israel Exploration Society.

2004 *The Manasseh Hill Country Survey I: The Shechem Syncline.* CHANE 21.1. Leiden: Brill.

Zettler, Richard L., ed.

1997a *Subsistence and Settlement in a Marginal Environment: Tell es-Sweyhat, 1989–1995 Preliminary Report.* MASCA Research Papers in Science and Archaeology 14. Philadelphia: The University Museum, University of Pennsylvania.

1997b Surface Collections and Excavations in the Lower Town and Lower Town South. In *Subsistence and Settlement in a Marginal Environment: Tell es-Sweyhat, 1989–1995 Preliminary Report*, ed. R. L. Zettler, 35–72. MASCA Research Papers in Science and Archaeology 14. Philadelphia: The University Museum, University of Pennsylvania.

Zohar, Mattanyah

1980 Tell Marjamah ('Ein Samiyeh). *IEJ* 30:219.

INDEX

Pages in **bold** denote primary discussions. *Italics* denote relevant figures. Akkadian terms are presented according to the conventions of the *CAD*, which include the omission of mimation. Fortification elements mentioned in the appendixes are not indexed; for occurrences, see appropriate tables in chapter 3.

aban waṣpim (sling stone), 32. *See also* sling
Abba'el (of Mari), 96
Abel Beth-Ma'akha (T. Abil al-Qamh), 99, *228*, **229**
Abou Danne, Tell, 52, 55, 62, 76, *186*, **187**
Abu Hawam, Tell, 30
Abu Hureira, Tell. *See* Haror
Abu Kharaz, Tell, 61, 82, 83, 99, *228*, **229**
abullu (gate), 110
Abu Schachat, 78, *79*
Abu Shusha Tell. *See* Shush
Abu Snesleh, 124
Abu Zarad, Tell (Sheikh), 99, 118, *228*
'Acharneh, Tell, 56, 57, 78. *See also* Tunip
Achzib, 49, 51, 52, 54, 55, 56, 57, 59, 62, 99, 101, **229–30**
Adams, Robert, 106, 107, 110, 116
adaššu: inner rampart, 108, 110; outer town, 108
Adda-danu (ruler of Lakiša), 138
Aegean Bronze Age fortifications, 15
Aeneas Tacticus: *Siegecraft*, 27, 37, 40, 41, 43, 103
Afis, Tell (Apsuna), 55, 62, 76, 79, 97, 98, 105, 123, 162, *186*, **187–88**; destruction, 98
Aghbariyeh, Tell el-. *See* Bar
Ahmose (Egyptian pharaoh), 30, 97
Aḫtiashna (city in EA), 136
Ai (et-Tell), 8, 57, 93
'Ain Hassan, Tell, 146, *186*, **188**;
'Ain Samiya. *See* Marjama
'Ajjul, Tell el-, 5, 7, 8, 39, 51, 56, 57, 64, 73, 81, 95, 99, 101, 123, 128, 129, *130*, 132, 133, 135, 136, 137, 138, 157, *228*, **230–33**; siege tunnels, 135, 232–33. *See also* Sharuhen
Akkadian empire, as military threat, 28, 46, 88, 89, 90–1, 92, 159–60. *See also* Naram-Sin, Sargon, Šar-kali-šarri
Akko, Tel (Tell el-Fukhar), 13, 49, 51, 52, 54, 55, 61, 62, 64, 67, 70, 71, 75, 80, 83, 96, 99, 101, 123, 126, 135, 156, 160, 161, *228*, **233-35**
Alaça Höyük, 35, 71, 84
Alalaḫ (Tell Atchana), 4, 5, 6, 8, 12, 30, 45, 49, 52, 68, 69, 72, 75, 84, 97, 98, 111, 112, 119, 158, 161, *186*, **189–90**
Albright, William F., 3, 4, **5–6**, 7, 8, 9, 19, 136
Aleppo. *See* Ḫalab
ālu (town), 104, 106–08, 110, 113, 114; *āl pāṭi*, 109, 110; *ālu elû*, 108, 110; fortified (*ālū dannu* or *ālānū dannūtu*), 107; unfortified (*ul dannū*), 110
ama (Indo-Aryan *wall*), 5
Amanus, region of, 90, 91, 97
Amarna texts, 120, 121, 122, 136, 137; EA 69, 72; EA 70, 115; EA 85, 120; EA 98, 120; EA 101, 120; EA 138, 120; EA 185, 115; EA 186, 115; EA 234, 115; EA 244, 100; EA 250, 100; EA 253, 136; EA 254, 136; EA 256, 115; EA 272, 136; EA 273, 136; EA 284, 136; EA 286, 120; EA 287, 120, 136; EA 288, 120, 136; EA 289, 120; EA 290, 136; EA 292, 136; EA 298, 136; EA 299, 136; EA 306, 136; EA 314, 136; EA 319, 136; EA 320–333, 136; EA 335, 136; EA 365, 142; EA 370, 136
ambushes, 27
Amenhotep II (Egyptian pharaoh), 34
Amman (Rabbath-Ammon), 49, 50, 52, 62, 73, 99, *228*, **235–36**
Amman airport structure, *123*, 124
Amorites: 14, 97; Amorite Hypothesis, 5, 92; cultural *koiné* of, 7, 160; defensive wall against, 10, 28, 58, 91, 97. *See also* Muriq-Tidnim; destruction of Levantine towns by, 7, 91, 160; earthen ramparts and, 6–7, 9–10; as threat, 28, 91–2; use of term, 37
amrummu (fosse), 107, 110
Amurru (of Amarna texts), 120, 213
Amutpi'el (of Qatna), 95
Anatolian fortifications, 15. *See also* Alalaḫ, Hattuša, Kültepe-Kaneš, Alaça Höyük, Kuşakli
ʿAnatu, 33
Aniba, 16
AN.ZA.GÀR, 110, 115, 116. *See also* dimtu, magdalu
Aphek, Tel (Tell Ras el-'Ain), 11, 18, 20, 50, 56, 62, 73, 75, 80, 99, 101, 126, 128, *228*, **236**
appāru (swamp), 105, 108, 110
Apsuna (Afis?), 98
apu (reed thicket), 89, 105
Arad, 57
Armānum: walls of, 60, 105; identification of, 90; siege of, 28, 60, 90, 97, 105, 159
Armi, 90
'Arqa, Tell (*Irqata*), 62, *186*, **190**
ʾarr (Ugaritic district), 113
Arrapḫe, kingdom of, 127; settlement patterns and types, 104, 108, 110, 112–13, 125

arrowheads, 32, 34, 35, 44. *See also* šiltāhu, mortuary evidence
Art of War. See Sun Tzu
Asawir, Tell. *See* Esur
Aseret, 129, *130*, 131, 133
ash, 33, 51, 53, 73–74
Ashara, Tell. *See* Terqa
Ashbee, Paul, 145
Ashdod, Tel, 61, 62, 70, 76, 82, 99, 101, 128, 129, *130*, 133, 136, 138, **236–37**
Ashkelon, 10, 13, 14, 15, 19, 20, 40, 48, 49, 50, 51, 54, 55, 56, 57, 59, 61, 62, 66, 67, 68, 69, 70, 71, 72, 73, 75, 76, 78, 80, 96, 99, 101, 103, 116, 117, 120, 122, 123, *130*, 154, 156, 157, 159, 160, 161, 162, *228*; **237–43**; Amorite kings of, 126, 136, 138. *See also* Yidya; identified as city-state, 116; in Amarna texts, 120, 126, 136, 138; kingdom of, 125–39, 155; settlements in MB kingdom, *130*; Ashkelon (b), 127, 129, *130*
Askut, 15, 16, 75
Ašqaluna (EA), 120, 126, 136, 138. *See* Ashkelon
Aššur, 57, 95
Assyria, as military threat, 45, 94–95, 96, 97, 101. *See also* Puzur-Assur, Šamši-Adad
Astour, Michael, 88, 89, 92
Atkinson, Richard, 145
Autobiography of Ahmose son of Abana, 36, 136, 231, 233
Avaris (Tell ed-Dab'a), 5, 7, 20, 29, 101, 158, 162, **320**; siege by Ahmose, 97
'Avdon, *228*, **243**
ax, **43**. See *pašu*
Ayanot (school), 129, *130*, 131
Azu. *See* Hadidi

bābu (gate), 110
Babylon, 27, 30, 98, 100, 107, 127; as military threat, 95–96, 97, 108, 109, 141, 143, 144, 155, 160, 162. *See also* Ḫammurapi (of Babylon), Samsuiluna (of Babylon)
Bal'ame, Khirbet. *See* Ibleam
Balatah, Tell el-. *See* Shechem
Bardawil, fosse of, 57
barrel vaults in gate construction, 70, 72
Bar, Tel (Tell el-Aghbariyeh), *228*, **243**
bastions: **65–67**; rectangular, 66, 68, 81, 84. *See also* towers
Batash, Tel. *See* Timnah
Batash, Telluliot. *See* Telluliot
battering ram: 5–6, 8, 10, 28, 38, **39–40**; 41, 45, 46, 55, 59, 66, 68, 72, 80, 102, 159; defense against, 40; Neo-Assyrian, *39*. *See also* GUD.SI.DILI, yašibu
battle(s): Kadesh, 27; Megiddo, 27; Qarqar, 27
bazaḫ(ā)tu (border fortress), 109, 110
Bderi, Tell, *79*, 91, *167*, **168**
Beck, Pirhiya, 11, 125, 128
Beirut. *See* Biruta
Beitin. *See* Beth-El
Beit Mirsim, Tell, 5, 6, 8, 50, 55, 62, 64, 66, 67, 70, 82, 99, 101, *228*, **243–45**
bēl dimti (as mayor), 125. *See also* ḫazannu
Beni Hasan, 6, 29, 35, 37, 39, 41, 43
Bergman, C. A., 33, 34
Beth-El (Beitin), 13, 55, 71, 73, 82, 83, 99, 101, 118, *228*, **245–47**
Beth-Shemesh (Tell Rumeileh), 62, 67, 68, 69, 75, 82, 83, 99, 101, 124, 129, *130*, 131, 134, 135, *228*, **247–48**
Beth-Zur (Khirbet et-Ṭubeiqa), 62, 82, 99, 101, 118, *228*, **248–49**
Beydar, Tell (Nabada), 17, 69, 78, *79*, 104, *167*, **168–69**
Bi'a, Tell. *See* Tuttul
Bira, Tel, 81, 99, 127, *228*, **249**
Biran, Avraham, 13, 64, 144, 148
birtu (fortress), 106, 108, 110
Biruta (Beirut), 30, 52, 55, 56, 62, 84, 123, *186*, **190**
Boas-Vedder, Diny, 151
Bosra, 49, 62, 73, 81, *228*, **249**
bows, 31, **33–36**, 46. *See also* composite bow, qaštu, simple bow, tilpānu
Brak, Tell. *See* Nagar
Braudel, Fernand, 87
breaching of walls, **38**. *See also* pilšu
Buccellati, Giorgio, 73
Bugha, 78, *79*
Buhen, 15, 16, 32, 60, 66, 75
Bullard, Reuben, 75
Bunimovitz, Shlomo, 12, 13, 117, 119, 120, 127, 128, 135, 155, 157, 158
Burak, Tell el-, 62, 67, 76, 123, 124, *186*, **192**
Burga, Tel, 11, 14, 49, 55, 56, 57, 59, 66, 80, 97, 98, 99, 124, 127, 160, *228*, **249–50**; MB IIA abandonment, 99
Burnah, Tell, 129, *130*, 132
Byblos (Gebal), 4, 8, 9, 12, 30, 49, 52, 54, 55, 56, 69, 70, 71, 72, 75, 79, 82, 84, 91, 92, 93, 94, 97, 102, 119, 120, 122, 123, 126, 156, 159, 160, *186*, **192–97**; destruction, 92

Canaanite, use of term, 37
canals, 58, 107, 144, 145, 155. *See also* E.KI.SUR.RA, nāru
Carthage, 114
Casana, Jesse, **111–12**, 116
chariot, 27, **29–31**, 45, 46, 71, 238, 240, 291; composite bow and, 36, 37, 43; introduction to

Egypt, 30; ramparts and, 4, 5–6, 7, 10; Tut'ankhamun's, 30
Charlton, Thomas, 120
chiefdoms and fortifications, 156
chronology: **18–20**; very low chronology, 19–20
Chuera, Tell, 17, 57, 58, 78, *79*, *167*, **170–71**
city-state: 118, **119–21**; as districts within kingdoms, 137–38, 161; criteria, 120; in southern coastal plain, 135–36; in southern Levant, 10, 101, 116–17, 137, 156; Ugarit as, 113
Clausewitz, Carl von, 29
clay: in glacis, 52; in earthen ramparts, 51
Cohen, Susan, 18, 98, 99, 100, 126, 127, 129, 131, 132
communication networks, 125
competitive emulation, building of ramparts for, 13, 158, 160
composite bow, 27, 31, **33–36**, 38, 41, 43, 45, 46, 102, 159; chariot and, 36, 37, 43; double-curved reflexed, 34–35; Hyksos and, 4
conspicuous consumption, ramparts as, 13, 117, 151, 157, 158, 161
construction: process for fortifications, 143–44; rate, 144–46, 153–55. *See also* earthen ramparts, fortifications walls, labor
corbel vaulting, 71, 84, 224, 235, 241
core walls, **54–55**
Cornwall, Ian W., 145
corvée labor. *See* labor
Cotterell, B., 145
Cremaschi, Mauro, 75
Crouwel, Joost, 29, 30, 31, 71
Cypriot fortifications, 15

Dab'a, Tell ed-. *See* Avaris
dagger, **43**. *See also* paṭru
Dalley, Stephanie, 27, 28, 35, 38, 102
Dan, Tel (Laiš), 13, 14, 42, 48, 49, 50, 52, 53, 54, 55, 56, 61, 62, 64, 66, 67, 68, 69, 71, 72, 73, 76, 81, 95, 99, 101, 117, 119, 123, 127, 143, 144, 146, 147, 148, 149, 151, 152, *228*, **250–54**
Dapur, siege of, 37
defensive settlement networks, 122–25
Debbeh, *228*, **255**
Deir 'Allā, Tell, 55, 62, 99, *228*, **255**
Deir el-Bahri, 44–45
Deir Khabiye, 49, 81, 95, 123, *186*, **197–98**
Der, Tell ed-, 16, 58
Deshashe, tomb at, 8, 38, 40
destruction processes: natural, 74. *See also* sebakh
Dever, William G., 3, 10, 13, 18, 19, 45, 53, 93, 98, 101, 116, 119, 120, 121, 155, 156
Dhaharat el-Humriya cemetery, *130*
Dhahrat Makmish. *See* Michal
Dimat, 116
Dimti-Dada, 122
Dimti-Enlil, 108, 122
dimtu: Old Babylonian, 28, 38, 46, 108, 110, 159
Diodorus Siculus, 32
districts, 110, 124–125. *See also* ḫalṣu
doors and door sockets, 68, 72
Dossin, Georges, 108, 125
Dothan, Moshe, 71, 101
Dothan, Trude, 131, 256
Dothan, Tell, 61, 62, 63, 82, 99, 118, *228*, **255–56**
drainage and drains, 53, 71
Drews, Robert, 29, 30
Duheisha, Khirbet ed-, 129, *130*, 131, 133
Dumatu, 115
Dumat(u)-'Agimi, 115
Dumatu-qidši, 115
Dunand, Maurice, 4, 12, 92
Dunnum, 107
Dūr-Samsuiluna (Khafajah Mound B), 17, 109, 155
dūru (wall or fortress), 110
Dūr-Yaḫdun-Lim, 109
Dūr-Yasmaḫ-Addu (Mari district), 111
Duweir, Tell ed-. *See* Lachish

E.KI.SUR.RA (boundary canal), 107
Eanatum (of Lagash), 88
Early Bronze Age warfare, 88–94
Early State Modules, 134
earthen ramparts, **48–54**; Akkadian terms for, 108; architectural evolution of, 78–84; beaten earth construction or *terre pisée*, 6, 7, 8, 48, 51, 53; composition of. *See also* clay, hamra, loess, marl, sand; construction rate, 144–45, 147–51; core walls within, 54–55; distinguished from glacis, 48, 55; elevation of, 54; endogenous origin hypothesis, 7, 9; freestanding type, 6, 10, 13, **48–49**, 50, 54, 55, 56, 57, 79, 80, 81, 84, 95, 128, 131, 132, 157, 159, 160, 161; functional interpretations of, 5, 10–11; layouts (elliptical vs. rectilinear), 49, 65, 66, 79–81, 84, 95, 96, 126, 161; slopes of, 52, 54; supplemental type, 7, 48, **49–50**, 51, 52, 55, 57, 80, 81, 82, 123, 157; 110; typologies of, 11. *See also* adaššum, glacis, kirḫu
EB IV fortified sites in southern Levant, 93. *See also* Iskander
Ebla (Tell Mardikh), 8, 12, 14, 25, 28, 32, 39, 42, 43, 45, 48, 49, 50, 52, 53, 54, 55, 56, 57, 59, 62, 65, 66, 67, 68, 69, 70, 71, 72, 75, 76, 79, 80, 81, 82, 87, 88, 89, 90, 91, 92, 95, 97, 98, 102, 105, 112, 123, 156, 159, 160, 161, 162, *186*, **198–204**; destruction by Pizikarra, 92, 97, 98; destruction by Sargon or Naram-Sin, 90, 92; outlying rampart, 52; Palace G, 28; settlement pattern of,

105–06; Tombeau des citernes, 42; Western Palace, 32, 66
Edzard, Dietz O., 88, 89
Egyptian fortifications: MK fortresses, 15. *See also individual site names*
Ekallatum, 127
Ekalte. *See* Yakaltum
Ekron (Tel Miqne), 56, 81, 95, 99, 123, 128, 129, *130*, 131, 133, 138, 157, *228*, **256**; identification with Tianna (city in EA), 136–37
Elam: campaigns into, 88, 90; Elamites, 26, 160
Emar (Tell Meskene), 78, *79*, 84, 89, 97, 122, 123, *186*, **205**
emūqu (fortress), 114
Engberg, Robert M., 4
Eninmar, 28
Enna-Dagan, letter of, 89, 97, 102, 105
eperu (siege ramp), 38
Epitome of the Military Science, 27, 33, 38, 47, 51, 58. *See also* Vegetius Renatus
Erez, 127, 129, *130*, 133
erosion, effects of, 74
escalade using siege ladders, **40**
Ešnunna, 95, 96, 97, 100, 127
Esur, Tel (Tell Asawir), 99, *228*, **256–57**
Eurasian theory of rampart origins, 4, 5, 6, 8, 30, 33
evolution of fortifications, **78–84**
Execration Texts (Egyptian), 94, 118, 126, 156, 160

Falconer, Steven, 100, 106, 124, 155, 156, 157
Far'ah North, Tell el-, 8, 50, 52, 55, 62, 66, 67, 70, 82, 93, 99, 101, 118, *228*, **257–58**
Far'ah South, Tell el-, 8, 52, 56, 57, 61, 62, 64, 68, 69, 77, 81, 82, 83, 99, 101, 129, *130*, 134, 135, *228*, **258–59**
Farani, Tell el-. *See* Poran
Ferzat, Tell (Tell es-Salihiyeh), 76, *186*, **205–206**
Fifteenth Dynasty (Egypt), 5, 127
Finkelstein, Israel, 12, 13, 117, 119, 135, 136, 137, 138, 144, 145, 147, 148, 149, 150, 153, 156, 157
Fleming, Daniel, 111
flint, use in rampart and glacis construction, 51, 55
fortification walls: **59–64**; casemate-style, 59, **61–63**, 64, 82–83, 84, 151; crenellations on, 60; depictions of, 59; exterior walls as, **64**; foundations for, 13, 48, 50, 54, 55, 63, 82, 148; offset-inset, 61; saw-tooth, 61; solid type, **60–61**, 62–63, 82
Fortin, Michel, 15
fortresses, 15, 16, 17, 28, 29, 49, 60, 65, 66, 103, 106, 107, 108, 109, 114, 118, 122, 124, 125, 138, 154, 155, 159; Middle Bronze Age Levant, 124; terms for, 108, 110, 114. *See also* birtu, dūru, Egyptian fortifications, emūqu
fosse: 3, 16, 20, 40, 47, 51, **56–59**, 75, 78, 79, 80, 81, 84, 95, 159; construction of, 143; terms for, 107, 110. *See also* ḫirītu, amrummu
freestanding ramparts. *See* earthen ramparts
Fukhar, Tell el- (Jordan), 83

ganūnu (Akk.), 116
Garstang, John, 5, 6, 10, 64, 67
gates: **67–72**; four-pier, **70–71**; miscellaneous types, **71–72**; outer, 68; placement of, 67; postern, **71**; six-pier, 6, 66, 67, **68–70**, 71, 72, 81, 83, 84, 95, 118, 131, 135, 161; terms for, 110; traffic and, 71–72. *See also* abullu, bābu, ṯġr
gateway, as political term, 119
Gat-Galon Road site, 129, *130*, 133
gath, 113, 116, 124
Gath (Tell es-Safi), 99, 124, 129, *130*, 131, 133, 135, 136, 137, 138, 156, *228*, **259**; *Gimtu*, 136, 138
Gaza, 122, 132, 136, 137, 138
Gazru (EA), 120, 136, 138. *See also* Gezer
Gebal. *See* Byblos
Gedera, 129, *130*, 131, 133
Genesis 23:2, 114
Gerar (biblical), 132, 134. *See also* Haror
Gerisa, Tel (Tell Jerishe), 7, 52, 55, 62, 73, 76, 80, 83, 97, 98, 99, 100, 101, 127, 160, *228*, **259–60**; MB IIA destruction, 98
Gezer (Tell el-Jezer), 11, 45, 49, 50, 52, 53, 55, 62, 65, 66, 67, 68, 69, 73, 75, 76, 81, 82, 99, 101, 120, 127, 128, 129, *130*, 131, 134, 135, 136, 137, 138, 139, 156, *228*, **260–63**
Ghiaur Kalah (Turkmenistan), 4
Gibson, McGuire, 16, 42, 58, 74, 145
Gindaris, Tell, *186*, **206**
gintu. *See* gittu
gittu (agricultural estate), 110, 113, 114, 115–16, 124, 129, 131, 133
Giv'at Ha-Parsa, 129, *130*, 131, 133
Giv'at Sharett, 64, 124, *228*, **263**
GL'IL, 33
glacis, 3, 5, 6, 7, 8, 9, 10, 14, 20, 47, 48, 49, 50, 51, 52, 53, 54, **55–56**, 59, 75, 77, 78, 80, 84, 118, 132; distinguished from earthen ramparts, 48, 55; type relative to precipitation, 56
Glea, 78
Glock, Albert, 25, 26, 27, 30, 83, 101, 106, 107, 109, 125
Goetze, Albrecht, 144
Goldberg, Paul, 74
Gonen, Rivka, 13, 25, 82, 83, 116
Gophna, Ram, 11, 12, 93, 124, 125
ǵr (Ugaritic district), 113
Gregori, Barbara, 10, 12, 47, 61, 67, 68, 70, 72, 81
Gubla. *See* Byblos
GUD.SI.DILI, 39, 46, 159. *See* battering ram, yašibu

Gungunum (of Larsa), 155

Habuba Kabira, Tell, 76, 80, *186*, **206–07**
Hadidi, Tell (Azu), 35, 55, 56, 84, 91, 92, 97, 160, *186*, **207–08**; destruction of, 92
Ḫalab (Aleppo), 79, 89, 90, 94, 97, 111, 122, 123, 159, 162, *186*, **208–10**; identification as Ḫalam, 90
Hall, Harold R., 4, 146
ḫalṣu (district), 110–11, 113, 121, 124–25, 137, 161
Hama, Tell, 19, 30, 91, 92, 97
Hamadiya, 124
Hammah, Tell el-, 62, 99, *228*, **263–64**
Hammam et-Turkman (Zalpah), 76, 91, 92, 97, 160, *167*, **171**
Ḫammurapi (of Babylon), 27, **96**, 97, 155, 160, 173
Ḫammurapi I (of Yamḫad), 96
Hamoukar, Tell, 58
hamra, for earthen rampart construction, 51
Haradum (Khirbet ed-Diniye), 17, 109
Harappa, 5
Harasim, Tel, 129, *130*
Haraz, Tell el-, 129, *130*, 131, 133
Harif, Amos, **11–12**
Hariri, Tell. *See* Mari
Haror, Tel (Tell Abu Hureira), 14, 49, 51, 55, 56, 57, 64, 81, 95, 99, 123, 128, 129, *130*, 132, 133, 135, 136, 137, 138, 144, 147, 150, 151, 152, 153, 154, 155, 157, *228*, **264–65**
Ḫarran (Mari district), 111
Hattuša: fortifications, 15, 35, 65, 71, 84; Phrygischer Tempel, 32
Ḫattušili I (of Ḫatti), 98
ḫazannu (mayor), 120, 121, 135, 136, 137, 161; military obligations of, 124–25
Ḫazzatu, 136, 137, 138. *See also* Gaza
Hazor (Tell el-Qedah), 5, 7, 8, 11, 13, 14, 15, 20, 39, 49, 50, 51, 52, 54, 55, 56, 57, 61, 62, 64, 68, 69, 72, 73, 75, 76, 81, 82, 83, 94, 95, 96, 98, 99, 100, 101, 102, 119, 120, 121, 122, 123, 126, 127, 137, 138, 144, 146, 147, 148, 149, 151, 152, 153, 154, 156, 159, 160, 162, *228*, **265–70**; identifications as city-state or kingdom, 116, 117
Hebron (Jebel Rumeidah), 13, 55, 56, 57, 62, 82, 83, 99, 114, 117, 118, 123, 149, 161, *228*, **270–71**
Hefer, Tel (Tell el-Ifshar), 20, 62, 97, 98, 99, 100, 126, 160, *228*, **271**; MB IIA destructions, 98
Heliopolis, 8, 49
Heltzer, Michael, 113, 115, 116
Herakleopolis, 44, 94, 97
Herzog, Ze'ev, 12, 13–14, 47, 60, 61, 67, 68, 71, 82, 101, 117, 122, 132, 141, 144, 147, 149
Hesi, Tell el-, 101, 127, 129, *130*, 132, 133, 135, 136
Hierakonpolis Tomb 100, 36
highlands (southern Levant): fortifications, 81–83; MB IIC polities, 117–19
hired labor. *See* labor
Hittites, destructions and conquests by, 98. *See also* Ḫattušili I and Muršili I
Ḫuraru (Ugaritic district), 113
Hurrians, **4–5**, 6, 30; *Epos der Freilassung*, 98
huwwar limestone, 51, 55
Hyksos, 8, 10, 11, 18, 19, 25, 29, 30, 36, 45, 82, 101, 132, 162; *ḥk3w ḫ3swt*, 26; rampart origins, 4–7

Iarmuti, 90, 97
Ibleam (Khirbet Bal'ame), 82, 99, *228*, **271**
Ida-Maraṣ (Mari district), 111
Ifshar, Tell el-. *See* Hefer
igāru or *igartu* (wall), 60, 107, 110
Ikkur, 16
Ilakabkabu, 95
ilku (service), 141, 142, 153. *See also* labor
Indo-Europeans, **4–5**
Intef, tomb of, 29, 36, 60
Iran, 9, 15, 146
Irbid, Tell, 81, 83, 99, 124, *228*, **271**
Irkabtum (of Mari), 96
Iron I, 75, 120, 138–39, 145
irrigation, 9, 10, 16, 58, 106, 142, 156
Išihi-Adad (of Qatna), 95
Iskander, Khirbet, 93, *228*, **272**
Išme-Dagan (of Mari), 38
itinnu (mason), 143, 158

Jaffa (Tel Yafo), 9, 10, 43, 55, 57, 72, 81, 99, 101, 120, *228*, **272–73**. *See also* Taking of Joppa
Jasmin, Michaël, 119, 136, 137
Jatt, 99, *228*, **273**
Jazira, 104–105, 106, 112
Jebel Bishri, 97
Jemmeh, Tell, 64, 77, 81, 95, 99, 128, 129, *130*, 132, 133, 135, 136, 138, *228*, **273–74**
Jenin, 82, 99, 118, *228*, **274**
Jerablus. *See* Kar-Kemiš
Jericho (Tell es-Sultan), 5, 6, 7, 8, 11, 13, 49, 52, 53, 54, 55, 57, 62, 64, 67, 73, 74, 75, 76, 80, 82, 83, 93, 99, 101, 142, *228*, **274–82**
Jerusalem, 62, 66, 67, 73, 83, 99, 100, 117, 118, 120, 123, 124, 135, 142, 149, 151, 156, 157, 161, *228*, **282–83**; MB kingdom, 118
Jisr, el- (cemetery), *130*
Joshua 11, 119; 12, 138; 15:15, 114; 15:49, 114; 15:60, 114; 15:9, 114; 9:17, 114

Kabri, Tell, 13, 15, 48, 49, 52, 62, 73, 99, 101, 119, 123, 127, 158, *228*, **283–85**; kingdom of, 116
Kadesh (Tell Nebi Mend): battle of, 27; destruction,

98; fortifications, 61, 97, 98, 162, *186*, **210–11**
Kaemheset, tomb of, 40
Kagan, Kimberly, 31
Kamid el-Loz. *See* Kumidi
Kamminga, J., 145
Kannas, Tell, 65, 67, 76, *186*, **211**
kaparu (Ug. village), 110, 113, 114, 123, 124, 129, 133
Kaplan, Jacob, 7, **9–10**, 11, 13, 48, 50, 58, 68, 72, 78, 80, 101, 131
kapru (OB village), 109, 110, 114
Karana, 127
Kar-Kemiš (Jerablus), 7, 8, 68, 84, 122, *186*, **211**
kāru (merchant enclave), 60, 94, 105, 108, 110
Katzenstein, Hannah, 48
Kazel, Tell (*Sumur*?), 52, 55, *186*, **211–12**
Kebara, el-, 62, 82, 99, 118, *228*, **285**
Kedesh, Tel, 117, *228*, **285**
Keegan, John, 31, 124
kefar (village), 113, 123
Kefar Menahem, 129, *130*, 131, 133
Kefar Ruppin, 124
Keisan, Tell, 52, 54, 55, 62, 81, 99, *228*, **285**
Kemp, Barry, 16, 74
Kenyon, Kathleen, 3, 5, 6, 7, 8, 10, 14, 31, 51, 64, 92, 93, 160
Kepinski-Lecomte, Christine, 17, 109
Kern, Paul, 27, 37
Khafajah (Mound B), 17, 109, 155. *See* also Dūr-Samsuiluna
Khan Sheikhoun, Tell, 7, *186*, **212**
Khanzir, 78, *79*
Kheibar, Khirbet, 82, 99, 118, *228*, **286**
Khoshi, Tell, 9
Khu-Sebek stele, 100
kilīlu (battlements), 107
kingdom, characteristics of, 119–121. *See also* mātu
Kinneret (Tell el-'Oreme), 82, 83, 117, *228*, **286–87**
kirḫu (or *kerḫu*): inner town, 108; rampart, 108, 110
Kirk Tepe (Turkmenistan), 4
Kirta Epic, 26, 31
kirû (fields), 104
Kiš, 107, 109, 141, 144, 155
Kisufim cemetery, *130*
Kitan, Tel (Tell Musa), 62, 99, 101, *228*, **287**
Klengel, Horst, 88, 94, 95, 96, 122, 142
Kochavi, Moshe, 11, 101
Koliński, Rafał, 108, 122
Kor, 15
Korfmann, Manfred, 25, 32, 33
Kôtaru-wa-Hasīsu, 33
Kotter, Wade, 116, 117, 149, 156
Kranzhügeln, 8, 11, 17, 49, 57, 58, **78–79**, *79*, 81, 84, 159; evolution of fortifications and, 78; sites identified as, *79*; Stager's hypothesis, 11. *See also site names*
Kubban, 16
Kültepe-Kaneš, 15, 35, 84, 94, 97, 160
Kumidi (Kamid el-Loz), 61, 62, 76, 84, 122, 151, *186*, **212–13**
Kumma, 15, 16
Kurdane, Tel (Aphek?), 99, *228*, **287**
kurkar sandstone, 51, 55, 148
Kuşakli (Sarissa), 65, 84

Lab'ayu (of Shechem), 100
labor: army, 141, 154–55; corvée, 141, 142, 143, 152, 153, 154, 157, 161; estimating, 146–52; hiring labor, 142, 143, 152, 153, 154, 157, 161; *ilku* service, 141, 142, 153; Old Babylonian, 141–42; specialized vs. unspecialized, 143; used in Jerusalem's wall, 142–43; women, children, and slaves as, 142. *See also* itinnu
Lachish (Tell ed-Duweir), 5, 7, 8, 33, 39, 43, 49, 50, 52, 55, 56, 57, 62, 72, 73, 81, 99, 101, 120, 123, 127, 128, 129, *130*, 131, 132, 133, 135, 136, 137, 138, 139, 156, 157, *228*, **287–88**
Lagarce, Jacques, 10
Lagaš, 35, 36, 42, 88
Laiš, 95, 119, 127. *See also* Dan
Lakiša (EA). *See* Lachish
Lamentation over the Destruction of Ur, 26, 91
landscape: fortifications and, 77; Landscapes of Tells, 104; signature landscape, 9, 78, 81
Larsa, 16, 60, 95, 100, 107, 108, 109, 110, 142, 155, 175
Late Bronze Age warfare and fortifications, 83–84; LB I destructions, 101. *See also* Gonen, Kh. Rabud, Kinneret
Lavee, Hanoch, 75
Leilan, Tell, 15, 17, 96. *See also* Šubat-Enlil
Li'mu, 34
Littauer, Mary, 29, 30, 31, 71
loess, 51
Lönnquist, Minna A, 92
loopholes, evidence for, 16, 75
Lugalzagesi (of Uruk), 88, 90, 97, 101, 159

Mabtuh (E) Sharqi, 78, *79*
Mabtuh (W) Gharbi, 78, *79*
mace, **42**
madgaltu (watchtower), 110, 115. *See also* watchtowers
madinatu (village), 110, 114
Magdala, 122
magdalu (watchtower), 110, 113, 114, 115, 116, 124, 125, 126, 129, 133. *See also* watchtowers
Maghr, 78

Mahrum, 78, *79*
májdal (Ar.), 115
Májdal (Ashkelon), cemetery, 115, 118, 126, 127, 129, *130*, 133
Májdal-Ba', Khirbet, 118
Májdal-Kikhia, 115
Májdal-Saleh, 115
Makhruq, Khirbet el-, 57, 93
Malat, Tel, 129, *130*, 131
Malhata, Tel (Tell el-Milh), 52, 54, 55, 56, 62, 77, 81, 99, 101, 124, *228*, **288**
Manahat (Nahal Rephaim), 64, 118, 123, 124, *228*, **288**
mantelet, 39
Mardikh, Tell. *See* Ebla
Margueron, Jean-Claude, 17, 32, 79, 92, 94, 96
Mari (Tell Hariri), 5, 6, 8, 9, 16, 20, 25, 27, 28, 30, 32, 33, 34, 35, 36, 37, 38, 39, 45, 46, 49, 50, 54, 56, 58, 60, 67, 70, 76, 78, *79*, 88, 89, 90, 91, 92, 94, 95, 96, 97, 98, 101, 102, 105, 106, 107, 108, 109, 110–11, 111, 117, 119, 120, 123, 125, 127, 134, 142, 156, 159, 160, *167*, **171–75**; destruction by Ḫammurapi, 96; districts of, 110; settlement patterns and types, 106–11; temple of Dagan, 32
Mari letters, 28, 32, 36, 38, 45, 46, 56, 67, 94, 96, 97, 98, 101, 106, 107, 108, 109, 110, 111, 117, 119, 120, 125, 127, 134, 156, 159, 160; ARM **I** 4, 38; 5, 26; 20, 108; 38, 35; 42, 125; 62, 125; 73, 111; 90, 107; 118, 39; 131, 6, 28, 38, 39; 135, 38, 39; 138, 6; 139, 107; 16, 111; ARM **II** 3, 107, 125; 7, 6, 38, 39; 22, 27; 30, 107; 35, 125; 67, 109; 72, 107; 88, 107, 109; 92, 109; 97, 125; 101, 107, 142, 143; 102, 109; 103, 109; 105, 109; 120, 111; 127, 32; ARM **III** 10, 67, 111; 12, 109; 17, 109; 27, 125; 30, 109, 125; 36, 107; 37, 39; 71, 107; 78, 67; ARM **IV** 26, 107; 31, 125; 32, 107; 44, 125; ARM **V** 2, 39; 16, 107, 110; 37, 125; ARM **VI** 14, 111; 23, 117, 127; 29, 107; 63, 39; 64, 109; 76, 125; 78, 117; ARM **VII** 16, 39; 63, 39; 69, 39; 236, 117; ARM **VIII** 88, 108; ARM **IX** 102, 32; ARM **XII** 747, 117; ARM **XIII** 117, 125; 118, 125; 146, 38, 39; ARM **XIV** 24, 107, 142; 45, 38, 39; ARM **XVIII** 5, 35; ARM **XXIII** 556, 117;ARM **XXV** 43, 117; 103, 117; ARM **XXVI** 375, 117
Marjama, Kh. ('Ain Samiya), 99, 118, *228*, **288–89**
marl, 51
Marrassini, Paolo, 9
Marsden, Eric, 32, 34
Mashash, Khirbet el-. *See* Masos
Mashkuta, Tell el-, 7
Mašin, Tell, 30, 76, *186*, **213**
maškanu (village): Old Babylonian, 109, 110, 114
Masos, Tel (Khirbet el-Mashash), 49, 52, 55, 56, 57, 64, 77, 81, 95, 99, *228*, **289–90**
Matthiae, Paolo, 12, 14, 32, 42, 43, 45, 67, 68, 72, 92, 98
mātu (land): designating territories, 120, 122, 137, 161; in Mari texts, 110–11
Mazar, Amihai, 3, 11, 15, 20, 49, 93, 116, 123, 124, 131, 144, 150
Mazar, Benjamin, 3, 61, 101
Mazkeret Batya (site east of), 129, *130*, 131
MB IIC destructions, 101
McLeod, Wallace, 34, 36
Me'amer, Tel (Tell el-'Amr), 99, *228*, **290**
medinah (village), 114
Medinet Habu, *59*, 60
Megiddo (Tell el-Mutesellim), 6, 8, 11, 20, 50, 52, 55, 56, 62, 64, 65, 66, 67, 68, 69, 70, 73, 75, 76, 80, 83, 97, 99, 100, *228*, **290–95**; Thutmose III's battle at, 27
Mejdel (near Bdama), 115
Mentuhotep II, Nebhepetre, 29, 36, 44, 60, 94, 97
Merneptah stele, 136
Meskene, Tell. *See* Emar
Mesopotamian fortifications: Old Babylonian, 15, 16; third millennium, 78–80. *See also* Kranzhügeln *and site names*
Me-Turan, 107
Mevorakh, Tel (Tell Mubarak), 13, 50, 64, 76, 99, 101, 124, 127, 144, 148, 149–50, 151, *228*, **295–96**
Meyer, Jan-Waalke, 16, 105, 123
Michal, Tel (Dhahrat Makmish), 12, 64, 78, 81, 82, 101, 124, 128, 144, 146, 147–48, 149, 151, *228*, **296–98**
Middle Bronze Age warfare: Egyptian involvement, 100–101; northern Levant, 94–98; southern Levant, 98–101
Middle Kingdom Egyptian fortresses, 15–16. *See also individual names of sites*
migdāl, 113, 124
Milh, Tell el-. *See* Malhata
Milha, Tel, 129, *130*, 132, 133
Milk-Ilu (ruler of Gazru), 138
Miller, Robert L., 10, 33, 34, 35, 36
Miqne, Tell. *See* Ekron
Mirgissa, 15, 16, 66
Mishrife, Tell. *See* Qatna
Mitanni, 120
moat, 56, 58, 107, 109, 110, 144, 155. *See also* fosse, appāru
Moorey, P. R. S., 29, 30, 31, 36
Moortgat-Correns, Ursula, 78
Mor, Tel, 82, 99, 101, 128, 129, *130*, 131, 133

morphology vs. construction, 48, 50–51
Morris, Ian, 121
mortar, 66, **74**; volume relative to mudbricks, 145–46
mortuary evidence, **44–45**
Mozan, Tell. *See* Urkeš
Muazzar, Tell, 78, *79*
Mubarak, Tell. *See* Mevorakh
mudbricks, 3, 4, 11, 13, 15, 16, 51, 52, 54, 55, 56, 59, 60, 61, 62, 63, 64, 66, 69, 70, 72, **73–74**, 75, **76**, 77, 78, 80, 81, 82, 84, 95, 98, 102, 104, 126, 143, 144, 147, 150, 151, 153, 155, 159, 160; dimensions, 76; production of, 145–46
Muḫḫazu (city in EA), 136
Muḫraštu (city in EA), 136
Mukiš: kingdom, 91, 97, 111–12, 113, 116, 117, 134
Munbaqa, Tell. *See* Yakaltum
Muriq-Tidnim, wall of, 10, 28, 58, 91, 97
Muršili I (of Ḫatti), 97, 98, 161
Musa, Tell. *See* Kitan
Mutesellim, Tell el-. *See* Megiddo
Mycenean fortifications, 15, 72

Na'an cemetery, *130*
Nabada, 17, 104. *See* Beydar
Nagar (Tell Brak), *167*, **175**; destruction, 89, 91, 92, 97; kingdom, 104
Nagila, Tel, 49, 52, 55, 56, 57, 62, 81, 95, 99, 101, 128, 129, *130*, 132, 133, 135, *228*, **298–99**
Nahal Rephaim. *See* Manahat
Nahariya, Tel, 62, 99, *228*, **299**
Naḫrima, 120
Najjar, Khirbet en-, 62, 82, 99, 118, *228*, **299**
Nami, Tel, 97, 98, 99, 100, 127, 160, *228*, **300**; MB IIA destruction, 98
Naram-Sin (of Akkad), 28, 35, 60, 89, **90–91**, 92, 97, 105, 106, 108, 159, 173
nāru (rivers or canals), 109
Naumann, Rudolf, 15, 68
Nebi Mend, Tell. *See* Kadesh
Nebi Rubin, 99, 127, 129, *130*, 131, 133, *228*, **300**
Neev, David, 78
Nehemiah 3:12, 143; 6:15, 142
Nichols, Deborah L., 120
Nilimmar (in Mari texts), 38
Nimrin, Tell, 62, 82, 99, *228*, **300**
Nippur, 16, 42, 57, 106, 107, 108, 110
Nuḫḫašše, kingdom, 122
Nubia, 16, 60, 66
Numbers 22:39, 114; 32:37, 114
Nur-Adad (of Larsa), 142

Odyssey, 43
open battle (pitched battle), 27, 28, 29, 30, 36, 45. *See also battles of* Kadesh, Megiddo, Qarqar
Opificius, Ruth, 16
Oredsson, Dag, 14, 47, 56, 57, 58, 78
'Oreme, Tell el-. *See* Kinneret
Oren, Eliezer, 3, 21, 116, 132, 144, 150

Par, Tel, 124
Parr, Peter J., **7–9**, 10, 12, 15, 20, 25, 48, 50, 78, 80
pašu (ax), 43. *See also* ax
Patrimonial Household Model (PHM), **121–22**, 125, 134
paṭru (dagger), 43. *See also* dagger
Pax Amoritica, **96**, 97, 100, 161
pdr (town), 110, 113, 114
Pella (Tabaqat Fahl), 62, 66, 67, 74, 76, 82, 99, *228*, **300–02**
Pennells, Ernest, 10, 47, 53
Petrie, Flinders, 4, 132, 135
Philip, Graham, 25, 35, 42, 121
Philistine pentapolis, 138
pilšu (tunneling or digging), 38–39, 159; defense against, 40–41
Pinnock, Frances, 12, 65, 66
pitched battle. *See* open battle
Pizikarra (of Nineveh), 97, 98
plaster: **74–75**; glacis, 48, 52, 55, 80; gypsum, 56, 74–75; lime, 48, 52, 53, 74–75; mud, 55, 56, 74–75; on walls, 16; rampart construction with, 51, 55
Poidebard, R. P. Antoine, 78
Poleg, Tel, 8, 9, 11, 50, 52, 63, 66, 67, 76, 80, 97, 98, 99, 124, 127, 160, *228*, **302**; East Fort. *See* Zurekiyeh; MB IIA abandonment, 99
political organization, 21, 103, 111, 122, 127, 135, 137, 141, 155, 156, 157, 161; northern vs. southern Levant, 119–21. *See also* city-state, gateway, kingdom, polity
polity, use of the term, 119
Poran, Tel (Tell el-Farani), 12, 52, 55, 63, 64, 99, 128, 129, *130*, 133, *228*, **302–03**; Poran (b), 128, 129, *130*; Poran (c) cemetery, *130*
Postgate, J. Nicholas, 43
post-processual approaches, **12–15**
Powell, Marvin A., 141, 144, 145
Prayer for a City Under Siege, 26
Pu-Baḫla (ruler of Yurza), 138
Pumpelly, Raphael, 4
pur (Indo-Aryan), 5
Puzur-Assur, 94, 97

Qaṭṭunan (Mari district), 111
Qana, Tel, 80, 99, *228*, **303**
qarītu (town), 110, 113, 114, 123, 129
Qarqaf, Khirbet, 82, 99, 118, *228*, **303**
Qarqar, battle of, 27

qart (town), 123. *See also* qarītu, Qiryat
Qashish, Tel (Tell Qasis), 63, 67, 99, *228*, **303–04**
qaštu (bow), 34. *See also* bow
Qatna (Tell Mishrife), 5, 7, 12, 30, 37, 38, 47, 48, 49, 50, 52, 53, 56, 57, 63, 64, 68, 69, 72, 73, 75, 77, 78, 81, 84, 94, 95, 96, 97, 98, 100, 101, 119, 122, 123, 127, 156, 159, 160, *186*, **213–17**
qīr (town), 114
Qîrḫadat, 38
qiryat (town), 113, 114
Qiryataim, 114
Qiryat-ʻArba, 114, 118
Qiryat-Baʻal, 114
Qiryat-Huzoth, 114
Qiryat-Sepher/Sanna, 114
Qiryat Shemona South, 80, 99, 124, *228*, **304**
Qitar, Tell el-, 55, 61, 63, 71, 84, *186*, **217**
Qumei, Khirbet, 82, 99, 118, *228*, **304**

rābiṣu, 121
Rabud, Khirbet, 82, 83
radiocarbon dates, 101
Randall-Maciver, David, 16
Ras Ibn Hani, 10, 84
Ras Shamra. *See* Ugarit
Rehov, Tel (Tell es-Sarem), 99, *228*, **304**
Reich, Ronny, 48, 73
Reisner, George, 16, 145, 146
Renfrew, Colin, 134, 149, 153
retaining walls, **54**
Revadim quarry site, 129, *130*, 131, 133
revetment walls, 3, 7, 54, **55**, 56, 59, 75, 82, 84, 118, 143
Richardson, Seth, 142
Ridan, Tel, 129, *130*, 132
Rigveda, 5
Rimah, Tell er-, 9
Rosen, Arlene, 56, 74, 75
Rowton, Michael, 117
Rujjum, er-, 57
Rukais, 63, 70, 76, 99, *228*, *228*, **304–05**
Rumeidah, Jebel. *See* Hebron.
Rumeileh, Tell. *See* Beth-Shemesh
rural agricultural estates, 104, 110, 114, 115–16, **124**, 133. *See also* gittu
ruse for entering cities, **43**. *See also* Taking of Joppa

Ṣabuma *or* Ṣapuma (city in EA), 136
Šaduppum (Tell Harmal), 109
Safut, Tell, 49, 83
Saggaratum, 67, 142, 143; Mari district, 111
Sahab, 81, 83, 99, *228*, **305**
sail, defensive use of, **37**
šalḫu (outer wall), 107
Salihiyeh, Tell es-. *See* Ferzat
Samanum, defeat of Yamḫad coalition at, 94–95, 97
Šamši-Adad I (Assyria), 45, 94, **95**, 96, 97, 101, 176
Samsuiluna (of Babylon), **96**, 97, 108, 109, 141, 143, 144, 155, 160, 162
sand: in plasters, 74; in ramparts, 51
šapīṭu (district governor), 111, 133; military obligations, 124
sapping, **38**; defense against, 40. *See also* pilšu
Ṣapunu (Ugaritic district), 113
Sargon (of Akkad), 28, 35, **90**, 92, 97, 159
Šar-kali-šarri (of Akkad), 91, 97
Sasson, Jack, 25, 32, 38, 106, 107, 125
Schaeffer, Claude, 4, 30, 92
Schloen, J. David, 71, 112, **121–22**, 156. *See also* Patrimonial Household Model
Schulman, Alan, 25, 27, 29, 38, 60, 94
Scurlock, Jo Anne, 39
sebakh, mining for, 64
Sefinat-Nouh, Tell, 47, 49, 52, 56, 57, 81, 95, *186*, **217–18**
Seger, Joe, 11
Seilun, Khirbet. *See* Shiloh
Selenkahiye, Tell es-, 17, 56, 57, 59, 63, 66, 67, 69, 70, 79, 91, 92, 97, 160, *186*, **218–19**
Semna, 15–16, 66
Semna South, 16
Semunieh. *See* Shimron.
Seqenenre Tao II, 43, 45
Seraʻ, Tel (Tell esh-Shariʻa), 81, 99, 129, *130*, 132, 133, *228*, **305**
Serangeli, Flavia, 111, 112
Serra, 16
Šeru (city in EA), 120
Sesostris I, 16
Sesostris II, 44, 94
Sesostris III, 16, 97, 100, 118
settlement patterns and types: Arrapḫe, 112–13; Ebla, 105–106; Hazor, 117; highlands, 117–19; Jazira, 104–105; Mari (Old Babylonian), 106–11; Mukiš, 111–12; Nabada, 104; typology for Levant, 122; Ugarit, 113–16. *See also individual settlement types*
Shalfak, 15, 16
Shang period fortifications (China), 51
Shariʻa Tell esh-. *See* Seraʻ
Sharuhen, 8, 29, 97, 116, 119, 129, 132, 135, 136, 137, 138; identification with Tell el-ʻAjjul, 231; kingdom of, 116. *See also* ʻAjjul
Shechem (Tell el-Balatah), 5, 6, 8, 10, 11, 13, 45, 49, 50, 52, 53, 55, 56, 61, 63, 64, 68, 69, 70, 71, 75, 76, 82, 83, 99, 100, 101, 117, 120, 123, 145, 149, 151, 156, 157, 161, *228*, **306–10**; MB kingdom, 118

shield, **36–37**
Shiloh (Khirbet Seilun), 12, 13, 15, 49, 50, 52, 54, 55, 61, 63, 64, 75, 77, 82, 99, 101, 117, 118, 144, 145, 146, 147, 148–49, 151, 152, 153, 157, 161, *228*, **310–12**
Shimron, Tel (Semunieh), 14, 49, 68, 80, 99, 160, *228*, **312–13**
Shush, Tel (Tell Abu Shusha), 99, *228*, **313**
Shuweiket er-Ras, Khirbet, 99, *228*, **313**
Siegecraft. See Aeneas Tacticus
siege ladder, **40**
siege machinery: **37–41**; countermeasures against, 37, **40–41**. *See also* battering ram, breaching, *pilšu* sapping, siege ramp, siege tower, tunneling
Siege of Uršu, 6
siege pole, 6, **39–40**, 45, 46. *See also* battering ram
siege ramp, **38**. *See also* eperu
siege scenes (Egyptian), 29. *See also* Beni Hasan
siege tower: **38**; defense against, 40. *See also* dimtu
siege warfare: **27–29**; *See also* mortuary evidence, siege machinery
simple bow, 31, **33**, 34, 36, 46
Sippar (T. Abu Habbah), 16, 58, 107–109, 141, 155
Šipṭi-Ba'la (ruler of Lakiša), 138
site morphological processes, **77**
site selection, **77**
sītu (battlemented parapet?), 107
Skmm, 97, 100, 118. *See also* Khu-Sebek stele
sling, **32–33**. *See also* aban waṣpim
Smith, Monica L., 134
social and political complexity, 21, 81, **155–58**, 161
soil micromorphology studies, **75–77**
Soldt, Wilfred H. van, 113, 115, 134
Ṣour, Tell aṣ- (Ṣourane), 81, 95, 123, *186*, **219–20**
spear, **42–43**
Stager, Lawrence, **10–11**, 14, 17, 19, 40, 58, 71, 79, 93, 126, 127, 134, 135, 138, 145
Standard of Ebla, 88
Steinkeller, Piotr, 8, 39
Stele of the Vultures, 36, 43
Stern, Ephraim, 51, 101, 150
stone, 52, 62, 63, **75**, 102. *See also* flint, huwwar, kurkar, travertine
strategic defenses, concept of, **124**, 125
structural walls, **54–55**. *See also* core walls, retaining walls, revetment walls
Šubat-Enlil (Tell Leilan), 15, 17, 60, 89, 96, 97, 123, 160, *167*, **175–76**
Šubat-Šamaš (Mari district), 111
Suhum (Mari district), 111
Šulgi (Ur III king), 20, 28, 91, 97, 106, 160
Sumu'epuh (of Yamḫad), 94, 95
Sumur, 122. *See also* Kazel
Sumuyamam, 95
Sun Tzu (*Art of War*), 7, 25, 27 29, 31, 141, 159
Šuppiluliuma (of Ḫatti), 122
supplemental ramparts. *See* earthen ramparts
Šu-Sin (Ur III king), 10, 28, 91, 92, 97, 160
Sweyhat, Tell es-, 17, 32, 66, 76, 79, 91, 92, 97, 144, 147, 150–51, 152, 154, 160, *167*, **176–77**; destruction, 92

Ta'anach (Tell Ta'annek), 55, 61, 82, 99, 101, *228*, **313–14**
Taking of Joppa, 43
Tale of Sinuhe, 26, 28
Tarrum (in Mari texts), 28
Taya, Tell, 68
Tell Dhurur, Khirbet et-. *See* Zeror
Telluliot Batash, 129, *130*, 131
Terqa (Tell Ashara), 16, 17, 49, 54, 56, 57, 58, 60, 76, 78, *79*, 105, 107, 109, 111, 142, 159, *167*, **177–79**
terre pisée: use in Shang period (China), 51
textual traditions about warfare, **25–27**; Canaanite, 43. *See also* Taking of Joppa, Tale of Sinuhe; Chinese. *See* Sun Tzu; Egyptian. *See also* Autobiography of Ahmose, Battle of Kadesh, Battle of Megiddo, Tale of Sinuhe, *and* Wisdom of Merikare; Greek. *See* Aeneas Tacticus, Odyssey; Hittite. *See* Siege of Uršu; Mari, 25; Mesopotamian, 26, 91. *See also* Lamentation; Roman, 27, 33, 38, 47, 51, 58. *See also* Vegetius Renatus; Ugaritic, 26, 31. *See also* Prayer, Kirta Epic
ṯǵr (gate), 110
Thiessen polygons, 116, 137
Thutmose III, 43, 45, 97; battle at Megiddo, 27
Tianna (city in EA), identification of, 136, 137, 138
Tidnum, 10, 28, 58, 91, 97. *See also* Muriq-Tidnim
tilpānu (bow), 34
Timnah (Tel Batash), 14, 15, 47, 48, 49, 51, 52, 53, 56, 57, 63, 76, 81, 95, 99, 101, 123, 128, 129, *130*, 131, 133, 135, 137, 144, 146, 147, 150, 151, 152, 153, *228*, **314–15**
towers, **64–67**; fields of fire from, *65*; rectangular, 84. *See also* bastions, madgaltu, magdalu, watchtowers
town: inner versus outer, 108; terms for, 110. *See also* adaššu, kirḫu
travertine limestone, 51, 55
Trojan horse, 43. *See also* ruse
Tuba. *See* Umm el-Marra
Tubb, Jonathan, **10**, 36, 42, 93
Tubeiqa, Khirbet et-. *See* Beth-Zur
Tunip (T. 'Acharneh), 56, 57, 78, 117, *186*, **220–21**
tunneling, **38–39**; defense against, 40; *See also* pilšu
tunnels, 38, 39, 40, 41, 73, 135, 232–33. *See also*

'Ajjul, Lachish, water systems
Tuqan, Tell, 12, 48, 49, 52, 56, 63, 65, 67, 68, 69, 72, 75, 76, 79, 80, 91, 92, 95, 97, 98, 105, 123, 160, 162, *186*, **221–23**; EB IVB destruction, 92; MB II destruction, 97, 98
Turkey, 6, 18, 146
Tuttul (Tell Bi'a), 17, 45, 60, 71, 76, *79*, 90, 91, 94, 95, 96, 97, 105, 160, *167*, **179–80**; Mari district, 111; Palace A destructions, 95
Tweini, Tell, *186*, **223**
typologies of fortifications: Egypt, 16; Levant, 11

'Udmu, 26
Ugarit (Ras Shamra), 4, 5, 6, 7, 8, 30, 44, 45, 52, 55, 56, 71, 81, 84, 91, 92, 94, 102, 104, 111, 117, 119, 120, 121, 122, 123, 134, 156, *186*, **223–25**; destruction, 92; districts of, 111, 113; RS 24.266, 26; settlement pattern and types, 110, 113–16. *See also* ʿamqu, dimtu, ḍmr, gittu, ḥl, kaparu, magdalu, maškanum, pdr, qarītu, ʿr
ugāru (field), 104
Ulišum, 90, 97
'Umayri, Tell el-, 50, 55, 56, 57, 82, 99, *228*, **315**
Umma, 28, 110
Umm el-Marra, 49, 56, 59, 63, 79, 80, *186*, **225–27**
unwalled settlements, **63–64**. *See also* villages
Ur, 10, 20, 26, 28, 91, 92, 97, 104, 106, 107, 108, 110, 160; destruction by Elamites, 20, 26, 91, 160; fortifications, 107. See also *Lamentation*
Ur III Dyn., as threat, **91**, 97. *See also* Šulgi, Šu-Sin
urbanization, fortifications and, 11, 13, 16, 17, 80, 93, 106, 156
Urkeš (Tell Mozan), 17, 73, *79*, 89, 91, 97, *167*, **180–81**
'Urma, Khirbet, 82, 99, 118, *228*, **315**
Ur-Nammu (Ur III king), 91, 106
Uronarti, 15, 16
Uršu (Warsuwa), siege of, 97
Uruk, 28, 33, 80, 88, 97, 101, 110
Urusalim (EA). *See* Jerusalem
Ussishkin, David, 12, 33, 101, 131

Van Beek, Gus W., 132
Vegetius Renatus, 27, 33, 38, 47, 51, 57. *See also* Epitome
Vercoutter, Jean, 15
village: terms for, 110, 116, 128, 132; unfortified settlements, 109, 114, 123. *See also* kaparu, kapru, maškanum, maškanatu, medinah
Villard, Pierre, 111
Voss, Ross, 19, 53, 74

waṣpu (sling), 32
Wall of the Land, 28, 91, 97, 106
Wall of Ur, 91
walls. *See* core walls, fortification walls, retaining walls, revetment walls, structural walls
Walls of the Ruler, 28
Walters, Stanley D., 142
Warring States Period (China), 27
warrior burials, 42, 43
watchtowers, 110, 113, 114, 115, 116, 124, 126, 129. *See also* magdalu, madgaltu, dimtu
water systems, **72–73**
weaponry, range, 37. *See also individual weapons*
Weiss, Harvey, 15, 92, 96
Weni, campaign of, 87, 88, 93, 97
Wheeler, Mortimer, 5
Wilkinson, Tony, 57, 58, 67, 78, 81, **104**, 106, 111–. 12, 113, 116, 122
Wisdom of Merikare, 125
wood, **75**. *See also* doors
Woolley, Sir Leonard, 4, 12, 16, 30, 45, 72, 112
Wright, George E., 3
Wright, George R. H., 4, **8–9**, 47, 48, 51, 53, 55, 67, 74, 75

Yabni-Ilu (ruler of Lakiša), 138
Yadin, Yigael, 5, **6–7**, 8, 10, 11, 14, 19, 25, 31, 32, 33, 34, 35, 36, 37, 40, 42, 50, 51, 73, 81, 83, 117, 127
Yaḫdun-Lim (of Mari), **94–95**, 97, 102, 109, 173
Yakaltum (Tell Munbaqa), 17, 56, 60, 61, 68, 69, 72, 76, 84, 109, 123, *167*, **181–83**
Yamḫad, kingdom, 94, 95, 96, 97, 98, 100, 101, 111, 119, 122, 127, 156, 160
Yapaḫu (ruler of Gazru), 138
Yarim-Lim I (of Yamḫad), **95–96**, 97
Yarim-Lim II (of Yamḫad), 96
Yarim-Lim III (of Yamḫad), 96, 97
Yarimuta, 120
Yarmuth, 57
yarru (pond or pool), 58, 104
yašibu (battering ram), 28, 38, 39, 159. *See also* battering ram, GUD.SI.DILI
Yasmaḫ-Adad (of Mari), 95, 97
Yavneh (coast site), 129, 131
Yavneh (coast), *130*
Yavneh (East), *130*, 131
Yavneh-Rehovot Road cemetery, *130*
Yavneh (Shikun Holot), 129, *130*, 131, 133
Yavneh (South) cemetery, *130*
Yavneh-Yam, 9, 10, 13, 48, 49, 52, 55, 61, 64, 68, 69, 70, 78, 81, 95, 99, 123, 127, 128, 129, *130*, 131, 133, 135, 137, 138, 157, *228*, **315–17**
year-names recording battles and sieges, 28, 91, 96, 105, 106, 107
Yehudiyeh, Tell el-, 4, 5, **7**, 8, 49
Yidya (ruler of Ašqaluna), 136, 138

Yoqne'am, Tel, 63, 99, *228*, **317**
Yurza (city in EA), as Haror, 138

Zaccagnini, Carlo, 104, 108, **112–13**, 125
Zalpah, 92. *See also* Hammam et-Turkman
Zeitah, Tell (Tel Zayit), 129, *130*, 132
Zeror (Khirbet et-Tell Dhurur), 11, 49, 56, 57, 59, 63, 66, 67, 76, 80, 97, 98, 99, 127, 160, *228*, **317–19**; MB IIA abandonment, 99
Zettler, Richard, 144, 150
Zilu (city in EA), 136
Zimr-Edda (ruler of Lakiša), 138
Zimri-Lim (of Mari), **95–96**, 97
Zuḫra (city in EA), 136
Zurekiyeh, 'Ain (Poleg East Fort), 50, 63, 80, 99, 124, 127, *228*, **319**

www.ingramcontent.com/pod-product-compliance
Lightning Source LLC
LaVergne TN
LVHW080016210926
848481LV00011B/3036